W9-BZR-053

Applied Linear Statistical Models

Applied Linear Statistical Models

Regression, Analysis of Variance, and Experimental Designs

John Neter
University of Georgia

and

William Wasserman
Syracuse University

1974

RICHARD D. IRWIN, INC. Homewood, Illinois 60430
IRWIN-DORSEY INTERNATIONAL Arundel, Sussex BN18 9AB
IRWIN-DORSEY LIMITED Georgetown, Ontario L7G 4B3

First Printing, January 1974
Second Printing, February 1975
Third Printing, August 1975
Fourth Printing, November 1975

ISBN 0–256–01498–1
Library of Congress Catalog Card No. 73–80099

Printed in the United States of America

To
Ronald, David
Christopher, Timothy, Randall

Preface

LINEAR STATISTICAL MODELS for regression, analysis of variance, and experimental designs are widely used today in management, management sciences, and social sciences. Successful application of these models requires a sound understanding of both the underlying theory and the practical problems which are encountered in using the models in real-life situations. While *Applied Linear Statistical Models* is basically an applied book, it seeks to effectively blend theory and applications, avoiding the extremes of presenting theory in isolation, or of giving elements of applications without the needed understanding of the theoretical foundations.

A key feature of this book is its unified approach to the application of linear statistical models in regression, analysis of variance, and experimental designs. Instead of treating these areas in isolated fashion, we seek to show the interrelationships between them. Use of a common notation for regression on the one hand and analysis of variance and experimental designs on the other facilitates a unified view. The notion of a general linear statistical model, which arises naturally in the context of regression models, is carried over to analysis of variance and experimental design models to bring out their relation to regression models. This unified approach also has the advantage of simplified presentation.

We have included in this book not only the more conventional topics in regression, analysis of variance, and basic experimental designs, but also have taken up topics that are frequently slighted but are important in practice. Thus, we devote a full chapter to indicator variables, covering both dependent and independent indicator variables. Another chapter takes up computerized selection procedures for obtaining a "best" set of independent variables. The use of residual analysis for examining the aptness of the model is a recurring theme throughout this book. So is the use of remedial measures that may be helpful when the model is not appropriate. In the analysis of the results of a study, we emphasize the use of estimation procedures, rather than tests, since estimation is often more meaningful in practice. Also, since practical problems seldom are concerned with a single estimate, we stress the use of multiple comparison estimation procedures.

Applications of linear statistical models frequently require extensive computations. In the past, this has proved to be a major deterrent. As a consequence, much attention was paid to computational procedures which would simplify the calculations, such as the Doolittle method. We take the point of view that in most applied work, an electronic computer is available. Further, almost every computer user has access to program packages for fitting linear statistical models. Hence, we explain the basic mathematical steps in fitting a linear statistical model, but do not dwell on computational details. This approach permits us to avoid many complex formulas and enables us to focus on basic principles. We make extensive use in this text of computer capabilities for performing computations, and illustrate a variety of computer printouts and explain how these are used for analysis.

A wide variety of case examples is presented, both to illustrate the great diversity of applications of linear statistical models and to show how analyses are carried out for different problems. Theoretical ideas are presented to the degree needed for good understanding in making sound applications. Proofs are given in those instances where we feel they serve to demonstrate an important method of approach. Emphasis is placed on a thorough understanding of the models, particularly the meaning of the model parameters, since such understanding is basic to proper applications.

We have used "Notes" and "Comments" sections in each chapter to present additional discussion and matters related to the main stream of development. In this way, the basic ideas in a chapter are presented concisely and without distraction. For the same reason, we have set up a number of optional "Topics" chapters following chapters containing the main development, which present a variety of supplementary topics. In most cases, these topics can be omitted without loss of continuity.

A selection of problems is provided at the end of each chapter (excepting Chapter 1). Here the reader can reinforce his understanding of the methodology and use the concepts he has learned to analyze data. We have been careful to supply data analysis problems that typify genuine applications. In some problems the calculations are best handled on a computer, and we urge that this avenue be used when possible.

We assume that the reader of *Applied Linear Statistical Models* has had an introductory course in statistical inference, covering the material outlined in Chapter 1. Should some gaps in the reader's background exist, he can read on his own the relevant portions of an introductory text, or the instructor of the class may use supplemental materials for covering the missing segments. Chapter 1 is not intended as a remedial chapter but rather as a reference chapter of basic statistical results, for continuing use as the reader progresses through the book.

Calculus is not required for reading *Applied Linear Statistical Models.* In a number of instances we use calculus to demonstrate how some important results are obtained, but these are confined to supplementary comments or notes and can be omitted without any loss of continuity. Readers who do know calculus will find these comments and notes in natural sequence so that the benefits of the mathematical developments are obtained in their immediate context. Some basic elements of matrix algebra are needed for multiple regression and related areas. Chapter 6 introduces these elements of matrix algebra in the context of simple regression for easy learning.

Applied Linear Statistical Models can be used in a variety of courses:

1. A two-quarter or two-semester course in regression, analysis of variance, and basic experimental designs might be based on the following chapters:
   Regression: 2, 3, 4, 5 (Sections 5.1–5.4 only), 6, 7, 9 (Sections 9.1–9.3 only).
   Analysis of variance: 13, 14, 15, 17, 18.
   Experimental designs: 21, 23 (omit Section 23.13), 24.
2. A one-quarter or one-semester course in regression analysis might be based on the following chapters:
   2, 3, 4, 5 (selected topics including Section 5.1), 6, 7, 8, 9, 10 (selected topics), 11,12.
3. A one-quarter or one-semester course in analysis of variance might be based on the following chapters:
   13, 14, 15, 16 (selected topics), 17, 18, 19 (selected topics), 20, 21.
4. A one-quarter or one-semester course in basic experimental designs might be based on the following chapters:
   21, 22, 23, 24.
   As time permits, the instructor could cover additional topics with supplementary materials.

This book can also be used by persons engaged in the fields of management, management sciences, and social sciences who desire to obtain competence in the application of linear statistical models by self-study.

A book such as this cannot be written without substantial assistance from others. We are indebted to the many contributors who have developed the theory and practice discussed in this book. We also would like to acknowledge appreciation to our students who helped us in a variety of ways to fashion the method of presentation contained herein. We are grateful to Ronald S. Koot, The Pennsylvania State University, who carefully reviewed the manuscript and made many valuable suggestions. Yu-tsung Lin assisted us diligently in the checking of the manuscript. Almost all of the typing was done by Diane M. Berube, who ably handled the preparation of a difficult manuscript. Thanks also go to Mrs. Loretta Scholten for her care in copyediting and seeing the manuscript through production. Finally, our families bore patiently the pressures caused by our commitment to complete this book. To all of them, we are most grateful.

We are also indebted to the Literary Executor of the late Sir Ronald A. Fisher, F.R.S., to Dr. Frank Yates, F.R.S., and to Oliver and Boyd, Edinburgh, for permission to reprint Table III from their book *Statistical Tables for Biological, Agricultural and Medical Research.*

December 1973

JOHN NETER
WILLIAM WASSERMAN

Contents

Part V
EXPERIMENTAL DESIGNS

1

Some Basic Results in Probability and Statistics

THIS CHAPTER contains some basic results in probability and statistics. It is intended as a reference chapter to which you may refer as you read this book. Sometimes, specific references to results in this chapter are made in the text. At other times, you may wish to refer on your own to particular results in this chapter as you feel the need.

You may prefer to scan the results on probability and statistical inference in this chapter before reading Chapter 2, or you may proceed directly to the next chapter.

1.1 SUMMATION AND PRODUCT OPERATORS

Summation Operator

The summation operator $\sum$ is defined as follows:

(1.1) $$\sum_{i=1}^{n} Y_i = Y_1 + Y_2 + \cdots + Y_n$$

Some important properties of this operator are:

(1.2a) $$\sum_{i=1}^{n} k = nk \qquad \text{where } k \text{ is a constant}$$

(1.2b) $$\sum_{i=1}^{n} (Y_i + Z_i) = \sum_{i=1}^{n} Y_i + \sum_{i=1}^{n} Z_i$$

(1.2c) $$\sum_{i=1}^{n} (a + cY_i) = na + c\sum_{i=1}^{n} Y_i \qquad \text{where } a \text{ and } c \text{ are constants}$$

The double summation operator $\sum\sum$ is defined as follows:

$$\sum_{i=1}^{n}\sum_{j=1}^{m} Y_{ij} = \sum_{i=1}^{n}(Y_{i1} + \cdots + Y_{im}) \tag{1.3}$$

$$= Y_{11} + \cdots + Y_{1m} + Y_{21} + \cdots + Y_{2m} + \cdots + Y_{nm}$$

An important property of the double summation operator is:

$$\sum_{i=1}^{n}\sum_{j=1}^{m} Y_{ij} = \sum_{j=1}^{m}\sum_{i=1}^{n} Y_{ij} \tag{1.4}$$

Product Operator

The product operator $\prod$ is defined as follows:

$$\prod_{i=1}^{n} Y_i = Y_1 \cdot Y_2 \cdot Y_3 \cdots Y_n \tag{1.5}$$

1.2 PROBABILITY

Addition Theorem

Let A_i and A_j be two events defined on a sample space. Then:

$$P(A_i \cup A_j) = P(A_i) + P(A_j) - P(A_i \cap A_j) \tag{1.6}$$

where $P(A_i \cup A_j)$ denotes the probability of either A_i or A_j or both occurring, $P(A_i)$ and $P(A_j)$ denote respectively the probability of A_i and the probability of A_j, and $P(A_i \cap A_j)$ denotes the probability of both A_i and A_j occurring.

Multiplication Theorem

Let $P(A_i|A_j)$ denote the conditional probability of A_i occurring, given that A_j has occurred. This conditional probability is defined as follows:

$$P(A_i|A_j) = \frac{P(A_i \cap A_j)}{P(A_j)} \qquad P(A_j) \neq 0 \tag{1.7}$$

The multiplication theorem states:

$$P(A_i \cap A_j) = P(A_i)P(A_j|A_i) \tag{1.8}$$

$$= P(A_j)P(A_i|A_j)$$

Complementary Events

The complementary event of A_i is denoted $\bar{A}_i$. The following results for complementary events are useful:

$$P(\bar{A}_i) = 1 - P(A_i) \tag{1.9}$$

$$P(\overline{A_i \cup A_j}) = P(\bar{A}_i \cap \bar{A}_j) \tag{1.10}$$

1.3 RANDOM VARIABLES

Throughout this section, we assume that the random variable Y assumes a finite number of outcomes. (If Y is a continuous random variable, the summation process is replaced by integration.)

Expected Value

Let the random variable Y assume the outcomes $Y_1, \ldots, Y_k$ with probabilities given by the probability function:

(1.11) $$f(Y_s) = P(Y = Y_s) \qquad s = 1, \ldots, k$$

The expected value of Y is defined:

(1.12) $$E(Y) = \sum_{s=1}^{k} Y_s f(Y_s)$$

An important property of the expectation operator E is:

(1.13) $$E(a + cY) = a + cE(Y) \qquad \text{where } a \text{ and } c \text{ are constants}$$

Special cases of this are:

(1.13a) $$E(a) = a$$

(1.13b) $$E(cY) = cE(Y)$$

(1.13c) $$E(a + Y) = a + E(Y)$$

Variance

The variance of the random variable Y is denoted $\sigma^2(Y)$, and is defined as follows:

(1.14) $$\sigma^2(Y) = E\{[Y - E(Y)]^2\}$$

An equivalent expression is:

(1.14a) $$\sigma^2(Y) = E(Y^2) - [E(Y)]^2$$

The variance of a linear function of Y is frequently encountered. We denote the variance of $a + cY$ by $\sigma^2(a + cY)$, and have:

(1.15) $$\sigma^2(a + cY) = c^2\sigma^2(Y) \qquad \text{where } a \text{ and } c \text{ are constants}$$

Special cases of this result are:

(1.15a) $$\sigma^2(a + Y) = \sigma^2(Y)$$

(1.15b) $$\sigma^2(cY) = c^2\sigma^2(Y)$$

Joint, Marginal, and Conditional Probability Distributions

Let the joint probability function for the two random variables Y and Z be denoted $g(Y, Z)$:

$$g(Y_s, Z_t) = P(Y = Y_s, Z = Z_t) \qquad s = 1, \ldots, k;\ t = 1, \ldots, m \tag{1.16}$$

The marginal probability function of Y, denoted $f(Y)$, is:

$$f(Y_s) = \sum_{t=1}^{m} g(Y_s, Z_t) \qquad s = 1, \ldots, k \tag{1.17a}$$

and the marginal probability function of Z, denoted $h(Z)$, is:

$$h(Z_t) = \sum_{s=1}^{k} g(Y_s, Z_t) \qquad t = 1, \ldots, m \tag{1.17b}$$

The conditional probability function of Y, given $Z = Z_t$, is:

$$f(Y_s | Z_t) = \frac{g(Y_s, Z_t)}{h(Z_t)} \qquad h(Z_t) \neq 0;\ s = 1, \ldots, k \tag{1.18a}$$

and the conditional probability function of Z, given $Y = Y_s$, is:

$$h(Z_t | Y_s) = \frac{g(Y_s, Z_t)}{f(Y_s)} \qquad f(Y_s) \neq 0;\ t = 1, \ldots, m \tag{1.18b}$$

Covariance

The covariance of Y and Z is denoted $\sigma(Y, Z)$, and is defined:

$$\sigma(Y, Z) = E\{[Y - E(Y)][Z - E(Z)]\} \tag{1.19}$$

An equivalent expression is:

$$\sigma(Y, Z) = E(YZ) - [E(Y)][E(Z)] \tag{1.19a}$$

The covariance of $a_1 + c_1 Y$ and $a_2 + c_2 Z$ is denoted $\sigma(a_1 + c_1 Y, a_2 + c_2 Z)$, and we have:

$$\sigma(a_1 + c_1 Y, a_2 + c_2 Z) = c_1 c_2 \sigma(Y, Z) \qquad \text{where } a_1, a_2, c_1, c_2 \text{ are constants} \tag{1.20}$$

Special cases of this are:

$$\sigma(c_1 Y, c_2 Z) = c_1 c_2 \sigma(Y, Z) \tag{1.20a}$$

$$\sigma(a_1 + Y, a_2 + Z) = \sigma(Y, Z) \tag{1.20b}$$

By definition, we have:

$$\sigma(Y, Y) = \sigma^2(Y) \tag{1.21}$$

where $\sigma^2(Y)$ is the variance of Y.

Independent Random Variables

(1.22) Random variables Y and Z are independent if and only if:

$$g(Y_s, Z_t) = f(Y_s)h(Z_t) \qquad s = 1, \ldots, k;\ t = 1, \ldots, m$$

If Y and Z are independent random variables:

(1.23) $$\sigma(Y, Z) = 0 \qquad \text{when } Y, Z \text{ are independent}$$

(In the special case where Y and Z are jointly normally distributed, $\sigma(Y, Z) = 0$ implies that Y and Z are independent.)

Functions of Random Variables

Let $Y_1, \ldots, Y_n$ be n random variables. Consider the function $\sum a_i Y_i$, where the a_i are constants. We then have:

(1.24a) $$E\left(\sum_{i=1}^{n} a_i Y_i\right) = \sum_{i=1}^{n} a_i E(Y_i) \qquad \text{where the } a_i \text{ are constants}$$

(1.24b) $$\sigma^2\left(\sum_{i=1}^{n} a_i Y_i\right) = \sum_{i=1}^{n} \sum_{j=1}^{n} a_i a_j \sigma(Y_i, Y_j) \qquad \text{where the } a_i \text{ are constants}$$

Specifically, we have for $n = 2$:

(1.25a) $$E(a_1 Y_1 + a_2 Y_2) = a_1 E(Y_1) + a_2 E(Y_2)$$

(1.25b) $$\sigma^2(a_1 Y_1 + a_2 Y_2) = a_1^2 \sigma^2(Y_1) + a_2^2 \sigma^2(Y_2) + 2a_1 a_2 \sigma(Y_1, Y_2)$$

If the random variables Y_i are independent, we have:

(1.26) $$\sigma^2\left(\sum_{i=1}^{n} a_i Y_i\right) = \sum_{i=1}^{n} a_i^2 \sigma^2(Y_i) \qquad \text{when the } Y_i \text{ are independent}$$

Special cases of this are:

(1.26a) $$\sigma^2(Y_1 + Y_2) = \sigma^2(Y_1) + \sigma^2(Y_2) \qquad \text{when } Y_1, Y_2 \text{ are independent}$$

(1.26b) $$\sigma^2(Y_1 - Y_2) = \sigma^2(Y_1) + \sigma^2(Y_2) \qquad \text{when } Y_1, Y_2 \text{ are independent}$$

The covariance of two linear functions $\sum a_i Y_i$ and $\sum c_i Y_i$ is, when the Y_i are independent random variables:

(1.27) $$\sigma\left(\sum_{i=1}^{n} a_i Y_i, \sum_{i=1}^{n} c_i Y_i\right) = \sum_{i=1}^{n} a_i c_i \sigma^2(Y_i) \qquad \text{when the } Y_i \text{ are independent}$$

Central Limit Theorem

(1.28) If $Y_1, \ldots, Y_n$ are independent random observations from a population with probability function $f(Y)$ for which $\sigma^2(Y)$ is finite, then, when the sample size n is reasonably large, the sample mean $\overline{Y}$:

$$\overline{Y} = \frac{\sum_{i=1}^{n} Y_i}{n}$$

is approximately normally distributed with mean $E(Y)$ and variance $\sigma^2(Y)/n$.

1.4 NORMAL PROBABILITY DISTRIBUTION AND RELATED DISTRIBUTIONS

Normal Probability Distribution

The density function for a normal random variable Y is:

(1.29) $$f(Y) = \frac{1}{\sqrt{2\pi}\,\sigma} \exp\left[-\frac{1}{2}\left(\frac{Y-\mu}{\sigma}\right)^2\right] \qquad -\infty < Y < +\infty$$

where μ and σ are the two parameters of the normal distribution.

The mean and variance of a normal random variable Y are:

(1.30a) $$E(Y) = \mu$$

(1.30b) $$\sigma^2(Y) = \sigma^2$$

Function of Normal Random Variable. A linear function of a normal random variable Y has the following property:

(1.31) If Y is a normal random variable, the transformed variable $Y' = a + cY$ (a and c are constants) is normally distributed, with mean $a + cE(Y)$ and variance $c^2\sigma^2(Y)$.

Standard Normal Variable. The standard normal variable z:

(1.32) $$z = \frac{Y-\mu}{\sigma} \qquad \text{where } Y \text{ is a normal random variable}$$

is normally distributed, with mean 0 and variance 1. We denote this as follows:

(1.33) $$z \text{ is } N(0, 1)$$

Mean Variance

Table A–1 in the Appendix contains the cumulative probabilities $1 - \alpha$ for percentiles $z(1 - \alpha)$, where:

(1.34) $$P\{z \leq z(1 - \alpha)\} = 1 - \alpha$$

For instance, when $z(1-\alpha) = 2.00$, $1-\alpha = .9772$. Because the normal distribution is symmetrical about 0, when $z(1-\alpha) = -2.00$, $1-\alpha = .0228$.

Functions of Independent Normal Random Variables. Let $Y_1, \ldots, Y_n$ be independent normal random variables. We then have:

(1.35) When $Y_1, \ldots, Y_n$ are independent normal random variables, the linear combination $a_1 Y_1 + a_2 Y_2 + \cdots + a_n Y_n$ is normally distributed, with mean $\sum a_i E(Y_i)$ and variance $\sum a_i^2 \sigma^2(Y_i)$.

χ^2 Distribution

Let $z_1, \ldots, z_\nu$ be ν independent standard normal variables. We then define:

(1.36) $$\chi^2(\nu) = z_1^2 + z_2^2 + \cdots + z_\nu^2 \qquad \text{where the } z_i \text{ are independent}$$

The χ^2 distribution has one parameter, ν, which is called the *degrees of freedom* (*df*). The mean of the χ^2 distribution with ν degrees of freedom is:

(1.37) $$E[\chi^2(\nu)] = \nu$$

Table A–3 in the Appendix contains percentiles of various χ^2 distributions. We define $\chi^2(1-\alpha; \nu)$ as follows:

(1.38) $$P\{\chi^2(\nu) \leq \chi^2(1-\alpha; \nu)\} = 1-\alpha$$

Suppose $\nu = 5$. The 90th percentile of the χ^2 distribution with 5 degrees of freedom is $\chi^2(.90; 5) = 9.24$.

t Distribution

Let z and $\chi^2(\nu)$ be independent random variables (standard normal and χ^2, respectively). We then define:

(1.39) $$t(\nu) = \frac{z}{\left[\dfrac{\chi^2(\nu)}{\nu}\right]^{1/2}} \qquad \text{where } z \text{ and } \chi^2(\nu) \text{ are independent}$$

The t distribution has one parameter, the *degrees of freedom* ν. The mean of the t distribution with ν degrees of freedom is:

(1.40) $$E[t(\nu)] = 0$$

Table A–2 in the Appendix contains percentiles of various t distributions. We define $t(1-\alpha; \nu)$ as follows:

(1.41) $$P\{t(\nu) \leq t(1-\alpha; \nu)\} = 1-\alpha$$

Suppose $\nu = 10$. The 90th percentile of the t distribution with 10 degrees of freedom is $t(.90; 10) = 1.372$. Because the t distribution is symmetrical about 0, we have $t(.10; 10) = -1.372$.

F Distribution

Let $\chi^2(\nu_1)$ and $\chi^2(\nu_2)$ be two independent χ^2 random variables. We then define:

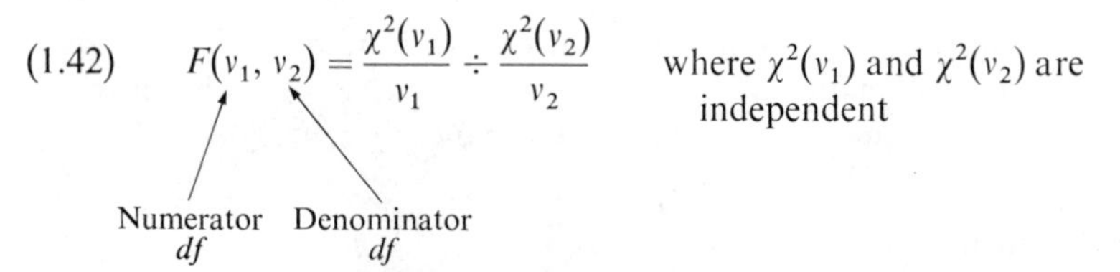

(1.42) $$F(\nu_1, \nu_2) = \frac{\chi^2(\nu_1)}{\nu_1} \div \frac{\chi^2(\nu_2)}{\nu_2}$$ where $\chi^2(\nu_1)$ and $\chi^2(\nu_2)$ are independent

The F distribution has two parameters, the *numerator degrees of freedom* ν_1 and the *denominator degrees of freedom* ν_2.

Table A–4 in the Appendix contains percentiles of various F distributions. We define $F(1-\alpha; \nu_1, \nu_2)$ as follows:

(1.43) $$P\{F(\nu_1, \nu_2) \leq F(1-\alpha; \nu_1, \nu_2)\} = 1 - \alpha$$

Suppose $\nu_1 = 2$, $\nu_2 = 3$. The 90th percentile of the F distribution with 2 and 3 degrees of freedom respectively in the numerator and denominator is $F(.90; 2, 3) = 5.46$.

Percentiles below 50 percent can be obtained by utilizing the relation:

(1.44) $$F(\alpha; \nu_2, \nu_1) = \frac{1}{F(1-\alpha; \nu_1, \nu_2)}$$

Thus, $F(.10; 3, 2) = 1/5.46 = .183$.

The following relation exists between the t and F random variables:

(1.45a) $$[t(\nu)]^2 = F(1, \nu)$$

and the percentiles of the t and F distributions are related as follows:

(1.45b) $$[t(1-\alpha/2; \nu)]^2 = F(1-\alpha; 1, \nu)$$

1.5 STATISTICAL ESTIMATION

Properties of Estimators

(1.46) An estimator $\hat{\theta}$ of the parameter θ is *unbiased* if:

$$E(\hat{\theta}) = \theta$$

(1.47) An estimator $\hat{\theta}$ is a *consistent estimator* of θ if:

$$\lim_{n \to \infty} P(|\hat{\theta} - \theta| \geq \varepsilon) = 0 \qquad \text{for any } \varepsilon > 0$$

(1.48) An estimator $\hat{\theta}$ is a *sufficient estimator* of θ if the conditional joint probability function of the sample observations, given $\hat{\theta}$, does not depend on the parameter θ.

(1.49) An estimator $\hat{\theta}$ is a *minimum variance estimator* of θ if, for any other estimator θ^*:

$$\sigma^2(\hat{\theta}) \le \sigma^2(\theta^*) \qquad \text{for all } \theta^*$$

Maximum Likelihood Estimators

The method of maximum likelihood is a general method of finding estimators. Suppose we are sampling a population whose probability function $f(Y; \theta)$ involves one parameter, θ. Given independent observations $Y_1, \ldots, Y_n$, the joint probability function of the sample observations is:

(1.50a)
$$g(Y_1, \ldots, Y_n) = \prod_{i=1}^{n} f(Y_i; \theta)$$

When this joint probability function is viewed as a function of θ, with the observations given, it is called the *likelihood function* $L(\theta)$.

(1.50b)
$$L(\theta) = \prod_{i=1}^{n} f(Y_i; \theta)$$

Maximizing $L(\theta)$ with respect to θ yields the maximum likelihood estimator of θ. Under quite general conditions, maximum likelihood estimators are consistent and sufficient.

Least Squares Estimators

The method of least squares is another general method of finding estimators. The sample observations are assumed to be of the form (for the case of a single parameter θ):

(1.51)
$$Y_i = f_i(\theta) + \varepsilon_i$$

where $f_i(\theta)$ is a known function of the parameter θ and the ε_i are random variables, usually assumed to have expectation $E(\varepsilon_i) = 0$.

With the method of least squares, the sum of squares:

(1.52)
$$Q = \sum_{i=1}^{n} [Y_i - f_i(\theta)]^2$$

is considered. The least squares estimator of θ is obtained by minimizing Q with respect to θ. In many instances, least squares estimators are unbiased and consistent.

1.6 INFERENCES ABOUT POPULATION MEAN—NORMAL POPULATION

We have a random sample of n observations $Y_1, \ldots, Y_n$ from a normal population with mean μ and standard deviation σ. The sample mean and sample standard deviation are:

(1.53a)
$$\bar{Y} = \frac{\sum_i Y_i}{n}$$

(1.53b)
$$s = \left[\frac{\sum_i (Y_i - \bar{Y})^2}{n-1}\right]^{1/2}$$

and the estimated standard deviation of the sampling distribution of $\bar{Y}$ is:

(1.53c)
$$s(\bar{Y}) = \frac{s}{\sqrt{n}}$$

We then have:

(1.54) $\frac{\bar{Y} - \mu}{s(\bar{Y})}$ is distributed as t with $n - 1$ degrees of freedom when the random sample is from a normal population.

Interval Estimation

The confidence interval for μ, with a confidence coefficient of $1 - \alpha$, is obtained by means of (1.54):

(1.55)
$$\bar{Y} - t(1 - \alpha/2; n - 1)s(\bar{Y}) \le \mu \le \bar{Y} + t(1 - \alpha/2; n - 1)s(\bar{Y})$$

Example 1. Obtain a 95 percent confidence interval for μ, when:

$$n = 10 \qquad \bar{Y} = 20 \qquad s = 4$$

We require:

$$s(\bar{Y}) = \frac{4}{\sqrt{10}} = 1.265 \qquad t(.975; 9) = 2.262$$

so that we find:

$$17.1 = 20 - (2.262)(1.265) \le \mu \le 20 + (2.262)(1.265) = 22.9$$

Tests

One-sided and two-sided tests concerning the population mean μ are constructed by means of (1.54). Table 1.1 contains the decision rules for each of three possible cases, with the risk of making a Type I error controlled at α.

Example 2. Choose among the alternatives:

$$C_1: \mu \le 20$$
$$C_2: \mu > 20$$

when α is to be controlled at .05 and:

$$n = 15 \qquad \bar{Y} = 24 \qquad s = 6$$

TABLE 1.1

Decision Rules for Tests concerning Mean μ of Normal Population

Alternatives	*Decision Rule*
	(a)
$C_1: \mu = \mu_o$ $C_2: \mu \neq \mu_o$	If $A_1 \leq \bar{Y} \leq A_2$, conclude C_1 Otherwise conclude C_2 where: $A_1 = \mu_o - t(1 - \alpha/2; n - 1)s(\bar{Y})$ $A_2 = \mu_o + t(1 - \alpha/2; n - 1)s(\bar{Y})$
	(b)
$C_1: \mu \geq \mu_o$ $C_2: \mu < \mu_o$	If $\bar{Y} \geq A$, conclude C_1 If $\bar{Y} < A$, conclude C_2 where: $A = \mu_o + t(\alpha; n - 1)s(\bar{Y})$
	(c)
$C_1: \mu \leq \mu_o$ $C_2: \mu > \mu_o$	If $\bar{Y} \leq A$, conclude C_1 If $\bar{Y} > A$, conclude C_2 where: $A = \mu_o + t(1 - \alpha; n - 1)s(\bar{Y})$

We require:

$$s(\bar{Y}) = \frac{6}{\sqrt{15}} = 1.549$$

$$t(.95; 14) = 1.761$$

$$A = 20 + (1.761)(1.549) = 22.73$$

so that the decision rule is:

$$\text{If } \bar{Y} \leq 22.73, \text{ conclude } C_1$$

$$\text{If } \bar{Y} > 22.73, \text{ conclude } C_2$$

Since $\bar{Y} = 24 > 22.73$, we conclude C_2.

Relation between Tests and Confidence Intervals. There is a direct relation between tests and confidence intervals. For example, the two-sided confidence interval (1.55) can be used for testing:

$$C_1: \mu = \mu_0$$

$$C_2: \mu \neq \mu_0$$

If μ_0 is contained within the $1 - \alpha$ confidence interval, then the two-sided decision rule in Table 1.1a, with level of significance α, will lead to conclusion

C_1, and vice versa. If μ_0 is not contained within the confidence interval, the decision rule will lead to C_2, and vice versa.

There are similar correspondences between one-sided confidence intervals and one-sided decision rules.

1.7 COMPARISONS OF TWO POPULATION MEANS—NORMAL POPULATIONS

Independent Samples

There are two normal populations, with means μ_1 and μ_2 respectively, and with the same standard deviation σ. The means μ_1 and μ_2 are to be compared on the basis of independent samples for each of the two populations:

$$\text{Sample 1: } Y_1, \ldots, Y_{n_1}$$
$$\text{Sample 2: } Z_1, \ldots, Z_{n_2}$$

Estimators of the two population means are the sample means:

$$\bar{Y} = \frac{\sum_i Y_i}{n_1} \tag{1.56a}$$

$$\bar{Z} = \frac{\sum_i Z_i}{n_2} \tag{1.56b}$$

and an estimator of $\mu_1 - \mu_2$ is $\bar{Y} - \bar{Z}$.

An estimator of the common variance σ^2 is:

$$s^2 = \frac{\sum_i (Y_i - \bar{Y})^2 + \sum_i (Z_i - \bar{Z})^2}{n_1 + n_2 - 2} \tag{1.57}$$

and an estimator of $\sigma^2(\bar{Y} - \bar{Z})$, the variance of the sampling distribution of $\bar{Y} - \bar{Z}$, is:

$$s^2(\bar{Y} - \bar{Z}) = s^2\left[\frac{1}{n_1} + \frac{1}{n_2}\right] \tag{1.58}$$

We have:

(1.59) $\dfrac{(\bar{Y} - \bar{Z}) - (\mu_1 - \mu_2)}{s(\bar{Y} - \bar{Z})}$ is distributed as t with $n_1 + n_2 - 2$ degrees of freedom when the two independent samples come from normal populations with the same standard deviation.

Interval Estimation. The confidence interval for $\mu_1 - \mu_2$, with confidence coefficient $1 - \alpha$, is obtained by means of (1.59):

$$(\bar{Y} - \bar{Z}) - t(1 - \alpha/2; n_1 + n_2 - 2)s(\bar{Y} - \bar{Z}) \leq \mu_1 - \mu_2 \leq (\bar{Y} - \bar{Z}) + t(1 - \alpha/2; n_1 + n_2 - 2)s(\bar{Y} - \bar{Z}) \tag{1.60}$$

Example 3. Obtain a 95 percent confidence interval for $\mu_1 - \mu_2$, when:

$$n_1 = 10 \qquad \bar{Y} = 14 \qquad \sum (Y_i - \bar{Y})^2 = 105$$
$$n_2 = 20 \qquad \bar{Z} = 8 \qquad \sum (Z_i - \bar{Z})^2 = 224$$

We require:

$$s^2 = \frac{105 + 224}{10 + 20 - 2} = 11.75$$

$$s^2(\bar{Y} - \bar{Z}) = 11.75\left(\frac{1}{10} + \frac{1}{20}\right) = 1.7625$$

$$s(\bar{Y} - \bar{Z}) = 1.328$$
$$t(.975; 28) = 2.048$$

$$3.3 = (14 - 8) - (2.048)(1.328) \leq \mu_1 - \mu_2 \leq (14 - 8) + (2.048)(1.328) = 8.7$$

Tests. One-sided and two-sided tests concerning $\mu_1 - \mu_2$ are constructed by means of (1.59). Table 1.2 contains the decision rules for each of three possible cases, with the risk of making a Type I error controlled at α.

Example 4. Choose among the alternatives:

$$C_1: \mu_1 = \mu_2$$
$$C_2: \mu_1 \neq \mu_2$$

when α is to be controlled at .10 and the data are those of Example 3.

TABLE 1.2

Decision Rules for Tests concerning Means μ_1 and μ_2 of Two Normal Populations ($\sigma_1 = \sigma_2 = \sigma$)

Alternatives	*Decision Rule*
	(a)
$C_1: \mu_1 = \mu_2$ $C_2: \mu_1 \neq \mu_2$	If $A_1 \leq \bar{Y} - \bar{Z} \leq A_2$, conclude C_1 Otherwise conclude C_2 where: $A_1 = -t(1 - \alpha/2; n_1 + n_2 - 2)s(\bar{Y} - \bar{Z})$ $A_2 = t(1 - \alpha/2; n_1 + n_2 - 2)s(\bar{Y} - \bar{Z})$
	(b)
$C_1: \mu_1 \geq \mu_2$ $C_2: \mu_1 < \mu_2$	If $\bar{Y} - \bar{Z} \geq A$, conclude C_1 If $\bar{Y} - \bar{Z} < A$, conclude C_2 where: $A = t(\alpha; n_1 + n_2 - 2)s(\bar{Y} - \bar{Z})$
	(c)
$C_1: \mu_1 \leq \mu_2$ $C_2: \mu_1 > \mu_2$	If $\bar{Y} - \bar{Z} \leq A$, conclude C_1 If $\bar{Y} - \bar{Z} > A$, conclude C_2 where: $A = t(1 - \alpha; n_1 + n_2 - 2)s(\bar{Y} - \bar{Z})$

We require:

$$t(.95; 28) = 1.701$$
$$A_1 = -1.701(1.328) = -2.26$$
$$A_2 = 1.701(1.328) = 2.26$$

so that the decision rule is:

If $-2.26 \leq \bar{Y} - \bar{Z} \leq 2.26$, conclude C_1
Otherwise conclude C_2

Since $\bar{Y} - \bar{Z} = 6 > 2.26$, we conclude C_2.

Paired Observations

When the observations in the two samples are paired (e.g., attitude scores Y_i and Z_i for the ith sample employee before and after a year's experience on the job), we use the differences:

(1.61) $$W_i = Y_i - Z_i \qquad i = 1, \ldots, n$$

in the fashion of a sample from a single population. Thus we have, when the W_i can be treated as observations from a normal population:

(1.62) $\dfrac{\bar{W} - (\mu_1 - \mu_2)}{s(\bar{W})}$ is distributed as t with $n - 1$ degrees of freedom when the differences W_i can be considered to be observations from a normal population, and:

$$\bar{W} = \frac{\sum_i W_i}{n}$$

$$s^2(\bar{W}) = \frac{\sum_i (W_i - \bar{W})^2}{n - 1} \div n$$

1.8 INFERENCES ABOUT POPULATION VARIANCE—NORMAL POPULATION

When sampling from a normal population, the following holds for the sample variance s^2, where s is defined in (1.53b):

(1.63) $\dfrac{(n - 1)s^2}{\sigma^2}$ is distributed as χ^2 with $n - 1$ degrees of freedom when the random sample is from a normal population.

Interval Estimation

The confidence interval for the population variance σ^2, with a confidence coefficient of $1 - \alpha$, is obtained by means of (1.63):

$$\frac{(n-1)s^2}{\chi^2(1-\alpha/2; n-1)} \leq \sigma^2 \leq \frac{(n-1)s^2}{\chi^2(\alpha/2; n-1)} \tag{1.64}$$

Example 5. Obtain a 98 percent confidence interval for σ^2, using the data of Example 1 ($n = 10$, $s = 4$).

We require:

$$s^2 = 16 \qquad \chi^2(.01; 9) = 2.09 \qquad \chi^2(.99; 9) = 21.67$$

$$6.6 = \frac{9(16)}{21.67} \leq \sigma^2 \leq \frac{9(16)}{2.09} = 68.9$$

Tests

One-sided and two-sided tests concerning the population variance σ^2 are constructed by means of (1.63). Table 1.3 contains the decision rule for each of three possible cases, with the risk of making a Type I error controlled at α.

TABLE 1.3

Decision Rules for Tests concerning Variance σ^2 of Normal Population

Alternatives	*Decision Rule*
	(a)
$C_1: \sigma^2 = \sigma_o^2$	If $\chi^2(\alpha/2; n-1) \leq \frac{(n-1)s^2}{\sigma_o^2} \leq \chi^2(1-\alpha/2; n-1)$, conclude C_1
$C_2: \sigma^2 \neq \sigma_o^2$	Otherwise conclude C_2
	(b)
$C_1: \sigma^2 \geq \sigma_o^2$	If $\frac{(n-1)s^2}{\sigma_o^2} \geq \chi^2(\alpha; n-1)$, conclude C_1
$C_2: \sigma^2 < \sigma_o^2$	If $\frac{(n-1)s^2}{\sigma_o^2} < \chi^2(\alpha; n-1)$, conclude C_2
	(c)
$C_1: \sigma^2 \leq \sigma_o^2$	If $\frac{(n-1)s^2}{\sigma_o^2} \leq \chi^2(1-\alpha; n-1)$, conclude C_1
$C_2: \sigma^2 > \sigma_o^2$	If $\frac{(n-1)s^2}{\sigma_o^2} > \chi^2(1-\alpha; n-1)$, conclude C_2

1.9 COMPARISONS OF TWO POPULATION VARIANCES—NORMAL POPULATIONS

Independent samples are selected from two normal populations, with means and variances of μ_1 and σ_1^2 and μ_2 and σ_2^2 respectively. Using the notation of Section 1.7, the two sample variances are:

(1.65a)
$$s_1^2 = \frac{\sum_i (Y_i - \bar{Y})^2}{n_1 - 1}$$

(1.65b)
$$s_2^2 = \frac{\sum_i (Z_i - \bar{Z})^2}{n_2 - 1}$$

We have:

(1.66) $\dfrac{s_1^2}{\sigma_1^2} \div \dfrac{s_2^2}{\sigma_2^2}$ is distributed as $F(n_1 - 1, n_2 - 1)$ when the two independent samples come from normal populations.

Interval Estimation

The confidence interval for σ_1^2/σ_2^2, with confidence coefficient $1 - \alpha$, is obtained by means of (1.66):

(1.67)
$$\frac{s_1^2}{s_2^2}\frac{1}{F(1 - \alpha/2; n_1 - 1, n_2 - 1)} \le \frac{\sigma_1^2}{\sigma_2^2} \le \frac{s_1^2}{s_2^2}\frac{1}{F(\alpha/2; n_1 - 1, n_2 - 1)}$$

Example 6. Obtain a 90 percent confidence interval for σ_1^2/σ_2^2 when the data are:

$$n_1 = 16 \qquad n_2 = 21$$
$$s_1^2 = 54.2 \qquad s_2^2 = 17.8$$

We require:

$$F(.05; 15, 20) = .430 \qquad F(.95; 15, 20) = 2.20$$

$$1.4 = \frac{54.2}{17.8}\frac{1}{2.20} \le \frac{\sigma_1^2}{\sigma_2^2} \le \frac{54.2}{17.8}\frac{1}{.430} = 7.1$$

Tests

One-sided and two-sided tests concerning σ_1^2/σ_2^2 are constructed by means of (1.66). Table 1.4 contains the decision rules for each of three possible cases, with the risk of making a Type I error controlled at α.

TABLE 1.4

Decision Rules for Tests concerning Variances σ_1^2 and σ_2^2 of Two Normal Populations

Alternatives	*Decision Rule*
	(a)
$C_1: \sigma_1^2 = \sigma_2^2$	If $F(\alpha/2; n_1 - 1, n_2 - 1) \leq \frac{s_1^2}{s_2^2} \leq F(1 - \alpha/2; n_1 - 1, n_2 - 1)$, conclude C_1
$C_2: \sigma_1^2 \neq \sigma_2^2$	Otherwise conclude C_2
	(b)
$C_1: \sigma_1^2 \geq \sigma_2^2$	If $\frac{s_1^2}{s_2^2} \geq F(\alpha; n_1 - 1, n_2 - 1)$, conclude C_1
$C_2: \sigma_1^2 < \sigma_2^2$	If $\frac{s_1^2}{s_2^2} < F(\alpha; n_1 - 1, n_2 - 1)$, conclude C_2
	(c)
$C_1: \sigma_1^2 \leq \sigma_2^2$	If $\frac{s_1^2}{s_2^2} \leq F(1 - \alpha; n_1 - 1, n_2 - 1)$, conclude C_1
$C_2: \sigma_1^2 > \sigma_2^2$	If $\frac{s_1^2}{s_2^2} > F(1 - \alpha; n_1 - 1, n_2 - 1)$, conclude C_2

Example 7. Choose among the alternatives:

$$C_1: \sigma_1^2 = \sigma_2^2$$
$$C_2: \sigma_1^2 \neq \sigma_2^2$$

when α is to be controlled at .02, and the data are those of Example 6.

We require:

$$F(.01; 15, 20) = .297 \qquad F(.99; 15, 20) = 3.09$$

so that the decision rule is:

$$\text{If } .297 \leq \frac{s_1^2}{s_2^2} \leq 3.09, \text{ conclude } C_1$$
$$\text{Otherwise conclude } C_2$$

Since $s_1^2/s_2^2 = 54.2/17.8 = 3.04$, we conclude C_1.

part I

Basic Regression Analysis

2

Linear Regression with One Independent Variable

REGRESSION ANALYSIS is a statistical tool which utilizes the relation between two or more quantitative variables so that one variable can be predicted from the other, or others. For example, if one knows the relation between advertising expenditures and sales, one can predict sales by means of regression analysis once the level of advertising expenditures has been set.

In Part I, we take up regression analysis when a single predictor variable is used for predicting the variable of interest. In this chapter specifically, we consider the basic ideas of regression analysis and discuss the estimation of the parameters of the regression model.

2.1 RELATIONS BETWEEN VARIABLES

The concept of a relation between two variables, such as between family income and family expenditures for housing, is a familiar one. We distinguish between a *functional* relation and a *statistical* relation, and consider each of these in turn.

Functional Relation between Two Variables

A functional relation between two variables is expressed by a mathematical formula. If X is the *independent* variable and Y the *dependent* variable, a functional relation is of the form:

$$Y = f(X)$$

Given a particular value of X, the function f indicates the corresponding value of Y.

Example 1. Consider the relation between dollar sales (Y) of a product, sold at a fixed price, and number of units sold (X). If the selling price is \$2 per unit, the relation is expressed by the equation:

$$Y = 2X$$

This functional relation is shown in Figure 2.1. Number of units sold and dollar sales during three recent periods (while the unit price remained constant at \$2) were as follows:

Period	*Number of Units Sold*	*Dollar Sales*
1	75	\$150
2	25	50
3	130	260

FIGURE 2.1

Example of Functional Relation

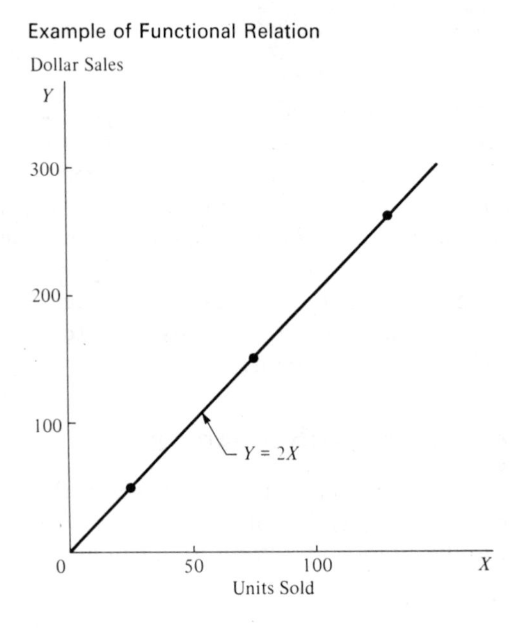

These observations are plotted also in Figure 2.1. Note that all fall directly on the line of functional relationship. This is characteristic of all functional relations.

Example 2. A car rental firm prices the rental of a car as follows:

$$Y = 8 + .15X$$

where:

Y is the rental fee (in dollars)
X is the distance traveled (in miles)
8 is the service charge (in dollars)
.15 is the rate per mile (in dollars)

This functional relation is shown in Figure 2.2. Two recent rentals were:

Distance	*Rental Fee*
20 miles	$11.00
51 miles	15.65

These observations are also plotted in Figure 2.2. Again, the observations fall directly on the line of functional relationship.

FIGURE 2.2

Example of Functional Relation

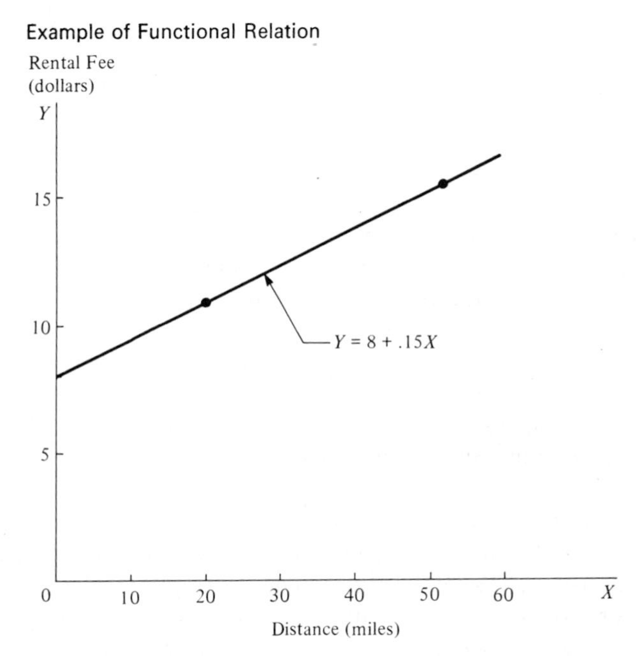

Statistical Relation between Two Variables

A statistical relation, unlike a functional relation, is not a perfect one. In general, the observations for a statistical relation do not fall directly on the curve of relationship.

Example 1. A certain spare part is manufactured by the Westwood Company once a month in lots which vary in size as demand fluctuates. Table 2.1,

Westwood Example

FIGURE 2.3

Statistical Relation between Lot Size and Number of Man-Hours

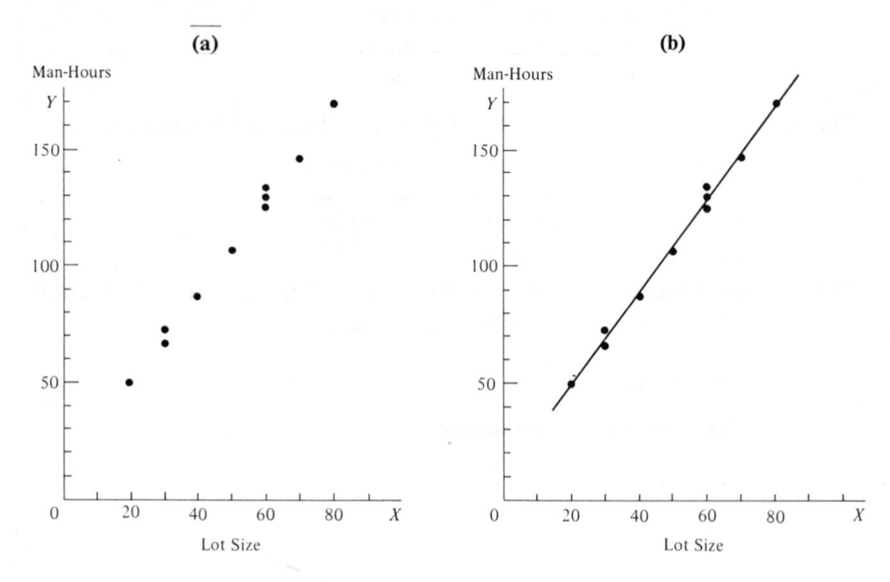

page 35, contains data on lot size and number of man-hours of labor for 10 recent production runs performed under similar production conditions. These data are plotted in Figure 2.3a. Man-hours are taken as the *dependent* or *response* variable Y, and lot size as the *independent* or *predictor* variable X. The plotting is done as for previous graphs. For instance, the first production run results are plotted as $X = 30$, $Y = 73$.

Figure 2.3a clearly suggests that there is a relation between lot size and number of man-hours, in the sense that the larger the lot size, the greater tends to be the number of man-hours. However, the relation is not a perfect one. There is a scattering of points, suggesting that some of the variation in man-hours is not accounted for by lot size. For instance, two production runs (1 and 8) consisted of 30 parts, yet they required somewhat different numbers of man-hours. Because of the scattering of points in a statistical relation, Figure 2.3a is called a *scatter diagram* or *scatter plot*. In statistical terminology, each point in the scatter diagram represents an *observation* or *trial*.

In Figure 2.3b, we have plotted a line of relationship which describes the statistical relation between man-hours and lot size. It indicates the general tendency by which man-hours vary with changes in lot size. Note that most of the points do not fall directly on the line of statistical relationship. This scattering of points around the line represents variation in man-hours which is not associated with the lot size, and which is usually considered to be of a random nature. Statistical relations can be highly useful, even though they do not have the exactitude of a functional relation.

Example 2. Figure 2.4 presents data on age and estimated labor force participation rate of males in industrialized countries in 1960. The data strongly suggest that the statistical relation is *curvilinear* (not linear). The curve of relationship has been drawn in Figure 2.4. It implies that as age becomes increasingly higher, the participation rate increases up to a point and then begins to decline. Note again the scattering of points around the curve of statistical relationship, typical of all statistical relations.

FIGURE 2.4

Curvilinear Statistical Relation between Age and Estimated Labor Force Participation Rate of Males in Industrialized Countries, 1960

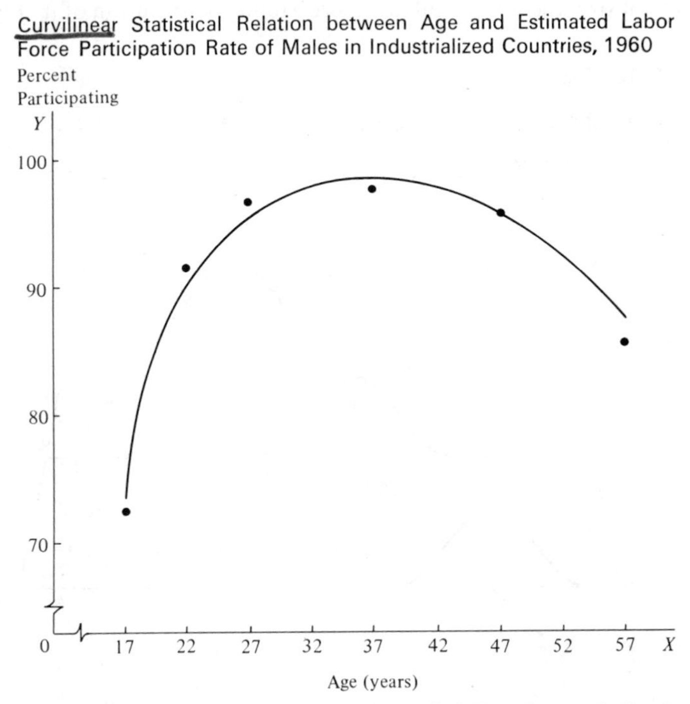

Source: Data adapted from I. Beller, "Latin America's Unemployment Problem," *Monthly Labor Review*, Vol. 93, November 1970, p. 5.

2.2 REGRESSION MODELS AND THEIR USES

Basic Concepts

A regression model is a formal means of expressing the two essential ingredients of a statistical relation:

1. A tendency of the dependent variable Y to vary with the independent variable or variables in a systematic fashion.
2. A scattering of observations around the curve of statistical relationship.

These two characteristics are embodied in a regression model by postulating that:

1. In the population of observations associated with the sampled process, there is a probability distribution of Y for each level of X.
2. The means of these probability distributions vary in some systematic fashion with X.

Example. Consider again the Westwood Company lot size example. The number of man-hours Y is treated in a regression model as a random variable. For each lot size, there is postulated a probability distribution of Y. Figure 2.5

FIGURE 2.5

Pictorial Representation of Linear Regression Model

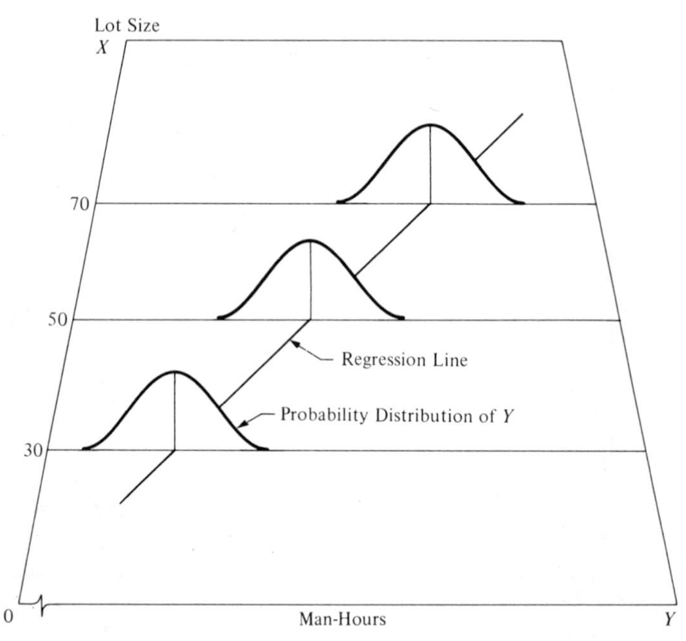

shows such a probability distribution for $X = 30$, which is the lot size for the first production run in Table 2.1. The actual number of man-hours Y (73, in our example in Table 2.1) is then viewed as a random selection from this probability distribution.

Figure 2.5 also shows probability distributions of Y for lot sizes $X = 50$ and $X = 70$. Note that the means of the probability distributions have a systematic relation to the level of X. This systematic relationship is called the *regression function of Y on X*. The graph of the regression function is called the *regression curve*. Note that in Figure 2.5, the regression function is linear. This would imply for our example that the expected (mean) number of man-hours varies linearly with lot size.

There is of course no a priori reason why man-hours need be linearly related to lot size. Figure 2.6 shows another possible regression model for our example. Here the regression function is curvilinear, with a shape reflecting economies of scale with larger lot sizes. Figure 2.6 differs in orientation

FIGURE 2.6

Pictorial Representation of Curvilinear Regression Model

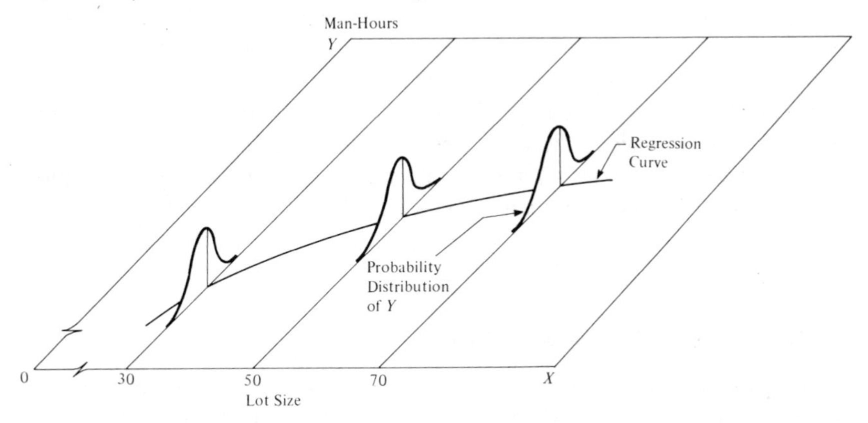

from Figure 2.5 in that the X and Y axes are plotted conventionally in Figure 2.6. While this makes it not quite as easy to view the probability distributions, the orientation of Figure 2.6 shows the regression curve in the perspective to be utilized from here on.

Regression models may differ in the form of the regression function, as in Figures 2.5 and 2.6. They may also differ with respect to the form of the probability distributions of the Y's, and in still other ways. Whatever the variation, the concept of a probability distribution of Y for given X is the formal counterpart to the empirical scatter in a statistical relation. Similarly, the regression curve, which describes the relation between the means of the probability distributions and X, is the counterpart to the general tendency of Y to vary with X systematically in a statistical relation.

Note

The expressions "independent variable" for X and "dependent variable" or "response variable" for Y in a regression model simply are conventional labels. There is no implication that Y causally depends on X in a given case. No matter how strong the statistical relation, no cause-and-effect pattern is necessarily implied by the regression model. In some applications an independent variable actually is dependent causally on the response variable, as when we estimate temperature (the response) from the height of mercury (the independent variable) in a thermometer.

Regression Models with More than One Independent Variable. Regression models may contain more than one independent variable. For example, in one application of regression analysis pertaining to 67 branch offices of a consumer finance chain, the regression model contained direct operating cost for the year just ended as the response variable and four independent variables—average size of loan outstanding during the year, average number of loans outstanding, total number of new loan applications processed, and office salary scale index. In a tractor purchase study the response variable was volume (in horsepower) of tractor purchases in each sales territory of a farm equipment firm. There were nine independent variables, including average age of tractors on farms in the territory, number of farms in the territory, and a quantity index of crop production in the territory.

The features represented in Figures 2.5 and 2.6 must be extended into further dimensions when there is more than one independent variable. With two independent variables X_1 and X_2, for instance, a probability distribution of Y for each (X_1, X_2) combination is assumed by the regression model. The systematic relation between the means of these probability distributions and the independent variables X_1 and X_2 is then given by a regression surface.

Construction of Regression Models

Selection of Independent Variables. Since reality must be reduced to manageable proportions whenever we construct models, only a limited number of independent or predictor variables can—or should—be included in the regression model for any situation of interest. A central problem therefore is that of choosing, for the regression model, a set of independent variables which is "best" in some sense for the purposes of the analysis. A major consideration in making this choice is the extent to which a chosen variable contributes to reducing the remaining variation in Y after allowance is made for the contributions of other independent variables which have tentatively been included in the regression model. Other considerations include the importance of the variable as a causal agent in the process under analysis; the degree to which observations on the variable can be obtained more accurately, or quickly, or economically than on competing variables; and the degree to which the variable can be preset by management. In Chapter 11, we shall discuss procedures and problems in choosing the independent variables to be included in a regression model.

Functional Form of Regression Equation. The choice of the functional form of the regression equation is related to the choice of the independent variables. Sometimes, relevant theory may indicate the appropriate functional form of the regression equation. Learning theory, for instance, may indicate that the regression function relating unit production costs to the number of previous times the item has been produced should have a specified shape with particular asymptotic properties.

More frequently, however, the functional form of the regression equation is not known in advance and must be decided upon once the data have been collected and analyzed. Thus, linear or quadratic regression functions are often used as satisfactory first approximations to regression functions of unknown nature. Indeed, these simple types of regression functions may be used even when theory provides the relevant functional form, notably when the known form is highly complex but can be reasonably approximated by a linear or quadratic regression function. Figure 2.7a illustrates a case

FIGURE 2.7

Uses of Linear Regression Function to Approximate Complex Regression Functions

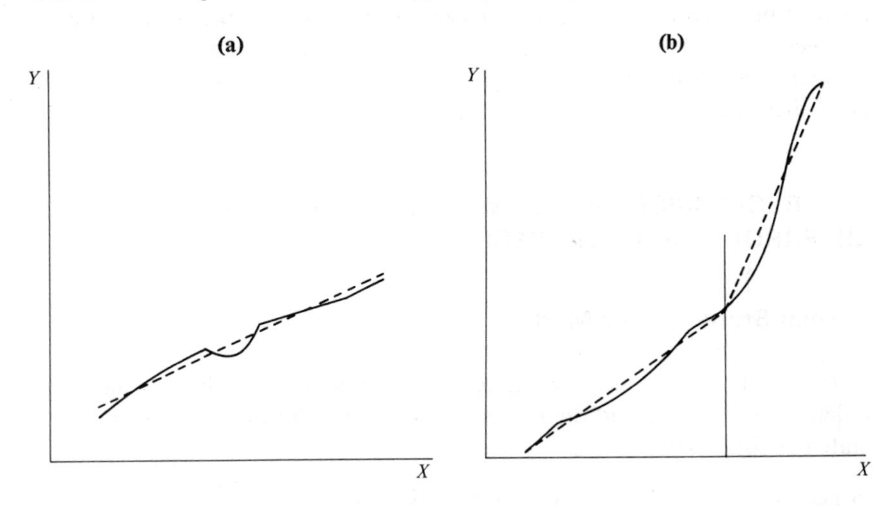

where a complex regression function may be reasonably approximated by a linear regression function. Figure 2.7b provides an example where two linear regression functions may be used "piecewise" to approximate a complex regression function.

Scope of Model. In formulating a regression model, we usually need to restrict the coverage of the model to some interval or region of values of the independent variable or variables. The scope is determined either by the design of the investigation or by the range of data at hand. For instance, a company studying the effect of price on sales volume investigated six price levels, ranging from \$4.95 to \$6.95. Here, the scope of the model would be limited to price levels ranging from near \$5 to near \$7. The shape of the regression function would be in serious doubt substantially outside this range because the investigation provided no evidence as to the nature of the statistical relation below \$4.95 or above \$6.95.

Uses of Regression Analysis

Regression analysis serves three major purposes: (1) description, (2) control, and (3) prediction. The tractor purchase study, cited earlier, served a descriptive purpose. In the study of branch office operating costs, also mentioned above, the purpose was administrative control; management, by developing a usable statistical relation between costs and independent variables in the system, was able to set cost standards for each branch office in the company chain. Many examples of the use of regression analysis for prediction are found in business, such as for estimating costs and forecasting sales.

The several purposes of regression analysis frequently overlap in practice. The situation in the Westwood Company lot size example provides a case in point. Knowledge of the relation between lot size and man-hours in past production runs enables management to predict the man-hour requirements for the next production run of given lot size, for purposes of cost estimation and production scheduling. After the run is completed, management may wish to compare the actual man-hours against the predicted hours for purposes of administrative control.

2.3 REGRESSION MODEL WITH DISTRIBUTION OF ERROR TERMS UNSPECIFIED

Formal Statement of Model

In Part I, we consider a basic regression model where there is only one independent variable and the regression function is linear. The model can be stated as follows:

$$Y_i = \beta_0 + \beta_1 X_i + \varepsilon_i \tag{2.1}$$

where:

- Y_i is the value of the response variable in the ith trial
- β_0 and β_1 are parameters
- X_i is a known constant, namely the value of the independent variable in the ith trial
- ε_i is a random error term with mean $E(\varepsilon_i) = 0$ and variance $\sigma^2(\varepsilon_i) = \sigma^2$; ε_i and ε_j are uncorrelated, so that the covariance $\sigma(\varepsilon_i, \varepsilon_j) = 0$ for all i, j; $i \neq j$
- $i = 1, \ldots, n$

Model (2.1) is said to be *simple*, *linear in the parameters*, and *linear in the independent variable*. It is "simple" in that there is only one independent variable, "linear in the parameters" because no parameter appears as an exponent or is multiplied or divided by another parameter, and "linear in the independent variable" because this variable appears only in the first power. A model which is linear in the parameters and the independent variable is also called a *first-order model*.

Important Features of Model

1. The observed value of Y in the ith trial is the sum of two components: (1) the constant term $\beta_0 + \beta_1 X_i$, and (2) the random term ε_i. Hence, Y_i is a random variable.

2. Since $E(\varepsilon_i) = 0$, it follows that:

$$E(Y_i) = E(\beta_0 + \beta_1 X_i + \varepsilon_i) = \beta_0 + \beta_1 X_i + E(\varepsilon_i) = \beta_0 + \beta_1 X_i$$

Thus, the response variable Y_i, when the level of X existing in the ith trial is X_i, comes from a probability distribution whose mean is:

$$E(Y_i) = \beta_0 + \beta_1 X_i \tag{2.2}$$

We therefore know that the regression function for model (2.1) is:

$$E(Y) = \beta_0 + \beta_1 X \tag{2.3}$$

since the regression function relates the means of the probability distributions of Y for any given X to the level of X.

3. The observed value of Y in the ith trial exceeds or falls short of the value of the regression function by the error term amount ε_i.

4. The error terms ε_i are assumed to have constant variance σ^2. It therefore follows that the variance of the response variable Y_i is:

$$\sigma^2(Y_i) = \sigma^2 \tag{2.4}$$

since, using theorem (1.15a), we have:

$$\sigma^2(\beta_0 + \beta_1 X_i + \varepsilon_i) = \sigma^2(\varepsilon_i) = \sigma^2$$

Note that $\beta_0 + \beta_1 X_i$ plays the role of the constant a in theorem (1.15a).

Thus, model (2.1) assumes that the probability distributions of Y have the same variance σ^2, regardless of the level of the independent variable X.

5. The error terms are assumed to be uncorrelated. Hence the outcome in any one trial has no effect on the error term for any other trial—as to whether it is positive or negative, small or large, and the like. Since the error terms ε_i and ε_j are uncorrelated, so are the response variables Y_i and Y_j.

6. In summary, model (2.1) implies that the response variable observations Y_i come from probability distributions whose means are $E(Y_i) = \beta_0 + \beta_1 X_i$ and whose variances are σ^2, the same for all levels of X. Further, any two observations Y_i and Y_j are uncorrelated.

Example

Suppose that regression model (2.1) is applicable for the Westwood Company lot size application and is as follows:

$$Y_i = 9.5 + 2.1 X_i + \varepsilon_i$$

Figure 2.8 contains a presentation of the regression function:

$$E(Y) = 9.5 + 2.1X$$

Suppose that in the ith trial a lot of $X_i = 45$ units is produced and the actual number of man-hours is $Y_i = 108$. In that case, the error term value is $\varepsilon_i = +4$, for we have:

$$E(Y_i) = 9.5 + 2.1(45) = 104$$

and:

$$108 = 104 + 4$$

Figure 2.8 displays the probability distribution of Y when $X_i = 45$, and indicates from where in this distribution the observation $Y_i = 108$ came. Note again that the error term ε_i is simply the deviation of Y_i from its mean value $E(Y_i)$.

FIGURE 2.8

Illustration of Linear Regression Model (2.1)

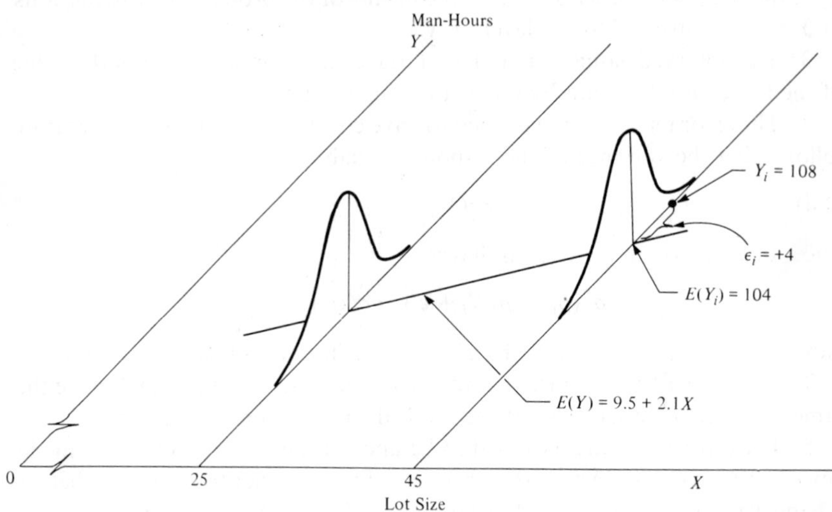

Figure 2.8 also shows the probability distribution of Y when $X = 25$. Note that this distribution exhibits the same variability as the probability distribution when $X = 45$, in conformance with the requirements of model (2.1).

Meaning of Regression Parameters

The parameters β_0 and β_1 in regression model (2.1) are called regression coefficients. β_1 is the slope of the regression line. It indicates the change in the mean of the probability distribution of Y per unit increase in X. The parameter β_0 is the Y intercept of the regression line. If the scope of the model includes $X = 0$, β_0 gives the mean of the probability distribution of Y at $X = 0$. When the scope of the model does not cover $X = 0$, β_0 does not have any particular meaning as a separate term in the regression model.

Example. Figure 2.9 shows the regression function:

$$E(Y) = 9.5 + 2.1X$$

for the previous Westwood Company lot size example. The slope $\beta_1 = 2.1$ indicates that an increase of one unit in lot size leads to an increase in the mean of the probability distribution of Y of 2.1 man-hours.

The intercept $\beta_0 = 9.5$ indicates the value of the regression function at $X = 0$. However, since the linear regression model was formulated to apply to lot sizes ranging from 20 to 80 units, β_0 does not have any intrinsic meaning of its own. In particular, it does not necessarily indicate the average setup time for the process (the average man-hours before actual output begins). A model with a curvilinear regression function, and some different value of β_0 than that in the linear model, might well be required if the scope of the model were to extend to lot sizes down to zero.

FIGURE 2.9

Meaning of Linear Regression Parameters

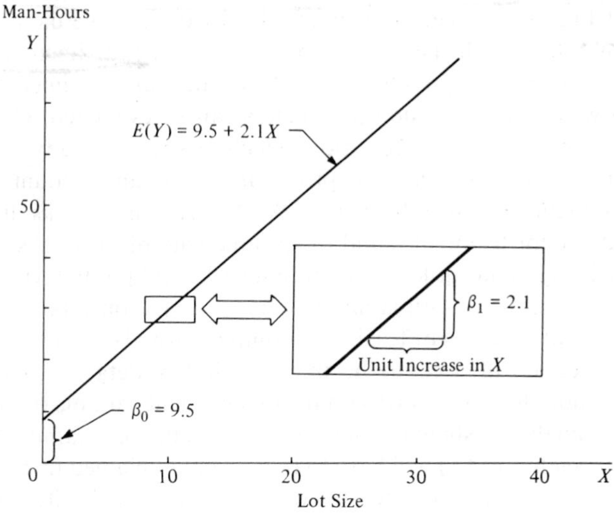

Alternative Versions of Model

Sometimes it is convenient to write model (2.1) in somewhat different, though equivalent, forms. Let X_0 be a *dummy variable* identically equal to one. Then, we can write (2.1) as follows:

(2.5) $$Y_i = \beta_0 X_0 + \beta_1 X_i + \varepsilon_i \qquad \text{where } X_0 \equiv 1$$

Another modification sometimes helpful is to use for the independent variable the deviation $(X_i - \bar{X})$ rather than X_i. To leave model (2.1) unchanged, we need to write:

$$\begin{aligned} Y_i &= \beta_0 + \beta_1(X_i - \bar{X}) + \beta_1\bar{X} + \varepsilon_i \\ &= (\beta_0 + \beta_1\bar{X}) + \beta_1(X_i - \bar{X}) + \varepsilon_i \\ &= \beta_0^* + \beta_1(X_i - \bar{X}) + \varepsilon_i \end{aligned}$$

Thus, an alternative model version is:

(2.6) $$Y_i = \beta_0^* + \beta_1(X_i - \bar{X}) + \varepsilon_i$$

where:

(2.6a) $$\beta_0^* = \beta_0 + \beta_1\bar{X}$$

We shall use models (2.1), (2.5), and (2.6) interchangeably as convenience dictates.

2.4 ESTIMATION OF REGRESSION FUNCTION

Obtaining Needed Sample Data

Ordinarily, we do not know the values of the regression parameters β_0 and β_1 in model (2.1) and need to estimate them from sample data. Two basic methods of obtaining the needed sample data are by experimentation and by survey. Sometimes, it is possible to conduct a formal experiment to provide data from which the regression parameters can be estimated. Consider, for instance, a health care organization which wishes to study the relation between productivity of its clerks in processing claims and amount of training. Five clerks might be trained for two weeks, five for three weeks, five for four weeks, and five for five weeks, and the productivity of the clerks would then be observed. The data on length of training (X) and productivity (Y) would then serve as a basis for estimating the regression parameters.

Often it is not practical or feasible to conduct formal experiments, in which case survey data will need to be utilized. Survey data are data that are obtained without formal controls on the independent variable of interest. For example, an analyst wishing to study the relation between family income (X) and food expenditures (Y) will have to rely on data obtained from a survey of families, since he would lack the controls needed to fix a family's income at a specified level. Similarly, the Westwood Company in our earlier lot size example needed to rely on survey data since the lot size at any given time was dictated by the demand for the product, which was not under the control of the company.

Once the data have been obtained, either by experiment or survey, they can be assembled in a table such as Table 2.1 for the Westwood Company example. We shall denote the (X, Y) observations for the first trial as (X_1, Y_1), for the second trial (X_2, Y_2), and in general for the ith trial (X_i, Y_i) where

TABLE 2.1

Data on Lot Size and Number of Man-Hours, Westwood Company Example

Production Run i	*Lot Size* X_i	*Man-Hours* Y_i
1	30	73
2	20	50
3	60	128
4	80	170
5	40	87
6	50	108
7	60	135
8	30	69
9	70	148
10	60	132

$i = 1, \ldots, n$. For the data in Table 2.1, $X_1 = 30$, $Y_1 = 73$, and so on, and $n = 10$.

Method of Least Squares

To find "good" estimators of the regression parameters β_0 and β_1, we shall employ the method of least squares. For each sample observation (X_i, Y_i), the method of least squares considers the deviation of Y_i from its expected value:

$$Y_i - (\beta_0 + \beta_1 X_i) = \varepsilon_i \tag{2.7}$$

In particular, the method of least squares requires that we consider the sum of the n squared deviations, denoted by Q:

$$Q = \sum_{i=1}^{n} (Y_i - \beta_0 - \beta_1 X_i)^2 \tag{2.8}$$

According to the method of least squares, the estimators of β_0 and β_1 are those values b_0 and b_1 respectively which minimize Q.

Example. Figure 2.10a contains the sample data of Table 2.1 for the Westwood Company example. In Figure 2.10b is plotted a regression line using the arbitrary estimates:

$$b_0 = 30$$

$$b_1 = 0$$

Also shown in Figure 2.10b are the deviations $Y_i - 30 - (0)X_i$. Note that each deviation corresponds to the vertical distance between Y_i and the fitted

FIGURE 2.10

Example of Deviations from Fitted Regression Lines

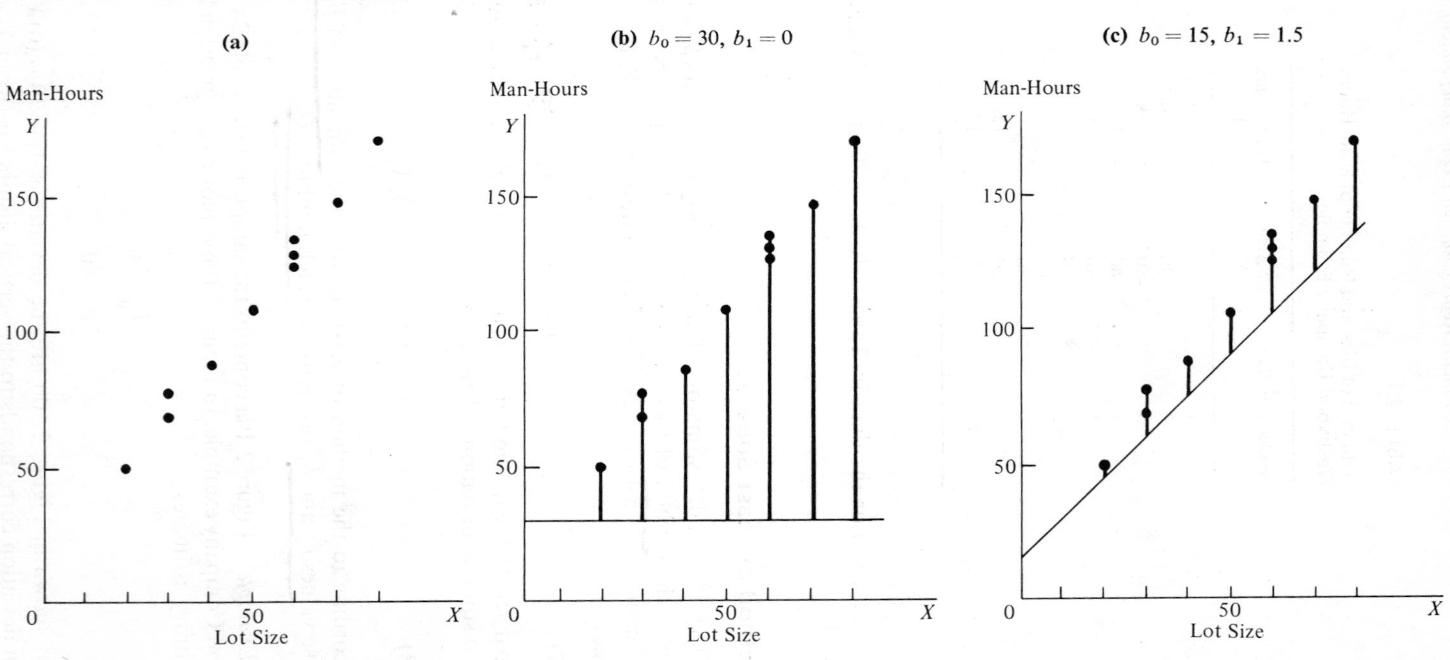

regression line. Clearly, the fit is poor, and hence the deviations are large and so are the squared deviations and their sum Q.

Figure 2.10c contains the deviations $Y_i - b_0 - b_1 X_i$ for the estimates $b_0 = 15$, $b_1 = 1.5$. Here, the fit is better (though still not good), the deviations are much smaller, and hence so is Q, the sum of the squared deviations. Thus, a better fit of the regression line to the data corresponds to a smaller sum Q.

The objective of the method of least squares is to find those numbers b_0 and b_1 for β_0 and β_1 respectively for which Q is a minimum. In a certain sense, to be discussed shortly, these numbers will provide a "good" fit of the linear regression function.

Least Squares Estimators. The estimators b_0 and b_1 which satisfy the least squares criterion could be found by a trial and error procedure. However, this is not necessary since it can be shown that the values b_0 and b_1 which minimize Q for any particular set of sample data are given by the following simultaneous equations:

$$\sum Y_i = nb_0 + b_1 \sum X_i \tag{2.9a}$$

$$\sum X_i Y_i = b_0 \sum X_i + b_1 \sum X_i^2 \tag{2.9b}$$

"Normal Equations"

The equations (2.9a) and (2.9b) are called *normal equations*; b_0 and b_1 are called *point estimators* of β_0 and β_1 respectively.

The quantities $\sum Y_i$, $\sum X_i$, and so on in (2.9) are calculated from the sample observations (X_i, Y_i). The equations then can be solved simultaneously for b_0 and b_1. Alternatively, b_0 and b_1 can be obtained directly as follows:

$$b_1 = \frac{\sum X_i Y_i - \frac{(\sum X_i)(\sum Y_i)}{n}}{\sum X_i^2 - \frac{(\sum X_i)^2}{n}} = \frac{\sum (X_i - \bar{X})(Y_i - \bar{Y})}{\sum (X_i - \bar{X})^2} \tag{2.10a}$$

$$b_0 = \frac{1}{n}\left(\sum Y_i - b_1 \sum X_i\right) = \bar{Y} - b_1 \bar{X} \tag{2.10b}$$

where $\bar{X}$ and $\bar{Y}$ have the usual meaning.

Note

The normal equations (2.9) can be derived by calculus. For given sample observations (X_i, Y_i), the quantity Q in (2.8) is a function of β_0 and β_1. The values of β_0 and β_1 which minimize Q can be derived by differentiating (2.8) with respect to β_0 and β_1. We obtain:

$$\frac{\partial Q}{\partial \beta_0} = -2 \sum (Y_i - \beta_0 - \beta_1 X_i)$$

$$\frac{\partial Q}{\partial \beta_1} = -2 \sum X_i (Y_i - \beta_0 - \beta_1 X_i)$$

We then set these partial derivatives equal to zero, using b_0 and b_1 to denote the particular values of β_0 and β_1 respectively which minimize Q:

$$-2 \sum (Y_i - b_0 - b_1 X_i) = 0$$
$$-2 \sum X_i(Y_i - b_0 - b_1 X_i) = 0$$

Simplifying, we obtain:

$$\sum_{i=1}^{n} (Y_i - b_0 - b_1 X_i) = 0$$

$$\sum_{i=1}^{n} X_i(Y_i - b_0 - b_1 X_i) = 0$$

Expanding out, we have:

$$\sum Y_i - nb_0 - b_1 \sum X_i = 0$$
$$\sum X_i Y_i - b_0 \sum X_i - b_1 \sum X_i^2 = 0$$

from which the normal equations (2.9) are obtained by rearranging terms.

Properties of Least Squares Estimators. An important theorem, called the *Gauss-Markov theorem*, states:

(2.11) Under the conditions of model (2.1), the least squares estimators b_0 and b_1 in (2.10) are unbiased and have minimum variance among all unbiased linear estimators.

This theorem states first that both b_0 and b_1 are unbiased estimators. Hence:

$$E(b_0) = \beta_0$$
$$E(b_1) = \beta_1$$

so that neither estimator tends to overestimate or underestimate systematically.

Second, the theorem states that the sampling distributions of b_0 and b_1 have smaller variability than those of any other estimators belonging to a particular class of estimators. Thus, the least squares estimators are more precise than any of these other estimators. The class of estimators for which the least squares estimators are "best" consists of all unbiased estimators which are linear functions of the observations $Y_1, \ldots, Y_n$. The estimators b_0 and b_1 are such linear functions of the Y's. Consider, for instance, b_1. We have from (2.10a):

$$b_1 = \frac{\sum (X_i - \bar{X})(Y_i - \bar{Y})}{\sum (X_i - \bar{X})^2}$$

It will be shown in (3.5) that this expression is equal to:

$$b_1 = \frac{\sum (X_i - \bar{X}) Y_i}{\sum (X_i - \bar{X})^2} = \sum k_i Y_i$$

where:

$$k_i = \frac{X_i - \bar{X}}{\sum (X_i - \bar{X})^2}$$

Since the k_i are known constants (because the X_i are known constants), b_1 is a linear combination of the Y_i and hence is a linear estimator.

In the same fashion, it can be shown that b_0 is a linear estimator.

Among all linear estimators which are unbiased then, b_0 and b_1 have the smallest variability in repeated samples in which the X levels remain unchanged.

Example. To illustrate the calculation of the least squares estimators b_0 and b_1, we will use the Westwood Company case discussed earlier. The sample data are given in Table 2.1 and plotted in Figure 2.10a. Table 2.2 gives the

TABLE 2.2

Basic Calculations to Obtain b_0 and b_1 for Westwood Company Example

	Y_i	X_i	$X_i Y_i$	X_i^2	(*for later use*) Y_i^2
	73	30	2,190	900	5,329
	50	20	1,000	400	2,500
	128	60	7,680	3,600	16,384
	170	80	13,600	6,400	28,900
	87	40	3,480	1,600	7,569
	108	50	5,400	2,500	11,664
	135	60	8,100	3,600	18,225
	69	30	2,070	900	4,761
	148	70	10,360	4,900	21,904
	132	60	7,920	3,600	17,424
Total	1,100	500	61,800	28,400	134,660

basic results required to calculate b_0 and b_1. We have: $\sum Y_i = 1{,}100$, $\sum X_i = 500$, $\sum X_i Y_i = 61{,}800$, $\sum X_i^2 = 28{,}400$, and $n = 10$. Using (2.10) we obtain:

$$b_1 = \frac{\sum X_i Y_i - \dfrac{(\sum X_i)(\sum Y_i)}{n}}{\sum X_i^2 - \dfrac{(\sum X_i)^2}{n}} = \frac{61{,}800 - \dfrac{(500)(1{,}100)}{10}}{28{,}400 - \dfrac{(500)^2}{10}} = 2.0$$

$$b_0 = \frac{1}{n}(\sum Y_i - b_1 \sum X_i) = \frac{1}{10}[1{,}100 - (2.0)(500)] = 10.0$$

Thus, we estimate that the mean number of man-hours increases by 2.0 hours for each unit increase in lot size.

Point Estimation of Mean Response

Estimated Regression Function. Given sample estimators b_0 and b_1 of the parameters in the regression function (2.3):

$$E(Y) = \beta_0 + \beta_1 X$$

it is natural that we estimate the regression function as follows:

$$\hat{Y} = b_0 + b_1 X \tag{2.12}$$

where $\hat{Y}$ (read Y hat) is the value of the estimated regression function at the level X of the independent variable.

We will call a *value* of the response variable a *response* and will call $E(Y)$ the *mean response*. Thus, the mean response is the mean of the probability distribution of Y corresponding to the level X of the independent variable. $\hat{Y}$ then is a point estimator of the mean response when the level of the independent variable is X. It can be shown as an extension of the Gauss-Markov theorem (2.11) that $\hat{Y}$ is an unbiased estimator of $E(Y)$, with minimum variance in the class of unbiased linear estimators.

For the observations in the sample, we will call $\hat{Y}_i$:

$$\hat{Y}_i = b_0 + b_1 X_i \tag{2.13}$$

the *fitted value* at the level X for the ith observation. Thus, the fitted value $\hat{Y}_i$ is to be viewed in distinction to the *observed value* Y_i.

Example. For the Westwood Company case, we found that the least squares estimates of the regression coefficients were:

$$b_0 = 10.0 \qquad b_1 = 2.0$$

Hence, the estimated regression equation is:

$$\hat{Y} = 10.0 + 2.0X$$

If we are interested in the mean number of man-hours when the lot size is $X = 55$, our point estimate would be:

$$\hat{Y} = 10.0 + 2.0(55) = 120$$

Thus, we would estimate that the mean number of man-hours for production runs of $X = 55$ units is 120. We interpret this to mean that if many runs of size 55 are produced under the conditions of the 10 runs in the sample, the mean labor time for these many runs is about 120 hours. Of course, the labor time for any one run of size 55 is likely to fall above or below the mean response because of inherent variability in the system, as represented by the error term in the model.

Figure 2.11 contains a plot of the estimated regression function $\hat{Y} = 10.0 + 2.0X$, as well as the original data.

FIGURE 2.11

Least Squares Regression Line and Residuals for Westwood Company Example (observed values and residuals not plotted to scale)

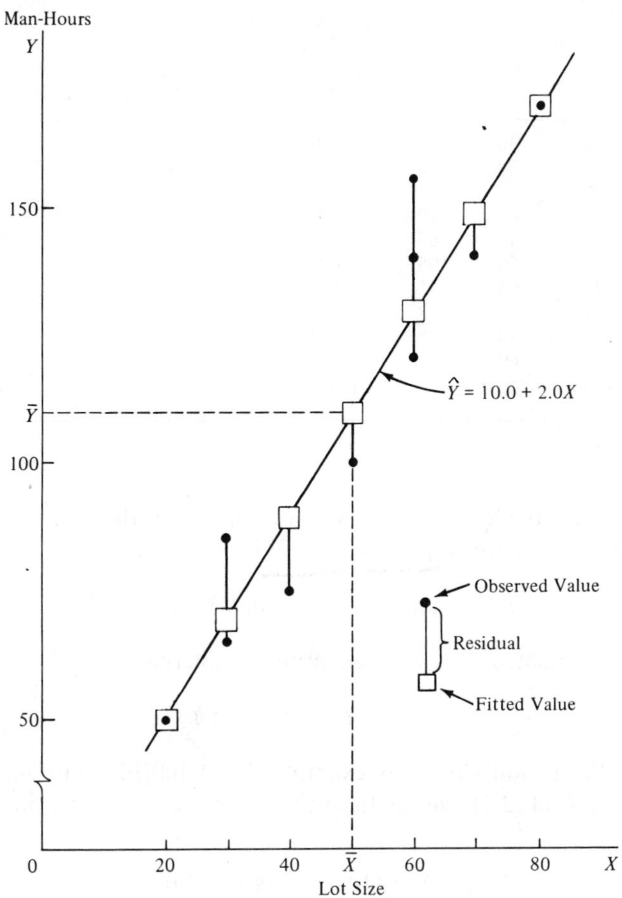

Fitted values for the sample data are obtained by substituting the X values in the sample into the estimated regression equation. For example, for our sample data $X_1 = 30$. Hence the fitted value is:

$$\hat{Y}_1 = 10.0 + 2.0(30) = 70$$

This compares with the actual man-hours of $Y_1 = 73$. Table 2.3 contains all the observed and fitted values for the Westwood Company data. The fitted values are also shown in Figure 2.11.

***Alternate Model* (2.6).** If the alternate regression model (2.6):

$$Y_i = \beta_0^* + \beta_1(X_i - \bar{X}) + \varepsilon_i$$

TABLE 2.3

Fitted Values, Residuals, and Squared Residuals for Westwood Company Example

Observation Number i	*Lot Size* X_i	*Man-Hours* Y_i	*Estimated Mean Response* $\hat{Y}_i$	*Residual* $(Y_i - \hat{Y}_i) = e_i$	*Squared Residual* $(Y_i - \hat{Y}_i)^2 = e_i^2$
1	30	73	70	+3	9
2	20	50	50	0	0
3	60	128	130	−2	4
4	80	170	170	0	0
5	40	87	90	−3	9
6	50	108	110	−2	4
7	60	135	130	+5	25
8	30	69	70	−1	1
9	70	148	150	−2	4
10	60	132	130	+2	4
Total	500	1,100	1,100	0	60

is to be utilized, the least squares estimator b_1 of β_1 is the same as before. The least squares estimator of $\beta_0^* = \beta_0 + \beta_1 \bar{X}$ is, using (2.10b):

$$b_0^* = b_0 + b_1 \bar{X} = (\bar{Y} - b_1 \bar{X}) + b_1 \bar{X} = \bar{Y} \tag{2.14}$$

Hence, the estimated regression equation for alternate model (2.6) is:

$$\hat{Y} = \bar{Y} + b_1(X - \bar{X}) \tag{2.15}$$

In our Westwood Company example, $\bar{Y} = 1{,}100/10 = 110$ and $\bar{X} = 500/10 = 50$ (see Table 2.2). Hence, the estimated regression equation in alternate form is:

$$\hat{Y} = 110.0 + 2.0(X - 50)$$

For our sample data, $X_1 = 30$; hence we estimate the mean response to be:

$$\hat{Y}_1 = 110.0 + 2.0(30 - 50) = 70$$

which, of course, is identical to our earlier result.

Residuals

The ith residual is the difference between the observed value Y_i and the corresponding fitted value $\hat{Y}_i$. Denoting this residual by e_i, we can write:

$$e_i = Y_i - \hat{Y}_i = Y_i - b_0 - b_1 X_i \tag{2.16}$$

Figure 2.11 shows the 10 residuals for the Westwood Company example. The magnitudes of the residuals are shown by the vertical lines between an observation and the fitted value on the estimated regression line. The residuals are calculated in Table 2.3.

We need to distinguish between the model error term value $\varepsilon_i = Y_i - E(Y_i)$ and the residual $e_i = Y_i - \hat{Y}_i$. The former involves the vertical deviation of Y_i from the unknown population regression line, and hence is unknown. On the other hand, the residual is the observed vertical deviation of Y_i from the fitted regression line.

Residuals are highly useful for studying whether a given regression model is appropriate for the data at hand. We shall discuss this use in Chapter 4.

Properties of Fitted Regression Line

The regression line fitted by the method of least squares has a number of properties worth noting:

1. The sum of the residuals is zero:

$$\sum_{i=1}^{n} e_i = 0 \tag{2.17}$$

This property can be proven easily. We have:

$$\begin{aligned}\sum e_i &= \sum (Y_i - b_0 - b_1 X_i) \\ &= \sum Y_i - nb_0 - b_1 \sum X_i = 0\end{aligned}$$

by the first normal equation (2.9a). Table 2.3 illustrates this property for our earlier example. Rounding errors may, of course, be present in any particular case.

2. The sum of the squared residuals, $\sum e_i^2$, is a minimum. This was the requirement to be satisfied in deriving the least squares estimators of the regression parameters.

3. The sum of the observed values Y_i equals the sum of the fitted values $\hat{Y}_i$:

$$\sum_{i=1}^{n} Y_i = \sum_{i=1}^{n} \hat{Y}_i \tag{2.18}$$

This condition is implicit in the first normal equation (2.9a):

$$\begin{aligned}\sum Y_i &= nb_0 + b_1 \sum X_i \\ &= \sum b_0 + \sum b_1 X_i = \sum (b_0 + b_1 X_i) = \sum \hat{Y}_i\end{aligned}$$

It follows from (2.18) that the mean of the $\hat{Y}_i$ is the same as the mean of the Y_i, namely $\bar{Y}$.

4. The sum of the weighted residuals is zero when the residual in the *i*th trial is weighted by the level of the independent variable in the *i*th trial:

$$\sum_{i=1}^{n} X_i e_i = 0 \tag{2.19}$$

This follows from the second normal equation (2.9b):

$$\begin{aligned}\sum X_i e_i &= \sum X_i(Y_i - b_0 - b_1 X_i)\\ &= \sum X_i Y_i - b_0 \sum X_i - b_1 \sum X_i^2 = 0\end{aligned}$$

5. The sum of the weighted residuals is zero when the residual in the ith trial is weighted by the fitted value of the response variable for the ith trial:

$$\sum_{i=1}^{n} \hat{Y}_i e_i = 0 \qquad (2.20)$$

The proof makes use of both normal equations.

6. The regression line always goes through the point $(\bar{X}, \bar{Y})$. This can be readily seen from the alternate form of the estimated regression line in (2.15). If $X = \bar{X}$, we have:

$$\hat{Y} = \bar{Y} + b_1(X - \bar{X}) = \bar{Y} + b_1(\bar{X} - \bar{X}) = \bar{Y}$$

Figure 2.11 demonstrates this property for our lot size example.

2.5 ESTIMATION OF ERROR TERMS VARIANCE σ^2

The variance σ^2 of the error terms ε_i in the regression model (2.1) needs to be estimated for a variety of purposes. Frequently, we would like to obtain an indication of the variability of the probability distributions of Y. Further, as we shall see in the next chapter, a variety of inferences concerning the regression function and the prediction of Y require an estimate of σ^2.

Point Estimator of σ^2

Single Population. To lay the basis for developing an estimator of σ^2 for the regression model (2.1), let us consider for a moment the simpler problem of sampling from a single population. In obtaining the sample variance s^2, we begin by considering the deviation of an observation Y_i from the estimated mean $\bar{Y}$, squaring it, and then summing all such deviations:

$$\sum_{i=1}^{n} (Y_i - \bar{Y})^2$$

Such a sum is called a *sum of squares.* The sum of squares is then divided by the degrees of freedom associated with it. This number is $n - 1$ here, because one degree of freedom is lost by using the estimate $\bar{Y}$ instead of the population mean μ. The resulting estimator is the usual sample variance:

$$s^2 = \frac{\sum_{i=1}^{n} (Y_i - \bar{Y})^2}{n - 1}$$

which is an unbiased estimator of the variance σ^2 of an infinite population. The sample variance is often called a *mean square,* because a sum of squares has been divided by the appropriate number of degrees of freedom.

Regression Model. The logic of developing an estimator of σ^2 for the regression model is the same as when sampling a single population. Recall in this connection from (2.4) that the variance of the observations Y_i is also σ^2, the same as that of the error terms ε_i. We again need to calculate a sum of squared deviations, but must recognize that the Y_i come from different probability distributions with different means, depending upon the level X_i. Thus, the deviation of an observation Y_i must be calculated around its own estimated mean $\hat{Y}_i$. Hence, the deviations are the residuals:

$$Y_i - \hat{Y}_i = e_i$$

and the appropriate sum of squares, denoted by *SSE*, is:

$$SSE = \sum_{i=1}^{n} (Y_i - \hat{Y}_i)^2 = \sum_{i=1}^{n} (Y_i - b_0 - b_1 X_i)^2 \tag{2.21}$$

where *SSE* stands for *error sum of squares* or *residual sum of squares*.

The sum of squares *SSE* has $n - 2$ degrees of freedom associated with it. Two degrees of freedom are lost because both β_0 and β_1 had to be estimated in obtaining $\hat{Y}_i$. Hence, the appropriate mean square, denoted by *MSE*, is:

$$s_e^2 = MSE = \frac{SSE}{n-2} = \frac{\sum (Y_i - \hat{Y}_i)^2}{n-2} = \frac{\sum (Y_i - b_0 - b_1 X_i)^2}{n-2} = \frac{\sum e_i^2}{n-2} \tag{2.22}$$

where *MSE* stands for *error mean square* or *residual mean square*.

It can be shown that *MSE* is an unbiased estimator of σ^2 for the regression model (2.1):

$$E(MSE) = \sigma^2 \tag{2.23}$$

An estimator of the standard deviation σ is simply the positive square root of *MSE*.

Alternative Computational Formulas

There are a number of alternative computational formulas for *SSE*. Three of these are as follows:

$$SSE = \sum Y_i^2 - b_0 \sum Y_i - b_1 \sum X_i Y_i \tag{2.24a}$$

$$SSE = \sum (Y_i - \bar{Y})^2 - \frac{[\sum (X_i - \bar{X})(Y_i - \bar{Y})]^2}{\sum (X_i - \bar{X})^2} \tag{2.24b}$$

$$SSE = \left[\sum Y_i^2 - \frac{(\sum Y_i)^2}{n} \right] - \frac{\left[\sum X_i Y_i - \frac{\sum X_i \sum Y_i}{n} \right]^2}{\sum X_i^2 - \frac{(\sum X_i)^2}{n}} \tag{2.24c}$$

Comments

1. Formula (2.24a) is useful if b_0 and b_1 have already been calculated. Otherwise, (2.24b) and (2.24c) are more direct.

2. In (2.24a), the estimates b_0 and b_1 should be carried to a large number of significant digits in order to yield reliable results for *SSE*.

3. To obtain (2.24a), recall that by (2.21) we have:

$$SSE = \sum (Y_i - b_0 - b_1 X_i)^2$$

Thus:

$$\begin{aligned} SSE &= \sum Y_i^2 - 2b_0 \sum Y_i - 2b_1 \sum X_i Y_i + nb_0^2 + 2b_0 b_1 \sum X_i + b_1^2 \sum X_i^2 \\ &= \sum Y_i^2 - 2b_0 \sum Y_i - 2b_1 \sum X_i Y_i + b_0(nb_0 + b_1 \sum X_i) \\ &\qquad + b_1(b_0 \sum X_i + b_1 \sum X_i^2) \end{aligned}$$

The expressions in parentheses are equal to $\sum Y_i$ and $\sum X_i Y_i$ respectively by the normal equations (2.9). Substituting these terms within the parentheses yields an expression which reduces directly to (2.24a).

4. None of the three alternative formulas explicitly provides the residuals e_i. As noted earlier, the residuals are useful for studying the appropriateness of the model.

Example

Returning to our Westwood Company lot size example, we will calculate *SSE* by (2.21). The residuals were obtained earlier, in Table 2.3. This table also shows the squared residuals. From these results, we obtain:

$$SSE = 60$$

Since $10 - 2 = 8$ degrees of freedom are associated with *SSE*, we find:

$$MSE = \frac{60}{8} = 7.5$$

Finally, a point estimate of σ, the standard deviation of the probability distribution of Y for any X, is $\sqrt{7.5} = 2.74$ man-hours.

Consider again the case where the lot size is $X = 55$ units. We estimated earlier that the probability distribution of Y for this lot size has a mean of 120 man-hours. Now, we have the additional information that the standard deviation of this distribution is estimated to be 2.74 man-hours.

If we wished to use, say, (2.24a) for calculating *SSE*, we would need $\sum Y_i^2$. This sum is calculated in Table 2.2. We then obtain, using the results in Table 2.2 and the estimates $b_0 = 10.0$, $b_1 = 2.0$:

$$\begin{aligned} SSE &= \sum Y_i^2 - b_0 \sum Y_i - b_1 \sum X_i Y_i \\ &= 134{,}660 - 10.0(1{,}100) - 2.0(61{,}800) = 60 \end{aligned}$$

which is, of course, the same result (except sometimes for rounding errors) as obtained earlier.

2.6 NORMAL ERROR REGRESSION MODEL

No matter what may be the functional form of the distribution of ε_i (and hence of Y_i), the least squares method provides unbiased point estimators of β_0 and β_1 which have minimum variance among all unbiased linear estimators. To set up interval estimates and make tests, however, we do need to make an assumption about the functional form of the distribution of the ε_i. The standard assumption is that the error terms are normally distributed, and we will adopt it here. A normal error term greatly simplifies the theory of regression analysis and is justifiable in many real world situations where regression analysis is applied.

Normal Error Model

The normal error model is as follows:

$$Y_i = \beta_0 + \beta_1 X_i + \varepsilon_i \tag{2.25}$$

where:

Y_i is the observed response in the ith trial
X_i is a known constant, the level of the independent variable in the ith trial
β_0 and β_1 are parameters
ε_i are independent $N(0, \sigma^2)$
$i = 1, \ldots, n$

Comments

1. The symbol $N(0, \sigma^2)$ stands for "normally distributed, with mean $\mu = 0$ and variance σ^2."

2. The normal error model (2.25) is the same as the regression model (2.1) with unspecified error distribution, except that model (2.25) assumes that the errors ε_i are normally distributed.

3. Because model (2.25) assumes that the errors are normally distributed, the assumption of uncorrelatedness of the ε_i in model (2.1) becomes one of independence in the normal error model.

4. Model (2.25) implies that the Y_i are independent normal random variables, with mean $E(Y_i) = \beta_0 + \beta_1 X_i$ and variance σ^2. Figure 2.5 (p. 26) pictures this normal error model. Each of the probability distributions of Y there is normally distributed, with constant variability, and the regression function is linear.

5. A major reason why the normality assumption for the error terms is justifiable in many situations is that the error terms frequently represent the effects of many factors omitted explicitly from the model, which do affect the response to some extent and which vary at random without reference to the independent variable X. For instance, in the lot size example, such factors as time lapse since the last production run, particular machines used, season of the year, and personnel employed, could vary more or less at random from run to run, independent of lot size. Also, there might be random measurement errors in recording Y. Insofar as

these random effects have a degree of mutual independence, the composite error term ε_i representing all these factors would tend to comply with the central limit theorem and the error term distribution would approach normality as the number of factor effects becomes large.

A second reason why the normality assumption for the error terms is frequently justifiable is that the estimation and testing procedures to be discussed in the next chapter are based on the t distribution, which is not sensitive to moderate departures from normality. Thus, unless the departures from normality are serious, particularly with respect to skewness, the actual confidence coefficients and risks of errors will be close to the levels for exact normality.

Estimation of Parameters by Method of Maximum Likelihood

When the functional form of the probability distribution of the error terms is specified, estimators of the parameters β_0, β_1, and σ^2 can be obtained by the *method of maximum likelihood.* This method utilizes the joint probability distribution of the sample observations. When this joint probability distribution is viewed as a function of the parameters, given the particular sample observations, it is called the *likelihood function.* The likelihood function for the normal error model (2.25), given the sample observations $Y_1, \ldots, Y_n$, is:

$$(2.26) \qquad L(\beta_0, \beta_1, \sigma^2) = \prod_{i=1}^{n} \frac{1}{(2\pi\sigma^2)^{1/2}} \exp\left[-\frac{1}{2\sigma^2}(Y_i - \beta_0 - \beta_1 X_i)^2\right]$$

$$= \frac{1}{(2\pi\sigma^2)^{n/2}} \exp\left[-\frac{1}{2\sigma^2}\sum(Y_i - \beta_0 - \beta_1 X_i)^2\right]$$

The values of β_0, β_1, and σ^2 which maximize this likelihood function are the maximum likelihood estimators. These are:

(2.27)

Parameter	*Maximum Likelihood Estimator*	
β_0	b_0	same as (2.10b)
β_1	b_1	same as (2.10a)
σ^2	$\hat{\sigma}^2$	$= \dfrac{\sum(Y_i - \hat{Y}_i)^2}{n}$

Thus, the maximum likelihood estimators of β_0 and β_1 are the same estimators as provided by the method of least squares. The maximum likelihood estimator $\hat{\sigma}^2$ is biased, and ordinarily the unbiased estimator MSE as given in (2.22) is used. Note that the unbiased estimator MSE differs but slightly from the maximum likelihood estimator $\hat{\sigma}^2$, especially if n is not small:

$$(2.28) \qquad MSE = \frac{n}{n-2}\hat{\sigma}^2$$

Comments

1. Since the maximum likelihood estimators b_0 and b_1 are the same as the least squares estimators, they have the properties of all least squares estimators:

a) They are unbiased.
b) They have minimum variance among all unbiased linear estimators.

In addition, the maximum likelihood estimators b_0 and b_1 for the normal error model (2.25) have other desirable properties:

c) They are consistent, as defined in (1.47).
d) They are sufficient, as defined in (1.48).
e) They are minimum variance unbiased; that is, they have minimum variance in the class of all unbiased estimators (linear or otherwise).

Thus, for the normal error model the estimators b_0 and b_1 have many desirable properties.

2. We find the values of β_0, β_1, and σ^2 which maximize the likelihood function L in (2.26) by taking partial derivatives of L with respect to β_0, β_1, and σ^2, equating each of the partials to zero, and solving the system of equations thus obtained. We can work with log L, rather than L, because both L and log L are maximized for the same values of β_0, β_1, and σ^2:

$$\log L = -\frac{n}{2}\log 2\pi - \frac{n}{2}\log \sigma^2 - \frac{1}{2\sigma^2}\sum (Y_i - \beta_0 - \beta_1 X_i)^2 \tag{2.29}$$

Partial differentiation of this logarithmic likelihood is much easier; it yields:

$$\frac{\partial(\log L)}{\partial \beta_0} = \frac{1}{\sigma^2}\sum (Y_i - \beta_0 - \beta_1 X_i)$$

$$\frac{\partial(\log L)}{\partial \beta_1} = \frac{1}{\sigma^2}\sum X_i(Y_i - \beta_0 - \beta_1 X_i)$$

$$\frac{\partial(\log L)}{\partial \sigma^2} = -\frac{n}{2\sigma^2} + \frac{1}{2\sigma^4}\sum (Y_i - \beta_0 - \beta_1 X_i)^2$$

We now set these partial derivatives equal to zero, replacing β_0, β_1, and σ^2 by the estimators b_0, b_1, and $\hat{\sigma}^2$. We obtain after some simplification:

$$\sum (Y_i - b_0 - b_1 X_i) = 0 \tag{2.30a}$$

$$\sum X_i(Y_i - b_0 - b_1 X_i) = 0 \tag{2.30b}$$

$$\frac{\sum (Y_i - b_0 - b_1 X_i)^2}{n} = \hat{\sigma}^2 \tag{2.30c}$$

Formulas (2.30a) and (2.30b) are identical to the earlier least squares normal equations (2.9), and (2.30c) is the biased estimator of σ^2 given earlier in (2.27).

2.7 COMPUTER INPUTS AND OUTPUTS

Regression calculations used to be tedious chores, especially when the number of observations was large and when there were several independent variables. Today electronic computers can be used quite easily to perform regression calculations, with one of many available packaged programs.

The inputting of data will vary from program to program. With some, only the X and Y observations are entered. For our Westwood Company data in Table 2.1, one form of data input would be:

X_i	Y_i
30	73
20	50
etc.	etc.

In some other cases, the inputting of the data would be in the form: X_1, Y_1, X_2, Y_2, etc. For our example, this input form would be: 30, 73, 20, 50, etc.

In still other cases, model (2.5) is utilized in a package, requiring an entry of $X_0 \equiv 1$. For our example, the appropriate data input for this approach might be:

FIGURE 2.12

Segment of Computer Output for Regression Run on Westwood Company Data

```
SIMPLE LINEAR REGRESSION

      X

 30        73
 20        50
 60       128
 80       170
 40        87
 50       108
 60       135
 30        69
 70       148
 60       132

X IS A MATRIX OF OBSERVATIONS WHERE THE COLUMNS CORRESPOND TO
VARIABLES AND THE ROWS TO OBSERVATIONS.  THE MATRIX BELOW GIVES
THE RESULTS OF THE BEST LEAST SQUARES FIT OF THE FUNCTION:
                X4 = A + BxX1

IN THE FOLLOWING FORMAT:

ROW 1:  2,A,ST. ERROR OF A, T-VALUE, 0
ROW 2:  1,B,ST. ERROR OF B, T-VALUE, 0
ROW 3:  0,DF FOR REGRESSION, SUM OF SQUARES, MEAN SQUARE, F-VALUE
ROW 4:  0,DF FOR ERROR, SUM OF SQUARES, MEAN SQUARE, 0
ROW 5:  0,DF FOR TOTAL, SUM OF SQUARES, ST. ERROR OF ESTIMATE,
                                  SQUARE OF MULTIPLE CORR. COEF.

2.00000   b0 -> 10.00000       2.50294            3.99530       0.00000
1.00000   b1 ->  2.00000       0.04697           42.58325       0.00000
0.00000          1.00000   13600.00000        13600.00000    1813.33333
0.00000          8.00000      60.00000            7.50000       0.00000
0.00000          9.00000   13660.00000   √MSE -> 2.73861        0.99561
```

X_0	X_i	Y_i
1	30	73
1	20	50
etc.	etc.	etc.

The computer output will vary from one program package to another. Figure 2.12 illustrates a typical output format when the linear regression model is fitted to the Westwood Company data in Table 2.1 by a computer program. The computed values of b_0, b_1, and $\sqrt{MSE}$ are annotated in Figure 2.12, and agree with our earlier results. In subsequent chapters, we shall explain the additional output in Figure 2.12.

PROBLEMS

2.1. Refer to example 2 on page 22. Suppose that mileage is measured accurately but the clerks frequently make computational errors in determining the rental fee. Would the relation between rental fee and distance traveled still be a functional one? Discuss.

2.2. Experience with a certain type of plastic indicates that a relation exists between the hardness (measured in Brinell units) of items molded from the plastic (Y) and the elapsed time since termination of the molding process (X). It is proposed to study this relation by means of regression analysis. A participant in the discussion objects. He points out that the hardening of the plastic "is the result of a natural chemical process that doesn't leave anything to chance, so the relation must be mathematical and regression analysis is not appropriate." Evaluate this objection.

2.3. A student, when asked to state the simple linear regression model, wrote it as follows: $E(Y_i) = \beta_0 + \beta_1 X_i + \varepsilon_i$. Do you agree?

2.4. A student in accounting enthusiastically declared: "Regression is a very powerful tool. We can isolate fixed and variable costs by fitting a linear regression, even when we have no data for small lots." Discuss.

2.5. Evaluate the following statement: "For the least squares method to be fully valid, it is required that the distributions of Y be normal."

2.6. Refer to model (2.1). Assume that $X = 0$ is within the scope of the model. What is the implication for the regression function if $\beta_0 = 0$ so that the model is $Y_i = \beta_1 X_i + \varepsilon_i$? How would the regression function plot on a graph such as Figure 2.2?

2.7. Refer to model (2.1). What is the implication for the regression function if $\beta_1 = 0$ so that the model is: $Y_i = \beta_0 + \varepsilon_i$? How would the regression function plot on a graph such as Figure 2.2?

2.8. Derive the expression for b_1 in (2.10a) from the normal equations in (2.9).

2.9. (Calculus needed.) Refer to the model $Y_i = \beta_0 + \varepsilon_i$ in Problem 2.7. Using the least squares method, derive the estimator of β_0 for this model. What population characteristic is β_0 equivalent to here? What does your result imply in general about the sample mean as an estimator?

2.10. Prove the result in (2.20) that the sum of the residuals weighted by the fitted values is zero.

2.11. Consider the normal error regression model (2.25). Suppose that the parameter values are $\beta_0 = 210$, $\beta_1 = 3.9$, $\sigma = 5$.

a) Plot the normal error regression model in the fashion of Figure 2.8. Show the distributions of Y for $X = 10, 20, 40$.

b) Explain the meaning of the parameters β_0 and β_1. Assume that the scope of the model includes $X = 0$.

2.12. A substance used in biological and medical research is shipped by air freight to users in cartons of 1,000 ampules. The data below, involving eight shipments, were collected on the number of times the carton was transferred from one aircraft to another (including the initial loading) over the shipment route (X) and the number of ampules found to be broken upon arrival (Y). Assume the first-order regression model (2.1) is appropriate.

i:	1	2	3	4	5	6	7	8
X_i:	2	1	3	1	4	2	1	2
Y_i:	16	9	17	12	22	13	8	15

a) Obtain the estimated regression equation. Plot the estimated regression function and the data. Does a linear regression function appear to give a good fit here?

b) Obtain the residuals. Do they sum to 0 in accord with (2.17)?

c) Obtain a point estimate of σ^2. In what units is your estimate expressed?

2.13. Refer to Problem 2.2. Twelve batches of the plastic were made, and from each batch one test item was molded and the hardness measured at some specific point in time. The results are shown below; X is elapsed time in hours and Y is hardness in Brinell units. Assume the first-order regression model (2.1) is appropriate.

i:	1	2	3	4	5	6	7	8	9	10	11	12
X_i:	32	48	72	64	48	16	40	48	48	24	80	56
Y_i:	220	262	323	298	255	199	236	279	267	214	369	305

a) Obtain the estimated regression equation. Plot the estimated regression function and the data. Does a linear regression function appear to give a good fit here?

b) Obtain the following: (1) a point estimate of the change in the mean response when X increases by one hour; (2) a point estimate of the mean response when $X = 40$; (3) the value of the residual for the fourth observation; and (4) a point estimate of σ^2.

c) Suppose one test item had been molded from a single batch of plastic and the hardness of this one item had been measured at 12 different points in time. Would the error term in the model for this case still reflect the same effects as for the experiment initially described? Would you expect the error terms for the different points in time to be uncorrelated now? Discuss.

3

Inferences in Regression Analysis

IN THIS CHAPTER, we first take up inferences concerning the regression parameters β_0 and β_1, considering both interval estimation of these parameters and tests about them. We then discuss interval estimation of the mean $E(Y)$ of the probability distribution of Y, for given X, and prediction intervals for a new observation Y, given X. Finally, we take up the analysis of variance approach to regression analysis.

Throughout this chapter, and in the remainder of Part I unless otherwise stated, we assume that the normal error model (2.25) *is applicable.* This model is:

(3.1) $$Y_i = \beta_0 + \beta_1 X_i + \varepsilon_i$$

where:

β_0 and β_1 are parameters
X_i are known constants
ε_i are independent $N(0, \sigma^2)$

3.1 INFERENCES CONCERNING β_1

Frequently, we are interested in drawing inferences about β_1, the slope of the regression line in model (3.1). For instance, a market research analyst studying the relation between sales (Y) and advertising expenditures (X) may wish to obtain an interval estimate of β_1 because it will provide him with information as to how many additional sales dollars, on the average, are generated by an additional dollar of advertising expenditure.

At times, tests concerning β_1 are of interest, particularly one of the form:

$$C_1: \beta_1 = 0$$
$$C_2: \beta_1 \neq 0$$

The reason for interest in testing whether or not $\beta_1 = 0$ is that $\beta_1 = 0$ implies that there is no relation between Y and X. Figure 3.1 illustrates the case when $\beta_1 = 0$ for the normal error model (3.1). Note that the means of the probability distributions of Y are then the same, namely:

$$E(Y) = \beta_0 + (0)X = \beta_0$$

Since model (3.1) assumes normality of the probability distributions of Y with constant variance, and since the means are the same when $\beta_1 = 0$, it follows that the probability distributions of Y are identical when $\beta_1 = 0$. This is shown in Figure 3.1. Thus, a test of whether or not $\beta_1 = 0$ for model (3.1) is the equivalent of a test of whether or not Y is related to X.

FIGURE 3.1

Model (3.1) When $\beta_1 = 0$

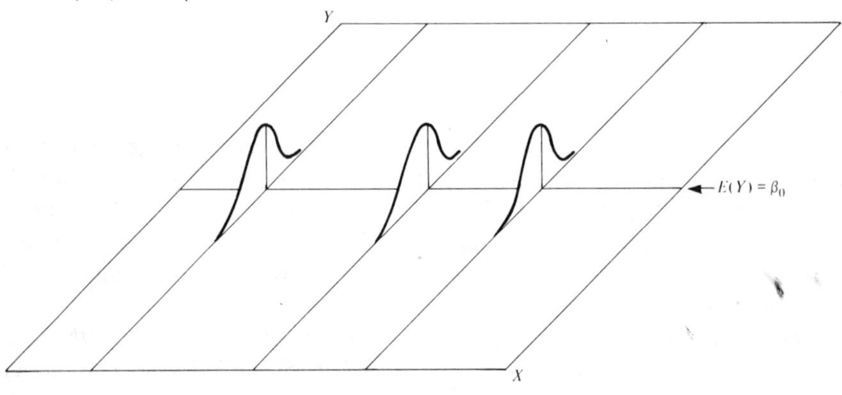

Before discussing inferences concerning β_1 further, we need to consider the sampling distribution of b_1, our point estimator of β_1.

Sampling Distribution of b_1

The point estimator b_1 was given in (2.10a) as follows:

$$b_1 = \frac{\sum (X_i - \bar{X})(Y_i - \bar{Y})}{\sum (X_i - \bar{X})^2} \tag{3.2}$$

The sampling distribution of b_1 refers to the different values of b_1 which would be obtained with repeated sampling, when the levels of the independent variable X are held constant from sample to sample.

(3.3) For model (3.1), the sampling distribution of b_1 is normal, with mean and variance:

(3.3a) $$E(b_1) = \beta_1$$

(3.3b) $$\sigma^2(b_1) = \frac{\sigma^2}{\sum(X_i - \bar{X})^2}$$

To show this, we need to recognize that b_1 is a linear combination of the observations Y_i.

b_1 as Linear Combination of the Y_i. We will show that b_1, as defined in (3.2), can be written:

$$b_1 = \sum k_i Y_i$$

Here the k_i are constants, hence b_1 is a linear combination of the Y_i. We first prove:

(3.4) $$\sum(X_i - \bar{X})(Y_i - \bar{Y}) = \sum(X_i - \bar{X})Y_i$$

This follows since:

$$\sum(X_i - \bar{X})(Y_i - \bar{Y}) = \sum(X_i - \bar{X})Y_i - \sum(X_i - \bar{X})\bar{Y}$$

But $\sum(X_i - \bar{X})\bar{Y} = \bar{Y}\sum(X_i - \bar{X})$ and $\sum(X_i - \bar{X}) = 0$. Hence, (3.4) holds.

We now express b_1, using (3.4):

$$b_1 = \frac{\sum(X_i - \bar{X})(Y_i - \bar{Y})}{\sum(X_i - \bar{X})^2} = \frac{\sum(X_i - \bar{X})Y_i}{\sum(X_i - \bar{X})^2}$$

We can rewrite this as follows:

(3.5) $$b_1 = \sum k_i Y_i$$

where:

(3.5a) $$k_i = \frac{X_i - \bar{X}}{\sum(X_i - \bar{X})^2}$$

Note that the k_i are fixed quantities, since the X_i are fixed.

Note

The quantities k_i have a number of interesting properties which will be used later:

1.

(3.6) $$\sum k_i = 0$$

because $\sum k_i = \sum(X_i - \bar{X})/\sum(X_i - \bar{X})^2 = 0/\sum(X_i - \bar{X})^2 = 0$.

2.

(3.7) $$\sum k_i X_i = 1$$

because $\sum k_i X_i = \sum(X_i - \bar{X})X_i/\sum(X_i - \bar{X})^2 = \sum(X_i - \bar{X})(X_i - \bar{X})/\sum(X_i - \bar{X})^2 = \sum(X_i - \bar{X})^2/\sum(X_i - \bar{X})^2 = 1$. The result $\sum(X_i - \bar{X})(X_i - \bar{X}) = \sum(X_i - \bar{X})X_i$ is obtained in the same way as (3.4).

3.

$$\sum k_i^2 = \frac{1}{\sum(X_i - \bar{X})^2} \tag{3.8}$$

because:

$$\sum k_i^2 = \sum\left[\frac{(X_i - \bar{X})}{\sum(X_i - \bar{X})^2}\right]^2 = \frac{1}{[\sum(X_i - \bar{X})^2]^2}\sum(X_i - \bar{X})^2$$

$$= \frac{1}{\sum(X_i - \bar{X})^2}$$

Normality. We return now to the sampling distribution of b_1 for the normal error model (3.1). The normality of the sampling distribution of b_1 follows at once from the fact that b_1 is a linear combination of the Y_i. The Y_i are independently, normally distributed according to model (3.1), and theorem (1.35) states that a linear combination of independent normal random variables is normally distributed.

Mean. The unbiasedness of the point estimator b_1, stated earlier in the Gauss-Markov theorem (2.11), is easy to show:

$$E(b_1) = E(\sum k_i Y_i) = \sum k_i E(Y_i) = \sum k_i(\beta_0 + \beta_1 X_i)$$
$$= \beta_0 \sum k_i + \beta_1 \sum k_i X_i$$

Hence by (3.6) and (3.7):

$$E(b_1) = \beta_1$$

Variance. The variance of b_1 can be derived readily. We need only remember that the Y_i are independent random variables, each with variance σ^2, and that the k_i are constants. Hence we obtain by (1.26):

$$\sigma^2(b_1) = \sigma^2(\sum k_i Y_i) = \sum k_i^2 \sigma^2(Y_i)$$
$$= \sum k_i^2 \sigma^2 = \sigma^2 \sum k_i^2$$
$$= \sigma^2 \frac{1}{\sum(X_i - \bar{X})^2}$$

The last step follows from (3.8).

Estimated Variance. We can estimate the variance of the sampling distribution of b_1:

$$\sigma^2(b_1) = \frac{\sigma^2}{\sum(X_i - \bar{X})^2}$$

by replacing the parameter σ^2 with the unbiased estimator of σ^2, namely MSE:

$$s^2(b_1) = \frac{MSE}{\sum(X_i - \bar{X})^2} = \frac{MSE}{\sum X_i^2 - \frac{(\sum X_i)^2}{n}} \tag{3.9}$$

The point estimator $s^2(b_1)$ is an unbiased estimator of $\sigma^2(b_1)$. Taking the square root, we obtain $s(b_1)$, our point estimator of $\sigma(b_1)$.

Note

We stated in theorem (2.11) that b_1 has minimum variance among all unbiased linear estimators of the form:

$$\hat{\beta}_1 = \sum c_i Y_i$$

where the c_i are arbitrary constants. We shall now prove this. Since $\hat{\beta}_1$ must be unbiased, the following must hold:

$$E(\hat{\beta}_1) = E(\sum c_i Y_i) = \sum c_i E(Y_i) = \beta_1$$

Now $E(Y_i) = \beta_0 + \beta_1 X_i$ by (2.2) so that the above condition becomes:

$$E(\hat{\beta}_1) = \sum c_i(\beta_0 + \beta_1 X_i) = \beta_0 \sum c_i + \beta_1 \sum c_i X_i = \beta_1$$

For the unbiasedness condition to hold, the c_i must follow the restrictions:

$$\sum c_i = 0 \qquad \sum c_i X_i = 1$$

Now the variance of $\hat{\beta}_1$ is, by (1.26):

$$\sigma^2(\hat{\beta}_1) = \sum c_i^2 \sigma^2(Y_i) = \sigma^2 \sum c_i^2$$

Let us define $c_i = k_i + d_i$, where the k_i are the least squares constants in (3.5) and the d_i are arbitrary constants. We can then write:

$$\sigma^2(\hat{\beta}_1) = \sigma^2 \sum c_i^2 = \sigma^2 \sum (k_i + d_i)^2 = \sigma^2[\sum k_i^2 + \sum d_i^2 + 2 \sum k_i d_i]$$

We know that $\sigma^2 \sum k_i^2 = \sigma^2(b_1)$ from our proof above. Further, $\sum k_i d_i = 0$; this follows from the restrictions on the k_i and the c_i. Hence, we have:

$$\sigma^2(\hat{\beta}_1) = \sigma^2(b_1) + \sigma^2 \sum d_i^2$$

Note that the smallest value of $\sum d_i^2$ is zero. Hence, the variance of $\hat{\beta}_1$ is at a minimum when $\sum d_i^2 = 0$. But this can only occur if all $d_i = 0$, which implies $c_i \equiv k_i$. Thus, the least squares estimator b_1 has minimum variance among all unbiased linear estimators.

Sampling Distribution of $(b_1 - \beta_1)/s(b_1)$

Since b_1 is normally distributed, we know that the standardized statistic $(b_1 - \beta_1)/\sigma(b_1)$ is a standard normal variable. Ordinarily, of course, we need to estimate $\sigma(b_1)$ by $s(b_1)$, and hence are interested in the distribution of the standardized statistic $(b_1 - \beta_1)/s(b_1)$. An important theorem in statistics states:

(3.10) $$\frac{(b_1 - \beta_1)}{s(b_1)} \text{ is distributed as } t(n-2) \text{ for model (3.1)}$$

Intuitively, this result should not be unexpected. We know that if the observations Y_i come from the same normal population, $(\bar{Y} - \mu)/s(\bar{Y})$ follows the t distribution with $n - 1$ degrees of freedom. The estimator b_1, like $\bar{Y}$, is a linear combination of the observations Y_i. The reason for the difference in the degrees of freedom is that two parameters (β_0 and β_1) need to be estimated for the regression model, hence two degrees of freedom are lost here.

Note

We can show that $(b_1 - \beta_1)/s(b_1)$ is distributed as t with $n-2$ degrees of freedom by relying on the following theorem:

(3.11) For model (3.1), SSE/σ^2 is distributed as χ^2, with $n-2$ degrees of freedom, and is independent of b_0 and b_1.

First, let us rewrite $(b_1 - \beta_1)/s(b_1)$ as follows:

$$\frac{b_1 - \beta_1}{\sigma(b_1)} \div \frac{s(b_1)}{\sigma(b_1)}$$

The numerator is a standard normal variable z. The nature of the denominator can be seen by first considering:

$$\frac{s^2(b_1)}{\sigma^2(b_1)} = \frac{\dfrac{MSE}{\sum (X_i - \bar{X})^2}}{\dfrac{\sigma^2}{\sum (X_i - \bar{X})^2}} = \frac{MSE}{\sigma^2} = \frac{\dfrac{SSE}{n-2}}{\sigma^2}$$

$$= \frac{SSE}{\sigma^2(n-2)} = \frac{\chi^2(n-2)}{n-2}$$

The last step follows from (3.11). Hence, we have:

$$\frac{b_1 - \beta_1}{s(b_1)} = \frac{z}{\sqrt{\dfrac{\chi^2(n-2)}{n-2}}}$$

But by theorem (3.11), z and χ^2 are independent, since z is a function of b_1 and b_1 is independent of $SSE/\sigma^2 = \chi^2$. Hence, by definition (1.39), it follows that:

$$\frac{b_1 - \beta_1}{s(b_1)} = t(n-2)$$

This result places us in a position to readily make inferences concerning β_1.

Confidence Interval for β_1

Since $(b_1 - \beta_1)/s(b_1)$ follows a t distribution, we can make the following probability statement:

(3.12) $$P\{t(\alpha/2; n-2) \leq (b_1 - \beta_1)/s(b_1) \leq t(1 - \alpha/2; n-2)\} = 1 - \alpha$$

Here, $t(\alpha/2; n-2)$ denotes the $(\alpha/2)100$ percentile of the t distribution with $n-2$ degrees of freedom. Because of the symmetry of the t distribution, it follows that:

(3.13) $$t(\alpha/2; n-2) = -t(1 - \alpha/2; n-2)$$

Rearranging the inequalities in (3.12) and using (3.13), we obtain:

$$P\{b_1 - t(1-\alpha/2; n-2)s(b_1) \le \beta_1 \le b_1 + t(1-\alpha/2; n-2)s(b_1)\} = 1-\alpha \tag{3.14}$$

Since (3.14) holds for all possible values of β_1, a $1-\alpha$ confidence interval for β_1 is:

$$b_1 - t(1-\alpha/2; n-2)s(b_1) \le \beta_1 \le b_1 + t(1-\alpha/2; n-2)s(b_1) \tag{3.15}$$

Example. Let us return to the Westwood Company lot size example of Chapter 2. Management wishes an estimate of β_1 with a 95 percent confidence coefficient. We summarize in Table 3.1 the needed results obtained earlier. First, we need to obtain $s(b_1)$:

$$s^2(b_1) = \frac{MSE}{\sum(X_i - \bar{X})^2} = \frac{7.5}{3{,}400} = .002206$$

and:

$$s(b_1) = .04697$$

TABLE 3.1

Results for Westwood Company Example Obtained in Chapter 2

$n = 10$	$\bar{X} = 50$
$b_0 = 10.0$	$b_1 = 2.0$
$\hat{Y} = 10.0 + 2.0\,X$	$SSE = 60$
$\sum X_i^2 = 28{,}400$	$MSE = 7.5$

$$\sum X_i^2 - \frac{(\sum X_i)^2}{n} = \sum(X_i - \bar{X})^2 = 3{,}400$$

$$\sum X_i Y_i - \frac{\sum X_i \sum Y_i}{n} = \sum(X_i - \bar{X})(Y_i - \bar{Y}) = 6{,}800$$

$$\sum(Y_i)^2 - \frac{(\sum Y_i)^2}{n} = \sum(Y_i - \bar{Y})^2 = 13{,}660$$

We require, for a 95 percent confidence coefficient, $t(.975; 8)$. From Table A–2 in the Appendix, we find $t(.975; 8) = 2.306$. The 95 percent confidence interval, by (3.15), then is:

$$2.0 - (2.306)(.04697) \le \beta_1 \le 2.0 + (2.306)(.04697)$$

$$1.89 \le \beta_1 \le 2.11$$

Thus, with confidence coefficient .95, we estimate that the mean number of man-hours increases by somewhere between 1.89 and 2.11 for each increase of one part in the lot size.

Note

In Chapter 2, we noted that the scope of a regression model is restricted ordinarily to some interval of values of the independent variable. This is particularly important to keep in mind when using estimates of the slope β_1. In our lot size example, a linear regression model appeared appropriate for lot sizes between 20 and 80, the range of the independent variable in the recent past. It may not be reasonable to use the estimate of the slope to infer the effect of lot size on number of man-hours far outside this range, since the regression relation may not be linear there.

Tests concerning β_1

Since $(b_1 - \beta_1)/s(b_1)$ is distributed as t with $n - 2$ degrees of freedom, tests concerning β_1 can be set up in ordinary fashion using the t distribution.

Suppose a cost analyst in the Westwood Company is interested in testing whether or not man-hours are related to lot size. Thus, the two alternatives would be:

$$\begin{aligned} &C_1: \beta_1 = 0 \\ &C_2: \beta_1 \neq 0 \end{aligned} \tag{3.16}$$

If the analyst wishes to control the risk of a Type I error at .05, he could indeed conclude C_2 at once by referring to the 95 percent confidence interval for β_1 constructed above, since the interval does not include 0.

If an explicit decision rule for testing (3.16) is desired, the two action limits of the decision rule for level of significance α would be:

$$\begin{aligned} A_1 &= 0 - t(1 - \alpha/2; n - 2)s(b_1) \\ A_2 &= 0 + t(1 - \alpha/2; n - 2)s(b_1) \end{aligned} \tag{3.17}$$

and the decision rule would be:

$$\begin{aligned} &\text{If } A_1 \leq b_1 \leq A_2, \text{ conclude } C_1 \\ &\text{Otherwise conclude } C_2 \end{aligned} \tag{3.17a}$$

Example. For the Westwood Company example, where $\alpha = .05$, we have $t(.975; 8) = 2.306$ and $s(b_1) = .04697$. Hence, the action limits are:

$$\begin{aligned} A_1 &= 0 - (2.306)(.04697) = -.11 \\ A_2 &= 0 + (2.306)(.04697) = +.11 \end{aligned}$$

and the decision rule is:

$$\begin{aligned} &\text{If } -.11 \leq b_1 \leq +.11, \text{ conclude } C_1 \\ &\text{Otherwise conclude } C_2 \end{aligned}$$

Since $b_1 = +2.0$, we conclude C_2, that $\beta_1 \neq 0$ or that there is a relation between lot size and man-hours.

Note

The test for deciding whether or not $\beta_1 = 0$, just discussed, uses b_1 as the test statistic. An equivalent test is based on the test statistic:

(3.18) $$t^* = \frac{b_1}{s(b_1)}$$

The decision rule with this test statistic, when controlling the level of significance at α, is:

(3.18a)
$$\begin{aligned} &\text{If } |t^*| \le t(1-\alpha/2; n-2), \text{ conclude } C_1 \\ &\text{If } |t^*| > t(1-\alpha/2; n-2), \text{ conclude } C_2 \end{aligned}$$

For the Westwood Company example, where $\alpha = .05$, we have: $t(.975; 8) = 2.306$, $b_1 = 2.0$, $s(b_1) = .04697$. Thus:

$$|t^*| = \left| \frac{2.0}{.04697} \right| = 42.6$$

Since $42.6 > 2.306$, we conclude C_2 as before, that $\beta_1 \neq 0$.

Computer program packages frequently furnish the test statistic t^* in (3.18) for easy testing whether or not $\beta_1 = 0$.

3.2 INFERENCES CONCERNING β_0

As noted in Chapter 2, there are only infrequent occasions when we wish to make inferences concerning β_0, the intercept of the regression line. These occur when the scope of the model includes $X = 0$.

Sampling Distribution of b_0

The point estimator b_0 was given in (2.10b) as follows:

(3.19) $$b_0 = \bar{Y} - b_1 \bar{X}$$

The sampling distribution of b_0 refers to the different values of b_0 which would be obtained with repeated sampling when the levels of the independent variable X are held constant from sample to sample.

(3.20) For model (3.1), the sampling distribution of b_0 is normal, with mean and variance:

(3.20a) $$E(b_0) = \beta_0$$

(3.20b) $$\sigma^2(b_0) = \sigma^2 \frac{\sum X_i^2}{n \sum (X_i - \bar{X})^2} = \sigma^2 \left[\frac{1}{n} + \frac{\bar{X}^2}{\sum (X_i - \bar{X})^2} \right]$$

The normality of the sampling distribution of b_0 follows because b_0, like b_1, is a linear combination of the observations Y_i. The results on the mean and variance of the sampling distribution of b_0 can be obtained in similar fashion as those for b_1.

An estimator of $\sigma^2(b_0)$ is obtained by replacing σ^2 by its point estimator MSE:

$$(3.21) \qquad s^2(b_0) = MSE \frac{\sum X_i^2}{n \sum (X_i - \bar{X})^2} = MSE\left[\frac{1}{n} + \frac{\bar{X}^2}{\sum (X_i - \bar{X})^2}\right]$$

The square root, $s(b_0)$, is an estimator of $\sigma(b_0)$.

Sampling Distribution of $(b_0 - \beta_0)/s(b_0)$

Analogous to theorem (3.10) for b_1, there is a theorem for b_0 which states:

$$(3.22) \qquad \frac{b_0 - \beta_0}{s(b_0)} \text{ is distributed as } t(n-2) \text{ for model (3.1)}$$

Hence, confidence intervals for β_0 and tests concerning β_0 can be set up in ordinary fashion, using the t distribution.

Confidence Interval for β_0

The $1 - \alpha$ confidence interval for β_0 is obtained in the same manner as that for β_1 derived earlier. It is:

$$(3.23) \qquad b_0 - t(1 - \alpha/2; n - 2)s(b_0) \leq \beta_0 \leq b_0 + t(1 - \alpha/2; n - 2)s(b_0)$$

Example. As noted earlier, the scope of the model for the Westwood Company example does not extend to lot sizes of $X = 0$. Hence, the regression parameter β_0 may not have intrinsic meaning. If, nevertheless, a 90 percent confidence interval for β_0 were desired, we would proceed by finding $t(.95; 8)$ and $s(b_0)$. From Table A–2, we find $t(.95; 8) = 1.860$. Using the earlier results summarized in Table 3.1, we obtain by (3.21):

$$s^2(b_0) = MSE \frac{\sum X_i^2}{n \sum (X_i - \bar{X})^2} = (7.5)\frac{28{,}400}{10(3{,}400)} = 6.26471$$

or:

$$s(b_0) = 2.50294$$

Hence, the 90 percent confidence interval for β_0 is:

$$10.0 - (1.860)(2.50294) \leq \beta_0 \leq 10.0 + (1.860)(2.50294)$$
$$5.34 \leq \beta_0 \leq 14.66$$

We caution again that this confidence interval does not necessarily provide meaningful information. For instance, it does not necessarily provide information about the "setup" costs of producing a lot of parts (costs incurred in setting up the production process, no matter what is the lot size), since we are not certain whether a linear regression model is appropriate when the scope of the model is extended to $X = 0$.

3.3 SOME CONSIDERATIONS ON MAKING INFERENCES CONCERNING β_0 AND β_1

Effect of Departures from Normality

If the probability distributions of Y are not exactly normal but do not depart seriously, the sampling distributions of b_0 and b_1 will be approximately normal, and the use of the t distribution will provide approximately the specified confidence coefficient or level of significance. Even if the distributions of Y are far from normal, the estimators b_0 and b_1 generally have the property of *asymptotic normality*—their distributions approach normality under very general conditions as the sample size increases. Thus, with sufficiently large samples, the confidence intervals and decision rules given earlier still apply even if the probability distributions of Y depart far from normality. For large samples, the t value is, of course, replaced by the z value for the standard normal distribution.

Interpretation of Confidence Coefficient and Risks of Errors

Since model (3.1) assumes that the X_i are known constants, the confidence coefficient and risks of errors are interpreted with respect to taking repeated samples in which the X observations are kept at the same levels as in the observed sample. For instance, we constructed a confidence interval for β_1 with a confidence coefficient of .95 in the Westwood Company example. This coefficient is interpreted to mean that if many independent samples are taken, where the levels of X (the lot sizes) in the first sample are repeated in these other samples, and a 95 percent confidence interval is constructed for each sample, 95 percent of the intervals will contain the true value of β_1.

Spacing of the X Levels

Inspection of formulas (3.3b) and (3.20b) for the variances of b_1 and b_0 respectively indicates that for given n and σ^2, these variances are affected by the spacing of the X levels in the observed data. For example, the more the spread in the X levels, the larger is the quantity $\sum (X_i - \bar{X})^2$ and the smaller is the variance of b_1. We will discuss in Section 5.8 how the X observations should be spaced in experiments, where spacing can be controlled.

Power of Tests

The power of tests on β_0 and β_1 can be obtained from Table A–5 in the Appendix, which contains charts of the power function of the t test. Consider, for example, the general decision problem:

$$C_1: \beta_1 = \beta_{10} \qquad (3.24)$$
$$C_2: \beta_1 \neq \beta_{10}$$

for which the test statistic to be employed is:

$$t^* = \frac{b_1 - \beta_{10}}{s(b_1)} \qquad (3.24a)$$

and the decision rule for level of significance α is:

$$\text{If } |t^*| \leq t(1 - \alpha/2; n - 2), \text{ conclude } C_1 \qquad (3.24b)$$
$$\text{If } |t^*| > t(1 - \alpha/2; n - 2), \text{ conclude } C_2$$

The power of the test is the probability that the decision rule will lead to conclusion C_2 when C_2 in fact holds. Specifically, the power is given by:

$$\text{Power} = P\{|t^*| > t(1 - \alpha/2; n - 2) | \delta\} \qquad (3.25)$$

where δ is a measure of *noncentrality*—that is, how far the true value of β_1 is from β_{10}:

$$\delta = \frac{|\beta_1 - \beta_{10}|}{\sigma(b_1)} \qquad (3.26)$$

Table A–5 presents the power of the two-sided t test (in percent) for $\alpha = .01$ and $\alpha = .05$, for various degrees of freedom *df.* To illustrate the use of this table, let us return to the Westwood Company example where we tested:

$$C_1: \beta_1 = \beta_{10} = 0$$
$$C_2: \beta_1 \neq \beta_{10} = 0$$

Suppose we wish to know the power of the test when $\beta_1 = .25$. Assume that the variance of the error terms is $\sigma^2 = 10.0$, so that $\sigma^2(b_1)$ for our example would be:

$$\sigma^2(b_1) = \frac{\sigma^2}{\sum (X_i - \bar{X})^2} = \frac{10.0}{3{,}400} = .002941$$

or $\sigma(b_1) = .05423$. Then $\delta = |.25 - 0| \div .05423 = 4.6$. We enter the graph for $\alpha = .05$ (the level of significance used in the test) and approximate visually the curve for 8 degrees of freedom. Reading the ordinate at $\delta = 4.6$, we obtain 97 percent approximately. Thus, if $\beta_1 = .25$, the probability would be about .97 that we would be led to conclude C_2 ($\beta_1 \neq 0$). In other words, if $\beta_1 = .25$, we would be almost certain to conclude that there is a relation between man-hours and lot size.

The power of tests concerning β_0 can be obtained in completely analogous fashion. For one-sided tests, Table A–5 should be entered so that one half the level of significance shown there is the desired level of significance.

3.4 INTERVAL ESTIMATION OF $E(Y_h)$

In regression analysis, one of the major goals usually is to estimate the mean for one or more probability distributions of Y. Consider, for example, a study of the relation between level of piecework pay (X) and worker productivity (Y). The mean productivity at high and medium levels of piecework pay may be of particular interest for purposes of analyzing the benefits obtained from an increase in the pay. As another example, the Westwood Company may be interested in the mean response (mean number of man-hours) for lot sizes of $X = 40$ parts, $X = 55$ parts, and $X = 70$ parts for purposes of choosing appropriate lot sizes for production.

Let X_h denote the level of X for which we wish to estimate the mean response. X_h may be a value which occurred in the sample, or it may be some other value of the independent variable within the scope of the model. The mean response when $X = X_h$ is denoted $E(Y_h)$. Formula (2.12) gives us the point estimator $\hat{Y}_h$ of $E(Y_h)$:

(3.27)
$$\hat{Y}_h = b_0 + b_1 X_h$$

We consider now the sampling distribution of $\hat{Y}_h$.

Sampling Distribution of $\hat{Y}_h$

The sampling distribution of $\hat{Y}_h$, like the earlier sampling distributions discussed, refers to the different values of $\hat{Y}_h$ which would be obtained if repeated samples were selected, each holding the levels of the independent variable X constant, and calculating $\hat{Y}_h$ for each sample.

(3.28) For model (3.1), the sampling distribution of $\hat{Y}_h$ is normal, with mean and variance:

(3.28a)
$$E(\hat{Y}_h) = E(Y_h)$$

(3.28b)
$$\sigma^2(\hat{Y}_h) = \sigma^2\left[\frac{1}{n} + \frac{(X_h - \bar{X})^2}{\sum (X_i - \bar{X})^2}\right]$$

Normality. The normality of the sampling distribution of $\hat{Y}_h$ follows directly from the fact that $\hat{Y}_h$ is a linear combination of the observations Y_i.

Mean. To prove that $\hat{Y}_h$ is an unbiased estimator of $E(Y_h)$, we proceed as follows:

$$\begin{aligned} E(\hat{Y}_h) &= E(b_0 + b_1 X_h) = E(b_0) + X_h E(b_1) \\ &= \beta_0 + \beta_1 X_h \end{aligned}$$

by (3.3a) and (3.20a).

Variance. First, we show that b_1 and $\bar{Y}$ are uncorrelated and hence, for model (3.1), independent:

$$\sigma(\bar{Y}, b_1) = 0 \tag{3.29}$$

where $\sigma(\bar{Y}, b_1)$ denotes the covariance between $\bar{Y}$ and b_1. We begin with the definitions:

$$\bar{Y} = \sum \left(\frac{1}{n}\right) Y_i$$

$$b_1 = \sum k_i Y_i$$

where k_i is as defined in (3.5a). We now use theorem (1.27), with $a_i = 1/n$, $c_i = k_i$; remember that the Y_i are independent random variables:

$$\sigma(\bar{Y}, b_1) = \sum \left(\frac{1}{n}\right) k_i \sigma^2(Y_i) = \frac{\sigma^2}{n} \sum k_i$$

But we know from (3.6) that $\sum k_i = 0$. Hence, the covariance is 0.

Now we are ready to find the variance of $\hat{Y}_h$. We shall use the estimator in the alternate form (2.15):

$$\sigma^2(\hat{Y}_h) = \sigma^2(\bar{Y} + b_1[X_h - \bar{X}])$$

Since $\bar{Y}$ and b_1 are independent and X_h and $\bar{X}$ are constants, we obtain:

$$\sigma^2(\hat{Y}_h) = \sigma^2(\bar{Y}) + (X_h - \bar{X})^2 \sigma^2(b_1)$$

Now $\sigma^2(b_1)$ is given in (3.3b), and:

$$\sigma^2(\bar{Y}) = \frac{\sigma^2(Y_i)}{n} = \frac{\sigma^2}{n}$$

Hence:

$$\sigma^2(\hat{Y}_h) = \frac{\sigma^2}{n} + (X_h - \bar{X})^2 \frac{\sigma^2}{\sum (X_i - \bar{X})^2}$$

which, upon a slight rearrangement of terms, yields (3.28b).

Note the effect of the term $(X_h - \bar{X})^2$ on $\sigma^2(\hat{Y}_h)$. The further X_h is from $\bar{X}$, the greater is the quantity $(X_h - \bar{X})^2$ and the larger is the variance of $\hat{Y}_h$. An intuitive explanation of this effect is found in Figure 3.2. Shown there are two sample regression lines, based on two samples for the same set of X values. The two regression lines are assumed to go through the same $(\bar{X}, \bar{Y})$ point to isolate the effect of interest, namely the effect of variation in the estimated slope b_1 from sample to sample. Note that at X_1, near $\bar{X}$, the fitted values $\hat{Y}_1$ for the two sample regression lines are close to each other. At X_2, which is far from $\bar{X}$, the situation is different. Here, the fitted values $\hat{Y}_2$ differ substantially. Thus, variation in the slope b_1 from sample to sample has a much more pronounced effect on $\hat{Y}_h$ for X levels far from the mean $\bar{X}$

FIGURE 3.2

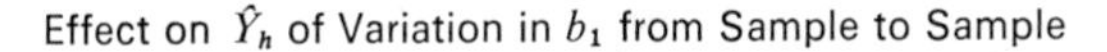
Effect on $\hat{Y}_h$ of Variation in b_1 from Sample to Sample

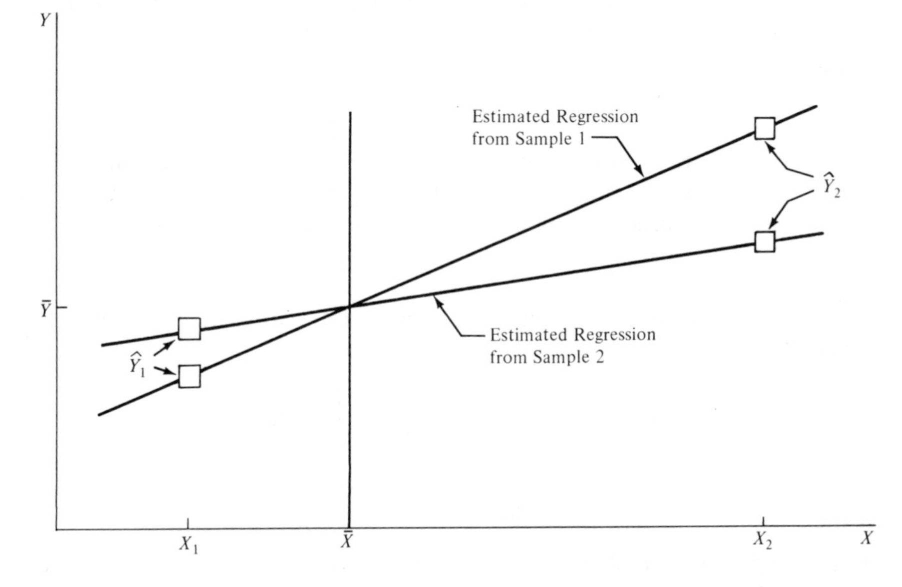

than for X levels near $\bar{X}$. Hence, the variation in the $\hat{Y}_h$ values from sample to sample will be greater when X_h is far from the mean than when X_h is near the mean.

When MSE is substituted for σ^2 in (3.28b), we obtain $s^2(\hat{Y}_h)$, the estimated variance of $\hat{Y}_h$:

(3.30)
$$s^2(\hat{Y}_h) = MSE\left[\frac{1}{n} + \frac{(X_h - \bar{X})^2}{\sum(X_i - \bar{X})^2}\right]$$

The estimated standard deviation of $\hat{Y}_h$ then is $s(\hat{Y}_h)$, the square root of $s^2(\hat{Y}_h)$.

Sampling Distribution of $[\hat{Y}_h - E(Y_h)]/s(\hat{Y}_h)$

Since we have encountered the t distribution in each type of inference for regression model (3.1) considered up to this point, it should not be surprising that:

(3.31)
$$\frac{\hat{Y}_h - E(Y_h)}{s(\hat{Y}_h)} \text{ is distributed as } t(n-2) \text{ for model (3.1)}$$

Hence, all inferences concerning $E(Y_h)$ are carried out in the usual fashion with the t distribution. We illustrate the construction of confidence intervals, since in practice these are more frequently used than tests.

Confidence Interval for $E(Y_h)$

A confidence interval for $E(Y_h)$ is constructed in the standard fashion, making use of the t distribution as indicated by theorem (3.31). The $1-\alpha$ confidence interval is:

$$\hat{Y}_h - t(1-\alpha/2; n-2)s(\hat{Y}_h) \leq E(Y_h) \leq \hat{Y}_h + t(1-\alpha/2; n-2)s(\hat{Y}_h) \tag{3.32}$$

Example 1. Returning to the Westwood Company lot size example, let us find a 90 percent confidence interval for $E(Y_h)$ when the lot size is $X_h = 55$ parts. Using the earlier results in Table 3.1, we find the point estimator $\hat{Y}_h$:

$$\hat{Y}_{55} = 10.0 + 2.0(55) = 120$$

Next, we need to find the estimated standard deviation $s(\hat{Y}_h)$. We obtain, using (3.30):

$$s^2(\hat{Y}_{55}) = 7.5\left[\frac{1}{10} + \frac{(55-50)^2}{3,400}\right] = .80515$$

so that:

$$s(\hat{Y}_{55}) = .89730$$

For a 90 percent confidence coefficient, we require $t(.95; 8) = 1.860$. Hence our confidence interval, with confidence coefficient .90, is by (3.32):

$$120 - (1.860)(.89730) \leq E(Y_{55}) \leq 120 + (1.860)(.89730)$$
$$118.3 \leq E(Y_{55}) \leq 121.7$$

We conclude, with confidence coefficient .90, that the mean number of man-hours required when lots of 55 parts are produced is somewhere between 118.3 and 121.7.

Example 2. Suppose the Westwood Company wishes to estimate $E(Y_h)$ when $X_h = 80$ parts, with a 90 percent confidence interval. We require:

$$\hat{Y}_{80} = 10.0 + 2.0(80) = 170$$
$$s^2(\hat{Y}_{80}) = 7.5\left[\frac{1}{10} + \frac{(80-50)^2}{3,400}\right] = 2.73529$$
$$s(\hat{Y}_{80}) = 1.65387$$
$$t(.95; 8) = 1.860$$

Hence, the 90 percent confidence interval is:

$$170 - (1.860)(1.65387) \leq E(Y_{80}) \leq 170 + (1.860)(1.65387)$$
$$166.9 \leq E(Y_{80}) \leq 173.1$$

Note that this confidence interval is somewhat wider than that for example 1, since the X_h level here ($X_h = 80$) is substantially farther from the mean $\bar{X} = 50$ than the X_h level for example 1 ($X_h = 55$).

Comments

1. Since the X_i are known constants in model (3.1), the interpretation of confidence intervals and risks of errors in inferences on the mean response is in terms of taking repeated samples in which the X observations are at the same levels as in the sample actually taken. We noted this same point earlier in connection with inferences on β_0 and β_1.

2. We see from formula (3.28b) that for given sample results, the variance of $\hat{Y}_h$ is smallest when $X_h = \bar{X}$. Thus in an experiment to estimate the mean response at a particular level X_h of the independent variable, the precision of the estimate will be greatest if (everything else remaining equal) the observations on X are spaced so that $\bar{X} = X_h$.

3. When the sample size is large, the t value in the confidence interval (3.32) may be replaced by the standard normal z value, since the t distribution approaches the standard normal distribution with increasing sample size.

4. The usual relationship between confidence intervals and tests applies in inferences concerning the mean response. Thus formula (3.32) can be utilized for two-sided tests concerning the mean response at X_h. Alternatively, a regular decision rule can be set up.

5. Confidence interval (3.32) applies when a single mean response is to be estimated from the sample. We discuss in Chapter 5 how to proceed when a number of mean responses are to be estimated from the same sample.

3.5 PREDICTION OF NEW OBSERVATION

We consider now the prediction of a new observation Y corresponding to a given level X of the independent variable. In our Westwood Company illustration, for instance, the next lot to be produced consists of 55 parts and management wishes to predict the number of man-hours for this particular lot. As another example, an economist has estimated the regression relation between company sales and number of persons 16 or more years old, based on data for the past 10 years. Given a reliable demographic projection of the number of persons 16 or more years old for next year, the economist wishes to predict next year's company sales.

The new observation on Y is viewed as the result of a new trial, independent of the trials on which the regression analysis is based. We shall denote the level of X for the new trial as X_h, and the new observation on Y as $Y_{h(\text{new})}$. Of course, we assume that the underlying regression model applicable for the basic sample data continues to be appropriate for the new observation.

The distinction between estimation of the mean response $E(Y_h)$, discussed in the preceding section, and prediction of a new response $Y_{h(\text{new})}$, discussed now, is basic. In the former case we estimate the *mean* of the distribution of Y. In the present case, we predict an individual outcome drawn from the distribution of Y. Of course, the great majority of individual outcomes deviate from the mean response, and this must be allowed for in the procedure for predicting $Y_{h(\text{new})}$.

Prediction Interval When Parameters Known

To illustrate the nature of a *prediction interval* for a new observation $Y_{h(new)}$ in as simple a fashion as possible, we shall first assume that all regression parameters are known. Later we shall drop this assumption and make appropriate modifications.

Suppose that the Westwood Company plans to produce a lot of $X_h = 40$ parts in a few weeks, and that the relevant parameters of the regression model are known to be:

$$\beta_0 = 9.5 \qquad \beta_1 = 2.1$$
$$E(Y) = 9.5 + 2.1X$$
$$\sigma^2 = 10.0$$

Thus, for $X_h = 40$ parts, we have:

$$E(Y_{40}) = 9.5 + 2.1(40) = 93.5$$

Figure 3.3 shows the probability distribution of Y for $X_h = 40$ parts. It has a mean $E(Y_{40}) = 93.5$, and a standard deviation $\sigma = \sqrt{10.0} = 3.162$. Further, the distribution is normal in accord with model (3.1).

FIGURE 3.3

Prediction of $Y_{h(new)}$ When Parameters Known

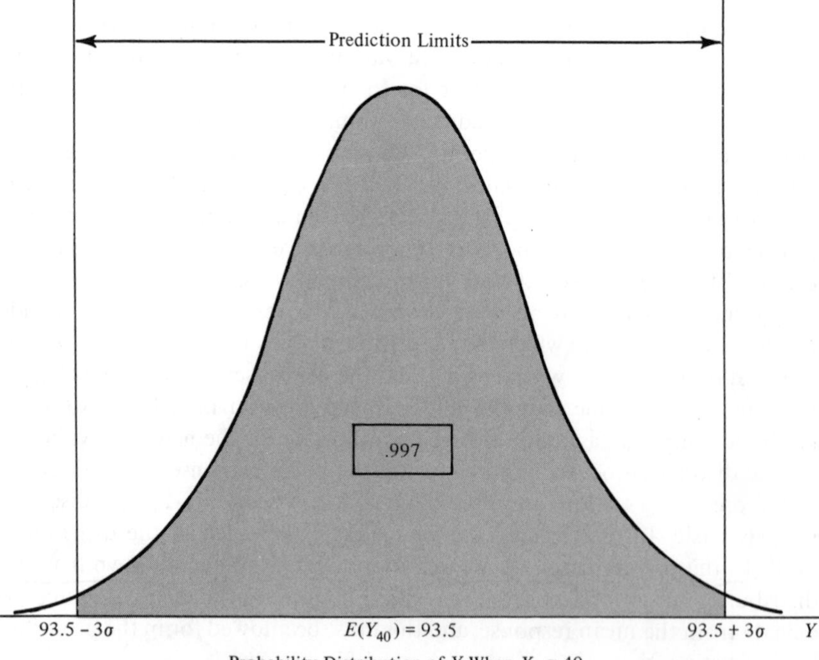

Suppose we were to predict that the number of man-hours for the next lot of $X_h = 40$ parts will be between:

$$E(Y_{40}) \pm 3\sigma$$
$$93.5 \pm 3(3.162)$$

so that the prediction interval would be:

$$84.0 \leq Y_{40(\text{new})} \leq 103.0$$

Since 99.7 percent of the area in a normal probability distribution falls within three standard deviations from the mean, the probability is .997 that this prediction interval will give a correct prediction for any given trial.

The basic idea of a prediction interval is thus to choose a range in the distribution of Y wherein most of the observations will fall, and to declare that the next observation will fall in this range. The usefulness of the prediction interval depends, as always, on the width of the interval and the needs for prediction by the user.

In general, the $1 - \alpha$ prediction interval for $Y_{h(\text{new})}$, when the regression parameters are known, is:

$$E(Y_h) - z(1-\alpha/2)\sigma \leq Y_{h(\text{new})} \leq E(Y_h) + z(1-\alpha/2)\sigma \tag{3.33}$$

In centering the interval around $E(Y_h)$, we obtain the narrowest interval consistent with the specified probability of a correct prediction.

Prediction Interval for $Y_{h(\text{new})}$ When Parameters Unknown

When the regression parameters are unknown, they must be estimated. The mean of the distribution of Y is estimated by $\hat{Y}_h$, as usual, and the variance of the distribution of Y is estimated by MSE. We cannot, however, simply use prediction interval (3.33) with the parameters replaced by the corresponding point estimators. The reason is illustrated intuitively in Figure 3.4. Shown there are two probability distributions of Y, corresponding to the upper and lower limits of a confidence interval for $E(Y_h)$. In other words, the distribution of Y could be located as far left as the one shown, as far right as the other one shown, or anywhere in between. Since we do not know the mean $E(Y_h)$ and only estimate it by a confidence interval, we cannot be certain of the location of the distribution of Y.

Figure 3.4 also shows the prediction limits for each of the two probability distributions of Y presented there. Since we cannot be certain of the location of the distribution of Y, prediction limits for $Y_{h(\text{new})}$ clearly must take account of two elements, as shown in Figure 3.4:

1. Variation in possible location of the distribution of Y.
2. Variation within the probability distribution of Y.

FIGURE 3.4

Prediction of $Y_{h(\text{new})}$ When Parameters Unknown

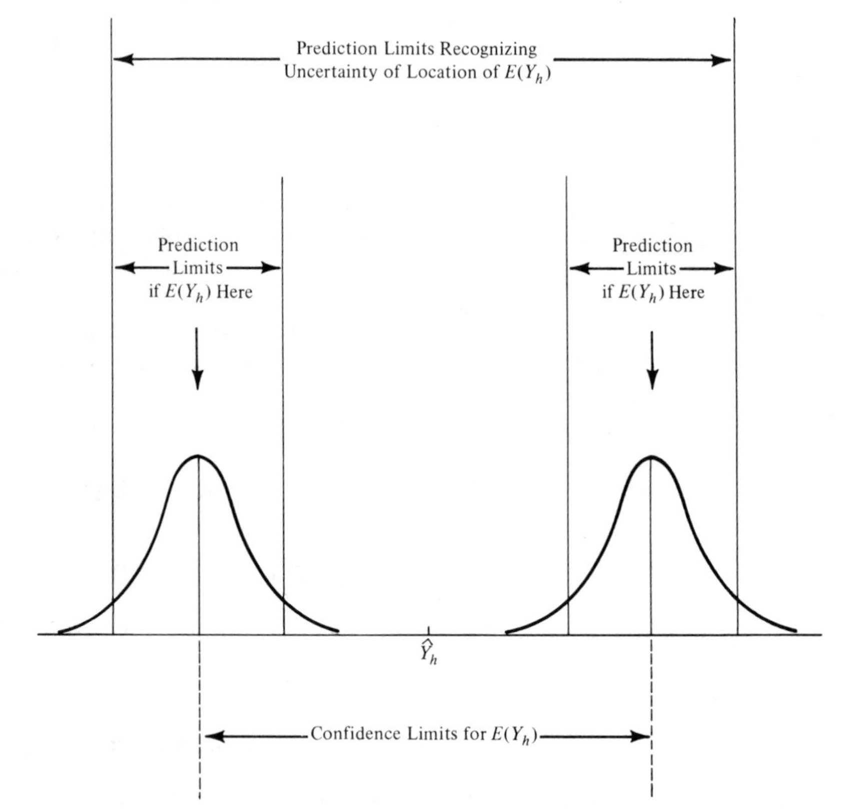

Since the new observation is assumed to be independent of the basic sample from which the regression estimates are obtained, it can be shown that the relevant variance, to be denoted $\sigma^2(Y_{h(\text{new})})$, is:

$$\sigma^2(Y_{h(\text{new})}) = \sigma^2(\hat{Y}_h) + \sigma^2 \tag{3.34}$$

Note that $\sigma^2(Y_{h(\text{new})})$ has two components:

1. The variance of the sampling distribution of $\hat{Y}_h$.
2. The variance of the distribution of Y.

An unbiased estimator of this variance is:

$$s^2(Y_{h(\text{new})}) = s^2(\hat{Y}_h) + MSE \tag{3.35}$$

which can be expressed, using (3.30), as follows:

$$s^2(Y_{h(\text{new})}) = MSE\left[1 + \frac{1}{n} + \frac{(X_h - \bar{X})^2}{\sum(X_i - \bar{X})^2}\right] \tag{3.35a}$$

It can be shown that the appropriate prediction interval can be obtained by means of the t distribution. The $1-\alpha$ prediction interval for $Y_{h(\text{new})}$ is:

$$(3.36)\qquad \hat{Y}_h - t(1-\alpha/2;\, n-2)s(Y_{h(\text{new})}) \leq Y_{h(\text{new})} \leq \hat{Y}_h + t(1-\alpha/2;\, n-2)s(Y_{h(\text{new})})$$

Example. Suppose that the Westwood Company wishes to predict the number of man-hours required in the forthcoming production run of size 55, with a 90 percent prediction interval, and that the parameter values are unknown. We require $t(.95; 8) = 1.860$. From earlier work, we have:

$$\hat{Y}_{55} = 120 \qquad s^2(\hat{Y}_{55}) = .80515$$
$$MSE = 7.5$$

Using (3.35), we obtain:

$$s^2(Y_{55(\text{new})}) = .80515 + 7.5 = 8.30515$$

so that:

$$s(Y_{55(\text{new})}) = 2.88187$$

Hence, the 90 percent prediction interval for $Y_{55(\text{new})}$ is by (3.36):

$$120 - (1.860)(2.88187) \leq Y_{55(\text{new})} \leq 120 + (1.860)(2.88187)$$
$$114.6 \leq Y_{55(\text{new})} \leq 125.4$$

With confidence coefficient .90, we predict that the number of man-hours for the next production run of 55 parts will be somewhere between 114.6 and 125.4.

Comments

1. The 90 percent prediction interval for $Y_{55(\text{new})}$ just obtained is wider than the 90 percent confidence interval for $E(Y_{55})$ obtained on page 68. The reason is that when predicting a new observation, we encounter both the variability in $\hat{Y}_h$ from sample to sample as well as the variation within the probability distribution of Y.

2. Formula (3.35a) indicates that the prediction interval is wider the further X_h is from $\bar{X}$. The reason for this is that the estimate of the mean $\hat{Y}_h$, as noted earlier, is less precise as X_h is located further away from $\bar{X}$.

3. The confidence coefficient for the prediction interval (3.36) refers to the taking of repeated samples based on the same set of X values, and calculating prediction limits for $Y_{h(\text{new})}$ for each sample.

4. Prediction limits lend themselves to statistical control uses. In our example, suppose that the new production run of 55 parts, for which the prediction limits were 114.6 and 125.4 hours, actually required 135 hours. Management here would have an indication that a change in the production process may have occurred, and may wish to initiate a search for the assignable cause.

5. When the sample size is large, the last two terms inside the brackets in (3.35a) are small compared to 1, the first term in the brackets. Also, of course, the t distribution is then approximately normal. Hence, an approximate $1-\alpha$ prediction interval for $Y_{h(\text{new})}$ when n is large is:

$$(3.37)\qquad \hat{Y}_h - z(1-\alpha/2)\sqrt{MSE} \leq Y_{h(\text{new})} \leq \hat{Y}_h + z(1-\alpha/2)\sqrt{MSE}$$

6. Prediction interval (3.36) applies for a single prediction based on the sample data. Below we discuss how to predict the mean of several new observations at a given X_h, and in Chapter 5 we take up how to make several predictions at different X_h values.

7. Prediction intervals resemble confidence intervals. However, they differ conceptually. A confidence interval represents an inference on a parameter, and is an interval which is intended to cover the value of the parameter. A prediction interval, on the other hand, is a statement about the value to be taken by a random variable.

Prediction of Mean of m New Observations for Given X_h

Occasionally, one would like to predict the mean of m new observations on Y for a given level of the independent variable. Suppose the Westwood Company plans to run $m = 3$ new runs, each consisting of $X_h = 55$ parts. Management then may wish to predict the mean man-hours per run for these three runs as a basis for predicting the total man-hours for the three runs. We shall denote the mean value of Y to be predicted as $\bar{Y}_{h(\text{new})}$. It can be shown that the appropriate $1 - \alpha$ prediction interval is:

$$\begin{aligned} \text{(3.38)} \qquad \hat{Y}_h - t(1 - \alpha/2; n - 2)s(\bar{Y}_{h(\text{new})}) &\leq \bar{Y}_{h(\text{new})} \\ &\leq \hat{Y}_h + t(1 - \alpha/2; n - 2)s(\bar{Y}_{h(\text{new})}) \end{aligned}$$

where:

$$\text{(3.38a)} \qquad s^2(\bar{Y}_{h(\text{new})}) = MSE\left[\frac{1}{m} + \frac{1}{n} + \frac{(X_h - \bar{X})^2}{\sum(X_i - \bar{X})^2}\right]$$

or alternatively:

$$\text{(3.38b)} \qquad s^2(\bar{Y}_{h(\text{new})}) = s^2(\hat{Y}_h) + \frac{MSE}{m}$$

Note from (3.38b) that the variance $s^2(\bar{Y}_{h(\text{new})})$ has two components:

1. The variance of the sampling distribution of $\hat{Y}_h$.
2. The variance of the mean of m observations from the probability distribution of Y.

Example. Let us find the 90 percent prediction interval for the mean $\bar{Y}_{h(\text{new})}$ in three new production runs, each for $X_h = 55$ parts, in the Westwood Company example. From previous work, we have:

$$\hat{Y}_{55} = 120 \qquad s^2(\hat{Y}_{55}) = .80515$$
$$MSE = 7.5 \qquad t(.95; 8) = 1.860$$

Hence, we obtain:

$$s^2(\bar{Y}_{55(\text{new})}) = .80515 + \frac{7.5}{3} = 3.30515$$

or:

$$s(\bar{Y}_{55(\text{new})}) = 1.81801$$

The prediction interval then is:

$$120 - (1.860)(1.81801) \leq \bar{Y}_{55(\text{new})} \leq 120 + (1.860)(1.81801)$$
$$116.6 \leq \bar{Y}_{55(\text{new})} \leq 123.4$$

Note that these prediction limits are somewhat narrower than those for predicting the man-hours for a single lot of 55 parts because they involve a prediction of the mean man-hours for three lots rather than the man-hours for a single new lot.

3.6 CONSIDERATIONS IN APPLYING REGRESSION ANALYSIS

We have now discussed the major uses of regression analysis—to make inferences about the regression parameters, to estimate the mean response for a given X, and to predict a new observation Y for a given X. It remains to make a few cautionary remarks about implementing applications of regression analysis.

1. Frequently, regression analysis is used to make inferences for the future. For instance, the Westwood Company may wish to estimate expected man-hours for given lot sizes for purposes of planning future production. In applications of this type, it is important to remember that the validity of the regression application depends upon whether basic causal conditions in the period ahead will be similar to those in existence during the period upon which the regression analysis is based. This caution applies whether mean responses are to be estimated, new observations predicted, or regression parameters estimated.

2. In predicting new observations on Y, the independent variable X itself often has to be predicted. For instance, we mentioned earlier the prediction of company sales for next year from a demographic projection of the number of persons 16 years of age or older next year. A prediction of company sales under these circumstances is a conditional prediction, dependent upon the correctness of the population projection. It is easy to forget the conditional nature of this type of prediction.

3. Another caution deals with inferences pertaining to levels of the independent variable which fall outside the range of observations. Unfortunately, this situation frequently occurs in practice. A company which predicts its sales from a regression relation of company sales to disposable personal income will often find the level of disposable personal income of interest (e.g., for the year ahead) to fall beyond the range of past data. If the X level does not fall far beyond this range, one may have reasonable confidence in the application of the regression analysis. On the other hand, if the X level falls far beyond the range of past data, extreme caution should be exercised since one cannot be sure that the regression function which fits the past data is appropriate over the wider range of the independent variable.

4. Finally, we should note again that special problems arise when one wishes to estimate the mean response or predict a new observation for a number of different levels of the independent variable, as is frequently the case. The confidence coefficients for the interval (3.32) for estimating $E(Y)$ and for the prediction interval (3.36) for a new observation apply for a single level of X for a given sample. In Chapter 5, we discuss how to make multiple inferences from a given sample.

3.7 CASE WHEN X IS RANDOM

The normal error model (3.1), which has been used throughout this chapter and will continue to be used, assumes that the X values are known constants. As a consequence of this, the confidence coefficients and risks of errors refer to repeated sampling when the X values are kept the same from sample to sample.

Frequently, it may not be appropriate to consider the X values as known constants. For instance, consider regressing daily bathing suit sales by a department store on mean daily temperature. Surely, the department store cannot control daily temperatures, so it would not be meaningful to think of repeated sampling when the temperature levels are the same from sample to sample.

In this type of situation, it may be preferable to consider both Y and X as random variables. Does this mean then that all of our earlier results are not applicable here? Not at all. It can be shown that all results on estimation, testing, and prediction obtained for model (3.1) still apply if the following conditions hold:

(3.39)
1. The conditional distributions of the Y_i, given X_i, are normal and independent, with conditional means $\beta_0 + \beta_1 X_i$ and conditional variance σ^2.
2. The X_i are independent random variables, whose probability distribution $g(X_i)$ does not involve the parameters $\beta_0, \beta_1, \sigma^2$.

These conditions require only that model (3.1) is appropriate for each *conditional* distribution of Y_i, and that the probability distribution of X_i does not involve the regression parameters. If these conditions are met, all earlier results on estimation, testing, and prediction still hold even though the X_i are now random variables. The major modification occurs in the interpretation of confidence coefficients and specified risks of error. These refer, when X is random, to repeated sampling of pairs of (X_i, Y_i) values, where the X_i values as well as the Y_i values change from sample to sample. Thus, in our bathing suit sales illustration, a confidence coefficient would refer to the proportion of correct interval estimates if repeated samples of n days' sales and temperatures were obtained and the confidence interval calculated for each sample. Another modification occurs in the power of tests which is different when X is a random variable.

3.8 ANALYSIS OF VARIANCE APPROACH TO REGRESSION ANALYSIS

We now have developed the basic regression model and demonstrated its major uses. At this point, we will view the relationships in a regression model in a somewhat different way than before. This new perspective will not enable us to do anything new with the basic regression model, but it will come into its own when we take up more complex regression models and additional types of linear statistical models.

Partitioning of Total Sum of Squares

Basic Notions. The analysis of variance approach is based on the partitioning of sums of squares and degrees of freedom associated with the response variable Y. To explain the motivation of this approach, consider again the Westwood Company lot size example. Figure 3.5a shows the man-hours required for the 10 production runs presented earlier in Table 2.1. There is variation in the number of man-hours, as in all statistical data. Indeed, if all observations Y_i were identically equal, in which case $Y_i \equiv \bar{Y}$, there would be no statistical problems. The variation of the Y_i is conventionally measured in terms of the deviations:

$$Y_i - \bar{Y} \tag{3.40}$$

These deviations are shown in Figure 3.5a, and one is labeled explicitly. The measure of total variation, denoted by $SSTO$, is the sum of the squared deviations (3.40):

$$SSTO = \sum (Y_i - \bar{Y})^2 \tag{3.41}$$

Here $SSTO$ stands for *total sum of squares*. If $SSTO = 0$, all observations are the same. The greater is $SSTO$, the greater is the variation among the Y observations.

When we utilize the regression approach, the variation reflecting the uncertainty in the data is that of the Y observations around the regression line:

$$Y_i - \hat{Y}_i \tag{3.42}$$

These deviations are shown in Figure 3.5b. The measure of variation in the data with the regression model is the sum of the squared deviations (3.42), which is the familiar SSE of (2.21):

$$SSE = \sum (Y_i - \hat{Y}_i)^2 \tag{3.43}$$

Again, SSE denotes *error sum of squares*. If $SSE = 0$, all observations fall on the fitted regression line. The larger is SSE, the greater is the variation of the Y observations around the regression line.

FIGURE 3.5

Partitioning of Deviations $Y_i - \overline{Y}$ (Y values not plotted to scale)

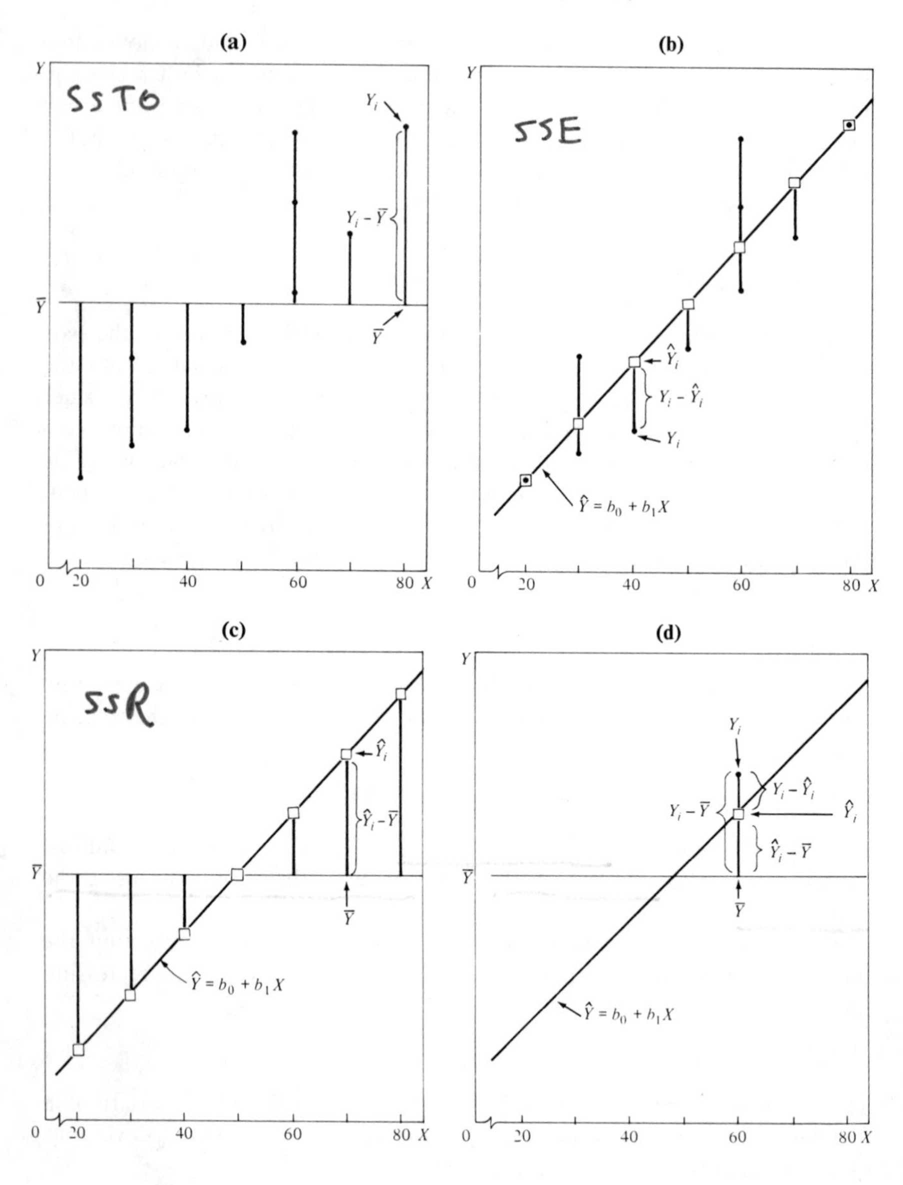

For the Westwood Company example, we know from earlier work (Table 3.1) that:

$$SSTO = 13{,}660$$
$$SSE = 60$$

What accounts for the substantial difference between these two sums of squares? The difference, as we shall show shortly, is another sum of squares:

$$SSR = \sum (\hat{Y}_i - \bar{Y})^2 \tag{3.44}$$

where SSR stands for *regression sum of squares*. Note that SSR is a sum of squared deviations, the deviations being:

$$\hat{Y}_i - \bar{Y} \tag{3.45}$$

These deviations are shown in Figure 3.5c. Each deviation is simply the difference between the fitted value on the regression line and the mean of the fitted values $\bar{Y}$. (Recall from (2.18) that the mean of the fitted values $\hat{Y}_i$ is $\bar{Y}$.) If the regression line is horizontal so that $\hat{Y}_i - \bar{Y} \equiv 0$, $SSR = 0$. Otherwise, SSR is positive.

SSR may be considered a measure of the variability of the Y's associated with the regression line. The larger is SSR in relation to $SSTO$, the greater is the effect of the regression relation in accounting for the total variation in the Y observations.

For our lot size example, we have:

$$SSR = SSTO - SSE = 13{,}660 - 60 = 13{,}600$$

which indicates that most of the total variability in man-hours is accounted for by the relation between lot size and man-hours.

Formal Development of Partitioning. Consider the deviation $Y_i - \bar{Y}$, the basic quantity measuring the variation of the observations Y_i. We can decompose this deviation as follows:

$$\underbrace{Y_i - \bar{Y}}_{\substack{\text{Total}\\ \text{deviation}}} = \underbrace{\hat{Y}_i - \bar{Y}}_{\substack{\text{Deviation}\\ \text{of fitted}\\ \text{regression}\\ \text{value}\\ \text{around mean}}} + \underbrace{Y_i - \hat{Y}_i}_{\substack{\text{Deviation}\\ \text{around}\\ \text{regression}\\ \text{line}}} \tag{3.46}$$

Thus, the total deviation $Y_i - \bar{Y}$ can be viewed as the sum of two components:

1. The deviation of the fitted value $\hat{Y}_i$ around the mean $\bar{Y}$.
2. The deviation of Y_i around the regression line.

Figure 3.5d shows this decomposition for one of the observations.

It is a remarkable property that the sums of these squared deviations have the same relationship:

$$\sum (Y_i - \bar{Y})^2 = \sum (\hat{Y}_i - \bar{Y})^2 + \sum (Y_i - \hat{Y}_i)^2 \tag{3.47}$$

or, using the notation in (3.41), (3.43), and (3.44):

$$SSTO = SSR + SSE \tag{3.47a}$$

To prove this basic result in the analysis of variance, we proceed as follows:

$$\begin{aligned}\sum (Y_i - \bar{Y})^2 &= \sum [(\hat{Y}_i - \bar{Y}) + (Y_i - \hat{Y}_i)]^2 \\ &= \sum [(\hat{Y}_i - \bar{Y})^2 + (Y_i - \hat{Y}_i)^2 + 2(\hat{Y}_i - \bar{Y})(Y_i - \hat{Y}_i)] \\ &= \sum (\hat{Y}_i - \bar{Y})^2 + \sum (Y_i - \hat{Y}_i)^2 + 2 \sum (\hat{Y}_i - \bar{Y})(Y_i - \hat{Y}_i)\end{aligned}$$

The last term on the right is zero, as can be seen by expanding it out:

$$2 \sum (\hat{Y}_i - \bar{Y})(Y_i - \hat{Y}_i) = 2 \sum \hat{Y}_i(Y_i - \hat{Y}_i) - 2\bar{Y} \sum (Y_i - \hat{Y}_i)$$

The first summation on the right is zero by (2.20), and the second is zero by (2.17). Hence, (3.47) follows.

Computational Formulas. The definitional formulas for *SSTO*, *SSR*, and *SSE* presented above are often not convenient for hand computation. Useful computational formulas for *SSTO* and *SSR*, which are algebraically equivalent to the definitional formulas, are:

$$SSTO = \sum Y_i^2 - \frac{(\sum Y_i)^2}{n} = \sum Y_i^2 - n\bar{Y}^2 \tag{3.48}$$

$$SSR = b_1\left[\sum X_i Y_i - \frac{\sum X_i \sum Y_i}{n}\right] = b_1[\sum (X_i - \bar{X})(Y_i - \bar{Y})] \tag{3.49a}$$

or:

$$SSR = b_1^2 \sum (X_i - \bar{X})^2 \tag{3.49b}$$

Computational formulas for *SSE* were given earlier in (2.24).

Using the results for the Westwood Company example summarized in Table 3.1, we obtain for *SSR* by (3.49a):

$$SSR = 2.0(6{,}800) = 13{,}600$$

This, of course, is the same result obtained previously by taking the difference $SSTO - SSE$, except sometimes for a slight difference due to rounding.

Breakdown of Degrees of Freedom

Corresponding to the partitioning of the total sum of squares *SSTO*, there is a partitioning of the associated degrees of freedom (abbreviated *df*). We have $n - 1$ degrees of freedom associated with *SSTO*. One degree of freedom is "lost" because of a constraint on the deviations $Y_i - \bar{Y}$, namely $\sum (Y_i - \bar{Y}) = 0$.

SSE, as noted earlier, has $n - 2$ degrees of freedom associated with it. Two degrees of freedom are lost because of two constraints on the deviations

$Y_i - \hat{Y}_i$, associated with estimating the two parameters β_0 and β_1. These constraints are:

$$\sum (Y_i - \hat{Y}_i) = 0$$
$$\sum X_i(Y_i - \hat{Y}_i) = 0$$

as implied by the normal equations (2.9).

SSR has 1 degree of freedom associated with it. There are two parameters in the regression equation, but the deviations $\hat{Y}_i - \bar{Y}$ are subject to the constraint $\sum (\hat{Y}_i - \bar{Y}) = 0$.

Note that the degrees of freedom are additive:

$$n - 1 = 1 + (n - 2)$$

For our Westwood Company example, these degrees of freedom are:

$$9 = 1 + 8$$

Mean Squares

A sum of squares divided by its associated degrees of freedom is called a *mean square* (abbreviated *MS*). For instance, an ordinary sample variance is a mean square since a sum of squares, $\sum (Y_i - \bar{Y})^2$, is divided by its associated degrees of freedom, $n - 1$. We are interested here in the *regression mean square*, denoted *MSR*:

$$MSR = \frac{SSR}{1} = SSR \tag{3.50}$$

and in the *error mean square*, *MSE*, defined earlier in (2.22):

$$MSE = \frac{SSE}{n - 2} \tag{3.51}$$

For our Westwood Company example, we have $SSR = 13{,}600$ and $SSE = 60$. Hence:

$$MSR = \frac{13{,}600}{1} = 13{,}600$$

Also, we obtained earlier:

$$MSE = \frac{60}{8} = 7.5$$

Note

The two mean squares *MSR* and *MSE* do not add to $SSTO \div n - 1 = 13{,}660 \div 9 = 1{,}518$. Thus, mean squares are not additive.

Analysis of Variance Table

Basic Table. The breakdowns of the total sum of squares and associated degrees of freedom are displayed in the form of an analysis of variance table (ANOVA table) in Table 3.2. Mean squares of interest also are shown.

TABLE 3.2

ANOVA Table for Simple Regression

Source of Variation	*SS*	*df*	*MS*	*E(MS)*
Regression	$SSR = \sum(\hat{Y}_i - \bar{Y})^2$	1	$MSR = \frac{SSR}{1}$	$\sigma^2 + \beta_1^2 \sum(X_i - \bar{X})^2$
Error	$SSE = \sum(Y_i - \hat{Y}_i)^2$	$n-2$	$MSE = \frac{SSE}{n-2}$	σ^2
Total	$SSTO = \sum(Y_i - \bar{Y})^2$	$n-1$		

In addition, there is a column of expected mean squares which will be utilized below. The ANOVA table for our Westwood Company example is shown in Table 3.3.

TABLE 3.3

ANOVA Table for Westwood Company Example

Source of Variation	*SS*	*df*	*MS*
Regression	13,600	1	13,600
Error	60	8	7.5
Total	13,660	9	

Modified Table. Sometimes, an ANOVA table showing one additional element of decomposition is utilized. Recall that by (3.48):

$$SSTO = \sum(Y_i - \bar{Y})^2 = \sum Y_i^2 - n\bar{Y}^2$$

In the modified ANOVA table, the *total uncorrected sum of squares*, denoted *SSTOU*, is defined as:

$$SSTOU = \sum Y_i^2 \tag{3.52}$$

and the *correction for the mean sum of squares*, denoted *SS*(correction for mean), is defined as:

$$SS(\text{correction for mean}) = n\bar{Y}^2 \tag{3.53}$$

Table 3.4 shows this modified ANOVA table. The general format is presented in part (a), and the Westwood Company results in part (b). Both types of ANOVA tables are widely used. Ordinarily we shall utilize the basic type of table, but occasionally we will use the modified table.

TABLE 3.4

Modified ANOVA Table for Simple Regression and Results for Westwood Company Example

(a)

Source of Variation	*SS*	*df*	*MS*
Regression	$SSR = \sum(\hat{Y}_i - \bar{Y})^2$	1	$MSR = \frac{SSR}{1}$
Error	$SSE = \sum(Y_i - \hat{Y}_i)^2$	$n-2$	$MSE = \frac{SSE}{n-2}$
Total	$SSTO = \sum(Y_i - \bar{Y})^2$	$n-1$	
Correction for mean	$SS(\text{correction for mean}) = n\bar{Y}^2$	1	
Total, uncorrected	$SSTOU = \sum Y_i^2$	n	

(b)

Source of Variation	*SS*	*df*	*MS*
Regression	13,600	1	13,600
Error	60	8	7.5
Total	13,660	9	
Correction for mean	121,000	1	
Total, uncorrected	134,660	10	

Expected Mean Squares

We now find the expected value of each of the mean squares in the ANOVA table, so that we can know what quantity each mean square estimates.

We stated earlier that *MSE* is an unbiased estimator of the error variance σ^2:

(3.54) $$E(MSE) = \sigma^2$$

This follows from theorem (3.11), which states that $SSE/\sigma^2 = \chi^2(n-2)$ for model (3.1). Hence it follows from property (1.37) of the chi-square distribution that:

$$E\left[\frac{SSE}{\sigma^2}\right] = n - 2$$

or that:

$$E\left[\frac{SSE}{n-2}\right] = E(MSE) = \sigma^2$$

To find the expected value of MSR, we begin with (3.49b):

$$SSR = b_1^2 \sum (X_i - \bar{X})^2$$

Now by (1.14a), we have:

$$\sigma^2(b_1) = E(b_1^2) - [E(b_1)]^2 \tag{3.55}$$

We know from (3.3a) that $E(b_1) = \beta_1$, and from (3.3b) that:

$$\sigma^2(b_1) = \frac{\sigma^2}{\sum (X_i - \bar{X})^2}$$

Hence, substituting into (3.55), we obtain:

$$E(b_1^2) = \frac{\sigma^2}{\sum (X_i - \bar{X})^2} + \beta_1^2$$

It now follows that:

$$E(SSR) = E(b_1^2) \sum (X_i - \bar{X})^2 = \sigma^2 + \beta_1^2 \sum (X_i - \bar{X})^2$$

Finally, $E(MSR)$ is:

$$E(MSR) = E\left(\frac{SSR}{1}\right) = \sigma^2 + \beta_1^2 \sum (X_i - \bar{X})^2 \tag{3.56}$$

Table 3.2 contains the expected mean squares which we have just derived.

Comments

1. The expectation of MSE is σ^2, whether or not X and Y are related, that is, whether or not $\beta_1 = 0$.

2. The expectation of MSR is also σ^2 when $\beta_1 = 0$. On the other hand, when $\beta_1 \neq 0$, $E(MSR)$ is greater than σ^2 since the term $\beta_1^2 \sum (X_i - \bar{X})^2$ in (3.56) then must be positive. Thus, for testing whether or not $\beta_1 = 0$, a comparison of MSR and MSE suggests itself. If MSR and MSE are of the same order of magnitude, this would suggest that $\beta_1 = 0$. On the other hand, if MSR is substantially greater than MSE, this would suggest that $\beta_1 \neq 0$. This indeed is the basic idea underlying the analysis of variance test to be discussed next.

F Test of $\beta_1 = 0$ versus $\beta_1 \neq 0$

The general analysis of variance approach provides us with a battery of highly useful tests for regression models (and other linear statistical models). For the simple regression case considered here, the analysis of variance provides us with a test for:

(3.57)
$$C_1: \beta_1 = 0$$
$$C_2: \beta_1 \neq 0$$

Test Statistic. The test statistic for the analysis of variance approach is denoted F^*. As just mentioned, it compares MSR and MSE, in the following fashion:

(3.58)
$$F^* = \frac{MSR}{MSE}$$

The earlier motivation, based on the expected mean squares in Table 3.2, suggests that large values of F^* support C_2, and values of F^* near 1 support C_1. In other words, the appropriate test is an upper-tail one.

Distribution of F^***.** In order to be able to construct a statistical decision rule and examine its properties, we need to know the sampling distribution of F^*. We begin by considering the sampling distribution of F^* when C_1 $(\beta_1 = 0)$ holds. The famous *Cochran's theorem* will be most helpful in this connection. For our purposes, this theorem can be put as follows:

(3.59) If all n observations Y_i come from the same normal distribution with mean μ and variance σ^2, and $SSTO$ is decomposed into k sums of squares SS_r, each with degrees of freedom df_r, then the SS_r/σ^2 terms are independent χ^2 variables with df_r degrees of freedom if:

$$\sum_{r=1}^{k} df_r = n - 1$$

Note from Table 3.2 that we have decomposed $SSTO$ into the two sums of squares SSR and SSE, and that their degrees of freedom are additive. Hence:

If $\beta_1 = 0$ so that all Y_i have the same mean $\mu = \beta_0$ and the same variance σ^2, $\dfrac{SSE}{\sigma^2}$ and $\dfrac{SSR}{\sigma^2}$ are independent χ^2 variables.

Now consider the test statistic F^*, which we can write as follows:

$$F^* = \frac{\dfrac{SSR}{\sigma^2}}{1} \div \frac{\dfrac{SSE}{\sigma^2}}{n-2} = \frac{MSR}{MSE}$$

But by Cochran's theorem, we have when C_1 holds:

$$F^* = \frac{\chi^2(1)}{1} \div \frac{\chi^2(n-2)}{n-2}$$

where the χ^2 variables are independent. Thus, when C_1 holds, F^* is the ratio of two independent χ^2 variables, each divided by its degrees of freedom. But this is the definition of an F distributed random variable in (1.42).

We have thus established that if C_1 holds, F^* follows the F distribution, specifically the $F(1, n-2)$ distribution.

When C_2 holds, it can be shown that F^* follows the noncentral F distribution, a complex distribution that we need not consider further at this time.

Note

SSR and SSE are independent and $SSE/\sigma^2 = \chi^2$ even if $\beta_1 \neq 0$. But that both SSR/σ^2 and SSE/σ^2 are χ^2 random variables requires that $\beta_1 = 0$.

Construction of Decision Rule. Since the test is upper-tailed and F^* is distributed as $F(1, n-2)$ when C_1 holds, the decision rule is as follows when the risk of a Type I error is to be controlled at α:

$$\begin{aligned} &\text{If } F^* \leq F(1-\alpha; 1, n-2), \text{ conclude } C_1 \\ &\text{If } F^* > F(1-\alpha; 1, n-2), \text{ conclude } C_2 \end{aligned} \tag{3.60}$$

where $F(1-\alpha; 1, n-2)$ is the $(1-\alpha)$ 100 percentile of the appropriate F distribution.

Example. Using our Westwood Company lot size example again, let us repeat the earlier test on β_1. This time we will use the F test. The alternative conclusions are:

$$\begin{aligned} C_1&: \beta_1 = 0 \\ C_2&: \beta_1 \neq 0 \end{aligned}$$

As before, let $\alpha = .05$. Since $n = 10$, we require $F(.95; 1, 8)$. We find from Table A–4 in the Appendix that $F(.95; 1, 8) = 5.32$. The decision rule is:

$$\begin{aligned} &\text{If } F^* \leq 5.32, \text{ conclude } C_1 \\ &\text{If } F^* > 5.32, \text{ conclude } C_2 \end{aligned}$$

We have from Table 3.3 that $MSR = 13{,}600$ and $MSE = 7.5$. Hence, F^* is:

$$F^* = \frac{13{,}600}{7.5} = 1{,}813$$

Since $F^* = 1{,}813 > 5.32$, we conclude C_2, that $\beta_1 \neq 0$, or that there is a relation between man-hours and lot size. This is the same result as when the t test was employed, as it must be according to our discussion below.

Equivalence of F Test and t Test. For a given α level, the F test of $\beta_1 = 0$ versus $\beta_1 \neq 0$ is equivalent algebraically to the two-tailed t test. To see this, recall from (3.49b) that:

$$SSR = b_1^2 \sum (X_i - \bar{X})^2$$

Thus we can write:

$$F^* = \frac{SSR \div 1}{SSE \div (n-2)} = \frac{b_1^2 \sum (X_i - \overline{X})^2}{MSE}$$

But since $s^2(b_1) = MSE / \sum (X_i - \overline{X})^2$, we obtain:

$$F^* = \frac{b_1^2}{s^2(b_1)} = \left[\frac{b_1}{s(b_1)}\right]^2 \tag{3.61}$$

Now, we know from earlier discussion that the t^* statistic for testing whether or not $\beta_1 = 0$ is, by (3.18):

$$t^* = \frac{b_1}{s(b_1)}$$

In squaring, we obtain the expression for F^* in (3.61). Thus:

$$(t^*)^2 = \left[\frac{b_1}{s(b_1)}\right]^2 = F^*$$

In our illustrative problem, we just calculated that $F^* = 1{,}813$. From earlier work we have: $b_1 = 2.0$, $s(b_1) = .04697$. Thus:

$$(t^*)^2 = \left[\frac{2.0}{.04697}\right]^2 = 1{,}813$$

Corresponding to the relation between t^* and F^*, we have the following relation between the required percentiles of the t and F distributions in the tests: $[t(1 - \alpha/2; n-2)]^2 = F(1 - \alpha; 1, n-2)$. In our tests on β_1 these percentiles were: $[t(.975; 8)]^2 = (2.306)^2 = 5.32 = F(.95; 1, 8)$. Remember that the t test is two-tailed while the F test is one-tailed.

Thus at a given α level, we can use either the t test or the F test for testing $\beta_1 = 0$ versus $\beta_1 \neq 0$. Whenever one test leads to C_1, so will the other, and correspondingly for C_2. The t test, however, is more flexible since it can be used for one-sided alternatives involving $\beta_1 (\leq \geq) 0$ versus $\beta_1 (> <) 0$, while the F test cannot.

General Linear Test. The analysis of variance test of $\beta_1 = 0$ versus $\beta_1 \neq 0$ is an example of a general test of a linear statistical model. We shall briefly explain this general test approach, which is based on a likelihood ratio test, in terms of our simple regression model. We do so at this time because of the generality of the approach and the wide use we shall make of it, and because of the simplicity of understanding the approach in terms of our present problem.

We begin with the model, which in this context is called the *full* or *unrestricted* model. For our simple regression case, the full model is:

$$Y_i = \beta_0 + \beta_1 X_i + \varepsilon_i \qquad \text{Full model} \tag{3.62}$$

We fit this full model by the method of least squares and obtain the error sum of squares SSE. In this context, we shall call this sum of squares $SSE(F)$

to indicate that it measures the variation of the Y's around the regression line for the full model.

Next we consider C_1. In this instance, we have:

$$C_1: \beta_1 = 0 \tag{3.63}$$

The model when C_1 holds is called the *reduced* or *restricted* model. Here, it is:

$$Y_i = \beta_0 + \varepsilon_i \qquad \text{Reduced model} \tag{3.64}$$

We fit this reduced model by the method of least squares and obtain the error sum of squares for this reduced model, denoted $SSE(R)$. When we fit the particular reduced model (3.64), it can be shown that the least squares estimator of β_0 is $\bar{Y}$. Hence, $\hat{Y}_i \equiv \bar{Y}$ and the error sum of squares for this reduced model is:

$$SSE(R) = \sum (Y_i - \bar{Y})^2 = SSTO \tag{3.65}$$

The logic now is to compare $SSE(F)$ and $SSE(R)$. It can be shown that $SSE(F)$ never is greater than $SSE(R)$. The reason is that the more parameters there are in the model, the better one can fit the data and the smaller are the deviations around the fitted regression line. If $SSE(F)$ is not much less than $SSE(R)$, using the full model does not account for much more of the variability of the Y's than the reduced model, in which case the data suggest C_1 holds. To put this another way, if $SSE(F)$ is close to $SSE(R)$, the variation of the observations around the regression line for the full model is almost as great as the variation around the regression line for the reduced model, in which case the added parameters in the full model really don't help to reduce the variation in the Y's. Thus, if $SSE(R) - SSE(F)$ is small, the evidence suggests that C_1 holds. On the other hand, if $SSE(R) - SSE(F)$ is large, this suggests that C_2 holds because the additional parameters in the model do help to reduce substantially the variation of the observations Y_i around the fitted regression line.

The actual test statistic used is a function of $SSE(R) - SSE(F)$, namely:

$$F^* = \frac{SSE(R) - SSE(F)}{df_R - df_F} \div \frac{SSE(F)}{df_F} \tag{3.66}$$

which follows the F distribution if C_1 holds. The degrees of freedom df_R and df_F are those associated with the reduced and full model error sums of squares respectively. Large values of F^* lead to C_2.

For our application, we have:

$$SSE(R) = SSTO \qquad SSE(F) = SSE$$
$$df_R = n - 1 \qquad df_F = n - 2$$

so that we obtain when substituting into (3.66):

$$F^* = \frac{SSTO - SSE}{(n-1) - (n-2)} \div \frac{SSE}{n-2} = \frac{SSR}{1} \div \frac{SSE}{n-2} = \frac{MSR}{MSE}$$

which is our old test statistic (3.58).

This general approach can be used for highly complex tests of linear statistical models, as well as for simpler tests. The basic steps again are:

1. Fit the full model and obtain the error sum of squares $SSE(F)$.
2. Fit the reduced model under C_1 and obtain the error sum of squares $SSE(R)$.
3. Use the test statistic (3.66).

3.9 DESCRIPTIVE MEASURES OF ASSOCIATION BETWEEN X AND Y IN REGRESSION MODEL

We have discussed the major uses of regression analysis—estimation of parameters and means and prediction of new observations—without mentioning the "degree of association" between X and Y, or similar terms. The reason is that the usefulness of estimates or predictions depends upon the width of the interval and the user's needs for precision, which vary from one application to another. Hence, no single descriptive measure of the "degree of association" can capture the essential information as to whether a given regression relation is useful in any particular application.

Nevertheless, two descriptive measures of association are frequently used in practice to describe the degree of relation between X and Y, and we discuss these briefly.

Coefficient of Determination

We saw earlier that $SSTO$ measures the variation in the observations Y_i when no account of the independent variable X is taken. Thus, $SSTO$ is a measure of the uncertainty in predicting Y when X is not considered. Similarly, SSE measures the variation in the Y_i, or the uncertainty in predicting Y, when a regression model utilizing the independent variable X is employed. A natural measure of the effect of X in reducing the variation in Y, that is, the uncertainty in predicting Y, is therefore:

$$r^2 = \frac{SSTO - SSE}{SSTO} = \frac{SSR}{SSTO} = 1 - \frac{SSE}{SSTO} \tag{3.67}$$

The measure r^2 is called the *coefficient of determination*. Since $0 \leq SSE \leq SSTO$, it follows that:

$$0 \leq r^2 \leq 1 \tag{3.68}$$

We may interpret r^2 as the proportionate reduction of total variation associated with the use of the independent variable X. Thus, the larger is r^2, the more is the total variation of Y reduced by introducing the independent variable X. The limiting values of r^2 occur as follows:

1. If all observations fall on the fitted regression line, $SSE = 0$ and $r^2 = 1$. This case is shown in Figure 3.6a. Here, the independent variable X accounts for all variation in the observations Y_i.

FIGURE 3.6

Scatter Plots When $r^2 = 0$ and $r^2 = 1$

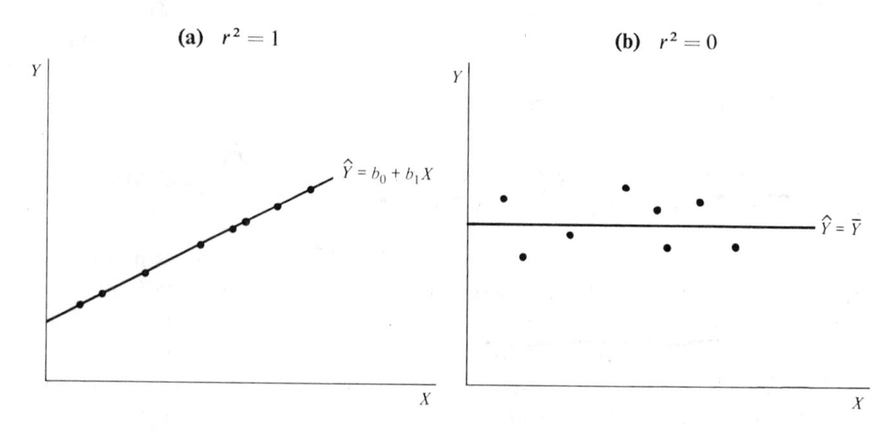

2. If the slope of the fitted regression line is $b_1 = 0$, so that $\hat{Y}_i \equiv \bar{Y}$, $SSE = SSTO$ and $r^2 = 0$. This case is shown in Figure 3.6b. Here, there is no relation between X and Y in the sample data, and the independent variable X is of no help in reducing the variation in the observations Y_i.

In practice, r^2 is not likely to be 0 or 1, but rather somewhere in between these limits. The closer it is to 1, the greater is said to be the degree of association between X and Y.

Coefficient of Correlation

The square root of r^2:

$$r = \pm\sqrt{r^2} \tag{3.69}$$

is called the *coefficient of correlation*. A plus or minus sign is attached to this measure according to whether the slope of the fitted regression line is positive or negative. Thus, the range of r is:

$$-1 \leq r \leq 1 \tag{3.70}$$

Whereas r^2 indicates the proportional reduction in the variability of Y attained by the use of information about X, the square root, r, does not have such a clear-cut operational interpretation. Nevertheless, there is a tendency to use r instead of r^2 in much applied work in business and economics. It is worth noting that since for any r^2 other than 0 or 1, $r^2 < |r|$, r may give the impression of a "closer" relationship between X and Y than does the corresponding r^2. For instance, $r^2 = .10$ indicates that the total variation in Y is reduced by only 10 percent when X is introduced, yet $|r| = .32$ may give an impression of greater association between X and Y.

Example

For the Westwood Company example, we obtained $SSTO = 13{,}660$ and $SSE = 60$. Hence:

$$r^2 = \frac{13{,}660 - 60}{13{,}660} = .996$$

Thus the variation in man-hours is reduced by 99.6 percent when lot size is considered.

The correlation coefficient in this example is:

$$r = +\sqrt{.996} = +.998$$

The plus sign is affixed since b_1 is positive.

Computational Formula for r

A direct computational formula for r, which automatically furnishes the proper sign, is:

$$(3.71) \qquad r = \frac{\sum (X_i - \bar{X})(Y_i - \bar{Y})}{[\sum (X_i - \bar{X})^2 \sum (Y_i - \bar{Y})^2]^{1/2}}$$

$$= \frac{\sum X_i Y_i - \frac{\sum X_i \sum Y_i}{n}}{\left[\left(\sum X_i^2 - \frac{(\sum X_i)^2}{n}\right)\left(\sum Y_i^2 - \frac{(\sum Y_i)^2}{n}\right)\right]^{1/2}}$$

Comments

1. The following relation between b_1 and r is interesting:

$$(3.72) \qquad b_1 = \left[\frac{\sum (Y_i - \bar{Y})^2}{\sum (X_i - \bar{X})^2}\right]^{1/2} r$$

Note that $b_1 = 0$ when $r = 0$ and vice versa. Thus, $r = 0$ implies a horizontal fitted regression line, and vice versa.

2. The value taken by r^2 in a given sample tends to be affected by the spacing of the X observations. This is implied in (3.67). SSE is not affected systematically by the spacing of the X's since for model (3.1), $\sigma^2(Y_i) = \sigma^2$ at all X levels. However, the wider the spacing is of the X's in the sample, the greater will tend to be the spread of the observed Y's around $\bar{Y}$ and hence the greater will be $SSTO$. Consequently, the wider the X's are spaced, the higher will tend to be r^2.

3. The regression sum of squares SSR is often called the "explained variation" in Y. The residual sum of squares SSE is then called the "unexplained variation," and the total sum of squares the "total variation." The coefficient r^2 then is inter-

preted in terms of the proportion of the total variation in Y which has been "explained" by X. Unfortunately this terminology frequently is taken literally, hence misunderstood. Remember that in a regression model there is no implication that Y necessarily depends on X in a causal or explanatory sense.

4. A value of r or r^2 relatively close to 1 sometimes is taken as an indication that sufficiently precise inferences on Y can be made from knowledge of X. As mentioned earlier, the usefulness of the regression relation depends upon the width of the confidence or prediction interval and the particular needs for precision, which vary from one application to another. Hence, no single measure is an adequate indicator of the usefulness of the regression relation.

5. Regression models do not contain any parameter to be estimated by r or r^2. These coefficients simply are descriptive measures of the degree of relationship between X and Y in the sample observations which may, or may not, be useful in any one instance. In a later chapter, we discuss correlation models which do contain a parameter for which r is an estimator.

3.10 COMPUTER OUTPUT

Figure 3.7 shows again the computer printout for the Westwood Company case presented in Figure 2.12. We referred to selected items in this printout in Chapter 2. Now, we are in a position to review the printout as a whole.

The 10 observations on lot size and man-hours are printed at the top. This enables us to verify that the observations were entered into the computer accurately. The variables are not labeled individually in the printout, but we have annotated them.

Next is an explanatory statement identifying the items given in the printout. The estimated regression equation happens to be shown in a different notation than is used in this book, hence we have annotated it.

The basic results for the Westwood Company example are given in the rectangular array of figures at the bottom of the printout. Referring to the explanatory statement, we note that row 1 contains the following items in sequence: the digit 2 (for identification); the value of b_0; the value of $s(b_0)$; the value of t^* in a test to decide between $C_1: \beta_0 = 0$ versus $C_2: \beta_0 \neq 0$; and the digit 0. This zero is entered because the program shows an entry for every cell in the array.

The second row contains: the digit 1; b_1; $s(b_1)$; t^* as obtained in the test to decide between $C_1: \beta_1 = 0$ versus $C_2: \beta_1 \neq 0$; and the digit 0.

Rows 3, 4, and 5 contain analysis of variance and related results. Row 3 gives: the digit 0; df associated with SSR; SSR; MSR; and the value of F^* obtained in the F test of $C_1: \beta_1 = 0$ versus $C_2: \beta_1 \neq 0$.

Row 4 gives: the digit 0; df associated with SSE; SSE; MSE; and the digit 0.

Row 5 gives: the digit 0; df associated with $SSTO$; $SSTO$; $\sqrt{MSE}$; and r^2.

We have annotated the ANOVA table in Figure 3.7 so that it can be seen clearly how the data fit into it. Computer printouts for regression analysis

FIGURE 3.7

Example of Computer Printout for Westwood Company Case

```
SIMPLE LINEAR REGRESSION

X_i     X      Y_i
 30      73
 20      50
 60     128
 80     170
 40      87
 50     108
 60     135
 30      69
 70     148
 60     132

X IS A MATRIX OF OBSERVATIONS WHERE THE COLUMNS CORRESPOND TO
VARIABLES AND THE ROWS TO OBSERVATIONS.  THE MATRIX BELOW GIVES
THE RESULTS OF THE BEST LEAST SQUARES FIT OF THE FUNCTION:
                   X4 = A + B×X1   ← Ŷ = b0 + b1 X

IN THE FOLLOWING FORMAT:

ROW 1:  2,A,ST.  ERROR OF A, T-VALUE, 0
ROW 2:  1,B,ST.  ERROR OF B, T-VALUE, 0
ROW 3:  0,DF FOR REGRESSION, SUM OF SQUARES, MEAN SQUARE, F-VALUE
ROW 4:  0,DF FOR ERROR, SUM OF SQUARES, MEAN SQUARE, 0
ROW 5:  0,DF FOR TOTAL, SUM OF SQUARES, ST.  ERROR OF ESTIMATE,
                                   SQUARE OF MULTIPLE CORR. COEF.
```

		Source d.f.	SS	MS	
2.00000		b_0 → 10.00000	$s(b_0)$ → 2.50294	t^* → 3.99530	0.00000
1.00000		b_1 → 2.00000	$s(b_1)$ → 0.04697	t^* → 42.58325	0.00000
0.00000	Regression	1.00000	13600.00000	13600.00000	F^* → 1813.33333
0.00000	Error	8.00000	60.00000	7.50000	0.00000
0.00000	Total	9.00000	13660.00000	$\sqrt{MSE}$ → 2.73861	r^2 → 0.99561

programs vary substantially in format from one program package to another. In some programs, for instance, the analysis of variance results are given in an explicit ANOVA table, with columns and rows labeled as in our annotation. In others, such as the one illustrated here, only the essential numbers are provided.

PROBLEMS

3.1. Suppose that the normal error, linear regression model (3.1) is applicable except that the error variance is not constant; rather it is larger, the larger X is. Does $\beta_1 = 0$ still imply that there is no relation between X and Y?

3.2. A student, working on a summer internship in the economic research office of a large corporation, studied the relation between sales of a product

(Y, in millions of dollars) and population (X, in millions of persons) in the firm's 50 marketing districts. Model (3.1) was employed. The student first wished to test whether or not a relation between Y and X existed. Using a time-sharing computer service available to the firm, the student accessed an interactive simple linear regression program and obtained the following information on the regression coefficients:

Parameter	*Estimated Value*	*95% Confidence Limits*	
Intercept	7.43119	−1.18518	16.0476
Slope	.755048	.452886	1.05721

a) The student concluded from these results that there is a relation between Y and X. Is the conclusion warranted? What is the implied level of significance?

b) Someone questioned the negative lower confidence limit for the intercept, pointing out that dollar sales cannot be negative even if the population in a district were zero. Discuss.

3.3. Show that b_0 as defined in (3.19) is an unbiased estimator of β_0.

3.4. Derive the expression in (3.20b) for the variance of b_0, making use of theorem (3.29).

3.5. Refer to Figure 3.7 for the Westwood Company case. A consultant has advised that an increase of one unit in lot size should require an increase of 1.8 in the expected number of man-hours for the given production item.

a) Conduct a test to decide whether or not the increase in the expected number of man-hours in the Westwood Company is equal to this standard. Use $\alpha = .05$. What is your conclusion?

b) Why is $t^* = 42.58325$, given in the printout, not relevant for the test in part (a)?

c) Obtain the power of your test if the consultant's standard actually is being exceeded by .1 hours. Assume $\sigma(b_1) = .05$.

3.6. The Tri-City Office Equipment Corporation sells an imported desk calculator on a franchise basis, and performs preventive maintenance and repair service on this calculator. The data below have been collected from 18 recent calls on users to perform routine preventive maintenance service; X is the number of machines serviced, and Y is the total number of minutes spent by the service man on the call. Assume model (3.1) is appropriate.

i:	1	2	3	4	5	6	7	8	9
X_i:	7	4	5	1	5	4	7	2	4
Y_i:	97	57	78	10	75	62	101	27	53

i:	10	11	12	13	14	15	16	17	18
X_i:	2	8	5	2	5	7	1	4	5
Y_i:	33	118	65	25	71	105	17	49	68

a) Estimate the change in the mean response when the number of machines serviced increases by one. Use a 95 percent confidence coefficient.

b) The manufacturer suggests that the mean required time should not increase by more than 13 minutes for each additional machine that is given routine preventive maintenance on a service call. Conduct a test to decide whether this standard is being satisfied in Tri-City. Control your risk of a Type I error at .05. What do you conclude?

c) Does b_0 give any information here on the "start-up" time on calls—that is, the time required before service work is begun on the machines?

3.7. Refer to Figure 3.7. A student, noting that $s(b_0)$ and $s(b_1)$ are given, asked why $s(\hat{Y})$ is not given. Discuss.

3.8. Refer to Figure 3.7 for the Westwood Company case.

a) Obtain an interval estimate of the mean number of man-hours in production runs where the lot size is 45 parts. Use a 90 percent confidence coefficient here and below.

b) Obtain a prediction interval for the number of man-hours required in the next production run for a lot of 45 parts.

c) Suppose that four new lots are to be produced, each for 45 parts, in separate production runs during the near future. Obtain: (1) a prediction interval for the mean number of man-hours required per run for these four runs; and (2) a prediction interval for the total number of man-hours required for these four runs.

d) Are your results in parts (a), (b), and (c) mutually consistent? Discuss, making reference to the locations and comparative widths of the intervals.

3.9. Refer to Problem 3.6.

a) Obtain an interval estimate of the mean service time on calls in which seven machines are serviced. Use a 90 percent confidence coefficient here and below.

b) Obtain a prediction interval for the service time on the next call in which seven machines are to be serviced. Is it reasonable that your interval here should be wider than in part (a)?

c) Interpret your confidence coefficients in parts (a) and (b).

d) Which type of interval—that in part (a) or in part (b)—would be more useful for statistical control purposes? Which would be more useful for ascertaining long-term manpower requirements in the firm? Discuss.

e) How would the meaning of the confidence coefficients in parts (a) and (b) change if the independent variable were considered a random variable and the conditions in (3.39) were applicable?

3.10. (Calculus needed.)

a) Obtain the likelihood function for the sample observations Y_i if the conditions in (3.39) apply.

b) Show that the maximum likelihood estimators of β_0, β_1, and σ^2 under the conditions in (3.39) are the same as in (2.27) when the X_i are fixed.

3.11. In a small-scale regression study, four observations on Y were obtained corresponding to $X = 2, 6, 10, 14$. Assume that $\sigma = .6$, $\beta_0 = 8$, $\beta_1 = 2$.

a) What are the expected values of MSR and MSE here?

b) For purposes of determining whether or not a regression relation exists, would it have been better or worse to have made observations for $X = 6, 7, 9, 10$? Why? Would the same answer apply if the principal purpose were to estimate the mean response for a given X?

3.12. Derive the expression in (3.49b) for SSR.

3.13. Refer to Problem 2.13. Plot the deviations $Y_i - \hat{Y}_i$ against X_i on a chart. On a separate chart plot the deviations $\hat{Y}_i - \bar{Y}$ against X_i. From your charts, can you tell which of the two components of *SSTO*—*SSR* or *SSE*—is the larger?

3.14. Refer to Figure 3.7 for the Westwood Company case.
   a) Explain the relation between $t^* = 42.58325$ and $F^* = 1813.33333$ here.
   b) For conducting statistical tests concerning the parameter β_1, why is the t test more versatile than the F test?

3.15. When testing whether or not $\beta_1 = 0$ by means of the F test, why is the F test one-sided even though the alternative possibilities are $\beta_1 < 0$ or $\beta_1 > 0$?

3.16. A value of r^2 near 1 is sometimes interpreted to imply that the relation between Y and X is sufficiently close so that suitably precise predictions of Y can be made from knowledge of X. Is this implication a necessary consequence of the definition of r^2?

3.17. Refer to Problem 3.6.
   a) Set up the ANOVA table. Which elements of your table are additive?
   b) Test by means of the F test whether or not there is a relation between time spent and number of machines serviced. Control the risk of Type I error at .05. What do you conclude?
   c) By how much, relatively, is the total variation in number of minutes spent per call reduced when the number of machines serviced is introduced into the analysis? Is this a relatively small or large reduction? What is the name of this measure?
   d) Calculate r and attach the appropriate sign.
   e) Which measure, r or r^2, has the more clear-cut operational interpretation?

3.18. Refer to Problem 2.13.
   a) Obtain the data for the ANOVA table and set it up using the format of Table 3.2; then set it up using the alternative format of Table 3.4a. How do the two tables differ?
   b) Calculate r^2 and interpret your result.

3.19. An analyst, in empirically developing a cost function from observed data, employed model (3.1) tentatively. For the process under study the intercept term in the model could be interpreted as representing fixed cost and the term $\beta_1 X_i$ as representing total variable cost. The analyst hypothesized that the fixed costs should be 10 thousand dollars, and wished to test this hypothesis by means of a general linear test.
   a) Indicate the alternative conclusions in the test.
   b) Specify the reduced model.
   c) Can you tell, without additional information, what the quantity $df_R - df_F$ in the test statistic (3.66) will equal in the analyst's test?

4

Aptness of Model and Remedial Measures

When a regression model, such as the simple linear regression model (3.1), is selected for an application, one can usually not be certain in advance that the model is appropriate for that application. Any one, or several, of the features of the model, such as linearity of the regression function or normality of the error terms, may not be appropriate for the particular data at hand. Hence, it is important to examine the aptness of the model for the data, before further analysis based on that model is undertaken. In this chapter, we discuss some simple graphic methods for studying the aptness of a model, as well as some formal statistical tests for doing so. We conclude with a consideration of some techniques whereby the simple regression model (3.1) can be made appropriate when the data do not accord with the conditions of the model.

While the discussion in this chapter is in terms of the aptness of the simple regression model (3.1), the basic principles apply to all statistical models discussed in this book.

4.1 RESIDUALS

A residual e_i, as defined in (2.16), is the difference between the observed value and the fitted value:

(4.1)
$$e_i = Y_i - \hat{Y}_i$$

As such, it may be regarded as the observed error, in distinction to the unknown true error ε_i in the regression model:

(4.2)
$$\varepsilon_i = Y_i - E(Y_i)$$

For regression model (3.1), the ε_i are assumed to be independent normal random variables, with mean 0 and constant variance σ^2. If the model is appropriate for the data at hand, the observed residuals e_i should then reflect the properties assumed for the ε_i. This is the basic idea underlying *residual analysis*, a highly useful means of examining the aptness of a model.

Properties of Residuals

The mean of the n residuals e_i is, by (2.17):

$$\bar{e} = \frac{\sum e_i}{n} = 0 \tag{4.3}$$

where $\bar{e}$ denotes the mean of the residuals. Thus, since $\bar{e}$ is always 0, it provides no information as to whether the true errors ε_i have expected value $E(\varepsilon_i) = 0$.

The variance of the n residuals e_i is defined as follows for model (3.1):

$$\frac{\sum (e_i - \bar{e})^2}{n-2} = \frac{\sum e_i^2}{n-2} = \frac{SSE}{n-2} = MSE \tag{4.4}$$

If the model is appropriate, MSE is, as noted earlier, an unbiased estimator of the variance of the error terms σ^2.

Standardized Residuals

For analytical convenience, the standardized residual:

$$\frac{e_i}{\sqrt{MSE}} \tag{4.5}$$

is sometimes used in residual analysis. We shall explain residual analysis mainly in terms of the residuals e_i, but occasionally will employ the standardized residuals.

Nonindependence of Residuals

The residuals e_i are not independent random variables. This is evident from (2.17), which constrains the sum of the e_i to 0. Thus, if we know $n - 1$ of the e_i, the last one is determined. Another constraint on the residuals is that of (2.19). The same lack of independence holds for the standardized residuals.

When the sample size is not small relative to the number of constraints on the e_i, however, the dependency effect is relatively unimportant and can be ignored for most purposes.

Departures from Model to Be Studied by Residuals

We shall consider the use of residuals for examining six important types of departures from model (3.1), the simple linear regression model with normal errors:

1. The regression function is not linear.
2. The error terms do not have constant variance.
3. The error terms are not independent.
4. The model fits all but one or a few outlier observations.
5. The error terms are not normally distributed.
6. One or several important independent variables have been omitted from the model.

4.2 GRAPHIC ANALYSIS OF RESIDUALS

We take up now some informal ways in which graphs of residuals can be analyzed to provide information on whether any of the six types of departures from the simple linear regression model (3.1) just mentioned are present.

Nonlinearity of Regression Function

Whether or not a linear regression function is appropriate for the data being analyzed can often be studied from a scatter diagram of the data, with the fitted regression function plotted on it. Figure 4.1a contains the data and the fitted regression line for a study of the relation between amount of transit information and bus ridership in eight comparable test cities, where X is the number of bus transit maps distributed free to residents of the city at the beginning of the test period and Y is the increase during the test period in average daily bus ridership during nonpeak hours. The original data and fitted values are given in Table 4.1. The graph suggests strongly that a linear regression function is not appropriate.

Figure 4.1b presents for this same example the residuals e_i, shown in Table 4.1, plotted against the independent variable X_i. The lack of fit of a linear regression function is also strongly suggested by the residual plot

TABLE 4.1

Number of Maps Distributed and Increase in Ridership

City i	*Increase in Ridership (thousands)* Y_i	*Maps Distributed (thousands)* X_i	*Fitted Value* $\hat{Y}_i$	*Residual* $Y_i - \hat{Y}_i = e_i$
1	.60	80	1.66	−1.06
2	6.70	220	7.75	−1.05
3	5.30	140	4.27	+1.03
4	4.00	120	3.40	+.60
5	6.55	180	6.01	+.54
6	2.15	100	2.53	−.38
7	6.60	200	6.88	−.28
8	5.75	160	5.14	+.61

$$\hat{Y} = -1.82 + .0435X$$

FIGURE 4.1

Scatter Plot and Residual Plot for Transit Example Illustrating Nonlinear Regression Function

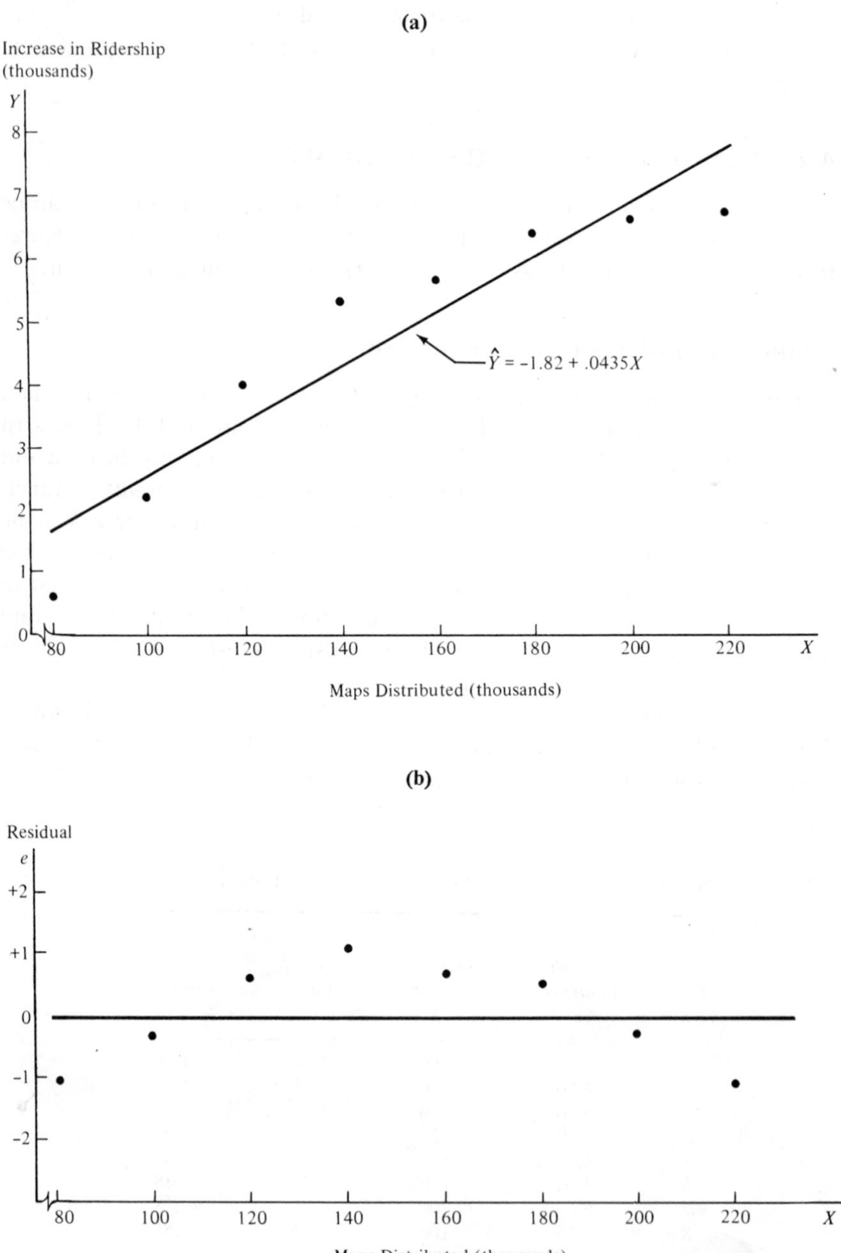

against the X_i in Figure 4.1b, since the residuals depart from 0 in a systematic fashion. Note that they are negative for smaller X values, positive for medium size X values, and negative again for large X values.

In this case, both Figures 4.1a and 4.1b are effective means of examining the appropriateness of the linearity of the regression function. Figure 4.1b, the residual plot, in general has some advantages over Figure 4.1a, the scatter plot. First, the residual plot can easily be used for examining other facets of the aptness of the model. Second, there are occasions when the scaling of the scatter plot places the Y_i observations close to the fitted values $\hat{Y}_i$, for instance, when there is a steep slope. It then becomes more difficult to study the appropriateness of a linear regression function from the scatter plot. A residual plot, on the other hand, could clearly show any systematic pattern in the deviations around the regression line under these conditions.

Figure 4.2a shows a prototype situation of the residual plot against X_i if

FIGURE 4.2

Prototype Residual Plots

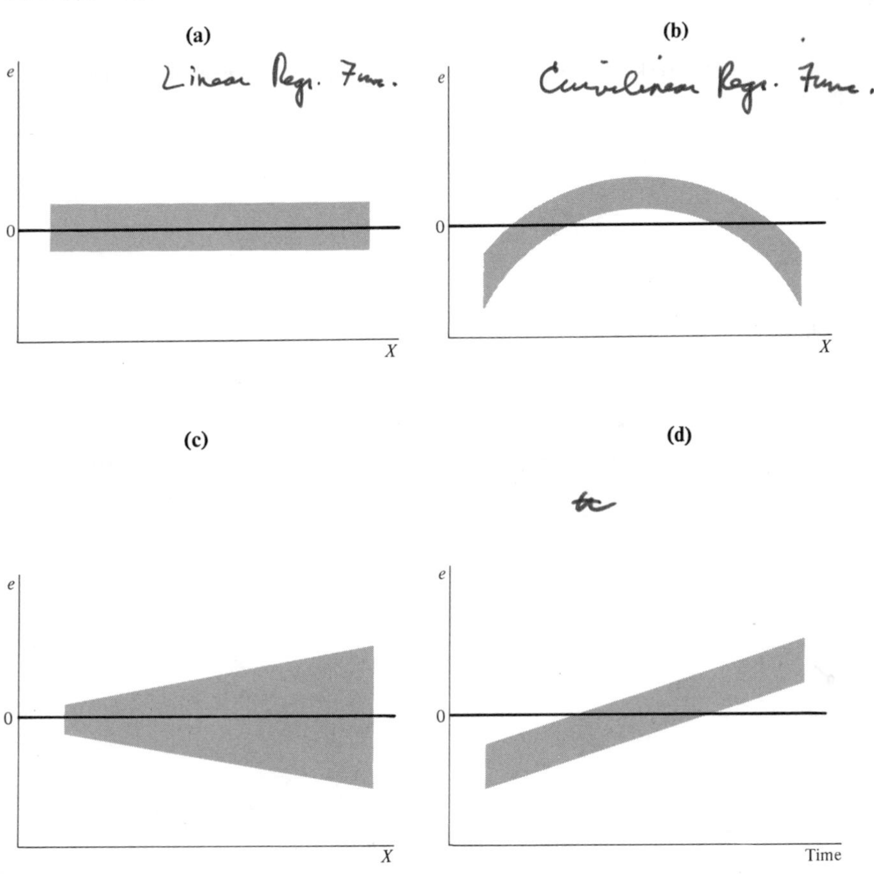

Source: Adapted, with permission, from N. R. Draper and H. Smith, *Applied Regression Analysis* (New York: John Wiley & Sons, Inc., 1966), p. 89.

the linear model is appropriate. The residuals should tend to fall within a horizontal band centered around 0, displaying no systematic tendencies to be positive and negative.

Figure 4.2b shows a prototype situation of a departure from the linear regression model indicating the need for a curvilinear regression function. Here the residuals tend to vary in a systematic fashion between being positive and negative. A different type of departure from linearity would, of course, lead to a different picture than the prototype pattern in Figure 4.2b.

Nonconstancy of Error Variance

A plot of the residuals against the independent variable is not only helpful to study whether a linear regression function is appropriate, but also to examine whether the variance of the error terms is constant. For instance, Figure 4.3 shows a residual plot against the independent variable X for a regression application involving time required to achieve a solution (Y) against

FIGURE 4.3

Residual Plot for Task Complexity Example Illustrating Nonconstant Error Variance

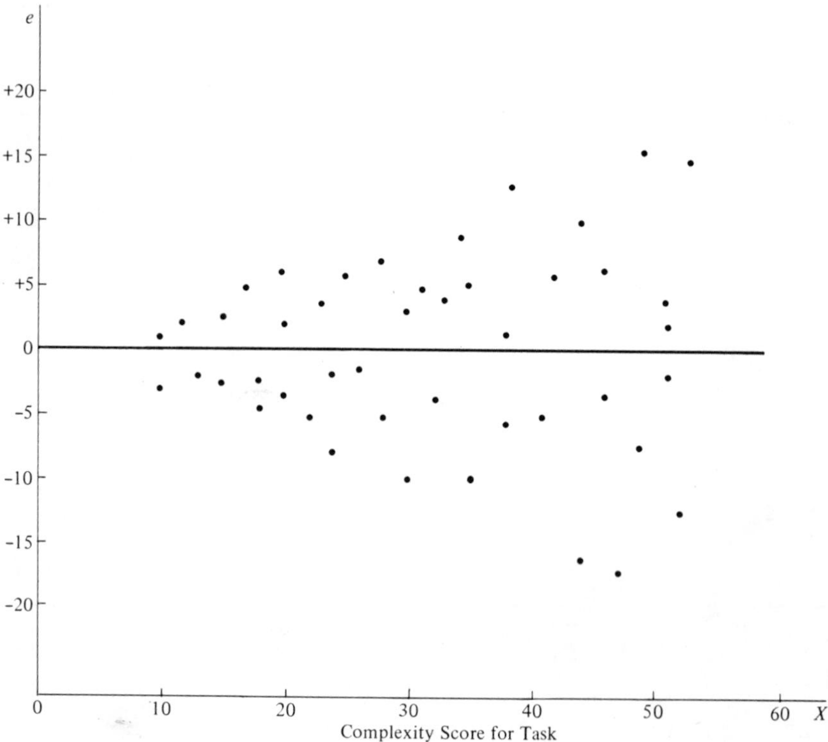

a measure of the complexity of the task (X). Note that the greater is the complexity of the task, the more spread out are the residuals. This suggests that the error variance is larger when X is large than when X is small.

Figure 4.2c shows a prototype picture of a residual plot when the error variance increases with X. In many business and economic applications, departures from constancy of the error variance are of the trapezoidal type shown in Figure 4.2c. One can also encounter error variances decreasing with increasing levels of the independent variable or varying in some other fashion with X.

A residual plot against the fitted values $\hat{Y}_i$ is also an effective means of studying the constancy of the error variance, particularly when the regression function is not linear or when a multiple regression model is employed. Figure 4.4 shows, for the same data as in Figure 4.3, a plot of the residuals e_i against the fitted values $\hat{Y}_i$. Again we see the prototype pattern of Figure 4.2c, suggesting that the error variance increases with X.

FIGURE 4.4

Residual Plot against Fitted Values $\hat{Y}_i$ for Task Complexity Example Illustrating Nonconstant Error Variance

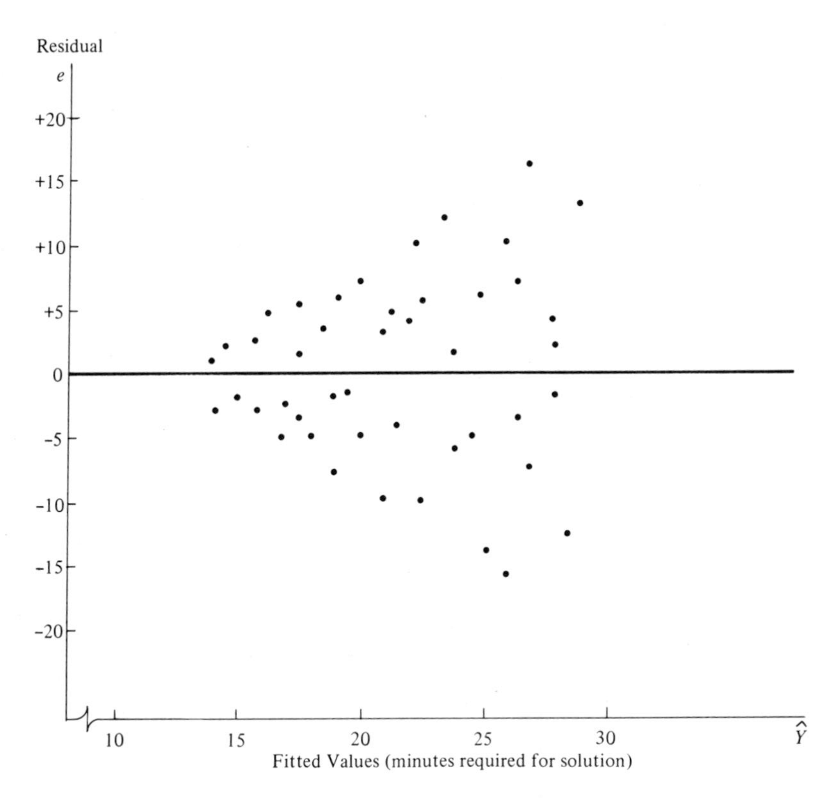

FIGURE 4.5

Residual Plots for Welding Example Illustrating Nonindependence of Error Terms

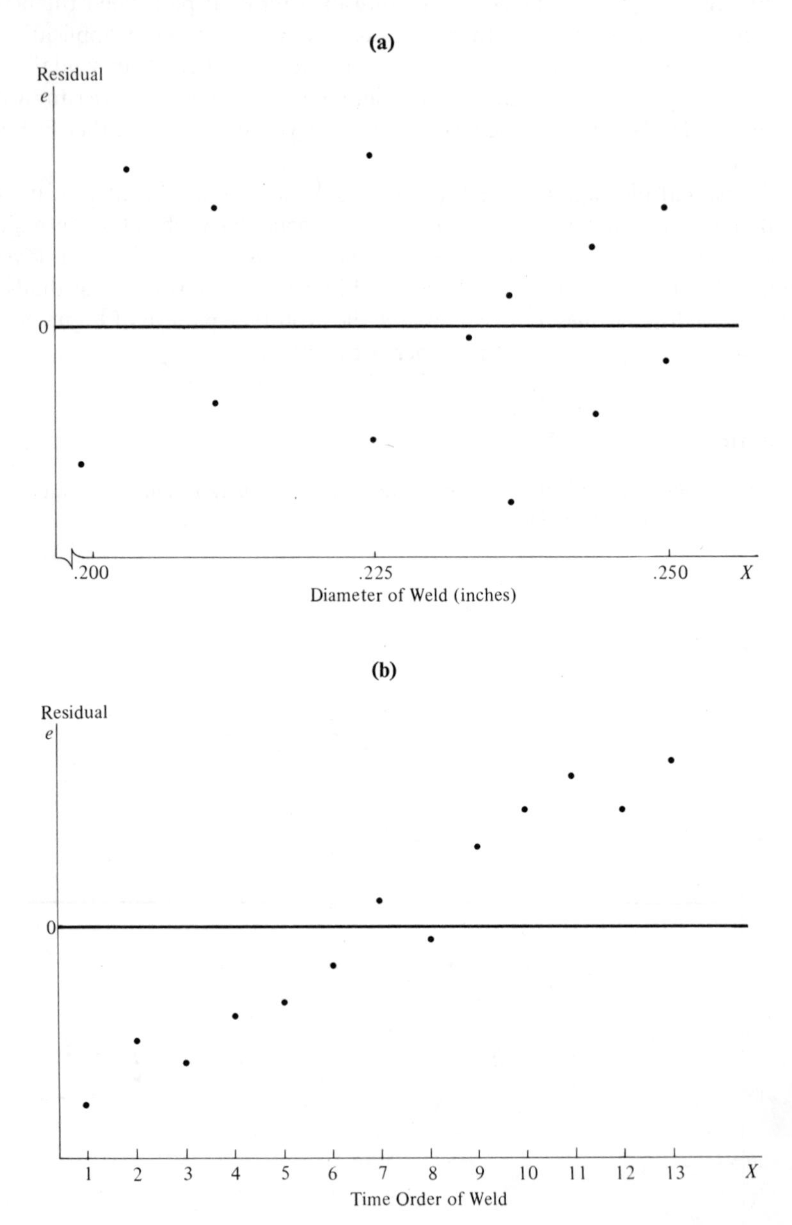

Nonindependence of Error Terms

Whenever data are obtained in a time sequence, it is a good idea to plot the residuals against time, even though time has not been explicitly incorporated as a variable into the model. The purpose is to see if there is any correlation between the error terms over time. In an experiment to study the relation between the diameter of a weld (X) and the shear strength of the weld (Y), the residual plot as shown in Figure 4.5a appears to indicate no departures from the simple regression model, either with respect to linearity or constancy of error variance. When the residuals are plotted in time order in which the welds were made in Figure 4.5b, however, an evident correlation between the error terms stands out. Negative residuals are associated mainly with the early trials, and positive residuals with the later trials. Apparently, some effect connected with time was present, such as learning by the welder or a gradual change in the welding equipment, so that the shear strength tended to be greater in the later welds on account of this effect.

A prototype of a time-related effect is shown in Figure 4.2d, which portrays a linear time-related effect. It is sometimes useful to view the problem of nonindependence of the error terms as one in which an important variable (in this case, time) has been omitted from the model. We shall discuss this type of problem shortly.

When the error terms are independent, we would expect the residuals to fluctuate in a more or less random pattern around the base line 0, such as the scattering shown in Figure 4.6. Lack of randomness can take the form of too

FIGURE 4.6

Residual Plot against Time Suggesting Independence of Error Terms

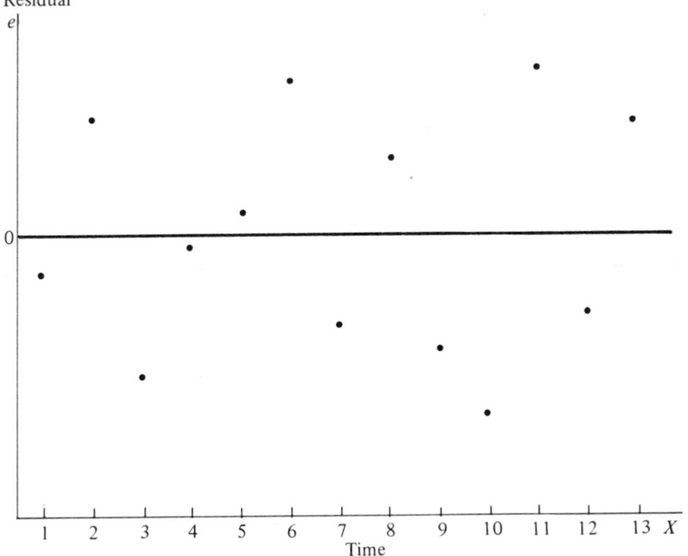

much alternation of points around the zero line, or too little alternation. In practice, there is little concern with the former case except in situations where the error term is subject to periodic or offsetting effects in observations at successive levels of X. Too little alternation, in contrast, frequently is a matter of concern, as in the welding application in Figure 4.5b.

Note

When the residuals are plotted against X, as in Figure 4.1b, the scatter may not appear to be random. For this plot, however, the basic problem is probably not lack of independence of the error terms, but rather a poorly fitting regression function. This, indeed, is the situation portrayed in Figure 4.1a.

Presence of Outliers

Outliers are extreme observations. In a residual plot, they are points that lie far beyond the scatter of the remaining residuals, perhaps four or more standard deviations from zero. Figure 4.7 shows a standardized residual plot with one outlier, which is circled. Note that this residual represents an observation almost six standard deviations from the fitted value.

FIGURE 4.7

Residual Plot with Outlier

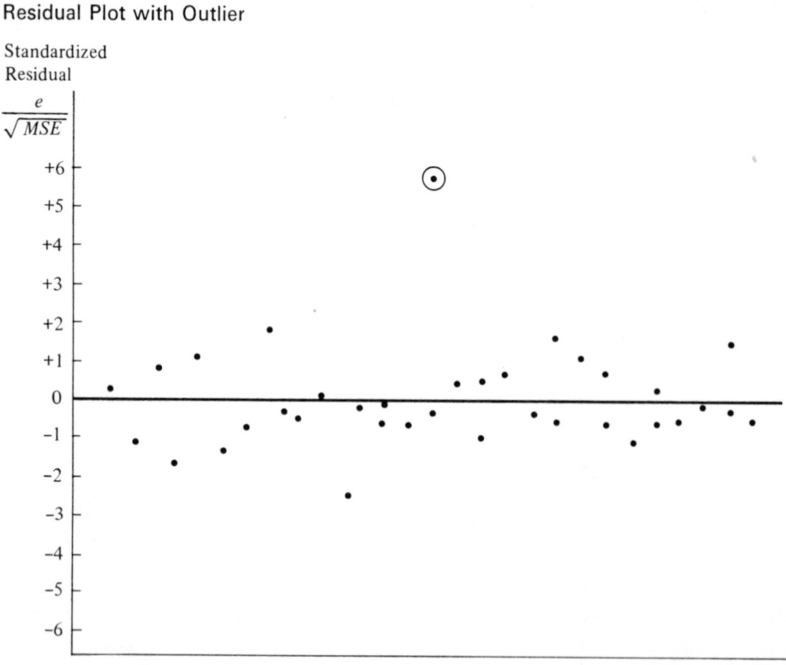

Outliers create great difficulty. When we encounter one, our first suspicion is that the observation resulted from a mistake or other extraneous effect, and hence should be discarded. A major reason for discarding it is that under the least squares method, a fitted line is pulled disproportionately toward an outlying observation because the sum of the *squared* deviations is minimized. This could cause a misleading fit if indeed the outlier observation resulted from a mistake or other extraneous cause. On the other hand, outliers may convey significant information, as when an outlier occurs because of an interaction with another independent variable omitted from the model. A safe rule frequently suggested is to discard an outlier only if there is direct evidence that it represents an error in recording, a miscalculation, a malfunctioning of equipment, or a similar type of circumstance.

Nonnormality of Error Terms

As we noted earlier, small departures from normality do not create any serious problems. Major departures, on the other hand, should be of concern. The normality of the error terms can be studied informally by examining the residuals in a variety of graphic ways. One can construct a histogram of the residuals and see if gross departures from normality are shown by it. Another possibility is to determine whether, say, about 68 percent of the standardized residuals $e_i/\sqrt{MSE}$ fall between -1 and $+1$, or about 90 percent between -1.64 and $+1.64$. (If the sample size is small, the corresponding t values would be used.)

Still another possibility is to plot the residuals on normal probability paper. Table 4.2 contains a cumulative frequency distribution of 21 standardized residuals from a study relating farm implement sales (Y) to total farm acreage (X), in each of 21 states in which this company operates. This cumulative frequency distribution is plotted in Figure 4.8 on normal probability paper.

TABLE 4.2

Cumulative Distribution of Residuals in Farm Implements Sales Study

Upper Class Limit of Residuals	*Number of Residuals*	*Percent of Residuals*
−3.0	0	0
−2.0	1	4.8
−1.0	2	9.5
−.5	7	33.3
0	10	47.6
+.5	13	61.9
+1.0	18	85.7
+1.5	19	90.5
+2.0	20	95.2
+3.0	21	100.0

FIGURE 4.8

Residual Plot on Normal Probability Paper for Farm Implements Example

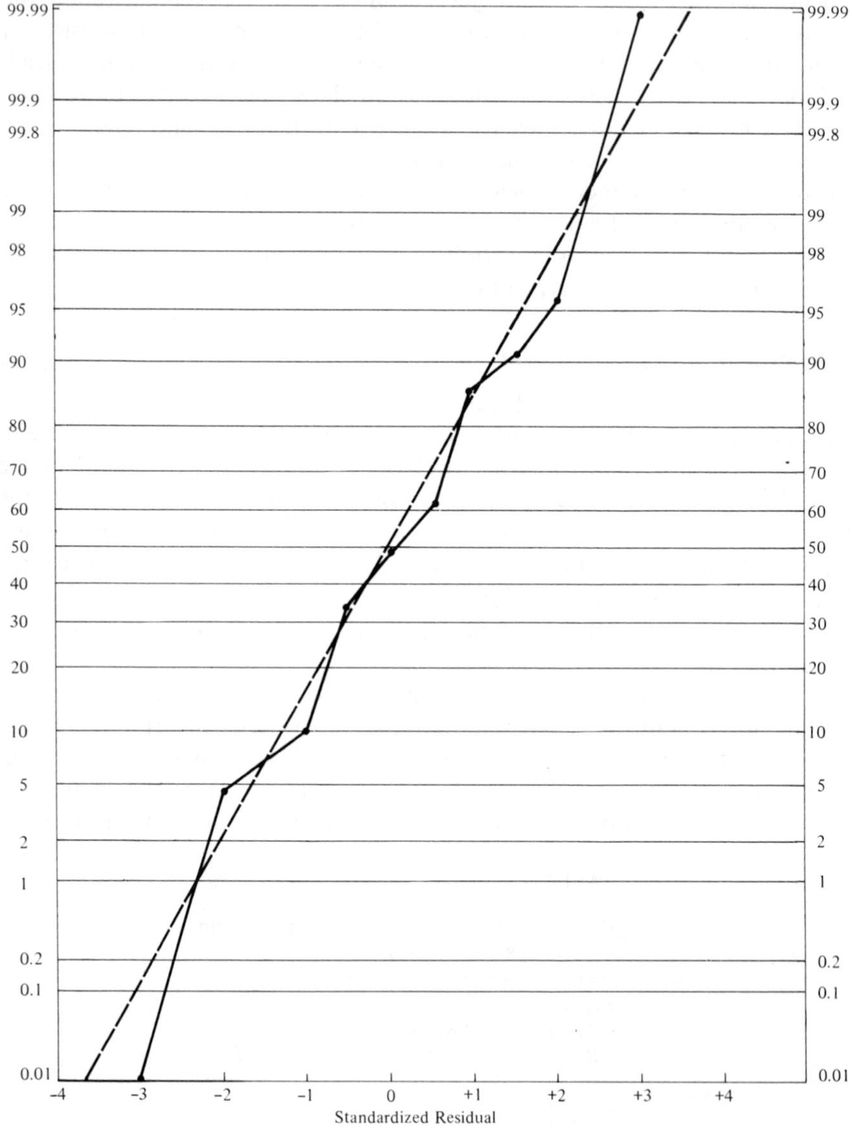

A characteristic of this type of graph paper is that a normal distribution plots as a straight line. Substantial departures from a straight line are grounds for suspecting that the distribution is not normal.

The horizontal scale in the paper used here is an ordinary arithmetic one. Since we are plotting standardized residuals, we have laid out this scale to range from -4 to $+4$. The vertical scale refers to percentages in the left tail of the distribution. The broken straight line drawn on the paper is the plot of the cumulative standard normal distribution. It runs through such points as $(-3, 0.13)$; $(-2, 2.28)$; $(-1, 15.87)$; $(0, 50.00)$; and so on. Remember that in the standard normal distribution, .13 percent of the area is to the left of $z = -3$, 2.28 percent of the area is to the left of $z = -2$, and so on. The purpose of plotting this straight line is simply to serve as a calibration guide for assessing the normality of the distribution of the error terms.

The cumulative frequency distribution of the standardized residuals in Table 4.2 can, for our purposes, be plotted on normal probability paper in straightforward fashion. For instance, the first cumulative percentage, 4.8, is plotted on the vertical scale corresponding to the z value -2.0. Other points are plotted in similar fashion in Figure 4.8. The plot of the cumulative frequency distribution of the standardized residuals, while not a perfect straight line, is reasonably close to that of the standard normal distribution to suggest that there is no strong evidence of any major departure from normality.

The analysis for model departures with respect to normality is, in many respects, more difficult than that for other types of departures. In the first place, random variation can be particularly mischievous when one studies the nature of a probability distribution unless the sample size is quite large. Even worse, other types of departures can and do affect the distribution of the residuals. For instance, residuals may appear to be not normally distributed because an inappropriate regression function is used or because the error variance is not constant. Hence, it is usually a good strategy to investigate these other types of departures first, before concerning oneself with the normality of the error terms.

Omission of Important Independent Variables

Residuals should be plotted against any variables omitted from the model which might have important effects on the response, data being available. The time variable cited earlier in the welding application is an example. The purpose of this additional analysis is to determine whether there are any other key independent variables which could provide important additional descriptive and predictive power to the model.

As another example, in a study to predict output by piece rate workers in an assembling operation, the relation between output (Y) and age (X) of worker was studied for a sample of employees. The plot of the residuals against X is shown in Figure 4.9a, and indicates no ground for suspecting the appropriateness of the linearity of the regression function or the constancy of the error

FIGURE 4.9

Residual Plots for Productivity Example

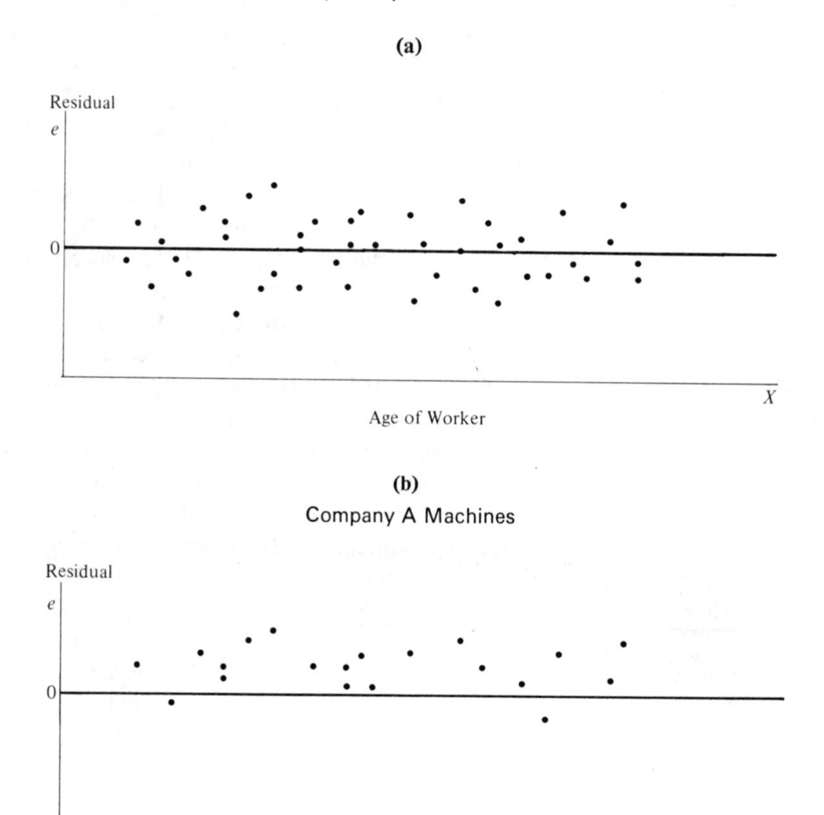

(c)

Company B Machines

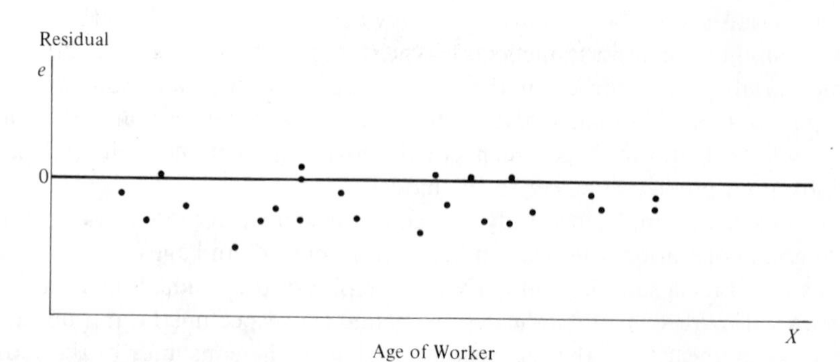

terms. Machines produced by two companies (denoted A and B) are used in the assembling operation. A residual plot by type of machine was undertaken, and is shown in Figures 4.9b and c. Note that the residuals for machines made by Company A tend to be positive, while those for machines made by Company B tend to be negative. Thus, type of machine appears to have a definite effect on productivity, and output predictions may turn out to be far superior when this independent variable is added to the model. While this example dealt with a classification variable (type of machine), the residual analysis for an additional quantitative variable is completely analogous. One would simply plot the residuals against the additional independent variable and see whether or not the residuals tend to vary systematically with the level of the additional independent variable.

Note

We do not say that the original model is "wrong" when it can be improved materially by adding one or more independent variables. Only a few of the factors operating on any dependent variable Y in real-world situations can be included explicitly in a regression model. The chief purpose of residual analysis in identifying other important independent variables is therefore to test the adequacy of the model and see whether it could be improved materially by adding one or a few independent variables.

Comments

1. We discussed the model departures one at a time. In actuality, several types of departures may occur together. For instance, a linear regression function may be a poor fit and the variance of the error terms may not be constant. In these cases, the prototype patterns of Figure 4.2 can still be useful, but they would need to be combined into composite patterns.
2. While graphic analysis of residuals is only an informal method of analysis, in many cases it suffices for examining the aptness of a model.
3. The basic approach to residual analysis applies not only to simple linear regression but also to more complex regression and other types of statistical models.
4. Most of the routine work in graphic analysis of residuals can be handled mechanically on computers. Some regression programs supply the fitted values and corresponding residuals in adjoining columns on the printout. Other programs supply a listing of the residuals identified by observation number. Routines are frequently available whereby the various types of residual plots can be obtained automatically on printout.

4.3 TESTS INVOLVING RESIDUALS

Graphic analysis of residuals is inherently subjective. Nevertheless, subjective analysis of a variety of interrelated residual plots will frequently reveal difficulties in the model more clearly than particular tests. There are occasions, however, when one wishes to put specific questions to a test. We now review some of the relevant tests briefly, and take up one new type of test.

Most statistical tests require independent observations. As we have seen, however, the residuals are dependent. Fortunately, the dependency becomes quite small for large samples, so that one can usually then ignore it.

Tests for Randomness

The run test is frequently used to test for lack of randomness in the residuals arranged in time order. Another test, specifically designed for lack of randomness in least squares residuals, is the Durbin-Watson test. This test will be discussed in Chapter 10.

Tests for Constancy of Variance

When a residual plot gives the impression that the variance may be increasing or decreasing in a systematic manner related to X or $E(Y)$, a simple test is to fit separate regression functions to each half of the observations arranged by level of X, calculate error mean squares for each, and test for equality of the error variances by an F test. Another simple test is by means of rank correlation between the absolute value of the residual and the value of the independent variable.

Tests for Outliers

A simple test for an outlier observation involves fitting a new regression line to the other $n - 1$ observations. The suspect observation, which was not used in fitting the new line, can now be regarded as a new observation. One can calculate the probability that in n observations, a deviation from the fitted line as great as that of the outlier will be obtained by chance. If this probability is sufficiently small, the outlier can be rejected as not having come from the same population as the other $n - 1$ observations. Otherwise, the outlier is retained.

Still other tests to aid in evaluating outliers have been developed. Some of these are discussed in References 4.1 and 4.2 (*see references at end of each chapter*).

Tests for Normality

Goodness-of-fit tests can be used for examining the normality of the error terms. For instance, the chi-square test or the Kolmogorov-Smirnov test can be employed for testing the normality of the error terms by analyzing the residuals.

Note

The run test, rank correlation, chi-square test, and Kolmogorov-Smirnov test, mentioned above, are commonly used statistical procedures which are discussed in basic statistics texts, such as Reference 4.3. The interested reader should refer to such a basic statistics text if he is not familiar with one or more of these procedures.

4.4 *F* TEST FOR LINEARITY

We will now take up in some detail a formal test for determining whether or not the regression function is linear. This test assumes that all of the conditions of model (3.1) except linearity of the regression function hold. Thus, the test assumes that the observations Y for any given X are (1) independent and (2) normally distributed, and (3) the distributions of Y have the same variance σ^2.

Replications

In order to be able to conduct the test for linearity, it is necessary that there be repeat observations at one or more X levels. In survey data, this may occur fortuitously, as when in a productivity study relating workers' output and age, several workers of the same age happen to be included in the study. In an experiment, one can assure by design that there are repeat observations. For instance, in an experiment on the effect of size of salesman bonus on sales, three salesmen can be offered a particular size of bonus, for each of six bonus sizes, and their sales then observed.

Repeated trials for the same level of the independent variable, of the type described, are called *replications.* The resulting observations are called *replicates.*

Example

In an experiment involving 12 similar but scattered suburban branch offices of a commercial bank, holders of checking accounts at the offices were offered gifts for setting up savings accounts at these same offices. The initial deposit in the new savings account had to be for a specified minimum amount to qualify for the gift. The value of the gift was directly proportional to the specified minimum deposit. Various levels of minimum deposit and related gift values were used in the experiment in order to ascertain the relation between the specified minimum deposit and gift value on the one hand and number of accounts opened at the office on the other. Altogether, six levels of minimum deposit and proportional gift value were used, with two of the branch offices assigned at random to each level. Table 4.3 contains the results, where X is the amount of minimum deposit and Y is the number of new savings accounts that were opened and qualified for the gift during the test period.

A linear regression function was fitted in the usual fashion; it is:

$$\hat{Y} = 50.85714 + .50286X$$

TABLE 4.3

Data for Bank Example

Observation i	*Size of Minimum Deposit (dollars)* X_i	*Number of New Accounts* Y_i	*Observation* i	*Size of Minimum Deposit (dollars)* X_i	*Number of New Accounts* Y_i
1	125	160	7	75	42
2	100	112	8	175	124
3	200	124	9	125	150
4	75	28	10	200	104
5	150	143	11	100	136
6	175	156	12	150	161

TABLE 4.4

ANOVA Tables for Bank Example

(a)

Source of Variation	*SS*	*df*	*MS*
Regression	$SSR = 5{,}531.4$	1	$MSR = 5{,}531.4$
Error	$SSE = 15{,}630.6$	10	$MSE = 1{,}563.1$
Total	$SSTO = 21{,}162.0$	11	

(b)

Source of Variation	*SS*	*df*	*MS*
Regression	$SSR = 5{,}531.4$	1	$MSR = 5{,}531.4$
Error	$SSE = 15{,}630.6$	10	$MSE = 1{,}563.1$
Lack of fit	$SSLF = 14{,}320.6$	4	$MSLF = 3{,}580.2$
Pure error	$SSPE = 1{,}310.0$	6	$MSPE = 218.3$
Total	$SSTO = 21{,}162.0$	11	

The analysis of variance table also was obtained, and is shown in Table 4.4a. Since no new principles are involved up to this point, we do not show details of the calculations.

A scatter plot, together with the fitted regression line, is shown in Figure 4.10. The indications are strong that a linear regression function is inappropriate. To test this formally, we need to perform a decomposition of the error sum of squares *SSE* in Table 4.4a.

FIGURE 4.10

Scatter Plot and Fitted Regression Line for Bank Example

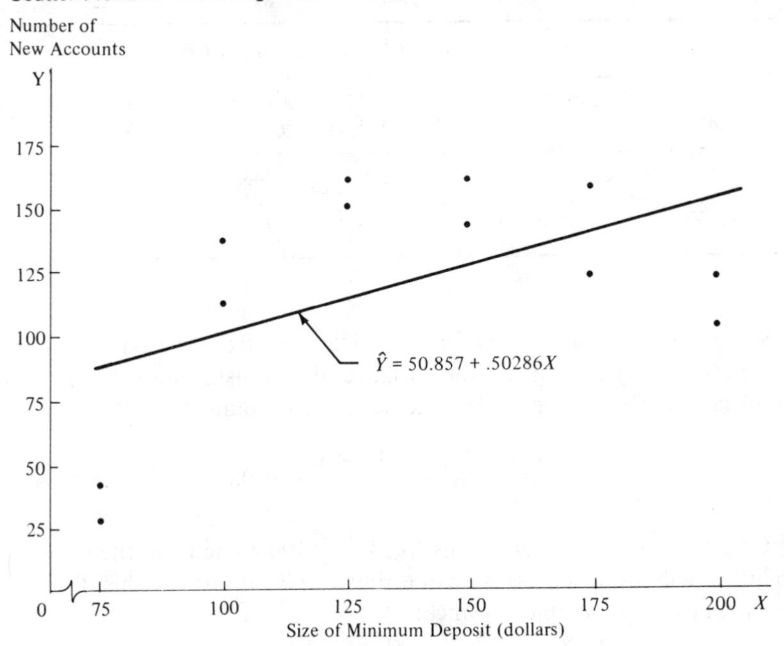

Decomposition of *SSE*

Pure Error Component. The basic idea for the first component of *SSE* rests on the fact that there are replications. In our example, for each minimum deposit size there are two observations. Let us denote the levels of X in the study as $X_1, \ldots, X_c$; for our example, $c = 6$ since there are six minimum deposit size levels in the study. Further we shall denote the number of observations for the jth level of X as n_j; for our example, $n_j \equiv 2$. Thus, the total number of observations n is given by:

$$n = \sum_{j=1}^{c} n_j \tag{4.6}$$

If we make no assumption about the nature of the regression function but assume all other elements of model (3.1), we can still estimate the error variance σ^2 because of the repeated observations. Table 4.5 presents the same data as in Table 4.3, but in a different arrangement. Table 4.5 also shows the mean response for each minimum deposit size. We shall denote the mean response when $X = X_j$ by $\bar{Y}_j$. Thus $\bar{Y}_1 = 35$ is the mean response in the study when the minimum deposit size is \$75, and so on.

TABLE 4.5

Data for Bank Example, Arranged by Observation Number and Minimum Deposit

	Size of Minimum Deposit (dollars)					
Observation	$X=75$ $j=1$	$X=100$ $j=2$	$X=125$ $j=3$	$X=150$ $j=4$	$X=175$ $j=5$	$X=200$ $j=6$
$i=1$	28	112	160	143	156	124
$i=2$	42	136	150	161	124	104
Mean $\bar{Y}_j$	35	124	155	152	140	114

Since the two observations for $X_1 = \$75$ come from the same probability distribution, we can estimate the variance of this distribution by calculating the usual sample variance, using the deviations around $\bar{Y}_1 = 35$:

$$\frac{(28-35)^2 + (42-35)^2}{2-1} = 98$$

Likewise, the two observations for $X_2 = \$100$ come from the same probability distribution, so that we can estimate the variance of this distribution by calculating the sample variance:

$$\frac{(112-124)^2 + (136-124)^2}{1} = 288$$

Similarly, we can estimate the variance of each of the other distributions. Since model (3.1) assumes that all probability distributions of Y have the same variance σ^2, we can combine the results for each of the X levels. The optimum way of combining is to add the numerators:

$$\begin{aligned}(28-35)^2 &+ (42-35)^2 + (112-124)^2 + (136-124)^2 \\ &+ (160-155)^2 + (150-155)^2 + (143-152)^2 + (161-152)^2 \\ &+ (156-140)^2 + (124-140)^2 + (124-114)^2 + (104-114)^2 = 1{,}310\end{aligned}$$

then add the denominators:

$$1+1+1+1+1+1 = 6$$

and finally take the ratio:

$$\frac{1{,}310}{6} = 218.3$$

To generalize, let us denote the ith observation for the jth level of X as Y_{ij}, where $i = 1, \ldots, n_j$; $j = 1, \ldots, c$. For our example (see Table 4.5), $Y_{11} = 28$, $Y_{21} = 42$, $Y_{12} = 112$, and so on. First, we calculate the sum of squares of the deviations from the mean at any given level of X. For $X = X_j$, this sum of squares is:

$$\sum_{i=1}^{n_j} (Y_{ij} - \bar{Y}_j)^2 \tag{4.7}$$

We then add these sums of squares over all levels of X, and denote this sum of the sums of squares $SSPE$:

$$SSPE = \sum_{j=1}^{c} \sum_{i=1}^{n_j} (Y_{ij} - \bar{Y}_j)^2 \tag{4.8}$$

Here $SSPE$ stands for *pure error sum of squares.*

The degrees of freedom associated with $SSPE$ is $n - c$. This is easy to see since there are as usual $n_1 - 1$ degrees of freedom associated with the sum of squares for X_1, $n_2 - 1$ degrees of freedom with the sum of squares for X_2, and so on. The sum of the degrees of freedom is:

$$\sum_{j=1}^{c} (n_j - 1) = \sum n_j - c = n - c \tag{4.9}$$

The *pure error mean square* $MSPE$ therefore is given by:

$$MSPE = \frac{SSPE}{n - c} \tag{4.10}$$

The reason for the term "pure error" is that $MSPE$ is an unbiased estimator of the error variance σ^2 no matter what is the nature of the regression function. $MSPE$ measures the variability of the distributions of Y without relying on any assumptions about the nature of the regression relation; hence it is a "pure" measure of the error variance.

Lack of Fit Component. The remaining component of SSE is:

$$SSLF = SSE - SSPE \tag{4.11}$$

where $SSLF$ denotes *lack of fit sum of squares.* It can be shown that:

$$SSLF = \sum_{j=1}^{c} n_j(\bar{Y}_j - \hat{Y}_j)^2 \tag{4.11a}$$

where $\hat{Y}_j$ denotes the fitted value when $X = X_j$. Thus, $SSLF$ is a weighted sum of squares (the weights are the sample sizes n_j) of the deviations:

$$\bar{Y}_j - \hat{Y}_j \tag{4.12}$$

Note that these deviations represent the difference between the mean response $\bar{Y}_j$ and the fitted value $\hat{Y}_j$ based on linear regression. The closer are the $\bar{Y}_j$ to the $\hat{Y}_j$, the greater is the evidence that the regression function is linear. The further the $\bar{Y}_j$ deviate from the $\hat{Y}_j$, the more the indication that a linear regression is inappropriate.

Figure 4.11 illustrates, for the case where $X = 150$ and $Y_{24} = 161$, the

FIGURE 4.11

Illustration of Decomposition of $Y_{ij} - \hat{Y}_j$ for $X = 150$

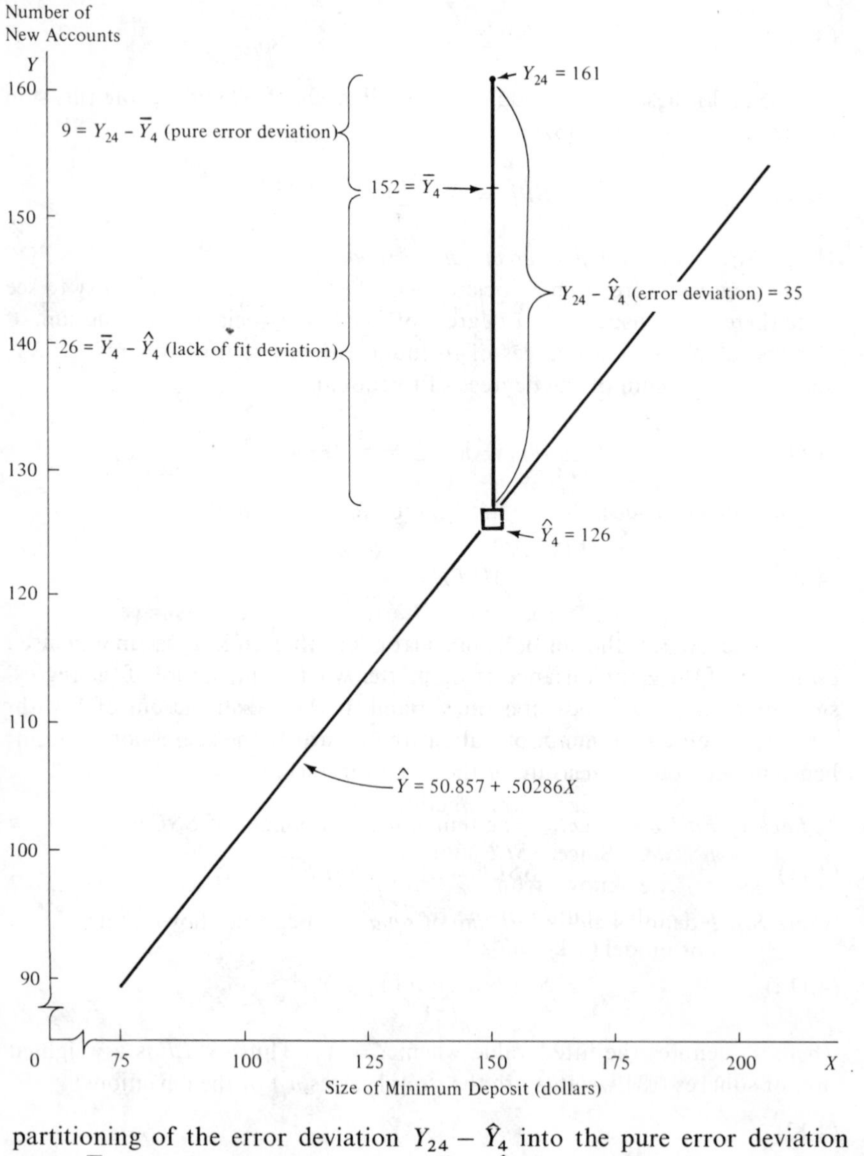

Number of New Accounts

$Y_{24} = 161$

$9 = Y_{24} - \overline{Y}_4$ (pure error deviation)

$152 = \overline{Y}_4$

$Y_{24} - \hat{Y}_4$ (error deviation) = 35

$26 = \overline{Y}_4 - \hat{Y}_4$ (lack of fit deviation)

$\hat{Y}_4 = 126$

$\hat{Y} = 50.857 + .50286X$

Size of Minimum Deposit (dollars)

partitioning of the error deviation $Y_{24} - \hat{Y}_4$ into the pure error deviation $Y_{24} - \overline{Y}_4$ and the lack of fit deviation $\overline{Y}_4 - \hat{Y}_4$.

There are $c - 2$ degrees of freedom associated with *SSLF*, since there are c levels of X and two degrees of freedom are lost because of the constraints on the fitted values $\hat{Y}_j$. Thus, the *lack of fit mean square MSLF* is:

$$MSLF = \frac{SSLF}{c - 2} \tag{4.13}$$

For our example, using (4.11) and the earlier results ($SSE = 15{,}630.6$ from Table 4.4a, $SSPE = 1{,}310.0$ from p. 116), we obtain:

$$SSLF = 15{,}630.6 - 1{,}310.0 = 14{,}320.6$$

and:

$$MSLF = \frac{14{,}320.6}{6 - 2} = 3{,}580.2$$

Table 4.4b contains for our example the decomposition of *SSE* just explained, as well as the mean squares of interest.

F Test

Test Statistic. For testing linearity of the regression function, the appropriate test statistic is:

$$F^* = \frac{MSLF}{MSPE} \tag{4.14}$$

We know that *MSPE* has expectation σ^2 no matter what is the nature of the regression function. It can be shown that:

$$E(MSLF) = \sigma^2 + \frac{\sum n_j[E(Y_j) - (\beta_0 + \beta_1 X_j)]^2}{c - 2} \tag{4.15}$$

where $E(Y_j)$ denotes the true mean of the distribution of Y when $X = X_j$, and $\beta_0 + \beta_1 X_j$ is the mean response indicated by the linear regression model. If the regression function is linear, the second term in (4.15) is 0, so that $E(MSLF) = \sigma^2$ then. On the other hand, if the regression function is not linear, $E(Y_j) \neq \beta_0 + \beta_1 X_j$ so that $E(MSLF)$ will be greater than σ^2. Hence, a value of F^* near 1 accords with a linear regression function; large values of F^* indicate that the regression function is not linear.

Decision Rule. Since *SSLF* and *SSPE* are additive, as are the degrees of freedom, we know from Cochran's theorem that F^* follows the $F(c - 2; n - c)$ distribution if the regression function is linear and all other conditions of model (3.1) hold. To decide between:

$$\begin{aligned} &C_1: E(Y) = \beta_0 + \beta_1 X \\ &C_2: E(Y) \neq \beta_0 + \beta_1 X \end{aligned} \tag{4.16}$$

we use the test statistic (4.14). The decision rule to control the risk of a Type I error at α is:

$$\begin{aligned} &\text{If } F^* \leq F(1 - \alpha; c - 2, n - c), \text{ conclude } C_1 \\ &\text{If } F^* > F(1 - \alpha; c - 2, n - c), \text{ conclude } C_2 \end{aligned} \tag{4.17}$$

Example. For our example, the test statistic can be constructed easily from the results in Table 4.4b:

$$F^* = \frac{3{,}580.2}{218.3} = 16.42$$

If the level of significance is to be $\alpha = .01$, we require $F(.99; 4, 6) = 9.15$. Hence, the decision rule is:

$$\text{If } F^* \leq 9.15, \text{ conclude } C_1$$
$$\text{If } F^* > 9.15, \text{ conclude } C_2$$

Since $F^* = 16.42 > 9.15$, we conclude C_2, that the regression function is not linear. This, of course, accords with our visual impression from Figure 4.10.

Comments

1. Not all levels of X need have repeat observations for the F test for linearity to be applicable. Repeat observations at only one or some levels of X are adequate. In that case, some n_j will equal 1, but all earlier formulas are still appropriate.

2. The F test for linearity fits into the framework of a general linear test discussed in Section 3.8. The full model is:

(4.18) For each X_j, Y is normal with mean $E(Y_j)$ and variance σ^2.

The least squares estimator of $E(Y_j)$ for the full model is $\bar{Y}_j$, so that the error sum of squares is:

$$SSE(F) = \sum_i \sum_j (Y_{ij} - \bar{Y}_j)^2 = SSPE \tag{4.19}$$

which has associated with it $n - c$ degrees of freedom.

Since C_1 states:

$$C_1: E(Y) = \beta_0 + \beta_1 X \tag{4.20}$$

the error sum of squares for the reduced model is:

$$SSE(R) = \sum_i \sum_j (Y_{ij} - \hat{Y}_j)^2 = SSE \tag{4.21}$$

which has associated with it $n - 2$ degrees of freedom. Substituting into (3.66) and utilizing (4.11), we obtain:

$$\frac{SSE - SSPE}{(n-2)-(n-c)} \div \frac{SSPE}{n-c} = \frac{SSLF}{c-2} \div \frac{SSPE}{n-c} = \frac{MSLF}{MSPE} \tag{4.22}$$

the same test statistic as in (4.14).

3. Suppose we had wished to test whether or not $\beta_1 = 0$ for the data underlying Table 4.4a, prior to any analysis of the aptness of the model. The test statistic would be:

$$F^* = \frac{MSR}{MSE} = \frac{5{,}531.4}{1{,}563.1} = 3.54$$

For $\alpha = .05$, $F(.95; 1, 10) = 4.96$, and we would conclude C_1, that $\beta_1 = 0$. A conclusion that there is no relation between minimum deposit size (and value of gift) and number of new accounts would be improper, however. Such an inference requires that model (3.1) is appropriate. Here it is not, as we have seen, because the regression function is not linear. There exists indeed a (curvilinear) relation between minimum deposit size and number of new accounts, and testing whether or not $\beta_1 = 0$ under these circumstances has entirely different implications. This illustrates the importance of always examining the aptness of a model before further inferences are drawn.

4. The F test approach just explained can be used to test the aptness of any regression function, not just a linear one. Only the degrees of freedom for $SSLF$ will need be modified.

5. The alternative C_2 in (4.16) includes all regression functions other than a linear one. For instance, it includes a quadratic regression function or a logarithmic one. If C_2 is concluded, a nonlinear regression function should be employed. A study of residuals can be helpful in identifying an appropriate function.

6. Clearly, repeat observations are most valuable whenever we are not certain of the nature of the regression function. If at all possible, provision should be made for some replications.

7. If we conclude that the employed model in C_1 is appropriate, the usual practice is to use the error mean square MSE as an estimator of σ^2 in preference to the pure error mean square $MSPE$, since the former contains more degrees of freedom.

8. Observations at the same level of X are genuine repeats only if they involve independent trials with respect to the error term. Suppose, in a regression analysis of the relation between hardness (Y) and amount of carbon (X) in specimens of an alloy, the error term in the model covers, among other things, random errors in the measurement of hardness by the analyst and effects of uncontrolled production factors which vary at random from specimen to specimen and affect hardness. If the analyst takes two readings on the hardness of a specimen, this will not provide genuine replication because the effects of random variation in the production factors are fixed in any given specimen. For genuine replication, different specimens with the same carbon content (X) would have to be measured by the analyst so that *all* the effects covered in the error term could vary at random from one repeated observation to the next.

4.5 REMEDIAL MEASURES

If the simple linear regression model (3.1) is not appropriate for the data at hand, there are two basic choices:

1. Abandon model (3.1) and search for a more appropriate model.
2. Use some transformation on the data so that model (3.1) is appropriate for the transformed data.

Each approach has advantages and disadvantages. The first approach may entail a more complex model which may yield better insights, but may also lead into serious difficulties in estimating the parameters. Successful use of transformations, on the other hand, leads to relatively simple methods of estimation, and may involve fewer parameters than a complex model, an advantage when the sample size is small. Yet transformations may obscure the fundamental interconnections between the variables, though at other times may illuminate them.

We shall consider both the use of more complex models and the use of transformations in this and other chapters. Here, we provide a brief overview of remedial measures.

Nonlinearity of Regression Function

If the regression function is not linear, a direct approach is to modify model (3.1) with respect to the nature of the regression function. For instance, a quadratic regression might be used:

$$E(Y) = \beta_0 + \beta_1 X + \beta_2 X^2 \tag{4.23}$$

or an exponential regression:

$$E(Y) = \beta_0 \beta_1^X \tag{4.24}$$

In Chapter 8, we discuss models where the regression function is a polynomial.

The transformation approach uses a transformation to linearize, at least approximately, a nonlinear regression function. For instance, the transformation:

$$Y' = \log Y \tag{4.25}$$

where Y' is the transformed variable, may be appropriate for the exponential regression function (4.24) above. We discuss the use of transformations to linearize regression functions in Section 4.6.

Nonconstancy of Error Variance

If the error variance is not constant but varies in a systematic fashion, a direct approach is to modify the model to allow for this and use, say, the method of *weighted least squares* to obtain the estimators of the parameters. We discuss the use of weighted least squares for this purpose in Section 4.7.

Transformations can also be effective in stabilizing the variance. Some of these are discussed in Chapter 15. For instance, the transformation:

$$Y' = \sqrt{Y} \tag{4.26}$$

is useful in a number of applications for stabilizing the variance. In Section 4.7, we discuss a transformation of both Y and X which achieves a constant variance in a frequently encountered case.

Nonindependence of Error Terms

If the error terms are correlated, a direct remedial measure is to work with a model which calls for correlated error terms. We discuss such a model in Chapter 10. A simple remedial transformation which is often helpful is to work with first differences, a topic also discussed in Chapter 10.

Nonnormality of Error Terms

Lack of normality and nonconstant error variances frequently go hand in hand. Fortunately, it is often the case that the same transformation which helps stabilize the variance, such as a logarithmic or a square root transformation, is also helpful in normalizing the error terms. It is therefore desirable that the transformation for stabilizing the error variance be utilized first, and then the residuals studied to see if serious departures from normality are still present. We discuss transformations to achieve normality in Chapter 15.

Omission of Important Independent Variables

When residual analysis indicates that an important independent variable has been omitted from the model, the solution is to modify the model. In Chapter 7 and following chapters of Part II, we discuss multiple regression analysis in which two or more independent variables are utilized.

4.6 TRANSFORMATIONS TO LINEARIZE REGRESSION FUNCTION

We now consider in more detail two remedial measures, namely, transformations to linearize the regression function and a transformation to stabilize the error term variance. First, we take up two basic types of transformations to linearize the regression function—logarithmic transformations and reciprocal transformations. Other transformations are also useful; see, for instance, Reference 4.1.

Logarithmic Transformations

Multiplicative Model. A logarithmic transformation is useful when the actual model needs to be multiplicative, say:

$$Y = \gamma_0 \gamma_1^X \varepsilon \tag{4.27}$$

where γ_0 and γ_1 are parameters and ε is a random variable with mean 1 and constant variance. Model (4.27) is neither linear in the parameters (γ_0 and γ_1 are multiplied together) nor is it linear in X (X appears as an exponent). Yet we say that model (4.27) is *intrinsically linear* since it can be expressed by a suitable transformation in the linear form of model (3.1). The transformation is the logarithmic one:

$$Y' = \log_{10} Y \tag{4.28}$$

Applying this transformation to (4.27), we obtain:

$$\log_{10} Y = Y' = \log_{10} \gamma_0 + X \log_{10} \gamma_1 + \log_{10} \varepsilon \tag{4.29}$$

Let:

$$\log_{10} \gamma_0 = \beta_0$$
$$\log_{10} \gamma_1 = \beta_1$$
$$\log_{10} \varepsilon = \varepsilon'$$

We can then write (4.27) in the standard form:

$$Y' = \beta_0 + \beta_1 X + \varepsilon' \tag{4.30}$$

Figure 4.12a shows two examples of regression functions for the multiplicative model (4.27), and Figure 4.12b shows the corresponding linear regression functions with the transformed variable Y'. Once the Y variable has been transformed by (4.28), all regression formulas previously developed are applicable, with the understanding that when Y is used in these formulas, the symbol stands for the transformed variable Y'.

FIGURE 4.12

Examples of Multiplicative Model (4.27) and Transformed Model

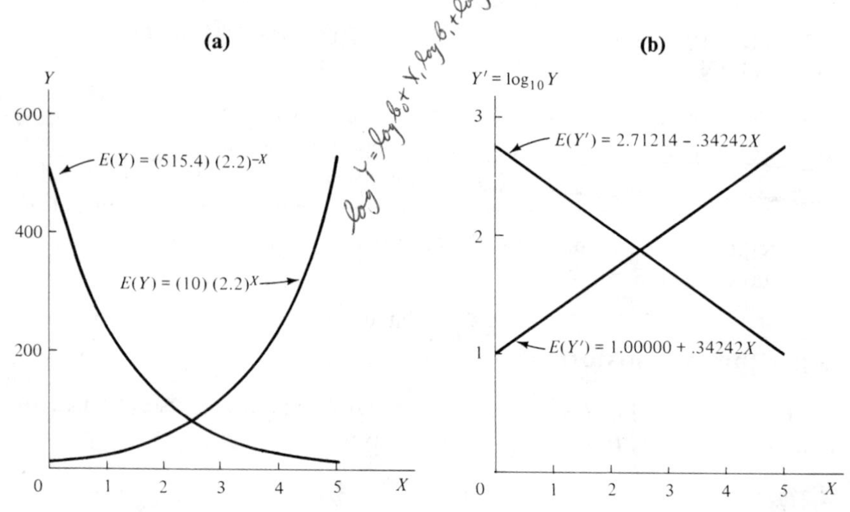

We should caution at once that use of model (3.1) with the transformed variable Y' requires that the transformed error terms ε' are independently normally distributed with mean 0 and constant variance σ^2. Residual analysis, as well as other methods at times, will need to be employed to check whether the transformed error terms meet these conditions.

Multiplicative models have been used in business and economics for various studies. For instance, growth in company sales (Y) as a function of time (X) may be described by the multiplicative model (4.27) if the expected growth rate is constant. Again, the time required to solve an optimization problem by iteration on a computer (Y) and the number of variables in the problem (X) may be related in such fashion that model (4.27) is appropriate.

Other Uses. The logarithmic transformation is also useful when no particular curvilinear model is suggested by theoretical or a priori considerations, but the scatter plot suggests that a logarithmic transformation would linearize the regression relation. Consider the data in Table 4.6, columns 1

TABLE 4.6

Regression Calculations with Logarithmic Transformation

	(1) *Days of Training* X	(2) *Performance Score* Y	(3) $\log_{10} Y = Y'$	(4) XY'	(5) X^2
	1	45	1.65321	1.65321	1
	1	40	1.60206	1.60206	1
	2	60	1.77815	3.55630	4
	2	62	1.79239	3.58478	4
	3	75	1.87506	5.62518	9
	3	81	1.90849	5.72547	9
	4	115	2.06070	8.24280	16
	5	150	2.17609	10.88045	25
	5	145	2.16137	10.80685	25
	5	148	2.17026	10.85130	25
Total	31	921	19.17778	62.52840	119

and 2. These show number of days of training (X) and performance score (Y) for 10 sales trainees in a battery of simulated sales situations in an experiment. The observations are plotted as a scatter diagram in Figure 4.13a. A nonlinear regression function is clearly required for a satisfactory fit. In Figure 4.13b, $\log_{10} Y$ is plotted against X. Here, the points tend to fall along a straight line, so that a linear function relating $\log_{10} Y$ and X appears to suffice. The model:

$$Y' = \beta_0 + \beta_1 X + \varepsilon' \tag{4.31}$$

where:

$$Y' = \log_{10} Y \tag{4.31a}$$

therefore should be studied to see if the other conditions of the basic simple linear regression model (3.1) are met.

Example of Regression Calculations. In our sales training example, the residual plots (not shown) suggest that model (3.1) is indeed appropriate when Y' is the independent variable. To illustrate that no new problems are encountered in the regression calculations, we present in Table 4.6 the basic

FIGURE 4.13

Scatter Plots and Fitted Regressions for Example of Logarithmic Transformation

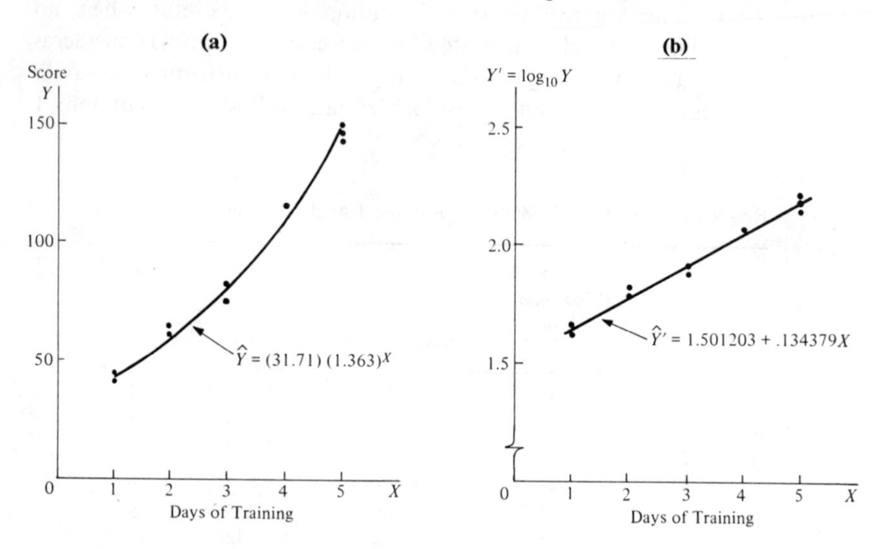

calculations required for the least squares estimators b_0 and b_1. Simply remember that Y' plays the role of Y in all earlier formulas. We obtain:

$$b_1 = \frac{\sum X_i Y_i' - \frac{\sum X_i \sum Y_i'}{n}}{\sum X_i^2 - \frac{(\sum X_i)^2}{n}} = \frac{62.52840 - \frac{(31)(19.17778)}{10}}{119 - \frac{(31)^2}{10}} = .134379$$

$$b_0 = \frac{1}{n}(\sum Y_i' - b_1 \sum X_i) = \frac{1}{10}[19.17778 - (.134379)(31)] = 1.501203$$

so that the fitted regression equation is:

$$\hat{Y}' = 1.501203 + .134379X$$

where $\hat{Y}'$ is the point estimator of $E(Y')$, the mean of the probability distribution of Y' for given X. This fitted regression equation is plotted in Figure 4.13b.

If we wish to obtain the fitted regression equation in the original units, we simply take antilogs:

$$\hat{Y} = \text{antilog}_{10}(1.501203 + .134379X)$$

or:

$$\hat{Y} = (31.71)(1.363)^X$$

This regression function in the original units is plotted in Figure 4.13a. For example, when $X = 3$, we have:

$$\hat{Y} = \text{antilog}_{10}[1.501203 + .134379(3)] = 80.2$$

Other Logarithmic Transformations. Other models which involve the logarithmic transformation are:

(4.32a) $$Y = \beta_0 + \beta_1 X' + \varepsilon \qquad \text{where } X' = \log_{10} X$$

(4.32b) $$Y' = \beta_0 + \beta_1 X' + \varepsilon' \qquad \text{where } Y' = \log_{10} Y,\ X' = \log_{10} X$$

Economists have used model (4.32b) when studying the relation between price of commodity (X) and quantity demanded (Y) at that price. In this model, the parameter β_1 stands for the price elasticity of demand. It is commonly interpreted as showing the percent change in quantity demanded per 1 percent change in price, where it is understood that the changes are in opposite directions.

FIGURE 4.14

Examples of Regression Functions for Model (4.32a)

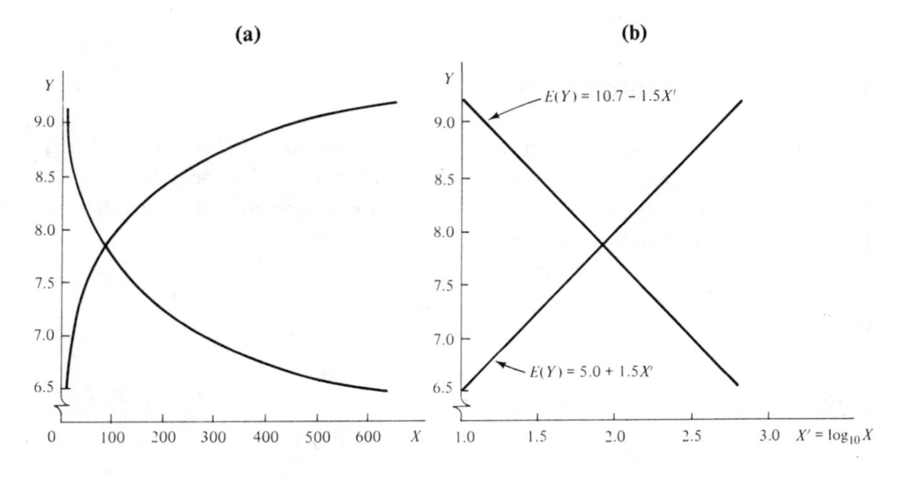

Figure 4.14 contains two examples of regression functions for model (4.32a) in both the original and transformed scales, and Figure 4.15 contains two examples for model (4.32b).

The choice of model is helped by plotting scatter diagrams of log Y against X, Y against log X, and log Y against log X, either manually or by computer. If the distributions of Y are nearly normal, an advantage of transforming X rather than Y is that this does not change the shape of the distributions. On the other hand, if the distributions of Y are skewed to the right, using log Y helps to normalize them.

FIGURE 4.15

Examples of Regression Functions for Model (4.32b)

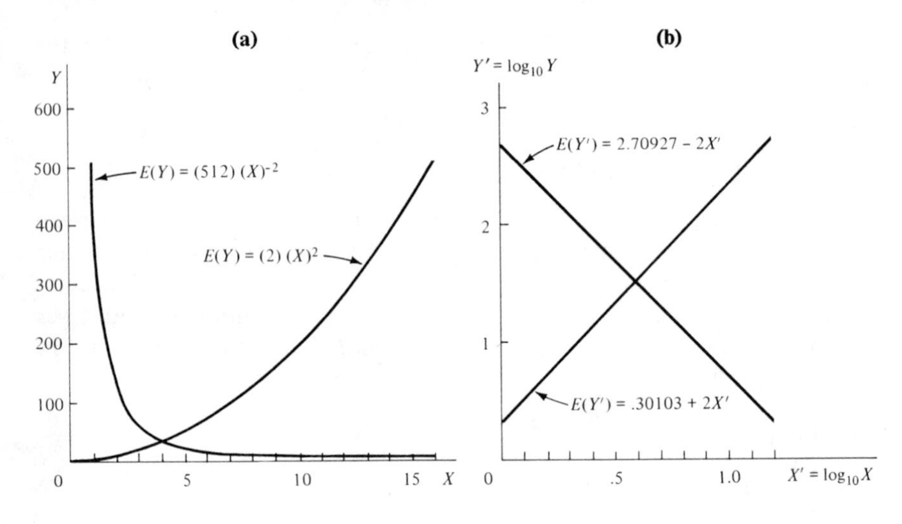

Reciprocal Transformations

A reciprocal transformation can be used to make certain intrinsically linear models linear, as well as when no particular model has been formulated but the scatter plot indicates the need for a curvilinear model. Consider the following model, which is not linear in X:

$$Y = \beta_0 + \frac{\beta_1}{X} + \varepsilon \tag{4.33}$$

If we use the transformation:

$$X' = \frac{1}{X} \tag{4.34}$$

model (4.33) becomes:

$$Y = \beta_0 + \beta_1 X' + \varepsilon \tag{4.35}$$

a simple linear model. Figure 4.16 shows two examples of regression functions for model (4.33) and the corresponding linear regression functions with the transformed variable X'. Note that for model (4.33), the regression function in the original units has an asymptote, and that the asymptote is $\beta_0 = 250$ in our examples.

FIGURE 4.16

Examples of Regression Functions for Model (4.33)

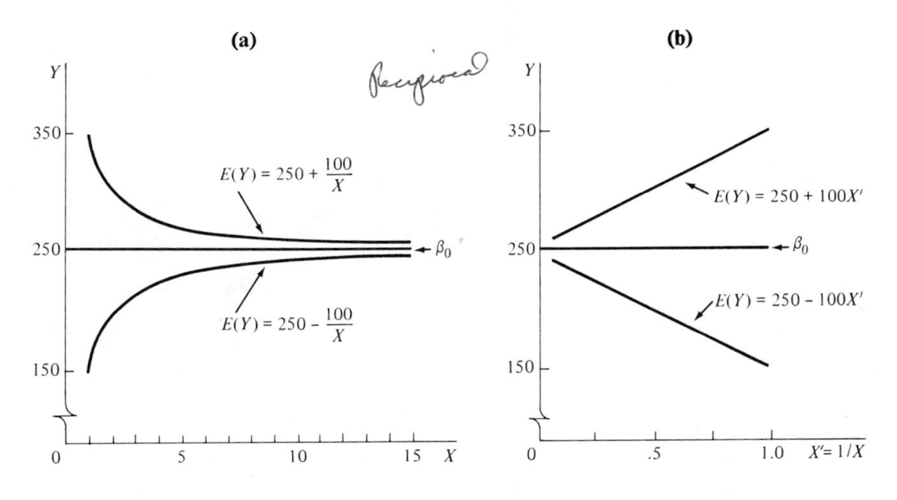

Other models involving the reciprocal transformation are:

$$Y' = \beta_0 + \beta_1 X + \varepsilon' \qquad \text{where } Y' = \frac{1}{Y} \tag{4.36a}$$

$$Y' = \beta_0 + \beta_1 X' + \varepsilon' \qquad \text{where } Y' = \frac{1}{Y},\ X' = \frac{1}{X} \tag{4.36b}$$

Example. In Figure 4.17a, we present a scatter plot of data on number of years experience (X) and current hourly earnings (Y) of five employees in a shop that makes hairpieces to order. The data suggest strongly that the regression function is curvilinear. A priori considerations indicate that an upper asymptote to hourly earnings is reasonable. We therefore investigate transformation (4.34). Figure 4.17b contains a scatter plot with the transformed variable X'. Note that this transformation has been successful since the points tend to fall in a linear pattern. The regression function is now computed in the usual way, with X in earlier formulas replaced by X'. The fitted regression function turns out to be:

$$\hat{Y} = 5.00 - .389X'$$

and the regression function for the original variables is:

$$\hat{Y} = 5.00 - \frac{.389}{X}$$

Each of these regression functions is plotted in Figure 4.17.

FIGURE 4.17

Scatter Plots for Example of Reciprocal Transformation

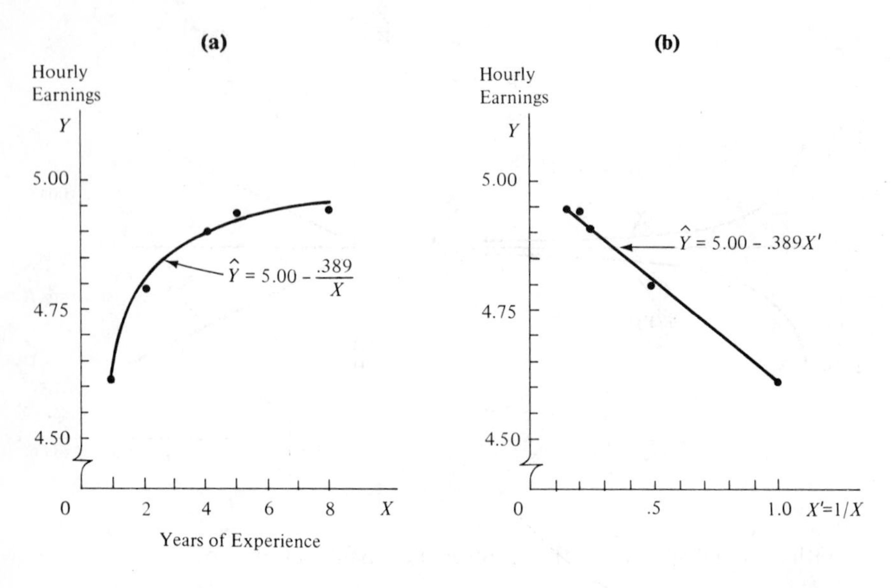

Comments

1. We reiterate the importance of checking the model (3.1) assumptions if a transformation on Y is employed. For instance, when the transformation $Y' = \log Y$ is employed with model (3.1), it is assumed that the distribution of $\log Y$ for given X is normal with constant variance. This needs to be checked after the transformation has been made.

2. When transformed models are employed, the estimators b_0 and b_1 obtained by least squares have the least squares properties with respect to the transformed observations, not the original ones.

3. The logarithmic and reciprocal transformations can be employed together. For instance, the following complex regression function (which has asymptotes both below and above):

$$Y = \frac{1}{1 + \exp(\beta_0 + \beta_1 X + \varepsilon)} \tag{4.37}$$

is transformed to the linear model by:

$$Y' = \log_e\left(\frac{1}{Y} - 1\right) \tag{4.38}$$

4.7 VARIANCE-STABILIZING TRANSFORMATIONS

Setting of Problem

We now take up the case where the error variance is not constant over all observations. This condition is called *heteroscedasticity*, in contrast to equal error variances or *homoscedasticity*.

When heteroscedasticity prevails but the other conditions of model (3.1) are met, the estimators b_0 and b_1 obtained by ordinary least squares procedures are still unbiased and consistent, but they are no longer minimum variance unbiased estimators.

Heteroscedasticity is inherent when the response in regression analysis follows a distribution in which the variance is functionally related to the mean. (Significant nonnormality in Y is encountered as well in most such cases.) Consider, in this connection, a regression analysis where X is the speed of a machine which puts a plastic coating on cable and Y is the number of blemishes in the coating per thousand feet drum of cable. If Y is Poisson distributed with a mean which increases as X increases, the distributions of Y cannot have constant variance at all levels of X since the variance of a Poisson variable equals the mean, which is increasing with X. Transformations to stabilize the variance of the error terms are discussed in Chapter 15. For instance, it is noted there that the transformation $Y' = \sqrt{Y}$ is highly useful when Y is a Poisson variable, both to stabilize the variance and to improve normality.

Nonconstancy of Error Terms

Standard Deviation Proportional to X

Here we consider a particular case which is frequently encountered in business and economic applications, namely where the standard deviation of the error terms is proportional to X, or equivalently, the variance of the error terms is proportional to X^2. In other words:

$$\sigma^2(\varepsilon_i) = kX_i^2 \tag{4.39}$$

where k is a proportionality factor. Studies of savings regressed on income, for instance, frequently entail this situation.

Appropriate Transformation. The proper transformation to obtain minimum variance unbiased estimators for the regression model with normal error terms:

$$Y_i = \beta_0 + \beta_1 X_i + \varepsilon_i \qquad \text{where } \sigma^2(\varepsilon_i) = kX_i^2 \tag{4.40}$$

is:

$$Y' = \frac{Y}{X} \tag{4.41}$$
$$X' = \frac{1}{X}$$

When these transformed variables are employed in the usual manner, minimum variance unbiased estimators of β_0 and β_1 are obtained once the fitted regression function is stated in the original variables again.

To see the effect of transformation (4.41), let us divide each side of equation (4.40) by X_i. We then obtain:

$$\frac{Y_i}{X_i} = \frac{\beta_0}{X_i} + \beta_1 + \frac{\varepsilon_i}{X_i} \tag{4.42}$$

Using transformation (4.41) where $Y_i' = Y_i/X_i$ and $X_i' = 1/X_i$, and letting:

$$\beta_0 = \beta_1' \tag{4.43a}$$

$$\beta_1 = \beta_0' \tag{4.43b}$$

$$\frac{\varepsilon_i}{X_i} = \varepsilon_i' \tag{4.43c}$$

we can write (4.42) as follows:

$$Y_i' = \beta_0' + \beta_1' X_i' + \varepsilon_i' \tag{4.44}$$

Note that (4.44) is in the usual simple linear regression form. Further, the variance of ε_i' is now a constant:

$$\sigma^2(\varepsilon_i') = \sigma^2\left(\frac{\varepsilon_i}{X_i}\right) = \frac{1}{X_i^2}\sigma^2(\varepsilon_i) = \frac{1}{X_i^2}(kX_i^2) = k \tag{4.45}$$

Thus, we can use the ordinary least squares estimators with the transformed variables Y' and X'. Let us denote the fitted regression equation so obtained as follows:

$$\hat{Y}' = b_0' + b_1' X' \tag{4.46}$$

where the primes on b_0 and b_1 remind us that they pertain to the transformed variables. To return to our original variables, we multiply both sides of (4.46) by X_i and obtain:

$$\hat{Y} = b_0' X + b_1' = b_0 + b_1 X \tag{4.47}$$

where:

$$b_0 = b_1' \tag{4.47a}$$

$$b_1 = b_0' \tag{4.47b}$$

To summarize, to fit model (4.40) we use the transformed variables Y' and X' defined in (4.41). The least squares estimators b_0' and b_1' so obtained are then respective estimators of β_1 and β_0 in the original model.

Example. The Nielsen Construction Company made a study of the relation between the size of a bid in million dollars (X) and the cost to the firm of preparing the bid in thousand dollars (Y), for 12 recent bids. The data are presented in Table 4.7, columns 1 and 2. Residual analysis suggested that the

TABLE 4.7

Regression Calculations for Unequal Variances Problem

	(1)	(2)	(3)	(4)	(5)	(6)
	Original Observations		*Transformed Observations*			
Observation Number i	X_i	Y_i	$X_i' = \frac{1}{X_i}$	$Y_i' = \frac{Y_i}{X_i}$	$X_i' Y_i'$	$(X_i')^2$
1	2.13	15.5	.4695	7.277	3.417	.2204
2	1.21	11.1	.8264	9.174	7.581	.6829
3	11.00	62.6	.0909	5.691	.517	.0083
4	6.00	35.4	.1667	5.900	.984	.0278
5	5.60	24.9	.1786	4.446	.794	.0319
6	6.91	28.1	.1447	4.067	.588	.0209
7	2.97	15.0	.3367	5.051	1.701	.1134
8	3.35	23.2	.2985	6.925	2.067	.0891
9	10.39	42.0	.0962	4.042	.389	.0093
10	1.10	10.0	.9091	9.091	8.265	.8265
11	4.36	20.0	.2294	4.587	1.052	.0526
12	8.00	47.5	.1250	5.938	.742	.0156
Total	63.02	335.3	3.8717	72.189	28.097	2.0987

error variance increases with size of bid (X), being approximately proportional to X^2. Hence model (4.40) is assumed to be applicable. We therefore use transformation (4.41):

$$Y_i' = \frac{Y_i}{X_i} \qquad X_i' = \frac{1}{X_i}$$

The transformed values are shown in columns 3 and 4 of Table 4.7. We now proceed with the least squares calculations in the usual way, and obtain from the results in Table 4.7:

$$b_1' = \frac{\sum X_i'Y_i' - \frac{\sum X_i' \sum Y_i'}{n}}{\sum (X_i')^2 - \frac{(\sum X_i')^2}{n}} = \frac{28.097 - \frac{(3.8717)(72.189)}{12}}{2.0987 - \frac{(3.8717)^2}{12}}$$

$$= 5.6570$$

$$b_0' = \frac{1}{n}(\sum Y_i' - b_1' \sum X_i') = \frac{1}{12}[72.189 - (5.6570)(3.8717)]$$

$$= 4.1906$$

L.S. calc. using Transformed Data

Hence, the fitted regression line in the transformed variables is:

$$\hat{Y}' = 4.1906 + 5.6570X'$$

Transforming into our original variables, we obtain from (4.47a) and (4.47b):

$$b_0 = b_1' = 5.6570$$
$$b_1 = b_0' = 4.1906$$

so that the fitted regression function in the original variables is:

$$\hat{Y} = 5.6570 + 4.1906X$$

Figure 4.18 contains a plot of the data, together with the fitted regression line. It appears from there that the fitted regression line is a reasonably good fit to the data.

FIGURE 4.18

Regression Function for Nielsen Construction Company Example

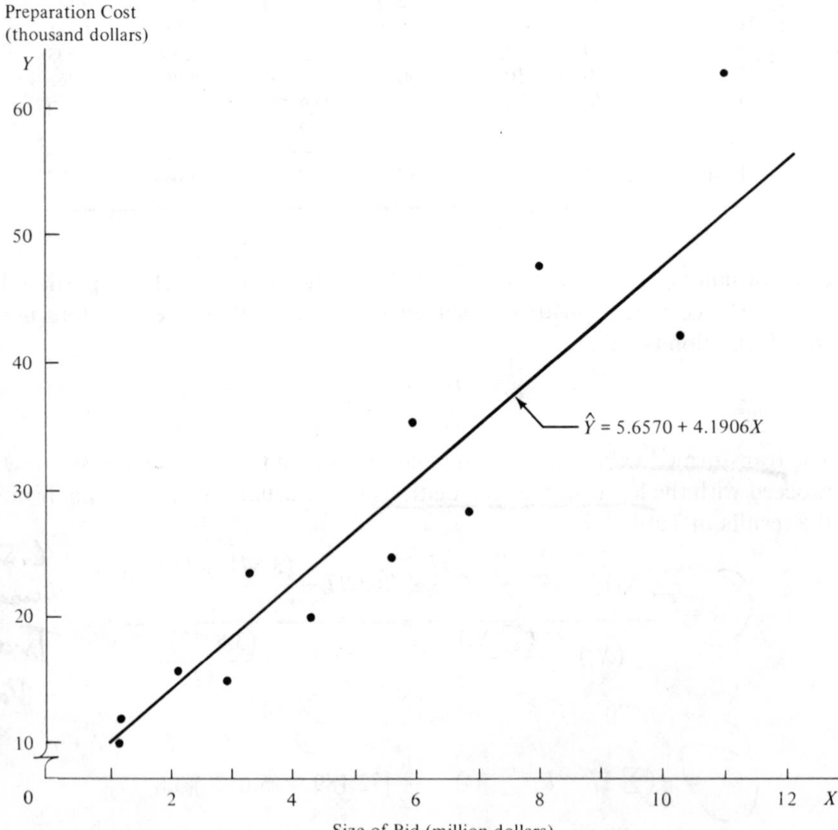

Comments

1. The use of transformation (4.41) to obtain constant error variances with model (4.40) and then applying ordinary least squares procedures is the equivalent of using *weighted least squares*. With ordinary least squares, the estimators of β_0 and β_1 are obtained by minimizing the quantity:

$$Q = \sum (Y_i - \beta_0 - \beta_1 X_i)^2 \tag{4.48}$$

In contrast, the quantity minimized with weighted least squares is:

$$Q = \sum w_i (Y_i - \beta_0 - \beta_1 X_i)^2 \tag{4.49}$$

where the w_i are weights. The weight w_i should be the inverse of the variance of ε_i. For model (4.40), the appropriate weights therefore are:

$$w_i = \frac{1}{kX_i^2}$$

When we minimize:

$$Q = \sum \frac{1}{kX_i^2} (Y_i - \beta_0 - \beta_1 X_i)^2$$

by differentiating with respect to β_0 and β_1 and setting the partial derivatives equal to zero, we obtain the normal equations:

$$\begin{aligned} \sum \frac{Y_i}{X_i^2} &= b_0 \sum \frac{1}{X_i^2} + b_1 \sum \frac{1}{X_i} \\ \sum \frac{Y_i}{X_i} &= b_0 \sum \frac{1}{X_i} + nb_1 \end{aligned} \tag{4.50}$$

Substituting into these normal equations with the data from Table 4.7 for our bidding cost example, we obtain:

$$\begin{aligned} 28.097 &= 2.0987b_0 + 3.8717b_1 \\ 72.189 &= 3.8717b_0 + 12b_1 \end{aligned}$$

When these two equations are solved, we find the same results as when transformed variables are used (except for possible slight rounding effects):

$$b_0 = 5.6570 \qquad b_1 = 4.1906$$

2. If we had fitted a straight line by unweighted least squares to the original data in our example, we would have obtained the regression function:

$$\hat{Y} = 4.2289 + 4.5153X$$

This differs from the weighted least squares results, as will generally be the case. The reason is that the weights $w_i = 1/kX_i^2$ give more emphasis to observations for smaller X (for which the distribution of Y has a smaller variance) and less emphasis to observations for larger X (for which the distribution of Y has a larger variance). While the estimates obtained by unweighted least squares are unbiased, as are the estimates obtained by weighted least squares, they are subject to greater sampling variation. This can be seen by comparing the estimated standard deviations of the regression coefficients for the two methods:

Unweighted Least Squares	*Weighted Least Squares*
$s(b_0) = 3.2517$	$s(b_0) = .9652$
$s(b_1) = .5285$	$s(b_1) = .4037$

3. Weighted least squares is appropriate whenever the error term variances are unequal. It can only be used, however, when the error variances are known completely or at least known up to a proportional constant such as:

$$\sigma^2(\varepsilon_i) = kX_i$$

$$\sigma^2(\varepsilon_i) = kX_i^2$$

and the like. The general normal equations which arise from minimizing (4.49) are:

$$\begin{aligned} \sum w_i Y_i &= b_0 \sum w_i + b_1 \sum w_i X_i \\ \sum w_i X_i Y_i &= b_0 \sum w_i X_i + b_1 \sum w_i X_i^2 \end{aligned} \tag{4.51}$$

where w_i, the weight for the ith observation, is the reciprocal of $\sigma^2(\varepsilon_i)$.

4. To obtain information as to the nature of the relation of the error variance to X, one may divide the residuals into, say, three or four groups of equal size, according to ascending or descending order of X, and calculate the sum of the squared residuals for each group. This rough analysis frequently will be an adequate guide to decide whether the assumption $\sigma^2(\varepsilon_i) = kX_i^2$ is reasonable or whether some other relation to X would be preferable.

PROBLEMS

4.1. Distinguish between: (1) residual and standardized residual; (2) $E(\varepsilon_i) = 0$ and $\bar{e} = 0$; and (3) error term and residual.

4.2. Prepare a prototype residual plot for each of the following cases: (1) error variance decreases with X; and (2) true regression function is U shaped but a linear regression function is fitted.

4.3. A student, in undertaking a case study, plotted the residuals e_i against Y_i and found a positive relation. When he plotted the residuals against the fitted values $\hat{Y}_i$, he found no relation. Why is there this difference, and which is the more meaningful plot?

4.4. Refer to Problem 2.12. Obtain the residuals and make all appropriate residual plots to ascertain whether any departures from model (3.1) are evident. The subscript i here indicates the time order of the observations. Summarize your conclusions.

4.5. Refer to Problem 3.6.

a) Obtain the residuals and make all appropriate residual plots to ascertain whether any departures from model (3.1) are evident. The subscript i here indicates the time order of the observations. Summarize your conclusions.

b) Information is given below on two variables not included in the model, namely mean operational age (in months) of machines serviced on the call (X_2) and amount of experience of the service man making the call (X_3). Make additional residual plots to ascertain whether the model can be improved by including either or both of these variables. What do you conclude?

i:	1	2	3	4	5	6	7	8	9
X_2:	12	21	36	16	25	32	18	16	12
X_3:	3	6	2	2	3	5	5	2	3

i:	10	11	12	13	14	15	16	17	18
X_2:	35	20	8	15	17	28	27	9	14
X_3:	6	5	2	5	6	3	5	3	6

4.6. Refer to Problem 4.5a. Make a residual plot on normal probability paper. What does the plot suggest?

4.7. A student states that he doesn't understand why the sum of squares defined in (4.8) is called a pure error sum of squares "since the formula looks like the one for an ordinary sum of squares." Explain.

4.8. Refer to Problem 2.13. Some preliminary results are: $SSR = 25{,}297.1$; $SSE = 1{,}519.8$.

a) Perform an F test to determine whether or not the regression function is linear. Use a level of significance of .01. What do you find?

b) Assuming the number of replications here was limited in advance to four, is there any advantage in conducting these all at the same level of X? Is there any disadvantage?

c) Does the test in part (a) indicate what regression function is appropriate when it leads to the conclusion that the regression function is not linear? How would you proceed?

4.9. Refer to Problem 3.6.

a) In an F test for linearity of the regression function, what are the alternative conclusions?

b) Perform the test indicated in part (a). Control your risk of Type I error at .05. What do you conclude?

c) Does your procedure in part (b) test for such departures from model (3.1) as lack of constant variance and lack of normality in the error terms? Could the results of the procedure be affected by such departures? Discuss.

4.10. Consider the following two models:

1. $Y = \beta_0 \beta_1^X + \varepsilon$ where ε is a random variable with mean 0 and constant variance
2. $Y = 1/(\beta_0 + \beta_1 X + \varepsilon)$ where ε is a random variable with mean 0 and constant variance

For each model, answer the following questions: (1) Is it linear in the parameters? (2) Is it linear in X? (3) Is it intrinsically linear—if so, how can it be expressed in the linear form of (3.1) by a suitable transformation?

4.11. An analyst used transformation (4.28) to linearize the multiplicative model (4.27). Which residuals are the relevant ones for study of the aptness of the fitted model?

4.12. Early in 1974 a marketing researcher used model (4.27) to study annual sales of a product that had been introduced in 1964. The data were as follows, where X is year (coded) and Y is sales in thousands of units:

Year:	1964	1965	1966	1967	1968	1969	1970	1971	1972	1973
X_i:	0	1	2	3	4	5	6	7	8	9
Y_i:	120	135	162	181	215	234	277	313	374	422

a) Linearize the researcher's model by a suitable transformation of the data, state the transformed model thus obtained, and calculate the estimated regression equation for the transformed model.
b) Plot the estimated regression line and the transformed data. Does the regression line appear to be a good fit?
c) Express the estimated regression equation in terms of the original data. What is the meaning of each of the estimated coefficients here?

4.13. Refer to Problem 4.12. Obtain the residuals and make all appropriate residual plots to evaluate the aptness of the model employed. Do your residuals involve the original or the transformed data? Summarize your findings.

4.14. If the error terms for model (4.33) are independent $N(0, \sigma^2)$, what can be said about the error terms after transformation (4.34) is used? Is the situation the same for model (4.27) and transformation (4.28)?

4.15. (Calculus needed.) An economist, studying the relation between personal income and personal savings in a developing country, found that the variance of the error terms is proportional to X—in other words, $\sigma^2(\varepsilon_i) = kX_i$.
a) Derive the optimal least squares normal equations for fitting a linear regression function here. Check your results by making appropriate substitutions in (4.51).
b) The program package for regression analysis available to the economist does not contain an option for weighted least squares. How may the economist proceed, using ordinary least squares?

4.16. The data below, from a study of computer-assisted learning by 10 students, show the total number of correct and incorrect responses in completing a lesson (X) and the cost of computer time (Y, in cents):

i:	1	2	3	4	5	6	7	8	9	10
X_i:	16	14	22	10	14	17	10	13	19	12
Y_i:	77	70	85	50	62	70	52	63	88	57

An analyst, employing model (3.1), obtained the following results:

i:	1	2	3	4	5	6	7	8	9	10
$\hat{Y}_i$:	71.5	65.2	90.7	52.4	65.2	74.7	52.4	62.0	81.1	58.8
e_i:	+5.5	+4.8	−5.7	−2.4	−3.2	−4.7	−.4	+1.0	+6.9	−1.8

$$\hat{Y}_i = 20.57047 + 3.18568X_i \qquad s(b_0) = 6.12691 \qquad s(b_1) = .40444$$

a) After graphic analysis of the residuals, the analyst concluded that model (4.40) and transformation (4.41) should be employed. Do you agree?

b) Obtain the estimated regression equation and residuals using (4.40) and (4.41). Does it appear that the purpose of the transformation has been attained? Discuss.

c) Restate the estimated regression equation obtained in part (b) so that it is now in terms of the original variables. Is the precision of the regression coefficients better than when model (3.1) was employed?

CITED REFERENCES

4.1. Natrella, Mary G. *Experimental Statistics.* National Bureau of Standards Handbook 91. Washington, D.C.: U.S. Government Printing Office, 1963.

4.2. Anscombe, F. J., and Tukey, John W. "The Examination and Analysis of Residuals," *Technometrics*, Vol. 5 (1963), pp. 141–60.

4.3. Dixon, Wilfrid J., and Massey, Frank J., Jr. *Introduction to Statistical Analysis.* 3d ed. New York: McGraw-Hill Book Co., 1969.

5

Topics in Regression Analysis—I

IN THIS CHAPTER, we take up a variety of topics in simple regression analysis. Several of the topics pertain to the problem of how to make a number of inferences from the same set of sample observations.

5.1 JOINT ESTIMATION OF β_0 AND β_1

Need for Joint Estimation

A market research analyst conducted a study of the relation between level of advertising (X) and sales (Y), in which there was no advertising ($X = 0$) for some observations while for other observations the level of advertising was varied. The scatter plot suggested a linear regression in the range of the advertising expenditures levels studied. The analyst wished to draw inferences about both the intercept β_0 and the slope β_1. To this end, he could construct, say, 95 percent confidence intervals for each parameter, by the methods of Chapter 3. Yet these would not provide him with 95 percent assurance that his conclusions for *both* β_0 and β_1 are correct. If the inferences were independent, the probability of both being correct would be $(.95)^2$, or only .9025. The inferences are not, however, independent, coming as they do from the same set of sample data, which makes the determination of the probability of both inferences being correct much more difficult.

Analysis of data frequently requires a series of estimates (or tests) based on the same sample data, such as the estimation of both β_0 and β_1, where the analyst would like to have an assurance about the correctness of the entire set of estimates (or tests). We shall call the set of estimates of interest the *family of estimates*. In our illustration, the *family* consists of the estimates of

β_0 and β_1. We then distinguish between a *statement confidence coefficient* and a *family confidence coefficient*. The former is the familiar type of confidence coefficient discussed earlier, which indicates the proportion of correct estimates that are obtained when repeated samples are selected and the specified confidence interval is calculated for each sample. A family confidence coefficient, on the other hand, indicates the proportion of correct families of estimates when repeated samples are selected and the specified confidence intervals for the entire family are calculated for each sample.

To illustrate the meaning of a family confidence coefficient further, let us return to the joint estimation of β_0 and β_1. A family confidence coefficient of, say, .95 would indicate for this situation that if repeated samples are selected and interval estimates for both β_0 and β_1 are calculated for each sample by specified procedures, 95 percent of the samples would lead to a family of estimates where *both* confidence intervals are correct. For 5 percent of the samples, either one or both of the interval estimates would be incorrect.

Clearly, a procedure which provides a family confidence coefficient is often highly desirable since it permits the analyst to weave the separate results together into an integrated set of conclusions, with an assurance that the entire set of estimates is correct. In our illustration for instance, the analyst could conclude, say, that mean sales with no advertising expenditures are somewhere near 500 cases per week and that the effect of each advertising dollar on sales is small compared to the sales obtained without advertising, in the range of advertising expenditures studied. With a family confidence coefficient, the analyst would have known assurance that both estimates which entered his integrated conclusion would be correct.

We turn now to the joint estimation of β_0 and β_1 with a family confidence coefficient which provides assurance that both estimates are correct.

Development of Joint Confidence Region

We shall initially work with the alternate regression model (2.6) for which the response function is:

$$E(Y) = \beta_0^* + \beta_1(X - \bar{X}) \tag{5.1}$$

where:

$$\beta_0^* = \beta_0 + \beta_1\bar{X} \tag{5.2}$$

For this model, we know from (2.14) and (2.15) that the estimated regression function is:

$$\hat{Y} = b_0^* + b_1(X - \bar{X}) \tag{5.3}$$

where:

$$b_0^* = b_0 + b_1\bar{X} = \bar{Y} \tag{5.4}$$

We know that b_0^* is normally distributed for model (3.1) since it is a linear combination of the observations Y_i. Further, the mean of b_0^* is:

$$E(b_0^*) = E(b_0 + b_1\bar{X}) = \beta_0 + \beta_1\bar{X} = \beta_0^* \tag{5.5}$$

and its variance is:

$$\sigma^2(b_0^*) = \sigma^2(\bar{Y}) = \frac{\sigma^2(Y_i)}{n} = \frac{\sigma^2}{n} \tag{5.6}$$

Hence, the standardized variable is a standard normal variable:

$$\frac{b_0^* - \beta_0^*}{\sigma(b_0^*)} = \frac{b_0^* - \beta_0^*}{\dfrac{\sigma}{\sqrt{n}}} = z \tag{5.7}$$

and its square, by (1.36), is a chi-square random variable with 1 degree of freedom:

$$\left[\frac{b_0^* - \beta_0^*}{\sigma(b_0^*)}\right]^2 = \frac{n(b_0^* - \beta_0^*)^2}{\sigma^2} = z^2 = \chi^2(1) \tag{5.8}$$

We also know from our earlier results that:

$$\frac{b_1 - \beta_1}{\sigma(b_1)} = \frac{b_1 - \beta_1}{\dfrac{\sigma}{\sqrt{\sum(X - \bar{X})^2}}} = z \tag{5.9}$$

so that:

$$\left[\frac{b_1 - \beta_1}{\sigma(b_1)}\right]^2 = \frac{\sum(X_i - \bar{X})^2(b_1 - \beta_1)^2}{\sigma^2} = \chi^2(1) \tag{5.10}$$

By theorem (3.29), $\bar{Y}$ and b_1 are independent. Since $\bar{Y} = b_0^*$, it follows that b_0^* and b_1 are independent. But then by definition (1.36) of $\chi^2(v)$, we obtain $\chi^2(2)$ when adding two independent $\chi^2(1)$ variables:

$$\frac{n(b_0^* - \beta_0^*)^2}{\sigma^2} + \frac{\sum(X_i - \bar{X})^2(b_1 - \beta_1)^2}{\sigma^2} = \chi^2(2) \tag{5.11}$$

We know also from (3.11) that:

1. $\dfrac{SSE}{\sigma^2} = \chi^2(n-2)$.

2. $\dfrac{SSE}{\sigma^2}$ is independent of b_0 and b_1.

Since b_0^* is a function of b_0 and b_1, SSE/σ^2 is independent of b_0^* and b_1. It therefore follows from definition (1.42) of an F random variable that $\chi^2(2)/2$ divided by an independent $\chi^2(n-2)/(n-2)$ is $F(2, n-2)$:

$$\frac{\dfrac{n(b_0^* - \beta_0^*)^2}{\sigma^2} + \dfrac{\sum(X_i - \bar{X})^2(b_1 - \beta_1)^2}{\sigma^2}}{2} \div \frac{\dfrac{SSE}{\sigma^2}}{n-2} = F(2, n-2) \tag{5.12}$$

Canceling the σ^2 terms, we obtain:

$$\frac{n(b_0^* - \beta_0^*)^2 + \sum(X_i - \bar{X})^2(b_1 - \beta_1)^2}{2MSE} = F(2, n-2) \tag{5.13}$$

If we now replace b_0^* and β_0^* by the equivalent expressions in (5.4) and (5.2) and then simplify, we find:

$$\frac{n(b_0 - \beta_0)^2 + 2(\sum X_i)(b_0 - \beta_0)(b_1 - \beta_1) + (\sum X_i^2)(b_1 - \beta_1)^2}{2MSE} = F(2, n-2) \tag{5.14}$$

Therefore, we can write the probability statement:

$$P\left\{\frac{n(b_0 - \beta_0)^2 + 2(\sum X_i)(b_0 - \beta_0)(b_1 - \beta_1) + (\sum X_i^2)(b_1 - \beta_1)^2}{2MSE} \leq F(1-\alpha; 2, n-2)\right\} = 1 - \alpha \tag{5.15}$$

Since the probability statement (5.15) holds for all values of β_0 and β_1, the $1 - \alpha$ confidence region for β_0 and β_1 is given by:

$$\frac{n(b_0 - \beta_0)^2 + 2(\sum X_i)(b_0 - \beta_0)(b_1 - \beta_1) + (\sum X_i^2)(b_1 - \beta_1)^2}{2MSE} \leq F(1-\alpha; 2, n-2) \tag{5.16}$$

The family confidence coefficient $1 - \alpha$ indicates that the confidence region (5.16), with repeated sampling, will contain both β_0 and β_1 in $(1-\alpha)100$ percent of the cases. The confidence region consists of all points (β_0, β_1) which satisfy the inequality (5.16). The boundary of the confidence region is obtained from the equality in (5.16):

$$\frac{n(b_0 - \beta_0)^2 + 2(\sum X_i)(b_0 - \beta_0)(b_1 - \beta_1) + (\sum X_i^2)(b_1 - \beta_1)^2}{2MSE} = F(1-\alpha; 2, n-2) \tag{5.17}$$

The boundary is an ellipse, as illustrated in Figure 5.1. We explain now by an example how the boundary is calculated.

FIGURE 5.1

Elliptical Joint 90 Percent Confidence Region for β_0 and β_1—Westwood Company Example

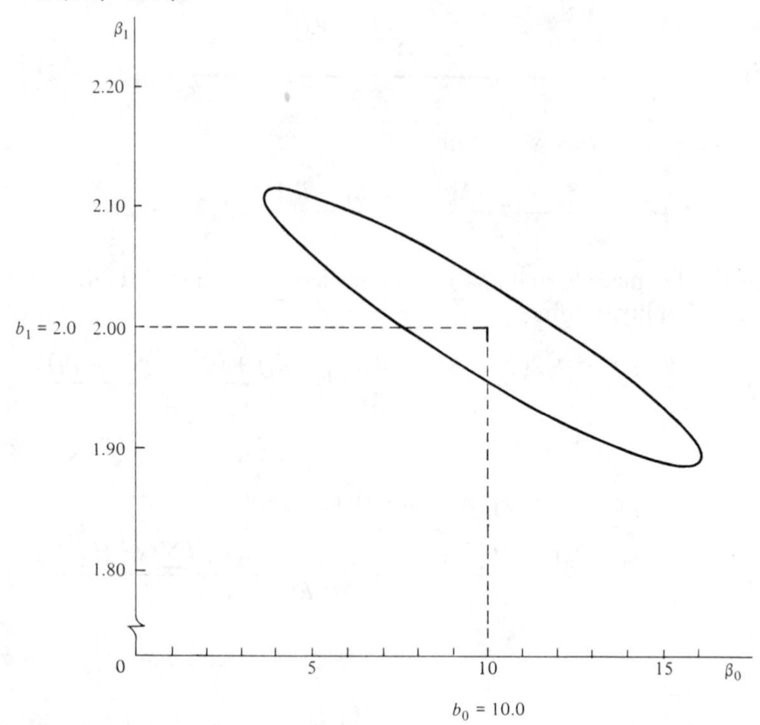

Example

Let us return to the Westwood Company lot size example of the previous chapters. Suppose that we require a joint confidence region for β_0 and β_1 with .90 family confidence coefficient. Then $1 - \alpha = .90$ and $F(.90; 2, 8) = 3.11$. We have, from previous work (see Table 3.1): $b_0 = 10.0$, $b_1 = 2.0$, $MSE = 7.5$, $\sum X_i = n\bar{X} = 500$, and $\sum X_i^2 = 28{,}400$. Substitution into (5.17) gives:

$$\frac{10(10.0 - \beta_0)^2 + 2(500)(10.0 - \beta_0)(2.0 - \beta_1) + 28{,}400(2.0 - \beta_1)^2}{2(7.5)} = 3.11$$

Boundary points are calculated by assigning a value to either β_0 or β_1 and finding corresponding values of the other unknown. For example, let $\beta_0 = 10.0$. The quantity $(10.0 - \beta_0)$ then equals 0, and the above expression reduces to:

$$\frac{28{,}400(2.0 - \beta_1)^2}{2(7.5)} = 3.11$$

Manipulation gives:

$$\beta_1^2 - 4.0\beta_1 + 3.998357 = 0$$

This quadratic equation has two roots, 1.95947 and 2.04053. Hence two boundary points are (10.0, 1.96) and (10.0, 2.04).

Additional boundary points are found in the same manner, by assigning a value to either β_0 or β_1 and solving for the other. The points can be plotted on a graph and connected to form the boundary of the elliptical confidence region, as shown in Figure 5.1. The region in our example is centered around $(b_0, b_1) = (10.0, 2.0)$. The joint confidence region indicates that β_1 is somewhere between 1.88 and 2.12, and that β_0 would be in the neighborhood of 4.0 if β_1 were near the upper limit and about 16.0 if β_1 were near its lower limit. Note carefully the interrelation between the estimates for β_0 and β_1: the larger is β_1, the smaller would be β_0, and vice versa. This interrelation is the result of the tilted position of the ellipse, with the major axis being sloped negatively.

Comments

1. The joint confidence region can be used directly for testing. To illustrate this use, suppose an industrial engineer working for the Westwood Company theorized that the regression function should have an intercept of 13.0 and a slope of 2.10. Since the point (13.0, 2.10) does not fall in the joint confidence region, we would conclude at the $\alpha = .10$ level of significance that either $\beta_0 \neq 13.0$ or $\beta_1 \neq 2.10$ or both. Note from Figure 5.1 that the engineer may be correct with respect to either the intercept or the slope, but that his particular *combination* is not supported by the data.

2. The tilt of the ellipse is a function of the covariance between b_0 and b_1. It can be shown that:

$$\sigma(b_0, b_1) = -\bar{X}\sigma^2(b_1) \tag{5.18}$$

Both the tilt of the ellipse and the covariance indicate the degree to which the point estimates of β_0 and β_1 obtained from the same sample are likely to err in a similar or an opposite direction because of sampling error. A positive covariance indicates a tendency for the values of b_0 and b_1 to be jointly too high or jointly too low, while a negative covariance means that the joint errors tend to be in opposite directions.

In our example, $\bar{X} = 50$; hence the covariance is negative and the ellipse's major axis is sloped negatively. This implies that the estimators b_0 and b_1 tend to err in opposite directions. We expect this intuitively. Since the observed points (X_i, Y_i) fall in the first quadrant, we anticipate that if the slope of the fitted regression line is too steep (b_1 overestimates β_1), the intercept is most likely to be too low (b_0 underestimates β_0), and vice versa.

When the independent variable is $X_i - \bar{X}$, we know that b_0^* and b_1 are independent and hence have zero covariance. Thus, the ellipse will in this case have axes parallel to the axes of the graph so that there is no tilt in the confidence region.

Bonferroni Joint Confidence Intervals

Another way to jointly estimate the regression parameters β_0 and β_1 is by the Bonferroni method. This method of joint estimation is a general method which can be applied in many cases, not just for the joint estimation of β_0 and β_1. Here, we explain the Bonferroni method as it applies for estimating β_0 and β_1 jointly.

Development of Joint Confidence Intervals. We start with ordinary confidence limits, with statement confidence coefficients $1 - \alpha$ each, for β_0 and β_1. These are:

$$b_0 \pm t(1 - \alpha/2; n - 2)s(b_0)$$
$$b_1 \pm t(1 - \alpha/2; n - 2)s(b_1)$$

We then ask what is the probability that both sets of limits are correct. Let A_1 denote the event that the first confidence interval does not cover β_0, and A_2 denote the event that the second confidence interval does not cover β_1. We know:

$$P(A_1) = \alpha \qquad P(A_2) = \alpha$$

Probability theorem (1.6) states:

$$P(A_1 \cup A_2) = P(A_1) + P(A_2) - P(A_1 \cap A_2)$$

and hence:

$$1 - P(A_1 \cup A_2) = 1 - P(A_1) - P(A_2) + P(A_1 \cap A_2) \tag{5.19}$$

Now by probability theorems (1.9) and (1.10), we have:

$$1 - P(A_1 \cup A_2) = P(\overline{A_1 \cup A_2}) = P(\bar{A}_1 \cap \bar{A}_2)$$

$P(\bar{A}_1 \cap \bar{A}_2)$ is the probability that both confidence intervals are correct. We thus have from (5.19):

$$P(\bar{A}_1 \cap \bar{A}_2) = 1 - P(A_1) - P(A_2) + P(A_1 \cap A_2) \tag{5.20}$$

Since $P(A_1 \cap A_2) \geq 0$, we obtain from (5.20) the Bonferroni inequality:

$$P(\bar{A}_1 \cap \bar{A}_2) \geq 1 - P(A_1) - P(A_2) \tag{5.21}$$

which for our situation is:

$$P(\bar{A}_1 \cap \bar{A}_2) \geq 1 - \alpha - \alpha = 1 - 2\alpha \tag{5.21a}$$

Thus, if β_0 and β_1 are separately estimated with, say, 95 percent confidence intervals, the Bonferroni inequality guarantees us a family confidence coefficient of at least 90 percent that both intervals based on the same sample are correct.

We can easily use the Bonferroni inequality (5.21a) to obtain a family confidence coefficient of at least $1 - \alpha$ for estimating β_0 and β_1. We do this by estimating β_0 and β_1 separately with statement confidence coefficients of $1 - \alpha/2$ each. Thus, the $1 - \alpha$ family confidence intervals for β_0 and β_1, often called a *confidence set*, are by the Bonferroni procedure:

(5.22)
$$\begin{aligned} b_0 - Bs(b_0) \leq \beta_0 \leq b_0 + Bs(b_0) \\ b_1 - Bs(b_1) \leq \beta_1 \leq b_1 + Bs(b_1) \end{aligned}$$

where:

(5.22a)
$$B = t(1 - \alpha/4; n - 2)$$

Example. For the Westwood Company lot size application, 90 percent family confidence intervals for β_0 and β_1 require $t(.975; 8) = 2.306$. We have from before:

$$b_0 = 10.0 \qquad s(b_0) = 2.50294$$
$$b_1 = 2.0 \qquad s(b_1) = .04697$$

Hence the joint confidence intervals are:

$$4.2282 = 10.0 - (2.306)(2.50294) \leq \beta_0 \leq 10.0 + (2.306)(2.50294) = 15.7718$$
$$1.8917 = 2.0 - (2.306)(.04697) \leq \beta_1 \leq 2.0 + (2.306)(.04697) = 2.1083$$

Thus, we conclude that β_0 is between 4.23 and 15.77 *and* β_1 is between 1.89 and 2.11. The family confidence coefficient is at least .90 that the procedure leads to correct pairs of interval estimates.

Comments

1. We reiterate that the Bonferroni $1 - \alpha$ family confidence coefficient is actually a lower bound on the true (but often unknown) family confidence coefficient. To the extent that incorrect interval estimates of β_0 and β_1 tend to pair up in the family (particularly when the covariance between b_0 and b_1 is large), the families of statements will tend to be correct more than $(1 - \alpha)100$ percent of the time.

2. The Bonferroni inequality (5.21a) can easily be extended to s simultaneous confidence intervals with family confidence coefficient $1 - \alpha$:

(5.23)
$$P\left(\bigcap_{i=1}^{s} \bar{A}_i\right) \geq 1 - s\alpha$$

Thus, if s interval estimates are desired with a family confidence coefficient $1 - \alpha$, constructing each interval estimate with statement confidence coefficient $1 - \alpha/s$ will suffice.

3. The larger the number of confidence intervals in the family, for a given family confidence coefficient, the greater becomes the multiple B, making some or all of the confidence intervals too wide to be helpful. Hence, the Bonferroni technique is ordinarily most useful when the number of simultaneous estimates is not too large.

4. It is not necessary with the Bonferroni procedure that the confidence intervals have the same statement confidence coefficient. Different statement confidence coefficients could be used, depending on the importance of each estimate. For instance, in our earlier illustration β_0 might be estimated with a 92 percent confidence interval

and β_1 with a 98 percent confidence interval. The family confidence coefficient by (5.21) will still be at least 90 percent.

Comparison of Two Approaches. Figure 5.2 contains the joint 90 percent confidence region by the Bonferroni approach for our example. Note that the region is a rectangle since the Bonferroni approach does not utilize the existing relationship between b_0 and b_1. The rectangle is centered at (b_0, b_1). Also shown in Figure 5.2 is the elliptical 90 percent confidence region which we obtained earlier. The elliptical region is more efficient in that it covers fewer (β_0, β_1) points for the same confidence coefficient. Nevertheless, the Bonferroni approach can be highly useful in particular cases. For instance, suppose the joint Bonferroni estimates in some application are:

$$3.4 \leq \beta_0 \leq 3.6$$
$$499 \leq \beta_1 \leq 503$$

These confidence intervals may not be as efficient as the elliptical confidence region, yet may be more than adequate for conveying information about the magnitudes of the two parameters.

FIGURE 5.2

Bonferroni and Elliptical Joint 90 Percent Confidence Regions for β_0 and β_1—Westwood Company Example

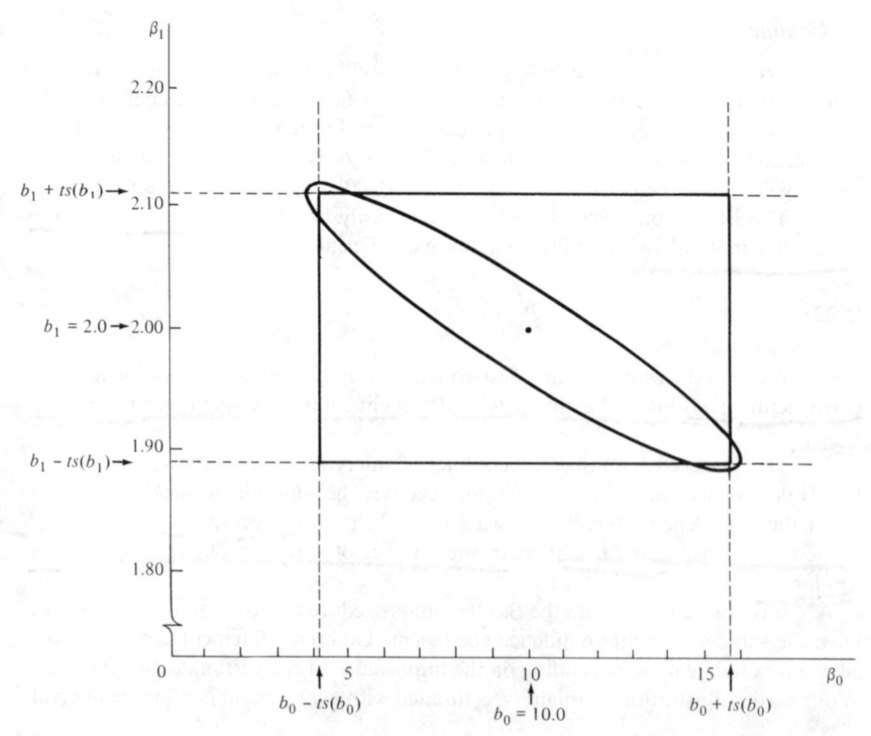

For the simple regression case, the elliptical confidence region is simple enough to obtain so that it probably should be used in preference to the Bonferroni region much of the time. In multiple regression cases, however, the Bonferroni method comes into its own because of the simplicity of obtaining and interpreting the joint estimates.

5.2 CONFIDENCE BAND FOR REGRESSION LINE

At times we would like to obtain a confidence band for the regression line $E(Y) = \beta_0 + \beta_1 X$, so that we can see in what region the entire regression line lies. This differs from estimating $E(Y_h) = \beta_0 + \beta_1 X_h$ for a particular value of X_h by an interval estimate, which we took up in Section 3.4.

To obtain a confidence band for the entire regression line, we essentially need to consider all possible (β_0, β_1) combinations in the elliptical joint confidence region (5.16) for β_0 and β_1. To each such combination corresponds a regression line. For our Westwood Company example, three possible (β_0, β_1) combinations (see Figure 5.1) and their corresponding regression lines are:

β_0	β_1	$E(Y) = \beta_0 + \beta_1 X$
4.00	2.11	$E(Y) = 4.00 + 2.11X$
9.00	2.00	$E(Y) = 9.00 + 2.00X$
16.00	1.89	$E(Y) = 16.00 + 1.89X$

Figure 5.3 contains a plot of these three possible regression lines. As additional lines for other possible (β_0, β_1) combinations in the confidence region in Figure 5.1 are plotted, we will fill out a confidence region or confidence band for the regression line. The boundaries of the region are sketched in Figure 5.3 by the broken lines, which are hyperbolas.

Working and Hotelling derived the formula for the $1 - \alpha$ hyperbolic confidence bounds for the regression line. At any level X_h, the bounds are:

$$\hat{Y}_h - Ws(\hat{Y}_h) \leq \beta_0 + \beta_1 X_h \leq \hat{Y}_h + Ws(\hat{Y}_h) \tag{5.24}$$

where:

$$W^2 = 2F(1 - \alpha; 2, n - 2) \tag{5.24a}$$

and $\hat{Y}_h$ and $s(\hat{Y}_h)$ are defined in (3.27) and (3.30) respectively.

Example

For our Westwood Company example, suppose that we wish to set up the 90 percent confidence band for the regression line. We developed earlier:

$$\hat{Y}_h = 10.0 + 2.0X_h \qquad \text{(p. 40)}$$

$$s^2(\hat{Y}_h) = 7.5\left[\frac{1}{10} + \frac{(X_h - 50)^2}{3{,}400}\right] \qquad \text{(p. 68)}$$

FIGURE 5.3

Plot of Three Possible Regression Lines for Joint Confidence Region in Figure 5.1 (*Y* values not plotted to scale)

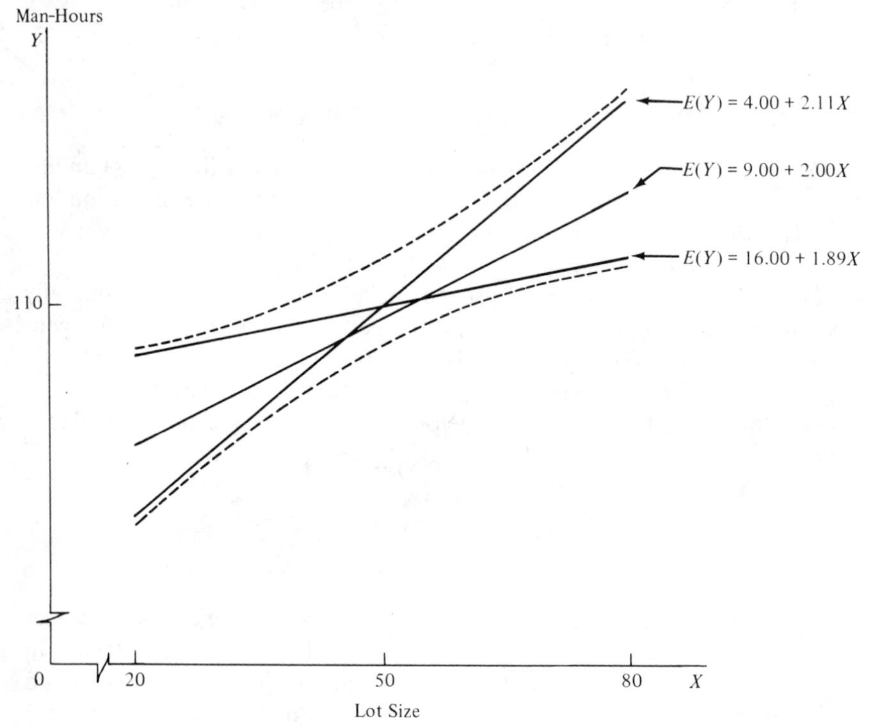

We need $F(.90; 2, 8) = 3.11$. Hence, we obtain:

$$W^2 = 2(3.11) = 6.22 \qquad \text{or} \qquad W = 2.494$$

Let us find the boundary points for the confidence band at $X_h = 55$:

$$\hat{Y}_{55} = 10.0 + 2.0(55) = 120.0$$

$$s^2(\hat{Y}_{55}) = 7.5\left[\frac{1}{10} + \frac{(55 - 50)^2}{3{,}400}\right] = .80515$$

or:

$$s(\hat{Y}_{55}) = .89730$$

Hence, the 90 percent boundary points for the regression line at $X_h = 55$ are:

$$120.0 - (2.494)(.89730) \leq \beta_0 + \beta_1 X_h \leq 120.0 + (2.494)(.89730)$$

or:

$$117.8 \leq \beta_0 + \beta_1 X_h \leq 122.2$$

FIGURE 5.4

90 Percent Confidence Band for Regression Line—Westwood Company Example

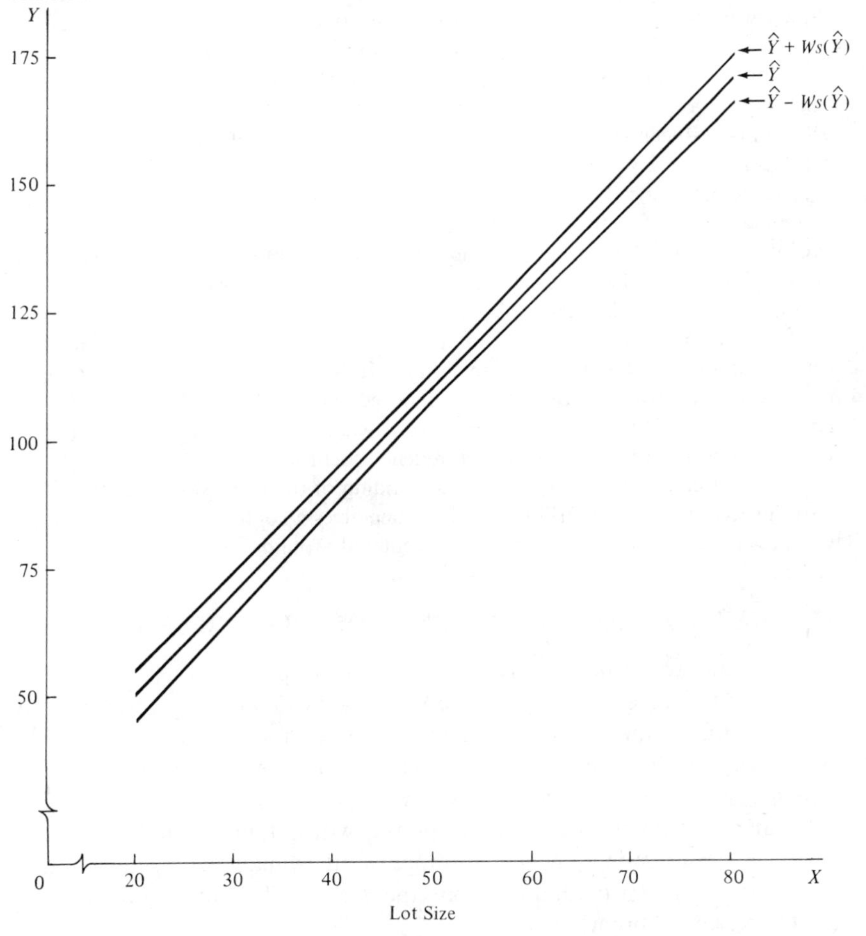

In similar fashion, the boundary points at a number of other values of X_h can be developed and the boundary curves then sketched in. This has been done in Figure 5.4, which contains the 90 percent confidence region for the regression line for the Westwood Company example.

Comments

1. The boundary points for the regression line are of exactly the same form as the confidence limits for a single mean response $E(Y_h)$ in (3.32) except that the t multiple has been replaced by the W multiple (named after Working).

2. If we had wished to estimate a single mean response with a 90 percent confidence coefficient, the required t value would have been $t(.95; 8) = 1.860$. Note that this t multiple, though smaller than the multiple $W = 2.494$ for the entire regression line, does not differ by any major extent. This is typically the case, so that the boundary points for the entire regression line usually are not very much further apart than the confidence limits for a single mean response $E(Y_h)$ at a given X_h value. With the somewhat wider limits for the entire regression line, one is able to draw conclusions about any and all mean responses for the entire regression line and not just about the mean response at a given X level. One use of this broader base for inferences will be explained in the next section.

3. The confidence band (5.24) applies to the entire regression line over all real-numbered values of X from $-\infty$ to $+\infty$. The confidence coefficient indicates the percent of time the estimating procedure will yield a band which covers the entire line, in a long series of samples in which the X observations are kept at the same level as in the sample actually taken.

In applications, the confidence band is ignored for that part of the regression line which is not of interest in the problem at hand. In the Westwood Company example, for instance, negative lot sizes would be ignored. The confidence coefficient for the limited segment of the band of interest is somewhat higher than $1 - \alpha$, so that $1 - \alpha$ serves then as a lower bound to the confidence coefficient.

4. Research continues on confidence banding of the regression line. One alternative procedure to the Working-Hotelling one gives a confidence band of uniform width over a finite interval on the X axis centered around $\bar{X}$ (Ref. 5.1).

5.3 SIMULTANEOUS ESTIMATION OF MEAN RESPONSES

Often one would like to estimate the mean responses at a number of X levels from the same sample data. The Westwood Company, for instance, may wish to estimate the mean number of man-hours for lots of 30, 55, and 80 parts. We already know how to do this for any one level of X with given statement confidence coefficient. Now we shall discuss two approaches for simultaneous estimation of mean responses with a family confidence coefficient, so that there is a known assurance of all estimates of mean responses being correct. The two approaches are the Working-Hotelling approach and the Bonferroni approach.

The reason for concern with a family confidence coefficient is that separate interval estimates of $E(Y_h)$ at various levels X_h need not all be correct or all be incorrect, even though they are all based on the same sample data and fitted regression line. The combination of sampling errors in b_0 and b_1 may be such that the interval estimates of $E(Y_h)$ will be correct over some range of X levels and incorrect elsewhere.

Working-Hotelling Approach

Since the Working-Hotelling confidence band for the entire regression line holds for all values of X, it certainly must hold for selected levels of the independent variable. Hence, to obtain with the Working-Hotelling approach a family of interval estimates of mean responses at different levels of X, with a $1 - \alpha$ family confidence coefficient, we simply use formula (5.24) repetitively to calculate boundary points of the confidence band at the various X levels. These boundary points serve then as the family confidence limits for the interval estimates of the mean responses of interest.

Example. Suppose we require, for the Westwood Company lot size example, a family of estimates of the mean number of man-hours at the following levels of lot size: 30, 55, 80. The family confidence coefficient is to be .90. We obtained earlier $\hat{Y}_h$ and $s(\hat{Y}_h)$ for $X_h = 55$, and found $W = 2.494$. In similar fashion, we can obtain the needed results for the other lot sizes. We summarize them here, without showing the calculations:

X_h	$\hat{Y}_h$	$s(\hat{Y}_h)$	$Ws(\hat{Y}_h)$
30	70.0	1.27764	3.1864
55	120.0	.89730	2.2379
80	170.0	1.65387	4.1248

Thus, the boundary limits for the regression line at $X_h = 30$, 55, and 80 are:

$$66.8 = 70.0 - 3.1864 \leq E(Y_{30}) \leq 70.0 + 3.1864 = 73.2$$

$$117.8 = 120.0 - 2.2379 \leq E(Y_{55}) \leq 120.0 + 2.2379 = 122.2$$

$$165.9 = 170.0 - 4.1248 \leq E(Y_{80}) \leq 170.0 + 4.1248 = 174.1$$

With family confidence coefficient .90, we conclude that the mean number of man-hours for lots of 30 parts is between 66.8 and 73.2, for lots of 55 parts between 117.8 and 122.2, and for lots of 80 parts between 165.9 and 174.1. The confidence coefficient .90 provides assurance that the procedure leads to all correct estimates in the family of estimates.

Bonferroni Approach

The Bonferroni approach, discussed earlier for simultaneous estimation of β_0 and β_1, is a completely general approach. To construct a family of confidence intervals for mean responses at different X levels, we construct the usual confidence interval for a single mean response $E(Y_h)$, given in (3.32), and adjust the statement confidence coefficient to yield the specified family confidence coefficient.

If $E(Y_h)$ is to be estimated for s levels of X, with a family confidence coefficient of $1 - \alpha$, the Bonferroni confidence intervals are:

(5.25) $$\hat{Y}_h - Bs(\hat{Y}_h) \leq E(Y_h) \leq \hat{Y}_h + Bs(\hat{Y}_h)$$

where:

(5.25a) $$B = t(1 - \alpha/2s; n - 2)$$

Example. The estimates of the mean man-hours for lot sizes of 30, 55, and 80 parts with a family confidence coefficient of .90 by the Bonferroni approach require the same data as the Working-Hotelling approach, presented on page 153. In addition, we require $t(1 - .10/(2)(3); 8) = t(.983; 8)$. We explain in Chapter 14 how one can interpolate for percentiles in the t distribution. Here, we simply state that $t(.983; 8) = 2.56$.

We thus obtain the confidence intervals, with a 90 percent family confidence coefficient:

$$66.7 = 70.0 - (2.56)(1.27764) \leq E(Y_{30}) \leq 70.0 + (2.56)(1.27764) = 73.3$$
$$117.7 = 120.0 - (2.56)(.89730) \leq E(Y_{55}) \leq 120.0 + (2.56)(.89730) = 122.3$$
$$165.8 = 170.0 - (2.56)(1.65387) \leq E(Y_{80}) \leq 170.0 + (2.56)(1.65387) = 174.2$$

Comments

1. In this instance the Working-Hotelling confidence limits are slightly tighter than the Bonferroni limits. In other cases where the number of statements is small, the Bonferroni limits may be tighter. For larger families, the Working-Hotelling confidence limits will always be the tighter, since W in (5.24a) stays the same for any number of statements in the family whereas B in (5.25a) becomes larger as the number of statements increases. In practice, once the family confidence coefficient has been decided upon one can calculate the W and B multiples to determine which procedure leads to tighter confidence limits.

2. Both the Working-Hotelling and Bonferroni approaches to multiple estimation of mean responses provide lower bounds to the actual family confidence coefficient. The reason why the Working-Hotelling approach furnishes a lower bound is that the confidence coefficient $1 - \alpha$ actually applies to the entire line from $-\infty$ to $+\infty$.

3. Sometimes, it is not known in advance for which levels of the independent variable to estimate the mean response. That is determined as the analysis proceeds. In such cases, it is better to use the Working-Hotelling approach. (See the discussion on data snooping in Chapter 14.)

5.4 SIMULTANEOUS PREDICTION INTERVALS FOR NEW OBSERVATIONS

Now we consider the simultaneous prediction of s new observations on Y in s independent trials at s different levels of X. To illustrate this type of application, let us suppose the Westwood Company plans to produce the next three lots in sizes of 30, 55, and 80 parts, and wishes to predict the man-hours for each of these lots with a family confidence coefficient of .95.

Two procedures will be considered here, the Scheffé procedure and the Bonferroni procedure. Both utilize the same type of interval as for predicting a single observation, given in (3.36), and only the multiple of the estimated standard deviation is changed. The Scheffé procedure uses the F distribution, while the Bonferroni procedure uses the t distribution. The simultaneous prediction limits for the s predictions with the Scheffé procedure, with family confidence coefficient $1 - \alpha$, are:

$$\hat{Y}_h - Ss(Y_{h(\text{new})}) \leq Y_{h(\text{new})} \leq \hat{Y}_h + Ss(Y_{h(\text{new})}) \tag{5.26}$$

where:

(5.26a) $$S^2 = sF(1 - \alpha; s, n - 2)$$

With the Bonferroni procedure, the $1 - \alpha$ simultaneous prediction limits are:

(5.27) $$\hat{Y}_h - Bs(Y_{h(\text{new})}) \leq Y_{h(\text{new})} \leq \hat{Y}_h + Bs(Y_{h(\text{new})})$$

where:

(5.27a) $$B = t(1 - \alpha/2s; n - 2)$$

We can evaluate the S and B multiples to see which procedure provides tighter prediction limits. For our example, we have:

$$S^2 = 3F(.95; 3, 8) = 3(4.07) = 12.21 \quad \text{or} \quad S = 3.49$$
$$B = t(1 - .05/(2)(3); 8) = t(.992; 8) = 3.04$$

so that the Bonferroni method will be used here. From earlier results, we obtain (calculations not shown):

X_h	$\hat{Y}_h$	$s(Y_{h(\text{new})})$	$Bs(Y_{h(\text{new})})$
30	70.0	3.02198	9.18682
55	120.0	2.88187	8.76088
80	170.0	3.19926	9.72575

and the simultaneous prediction limits are:

$$60.8 = 70.0 - 9.18682 \leq Y_{30(\text{new})} \leq 70.0 + 9.18682 = 79.2$$
$$111.2 = 120.0 - 8.76088 \leq Y_{55(\text{new})} \leq 120.0 + 8.76088 = 128.8$$
$$160.3 = 170.0 - 9.72575 \leq Y_{80(\text{new})} \leq 170.0 + 9.72575 = 179.7$$

With family confidence coefficient at least .95, we can predict that the man-hours for the next three production runs all will be within the above limits.

Comments

1. Simultaneous prediction intervals for s new observations on Y at s different levels of X with a $1 - \alpha$ family confidence coefficient are wider than the corresponding single prediction interval of (3.36). When the number of simultaneous predictions is not large, however, the difference in the width is only moderate. For instance, a single 95 percent prediction interval for our example would have utilized the t multiple $t(.975; 8) = 2.306$, which is only moderately smaller than the multiple $B = 3.04$ for three simultaneous predictions.

2. Note that both B and S become larger as s increases. This contrasts with simultaneous estimation of mean responses where only B becomes larger, but not W. When s is large, both the B and S multiples may become so large that the prediction intervals will be too wide to be useful. Other simultaneous estimation techniques could then be considered, as discussed in Reference 5.7.

5.5 REGRESSION THROUGH THE ORIGIN

Sometimes the regression line is known to go through the origin at (0, 0). This occurs, for instance, when X is units of output and Y is variable cost, so Y is zero by definition when X is zero. Another example is where X is the number of brands of cigarettes stocked in a supermarket in an experiment (including some supermarkets with no brands stocked), and Y is the volume of cigarette sales in the supermarket. The normal error model for these cases is the same as the general model (3.1) except $\beta_0 = 0$:

(5.28)
$$Y_i = \beta_1 X_i + \varepsilon_i$$

where:

β_1 is a parameter
X_i are known constants
ε_i are independent $N(0, \sigma^2)$

The regression function for model (5.28) is:

(5.29)
$$E(Y) = \beta_1 X$$

The least squares estimator of β_1 is obtained by minimizing:

(5.30)
$$Q = \sum (Y_i - \beta_1 X_i)^2$$

with respect to β_1. The resulting normal equation is:

(5.31)
$$\sum X_i(Y_i - b_1 X_i) = 0$$

leading to the point estimator:

(5.32)
$$b_1 = \frac{\sum X_i Y_i}{\sum X_i^2}$$

b_1 as given in (5.32) is also the maximum likelihood estimator.

An unbiased estimator of $E(Y)$ is:

(5.33)
$$\hat{Y} = b_1 X$$

Also, an unbiased estimator of σ^2 is:

(5.34)
$$MSE = \frac{\sum (Y_i - \hat{Y}_i)^2}{n - 1} = \frac{\sum (Y_i - b_1 X_i)^2}{n - 1}$$

The reason for the denominator $n - 1$ is that only one degree of freedom is lost in estimating the single parameter of the regression equation (5.29).

Confidence intervals for β_1, $E(Y_h)$, and a new observation $Y_{h(\text{new})}$ are shown in Table 5.1. Note that the t multiple has $n - 1$ degrees of freedom here, the degrees of freedom associated with MSE. The results in Table 5.1 are derived in analogous fashion to the earlier results for our general model (3.1). Whereas for the general case, we encounter terms $(X_i - \bar{X})^2$ or $(X_h - \bar{X})^2$, here we find X_i^2 and X_h^2 because of the regression through the origin.

TABLE 5.1

Confidence Intervals for Regression through Origin

Estimate of—	Estimated Variance		Confidence Interval
β_1	$s^2(b_1) = \frac{MSE}{\sum X_i^2}$	(5.35)	$b_1 - ts(b_1) \leq \beta_1 \leq b_1 + ts(b_1)$
$E(Y_h)$	$s^2(\hat{Y}_h) = \frac{X_h^2 MSE}{\sum X_i^2}$	(5.36)	$\hat{Y}_h - ts(\hat{Y}_h) \leq E(Y_h) \leq \hat{Y}_h + ts(\hat{Y}_h)$
$Y_{h(\text{new})}$	$s^2(Y_{h(\text{new})}) = MSE\left[1 + \frac{X_h^2}{\sum X_i^2}\right]$	(5.37)	$\hat{Y}_h - ts(Y_{h(\text{new})}) \leq Y_{h(\text{new})} \leq \hat{Y}_h + ts(Y_{h(\text{new})})$
		where:	$t = t(1-\alpha/2; n-1)$

Example

The Charles Plumbing Supply Company operates 12 warehouses. In an attempt to tighten procedures for planning and control, a consultant studied the relation between number of work units performed (X) and total variable labor cost (Y) in the warehouses during a test period. The data are given in Table 5.2, and the observations are shown as a scatter plot in Figure 5.5.

Model (5.28) for regression through the origin was employed since Y involved variable costs only and the other conditions of the model appeared

TABLE 5.2

Data for Example of Regression through Origin

Warehouse i	Work Units Performed X_i	Variable Labor Cost (dollars) Y_i	$X_i Y_i$	X_i^2
1	20	114	2,280	400
2	196	921	180,516	38,416
3	115	560	64,400	13,225
4	50	245	12,250	2,500
5	122	575	70,150	14,884
6	100	475	47,500	10,000
7	33	138	4,554	1,089
8	154	727	111,958	23,716
9	80	375	30,000	6,400
10	147	670	98,490	21,609
11	182	828	150,696	33,124
12	160	762	121,920	25,600
Total	1,359	6,390	894,714	190,963

FIGURE 5.5

Data and Fitted Regression through Origin—Warehouse Example

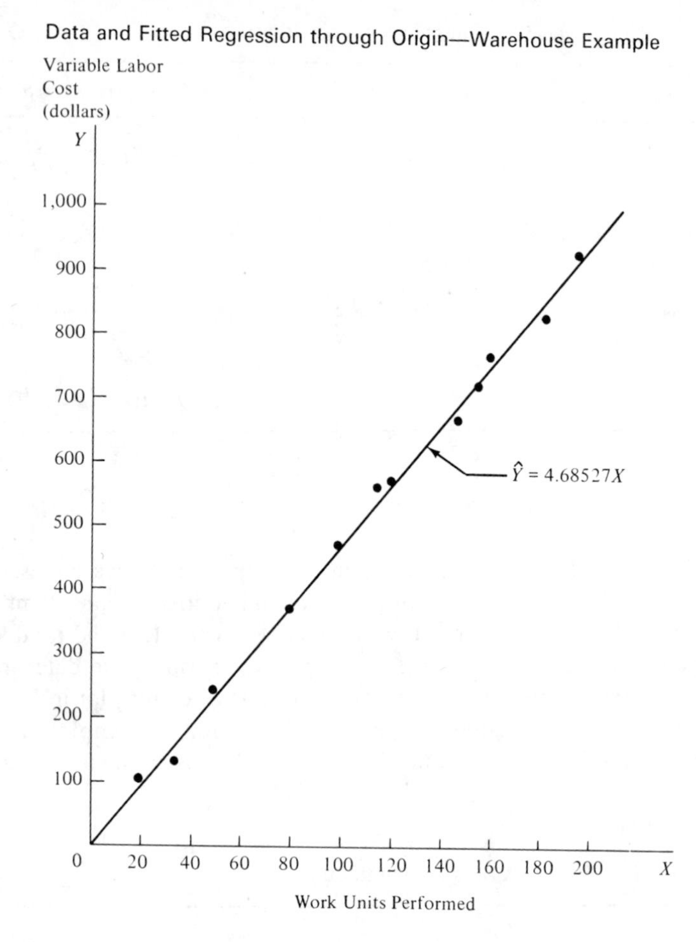

to be satisfied as well. From Table 5.2 we have: $\sum X_i Y_i = 894{,}714$ and $\sum X_i^2 = 190{,}963$. Hence:

$$b_1 = \frac{\sum X_i Y_i}{\sum X_i^2} = \frac{894{,}714}{190{,}963} = 4.68527$$

and the estimated regression equation is:

$$\hat{Y} = 4.68527X$$

The fitted regression line is plotted in Figure 5.5.

To illustrate inferences for regression through the origin, suppose an interval estimate of β_1 is desired, with a .95 confidence coefficient. We obtain (calculations not shown):

$$MSE = \frac{\sum (Y_i - b_1 X_i)^2}{n-1} = \frac{2{,}457.66}{11} = 223.42$$

From Table 5.2 we have: $\sum X_i^2 = 190{,}963$. Hence:

$$s^2(b_1) = \frac{MSE}{\sum X_i^2} = \frac{223.42}{190{,}963} = .0011700$$

and:

$$s(b_1) = .034205$$

For a .95 confidence coefficient, we require $t(.975; 11) = 2.201$. The interval estimate, by (5.35), is:

$$b_1 - ts(b_1) \leq \beta_1 \leq b_1 + ts(b_1)$$

or:

$$4.61 = 4.68527 - (2.201)(.034205) \leq \beta_1 \leq 4.68527 + (2.201)(.034205) = 4.76$$

Thus it is estimated, with 95 percent confidence coefficient, that the mean of the distribution of total variable labor costs increases by somewhere between \$4.61 and \$4.76 for each additional work unit performed.

Comments

1. In linear regression through the origin, there is no constraint of the form $\sum (Y_i - b_1 X_i) = \sum e_i = 0$. Consequently, the residuals usually will not sum to zero here. The only constraint on the residuals comes from the normal equation (5.31), namely $\sum X_i e_i = 0$.

2. When it is known that $\beta_0 = 0$, one could still use the general model (3.1), anticipating that b_0 will differ from 0 only by a small sampling error. However, it is more efficient to incorporate the knowledge $\beta_0 = 0$ into the model, thereby gaining a degree of freedom and simplifying the calculations.

3. In interval estimation of $E(Y_h)$ or $Y_{h(\text{new})}$, note that the intervals (5.36) and (5.37) in Table 5.1 widen the further X_h is from the origin. The reason, of course, is that the value of the true regression function is known precisely at the origin, so that the effect of the sampling error in the slope b_1 becomes increasingly important, the farther X_h is from the origin.

4. Since only one regression parameter, β_1, must be estimated for the regression function (5.29), simultaneous estimation methods are not required to make a family of statements about several mean responses. For a given confidence coefficient $1 - \alpha$, formula (5.36) can be used repetitively with the given sample results to generate a family of statements for which the family confidence coefficient is still $1 - \alpha$.

5. Like any other model, model (5.28) should be evaluated for aptness. Even though it is known that the regression function must go through the origin, the function might not be linear, for instance, or the variance of the error terms might not be constant.

5.6 COMPARISON OF TWO REGRESSION LINES

Frequently we encounter regressions for two or more populations and wish to examine their similarities and differences. In Parts III and IV, we will meet this situation repeatedly. Here we introduce the topic by considering the comparison of two regression lines. In Chapter 9, we shall take up a more general approach.

A comparison of two regression lines may be desired to obtain information as to the nature of the differences, if any, between them. For instance, an economist has studied for a sample of urban families and a sample of rural families the relation between amount of saving and level of income. He may then wish to compare whether, at given income levels, urban and rural families tend to save the same amount—that is, whether the two regression lines are the same. If they are not, he may wish to explore whether at least the amounts of saving out of an additional dollar of income are the same for the two groups—that is, whether the slopes of the two regression lines are the same.

Another reason for interest in comparing two regression lines is to determine whether they can be pooled. As a case in point, a company has two instruments constructed to identical specifications to measure pressure in an industrial process. A study has been made for each instrument of the relation between its gauge readings and actual pressures as determined by an almost exact but slow and costly method. If the two regression lines are the same, a single calibration schedule can be developed for the two instruments; otherwise, two different calibration schedules would be required. A single pooled regression would also have the advantage of containing greater precision than two different regressions.

The methods which we shall take up in this section all require that the error terms in the two regressions have equal variances. It is usually wise to test this before proceeding to employ the methods to be described. If the error variances for the two regressions are not equal, a transformation may be tried to see if it will equalize the variances, at least approximately.

Test Whether Two Regression Lines Identical

Example. A company operates two production lines for making soap bars. For each line, the relation between the speed of the line and the amount of scrap for the day was studied. The basic data are shown, in coded form, in Table 5.3. A scatter diagram of the data for the two production lines is presented in Figure 5.6. This plot suggests that the regression relation between production line speed and amount of scrap is not the same for the two production lines. The slopes appear to be about the same, but the heights of the regression lines seem to differ. A formal test is desired whether or not the two regression lines are identical.

TABLE 5.3

Data for Soap Production Lines Example (all data are coded)

Production Line 1			*Production Line 2*		
Observation i	*Line Speed* X_{i1}	*Amount of Scrap* Y_{i1}	*Observation* i	*Line Speed* X_{i2}	*Amount of Scrap* Y_{i2}
1	100	218	1	105	140
2	125	248	2	215	277
3	220	360	3	270	384
4	205	351	4	255	341
5	300	470	5	175	215
6	255	394	6	135	180
7	225	332	7	200	260
8	175	321	8	275	361
9	270	410	9	155	252
10	170	260	10	320	422
11	155	241	11	190	273
12	190	331	12	295	410
13	140	275			
14	290	425			
15	265	367			
	$n_1 = 15$			$n_2 = 12$	

General Linear Test Approach. We shall use the general linear test approach of Chapter 3 for testing the equality of two regression lines. Recall that the basic steps are:

1. Fit the full, or unrestricted, model and obtain the error sum of squares $SSE(F)$.
2. Obtain the reduced, or restricted, model under C_1, fit it, and determine the error sum of squares $SSE(R)$ for the reduced model.
3. Calculate the F^* statistic (3.66), which involves the difference $SSE(R) - SSE(F)$. The greater the difference, the more the data support C_2; the smaller the difference, the more the data support C_1.

Full Model. Let us denote the observations from production line 1 (X_{i1}, Y_{i1}) and those from production line 2 (X_{i2}, Y_{i2}). The full model, which provides separate regressions for each production line, can then be expressed as follows:

$$Y_{ij} = \beta_{0j} + \beta_{1j} X_{ij} + \varepsilon_{ij} \qquad \text{Full model} \tag{5.38}$$

where:

ε_{ij} are independent $N(0, \sigma^2)$

$i = 1, \ldots, n_j; j = 1, 2$

FIGURE 5.6

Scatter Plot for Soap Production Lines Example

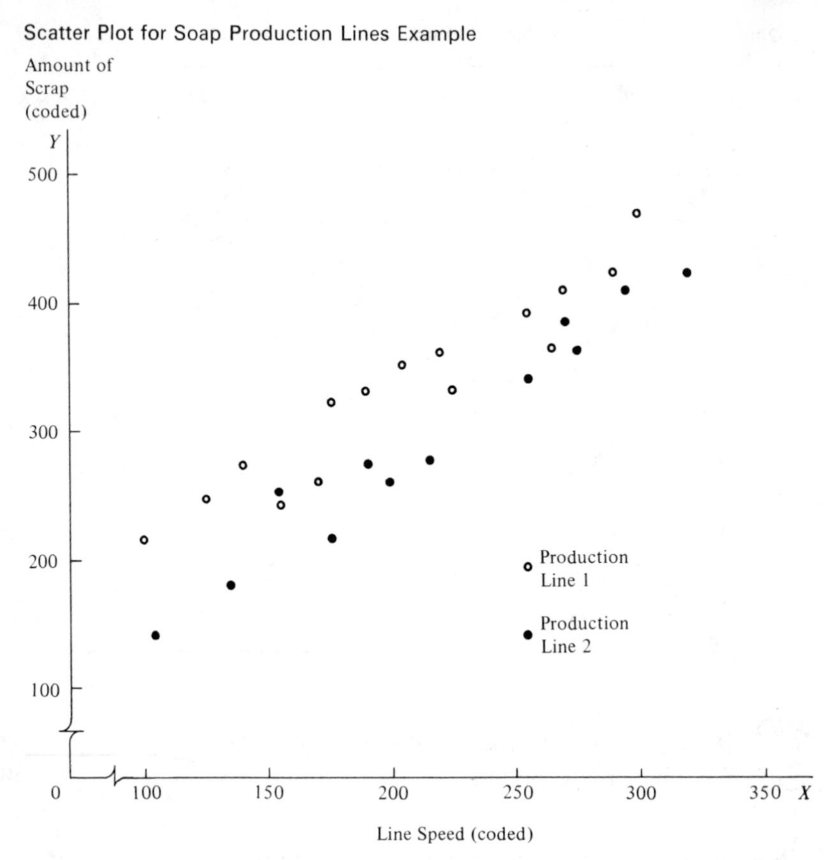

Note several important features of this model:

1. Observations from each production line have their own regression function. The expected value of an observation from production line 1 is:

$$E(Y_{i1}) = \beta_{01} + \beta_{11}X_{i1} \tag{5.39}$$

where the second subscript identifies production line 1.

The expected value of an observation from production line 2 is:

$$E(Y_{i2}) = \beta_{02} + \beta_{12}X_{i2} \tag{5.40}$$

2. All error terms are independent. Thus, the errors for each production line are independent, as are the errors for one production line and the other.

3. All error terms have constant variance. It is thus assumed that the error variances for the two production lines are the same.

It can easily be shown that fitting the full model (5.38) is equivalent to fitting two separate regression lines, one for each production line. Since by now the reader is thoroughly familiar with fitting a straight line regression function to a set of data, we shall not show detailed calculations but will simply present the relevant results from the computer run for each production line. We do this in Table 5.4.

TABLE 5.4

Computer Outputs for Separate Regressions for Two Production Lines

Production Line 1			*Production Line 2*		
Regression coefficients:			Regression coefficients:		
$b_{01} = 97.96533$			$b_{02} = 7.57446$		
$b_{11} = 1.14539$			$b_{12} = 1.32205$		
Analysis of variance:			Analysis of variance:		
Source of Variation	*SS*	*df*	*Source of Variation*	*SS*	*df*
Regression	$SSR_1 = 70{,}440.94$	1	Regression	$SSR_2 = 87{,}725.65$	1
Error	$SSE_1 = 6{,}402.79$	13	Error	$SSE_2 = 3{,}501.27$	10
Total	$SSTO_1 = 76{,}843.73$	14	Total	$SSTO_2 = 91{,}226.92$	11
Other:			Other:		
$\sum (X_{i1} - \bar{X}_1)^2 = 53{,}693.33$			$\sum (X_{i2} - \bar{X}_2)^2 = 50{,}191.67$		

Let us denote the error sum of squares for the regression for production line 1 SSE_1, and that for production line 2 SSE_2. It can readily be shown that the error sum of squares for fitting the full model (5.38) to the entire set of data (which, as noted, is equivalent to fitting separate regressions) is:

$$SSE(F) = SSE_1 + SSE_2 \tag{5.41}$$

For our example, we have:

$$SSE(F) = 6{,}402.79 + 3{,}501.27 = 9{,}904.06$$

Reduced Model. We wish to test whether the two regression lines are the same. Thus, the two alternative conclusions are:

$$\begin{aligned} &C_1\text{: } \beta_{01} = \beta_{02} \text{ and } \beta_{11} = \beta_{12} \\ &C_2\text{: Either } \beta_{01} \neq \beta_{02} \text{ or } \beta_{11} \neq \beta_{12} \text{ or both} \end{aligned} \tag{5.42}$$

Test

If the two regression lines are the same (C_1), both the intercept and slope terms must be equal. If the regression lines are not the same (C_2), they must differ with respect to either the intercept or the slope or with respect to both.

Model (5.38), when restricted by C_1 in (5.42), becomes the reduced model:

$$Y_{ij} = \beta_0 + \beta_1 X_{ij} + \varepsilon_{ij} \qquad \text{Reduced model} \tag{5.43}$$

where β_0 and β_1 denote the common parameters for both production lines. Fitting the reduced model (5.43) to the data implies fitting one regression line to the combined data ($n_1 + n_2 = 15 + 12 = 27$) for the two production lines. We have fitted a straight line regression to the combined 27 observations by computer, and the results are presented in Table 5.5. The error sum of squares for the reduced model is, according to Table 5.5:

$$SSE(R) = 29{,}407.76$$

Reduced Model

TABLE 5.5

Computer Output for Regression Fit to Combined Production Lines—Reduced Model

Regression coefficients:

$b_0 = 64.03568$
$b_1 = 1.19631$

Analysis of variance:

Source of Variation	SS	df
Regression	$SSR(R) = 149{,}660.98$	1
Error	$SSE(R) = 29{,}407.76$	25
Total	$SSTO = 179{,}068.74$	26

Test Statistic. The degrees of freedom associated with $SSE(R)$, based on $n_1 + n_2$ observations, are $n_1 + n_2 - 2$. Since $n_1 - 2$ degrees of freedom are associated with the error sum of squares SSE_1 for production line 1, and $n_2 - 2$ with SSE_2, the number of degrees of freedom associated with $SSE(F) = SSE_1 + SSE_2$ are $(n_1 - 2) + (n_2 - 2) = n_1 + n_2 - 4$.

We can now substitute into the general test statistic (3.66), and obtain for testing the equality of two regression lines:

$$\frac{SSE(R) - SSE(F)}{(n_1 + n_2 - 2) - (n_1 + n_2 - 4)} \div \frac{SSE(F)}{n_1 + n_2 - 4}$$

which simplifies to:

$$F^* = \frac{SSE(R) - SSE(F)}{2} \div \frac{SSE(F)}{n_1 + n_2 - 4} \tag{5.44}$$

If C_1 holds, F^* follows the $F(2, n_1 + n_2 - 4)$ distribution. Large values of F^* lead to C_2. Thus, the decision rule for controlling the risk of a Type I error at α is:

$$\text{(5.45)} \qquad \begin{aligned} &\text{If } F^* \leq F(1-\alpha; 2, n_1 + n_2 - 4), \text{ conclude } C_1 \\ &\text{If } F^* > F(1-\alpha; 2, n_1 + n_2 - 4), \text{ conclude } C_2 \end{aligned}$$

Example. For our soap production lines illustration, the test statistic is:

$$F^* = \frac{29{,}407.76 - 9{,}904.06}{2} \div \frac{9{,}904.06}{23} = 22.65$$

If we wish to control α at the level .05, we require $F(.95; 2, 23) = 3.42$, so that the decision rule is:

$$\begin{aligned} &\text{If } F^* \leq 3.42, \text{ conclude } C_1 \\ &\text{If } F^* > 3.42, \text{ conclude } C_2 \end{aligned}$$

Since $F^* = 22.65 > 3.42$, we conclude C_2, that the linear regression functions for the two production lines are not the same. The next step in the analysis usually will be to investigate how the two regression lines differ, and we shall take up this topic below.

Comments

1. The approach just presented is completely general, and can be used to test the equality of two curvilinear regression functions or two multiple regression functions. Only the degrees of freedom would need to be modified.

2. The extension of the test for the equality of two regression lines to testing the equality of three or more regression lines is straightforward. $SSE(F)$ in (5.41) simply would be the sum of the three or more error sums of squares for each of the separate regression lines, and the degrees of freedom would need to be modified accordingly.

3. The equality of the error variances for the two production lines, which is assumed by model (5.38), can be tested by the usual F test in Table 1.4a. The estimates of the error variances for the two production lines are, using the results in Table 5.4:

$$\frac{SSE_1}{n_1 - 2} = \frac{6{,}402.79}{13} = 492.52$$

$$\frac{SSE_2}{n_2 - 2} = \frac{3{,}501.27}{10} = 350.13$$

so that:

$$F^* = \frac{492.52}{350.13} = 1.41$$

Specifying the level of significance at .10, we require $F(.05; 13, 10) = .374$ and $F(.95; 13, 10) = 2.89$. Since F^* falls between these two action limits, we conclude that the two production line regressions have equal error variances.

Comparisons of Regression Parameters

Once it has been established that two regression lines differ, the next step in the analysis is usually to study how they differ. One may test, for example, whether the two slopes are equal. If so, the regression lines would be parallel and would differ only in height. In Chapter 22, we discuss how to test whether or not the slopes of two or more regression lines are equal. This test follows the general approach utilized for testing the equality of two regression lines.

When analyzing only two regression lines, a straightforward method of comparing the slopes of the regression functions is to construct an interval estimate. The confidence interval for the difference in the slopes, $\beta_{11} - \beta_{12}$, is:

$$(5.46) \qquad (b_{11} - b_{12}) - t(1 - \alpha/2; n_1 + n_2 - 4)s(b_{11} - b_{12}) \leq \beta_{11} - \beta_{12} \\ \leq (b_{11} - b_{12}) + t(1 - \alpha/2; n_1 + n_2 - 4)s(b_{11} - b_{12})$$

Here, b_{11} and b_{12} are the slopes from the two separately fitted regression lines, and:

$$(5.47) \qquad s^2(b_{11} - b_{12}) = MSE(F)\left[\frac{1}{\sum(X_{i1} - \bar{X}_1)^2} + \frac{1}{\sum(X_{i2} - \bar{X}_2)^2}\right]$$

where:

$$(5.47a) \qquad MSE(F) = \frac{SSE(F)}{n_1 + n_2 - 4}$$

Note that $s^2(b_{11} - b_{12})$ in (5.47) is essentially the sum $s^2(b_{11}) + s^2(b_{12})$, where each of these estimated variances uses $MSE(F)$ from the full model as the estimator of σ^2.

To construct 95 percent confidence limits for $\beta_{11} - \beta_{12}$ for our soap production lines example, we require $t(.975; 23) = 2.069$. We know from Table 5.4 that:

$$b_{11} = 1.14539 \qquad\qquad b_{12} = 1.32205$$
$$\sum(X_{i1} - \bar{X}_1)^2 = 53{,}693.33 \qquad \sum(X_{i2} - \bar{X}_2)^2 = 50{,}191.67$$

We found earlier that $SSE(F) = 9{,}904.06$, so that $MSE(F) = 9{,}904.06 \div 23 = 430.61$. Hence:

$$s^2(b_{11} - b_{12}) = 430.61\left[\frac{1}{53{,}693.33} + \frac{1}{50{,}191.67}\right] = .016599$$

or:

$$s(b_{11} - b_{12}) = .128838$$

From above we obtain $b_{11} - b_{12} = -.17666$. The 95 percent confidence limits for $\beta_{11} - \beta_{12}$ therefore are:

$$-.443 = -.17666 - (2.069)(.128838) \leq \beta_{11} - \beta_{12} \leq -.17666 + (2.069)(.128838) = +.090$$

These confidence limits suggest that the two slopes are the same or else do not differ much. This is in accord with our visual interpretation of Figure 5.6, which indicated that the main difference in the two regression lines is in their elevation. Indeed, a test via the above confidence interval, with a 5 percent level of significance, would lead to the conclusion that the two slopes are the same since the confidence interval covers $\beta_{11} - \beta_{12} = 0$.

Pooling of Regression Lines

As we mentioned earlier, one reason for comparing two regression lines is to determine whether the regression lines can be pooled into one, for purposes of increasing the precision of predictions and the like. If the statistical test indicates that the two regression lines are the same, pooling consists simply of forgetting about the original distinction (two production lines, in our example), combining all the data and fitting a regression function to the combined data.

If the statistical analysis indicates that the two regression lines have common slope but different intercepts, the data can be pooled for purposes of estimating the common slope. The proper method of pooling for this case is explained in Chapter 22.

5.7 EFFECT OF MEASUREMENT ERRORS

In our discussion of the regression model up to this point, we have not explicitly considered the presence of measurement errors in either X or Y. We now examine briefly the effect of measurement errors.

Measurement Errors in Y

If random measurement errors are present in the dependent variable Y, no new problems are created if these errors are uncorrelated and not biased (positive and negative measurement errors tend to cancel out). Consider, for example, a study of the relation between the time required to complete a task (Y) and the complexity of the task (X). The time to complete the task may not be measured accurately because the person activating and stopping the stopwatch may not do so at the precise instant called for. As long as such measurement errors are of a random nature, uncorrelated, and not biased, these measurement errors can simply be absorbed in the model error term ε_i. The model error term reflects the composite effects of a large number of factors not considered in the model, one of which simply would be random errors due to inaccuracy in the process of measuring Y.

Measurement Errors in X

Unfortunately, a different situation holds if the independent variable X is known only with measurement error. Frequently, to be sure, X is known without measurement error, as when the independent variable is price of a product, number of variables in an optimization problem, or wage rate for a class of employees. At other times, however, measurement errors may enter the value observed for the independent variable, for instance, when it is pressure, temperature, production line speed, or person's age. Let us use the latter illustration in our development of the nature of the problem. Suppose we are regressing employees' piecework earnings on age. Let X_i denote the true age of the ith employee and X_i^* the age given by the employee on his employment records. Needless to say, the two are not always the same. We define the measurement error δ_i as follows:

(5.48) $$\delta_i = X_i^* - X_i$$

The regression model we would like to study is:

(5.49) $$Y_i = \beta_0 + \beta_1 X_i + \varepsilon_i$$

Since, however, we only observe X_i^*, model (5.49) becomes:

(5.50) $$Y_i = \beta_0 + \beta_1(X_i^* - \delta_i) + \varepsilon_i$$

where we make use of (5.48) in replacing X_i. We can rewrite (5.50) as follows:

(5.51) $$Y_i = \beta_0 + \beta_1 X_i^* + (\varepsilon_i - \beta_1 \delta_i)$$

Model (5.51) may appear like an ordinary regression model, with independent variable X_i^* and error term $\varepsilon_i - \beta_1 \delta_i$, but it is not. The independent variable X_i^* is a random variable, which is correlated with the error term $(\varepsilon_i - \beta_1 \delta_i)$. Theorem (3.39) for the case of random independent variables requires that the error term be independent of the independent variable. Hence, the standard regression results are not applicable for model (5.51).

Intuitively, we know that $\varepsilon_i - \beta_1 \delta_i$ is not independent of X_i^* since (5.48) constrains $X_i^* - \delta_i$ to equal X_i. To determine the dependence formally, let us assume:

(5.52a) $$E(\delta_i) = 0$$

(5.52b) $$E(\varepsilon_i) = 0$$

(5.52c) $$E(\delta_i \varepsilon_i) = 0$$

so that $E(X_i^*) = E(X_i + \delta_i) = X_i$, and the measurement error δ_i is not correlated with the model error ε_i. We now wish to find the covariance:

$$\begin{aligned}\sigma(X_i^*, \varepsilon_i - \beta_1 \delta_i) &= E\{[X_i^* - E(X_i^*)][(\varepsilon_i - \beta_1 \delta_i) - E(\varepsilon_i - \beta_1 \delta_i)]\} \\ &= E[(X_i^* - X_i)(\varepsilon_i - \beta_1 \delta_i)] \\ &= E[\delta_i(\varepsilon_i - \beta_1 \delta_i)] = E(\delta_i \varepsilon_i - \beta_1 \delta_i^2)\end{aligned}$$

so that we obtain:

$$\sigma(X_i^*, \varepsilon_i - \beta_1\delta_i) = -\beta_1\sigma^2(\delta_i) \tag{5.53}$$

This covariance is not zero if there is a regression relation between X and Y.

If standard least squares procedures are applied to model (5.51), the estimators b_0 and b_1 are biased and also lack the property of consistency. Great difficulties are encountered in developing unbiased estimators when there are measurement errors in X. One approach is to impose severe conditions on the problem—for example, to make fairly strong assumptions about the properties of the distributions of δ_i, the covariance of δ_i and ε_j, and so on. Another approach is to use additional variables which are known to be related to the true value of X but not with the errors of measurement δ. Such variables are called "instrumental" variables because they are used as an instrument in studying the relation between X and Y. Instrumental variables make it possible to obtain consistent estimators of the regression parameters.

Discussions of approaches which can be taken, and further references, will be found in References 5.2, 5.3, and 5.4.

Note

It may be asked what is the distinction between the case when X is a random variable, considered in Chapter 3, and the case when X is subject to random measurement errors, and why are there special problems with the latter. When X is a random variable, it is not under the control of the analyst and will vary at random from trial to trial, as when X is the number of persons entering a store in a day. If this random variable X is not subject to measurement errors, however, it can be accurately ascertained for a given trial. Thus, if there are no measurement errors in counting the number of persons entering a store in a day, the analyst has accurate information to study the relation between number of persons entering the store and sales, even though he cannot control the levels of number of customers which actually occur. If, on the other hand, measurement errors are present in the number of persons entering the store, a distorted picture of the relation between number of persons and sales occurs because the sales observations will frequently be matched against an incorrect number of customers. This distorting effect of measurement errors is present whether X is fixed or random.

Berkson Model

There is one situation where measurement errors in X are no problem. This case was first noted by Berkson (Ref. 5.5). Frequently, the independent variable in an experiment is set at a target value. For instance, in an experiment on the effect of temperature on typist productivity, the temperature may be set at target levels of 68 degrees, 70 degrees, and so on, according to the temperature control on the thermostat. The observed temperature X_i^* is fixed here, while the actual temperature X_i is a random variable since the thermostat may not be completely accurate. Similar situations exist when water pressure is set according to a gauge, or employees of specified ages according to their employment records are selected for a study.

In all of these cases, the observed X_i^* is a fixed quantity, while the unobserved true value X_i is a random variable. The measurement error is, as before:

$$\delta_i = X_i^* - X_i \tag{5.54}$$

Here, however, there is no constraint on the relation between X_i^* and δ_i since X_i^* is a fixed quantity. Again, we assume that $E(\delta_i) = 0$.

Model (5.51), which we obtained when replacing X_i by $X_i^* - \delta_i$, is still applicable for the Berkson case:

$$Y_i = \beta_0 + \beta_1 X_i^* + (\varepsilon_i - \beta_1 \delta_i) \tag{5.55}$$

The expected value of the error term, $E(\varepsilon_i - \beta_1 \delta_i)$, is zero as before, since $E(\varepsilon_i) = 0$ and $E(\delta_i) = 0$. Further, $\varepsilon_i - \beta_1 \delta_i$ is now independent of X_i^* since X_i^* is a constant for the Berkson case. Hence, the conditions of an ordinary regression model are met:

1. The error terms have expectation 0.
2. The independent variable is a constant, and hence the error terms are independent of it.

Thus, standard least squares procedures can be applied for the Berkson case without modification, and the estimators b_0 and b_1 will be unbiased. If we can make the standard normality and constant variance assumptions for the errors $(\varepsilon_i - \beta_1 \delta_i)$, the usual tests and interval estimates can be utilized.

5.8 CHOICE OF X LEVELS

When regression data are obtained by experiment, the levels of X at which observations on Y are to be taken are under the control of the experimenter. Among other things, the experimenter will have to consider:

1. How many levels of X should be investigated?
2. What shall the two extreme levels be?
3. How shall the other levels of X, if any, be spaced?
4. How many observations should be taken at each level of X?

There is no single answer to these questions, since different purposes of the regression analysis lead to different answers. The possible objectives in regression analysis are varied, as we have noted earlier. The main objective may be to estimate the slope of the regression line, or in some cases to estimate the intercept. In many cases, the main objective is to predict one or more new observations or to estimate one or more mean responses. When the regression function is curvilinear, the main objective may be to locate the maximum or minimum mean response. At still other times, the main purpose is to determine the nature of the regression function.

To illustrate how the purpose affects the design, consider the variances of b_0, b_1, $\hat{Y}_h$, and $Y_{h(\text{new})}$ which were developed earlier:

$$\sigma^2(b_0) = \sigma^2 \frac{\sum X_i^2}{n \sum (X_i - \bar{X})^2} = \sigma^2 \left[\frac{1}{n} + \frac{\bar{X}^2}{\sum (X_i - \bar{X})^2} \right] \tag{5.56}$$

$$\sigma^2(b_1) = \frac{\sigma^2}{\sum (X_i - \bar{X})^2} \tag{5.57}$$

$$\sigma^2(\hat{Y}_h) = \sigma^2 \left[\frac{1}{n} + \frac{(X_h - \bar{X})^2}{\sum (X_i - \bar{X})^2} \right] \tag{5.58}$$

$$\sigma^2(Y_{h(\text{new})}) = \sigma^2 \left[1 + \frac{1}{n} + \frac{(X_h - \bar{X})^2}{\sum (X_i - \bar{X})^2} \right] \tag{5.59}$$

The variance of the slope is minimized if $\sum (X_i - \bar{X})^2$ is maximized. This is accomplished by using two levels of X, at the two extremes for the scope of the model, and placing half of the observations at each of the two levels. One would be hesitant to use only two levels, incidentally, if one were not sure of the linearity of the regression function, since two levels provide no information about possible departures from linearity.

If the main purpose is to estimate β_0, the number and placement of levels does not matter as long as $\bar{X} = 0$. On the other hand, to estimate the mean response or predict a new observation at X_h, it is best to use X levels so that $\bar{X} = X_h$. If a number of mean responses are to be estimated or a number of new observations are to be predicted, it would be best to spread out the X levels such that $\bar{X}$ is in the center of the X_h levels of interest.

Although the number and spacing of X levels depends very much on the major purpose of the regression analysis, some general advice can be given, at least to be used as a point of departure. D. R. Cox, a well-known statistician, suggests as follows:

> Use two levels when the object is primarily to examine whether or not . . . (the independent variable) . . . has an effect and in which direction that effect is. Use three levels whenever a description of the response curve by its slope and curvature is likely to be adequate; this should cover most cases. Use four levels if further examination of the shape of the response curve is important. Use more than four levels when it is required to estimate the detailed shape of the response curve, or when the curve is expected to rise to an asymptotic value, or in general to show features not adequately described by slope and curvature. Except in these last cases it is generally satisfactory to use equally spaced levels with equal numbers of observations per level [Ref. 5.6].

PROBLEMS

5.1. When joint confidence intervals for β_0 and β_1 are developed by the Bonferroni method with a family confidence coefficient of 90 percent, does this imply that 10 percent of the time the confidence intervals for β_0 and β_1 will be incorrect?

5.2. What simplification takes place in the boundary of the joint confidence region (5.17) for β_0 and β_1 if the independent variable is so coded that $\bar{X}=0$? What happens to the shape of the boundary?

5.3. If the independent variable is so coded that $\bar{X}=0$ and the normal error model (3.1) applies, are b_0 and b_1 independent? Are the confidence intervals for b_0 and b_1 independent?

5.4. Refer to Problem 3.2. Suppose the student combines the two confidence intervals into a confidence set. What can you say about the family confidence coefficient for this set?

5.5. An analyst employed the linear regression model (3.1) to study the relation in 20 sales territories between target population (X, in thousands of persons) and sales of a breakfast cereal (Y, in thousands of cartons). Some results were:

$b_0 = 3.475$ $\quad s(b_0) = 1.771$ $\quad \bar{X} = 170.4$ $\quad \sum(X_i - \bar{X})^2 = 108{,}131$
$b_1 = .3045$ $\quad s(b_1) = .00954$ $\quad \sum X_i = 3{,}408$ $\quad \sum X_i^2 = 688{,}854$
$MSE = 9.85$ $\quad \bar{Y} = 55.35$

a) Will b_0 and b_1 tend to err in the same direction or in opposite directions here? What does this imply about the tilt of the elliptical joint confidence region for β_0 and β_1 here?
b) Obtain a few boundary points and plot the joint confidence region for β_0 and β_1. Use a 95 percent family confidence coefficient. Interpret your joint confidence region.
c) From experience with other breakfast cereals, the analyst hypothesized that the regression function for the cereal under study should have an intercept of 5.50 together with a slope of .320. Use your joint confidence region to test this hypothesis. What is the implicit level of significance? What conclusion do you reach?

5.6. Refer to Problem 5.5.
a) Obtain Bonferroni joint confidence intervals for β_0 and β_1, using a 95 percent family confidence coefficient.
b) Plot the Bonferroni joint confidence region. Is it adequate here, or is it substantially less efficient than the joint confidence ellipse obtained in Problem 5.5b?

5.7. Refer to Problem 3.6.
a) Obtain a few boundary points and plot the joint confidence region for β_0 and β_1. Use a 90 percent family confidence coefficient. Interpret your results.
b) What is the meaning of the family confidence coefficient in part (a)?
c) Obtain Bonferroni joint confidence intervals for β_0 and β_1, with a 90 percent family confidence coefficient. Plot the Bonferroni joint confidence region and compare it with that obtained in part (a). Which is more efficient?

5.8. Refer to Problem 3.6.
a) Obtain the 90 percent confidence band for the regression line. Plot it, together with the estimated regression equation. How precisely do we know the location of the true regression line here?
b) Estimate the expected number of minutes spent when there are 3, 5, and 7 machines to be serviced respectively. Use interval estimates, with a

90 percent family confidence coefficient, based on the Working-Hotelling approach.

c) Two service calls for preventive maintenance are scheduled, in which the numbers of machines to be serviced are 4 and 7 respectively. A family of predictions of the times to be spent on these calls is desired, with a 90 percent family confidence coefficient. Which procedure, Scheffé or Bonferroni, will provide tighter prediction limits here?

d) Obtain the family of prediction limits required in part (c), using the more efficient procedure.

5.9. Refer to Problem 5.5.

a) Management requires interval estimates of mean sales in territories containing target populations of 200 and 300 thousand persons respectively, with a 95 percent family confidence coefficient. It is possible that several interval estimates will be added later to the set, with the family confidence coefficient to be held at 95 percent. The manager of the Economic Research Department indicates that it would be highly desirable to employ a procedure in which the first two interval estimates would not have to be revised if the set is augmented later. Which procedure do you recommend?

b) Use your recommended procedure to obtain the estimates specified in part (a), and explain how your results and confidence coefficient are interpreted.

5.10. Refer to Problem 5.5. Management wishes to predict sales in three territories with target populations of 100, 200, and 250 thousand persons respectively, with a 90 percent family confidence coefficient.

a) Which procedure, the Scheffé or Bonferroni, will be more efficient here?

b) Use your recommended procedure to obtain the desired prediction intervals, and explain how your results and confidence coefficient are interpreted.

c) If management later decides to make a prediction for another territory, with target population 270 thousand, and wishes to add it to the previous confidence set with family confidence coefficient of 90 percent for the entire set, would the earlier confidence intervals have to be recalculated? Would this be true if the other of the two methods had been employed?

d) Would your answers in part (c) be affected if the fourth territory had a target population of 250 thousand?

5.11. A behavioral scientist stated recently: "I am never sure whether the regression line goes through the origin. Hence I will not use such a model." Comment.

5.12. Show that for the fitted least squares regression line (5.33) through the origin, $\sum X_i e_i = 0$.

5.13. Show that $\hat{Y}$ as defined in (5.33) for linear regression through the origin is an unbiased estimator of $E(Y)$.

5.14. Derive the formula for $s^2(\hat{Y}_h)$ given in Table 5.1 for linear regression through the origin.

5.15. Shown below are the number of galleys of type set (X) and the dollar cost of correcting typographical errors (Y) in a random sample of recent printing orders handled by a firm specializing in technical printing. Since Y involves variable costs only, an analyst wished to determine whether the regression through the origin model (5.28) was apt for studying the relation between the two variables.

i:	1	2	3	4	5	6	7	8	9	10	11	12
X_i:	7	12	10	10	14	25	30	25	18	10	4	6
Y_i:	128	213	186	178	250	446	540	457	324	177	75	103

a) Fit model (5.28).
b) Prepare appropriate residual plots to study the aptness of model (5.28). Summarize your findings.
c) Conduct a formal test of the appropriateness of linear regression through the origin. Use a level of significance of .01. What do you conclude?

5.16. Refer to Problem 5.15. Assume model (5.28) applies.
a) In estimating costs of handling prospective orders, management has used a standard of $17.50 per galley for the cost of correcting typographical errors. Do the results in the sample give any indication that this standard should be revised?
b) Obtain a prediction interval for the correction cost on a forthcoming job involving 10 galleys. Use a confidence coefficient of 95 percent.

5.17. A testing laboratory, using equipment that simulates highway driving, studied for two makes of a certain type of truck tire the relation between operating cost per mile (Y_1, Y_2) and cruising speed (X). The observations are shown below (all data are coded), together with key results from computer printouts. An engineer now wishes to decide whether or not the regression of cost on cruising speed is the same for the two makes. Assume that model (5.38) is appropriate.

i:	1	2	3	4	5	6	7	8	9	10
X_i:	10	20	20	30	40	40	50	60	60	70
Y_{i1}:	9.8	12.5	14.2	14.9	19.0	16.5	20.9	22.4	24.1	25.8
Y_{i2}:	15.0	14.5	16.1	16.5	16.4	19.1	20.9	22.3	19.8	21.4

	Full Model Separate Regressions		*Reduced Model Pooled Regression*
	Make 1	*Make 2*	
b_0:	7.61	13.02	10.32
$s(b_0)$:	.7024	.8984	.8693
b_1:	.260	.129	.195
$s(b_1)$:	.0159	.0203	.0196
SSE:	7.249	11.859	49.969

a) Plot the data. Does the relation between speed and cost appear to be the same for the two makes?
b) Conduct a formal test to decide whether or not the regression functions are the same for the two makes. Control your risk of Type I error at .05. What do you conclude?
c) Suppose the question of interest had been simply whether the two regression lines have equal slopes. Answer this question by setting

up a 95 percent confidence interval for the difference between the two slopes. What do you find?

d) Test whether the two populations have equal error variances, using a level of significance of .01. What does your test indicate? Was any assumption about the error variances implicit in your procedure in part (b)? In part (c)?

5.18. Refer to Problem 2.13. Suppose that errors arise in X here because the laboratory technician is instructed to measure the hardness of the ith specimen (Y_i) at a pre-recorded elapsed time (X_i), but his timing is imperfect so the true elapsed time varies at random from the pre-recorded elapsed time. Will ordinary least squares estimates be biased here?

CITED REFERENCES

5.1. Gafarian, A. V. "Confidence Bands in Straight Line Regression," *Journal of the American Statistical Association*, Vol. 59 (1964), pp. 182–213.

5.2. Johnston, J. *Econometric Methods*. New York: McGraw-Hill Book Co., 1963.

5.3. Madansky, Albert. "The Fitting of Straight Lines When Both Variables Are Subject to Error," *Journal of the American Statistical Association*, Vol. 54 (1959), pp. 173–205.

5.4. Halperin, Max. "Fitting of Straight Lines and Prediction When Both Variables Are Subject to Error," *Journal of the American Statistical Association*, Vol. 56 (1961), pp. 657–69.

5.5. Berkson, J. "Are There Two Regressions?" *Journal of the American Statistical Association*, Vol. 45 (1950), pp. 164–80.

5.6. Cox, D. R. *Planning of Experiments*, pp. 141–42. New York: John Wiley & Sons, Inc., 1958.

5.7. Miller, Rupert G., Jr. *Simultaneous Statistical Inference*, pp. 114–16. New York: McGraw-Hill Book Co., 1966.

part II

General Regression and Correlation Analysis

6

Matrix Approach to Simple Regression Analysis

MATRIX ALGEBRA is used increasingly for mathematical and statistical analysis. The matrix approach is practically a necessity in multiple regression analysis, since it permits extensive systems of equations and large arrays of data to be denoted compactly and operated upon efficiently.

In this chapter, we first take up a brief introduction to matrix algebra. Then we apply matrix methods to the simple linear regression model discussed in Part I. While matrix algebra is not really required for simple regression with one independent variable, the application of matrix methods to this case will provide a transition to multiple regression which will be taken up in succeeding chapters.

Readers who are familiar with matrix algebra may wish to scan those parts of this chapter introducing matrix methods.

6.1 MATRICES

Definition of Matrix

A matrix is a rectangular array of elements arranged in rows and columns. An example of a matrix is:

$$\begin{array}{r} \\ \text{Row 1} \\ \text{Row 2} \\ \text{Row 3} \end{array} \begin{array}{c} \begin{array}{cc} \text{Column 1} & \text{Column 2} \end{array} \\ \begin{bmatrix} 6{,}000 & 23 \\ 13{,}000 & 47 \\ 11{,}000 & 35 \end{bmatrix} \end{array}$$

The *elements* of this particular matrix are numbers representing income (column 1) and age (column 2) of three persons. The elements are arranged by row (person) and column (characteristic of person). Thus, the element in the first row and first column (6,000) represents the income of the first person. The element in the first row and second column (23) represents the age of the first person. The *dimension* of the matrix is 3×2, that is, 3 rows by 2 columns. If we wanted to present income and age for 1,000 persons in a matrix with the same format as the one above, we would require a $1{,}000 \times 2$ matrix.

Other examples of matrices are:

$$\begin{bmatrix} 1 & 0 \\ 5 & 10 \end{bmatrix} \quad \begin{bmatrix} 4 & 7 & 12 & 16 \\ 3 & 15 & 9 & 8 \end{bmatrix}$$

These two matrices have dimensions of 2×2 and 2×4 respectively. Note that we always specify the number of rows first and then the number of columns in giving the dimensions of a matrix.

As in ordinary algebra, we may use symbols to identify the elements of a matrix:

$$\begin{array}{c} \\ i=1 \\ i=2 \end{array} \begin{array}{ccc} j=1 & j=2 & j=3 \\ \left[a_{11} \right. & a_{12} & \left. a_{13} \right] \\ \left[a_{21} \right. & a_{22} & \left. a_{23} \right] \end{array}$$

Note that the first subscript identifies the row number and the second the column number. We shall use the general notation a_{ij} for the element in the ith row and the jth column. In our above example, $i = 1, 2$ and $j = 1, 2, 3$.

A matrix may be denoted by a symbol such as **A**, **X**, or **Z**. The symbol is in **boldface** to identify that it refers to a matrix. Thus, we might define for the above matrix:

$$\mathbf{A} = \begin{bmatrix} a_{11} & a_{12} & a_{13} \\ a_{21} & a_{22} & a_{23} \end{bmatrix}$$

Reference to the matrix **A** then implies reference to the 2×3 array just given.

Another notation for the matrix **A** just given is:

$$\mathbf{A} = [a_{ij}] \qquad i = 1, 2; j = 1, 2, 3$$

which avoids the need for writing out all elements of the matrix by stating only the general element. This notation can only be used, of course, when the elements of a matrix are symbols.

To summarize, a matrix with r rows and c columns will be represented either in full:

$$(6.1) \qquad \mathbf{A} = \begin{bmatrix} a_{11} & a_{12} & \cdots & a_{1j} & \cdots & a_{1c} \\ a_{21} & a_{22} & \cdots & a_{2j} & \cdots & a_{2c} \\ \vdots & \vdots & & \vdots & & \vdots \\ a_{i1} & a_{i2} & \cdots & a_{ij} & \cdots & a_{ic} \\ \vdots & \vdots & & \vdots & & \vdots \\ a_{r1} & a_{r2} & \cdots & a_{rj} & \cdots & a_{rc} \end{bmatrix}$$

or in the abbreviated form:

$$\mathbf{A} = [a_{ij}] \qquad i = 1, \ldots, r; j = 1, \ldots, c \tag{6.2}$$

or simply by a boldface symbol, such as **A**.

Comments

1. Do not think of a matrix as a number. It is a set of elements arranged in an array. Only when the matrix has dimension 1×1 is there a single number in the matrix, in which case one *can* think of it interchangeably as a matrix or as a number.
2. The following is *not* a matrix:

$$\begin{bmatrix} & 14 & \\ & 8 & \\ 10 & & 15 \\ 9 & & 16 \end{bmatrix}$$

since the numbers are not arranged in columns and rows.

Square Matrix

A matrix is said to be square if the number of rows equals the number of columns. Two examples are:

$$\begin{bmatrix} 4 & 7 \\ 3 & 9 \end{bmatrix} \qquad \begin{bmatrix} a_{11} & a_{12} & a_{13} \\ a_{21} & a_{22} & a_{23} \\ a_{31} & a_{32} & a_{33} \end{bmatrix}$$

Vector

A matrix containing only one column is called a *column vector* or simply a *vector*. Two examples are:

$$\mathbf{A} = \begin{bmatrix} 4 \\ 7 \\ 10 \end{bmatrix} \qquad \mathbf{C} = \begin{bmatrix} c_1 \\ c_2 \\ c_3 \\ c_4 \\ c_5 \end{bmatrix}$$

The vector **A** is a 3×1 matrix, and the vector **C** is a 5×1 matrix.

A matrix containing only one row is called a *row vector*. Two examples are:

$$\mathbf{B}' = [15 \quad 25 \quad 50] \qquad \mathbf{D}' = [c_1 \quad c_2]$$

We use the prime symbol for row vectors for reasons to be seen shortly. Note that the row vector $\mathbf{B}'$ is a 1×3 matrix and the row vector $\mathbf{D}'$ is a 1×2 matrix.

A single subscript suffices to identify the elements of a vector.

Transpose

The transpose of a matrix $\mathbf{A}$ is another matrix, denoted $\mathbf{A}'$, which is obtained by interchanging corresponding columns and rows of the matrix $\mathbf{A}$.

For example, if:

$$\underset{3\times 2}{\mathbf{A}} = \begin{bmatrix} 2 & 5 \\ 7 & 10 \\ 3 & 4 \end{bmatrix}$$

then the transpose $\mathbf{A}'$ is:

$$\underset{2\times 3}{\mathbf{A}'} = \begin{bmatrix} 2 & 7 & 3 \\ 5 & 10 & 4 \end{bmatrix}$$

Note that the first column of $\mathbf{A}$ is the first row of $\mathbf{A}'$, and similarly the second column of $\mathbf{A}$ is the second row of $\mathbf{A}'$. Correspondingly, the first row of $\mathbf{A}$ has become the first column of $\mathbf{A}'$, and so on. Note that the dimension of $\mathbf{A}$, indicated under the symbol $\mathbf{A}$, becomes reversed for the dimension of $\mathbf{A}'$.

As another example, consider:

$$\underset{3\times 1}{\mathbf{C}} = \begin{bmatrix} 4 \\ 7 \\ 10 \end{bmatrix} \qquad \underset{1\times 3}{\mathbf{C}'} = [4 \quad 7 \quad 10]$$

Thus, the transpose of a column vector is a row vector, and vice versa. This is the reason why we used the symbol $\mathbf{B}'$ earlier to identify a row vector, since it may be thought of as the transpose of a column vector $\mathbf{B}$.

In general, we have:

$$\underset{r\times c}{\mathbf{A}} = \begin{bmatrix} a_{11} & \cdots & a_{1c} \\ \cdot & & \cdot \\ \cdot & & \cdot \\ \cdot & & \cdot \\ a_{r1} & \cdots & a_{rc} \end{bmatrix} = \underset{\substack{\nearrow \quad \nwarrow \\ \text{Row index} \quad \text{Column index}}}{[a_{ij}]} \qquad i = 1, \ldots, r;\ j = 1, \ldots, c$$

(6.3)

$$\underset{c\times r}{\mathbf{A}'} = \begin{bmatrix} a_{11} & \cdots & a_{r1} \\ \cdot & & \cdot \\ \cdot & & \cdot \\ \cdot & & \cdot \\ a_{1c} & \cdots & a_{rc} \end{bmatrix} = \underset{\substack{\nearrow \quad \nwarrow \\ \text{Row index} \quad \text{Column index}}}{[a_{ji}]} \qquad j = 1, \ldots, c;\ i = 1, \ldots, r$$

Thus, the element in the ith row and jth column in $\mathbf{A}$ is found in the jth row and ith column in $\mathbf{A}'$.

Equality of Matrices

Two matrices $\mathbf{A}$ and $\mathbf{B}$ are said to be equal if they have the same dimensions and if all corresponding elements are equal. Conversely, if two matrices are equal their corresponding elements are equal. For example, if:

$$\mathbf{A} = \begin{bmatrix} a_1 \\ a_2 \\ a_3 \end{bmatrix} \qquad \mathbf{B} = \begin{bmatrix} 4 \\ 7 \\ 3 \end{bmatrix}$$

then $\mathbf{A} = \mathbf{B}$ implies:

$$a_1 = 4$$
$$a_2 = 7$$
$$a_3 = 3$$

Similarly, if:

$$\mathbf{A} = \begin{bmatrix} a_{11} & a_{12} \\ a_{21} & a_{22} \\ a_{31} & a_{32} \end{bmatrix} \qquad \mathbf{B} = \begin{bmatrix} 17 & 2 \\ 14 & 5 \\ 13 & 9 \end{bmatrix}$$

then $\mathbf{A} = \mathbf{B}$ implies:

$$a_{11} = 17 \qquad a_{12} = 2$$
$$a_{21} = 14 \qquad a_{22} = 5$$
$$a_{31} = 13 \qquad a_{32} = 9$$

Regression Examples

In regression analysis, one basic matrix is the vector $\mathbf{Y}$, consisting of the n observations on the dependent variable:

$$\underset{n \times 1}{\mathbf{Y}} = \begin{bmatrix} Y_1 \\ Y_2 \\ \cdot \\ \cdot \\ \cdot \\ Y_n \end{bmatrix} \tag{6.4}$$

Note that the transpose $\mathbf{Y}'$ is the row vector:

$$\underset{1 \times n}{\mathbf{Y}'} = [Y_1 \quad Y_2 \quad \cdots \quad Y_n] \tag{6.5}$$

Another basic matrix in regression analysis is the $\mathbf{X}$ matrix, which is defined as follows for simple regression analysis:

$$\underset{n \times 2}{\mathbf{X}} = \begin{bmatrix} 1 & X_1 \\ 1 & X_2 \\ \cdot & \cdot \\ \cdot & \cdot \\ \cdot & \cdot \\ 1 & X_n \end{bmatrix} \tag{6.6}$$

The matrix **X** consists of a column of 1's and a column containing the n values of the independent variable X_i. Note that the transpose of **X** is:

(6.7)
$$\underset{2\times n}{\mathbf{X}'} = \begin{bmatrix} 1 & 1 & \cdots & 1 \\ X_1 & X_2 & \cdots & X_n \end{bmatrix}$$

For the Westwood Company lot size example, the **Y** and **X** matrices are (see Table 2.1):

$$\mathbf{Y} = \begin{bmatrix} 73 \\ 50 \\ \cdot \\ \cdot \\ \cdot \\ 132 \end{bmatrix} \qquad \mathbf{X} = \begin{bmatrix} 1 & 30 \\ 1 & 20 \\ \cdot & \cdot \\ \cdot & \cdot \\ \cdot & \cdot \\ 1 & 60 \end{bmatrix}$$

6.2 MATRIX ADDITION AND SUBTRACTION

Adding or subtracting two matrices requires that they have the same dimension. The sum, or difference, of two matrices is another matrix whose elements each consists of the sum, or difference, of the corresponding elements of the two matrices. Suppose:

$$\underset{3\times 2}{\mathbf{A}} = \begin{bmatrix} 1 & 4 \\ 2 & 5 \\ 3 & 6 \end{bmatrix} \qquad \underset{3\times 2}{\mathbf{B}} = \begin{bmatrix} 1 & 2 \\ 2 & 3 \\ 3 & 4 \end{bmatrix}$$

then:

$$\underset{3\times 2}{\mathbf{A}+\mathbf{B}} = \begin{bmatrix} 1+1 & 4+2 \\ 2+2 & 5+3 \\ 3+3 & 6+4 \end{bmatrix} = \begin{bmatrix} 2 & 6 \\ 4 & 8 \\ 6 & 10 \end{bmatrix}$$

Similarly:

$$\underset{3\times 2}{\mathbf{A}-\mathbf{B}} = \begin{bmatrix} 1-1 & 4-2 \\ 2-2 & 5-3 \\ 3-3 & 6-4 \end{bmatrix} = \begin{bmatrix} 0 & 2 \\ 0 & 2 \\ 0 & 2 \end{bmatrix}$$

In general, if:

$$\underset{r\times c}{\mathbf{A}} = [a_{ij}] \qquad \underset{r\times c}{\mathbf{B}} = [b_{ij}] \qquad i = 1, \ldots, r; j = 1, \ldots, c$$

then:

(6.8)
$$\underset{r\times c}{\mathbf{A}+\mathbf{B}} = [a_{ij} + b_{ij}] \quad \text{and} \quad \underset{r\times c}{\mathbf{A}-\mathbf{B}} = [a_{ij} - b_{ij}]$$

Formula (6.8) generalizes in an obvious way to addition and subtraction of more than two matrices. Note also that $\mathbf{A}+\mathbf{B} = \mathbf{B}+\mathbf{A}$, as in ordinary algebra.

Regression Example

The regression model:

$$Y_i = E(Y_i) + \varepsilon_i \qquad i = 1, \ldots, n$$

can be written in matrix notation as follows:

$$\begin{bmatrix} Y_1 \\ Y_2 \\ \cdot \\ \cdot \\ \cdot \\ Y_n \end{bmatrix} = \begin{bmatrix} E(Y_1) \\ E(Y_2) \\ \cdot \\ \cdot \\ \cdot \\ E(Y_n) \end{bmatrix} + \begin{bmatrix} \varepsilon_1 \\ \varepsilon_2 \\ \cdot \\ \cdot \\ \cdot \\ \varepsilon_n \end{bmatrix}$$

Thus, the matrix $\mathbf{Y}$ equals the sum of two matrices, a matrix containing the expected values and another containing the error terms.

6.3 MATRIX MULTIPLICATION

Multiplication of a Matrix by a Scalar

A scalar is an ordinary number or a symbol representing a number. Thus, a scalar is not a matrix. We frequently encounter multiplication of a matrix by a scalar. In this, every element of the matrix is multiplied by the scalar. For example, suppose the matrix $\mathbf{A}$ is given by:

$$\mathbf{A} = \begin{bmatrix} 2 & 7 \\ 9 & 3 \end{bmatrix}$$

Then $4\mathbf{A}$, where 4 is the scalar, equals:

$$4\mathbf{A} = 4\begin{bmatrix} 2 & 7 \\ 9 & 3 \end{bmatrix} = \begin{bmatrix} 8 & 28 \\ 36 & 12 \end{bmatrix}$$

Similarly, $\lambda\mathbf{A}$ equals:

$$\lambda\mathbf{A} = \lambda\begin{bmatrix} 2 & 7 \\ 9 & 3 \end{bmatrix} = \begin{bmatrix} 2\lambda & 7\lambda \\ 9\lambda & 3\lambda \end{bmatrix}$$

where λ denotes the scalar.

If every element of a matrix has a common factor, this factor can be taken outside the matrix and treated as a scalar. For example:

$$\begin{bmatrix} 9 & 27 \\ 15 & 18 \end{bmatrix} = 3\begin{bmatrix} 3 & 9 \\ 5 & 6 \end{bmatrix}$$

Similarly:

$$\begin{bmatrix} \frac{5}{\lambda} & \frac{2}{\lambda} \\ \frac{3}{\lambda} & \frac{8}{\lambda} \end{bmatrix} = \frac{1}{\lambda}\begin{bmatrix} 5 & 2 \\ 3 & 8 \end{bmatrix}$$

In general, if $\mathbf{A} = [a_{ij}]$ and λ is a scalar, we have:

$$\lambda\mathbf{A} = \mathbf{A}\lambda = [\lambda a_{ij}] \tag{6.9}$$

Multiplication of a Matrix by a Matrix

Multiplication of a matrix by a matrix may appear somewhat complicated at first, but a little practice will make it into a routine operation.

Consider the two matrices:

$$\underset{2\times 2}{\mathbf{A}} = \begin{bmatrix} 2 & 5 \\ 4 & 1 \end{bmatrix} \qquad \underset{2\times 2}{\mathbf{B}} = \begin{bmatrix} 4 & 6 \\ 5 & 8 \end{bmatrix}$$

The product **AB** will be a 2×2 matrix, whose elements are obtained by finding the cross products of rows of **A** with columns of **B** and summing the cross products. For instance, to find the element in the first row and first column of the product **AB**, we work with the first row of **A** and the first column of **B**, as follows:

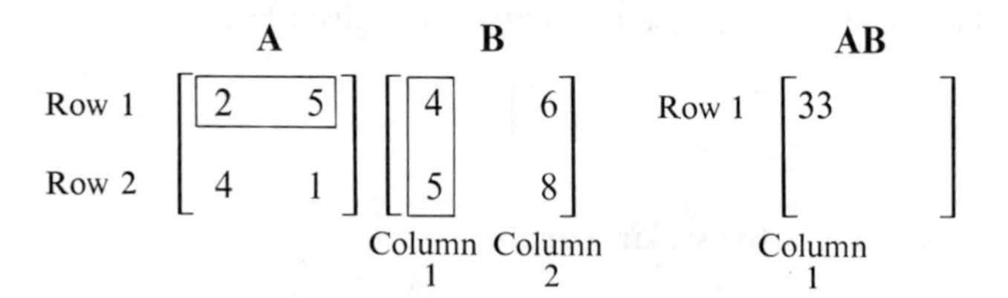

We take the cross products and sum:

$$(2)(4) + (5)(5) = 33$$

The number 33 is the element in the first row and first column of the matrix **AB**.

To find the element in the first row and second column of **AB**, we work with the first row of **A** and the second column of **B**:

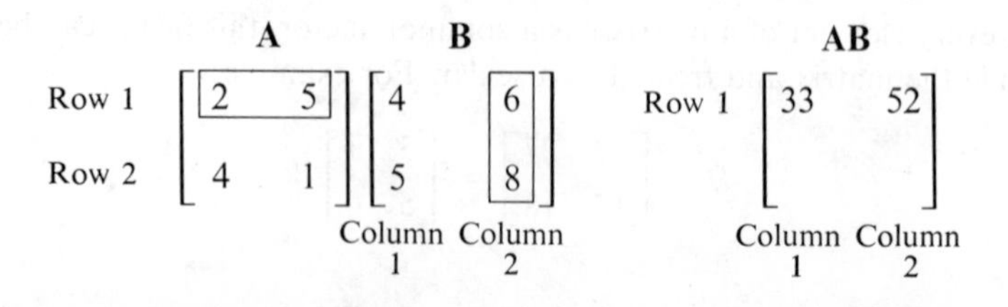

The cross products and sum are:

$$(2)(6) + (5)(8) = 52$$

Continuing this process, we find the product **AB** to be:

$$\underset{2\times 2}{\mathbf{AB}} = \begin{bmatrix} 2 & 5 \\ 4 & 1 \end{bmatrix} \begin{bmatrix} 4 & 6 \\ 5 & 8 \end{bmatrix} = \begin{bmatrix} 33 & 52 \\ 21 & 32 \end{bmatrix}$$

Let us consider another example:

$$\underset{2\times 3}{\mathbf{A}} = \begin{bmatrix} 1 & 3 & 4 \\ 0 & 5 & 8 \end{bmatrix} \qquad \underset{3\times 1}{\mathbf{B}} = \begin{bmatrix} 3 \\ 5 \\ 2 \end{bmatrix}$$

$$\underset{2\times 1}{\mathbf{AB}} = \begin{bmatrix} 1 & 3 & 4 \\ 0 & 5 & 8 \end{bmatrix} \begin{bmatrix} 3 \\ 5 \\ 2 \end{bmatrix} = \begin{bmatrix} 26 \\ 41 \end{bmatrix}$$

When obtaining the product **AB**, we say that **A** is *postmultiplied* by **B** or **B** is *premultiplied* by **A**. The reason for this careful terminology is that multiplication rules for ordinary algebra do not apply to matrix algebra. In ordinary algebra, $xy = yx$. In matrix algebra, $\mathbf{AB} \neq \mathbf{BA}$ usually. In fact, even though the product **AB** may be defined, the product **BA** may not be defined at all.

In general, the product **AB** is only defined when the number of columns in **A** equals the number of rows in **B** so that there will be corresponding terms in the cross products. Thus, in our previous two examples, we had:

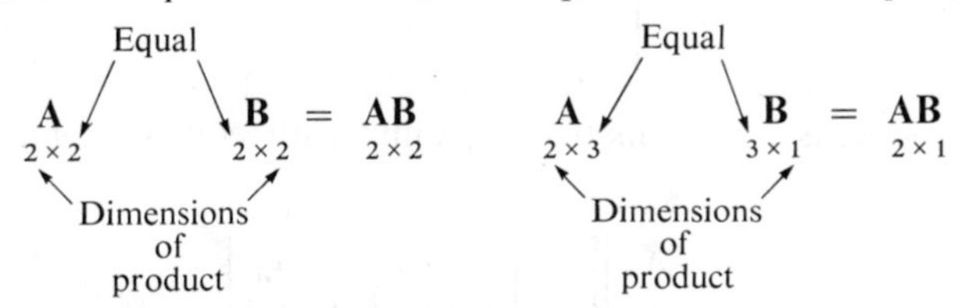

Note that the dimensions of the product **AB** are given by the number of rows in **A** and the number of columns in **B**. Note also that in the second case the product **BA** would not be defined since the number of columns in **B** is not equal to the number of rows in **A**:

$$\underset{3\times 1}{\mathbf{B}} \quad \underset{2\times 3}{\mathbf{A}} \qquad \text{(Unequal)}$$

Here is another example of matrix multiplication:

$$\mathbf{AB} = \begin{bmatrix} a_{11} & a_{12} & a_{13} \\ a_{21} & a_{22} & a_{23} \end{bmatrix} \begin{bmatrix} b_{11} & b_{12} \\ b_{21} & b_{22} \\ b_{31} & b_{32} \end{bmatrix}$$

$$= \begin{bmatrix} a_{11}b_{11} + a_{12}b_{21} + a_{13}b_{31} & a_{11}b_{12} + a_{12}b_{22} + a_{13}b_{32} \\ a_{21}b_{11} + a_{22}b_{21} + a_{23}b_{31} & a_{21}b_{12} + a_{22}b_{22} + a_{23}b_{32} \end{bmatrix}$$

In general, if **A** has dimension $r \times c$ and **B** has dimension $c \times s$, the product **AB** is a matrix of dimension $r \times s$, whose element in the ith row and jth column is:

$$\sum_{k=1}^{c} a_{ik} b_{kj}$$

so that:

$$\mathbf{AB} = \left[\sum_{k=1}^{c} a_{ik} b_{kj} \right] \qquad i = 1, \ldots, r; j = 1, \ldots, s \tag{6.10}$$

Thus, in the foregoing example, the element in, say, the first row and second column of the product **AB** is:

$$\sum_{k=1}^{3} a_{1k} b_{k2} = a_{11} b_{12} + a_{12} b_{22} + a_{13} b_{32}$$

as indeed we found by taking the cross products of the elements in the first row of **A** and second column of **B** and summing.

Additional Examples

1.

$$\begin{bmatrix} 4 & 2 \\ 5 & 8 \end{bmatrix} \begin{bmatrix} a_1 \\ a_2 \end{bmatrix} = \begin{bmatrix} 4a_1 + 2a_2 \\ 5a_1 + 8a_2 \end{bmatrix}$$

2.

$$[2 \quad 3 \quad 5] \begin{bmatrix} 2 \\ 3 \\ 5 \end{bmatrix} = [2^2 + 3^2 + 5^2] = [38]$$

Here, the product is a 1×1 matrix, or equivalently equals the number 38.

3.

$$\begin{bmatrix} 1 & X_1 \\ 1 & X_2 \\ 1 & X_3 \end{bmatrix} \begin{bmatrix} \beta_0 \\ \beta_1 \end{bmatrix} = \begin{bmatrix} \beta_0 + \beta_1 X_1 \\ \beta_0 + \beta_1 X_2 \\ \beta_0 + \beta_1 X_3 \end{bmatrix}$$

Regression Examples. Let us define the vector $\boldsymbol{\beta}$ as follows:

$$\boldsymbol{\beta} = \begin{bmatrix} \beta_0 \\ \beta_1 \end{bmatrix} \tag{6.11}$$

Then the product $\mathbf{X}\boldsymbol{\beta}$, where **X** is defined in (6.6), is an $n \times 1$ matrix:

$$\underset{n \times 1}{\mathbf{X}\boldsymbol{\beta}} = \begin{bmatrix} 1 & X_1 \\ 1 & X_2 \\ \cdot & \cdot \\ \cdot & \cdot \\ \cdot & \cdot \\ 1 & X_n \end{bmatrix} \begin{bmatrix} \beta_0 \\ \beta_1 \end{bmatrix} = \begin{bmatrix} \beta_0 + \beta_1 X_1 \\ \beta_0 + \beta_1 X_2 \\ \cdot \\ \cdot \\ \cdot \\ \beta_0 + \beta_1 X_n \end{bmatrix} \tag{6.12}$$

Since $\beta_0 + \beta_1 X_i = E(Y_i)$, we see that $\mathbf{X\beta}$ is the vector of expected values $E(Y_i)$ for the simple linear regression model.

Another product frequently needed is $\mathbf{Y'Y}$, where $\mathbf{Y}$ is the vector of observations on the dependent variable as defined in (6.4):

$$\underset{1\times 1}{\mathbf{Y'Y}} = [Y_1 \quad Y_2 \quad \cdots \quad Y_n]\begin{bmatrix} Y_1 \\ Y_2 \\ \cdot \\ \cdot \\ \cdot \\ Y_n \end{bmatrix} = [Y_1^2 + Y_2^2 + \cdots + Y_n^2] = [\textstyle\sum Y_i^2] \tag{6.13}$$

Note that $\mathbf{Y'Y}$ is a 1×1 matrix, or a scalar. We thus have a compact way of writing a sum of squares: $\mathbf{Y'Y} = \sum Y_i^2$.

We also will need $\mathbf{X'X}$, which is a 2×2 matrix:

$$\underset{2\times 2}{\mathbf{X'X}} = \begin{bmatrix} 1 & 1 & \cdots & 1 \\ X_1 & X_2 & \cdots & X_n \end{bmatrix}\begin{bmatrix} 1 & X_1 \\ 1 & X_2 \\ \cdot & \cdot \\ \cdot & \cdot \\ \cdot & \cdot \\ 1 & X_n \end{bmatrix} = \begin{bmatrix} n & \sum X_i \\ \sum X_i & \sum X_i^2 \end{bmatrix} \tag{6.14}$$

and $\mathbf{X'Y}$, which is a 2×1 matrix:

$$\underset{2\times 1}{\mathbf{X'Y}} = \begin{bmatrix} 1 & 1 & \cdots & 1 \\ X_1 & X_2 & \cdots & X_n \end{bmatrix}\begin{bmatrix} Y_1 \\ Y_2 \\ \cdot \\ \cdot \\ \cdot \\ Y_n \end{bmatrix} = \begin{bmatrix} \sum Y_i \\ \sum X_i Y_i \end{bmatrix} \tag{6.15}$$

6.4 SPECIAL TYPES OF MATRICES

Certain special types of matrices arise regularly in regression analysis. We shall consider the most important of these.

Symmetric Matrix

If $\mathbf{A} = \mathbf{A}'$, $\mathbf{A}$ is said to be symmetric. Thus $\mathbf{A}$ below is symmetric:

$$\mathbf{A} = \begin{bmatrix} 1 & 4 & 6 \\ 4 & 2 & 5 \\ 6 & 5 & 3 \end{bmatrix} \qquad \mathbf{A}' = \begin{bmatrix} 1 & 4 & 6 \\ 4 & 2 & 5 \\ 6 & 5 & 3 \end{bmatrix}$$

Clearly, a symmetric matrix necessarily is square. Symmetric matrices arise typically in regression analysis when we premultiply a matrix, say $\mathbf{X}$, by its transpose, $\mathbf{X}'$. The resulting matrix, $\mathbf{X'X}$, is symmetric, as can readily be seen from (6.14).

Diagonal Matrix

A diagonal matrix **A** is a square matrix whose off-diagonal elements a_{ij} $(i \neq j)$ are all zeros, such as:

$$\mathbf{A} = \begin{bmatrix} a_{11} & 0 & 0 \\ 0 & a_{22} & 0 \\ 0 & 0 & a_{33} \end{bmatrix} \qquad \mathbf{B} = \begin{bmatrix} 4 & 0 & 0 & 0 \\ 0 & 1 & 0 & 0 \\ 0 & 0 & 10 & 0 \\ 0 & 0 & 0 & 5 \end{bmatrix}$$

Two important types of diagonal matrices are the identity matrix and the scalar matrix.

Identity Matrix

The identity matrix or unit matrix is denoted **I**. It is a square matrix whose elements on the main diagonal are all ones and whose off-diagonal elements are all zeros. Premultiplying or postmultiplying any $r \times r$ matrix **A** by the $r \times r$ identity matrix **I** leaves **A** unchanged. For example:

$$\mathbf{IA} = \begin{bmatrix} 1 & 0 & 0 \\ 0 & 1 & 0 \\ 0 & 0 & 1 \end{bmatrix} \begin{bmatrix} a_{11} & a_{12} & a_{13} \\ a_{21} & a_{22} & a_{23} \\ a_{31} & a_{32} & a_{33} \end{bmatrix} = \begin{bmatrix} a_{11} & a_{12} & a_{13} \\ a_{21} & a_{22} & a_{23} \\ a_{31} & a_{32} & a_{33} \end{bmatrix}$$

Similarly, we have:

$$\mathbf{AI} = \begin{bmatrix} a_{11} & a_{12} & a_{13} \\ a_{21} & a_{22} & a_{23} \\ a_{31} & a_{32} & a_{33} \end{bmatrix} \begin{bmatrix} 1 & 0 & 0 \\ 0 & 1 & 0 \\ 0 & 0 & 1 \end{bmatrix} = \begin{bmatrix} a_{11} & a_{12} & a_{13} \\ a_{21} & a_{22} & a_{23} \\ a_{31} & a_{32} & a_{33} \end{bmatrix}$$

Note that the identity matrix **I** therefore corresponds to the number 1 in ordinary algebra, since we have there that $1 \cdot x = x \cdot 1 = x$.

In general, we have for any $r \times r$ matrix **A**:

(6.16) $$\mathbf{AI} = \mathbf{IA} = \mathbf{A}$$

Thus, the identity matrix can be inserted or dropped from a matrix expression whenever it is convenient to do so.

Scalar Matrix

A scalar matrix is a square matrix whose main-diagonal elements are a scalar quantity and whose off-diagonal elements are zeros. Two examples of scalar matrices are:

$$\begin{bmatrix} 2 & 0 \\ 0 & 2 \end{bmatrix} \qquad \begin{bmatrix} \lambda & 0 & 0 \\ 0 & \lambda & 0 \\ 0 & 0 & \lambda \end{bmatrix}$$

A scalar matrix can be denoted $\lambda\mathbf{I}$ where λ is the scalar. For instance:

$$\begin{bmatrix} 2 & 0 \\ 0 & 2 \end{bmatrix} = 2\begin{bmatrix} 1 & 0 \\ 0 & 1 \end{bmatrix}$$

$$\begin{bmatrix} \lambda & 0 & 0 \\ 0 & \lambda & 0 \\ 0 & 0 & \lambda \end{bmatrix} = \lambda\begin{bmatrix} 1 & 0 & 0 \\ 0 & 1 & 0 \\ 0 & 0 & 1 \end{bmatrix}$$

Multiplying an $r \times r$ matrix $\mathbf{A}$ by the $r \times r$ scalar matrix $\lambda\mathbf{I}$ is equivalent to multiplying $\mathbf{A}$ by the scalar λ.

Zero Vector

A zero vector is a column vector containing only zeros. It will be denoted $\mathbf{0}$. For instance, the 3×1 zero vector is:

$$\mathbf{0} = \begin{bmatrix} 0 \\ 0 \\ 0 \end{bmatrix}$$

We shall make use of this type of vector shortly.

6.5 LINEAR DEPENDENCE AND RANK OF MATRIX

Linear Dependence

Consider the following matrix:

$$\mathbf{A} = \begin{bmatrix} 1 & 2 & 5 & 1 \\ 2 & 2 & 10 & 6 \\ 3 & 4 & 15 & 1 \end{bmatrix} \tag{6.17}$$

Let us think now of the columns of this matrix as vectors. Thus, we view $\mathbf{A}$ as being made up of four column vectors. It happens in (6.17) that the columns are interrelated in a special manner. Note that the third column vector is a multiple of the first column vector:

$$\begin{bmatrix} 5 \\ 10 \\ 15 \end{bmatrix} = 5\begin{bmatrix} 1 \\ 2 \\ 3 \end{bmatrix}$$

We say that the columns of $\mathbf{A}$ are linearly dependent. They contain redundant information, so to speak, since one column can be obtained as a linear combination of the others.

We define a set of column vectors to be linearly dependent if one vector can be expressed as a linear combination of some of the others. If no vector in the set can be so expressed, we define the set of vectors to be linearly independent. A more general, though equivalent, definition for the c column vectors $\mathbf{C}_1, \ldots, \mathbf{C}_c$ in an $r \times c$ matrix is:

(6.18) When c scalars $\lambda_1, \ldots, \lambda_c$, not all zero, can be found such that:

$$\lambda_1 \mathbf{C}_1 + \lambda_2 \mathbf{C}_2 + \cdots + \lambda_c \mathbf{C}_c = \mathbf{0}$$

where $\mathbf{0}$ denotes the zero column vector, the c column vectors are linearly dependent. If the only set of scalars for which the equality holds is $\lambda_1 = 0, \ldots, \lambda_c = 0$, the set of c column vectors is linearly independent.

To illustrate for our example in (6.17), $\lambda_1 = 5$, $\lambda_2 = 0$, $\lambda_3 = -1$, $\lambda_4 = 0$ leads to:

$$5\begin{bmatrix} 1 \\ 2 \\ 3 \end{bmatrix} + 0\begin{bmatrix} 2 \\ 2 \\ 4 \end{bmatrix} - 1\begin{bmatrix} 5 \\ 10 \\ 15 \end{bmatrix} + 0\begin{bmatrix} 1 \\ 6 \\ 1 \end{bmatrix} = \begin{bmatrix} 0 \\ 0 \\ 0 \end{bmatrix}$$

Hence, the column vectors are linearly dependent. Note that some of the $\lambda_j = 0$ here. It is only required for linear dependence that at least one λ_j is not zero.

Rank of a Matrix

The rank of a matrix is defined to be the maximum number of linearly independent columns in the matrix. We know that the rank of $\mathbf{A}$ in (6.17) cannot be 4, since the four columns are linearly dependent. We can, however, find 3 columns (1, 2, and 4) which are linearly independent. There are no scalars $\lambda_1, \lambda_2, \lambda_4$ such that $\lambda_1 \mathbf{C}_1 + \lambda_2 \mathbf{C}_2 + \lambda_4 \mathbf{C}_4 = \mathbf{0}$ other than $\lambda_1 = \lambda_2 = \lambda_4 = 0$. Thus, the rank of $\mathbf{A}$ in (6.17) is 3.

The rank of a matrix is unique and can equivalently be defined as the maximum number of linearly independent rows. It follows that the rank of an $r \times c$ matrix cannot exceed $\min(r, c)$, the minimum of the two values r and c.

6.6 INVERSE OF A MATRIX

In ordinary algebra, the inverse of a number is its reciprocal. Thus, the inverse of 6 is $\frac{1}{6}$. A number multiplied by its inverse always equals 1:

$$6 \cdot \tfrac{1}{6} = 1$$

$$x \cdot \frac{1}{x} = x \cdot x^{-1} = x^{-1} \cdot x = 1$$

In matrix algebra, the inverse of a matrix $\mathbf{A}$ is another matrix, denoted $\mathbf{A}^{-1}$, such that:

(6.19) $$\mathbf{A}^{-1}\mathbf{A} = \mathbf{A}\mathbf{A}^{-1} = \mathbf{I}$$

where **I** is the identity matrix. Thus again, the identity matrix **I** plays the same role as the number 1 in ordinary algebra. An inverse of a matrix is defined only for square matrices. Even so, many square matrices do not have an inverse. If a square matrix does have an inverse, the inverse is unique.

Examples

1. The inverse of the matrix:

$$\mathbf{A} = \begin{bmatrix} 2 & 4 \\ 3 & 1 \end{bmatrix}$$

is:

$$\mathbf{A}^{-1} = \begin{bmatrix} -.1 & .4 \\ .3 & -.2 \end{bmatrix}$$

since:

$$\mathbf{A}^{-1}\mathbf{A} = \begin{bmatrix} -.1 & .4 \\ .3 & -.2 \end{bmatrix}\begin{bmatrix} 2 & 4 \\ 3 & 1 \end{bmatrix} = \begin{bmatrix} 1 & 0 \\ 0 & 1 \end{bmatrix}$$

or:

$$\mathbf{A}\mathbf{A}^{-1} = \begin{bmatrix} 2 & 4 \\ 3 & 1 \end{bmatrix}\begin{bmatrix} -.1 & .4 \\ .3 & -.2 \end{bmatrix} = \begin{bmatrix} 1 & 0 \\ 0 & 1 \end{bmatrix}$$

2. The inverse of the matrix:

$$\mathbf{A} = \begin{bmatrix} 3 & 0 & 0 \\ 0 & 4 & 0 \\ 0 & 0 & 2 \end{bmatrix}$$

is:

$$\mathbf{A}^{-1} = \begin{bmatrix} \frac{1}{3} & 0 & 0 \\ 0 & \frac{1}{4} & 0 \\ 0 & 0 & \frac{1}{2} \end{bmatrix}$$

since:

$$\mathbf{A}^{-1}\mathbf{A} = \begin{bmatrix} \frac{1}{3} & 0 & 0 \\ 0 & \frac{1}{4} & 0 \\ 0 & 0 & \frac{1}{2} \end{bmatrix}\begin{bmatrix} 3 & 0 & 0 \\ 0 & 4 & 0 \\ 0 & 0 & 2 \end{bmatrix} = \begin{bmatrix} 1 & 0 & 0 \\ 0 & 1 & 0 \\ 0 & 0 & 1 \end{bmatrix}$$

Note that the inverse of a diagonal matrix is a diagonal matrix consisting simply of the reciprocals of the elements on the diagonal.

Finding the Inverse

Up to this point, the inverse of a matrix $\mathbf{A}$ has been given, and we have only checked to make sure it is the inverse by seeing whether or not $\mathbf{A}^{-1}\mathbf{A} = \mathbf{I}$. But how does one find the inverse, and when does it exist?

An inverse of a square $r \times r$ matrix exists if the rank of the matrix is r. Such a matrix is said to be *nonsingular*. An $r \times r$ matrix with rank less than r is said to be *singular*, and does not have an inverse.

Finding the inverse of a matrix can often require a tremendous amount of computing. We shall take the approach in this book that the inverse of a 2×2 matrix and a 3×3 matrix can be calculated by hand. For any larger matrix, one is likely to use an electronic computer to find the inverse unless the matrix is of a special form such as a diagonal matrix. It can be shown that the inverses for 2×2 and 3×3 matrices are as follows:

1. If:

$$\mathbf{A} = \begin{bmatrix} a & b \\ c & d \end{bmatrix}$$

then:

$$\mathbf{A}^{-1} = \begin{bmatrix} a & b \\ c & d \end{bmatrix}^{-1} = \begin{bmatrix} \frac{d}{D} & \frac{-b}{D} \\ \frac{-c}{D} & \frac{a}{D} \end{bmatrix} \tag{6.20}$$

where:

$$D = ad - bc$$

D is called the *determinant* of the matrix $\mathbf{A}$. If $\mathbf{A}$ were singular, its determinant would equal zero and no inverse of $\mathbf{A}$ would exist.

2. If:

$$\mathbf{B} = \begin{bmatrix} a & b & c \\ d & e & f \\ g & h & k \end{bmatrix}$$

then:

$$\mathbf{B}^{-1} = \begin{bmatrix} a & b & c \\ d & e & f \\ g & h & k \end{bmatrix}^{-1} = \begin{bmatrix} A & B & C \\ D & E & F \\ G & H & K \end{bmatrix} \tag{6.21}$$

where:

$$\begin{array}{lll} A = (ek - fh)/Z & B = -(bk - ch)/Z & C = (bf - ce)/Z \\ D = -(dk - fg)/Z & E = (ak - cg)/Z & F = -(af - cd)/Z \\ G = (dh - eg)/Z & H = -(ah - bg)/Z & K = (ae - bd)/Z \end{array}$$

and:

$$Z = a(ek - fh) - b(dk - fg) + c(dh - eg)$$

Z is called the determinant of the matrix $\mathbf{B}$.

Let us use (6.20) to find the inverse of:

$$\mathbf{A} = \begin{bmatrix} 2 & 4 \\ 3 & 1 \end{bmatrix}$$

We have:

$$a = 2 \qquad b = 4$$
$$c = 3 \qquad d = 1$$
$$D = ad - bc = (2)(1) - (4)(3) = -10$$

Hence:

$$\mathbf{A}^{-1} = \begin{bmatrix} \dfrac{1}{-10} & \dfrac{-4}{-10} \\ \dfrac{-3}{-10} & \dfrac{2}{-10} \end{bmatrix} = \begin{bmatrix} -.1 & .4 \\ .3 & -.2 \end{bmatrix}$$

as was given in an earlier example.

When an inverse $\mathbf{A}^{-1}$ has been obtained, either by hand calculations or from a computer run, it is usually wise to compute $\mathbf{A}^{-1}\mathbf{A}$ to check whether the product equals the identity matrix, allowing for minor rounding departures from 0 and 1.

Regression Example

The principal inverse matrix encountered in regression analysis is the inverse of the matrix $\mathbf{X'X}$ in (6.14):

$$\mathbf{X'X} = \begin{bmatrix} n & \sum X_i \\ \sum X_i & \sum X_i^2 \end{bmatrix}$$

Using rule (6.20), we have:

$$a = n \qquad b = \sum X_i$$
$$c = \sum X_i \qquad d = \sum X_i^2$$

so that:

$$D = n \sum X_i^2 - \left(\sum X_i\right)\left(\sum X_i\right) = n\left(\sum X_i^2 - \frac{\left(\sum X_i\right)^2}{n}\right) = n \sum (X_i - \bar{X})^2$$

Hence:

$$(6.22)\qquad (\mathbf{X'X})^{-1} = \begin{bmatrix} \dfrac{\sum X_i^2}{n\sum(X_i - \bar{X})^2} & \dfrac{-\sum X_i}{n\sum(X_i - \bar{X})^2} \\ \dfrac{-\sum X_i}{n\sum(X_i - \bar{X})^2} & \dfrac{n}{n\sum(X_i - \bar{X})^2} \end{bmatrix}$$

Since $\sum X_i = n\bar{X}$, we can simplify (6.22):

$$(6.23)\qquad (\mathbf{X'X})^{-1} = \begin{bmatrix} \dfrac{\sum X_i^2}{n\sum(X_i - \bar{X})^2} & \dfrac{-\bar{X}}{\sum(X_i - \bar{X})^2} \\ \dfrac{-\bar{X}}{\sum(X_i - \bar{X})^2} & \dfrac{1}{\sum(X_i - \bar{X})^2} \end{bmatrix}$$

Uses of Inverse Matrix

In ordinary algebra, we solve an equation of the type:

$$5x = 20$$

by multiplying both sides of the equation by the inverse of 5, namely:

$$\tfrac{1}{5}(5x) = \tfrac{1}{5}(20)$$

We obtain:

$$x = \tfrac{1}{5}(20) = 4$$

In matrix algebra, if we have an equation:

$$\mathbf{AY} = \mathbf{C}$$

we correspondingly premultiply both sides by $\mathbf{A}^{-1}$, assuming $\mathbf{A}$ has an inverse, and obtain:

$$\mathbf{A}^{-1}\mathbf{AY} = \mathbf{A}^{-1}\mathbf{C}$$

Since $\mathbf{A}^{-1}\mathbf{AY} = \mathbf{IY} = \mathbf{Y}$, we obtain:

$$\mathbf{Y} = \mathbf{A}^{-1}\mathbf{C}$$

To illustrate this use, suppose we have two simultaneous equations:

$$2x + 4y = 20$$
$$3x + y = 10$$

which can be written as follows in matrix notation:

$$\begin{bmatrix} 2 & 4 \\ 3 & 1 \end{bmatrix}\begin{bmatrix} x \\ y \end{bmatrix} = \begin{bmatrix} 20 \\ 10 \end{bmatrix}$$

The solution of these equations then is:

$$\begin{bmatrix} x \\ y \end{bmatrix} = \begin{bmatrix} 2 & 4 \\ 3 & 1 \end{bmatrix}^{-1} \begin{bmatrix} 20 \\ 10 \end{bmatrix}$$

Earlier we found the required inverse, so we obtain:

$$\begin{bmatrix} x \\ y \end{bmatrix} = \begin{bmatrix} -.1 & .4 \\ .3 & -.2 \end{bmatrix} \begin{bmatrix} 20 \\ 10 \end{bmatrix} = \begin{bmatrix} 2 \\ 4 \end{bmatrix}$$

Hence, $x = 2$ and $y = 4$ satisfy these two equations.

6.7 SOME BASIC THEOREMS FOR MATRICES

We list here, without proof, some basic theorems for matrices which we will utilize in later work.

(6.24) $$\mathbf{A} + \mathbf{B} = \mathbf{B} + \mathbf{A}$$

(6.25) $$(\mathbf{A} + \mathbf{B}) + \mathbf{C} = \mathbf{A} + (\mathbf{B} + \mathbf{C})$$

(6.26) $$(\mathbf{AB})\mathbf{C} = \mathbf{A}(\mathbf{BC})$$

(6.27) $$\mathbf{C}(\mathbf{A} + \mathbf{B}) = \mathbf{CA} + \mathbf{CB}$$

(6.28) $$\lambda(\mathbf{A} + \mathbf{B}) = \lambda\mathbf{A} + \lambda\mathbf{B}$$

(6.29) $$(\mathbf{A}')' = \mathbf{A}$$

(6.30) $$(\mathbf{A} + \mathbf{B})' = \mathbf{A}' + \mathbf{B}'$$

(6.31) $$(\mathbf{AB})' = \mathbf{B}'\mathbf{A}'$$

(6.32) $$(\mathbf{AB})^{-1} = \mathbf{B}^{-1}\mathbf{A}^{-1}$$

(6.33) $$(\mathbf{ABC})^{-1} = \mathbf{C}^{-1}\mathbf{B}^{-1}\mathbf{A}^{-1}$$

(6.34) $$(\mathbf{A}^{-1})^{-1} = \mathbf{A}$$

(6.35) $$(\mathbf{A}')^{-1} = (\mathbf{A}^{-1})'$$

6.8 RANDOM VECTORS AND MATRICES

A random vector or a random matrix contains elements which are random variables. Thus, the observation vector $\mathbf{Y}$ in (6.4) is a random vector since the Y_i elements are random variables.

Expectation of Random Vector or Matrix

Suppose we have $n = 3$ observations and are concerned with the observation vector:

$$\mathbf{Y} = \begin{bmatrix} Y_1 \\ Y_2 \\ Y_3 \end{bmatrix}$$

By $\boldsymbol{E}(\mathbf{Y})$, we refer to the vector:

$$\boldsymbol{E}(\mathbf{Y}) = \begin{bmatrix} E(Y_1) \\ E(Y_2) \\ E(Y_3) \end{bmatrix}$$

Thus, the expectation of a random vector or matrix is a vector or matrix whose elements are the expected values of the corresponding elements in the original vector or matrix.

In general, for a random vector $\mathbf{Y}$ the expectation is:

$$\text{(6.36)} \qquad \boldsymbol{E}(\mathbf{Y}) = [E(Y_i)] \qquad i = 1, \ldots, n$$

and for a random matrix $\mathbf{Y}$ the expectation is:

$$\text{(6.37)} \qquad \boldsymbol{E}(\mathbf{Y}) = [E(Y_{ij})] \qquad i = 1, \ldots, n; j = 1, \ldots, p$$

Regression Example. Suppose the number of observations in a regression application is $n = 3$. The three error terms ε_1, ε_2, ε_3 each have expectation zero. If we define the error vector:

$$\boldsymbol{\varepsilon} = \begin{bmatrix} \varepsilon_1 \\ \varepsilon_2 \\ \varepsilon_3 \end{bmatrix}$$

we have:

$$\boldsymbol{E}(\boldsymbol{\varepsilon}) = \mathbf{0}$$

since:

$$\boldsymbol{E}(\boldsymbol{\varepsilon}) = \begin{bmatrix} E(\varepsilon_1) \\ E(\varepsilon_2) \\ E(\varepsilon_3) \end{bmatrix} = \begin{bmatrix} 0 \\ 0 \\ 0 \end{bmatrix}$$

Variance-Covariance Matrix of a Random Vector

Consider again the random vector $\mathbf{Y}$ consisting of three observations Y_1, Y_2, Y_3. Each random variable has a variance, $\sigma^2(Y_i)$, and any two random variables have a covariance, $\sigma(Y_i, Y_j)$. We can assemble these in a matrix called the *variance-covariance matrix of* $\mathbf{Y}$, and denoted $\boldsymbol{\sigma}^2(\mathbf{Y})$:

$$\text{(6.38)} \qquad \boldsymbol{\sigma}^2(\mathbf{Y}) = \begin{bmatrix} \sigma^2(Y_1) & \sigma(Y_1, Y_2) & \sigma(Y_1, Y_3) \\ \sigma(Y_2, Y_1) & \sigma^2(Y_2) & \sigma(Y_2, Y_3) \\ \sigma(Y_3, Y_1) & \sigma(Y_3, Y_2) & \sigma^2(Y_3) \end{bmatrix}$$

Note that the variances are on the main diagonal and the covariance $\sigma(Y_i, Y_j)$ is found in the ith row and jth column of the matrix. Thus $\sigma(Y_2, Y_1)$ is found in the second row, first column, and $\sigma(Y_1, Y_2)$ is found in the first row, second column. Remember, of course, that $\sigma(Y_2, Y_1) = \sigma(Y_1, Y_2)$. Hence it is clear that $\boldsymbol{\sigma}^2(\mathbf{Y})$ is a symmetric matrix containing redundant elements in it since the covariances below the main diagonal are the same as those above the main diagonal. However, the full matrix is obtained quite readily by matrix methods, and the matrix as defined turns out to be extremely useful.

It follows readily that:

$$\boldsymbol{\sigma}^2(\mathbf{Y}) = \boldsymbol{E}[\mathbf{Y} - \boldsymbol{E}(\mathbf{Y})][\mathbf{Y} - \boldsymbol{E}(\mathbf{Y})]' \tag{6.39}$$

For our illustration, we have:

$$\boldsymbol{\sigma}^2(\mathbf{Y}) = \boldsymbol{E}\begin{bmatrix} Y_1 - E(Y_1) \\ Y_2 - E(Y_2) \\ Y_3 - E(Y_3) \end{bmatrix}[Y_1 - E(Y_1) \quad Y_2 - E(Y_2) \quad Y_3 - E(Y_3)]$$

Multiplying the two matrices and then taking expectations, we obtain:

Location in Product	*Term*	*Expected Value*
Row 1, column 1	$[Y_1 - E(Y_1)]^2$	$\sigma^2(Y_1)$
Row 1, column 2	$[Y_1 - E(Y_1)][Y_2 - E(Y_2)]$	$\sigma(Y_1, Y_2)$
Row 1, column 3	$[Y_1 - E(Y_1)][Y_3 - E(Y_3)]$	$\sigma(Y_1, Y_3)$
Row 2, column 1	$[Y_2 - E(Y_2)][Y_1 - E(Y_1)]$	$\sigma(Y_2, Y_1)$
etc.	etc.	etc.

This, of course, leads to the variance-covariance matrix in (6.38).

To generalize, the variance-covariance matrix for an $n \times 1$ random vector $\mathbf{Y}$ is:

$$\boldsymbol{\sigma}^2(\mathbf{Y}) = \begin{bmatrix} \sigma^2(Y_1) & \sigma(Y_1, Y_2) & \cdots & \sigma(Y_1, Y_n) \\ \sigma(Y_2, Y_1) & \sigma^2(Y_2) & \cdots & \sigma(Y_2, Y_n) \\ \vdots & \vdots & & \vdots \\ \sigma(Y_n, Y_1) & \sigma(Y_n, Y_2) & \cdots & \sigma^2(Y_n) \end{bmatrix} \tag{6.40}$$

Note again that $\boldsymbol{\sigma}^2(\mathbf{Y})$ is a symmetric matrix.

Regression Example. Let us return to the example based on $n = 3$ observations. Suppose that the three error terms have constant variance, $\sigma^2(\varepsilon_i) = \sigma^2$, and are uncorrelated, $\sigma(\varepsilon_i, \varepsilon_j) = 0$ for $i \neq j$. We can then write the variance-covariance matrix for the random vector $\boldsymbol{\varepsilon}$ of the previous example as follows:

$$\boldsymbol{\sigma}^2(\boldsymbol{\varepsilon}) = \sigma^2\mathbf{I}$$

since:

$$\sigma^2\mathbf{I} = \sigma^2\begin{bmatrix} 1 & 0 & 0 \\ 0 & 1 & 0 \\ 0 & 0 & 1 \end{bmatrix} = \begin{bmatrix} \sigma^2 & 0 & 0 \\ 0 & \sigma^2 & 0 \\ 0 & 0 & \sigma^2 \end{bmatrix}$$

Note that all variances are σ^2 and all covariances are zero.

Some Basic Theorems

Frequently, we shall encounter a random vector $\mathbf{W}$ which is obtained by premultiplying the random vector $\mathbf{Y}$ by a constant matrix $\mathbf{A}$ (a matrix whose elements are fixed):

$$\mathbf{W} = \mathbf{AY} \tag{6.41}$$

Some basic theorems for this case are:

$$E(\mathbf{A}) = \mathbf{A} \tag{6.42}$$

$$E(\mathbf{W}) = E(\mathbf{AY}) = \mathbf{A}E(\mathbf{Y}) \tag{6.43}$$

$$\boldsymbol{\sigma}^2(\mathbf{W}) = \boldsymbol{\sigma}^2(\mathbf{AY}) = \mathbf{A}[\boldsymbol{\sigma}^2(\mathbf{Y})]\mathbf{A}' \tag{6.44}$$

where $\boldsymbol{\sigma}^2(\mathbf{Y})$ is the variance-covariance matrix of $\mathbf{Y}$.

Example. As a simple illustration of the use of these theorems, consider:

$$\underset{2\times 1}{\mathbf{W}} \begin{bmatrix} W_1 \\ W_2 \end{bmatrix} = \underset{2\times 2}{\mathbf{A}} \begin{bmatrix} 1 & -1 \\ 1 & 1 \end{bmatrix} \underset{2\times 1}{\mathbf{Y}} \begin{bmatrix} Y_1 \\ Y_2 \end{bmatrix} = \begin{bmatrix} Y_1 - Y_2 \\ Y_1 + Y_2 \end{bmatrix}$$

We then have by (6.43):

$$E(\mathbf{W}) = \begin{bmatrix} 1 & -1 \\ 1 & 1 \end{bmatrix} \begin{bmatrix} E(Y_1) \\ E(Y_2) \end{bmatrix} = \begin{bmatrix} E(Y_1) - E(Y_2) \\ E(Y_1) + E(Y_2) \end{bmatrix}$$

and by (6.44):

$$\boldsymbol{\sigma}^2(\mathbf{W}) = \begin{bmatrix} 1 & -1 \\ 1 & 1 \end{bmatrix} \begin{bmatrix} \sigma^2(Y_1) & \sigma(Y_1, Y_2) \\ \sigma(Y_2, Y_1) & \sigma^2(Y_2) \end{bmatrix} \begin{bmatrix} 1 & 1 \\ -1 & 1 \end{bmatrix}$$

$$= \begin{bmatrix} \sigma^2(Y_1) + \sigma^2(Y_2) - 2\sigma(Y_1, Y_2) & \sigma^2(Y_1) - \sigma^2(Y_2) \\ \sigma^2(Y_1) - \sigma^2(Y_2) & \sigma^2(Y_1) + \sigma^2(Y_2) + 2\sigma(Y_1, Y_2) \end{bmatrix}$$

Thus:

$$\sigma^2(W_1) = \sigma^2(Y_1 - Y_2) = \sigma^2(Y_1) + \sigma^2(Y_2) - 2\sigma(Y_1, Y_2)$$
$$\sigma^2(W_2) = \sigma^2(Y_1 + Y_2) = \sigma^2(Y_1) + \sigma^2(Y_2) + 2\sigma(Y_1, Y_2)$$
$$\sigma(W_1, W_2) = \sigma(Y_1 - Y_2, Y_1 + Y_2) = \sigma^2(Y_1) - \sigma^2(Y_2)$$

6.9 SIMPLE LINEAR REGRESSION MODEL IN MATRIX TERMS

We are now ready to develop simple linear regression in matrix terms. Remember again that we will not present any new results, but shall only state in matrix terms the results obtained earlier. We shall begin with the regression model (3.1):

$$Y_i = \beta_0 + \beta_1 X_i + \varepsilon_i \qquad i = 1, \ldots, n \tag{6.45}$$

This implies:

(6.46)
$$\begin{aligned} Y_1 &= \beta_0 + \beta_1 X_1 + \varepsilon_1 \\ Y_2 &= \beta_0 + \beta_1 X_2 + \varepsilon_2 \\ &\vdots \\ Y_n &= \beta_0 + \beta_1 X_n + \varepsilon_n \end{aligned}$$

We defined earlier the observation vector **Y** in (6.4), the **X** matrix in (6.6), and the **β** vector in (6.11). Let us repeat these definitions and also define the error term vector **ε**:

(6.47)
$$\mathbf{Y} = \begin{bmatrix} Y_1 \\ Y_2 \\ \vdots \\ Y_n \end{bmatrix} \qquad \mathbf{X} = \begin{bmatrix} 1 & X_1 \\ 1 & X_2 \\ \vdots & \vdots \\ 1 & X_n \end{bmatrix} \qquad \boldsymbol{\beta} = \begin{bmatrix} \beta_0 \\ \beta_1 \end{bmatrix} \qquad \boldsymbol{\varepsilon} = \begin{bmatrix} \varepsilon_1 \\ \varepsilon_2 \\ \vdots \\ \varepsilon_n \end{bmatrix}$$

Now we can write (6.46) in matrix terms compactly as follows:

(6.48)
$$\underset{n\times 1}{\mathbf{Y}} = \underset{n\times 2}{\mathbf{X}}\ \underset{2\times 1}{\boldsymbol{\beta}} + \underset{n\times 1}{\boldsymbol{\varepsilon}}$$

since:

$$\begin{bmatrix} Y_1 \\ Y_2 \\ \vdots \\ Y_n \end{bmatrix} = \begin{bmatrix} 1 & X_1 \\ 1 & X_2 \\ \vdots & \vdots \\ 1 & X_n \end{bmatrix} \begin{bmatrix} \beta_0 \\ \beta_1 \end{bmatrix} + \begin{bmatrix} \varepsilon_1 \\ \varepsilon_2 \\ \vdots \\ \varepsilon_n \end{bmatrix} = \begin{bmatrix} \beta_0 + \beta_1 X_1 + \varepsilon_1 \\ \beta_0 + \beta_1 X_2 + \varepsilon_2 \\ \vdots \\ \beta_0 + \beta_1 X_n + \varepsilon_n \end{bmatrix}$$

The column of 1's in the **X** matrix may be viewed as consisting of the dummy variable $X_0 \equiv 1$ in the alternate regression model (2.5):

$$Y_i = \beta_0 X_0 + \beta_1 X_i + \varepsilon_i \qquad \text{where } X_0 \equiv 1$$

Thus, the **X** matrix may be considered to contain a column vector of the dummy variable X_0 and another column vector consisting of the values of the independent variable X_i.

With respect to the error terms, model (3.1) assumes that $E(\varepsilon_i) = 0$, $\sigma^2(\varepsilon_i) = \sigma^2$, and that the ε_i are independent normal random variables. The condition $E(\varepsilon_i) = 0$ in matrix terms is:

(6.49)
$$E(\boldsymbol{\varepsilon}) = \mathbf{0}$$

since (6.49) states:

$$\mathbf{E}\begin{bmatrix} \varepsilon_1 \\ \varepsilon_2 \\ \cdot \\ \cdot \\ \cdot \\ \varepsilon_n \end{bmatrix} = \begin{bmatrix} E(\varepsilon_1) \\ E(\varepsilon_2) \\ \cdot \\ \cdot \\ \cdot \\ E(\varepsilon_n) \end{bmatrix} = \begin{bmatrix} 0 \\ 0 \\ \cdot \\ \cdot \\ \cdot \\ 0 \end{bmatrix}$$

The condition that the error terms have constant variance σ^2 and that all covariances $\sigma(\varepsilon_i, \varepsilon_j)$ for $i \neq j$ are zero (since the ε_i are independent) is expressed in matrix terms through the variance-covariance matrix:

(6.50)
$$\underset{n \times n}{\boldsymbol{\sigma}^2(\boldsymbol{\varepsilon})} = \sigma^2 \underset{n \times n}{\mathbf{I}}$$

since (6.50) states:

$$\boldsymbol{\sigma}^2(\boldsymbol{\varepsilon}) = \sigma^2 \begin{bmatrix} 1 & 0 & 0 & \cdots & 0 \\ 0 & 1 & 0 & \cdots & 0 \\ \cdot & \cdot & \cdot & & \cdot \\ \cdot & \cdot & \cdot & & \cdot \\ \cdot & \cdot & \cdot & & \cdot \\ 0 & 0 & 0 & \cdots & 1 \end{bmatrix} = \begin{bmatrix} \sigma^2 & 0 & 0 & \cdots & 0 \\ 0 & \sigma^2 & 0 & \cdots & 0 \\ \cdot & \cdot & \cdot & & \cdot \\ \cdot & \cdot & \cdot & & \cdot \\ \cdot & \cdot & \cdot & & \cdot \\ 0 & 0 & 0 & \cdots & \sigma^2 \end{bmatrix}$$

Thus, the normal error model (3.1) in matrix terms is:

(6.51)
$$\mathbf{Y} = \mathbf{X}\boldsymbol{\beta} + \boldsymbol{\varepsilon}$$

where:

> $\boldsymbol{\varepsilon}$ is a vector of independent normal random variables with $\mathbf{E}(\boldsymbol{\varepsilon}) = \mathbf{0}$ and $\boldsymbol{\sigma}^2(\boldsymbol{\varepsilon}) = \sigma^2\mathbf{I}$

6.10 LEAST SQUARES ESTIMATION OF REGRESSION PARAMETERS

Normal Equations

The normal equations (2.9):

(6.52)
$$\begin{aligned} nb_0 + b_1 \sum X_i &= \sum Y_i \\ b_0 \sum X_i + b_1 \sum X_i^2 &= \sum X_i Y_i \end{aligned}$$

in matrix terms are:

(6.53)
$$\mathbf{X'Xb} = \mathbf{X'Y}$$

where:

(6.53a)
$$\mathbf{b} = \begin{bmatrix} b_0 \\ b_1 \end{bmatrix}$$

To see this, recall that we obtained **X′X** in (6.14) and **X′Y** in (6.15). Equation (6.53) thus states:

$$\begin{bmatrix} n & \sum X_i \\ \sum X_i & \sum X_i^2 \end{bmatrix} \begin{bmatrix} b_0 \\ b_1 \end{bmatrix} = \begin{bmatrix} \sum Y_i \\ \sum X_i Y_i \end{bmatrix}$$

or:

$$\begin{bmatrix} nb_0 + b_1 \sum X_i \\ b_0 \sum X_i + b_1 \sum X_i^2 \end{bmatrix} = \begin{bmatrix} \sum Y_i \\ \sum X_i Y_i \end{bmatrix}$$

These are precisely the normal equations in (6.52).

Estimated Regression Coefficients

To obtain the estimated regression coefficients from the normal equations:

$$\mathbf{X'Xb} = \mathbf{X'Y}$$

by matrix methods, we premultiply both sides by the inverse of **X′X** (we assume this exists):

$$(\mathbf{X'X})^{-1}\mathbf{X'Xb} = (\mathbf{X'X})^{-1}\mathbf{X'Y}$$

so that we find, since $(\mathbf{X'X})^{-1}\mathbf{X'X} = \mathbf{I}$:

$$\mathbf{b} = (\mathbf{X'X})^{-1}\mathbf{X'Y} \tag{6.54}$$

The estimators b_0 and b_1 in **b** are the same as those given earlier in (2.10a) and (2.10b). We shall demonstrate this by an example.

Example. Let us find the estimated regression coefficients for the Westwood Company lot size example by matrix methods. From earlier work, we have (see Table 2.2):

$$n = 10 \qquad \sum Y_i = 1{,}100 \qquad \sum X_i = 500 \qquad \sum X_i^2 = 28{,}400$$
$$\sum X_i Y_i = 61{,}800$$

Let us now use (6.22) to evaluate $(\mathbf{X'X})^{-1}$. We have:

$$n \sum (X_i - \bar{X})^2 = n\left[\sum X_i^2 - \frac{(\sum X_i)^2}{n}\right] = 10\left[28{,}400 - \frac{(500)^2}{10}\right] = 34{,}000$$

Therefore:

$$(\mathbf{X'X})^{-1} = \begin{bmatrix} \dfrac{\sum X_i^2}{n \sum (X_i - \bar{X})^2} & \dfrac{-\sum X_i}{n \sum (X_i - \bar{X})^2} \\ \dfrac{-\sum X_i}{n \sum (X_i - \bar{X})^2} & \dfrac{n}{n \sum (X_i - \bar{X})^2} \end{bmatrix} = \begin{bmatrix} \dfrac{28{,}400}{34{,}000} & \dfrac{-500}{34{,}000} \\ \dfrac{-500}{34{,}000} & \dfrac{10}{34{,}000} \end{bmatrix}$$

$$= \begin{bmatrix} .83529412 & -.01470588 \\ -.01470588 & .00029412 \end{bmatrix}$$

We also wish to make use of (6.15) to evaluate $\mathbf{X'Y}$:

$$\mathbf{X'Y} = \begin{bmatrix} \sum Y_i \\ \sum X_i Y_i \end{bmatrix} = \begin{bmatrix} 1{,}100 \\ 61{,}800 \end{bmatrix}$$

Hence, by (6.54):

$$\mathbf{b} = \begin{bmatrix} b_0 \\ b_1 \end{bmatrix} = (\mathbf{X'X})^{-1}\mathbf{X'Y} = \begin{bmatrix} .83529412 & -.01470588 \\ -.01470588 & .00029412 \end{bmatrix} \begin{bmatrix} 1{,}100 \\ 61{,}800 \end{bmatrix} = \begin{bmatrix} 10.0 \\ 2.0 \end{bmatrix}$$

or $b_0 = 10.0$ and $b_1 = 2.0$. This agrees with the results in Chapter 2. Any difference would have been due to rounding errors.

To reduce the effect of rounding errors when obtaining the vector $\mathbf{b}$ by hand calculations, it is often desirable to move the constant in the denominator of the elements of $(\mathbf{X'X})^{-1}$ outside the matrix, and do the division as the last step. For our example, this would lead to:

$$(\mathbf{X'X})^{-1} = \frac{1}{n \sum (X_i - \bar{X})^2} \begin{bmatrix} \sum X_i^2 & -\sum X_i \\ -\sum X_i & n \end{bmatrix} = \frac{1}{34{,}000} \begin{bmatrix} 28{,}400 & -500 \\ -500 & 10 \end{bmatrix}$$

$$\mathbf{b} = \frac{1}{34{,}000} \begin{bmatrix} 28{,}400 & -500 \\ -500 & 10 \end{bmatrix} \begin{bmatrix} 1{,}100 \\ 61{,}800 \end{bmatrix} = \frac{1}{34{,}000} \begin{bmatrix} 340{,}000 \\ 68{,}000 \end{bmatrix} = \begin{bmatrix} 10.0 \\ 2.0 \end{bmatrix}$$

In this instance, the two methods of calculation lead to identical results. Often, however, postponing division by $n \sum (X_i - \bar{X})^2$ until the end yields more accurate results.

Comments

1. To derive the normal equations by the method of least squares, we minimize the quantity:

$$Q = \sum [Y_i - (\beta_0 + \beta_1 X_i)]^2 = \sum \varepsilon_i^2$$

In matrix notation:

$$Q = (\mathbf{Y} - \mathbf{X\boldsymbol{\beta}})'(\mathbf{Y} - \mathbf{X\boldsymbol{\beta}}) \tag{6.55}$$

Expanding out, we obtain:

$$Q = \mathbf{Y'Y} - \boldsymbol{\beta}'\mathbf{X'Y} - \mathbf{Y'X}\boldsymbol{\beta} + \boldsymbol{\beta}'\mathbf{X'X}\boldsymbol{\beta}$$

since $(\mathbf{X}\boldsymbol{\beta})' = \boldsymbol{\beta}'\mathbf{X}'$ by (6.31). Note now that $\mathbf{Y'X}\boldsymbol{\beta}$ is 1×1, hence is equal to its transpose $\boldsymbol{\beta}'\mathbf{X'Y}$. Thus, we find:

$$Q = \mathbf{Y'Y} - 2\boldsymbol{\beta}'\mathbf{X'Y} + \boldsymbol{\beta}'\mathbf{X'X}\boldsymbol{\beta} \tag{6.56}$$

To find the value of $\boldsymbol{\beta}$ which minimizes Q, we differentiate with respect to β_0 and β_1. Let:

$$\frac{\partial}{\partial \boldsymbol{\beta}}(Q) = \begin{bmatrix} \dfrac{\partial Q}{\partial \beta_0} \\ \dfrac{\partial Q}{\partial \beta_1} \end{bmatrix} \tag{6.57}$$

Then it follows that:

$$\frac{\partial}{\partial \boldsymbol{\beta}}(Q) = -2\mathbf{X}'\mathbf{Y} + 2\mathbf{X}'\mathbf{X}\boldsymbol{\beta} \tag{6.58}$$

Equating to zero and substituting **b** for $\boldsymbol{\beta}$ gives the matrix form of the least squares normal equations:

$$\mathbf{X}'\mathbf{X}\mathbf{b} = \mathbf{X}'\mathbf{Y}$$

2. A comparison of the normal equations and **X′X** shows that whenever the columns of **X′X** are linearly dependent, the normal equations will be linearly dependent also. No unique solutions can be obtained for b_0 and b_1 in that case. Fortunately, in most regression applications, the columns of **X′X** are linearly independent, leading to unique solutions for b_0 and b_1.

6.11 ANALYSIS OF VARIANCE RESULTS

Let the vector of the fitted values $\hat{Y}_i$ be denoted $\hat{\mathbf{Y}}$:

$$\hat{\mathbf{Y}} = \begin{bmatrix} \hat{Y}_1 \\ \hat{Y}_2 \\ \cdot \\ \cdot \\ \cdot \\ \hat{Y}_n \end{bmatrix} \tag{6.59}$$

and the vector of the residuals $e_i = Y_i - \hat{Y}_i$ be denoted **e**:

$$\mathbf{e} = \begin{bmatrix} e_1 \\ e_2 \\ \cdot \\ \cdot \\ \cdot \\ e_n \end{bmatrix} \tag{6.60}$$

In matrix notation, we then have:

$$\hat{\mathbf{Y}} = \mathbf{X}\mathbf{b} \tag{6.61}$$

because:

$$\begin{bmatrix} \hat{Y}_1 \\ \hat{Y}_2 \\ \cdot \\ \cdot \\ \cdot \\ \hat{Y}_n \end{bmatrix} = \begin{bmatrix} 1 & X_1 \\ 1 & X_2 \\ \cdot & \cdot \\ \cdot & \cdot \\ \cdot & \cdot \\ 1 & X_n \end{bmatrix} \begin{bmatrix} b_0 \\ b_1 \end{bmatrix} = \begin{bmatrix} b_0 + b_1 X_1 \\ b_0 + b_1 X_2 \\ \cdot \\ \cdot \\ \cdot \\ b_0 + b_1 X_n \end{bmatrix}$$

Similarly:

$$\mathbf{e} = \mathbf{Y} - \hat{\mathbf{Y}} \tag{6.62}$$

The sums of squares for the analysis of variance are as follows in matrix notation:

(6.63)
$$SSTO = \mathbf{Y'Y} - n\bar{Y}^2$$

(6.64)
$$SSR = \mathbf{b'X'Y} - n\bar{Y}^2$$

(6.65)
$$SSE = \mathbf{e'e} = \mathbf{Y'Y} - \mathbf{b'X'Y}$$

Example

Let us find SSE for the Westwood Company lot size example by matrix methods. We know from earlier results:

$$\mathbf{Y'Y} = \sum Y_i^2 = 134{,}660$$

We also know from earlier:

$$\mathbf{b} = \begin{bmatrix} 10.0 \\ 2.0 \end{bmatrix} \qquad \mathbf{X'Y} = \begin{bmatrix} 1{,}100 \\ 61{,}800 \end{bmatrix}$$

Hence:

$$\mathbf{b'X'Y} = [10.0 \quad 2.0]\begin{bmatrix} 1{,}100 \\ 61{,}800 \end{bmatrix} = 134{,}600$$

and:

$$SSE = \mathbf{Y'Y} - \mathbf{b'X'Y} = 134{,}660 - 134{,}600 = 60$$

which is the same result as that obtained in Chapter 2. Any difference would have been due to rounding errors.

Note

To illustrate the derivation of the sums of squares expressions in matrix notation, consider SSE:

$$SSE = \mathbf{e'e} = (\mathbf{Y} - \mathbf{Xb})'(\mathbf{Y} - \mathbf{Xb}) = \mathbf{Y'Y} - 2\mathbf{b'X'Y} + \mathbf{b'X'Xb}$$

In substituting for the right-most $\mathbf{b}$ we obtain, by (6.54):

$$\begin{aligned} SSE &= \mathbf{Y'Y} - 2\mathbf{b'X'Y} + \mathbf{b'X'X(X'X)^{-1}X'Y} \\ &= \mathbf{Y'Y} - 2\mathbf{b'X'Y} + \mathbf{b'IX'Y} \end{aligned}$$

In dropping $\mathbf{I}$ and subtracting, we obtain:

$$SSE = \mathbf{Y'Y} - \mathbf{b'X'Y}$$

6.12 INFERENCES IN REGRESSION ANALYSIS

As we saw in earlier chapters, all interval estimates are of the form: point estimator plus and minus a certain number of estimated standard deviations of the point estimator. Similarly, all tests require the point estimator and the estimated standard deviation of the point estimator or, in the case of analysis of variance tests, various sums of squares. Matrix algebra is of principal help in inference making when obtaining the estimated standard deviations and sums of squares. We have already given the matrix equivalents of the sums of squares for the analysis of variance. Hence, we focus here chiefly on the matrix expressions for the estimated standard deviations of point estimators of interest.

Regression Coefficients

The variance-covariance matrix of **b**:

$$\boldsymbol{\sigma}^2(\mathbf{b}) = \begin{bmatrix} \sigma^2(b_0) & \sigma(b_0, b_1) \\ \sigma(b_1, b_0) & \sigma^2(b_1) \end{bmatrix} \tag{6.66}$$

is:

$$\boldsymbol{\sigma}^2(\mathbf{b}) = \sigma^2(\mathbf{X}'\mathbf{X})^{-1} \tag{6.67}$$

or, using (6.23):

$$\boldsymbol{\sigma}^2(\mathbf{b}) = \begin{bmatrix} \dfrac{\sigma^2 \sum X_i^2}{n \sum (X_i - \bar{X})^2} & \dfrac{-\bar{X}\sigma^2}{\sum (X_i - \bar{X})^2} \\ \dfrac{-\bar{X}\sigma^2}{\sum (X_i - \bar{X})^2} & \dfrac{\sigma^2}{\sum (X_i - \bar{X})^2} \end{bmatrix} \tag{6.67a}$$

When MSE is substituted for σ^2 in (6.67a) we have:

$$\mathbf{s}^2(\mathbf{b}) = MSE(\mathbf{X}'\mathbf{X})^{-1} = \begin{bmatrix} \dfrac{MSE \sum X_i^2}{n \sum (X_i - \bar{X})^2} & \dfrac{-\bar{X}\,MSE}{\sum (X_i - \bar{X})^2} \\ \dfrac{-\bar{X}\,MSE}{\sum (X_i - \bar{X})^2} & \dfrac{MSE}{\sum (X_i - \bar{X})^2} \end{bmatrix} \tag{6.68}$$

where $\mathbf{s}^2(\mathbf{b})$ is the estimated variance-covariance matrix of **b**. In (6.67a), you will recognize the variances of b_0 (3.20b) and b_1 (3.3b) and the covariance of b_0 and b_1 (5.18). Likewise, the estimated variances in (6.68) are familiar from earlier chapters.

Joint Confidence Region for β_0 and β_1

The boundary for the joint confidence region for β_0 and β_1, given in (5.17), is expressed in matrix terms as follows:

$$\frac{(\mathbf{b} - \boldsymbol{\beta})'\mathbf{X}'\mathbf{X}(\mathbf{b} - \boldsymbol{\beta})}{2MSE} = F(1 - \alpha; 2, n - 2) \tag{6.69}$$

Mean Response

To estimate the mean response at X_h, let us define the vector:

$$\mathbf{X}_h = \begin{bmatrix} 1 \\ X_h \end{bmatrix} \quad \text{or} \quad \mathbf{X}'_h = [1 \quad X_h] \tag{6.70}$$

The fitted value in matrix notation then is:

$$\hat{Y}_h = \mathbf{X}'_h \mathbf{b} \tag{6.71}$$

since:

$$\mathbf{X}'_h \mathbf{b} = [1 \quad X_h]\begin{bmatrix} b_0 \\ b_1 \end{bmatrix} = [b_0 + b_1 X_h] = [\hat{Y}_h] = \hat{Y}_h$$

Note that $\mathbf{X}'_h \mathbf{b}$ is a 1×1 matrix, hence we can write the final result as a scalar.

The variance of $\hat{Y}_h$, given earlier in (3.28b), is in matrix notation:

$$\sigma^2(\hat{Y}_h) = \sigma^2 \mathbf{X}'_h (\mathbf{X}'\mathbf{X})^{-1} \mathbf{X}_h \tag{6.72}$$

and the estimated variance of $\hat{Y}_h$, given earlier in (3.30), is in matrix notation:

$$s^2(\hat{Y}_h) = MSE(\mathbf{X}'_h (\mathbf{X}'\mathbf{X})^{-1} \mathbf{X}_h) \tag{6.73}$$

Prediction of New Observation

The estimated variance $s^2(Y_{h(\text{new})})$, given earlier in (3.35a), is in matrix notation:

$$s^2(Y_{h(\text{new})}) = MSE(1 + \mathbf{X}'_h (\mathbf{X}'\mathbf{X})^{-1} \mathbf{X}_h) \tag{6.74}$$

Examples

1. We wish to find $s^2(b_0)$ and $s^2(b_1)$ for the Westwood Company lot size example by matrix methods. We found earlier that $MSE = 7.5$ and:

$$(\mathbf{X}'\mathbf{X})^{-1} = \begin{bmatrix} .83529412 & -.01470588 \\ -.01470588 & .00029412 \end{bmatrix}$$

Hence by (6.68):

$$\mathbf{s}^2(\mathbf{b}) = MSE(\mathbf{X}'\mathbf{X})^{-1} = 7.5\begin{bmatrix} .83529412 & -.01470588 \\ -.01470588 & .00029412 \end{bmatrix}$$

$$= \begin{bmatrix} 6.26471 & -.110294 \\ -.110294 & .002206 \end{bmatrix}$$

Thus, $s^2(b_0) = 6.26471$ and $s^2(b_1) = .002206$. These are the same as the results obtained in Chapter 3.

Note how simple it is to find the estimated variances of the regression coefficients as soon as $(\mathbf{X}'\mathbf{X})^{-1}$ has been obtained. This inverse is needed in the first place to find the regression coefficients, so that practically no extra work is required to obtain the estimated variances of the regression coefficients.

2. We wish to find $s^2(\hat{Y}_h)$ for the Westwood Company example when $X_h = 55$. We define:

$$\mathbf{X}_h' = [1 \quad 55]$$

and obtain by (6.73):

$$s^2(\hat{Y}_{55}) = MSE(\mathbf{X}_h'(\mathbf{X}'\mathbf{X})^{-1}\mathbf{X}_h)$$

$$= 7.5[1 \quad 55]\begin{bmatrix} .83529412 & -.01470588 \\ -.01470588 & .00029412 \end{bmatrix}\begin{bmatrix} 1 \\ 55 \end{bmatrix} = .80520$$

This is the same result as that obtained in Chapter 3, except for a minor difference due to rounding.

Comments

1. One of the major advantages of the matrix approach to regression is that programming a computer in terms of matrix operations simplifies the regression calculations greatly. Thus, the steps to find the regression coefficients by means of (6.54) are:

a) Input the **Y** vector and the **X** matrix.
b) Obtain the transpose of **X**.
c) Premultiply **Y** by **X′**.
d) Premultiply **X** by **X′**.
e) Obtain the inverse of **X′X**.
f) Premultiply **X′Y** by $(\mathbf{X}'\mathbf{X})^{-1}$.
g) Output the **b** vector.

Another major advantage, not fully evident until we discuss multiple regression analysis, is that highly complex formulas can be expressed in compact fashion with matrix methods.

2. To illustrate a derivation in matrix terms, let us find the variance-covariance matrix of **b**. Recall that:

$$\mathbf{b} = (\mathbf{X}'\mathbf{X})^{-1}\mathbf{X}'\mathbf{Y} = \mathbf{A}\mathbf{Y}$$

where **A** is a constant matrix:

$$\mathbf{A} = (\mathbf{X}'\mathbf{X})^{-1}\mathbf{X}'$$

Hence, by (6.44), we have:

$$\boldsymbol{\sigma}^2(\mathbf{b}) = \mathbf{A}[\boldsymbol{\sigma}^2(\mathbf{Y})]\mathbf{A}'$$

Now $\boldsymbol{\sigma}^2(\mathbf{Y}) = \sigma^2\mathbf{I}$. Further, by (6.31) and (6.29) and the fact that $(\mathbf{X}'\mathbf{X})^{-1}$ is symmetric so that it remains unchanged when transposed, we obtain:

$$\mathbf{A}' = \mathbf{X}(\mathbf{X}'\mathbf{X})^{-1}$$

We find therefore:

$$\begin{aligned}\sigma^2(\mathbf{b}) &= (\mathbf{X}'\mathbf{X})^{-1}\mathbf{X}'\sigma^2\mathbf{I}\mathbf{X}(\mathbf{X}'\mathbf{X})^{-1}\\ &= \sigma^2(\mathbf{X}'\mathbf{X})^{-1}\mathbf{X}'\mathbf{X}(\mathbf{X}'\mathbf{X})^{-1}\\ &= \sigma^2(\mathbf{X}'\mathbf{X})^{-1}\mathbf{I}\\ &= \sigma^2(\mathbf{X}'\mathbf{X})^{-1}\end{aligned}$$

3. Since $\hat{Y}_h = \mathbf{X}_h'\mathbf{b}$, it follows at once from (6.44) that:

$$\sigma^2(\hat{Y}_h) = \mathbf{X}_h'[\sigma^2(\mathbf{b})]\mathbf{X}_h$$

Hence:

$$\sigma^2(\hat{Y}_h) = [1 \quad X_h]\begin{bmatrix}\sigma^2(b_0) & \sigma(b_0, b_1)\\ \sigma(b_1, b_0) & \sigma^2(b_1)\end{bmatrix}\begin{bmatrix}1\\ X_h\end{bmatrix}$$

or:

$$\sigma^2(\hat{Y}_h) = \sigma^2(b_0) + 2X_h\sigma(b_0, b_1) + X_h^2\sigma^2(b_1) \tag{6.75}$$

Using the results from (6.67a), we obtain:

$$\sigma^2(\hat{Y}_h) = \frac{\sigma^2 \sum X_i^2}{n\sum(X_i - \bar{X})^2} + \frac{2X_h(-\bar{X})\sigma^2}{\sum(X_i - \bar{X})^2} + \frac{X_h^2\sigma^2}{\sum(X_i - \bar{X})^2}$$

which reduces to the familiar expression:

$$\sigma^2(\hat{Y}_h) = \sigma^2\left[\frac{1}{n} + \frac{(X_h - \bar{X})^2}{\sum(X_i - \bar{X})^2}\right] \tag{6.76}$$

We thus see that the variance expression in (6.76) actually contains contributions from $\sigma^2(b_0)$, $\sigma^2(b_1)$, and $\sigma(b_0, b_1)$, which it must according to theorem (1.25b) since $\hat{Y}_h$ is a linear combination of b_0 and b_1:

$$\hat{Y}_h = b_0 + b_1 X_h$$

4. We do not show the results in matrix terms for other types of inferences, such as simultaneous prediction of several new observations on Y at different X_h levels, since these are based on results we have developed.

PROBLEMS

6.1. For the matrices below, obtain: (1) $\mathbf{A} + \mathbf{B}$; (2) $\mathbf{A} - \mathbf{B}$; (3) $\mathbf{AC}$; (4) $\mathbf{AB}'$; (5) $\mathbf{B}'\mathbf{A}$.

$$\mathbf{A} = \begin{bmatrix}1 & 4\\ 2 & 5\\ 3 & 6\end{bmatrix} \qquad \mathbf{B} = \begin{bmatrix}1 & 5\\ 1 & 4\\ 2 & 3\end{bmatrix} \qquad \mathbf{C} = \begin{bmatrix}3 & 6 & 1\\ 5 & 2 & 0\end{bmatrix}$$

State the dimension of each resulting matrix.

6.2. For the matrices below, obtain: (1) $\mathbf{A}+\mathbf{C}$; (2) $\mathbf{A}-\mathbf{C}$; (3) $\mathbf{B}'\mathbf{A}$; (4) $\mathbf{AC}'$; (5) $\mathbf{C}'\mathbf{A}$.

$$\mathbf{A}=\begin{bmatrix}2 & 1\\ 3 & 5\\ 1 & 7\\ 4 & 6\end{bmatrix} \quad \mathbf{B}=\begin{bmatrix}6\\ 9\\ 5\\ 1\end{bmatrix} \quad \mathbf{C}=\begin{bmatrix}3 & 8\\ 4 & 6\\ 5 & 1\\ 2 & 0\end{bmatrix}$$

State the dimension of each resulting matrix.

6.3. Show how the following expressions are written in terms of matrices: (1) $Y_i - \hat{Y}_i = e_i$; (2) $\sum X_i e_i = 0$. Assume $i = 1, \ldots, 4$.

6.4. The results shown below were obtained in a small-scale experiment to study the relation between °F of storage temperature (X) and number of weeks before flavor deterioration begins to occur (Y), for a certain food product.

i:	1	2	3	4	5
X_i:	+10	+5	0	−5	−10
Y_i:	7.9	9.0	10.2	11.0	11.9

Assume that the first-order regression model (3.1) is applicable. Using matrix methods, find: (1) $\mathbf{Y}'\mathbf{Y}$; (2) $\mathbf{X}'\mathbf{X}$; (3) $\mathbf{X}'\mathbf{Y}$.

6.5. The data below show, for a consumer finance company operating in six cities, the number of competing loan companies operating in the city (X) and the number per thousand of the company's loans made in that city that are currently delinquent (Y).

i:	1	2	3	4	5	6
X_i:	3	0	1	2	2	3
Y_i:	18	5	10	16	13	21

Assume that the first-order regression model (3.1) is applicable. Using matrix methods, find: (1) $\mathbf{Y}'\mathbf{Y}$; (2) $\mathbf{X}'\mathbf{X}$; (3) $\mathbf{X}'\mathbf{Y}$.

6.6. Let **A** be defined as follows:

$$\mathbf{A}=\begin{bmatrix}0 & 1 & 7\\ 0 & 3 & 1\\ 0 & 5 & 5\end{bmatrix}$$

a) Are the column vectors of **A** linearly dependent?
b) Restate definition (6.18) in terms of row vectors. Are the row vectors of **A** linearly dependent?
c) What is the rank of **A**?
d) Calculate the determinant of **A**. Could you have anticipated your result from your answers in parts (a) through (c)?

6.7. Find the inverse of each of the following matrices:

$$\mathbf{A}=\begin{bmatrix}2 & 5\\ 3 & 1\end{bmatrix} \quad \mathbf{B}=\begin{bmatrix}4 & 3 & 2\\ 6 & 5 & 8\\ 10 & 1 & 6\end{bmatrix}$$

Check in each case that the resulting matrix is indeed the inverse.

6.8. Refer to Problem 6.4. Find $(\mathbf{X}'\mathbf{X})^{-1}$.

6.9. Refer to Problem 6.5. Find $(\mathbf{X}'\mathbf{X})^{-1}$.

6.10. Consider the simultaneous equations:

$$4x + 7y = 20$$
$$2x + 3y = 14$$

a) Write these equations in matrix notation.
b) Using matrix methods, find the solutions for x and y.

6.11. Consider the simultaneous equations:

$$6c + 2d = 8$$
$$21c + 7d = 28$$

a) Write these equations in matrix notation.
b) Can you obtain explicit solutions for c and d by matrix methods? Can you obtain explicit solutions by ordinary algebraic methods? What generalization is suggested by your answers?

6.12. Refer to regression model (4.27). Set up the expectation vector for $\boldsymbol{\varepsilon}$. Assume $i = 1, \ldots, 4$.

6.13. Refer to regression model (4.40). Set up the variance-covariance matrix for the error terms. Assume $\sigma(\varepsilon_i, \varepsilon_j) = 0$ for $i \neq j$, and that $i = 1, \ldots, 4$.

6.14. Consider the estimated linear regression equation in the form of (2.15). Write this expression in matrix terms, for $i = 1, \ldots, 5$.

6.15. Consider model (5.28) for regression through the origin, and the estimator b_1 given in (5.32). Obtain (5.32) by utilizing (6.54) with $\mathbf{X}$ suitably defined.

6.16. Consider the least squares estimator $\mathbf{b}$ given in (6.54). Using matrix methods, show that $\mathbf{b}$ is an unbiased estimator.

6.17. Consider the following functions of the random variables Y_1, Y_2, Y_3:

$$W_1 = Y_1 + Y_2 + Y_3$$
$$W_2 = Y_1 + Y_2$$
$$W_3 = Y_1 - Y_2 - Y_3$$

a) State the above in matrix notation.
b) Find the expectation of the random vector $\mathbf{W}$.
c) Find the variance-covariance matrix of $\mathbf{W}$.

6.18. Refer to Problems 6.4 and 6.8.

a) Using matrix methods, obtain the following: (1) vector of estimated regression coefficients; (2) vector of residuals; (3) SSR; (4) SSE; (5) estimated variance-covariance matrix of $\mathbf{b}$; (6) point estimate of $E(Y_h)$ when $X_h = -7$; (7) estimated variance of $\hat{Y}_h$ when $X_h = -7$.
b) What simplifications arose from the spacing of the X levels in the experiment?

6.19. Refer to Problem 6.18. Using matrix methods, obtain the numerator of (6.69).

6.20. Refer to Problems 6.5 and 6.9.

a) Using matrix methods, obtain the following: (1) vector of estimated regression coefficients; (2) vector of residuals; (3) *SSR*; (4) *SSE*; (5) estimated variance-covariance matrix of **b**; (6) point estimate of $E(Y_h)$ when $X_h = 2$; (7) estimated variance of $\hat{Y}_h$ when $X_h = 2$.

b) From your estimated variance-covariance matrix in part (a5), obtain the following: (1) $s(b_0, b_1)$; (2) $s^2(b_0)$; (3) $s^2(b_1)$.

7

Multiple Regression

MULTIPLE REGRESSION analysis is one of the most widely used of all statistical tools. In this chapter, we first discuss some important issues concerning multiple regression models. Then we present the basic statistical results for multiple regression in matrix form. Since the matrix expressions for multiple regression are the same as for simple regression, we state the results without much discussion. We then give an example, illustrating a variety of inferences in multiple regression analysis. Finally, we take up a series of topics dealing with special problems that arise in analyzing multiple regression models.

7.1 MULTIPLE REGRESSION MODELS

Need for Several Independent Variables

When we first introduced regression analysis in Chapter 2, we spoke of regression models containing a number of independent variables. We mentioned a regression model where the dependent variable was direct operating cost for a branch office of a consumer finance chain, and four independent variables were considered, including average number of loans outstanding at the branch and total number of new loan applications processed by the branch. We also mentioned a tractor purchase study where the response variable was volume of tractor purchases in a sales territory, and the nine independent variables included number of farms in the territory and quantity of crop production in the territory. In both of these examples, one independent variable in the model would have provided an inadequate description since a number of key independent variables affect the response variable in important and distinctive ways. Furthermore, in situations of this type, one will frequently find that predictions of the response variable based on a model containing only a single independent variable are too imprecise to be useful. A more complex model, containing additional independent variables, typically is more helpful in providing sufficiently precise predictions of the response variable.

In both examples mentioned, the analysis is based on survey data because some or all of the independent variables are not susceptible to direct control. Multiple regression analysis is also highly useful in experimental situations where the experimenter can control the independent variables. An experimenter typically will wish to investigate a number of independent variables simultaneously because almost always more than one key independent variable influences the response. For example, in a study on productivity of work crews, the experimenter may wish to control both the size of the crew and the level of bonus pay.

First-Order Model with Two Independent Variables

The model:

$$Y_i = \beta_0 + \beta_1 X_{i1} + \beta_2 X_{i2} + \varepsilon_i \tag{7.1}$$

is called a first-order model with two independent variables. A first-order model, it will be recalled from Chapter 2, is linear in the parameters and linear in the independent variables. Y_i denotes as usual the response in the ith trial, and X_{i1} and X_{i2} are the values of the two independent variables in the ith trial. The parameters of the model are β_0, β_1, and β_2, and the error term is ε_i.

Assuming that $E(\varepsilon_i) = 0$, the regression function for model (7.1) is:

$$E(Y) = \beta_0 + \beta_1 X_1 + \beta_2 X_2 \tag{7.2}$$

Analogous to simple linear regression, where the regression function $E(Y) = \beta_0 + \beta_1 X$ is a line, the regression function (7.2) is a plane. Figure 7.1 contains a representation of a portion of the response plane:

$$E(Y) = 20.0 + .95X_1 - .50X_2 \tag{7.3}$$

Frequently the regression function in multiple regression is called a *regression surface* or a *response surface*. In Figure 7.1, the response surface is just a simple plane, but in other cases the response surface may be complex in nature. Note that a point on the response plane (7.3) gives the mean response $E(Y)$ at the given combination of levels of X_1 and X_2.

Let us now consider the meaning of the regression parameters in the multiple regression function (7.2). The parameter β_0 is the Y intercept of the regression plane. If the scope of the model includes $X_1 = 0$, $X_2 = 0$, β_0 gives the mean response at $X_1 = 0$, $X_2 = 0$. Otherwise, β_0 does not have any particular meaning as a separate term in the regression model.

FIGURE 7.1
Example of Response Surface—A Response Plane

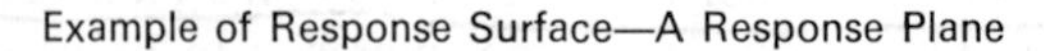

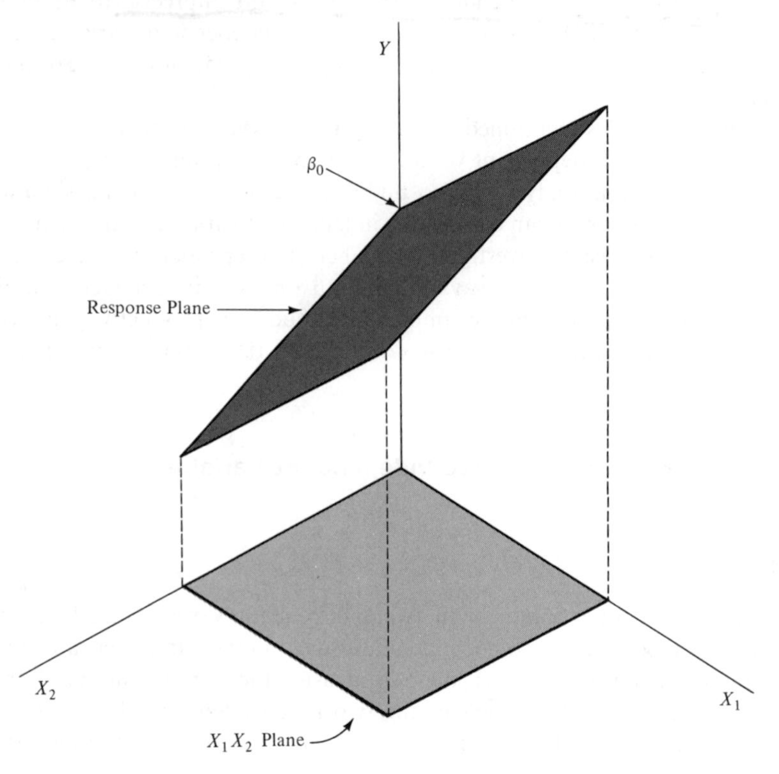

The parameter β_1 indicates the change in the mean response per unit increase in X_1 when X_2 is held constant. Likewise, β_2 indicates the change in the mean response per unit increase in X_2 when X_1 is held constant. To see this for our example, suppose X_2 is held at the level $X_2 = 20$. The regression function (7.3) now is:

$$E(Y) = 20.0 + .95X_1 - .50(20) = (20.0 - 10.0) + .95X_1 \tag{7.4}$$

The first term on the right, in parentheses, is a constant. Hence, (7.4) is the formula for a straight line. Changing the level at which X_2 is held constant affects the intercept but not the slope. Hence β_1 indicates the change in the mean response with a unit increase in X_1 when X_2 is constant, no matter at what level X_2 is held. More loosely speaking, we state that β_1 indicates the change in $E(Y)$ with a unit increase in X_1 when X_2 is held constant.

When the effect of X_1 on the mean response does not depend on the level of X_2, and correspondingly the effect of X_2 does not depend on the level of X_1, the two independent variables are said to have *additive effects* or *not to interact*. Thus, the first-order model (7.1) is designed for independent variables whose effects on the mean response are additive or do not interact.

The parameters β_1 and β_2 are frequently called *partial regression coefficients*, because they reflect the partial effect of one independent variable when the other independent variable is included in the model and is held constant.

Example. Suppose that the response surface in (7.3) pertains to the urban service stations of a major oil producer and shows the effect of variety and adequacy of services (X_1) and average time taken to reach car (X_2) on the ratio of actual gallonage of gasoline sold to potential gallonage (Y), where X_1 is expressed as an index with $100 =$ average, X_2 is in seconds, and Y is given as a percent. Increasing the index of adequacy of services by one point, while holding average time to reach car constant, leads to an increase of .95 percent points in the expected ratio of actual to potential gallonage. If the index of adequacy of services is held constant and the average time to reach car is increased by one second, the expected ratio of actual to potential gallonage decreases by .50 percent points.

Comments

1. A regression model for which the response surface is a plane can be used either in its own right when it is appropriate, or as an approximation to a more complex response surface. Many complex response surfaces can be approximated well by a plane for limited ranges of X_1 and X_2.

2. We can readily establish the meaning of β_1 and β_2 by calculus, taking partial derivatives of the response surface (7.2) with respect to X_1 and X_2 in turn:

$$\frac{\partial E(Y)}{\partial X_1} = \beta_1 \qquad \frac{\partial E(Y)}{\partial X_2} = \beta_2$$

The partial derivatives measure the rate of change in $E(Y)$ with respect to one independent variable, when the other is held constant.

First-Order Model with More Than Two Independent Variables

We consider now the case where there are $p - 1$ independent variables $X_1, \ldots, X_{p-1}$. The model:

(7.5) $$Y_i = \beta_0 + \beta_1 X_{i1} + \beta_2 X_{i2} + \cdots + \beta_{p-1} X_{i,p-1} + \varepsilon_i$$

is called a first-order model with $p - 1$ independent variables. It can also be written:

(7.5a) $$Y_i = \beta_0 + \sum_{k=1}^{p-1} \beta_k X_{ik} + \varepsilon_i$$

or, if we let $X_{i0} \equiv 1$, it can be written as:

(7.5b) $$Y_i = \sum_{k=0}^{p-1} \beta_k X_{ik} + \varepsilon_i \qquad \text{where } X_{i0} \equiv 1$$

Assuming that $E(\varepsilon_i) = 0$, the response function for model (7.5) is:

$$E(Y) = \beta_0 + \beta_1 X_{i1} + \beta_2 X_{i2} + \cdots + \beta_{p-1} X_{i,p-1} \tag{7.6}$$

This response function is a *hyperplane*, which is a plane in more than two dimensions. It is no longer possible to picture this response surface, as we were able to do in Figure 7.1 for the case of two independent variables. Nevertheless, the meaning of the parameters is analogous to the two independent variables case. The parameter β_k indicates the change in the mean response $E(Y)$ with a unit increase in the independent variable X_k, when all other independent variables X_1, X_2, etc. included in the model are held constant. Note again that the effect of any independent variable on the mean response is the same for model (7.5), no matter what are the levels at which the other independent variables are held. Hence, the first-order model (7.5) is designed for independent variables whose effects on the mean response are additive and therefore do not interact.

Note

If $p - 1 = 1$, model (7.5) reduces to:

$$Y_i = \beta_0 + \beta_1 X_{i1} + \varepsilon_i$$

which is the simple linear regression model considered in earlier chapters.

General Linear Regression Model

The general linear regression model, with normal error terms, is:

$$Y_i = \beta_0 + \beta_1 X_{i1} + \beta_2 X_{i2} + \cdots + \beta_{p-1} X_{i,p-1} + \varepsilon_i \tag{7.7}$$

where:

$\beta_0, \beta_1, \ldots, \beta_{p-1}$ are parameters
$X_{i1}, \ldots, X_{i,p-1}$ are known constants
ε_i are independent $N(0, \sigma^2)$
$i = 1, \ldots, n$

If we let $X_{i0} \equiv 1$, model (7.7) can be written as follows:

$$Y_i = \beta_0 X_{i0} + \beta_1 X_{i1} + \beta_2 X_{i2} + \cdots + \beta_{p-1} X_{i,p-1} + \varepsilon_i \qquad \text{where } X_{i0} \equiv 1 \tag{7.7a}$$

or:

$$Y_i = \sum_{k=0}^{p-1} \beta_k X_{ik} + \varepsilon_i \qquad \text{where } X_{i0} = 1 \tag{7.7b}$$

The response function for model (7.7) is, since $E(\varepsilon_i) = 0$:

$$E(Y) = \beta_0 + \beta_1 X_1 + \beta_2 X_2 + \cdots + \beta_{p-1} X_{p-1} \tag{7.8}$$

Thus, the general linear regression model implies that the observations Y_i are independent normal variables, with mean $E(Y_i)$ as given by (7.8) and with constant variance σ^2.

This general linear model encompasses a vast variety of situations. We shall consider a few of these now:

1.) $p - 1$ ***Independent Variables.*** When there are $p - 1$ different independent variables, the general linear model (7.7) represents, as we have seen, a first-order model in which there are no interacting effects between the independent variables.

2.) ***Polynomial Regression.*** Consider the curvilinear regression model with one independent variable:

$$Y_i = \beta_0 + \beta_1 X_i + \beta_2 X_i^2 + \varepsilon_i \tag{7.9}$$

If we let $X_{i1} = X_i$ and $X_{i2} = X_i^2$, we can write (7.9) as follows:

$$Y_i = \beta_0 + \beta_1 X_{i1} + \beta_2 X_{i2} + \varepsilon_i$$

so that model (7.9) is a particular case of the general linear regression model. While (7.9) illustrates a curvilinear model where the response function is quadratic, models with higher degree polynomial response functions are also particular cases of the general linear regression model.

3.) ***Transformed Variables.*** Consider the model:

$$\log Y_i = \beta_0 + \beta_1 X_{i1} + \beta_2 X_{i2} + \beta_3 X_{i3} + \varepsilon_i \tag{7.10}$$

Here, the response surface is a highly complex one, yet model (7.10) can be treated as a general linear regression model. If we let $Y_i' = \log Y_i$, we can write model (7.10) as follows:

$$Y_i' = \beta_0 + \beta_1 X_{i1} + \beta_2 X_{i2} + \beta_3 X_{i3} + \varepsilon_i$$

which is in the form of the general linear regression model. The dependent variable just happens to be measured in logarithms of Y.

Models that are intrinsically linear, that is, models that can be linearized by a transformation, can be treated as general linear regression models after the transformation. Thus, the model:

$$Y_i = \frac{1}{\beta_0 + \beta_1 X_{i1} + \beta_2 X_{i2} + \varepsilon_i} \tag{7.11}$$

can be transformed to a general linear regression model by letting $Y_i' = 1/Y_i$. We then have:

$$Y_i' = \beta_0 + \beta_1 X_{i1} + \beta_2 X_{i2} + \varepsilon_i$$

Interaction Effects. Consider the model:

$$Y_i = \beta_0 + \beta_1 X_{i1} + \beta_2 X_{i2} + \beta_3 X_{i1} X_{i2} + \varepsilon_i \tag{7.12}$$

The meaning of β_1 and β_2 here is not the same as that given earlier because of the cross-product term $\beta_3 X_{i1} X_{i2}$. It can be shown that the change in the mean response with a unit increase in X_1, when X_2 is held constant, is:

$$\beta_1 + \beta_3 X_2 \tag{7.13}$$

Similarly, the change in the mean response with a unit change in X_2, when X_1 is held constant, is:

$$\beta_2 + \beta_3 X_1 \tag{7.14}$$

Note that both the effect of X_1, for given level of X_2, and the effect of X_2, for given level of X_1, in model (7.12) depend on the level of the other independent variable. Model (7.12) is designed for independent variables which *interact*. The cross-product term $\beta_3 X_{i1} X_{i2}$ is called an *interaction term*. While the mean response in model (7.12) is still a linear function of X_1, when X_2 is constant, now both the intercept and the slope change as the level at which X_2 is held constant is varied. The same holds when the mean response is regarded as a function of X_2, with X_1 constant.

Despite these complexities of model (7.12), it can still be regarded as a general linear regression model. Let $X_{i3} = X_{i1}X_{i2}$. We can then write (7.12) as follows:

$$Y_i = \beta_0 + \beta_1 X_{i1} + \beta_2 X_{i2} + \beta_3 X_{i3} + \varepsilon_i$$

which is in the form of the general linear regression model.

Note

To derive (7.13) and (7.14), we differentiate:

$$E(Y) = \beta_0 + \beta_1 X_1 + \beta_2 X_2 + \beta_3 X_1 X_2$$

with respect to X_1 and X_2 respectively:

$$\frac{\partial E(Y)}{\partial X_1} = \beta_1 + \beta_3 X_2 \qquad \frac{\partial E(Y)}{\partial X_2} = \beta_2 + \beta_3 X_1$$

Combination of Cases. A regression model may combine a number of the elements we have just noted and still can be treated as a general linear regression model. Consider a model with two independent variables, each in quadratic form, with an interaction term:

$$Y_i = \beta_0 + \beta_1 X_{i1} + \beta_2 X_{i1}^2 + \beta_3 X_{i2} + \beta_4 X_{i2}^2 + \beta_5 X_{i1} X_{i2} + \varepsilon_i \tag{7.15}$$

Let us define:

$$Z_{i1} = X_{i1} \qquad Z_{i2} = X_{i1}^2 \qquad Z_{i3} = X_{i2} \qquad Z_{i4} = X_{i2}^2 \qquad Z_{i5} = X_{i1}X_{i2}$$

We can then write model (7.15) as follows:

$$Y_i = \beta_0 + \beta_1 Z_{i1} + \beta_2 Z_{i2} + \beta_3 Z_{i3} + \beta_4 Z_{i4} + \beta_5 Z_{i5} + \varepsilon_i$$

which is in the form of the general linear regression model.

Comments

1. It should be clear from the various examples that the general linear regression model (7.7) is not restricted to linear response surfaces. The term "linear" refers to the fact that (7.7) is linear in the parameters, not to the shape of the response surface.

2. Figure 7.2 illustrates some complex response surfaces that may be encountered when there are two independent variables.

FIGURE 7.2

Additional Examples of Response Functions

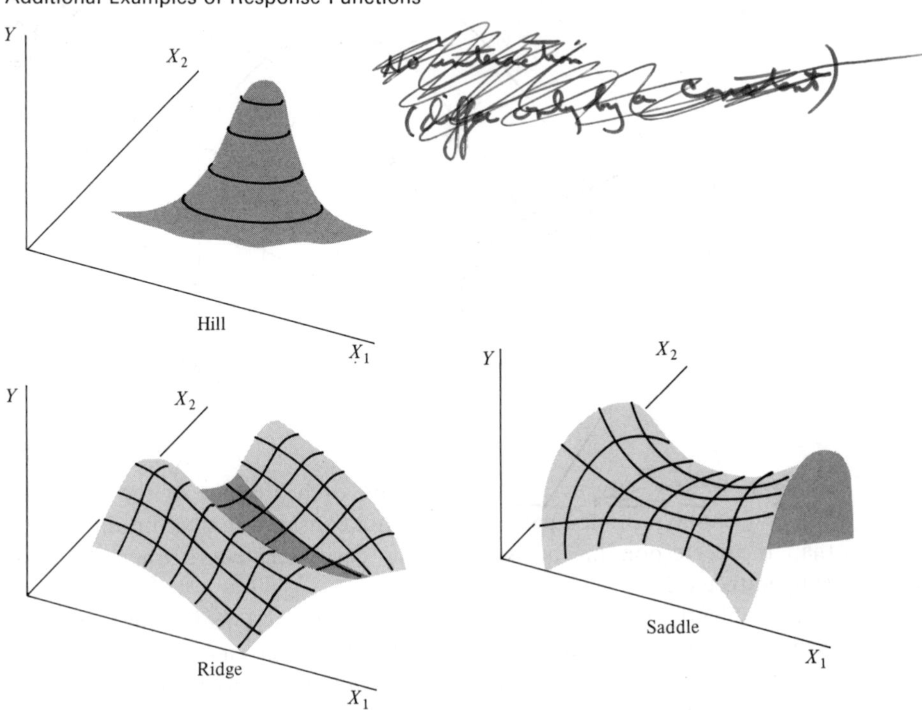

Source: Reprinted, with permission, from K. C. Peng, *The Design and Analysis of Scientific Experiments* (Reading, Mass.: Addison-Wesley Publishing Co., Inc., 1967), p. 150.

Interactions and Nature of Response Surface

The concept of interacting independent variables has been mentioned earlier. Here we shall discuss how the response surface differs when the independent variables do not interact and when they do interact. To simplify the discussion, we consider only the case of two independent variables.

Figure 7.3a contains a representation of a response surface in which the two independent variables (mean season temperature, amount of rainfall) do not interact on the dependent variable (corn yield). The absence of interactions can be seen by considering the corn yield curves as a function of rainfall, for any given mean season temperature. These curves all have the same

FIGURE 7.3

Response Surfaces for Additive and Interacting Independent Variables

(a) Independent Variables Do Not Interact
Yield of corn as function of season rainfall and mean temperature

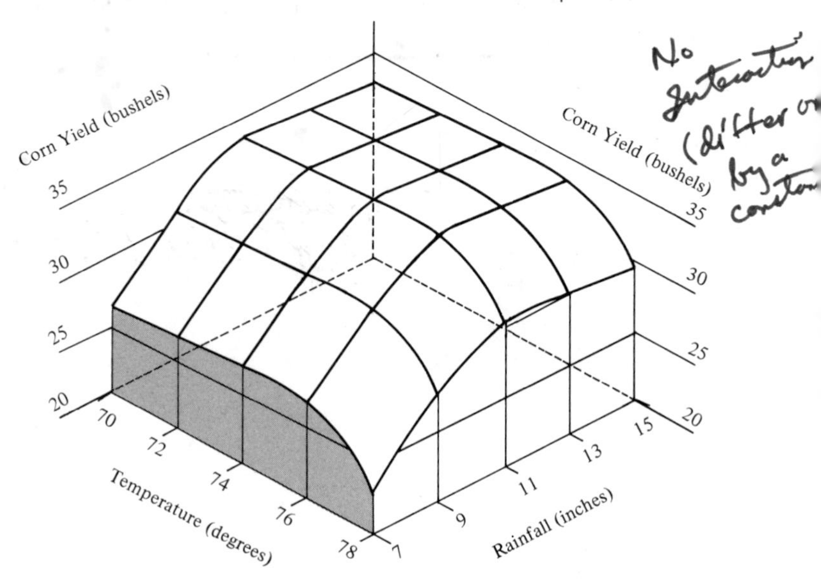

shape and differ only by a constant. Thus, each ordinate of the corn yield curve when the mean temperature is 70° is a constant number of units higher than the corresponding ordinate for the corn yield curve when the mean temperature is 78°.

Equivalently, one can note the absence of interactions by considering the corn yield curves as a function of temperature, for given amounts of rainfall. Again, these curves are the same in shape and differ only by a constant.

Absence of interactions therefore implies that the mean response $E(Y)$ can be expressed in the form:

$$E(Y) = f_1(X_1) + f_2(X_2) \tag{7.16}$$

where f_1 and f_2 can be any functions, not necessarily simple ones.

Figure 7.3b illustrates a case where the two independent variables (age, percent of normal weight) interact on the dependent variable (mortality ratio). Here, the shape of the mortality ratio curve as a function of percent of normal weight varies for different ages. For men 22 years old, both underweight and overweight persons have higher mortality rates than normal (normal = 100) for that age. On the other hand, for men 52 years old, the mortality rate is above normal for that age for overweight persons but not for underweight persons. Similarly, the mortality ratio curves as a function of age vary in shape for different weights.

FIGURE 7.3 (continued)

(b) Independent Variables Interact

Mortality ratio for men as function of age and percent of normal weight

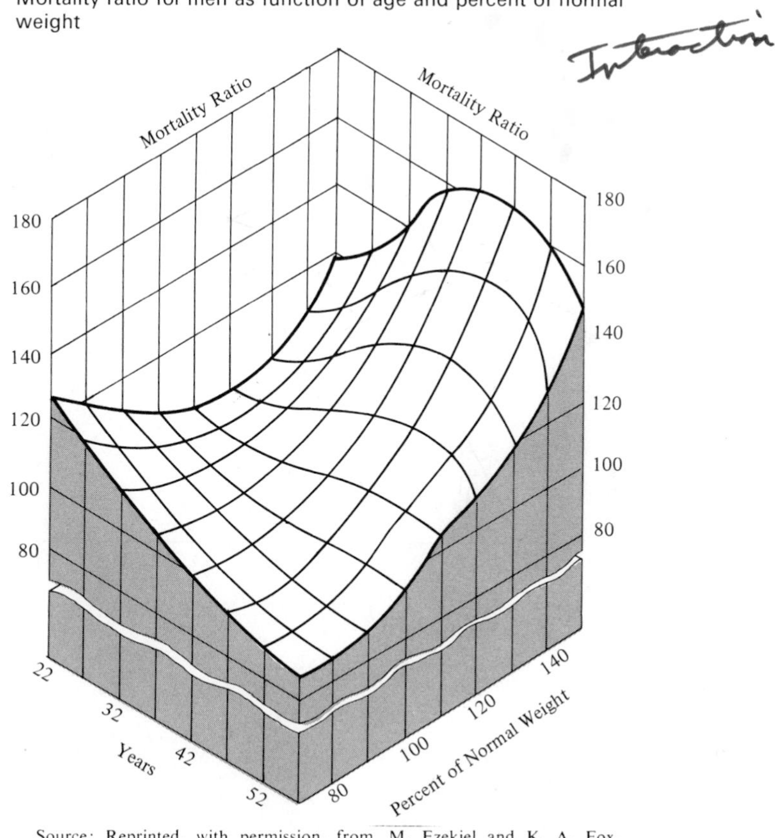

Source: Reprinted, with permission, from M. Ezekiel and K. A. Fox, *Methods of Correlation and Regression Analysis* (3d ed.; New York: John Wiley & Sons, Inc., 1959), pp. 349, 350.

One can represent response surfaces also by means of contour diagrams, which show the various combinations of the two independent variables which yield the same level of response, for a number of different response levels. Figure 7.4a shows a contour diagram for the response surface portrayed in Figure 7.1:

$$E(Y) = 20.0 + .95X_1 - .50X_2$$

Note that the independent variables do not interact in this response function. Figure 7.4b shows a contour diagram for the response function:

$$E(Y) = 5X_1 + 7X_2 + 3X_1X_2$$

where the two independent variables interact.

FIGURE 7.4

Response Contour Diagrams

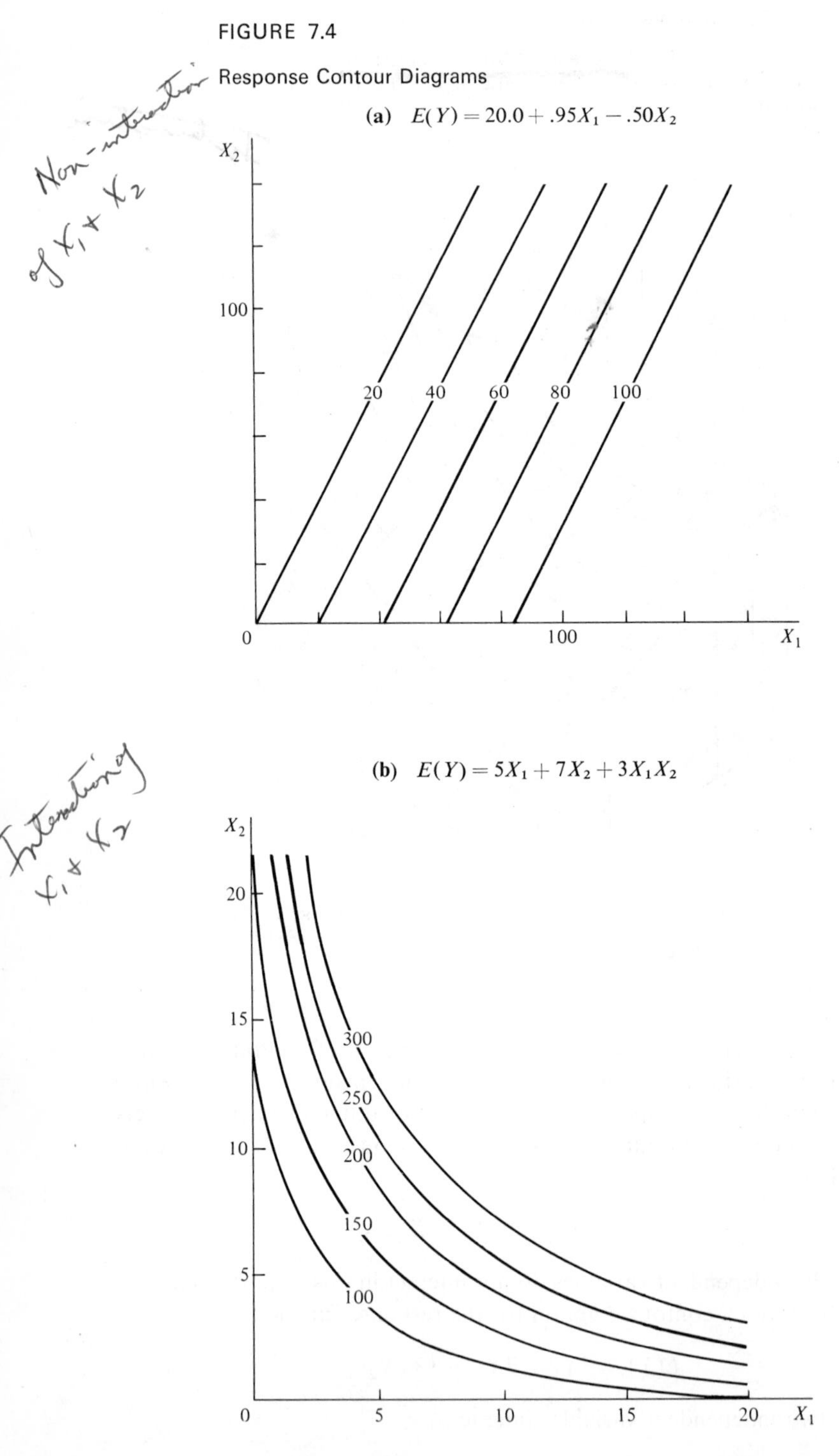

7.2 GENERAL LINEAR REGRESSION MODEL IN MATRIX TERMS

We shall now present the principal results for the general linear regression model (7.7) in matrix terms. This model, as we have noted, encompasses a wide variety of particular cases. The results to be presented are applicable to all of these.

It is a remarkable property of matrix algebra that the results for the general linear regression model (7.7) appear exactly the same in matrix notation as those for the simple linear regression model (6.51). Only the degrees of freedom and other constants related to the number of independent variables, and the dimensions of some matrices, will be different. Hence, we shall be able to present the results very concisely.

The matrix notation, to be sure, may hide enormous computational complexities. The inverse of a 10×10 matrix $\mathbf{A}$ requires tremendous amounts of computation, yet is simply represented as $\mathbf{A}^{-1}$. Our reason for emphasizing matrix algebra is that it indicates the essential conceptual steps in the solution. The actual computations will in all but the very simplest cases be done by electronic computer. Hence, it does not matter for us whether $(\mathbf{X}'\mathbf{X})^{-1}$ represents finding the inverse of a 2×2 or a 10×10 matrix. The important point is to know what the inverse of the matrix represents.

To express the general linear regression model (7.7):

$$Y_i = \beta_0 + \beta_1 X_{i1} + \beta_2 X_{i2} + \cdots + \beta_{p-1} X_{i,p-1} + \varepsilon_i$$

in matrix terms, we need to define the following matrices:

(7.17)

(7.17a)
$$\underset{n\times 1}{\mathbf{Y}} = \begin{bmatrix} Y_1 \\ Y_2 \\ \cdot \\ \cdot \\ \cdot \\ Y_n \end{bmatrix}$$

(7.17b)
$$\underset{n\times p}{\mathbf{X}} = \begin{bmatrix} 1 & X_{11} & X_{12} & \cdots & X_{1,p-1} \\ 1 & X_{21} & X_{22} & \cdots & X_{2,p-1} \\ \cdot & \cdot & \cdot & & \cdot \\ \cdot & \cdot & \cdot & & \cdot \\ \cdot & \cdot & \cdot & & \cdot \\ 1 & X_{n1} & X_{n2} & \cdots & X_{n,p-1} \end{bmatrix}$$

(7.17c)
$$\underset{p\times 1}{\boldsymbol{\beta}} = \begin{bmatrix} \beta_0 \\ \beta_1 \\ \cdot \\ \cdot \\ \cdot \\ \beta_{p-1} \end{bmatrix}$$

(7.17d)
$$\underset{n\times 1}{\boldsymbol{\varepsilon}} = \begin{bmatrix} \varepsilon_1 \\ \varepsilon_2 \\ \cdot \\ \cdot \\ \cdot \\ \varepsilon_n \end{bmatrix}$$

Note that the $\mathbf{Y}$ and $\boldsymbol{\varepsilon}$ vectors are the same as for simple regression. The $\boldsymbol{\beta}$ vector contains additional regression parameters, and the $\mathbf{X}$ matrix contains a column of 1's as well as a column of the n observations on each of the $p-1$ X variables in the regression model. The row subscript for each element X_{ik} in the $\mathbf{X}$ matrix identifies the trial, and the column subscript identifies the X variable.

In matrix terms, the general linear regression model (7.7) is:

(7.18)
$$\underset{n\times 1}{\mathbf{Y}} = \underset{n\times p}{\mathbf{X}}\ \underset{p\times 1}{\boldsymbol{\beta}} + \underset{n\times 1}{\boldsymbol{\varepsilon}}$$

where:

$\mathbf{Y}$ is a vector of observations
$\boldsymbol{\beta}$ is a vector of parameters
$\mathbf{X}$ is a matrix of constants
$\boldsymbol{\varepsilon}$ is a vector of independent normal random variables with expectation $\mathbf{E}(\boldsymbol{\varepsilon}) = \mathbf{0}$ and variance-covariance matrix $\boldsymbol{\sigma}^2(\boldsymbol{\varepsilon}) = \sigma^2\mathbf{I}$

Consequently, the random vector $\mathbf{Y}$ has expectation:

(7.18a)
$$\mathbf{E}(\mathbf{Y}) = \mathbf{X}\boldsymbol{\beta}$$

and the variance-covariance matrix of $\mathbf{Y}$ is:

(7.18b)
$$\boldsymbol{\sigma}^2(\mathbf{Y}) = \sigma^2\mathbf{I}$$

7.3 LEAST SQUARES ESTIMATORS

Let us denote the vector of estimated regression coefficients $b_0, b_1, \ldots, b_{p-1}$ as $\mathbf{b}$:

(7.19)
$$\underset{p\times 1}{\mathbf{b}} = \begin{bmatrix} b_0 \\ b_1 \\ b_2 \\ \cdot \\ \cdot \\ \cdot \\ b_{p-1} \end{bmatrix}$$

The least squares normal equations for the general linear regression model (7.18) are:

(7.20)
$$\underset{p\times p}{(\mathbf{X}'\mathbf{X})}\ \underset{p\times 1}{\mathbf{b}} = \underset{p\times n}{\mathbf{X}'}\ \underset{n\times 1}{\mathbf{Y}}$$

and the least squares estimators are:

(7.21)
$$\underset{p\times 1}{\mathbf{b}} = \underset{p\times p}{(\mathbf{X}'\mathbf{X})^{-1}}\ \underset{p\times 1}{\mathbf{X}'\mathbf{Y}}$$

For model (7.18), these least squares estimators are also maximum likelihood estimators and have all the properties mentioned in Chapter 2: they are unbiased, minimum variance unbiased estimators, consistent, sufficient.

7.4 ANALYSIS OF VARIANCE RESULTS

Let the vector of the fitted values $\hat{Y}_i$ be denoted $\hat{\mathbf{Y}}$ and the vector of the residual terms $e_i = Y_i - \hat{Y}_i$ be denoted $\mathbf{e}$:

(7.22) (7.22a) $$\underset{n \times 1}{\hat{\mathbf{Y}}} = \begin{bmatrix} \hat{Y}_1 \\ \hat{Y}_2 \\ \cdot \\ \cdot \\ \cdot \\ \hat{Y}_n \end{bmatrix}$$ (7.22b) $$\underset{n \times 1}{\mathbf{e}} = \begin{bmatrix} e_1 \\ e_2 \\ \cdot \\ \cdot \\ \cdot \\ e_n \end{bmatrix}$$

The fitted values are represented by:

(7.23) $$\hat{\mathbf{Y}} = \mathbf{Xb}$$

and the residual terms by:

(7.24) $$\mathbf{e} = \mathbf{Y} - \hat{\mathbf{Y}}$$

Sums of Squares and Mean Squares

The sums of squares for the analysis of variance are:

(7.25) $$SSTO = \mathbf{Y'Y} - n\bar{Y}^2$$

(7.26) $$SSR = \mathbf{b'X'Y} - n\bar{Y}^2$$

(7.27) $$SSE = \mathbf{e'e} = \mathbf{Y'Y} - \mathbf{b'X'Y}$$

SSTO, as usual, has $n - 1$ degrees of freedom associated with it. *SSE* has $n - p$ degrees of freedom associated with it since p parameters need to be estimated in the regression function for model (7.18). Finally, *SSR* has $p - 1$ degrees of freedom associated with it, representing the number of X variables $X_1, \ldots, X_{p-1}$.

Table 7.1 shows these analysis of variance results, as well as the mean squares *MSR* and *MSE*:

(7.28) $$MSR = \frac{SSR}{p - 1}$$

(7.29) $$MSE = \frac{SSE}{n - p}$$

The expectation of *MSE* is σ^2, as for simple regression. The expectation of *MSR* is σ^2 plus a quantity which is positive if any of the β_k $(k = 1, \ldots, p - 1)$ coefficients is not zero. For instance, when $p - 1 = 2$, we have:

$$E(MSR) = \sigma^2 + [\beta_1^2 \sum (X_{i1} - \bar{X}_1)^2 + \beta_2^2 \sum (X_{i2} - \bar{X}_2)^2 + 2\beta_1\beta_2 \sum (X_{i1} - \bar{X}_1)(X_{i2} - \bar{X}_2)]/2$$

Thus, if both β_1 and β_2 equal zero, $E(MSR) = \sigma^2$. Otherwise $E(MSR) > \sigma^2$.

TABLE 7.1

ANOVA Table for General Linear Regression Model (7.18)

Source of Variation	*SS*	*df*	*MS*
Regression	$SSR = \mathbf{b}'\mathbf{X}'\mathbf{Y} - n\bar{Y}^2$	$p-1$	$MSR = \frac{SSR}{p-1}$
Error	$SSE = \mathbf{Y}'\mathbf{Y} - \mathbf{b}'\mathbf{X}'\mathbf{Y}$	$n-p$	$MSE = \frac{SSE}{n-p}$
Total	$SSTO = \mathbf{Y}'\mathbf{Y} - n\bar{Y}^2$	$n-1$	

F Test for Regression Relation

To test whether there is a relation between the dependent variable Y and the set of X variables $X_1, \ldots, X_{p-1}$, that is, to choose between the alternatives:

(7.30a)
$$\begin{aligned} &C_1: \beta_1 = \beta_2 \cdots = \beta_{p-1} = 0 \\ &C_2: \text{not all } \beta_k \ (k = 1, \ldots, p-1) \text{ equal } 0 \end{aligned}$$

we use the test statistic:

(7.30b)
$$F^* = \frac{MSR}{MSE}$$

The decision rule to control the Type I error at α is:

(7.30c)
$$\begin{aligned} &\text{If } F^* \leq F(1-\alpha; p-1, n-p), \text{ conclude } C_1 \\ &\text{If } F^* > F(1-\alpha; p-1, n-p), \text{ conclude } C_2 \end{aligned}$$

The existence of a regression relation, by itself, does not of course assure that useful predictions can be made by using it.

Coefficient of Multiple Determination

The coefficient of multiple determination, denoted by R^2, is defined as follows:

(7.31)
$$R^2 = \frac{SSR}{SSTO} = 1 - \frac{SSE}{SSTO}$$

It measures the proportionate reduction of total variation in Y associated with the use of the set of X variables $X_1, \ldots, X_{p-1}$. The coefficient of multiple determination R^2 reduces to the coefficient of simple determination r^2 in (3.67) when $p - 1 = 1$, that is, when one independent variable is in model (7.18). As for r^2, we have:

(7.32)
$$0 \leq R^2 \leq 1$$

R^2 assumes the value 0 when all $b_k = 0$ $(k = 1, \ldots, p-1)$. R^2 takes on the value 1 when all observations fall directly on the fitted response surface, that is, when $Y_i = \hat{Y}_i$ for all i.

Comments

1. It can be shown that the coefficient of multiple determination R^2 can be viewed as a coefficient of simple determination r^2 between the responses Y_i and the fitted values $\hat{Y}_i$.

2. A large R^2 does not necessarily imply that the fitted model is a useful one. For instance, observations may have been taken at only a few levels of the independent variables. Despite a high R^2 in this case, the fitted model may not be useful because most predictions would require extrapolations outside the region of observations. Again, even though R^2 is large, MSE may still be too large for inferences to be useful in a case where high precision is required.

3. Adding more independent variables to the model can only increase R^2 and never reduce it, because SSE can never become larger with more independent variables and $SSTO$ is always the same for a given set of responses. Since R^2 often can be made large by including a large number of independent variables, it is sometimes suggested that a modified measure be used which recognizes the number of independent variables in the model. This *adjusted coefficient of multiple determination*, denoted R_a^2, is defined:

$$R_a^2 = 1 - \left(\frac{n-1}{n-p}\right)\frac{SSE}{SSTO} \tag{7.33}$$

This adjusted coefficient of multiple determination may actually become smaller when another independent variable is introduced into the model, because the decrease in SSE may be more than offset by the loss of a degree of freedom in the denominator $n-p$.

Coefficient of Multiple Correlation

The coefficient of multiple correlation R is the positive square root of R^2:

$$R = \sqrt{R^2} \tag{7.34}$$

It equals in absolute value the simple correlation coefficient r in (3.71) when $p-1=1$, that is, when there is one independent variable in model (7.18).

7.5 INFERENCES ABOUT REGRESSION PARAMETERS

The least squares estimators in $\mathbf{b}$ are unbiased:

$$E(\mathbf{b}) = \boldsymbol{\beta} \tag{7.35}$$

The variance-covariance matrix $\boldsymbol{\sigma}^2(\mathbf{b})$:

$$\boldsymbol{\sigma}^2(\mathbf{b}) = \begin{bmatrix} \sigma^2(b_0) & \sigma(b_0, b_1) & \cdots & \sigma(b_0, b_{p-1}) \\ \sigma(b_1, b_0) & \sigma^2(b_1) & \cdots & \sigma(b_1, b_{p-1}) \\ \vdots & \vdots & & \vdots \\ \sigma(b_{p-1}, b_0) & \sigma(b_{p-1}, b_1) & \cdots & \sigma^2(b_{p-1}) \end{bmatrix} \tag{7.36}$$

is given by:

$$\boldsymbol{\sigma}^2(\mathbf{b}) = \sigma^2(\mathbf{X}'\mathbf{X})^{-1} \tag{7.37}$$

The estimated variance-covariance matrix $\mathbf{s}^2(\mathbf{b})$:

$$\mathbf{s}^2(\mathbf{b}) = \begin{bmatrix} s^2(b_0) & s(b_0, b_1) & \cdots & s(b_0, b_{p-1}) \\ s(b_1, b_0) & s^2(b_1) & \cdots & s(b_1, b_{p-1}) \\ \vdots & \vdots & & \vdots \\ s(b_{p-1}, b_0) & s(b_{p-1}, b_1) & \cdots & s^2(b_{p-1}) \end{bmatrix} \tag{7.38}$$

is given by:

$$\mathbf{s}^2(\mathbf{b}) = MSE(\mathbf{X}'\mathbf{X})^{-1} \tag{7.39}$$

From $\mathbf{s}^2(\mathbf{b})$, one can obtain $s^2(b_0)$, $s^2(b_1)$ or whatever other variance is needed, or any needed covariances.

Interval Estimation of β_k

For the normal error model (7.18), we have:

$$\frac{b_k - \beta_k}{s(b_k)} = t(n-p) \qquad k = 0, 1, \ldots, p-1 \tag{7.40}$$

Hence, the confidence interval for β_k, with $1 - \alpha$ confidence coefficient, is:

$$b_k - t(1 - \alpha/2; n-p)s(b_k) \leq \beta_k \leq b_k + t(1 - \alpha/2; n-p)s(b_k) \tag{7.41}$$

Tests for β_k

Tests for β_k are set up in the usual fashion. To test:

$$\begin{aligned} C_1&: \beta_k = 0 \\ C_2&: \beta_k \neq 0 \end{aligned} \tag{7.42a}$$

we may use the test statistic:

$$t^* = \frac{b_k}{s(b_k)} \tag{7.42b}$$

and the decision rule:

$$\begin{aligned} &\text{If } |t^*| \leq t(1 - \alpha/2; n-p), \text{ conclude } C_1 \\ &\text{Otherwise conclude } C_2 \end{aligned} \tag{7.42c}$$

The power of the t test can be obtained as explained in Chapter 3, with the degrees of freedom modified to $n - p$.

Joint Inferences

The boundary of the joint confidence region for all p of the β_k regression parameters ($k = 0, 1, \ldots, p-1$), with confidence coefficient $1 - \alpha$, is:

$$\frac{(\mathbf{b} - \boldsymbol{\beta})'\mathbf{X}'\mathbf{X}(\mathbf{b} - \boldsymbol{\beta})}{pMSE} = F(1 - \alpha; p, n - p) \tag{7.43}$$

The region defined by this boundary is generally difficult to obtain and interpret.

The Bonferroni joint confidence intervals, on the other hand, are easy to obtain and interpret. If s parameters are to be estimated jointly (where $s \leq p$), the confidence intervals with family confidence coefficient $1 - \alpha$ are:

$$b_k - Bs(b_k) \leq \beta_k \leq b_k + Bs(b_k) \tag{7.44}$$

where:

$$B = t(1 - \alpha/2s; n - p) \tag{7.44a}$$

In Section 7.10 we discuss tests concerning a subset of the regression parameters.

7.6 INFERENCES ABOUT MEAN RESPONSE

Interval Estimation of $E(Y_h)$

For given values of $X_1, \ldots, X_{p-1}$, denoted $X_{h1}, \ldots, X_{h,p-1}$, the mean response is denoted $E(Y_h)$. To estimate this mean response, let us define the vector $\mathbf{X}_h$ as follows:

$$\mathbf{X}_h = \begin{bmatrix} 1 \\ X_{h1} \\ X_{h2} \\ \cdot \\ \cdot \\ \cdot \\ X_{h,p-1} \end{bmatrix} \tag{7.45}$$

The estimated mean response corresponding to $\mathbf{X}_h$ is denoted $\hat{Y}_h$:

$$\hat{Y}_h = \mathbf{X}_h'\mathbf{b} \tag{7.46}$$

This estimator is unbiased:

$$E(\hat{Y}_h) = E(Y_h) = \mathbf{X}_h'\boldsymbol{\beta} \tag{7.47}$$

and its variance is:

$$\sigma^2(\hat{Y}_h) = \sigma^2\mathbf{X}_h'(\mathbf{X}'\mathbf{X})^{-1}\mathbf{X}_h \tag{7.48}$$

The estimated variance $s^2(\hat{Y}_h)$ is given by:

(7.49) $$s^2(\hat{Y}_h) = MSE(\mathbf{X}_h'(\mathbf{X}'\mathbf{X})^{-1}\mathbf{X}_h)$$

The $1 - \alpha$ confidence interval for $E(Y_h)$ is:

(7.50) $$\hat{Y}_h - t(1 - \alpha/2; n - p)s(\hat{Y}_h) \leq E(Y_h) \leq \hat{Y}_h + t(1 - \alpha/2; n - p)s(\hat{Y}_h)$$

Confidence Region for Regression Surface

The $1 - \alpha$ confidence region for the entire regression surface is an extension of the Working-Hotelling confidence band for the regression line when there is one independent variable. Boundary points of the confidence region at $\mathbf{X}_h$ are obtained from:

(7.51) $$\hat{Y}_h - Ws(\hat{Y}_h) \leq \mathbf{X}_h'\boldsymbol{\beta} \leq \hat{Y}_h + Ws(\hat{Y}_h)$$

where:

(7.51a) $$W^2 = pF(1 - \alpha; p, n - p)$$

The confidence coefficient is $1 - \alpha$ that the region contains the entire regression surface over all combinations of real-numbered values of the X variables.

Simultaneous Confidence Intervals for Several Mean Responses

When it is desired to estimate a number of mean responses $E(Y_h)$ corresponding to different $\mathbf{X}_h$ vectors, one can employ two basic approaches:

1. Use the Working-Hotelling type confidence region bounds from (7.51) for the several $\mathbf{X}_h$ vectors of interest. Since these bounds cover the mean responses for all possible $\mathbf{X}_h$ vectors with confidence coefficient $1 - \alpha$, they will cover the mean responses for selected $\mathbf{X}_h$ vectors with confidence coefficient greater than $1 - \alpha$.
2. Use Bonferroni simultaneous confidence intervals. These are, when s statements are to be made with family confidence coefficient $1 - \alpha$:

(7.52) $$\hat{Y}_h - Bs(\hat{Y}_h) \leq E(Y_h) \leq \hat{Y}_h + Bs(\hat{Y}_h)$$

where:

(7.52a) $$B = t(1 - \alpha/2s; n - p)$$

For any particular application, one should compare W and B to see which procedure will lead to narrower confidence intervals. If the $\mathbf{X}_h$ levels are not specified in advance but are determined as the analysis proceeds, it is better to use the Working-Hotelling type intervals (7.51).

F Test for Lack of Fit

To test whether the response function:

(7.53) $$E(Y) = \beta_0 + \beta_1 X_1 + \cdots + \beta_{p-1} X_{p-1}$$

is an appropriate response surface for the data at hand requires repeat observations, as for simple regression analysis. Repeat observations in multiple regression are replicate observations on Y corresponding to levels of each of the X variables which are constant from trial to trial. Thus, with two independent variables repeat observations require that X_1 and X_2 each remain at given levels from trial to trial.

The procedures described in Chapter 4 for the F test for lack of fit are applicable to multiple regression. Once the ANOVA table, shown in Table 7.1, has been obtained, SSE is decomposed into pure error and lack of fit components. The pure error sum of squares $SSPE$ is obtained by first calculating the sum of squared deviations of the observations around the group mean, where a group has the same values for the X variables. Suppose there are c groups for which there are distinct sets of levels for the X variables, and let the mean for the jth group be denoted $\bar{Y}_j$. Then the sum of squares for the jth group is given by (4.7), and the pure error sum of squares is the sum of these sums of squares, as shown by (4.8). The lack of fit sum of squares $SSLF$ equals the difference $SSE - SSPE$, as indicated by (4.11).

The number of degrees of freedom associated with $SSPE$ is $n - c$, and the number of degrees of freedom associated with $SSLF$ is $(n - p) - (n - c) = c - p$.

The F test is conducted as described in Chapter 4, but with the degrees of freedom modified to those just stated.

7.7 PREDICTIONS OF NEW OBSERVATIONS

Prediction of New Observation $Y_{h(\text{new})}$

A prediction interval with $1 - \alpha$ confidence coefficient for a new observation $Y_{h(\text{new})}$ corresponding to $\mathbf{X}_h$, the specified values of the X variables, is:

$$\hat{Y}_h - t(1 - \alpha/2; n - p)s(Y_{h(\text{new})}) \leq Y_{h(\text{new})} \leq \hat{Y}_h + t(1 - \alpha/2; n - p)s(Y_{h(\text{new})}) \tag{7.54}$$

where:

$$s^2(Y_{h(\text{new})}) = MSE(1 + \mathbf{X}_h'(\mathbf{X}'\mathbf{X})^{-1}\mathbf{X}_h) \tag{7.54a}$$

Prediction of Mean of m New Observations at $\mathbf{X}_h$

When m new observations are to be selected at $\mathbf{X}_h$ and their mean $\bar{Y}_{h(\text{new})}$ is to be predicted, the $1 - \alpha$ prediction interval is:

$$\hat{Y}_h - t(1 - \alpha/2; n - p)s(\bar{Y}_{h(\text{new})}) \leq \bar{Y}_{h(\text{new})} \leq \hat{Y}_h + t(1 - \alpha/2; n - p)s(\bar{Y}_{h(\text{new})}) \tag{7.55}$$

where:

$$s^2(\bar{Y}_{h(\text{new})}) = MSE\left(\frac{1}{m} + \mathbf{X}_h'(\mathbf{X}'\mathbf{X})^{-1}\mathbf{X}_h\right) \tag{7.55a}$$

Predictions of s New Observations

Simultaneous prediction intervals for s new observations at s different levels of $\mathbf{X}_h$, with family confidence coefficient $1 - \alpha$, are given by:

$$\hat{Y}_h - Ss(Y_{h(\text{new})}) \le Y_{h(\text{new})} \le \hat{Y}_h + Ss(Y_{h(\text{new})}) \tag{7.56}$$

where:

$$S^2 = sF(1 - \alpha; s, n - p) \tag{7.56a}$$

and $s^2(Y_{h(\text{new})})$ is given by (7.54a).

Alternatively, the Bonferroni simultaneous prediction intervals can be used. They are, for s predictions with a $1 - \alpha$ family confidence coefficient:

$$\hat{Y}_h - Bs(Y_{h(\text{new})}) \le Y_{h(\text{new})} \le \hat{Y}_h + Bs(Y_{h(\text{new})}) \tag{7.57}$$

where:

$$B = t(1 - \alpha/2s; n - p) \tag{7.57a}$$

A comparison of S and B in advance of any particular use will indicate which procedure will lead to narrower prediction intervals.

7.8 AN EXAMPLE—MULTIPLE REGRESSION WITH TWO INDEPENDENT VARIABLES

In this section, we shall develop a multiple regression application with two independent variables. We shall illustrate a number of different types of inferences which might be made for this application but will not take up every possible type of inference.

Setting

The Zarthan Company sells a special skin cream through drugstores exclusively. It operates in 15 marketing districts and is interested in predicting district sales. Table 7.2 contains data on sales by district, as well as district data on target population and per capita income. Sales are to be treated as the dependent variable Y, and target population and per capita income as independent variables X_1 and X_2 respectively, in an exploration of the feasibility of predicting district sales from target population and per capita income. The first-order model:

$$Y_i = \beta_0 + \beta_1 X_{i1} + \beta_2 X_{i2} + \varepsilon_i \tag{7.58}$$

with normal error terms is expected to be appropriate.

TABLE 7.2

Basic Data for Zarthan Company Example

District i	*Sales (gross of jars; 1 gross = 12 dozen)* Y_i	*Target Population (thousands of persons)* X_{i1}	*Per Capita Income (dollars)* X_{i2}
1	162	274	2,450
2	120	180	3,254
3	223	375	3,802
4	131	205	2,838
5	67	86	2,347
6	169	265	3,782
7	81	98	3,008
8	192	330	2,450
9	116	195	2,137
10	55	53	2,560
11	252	430	4,020
12	232	372	4,427
13	144	236	2,660
14	103	157	2,088
15	212	370	2,605

Basic Calculations

The **Y** and **X** matrices for the Zarthan Company illustration are shown in Table 7.3. We shall require:

1.

$$\mathbf{X'X} = \begin{bmatrix} 1 & 1 & \cdots & 1 \\ 274 & 180 & \cdots & 370 \\ 2{,}450 & 3{,}254 & \cdots & 2{,}605 \end{bmatrix} \begin{bmatrix} 1 & 274 & 2{,}450 \\ 1 & 180 & 3{,}254 \\ \vdots & \vdots & \vdots \\ 1 & 370 & 2{,}605 \end{bmatrix}$$

which yields:

$$\mathbf{X'X} = \begin{bmatrix} 15 & 3{,}626 & 44{,}428 \\ 3{,}626 & 1{,}067{,}614 & 11{,}419{,}181 \\ 44{,}428 & 11{,}419{,}181 & 139{,}063{,}428 \end{bmatrix} \tag{7.59}$$

2.

$$\mathbf{X'Y} = \begin{bmatrix} 1 & 1 & \cdots & 1 \\ 274 & 180 & \cdots & 370 \\ 2{,}450 & 3{,}254 & \cdots & 2{,}605 \end{bmatrix} \begin{bmatrix} 162 \\ 120 \\ \vdots \\ 212 \end{bmatrix}$$

TABLE 7.3

X and **Y** Matrices for Zarthan Company Example

$$\mathbf{Y} = \begin{bmatrix} 162 \\ 120 \\ 223 \\ 131 \\ 67 \\ 169 \\ 81 \\ 192 \\ 116 \\ 55 \\ 252 \\ 232 \\ 144 \\ 103 \\ 212 \end{bmatrix} \qquad \mathbf{X} = \begin{bmatrix} 1 & 274 & 2{,}450 \\ 1 & 180 & 3{,}254 \\ 1 & 375 & 3{,}802 \\ 1 & 205 & 2{,}838 \\ 1 & 86 & 2{,}347 \\ 1 & 265 & 3{,}782 \\ 1 & 98 & 3{,}008 \\ 1 & 330 & 2{,}450 \\ 1 & 195 & 2{,}137 \\ 1 & 53 & 2{,}560 \\ 1 & 430 & 4{,}020 \\ 1 & 372 & 4{,}427 \\ 1 & 236 & 2{,}660 \\ 1 & 157 & 2{,}088 \\ 1 & 370 & 2{,}605 \end{bmatrix}$$

which yields:

$$\mathbf{X'Y} = \begin{bmatrix} 2{,}259 \\ 647{,}107 \\ 7{,}096{,}619 \end{bmatrix} \tag{7.60}$$

3.

$$(\mathbf{X'X})^{-1} = \begin{bmatrix} 15 & 3{,}626 & 44{,}428 \\ 3{,}626 & 1{,}067{,}614 & 11{,}419{,}181 \\ 44{,}428 & 11{,}419{,}181 & 139{,}063{,}428 \end{bmatrix}^{-1}$$

Using (6.21), we define:

$$\begin{array}{lll} a = 15 & b = 3{,}626 & c = 44{,}428 \\ d = 3{,}626 & e = 1{,}067{,}614 & f = 11{,}419{,}181 \\ g = 44{,}428 & h = 11{,}419{,}181 & k = 139{,}063{,}428 \end{array}$$

so that:

$$Z = 14{,}497{,}044{,}060{,}000$$
$$A = 1.246348416$$
$$B = .0002129664176$$

and so on. We obtain:

$$(\mathbf{X'X})^{-1} = \tag{7.61}$$

$$\begin{bmatrix} 1.2463484 & 2.1296642 \times 10^{-4} & -4.1567125 \times 10^{-4} \\ 2.1296642 \times 10^{-4} & 7.7329030 \times 10^{-6} & -7.0302518 \times 10^{-7} \\ -4.1567125 \times 10^{-4} & -7.0302518 \times 10^{-7} & 1.9771851 \times 10^{-7} \end{bmatrix}$$

For hand calculations, it is often preferable to take the determinant Z outside the matrix and postpone the division until the end of the computations:

$$(\mathbf{X}'\mathbf{X})^{-1} = \frac{1}{14.49704406 \times 10^{12}}$$

$$\times \begin{bmatrix} 1.8068368 \times 10^{13} & 3.0873835 \times 10^{9} & -6.0260045 \times 10^{9} \\ 3.0873835 \times 10^{9} & 1.1210424 \times 10^{8} & -1.0191787 \times 10^{7} \\ -6.0260045 \times 10^{9} & -1.0191787 \times 10^{7} & 2.866334 \times 10^{6} \end{bmatrix}$$

Algebraic Equivalents. Note that $\mathbf{X}'\mathbf{X}$ for the first-order model (7.58) with two independent variables is:

$$\mathbf{X}'\mathbf{X} = \begin{bmatrix} 1 & 1 & \cdots & 1 \\ X_{11} & X_{21} & \cdots & X_{n1} \\ X_{12} & X_{22} & \cdots & X_{n2} \end{bmatrix} \begin{bmatrix} 1 & X_{11} & X_{12} \\ 1 & X_{21} & X_{22} \\ \cdot & \cdot & \cdot \\ \cdot & \cdot & \cdot \\ \cdot & \cdot & \cdot \\ 1 & X_{n1} & X_{n2} \end{bmatrix}$$

or:

$$\mathbf{X}'\mathbf{X} = \begin{bmatrix} n & \sum X_{i1} & \sum X_{i2} \\ \sum X_{i1} & \sum X_{i1}^2 & \sum X_{i1}X_{i2} \\ \sum X_{i2} & \sum X_{i2}X_{i1} & \sum X_{i2}^2 \end{bmatrix} \tag{7.62}$$

Thus, for our example:

$$n = 15$$

$$\sum X_{i1} = 274 + 180 + \cdots = 3{,}626$$

$$\sum X_{i1}X_{i2} = (274)(2{,}450) + (180)(3{,}254) + \cdots = 11{,}419{,}181$$

etc.

These elements are found in (7.59).

Also note that $\mathbf{X}'\mathbf{Y}$ for the first-order model with two independent variables is:

$$\mathbf{X}'\mathbf{Y} = \begin{bmatrix} 1 & 1 & \cdots & 1 \\ X_{11} & X_{21} & \cdots & X_{n1} \\ X_{12} & X_{22} & \cdots & X_{n2} \end{bmatrix} \begin{bmatrix} Y_1 \\ Y_2 \\ \cdot \\ \cdot \\ \cdot \\ Y_n \end{bmatrix} = \begin{bmatrix} \sum Y_i \\ \sum X_{i1}Y_i \\ \sum X_{i2}Y_i \end{bmatrix} \tag{7.63}$$

For our example, we have:

$$\sum Y_i = 162 + 120 + \cdots = 2{,}259$$

$$\sum X_{i1}Y_i = (274)(162) + (180)(120) + \cdots = 647{,}107$$

$$\sum X_{i2}Y_i = (2{,}450)(162) + (3{,}254)(120) + \cdots = 7{,}096{,}619$$

These are the elements found in (7.60).

Estimated Regression Function

The least squares estimates **b** are readily obtained by (7.21), given our basic calculations in (7.60) and (7.61):

$$\mathbf{b} = (\mathbf{X'X})^{-1}\mathbf{X'Y}$$

$$= \begin{bmatrix} 1.2463484 & 2.1296642 \times 10^{-4} & -4.1567125 \times 10^{-4} \\ 2.1296642 \times 10^{-4} & 7.7329030 \times 10^{-6} & -7.0302518 \times 10^{-7} \\ -4.1567125 \times 10^{-4} & -7.0302518 \times 10^{-7} & 1.9771851 \times 10^{-7} \end{bmatrix} \times \begin{bmatrix} 2{,}259 \\ 647{,}107 \\ 7{,}096{,}619 \end{bmatrix}$$

$$= \begin{bmatrix} 3.4526127900 \\ .4960049761 \\ .009199080867 \end{bmatrix}$$

Thus:

$$\begin{bmatrix} b_0 \\ b_1 \\ b_2 \end{bmatrix} = \begin{bmatrix} 3.4526127900 \\ .4960049761 \\ .009199080867 \end{bmatrix}$$

and the estimated regression function is:

$$\hat{Y} = 3.45 + .496X_1 + .00920X_2$$

This estimated regression function indicates that mean sales are expected to increase by .496 gross when the target population increases by one thousand, holding per capita income constant, and that mean sales are expected to increase by .0092 gross when per capita income increases by one dollar, holding population constant.

Algebraic Version of Normal Equations. The normal equations in algebraic form for the case of two independent variables can be obtained readily from (7.62) and (7.63). We have:

$$(\mathbf{X'X})\mathbf{b} = \mathbf{X'Y}$$

$$\begin{bmatrix} n & \sum X_{i1} & \sum X_{i2} \\ \sum X_{i1} & \sum X_{i1}^2 & \sum X_{i1}X_{i2} \\ \sum X_{i2} & \sum X_{i2}X_{i1} & \sum X_{i2}^2 \end{bmatrix} \begin{bmatrix} b_0 \\ b_1 \\ b_2 \end{bmatrix} = \begin{bmatrix} \sum Y_i \\ \sum X_{i1}Y_i \\ \sum X_{i2}Y_i \end{bmatrix}$$

from which we obtain the normal equations:

$$\begin{aligned} \sum Y_i &= nb_0 + b_1 \sum X_{i1} + b_2 \sum X_{i2} \\ \sum X_{i1}Y_i &= b_0 \sum X_{i1} + b_1 \sum X_{i1}^2 + b_2 \sum X_{i1}X_{i2} \\ \sum X_{i2}Y_i &= b_0 \sum X_{i2} + b_1 \sum X_{i1}X_{i2} + b_2 \sum X_{i2}^2 \end{aligned} \tag{7.64}$$

Aptness of Model

To examine the aptness of regression model (7.58) with independent variables X_1 and X_2 for the data at hand, we require the fitted values $\hat{Y}_i$ and the residuals $e_i = Y_i - \hat{Y}_i$. We obtain by (7.23):

$$\hat{\mathbf{Y}} = \mathbf{X}\mathbf{b}$$

$$\begin{bmatrix} \hat{Y}_1 \\ \hat{Y}_2 \\ \cdot \\ \cdot \\ \cdot \\ \hat{Y}_{15} \end{bmatrix} = \begin{bmatrix} 1 & 274 & 2{,}450 \\ 1 & 180 & 3{,}254 \\ \cdot & \cdot & \cdot \\ \cdot & \cdot & \cdot \\ \cdot & \cdot & \cdot \\ 1 & 370 & 2{,}605 \end{bmatrix} \begin{bmatrix} 3.4526127900 \\ .4960049761 \\ .009199080867 \end{bmatrix} = \begin{bmatrix} 161.896 \\ 122.667 \\ \cdot \\ \cdot \\ \cdot \\ 210.938 \end{bmatrix}$$

FIGURE 7.5

Residual Plot against $\hat{Y}$— Zarthan Company Example

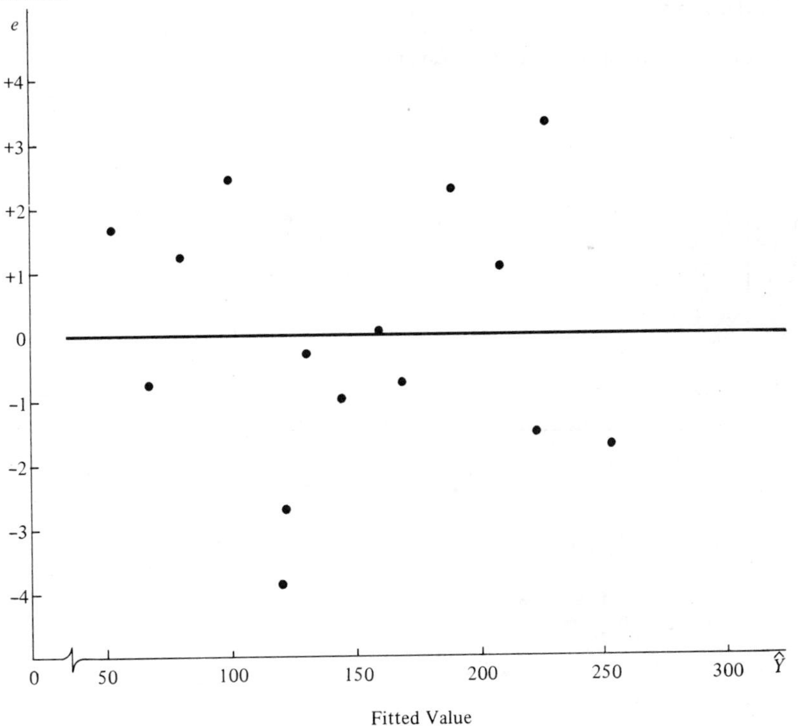

Further, by (7.24) we find:

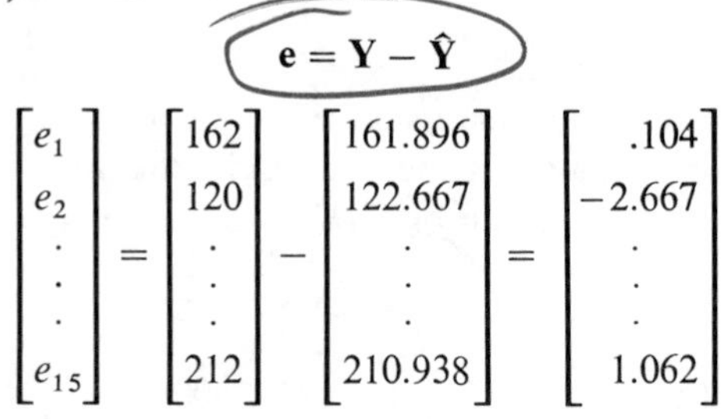

$$\mathbf{e} = \mathbf{Y} - \hat{\mathbf{Y}}$$

$$\begin{bmatrix} e_1 \\ e_2 \\ \cdot \\ \cdot \\ \cdot \\ e_{15} \end{bmatrix} = \begin{bmatrix} 162 \\ 120 \\ \cdot \\ \cdot \\ \cdot \\ 212 \end{bmatrix} - \begin{bmatrix} 161.896 \\ 122.667 \\ \cdot \\ \cdot \\ \cdot \\ 210.938 \end{bmatrix} = \begin{bmatrix} .104 \\ -2.667 \\ \cdot \\ \cdot \\ \cdot \\ 1.062 \end{bmatrix}$$

Figure 7.5 contains a plot of the residuals e_i against the fitted values $\hat{Y}_i$. Figure 7.6 contains a plot of the residuals against X_{i1}, and Figure 7.7 contains a plot of the residuals against X_{i2}. There are no suggestions in any of these plots that systematic deviations from the fitted response plane are present, nor that the error variance varies either with the level of $\hat{Y}$ or with the levels of X_1 or X_2. We do not show a plot on normal probability paper, but it does not indicate any major departure from normality. Hence, model (7.58) appears to be apt for this application.

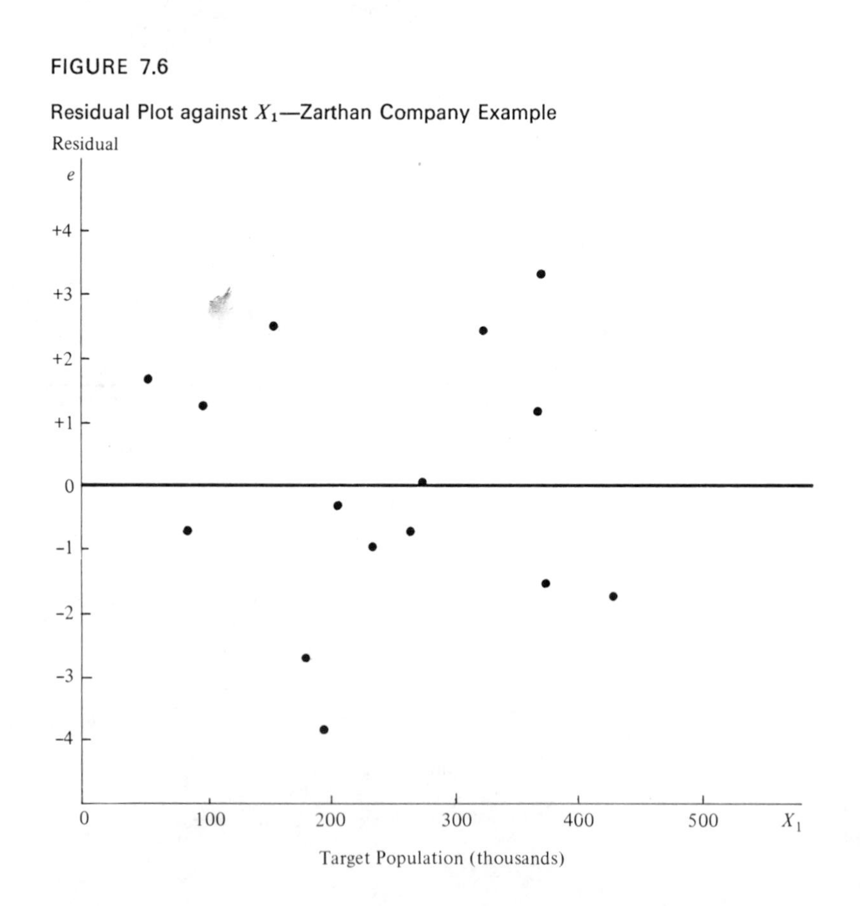

FIGURE 7.6

Residual Plot against X_1—Zarthan Company Example

FIGURE 7.7

Residual Plot against X_2—Zarthan Company Example

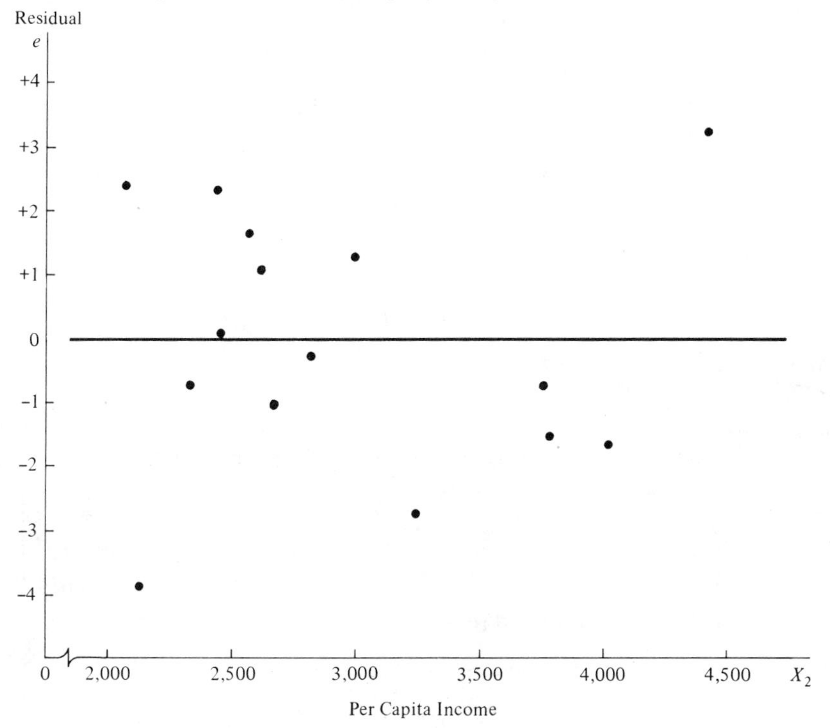

Analysis of Variance

To test whether sales are related to population and per capita income, we construct the ANOVA table in Table 7.4. The basic quantities needed there are:

$$\mathbf{Y'Y} = [162 \quad 120 \quad \cdots \quad 212]\begin{bmatrix} 162 \\ 120 \\ \cdot \\ \cdot \\ \cdot \\ 212 \end{bmatrix}$$

$$= (162)^2 + (120)^2 + \cdots + (212)^2$$

$$= 394{,}107.000$$

$$n\overline{Y}^2 = \frac{(\sum Y_i)^2}{n} = \frac{(2{,}259)^2}{15} = 340{,}205.400$$

TABLE 7.4

ANOVA Table for Zarthan Company Example

Source of Variation	*SS*	*df*	*MS*
Regression	$SSR = 53{,}844.716$	2	$MSR = 26{,}922.358$
Error	$SSE = 56.884$	12	$MSE = 4.740$
Total	$SSTO = 53{,}901.600$	14	

Thus:

$$SSTO = \mathbf{Y}'\mathbf{Y} - n\bar{Y}^2 = 394{,}107.000 - 340{,}205.400 = 53{,}901.600$$

and, using our result in (7.60):

$$\begin{aligned} SSE &= \mathbf{Y}'\mathbf{Y} - \mathbf{b}'\mathbf{X}'\mathbf{Y} \\ &= 394{,}107.000 - [3.4526127900 \quad .4960049761 \quad .009199080867] \times \begin{bmatrix} 2{,}259 \\ 647{,}107 \\ 7{,}096{,}619 \end{bmatrix} \\ &= 394{,}107.000 - 394{,}050.116 = 56.884 \end{aligned}$$

Finally, we obtain by subtraction:

$$SSR = SSTO - SSE = 53{,}901.600 - 56.884 = 53{,}844.716$$

The degrees of freedom and mean squares are entered in Table 7.4. Note that three regression parameters had to be estimated, hence $15 - 3 = 12$ degrees of freedom are associated with SSE. Also, the number of degrees of freedom associated with SSR are two—the number of X variables in the model.

Test of Regression Relation. To test whether sales are related to population and per capita income:

$$\begin{aligned} &C_1: \beta_1 = 0 \text{ and } \beta_2 = 0 \\ &C_2: \text{not both } \beta_1 \text{ and } \beta_2 \text{ equal } 0 \end{aligned}$$

we use test statistic (7.30b):

$$F^* = \frac{MSR}{MSE} = \frac{26{,}922.358}{4.740} = 5{,}680$$

Assuming α is to be held at .05, we require $F(.95; 2, 12) = 3.89$. Since F^* exceeds this action limit, we conclude C_2, that sales are related to population and per capita income. Whether this relation is useful for making predictions of sales or estimates of mean sales still remains to be seen.

Coefficient of Multiple Determination. For our example, we have by (7.31):

$$R^2 = \frac{SSR}{SSTO} = \frac{53{,}844.716}{53{,}901.600} = .9989$$

Thus, when the two independent variables population and per capita income are considered, the variation in sales is reduced by 99.9 percent.

Algebraic Expression for SSE. The error sum of squares for the case of two independent variables in algebraic terms is:

$$SSE = \mathbf{Y'Y} - \mathbf{b'X'Y} = \sum Y_i^2 - [b_0 \quad b_1 \quad b_2]\begin{bmatrix} \sum Y_i \\ \sum X_{i1}Y_i \\ \sum X_{i2}Y_i \end{bmatrix}$$

or:

(7.65)
$$SSE = \sum Y_i^2 - b_0 \sum Y_i - b_1 \sum X_{i1}Y_i - b_2 \sum X_{i2}Y_i$$

Note how this expression is a straightforward extension of (2.24a) for the case of one independent variable.

Estimation of Regression Parameters

The Zarthan Company is not interested in the parameter β_0 since it falls far outside the scope of the model. It is desired to estimate β_1 and β_2 jointly, with a family confidence coefficient .90. We shall use the simultaneous Bonferroni confidence intervals in (7.44), since these are easy to develop and interpret.

First, we need the estimated variance-covariance matrix $\mathbf{s}^2(\mathbf{b})$:

$$\mathbf{s}^2(\mathbf{b}) = MSE(\mathbf{X'X})^{-1}$$

MSE is given in Table 7.4, and $(\mathbf{X'X})^{-1}$ was obtained in (7.61). Hence:

(7.66)

$$\mathbf{s}^2(\mathbf{b}) = 4.7403$$

$$\times \begin{bmatrix} 1.2463484 & 2.1296642 \times 10^{-4} & -4.1567125 \times 10^{-4} \\ 2.1296642 \times 10^{-4} & 7.7329030 \times 10^{-6} & -7.0302518 \times 10^{-7} \\ -4.1567125 \times 10^{-4} & -7.0302518 \times 10^{-7} & 1.9771851 \times 10^{-7} \end{bmatrix}$$

$$= \begin{bmatrix} 5.9081 & .0010095 & -.0019704 \\ .0010095 & .000036656 & -.0000033326 \\ -.0019704 & -.0000033326 & .00000093725 \end{bmatrix}$$

The two elements we require are:

$$s^2(b_1) = .000036656 \qquad \text{or} \qquad s(b_1) = .006054$$
$$s^2(b_2) = .00000093725 \qquad \text{or} \qquad s(b_2) = .0009681$$

Next, we require:

$$B = t(1 - .10/2(2); 12) = t(.975; 12) = 2.179$$

Now we are ready to obtain the two simultaneous confidence intervals:

$$.4960 - (2.179)(.006054) \leq \beta_1 \leq .4960 + (2.179)(.006054)$$

or:

$$.483 \leq \beta_1 \leq .509$$

$$.009199 - (2.179)(.0009681) \leq \beta_2 \leq .009199 + (2.179)(.0009681)$$

or:

$$.0071 \leq \beta_2 \leq .0113$$

With family confidence coefficient .90, we conclude that β_1 falls between .483 and .509 and that β_2 falls between .0071 and .0113.

Note that the simultaneous confidence intervals suggest that both β_1 and β_2 are positive, which is in accord with theoretical expectations that sales should increase with either higher target population or higher per capita income, the other variable being held constant.

Estimation of Mean Response

Suppose the Zarthan Company would like to estimate expected (mean) sales in a district with target population $X_{h1} = 220$ thousand persons and per capita income $X_{h2} = 2{,}500$ dollars. We define:

$$\mathbf{X}_h = \begin{bmatrix} 1 \\ 220 \\ 2{,}500 \end{bmatrix}$$

The point estimate of mean sales is, by (7.46):

$$\hat{Y}_h = \mathbf{X}_h'\mathbf{b} = [1 \quad 220 \quad 2{,}500]\begin{bmatrix} 3.4526 \\ .4960 \\ .009199 \end{bmatrix} = 135.57$$

The estimated variance is, using (7.49) and earlier results:

$$s^2(\hat{Y}_h) = MSE(\mathbf{X}_h'(\mathbf{X}'\mathbf{X})^{-1}\mathbf{X}_h)$$

$$= 4.740[1 \quad 220 \quad 2{,}500]$$

$$\times \begin{bmatrix} 1.2463 & 2.1297 \times 10^{-4} & -4.1567 \times 10^{-4} \\ 2.1297 \times 10^{-4} & 7.7329 \times 10^{-6} & -7.0303 \times 10^{-7} \\ -4.1567 \times 10^{-4} & -7.0303 \times 10^{-7} & 1.9772 \times 10^{-7} \end{bmatrix} \begin{bmatrix} 1 \\ 220 \\ 2{,}500 \end{bmatrix}$$

$$= 4.740(.09835) = .46618$$

or:

$$s(\hat{Y}_h) = .68277$$

Assume that the confidence coefficient for the interval estimate of $E(Y_h)$ is to be .95. We then need $t(.975; 12) = 2.179$, and obtain by (7.50):

$$135.57 - (2.179)(.68277) \leq E(Y_h) \leq 135.57 + (2.179)(.68277)$$

or:

$$134.1 \leq E(Y_h) \leq 137.1$$

Thus, with confidence coefficient .95, we estimate that mean sales in a district with target population of 220 thousand and per capita income of $2,500 are somewhere between 134.1 and 137.1 gross.

Algebraic Version of Estimated Variance $s^2(\hat{Y}_h)$. Since we can write:

$$s^2(\hat{Y}_h) = MSE(\mathbf{X}_h'(\mathbf{X}'\mathbf{X})^{-1}\mathbf{X}_h) = \mathbf{X}_h'\mathbf{s}^2(\mathbf{b})\mathbf{X}_h$$

it follows for the case of two independent variables:

$$s^2(\hat{Y}_h) = s^2(b_0) + X_{h1}^2 s^2(b_1) + X_{h2}^2 s^2(b_2) + 2X_{h1}s(b_0, b_1) + 2X_{h2}s(b_0, b_2) + 2X_{h1}X_{h2}s(b_1, b_2) \tag{7.67}$$

When we substitute in (7.67), utilizing the estimated variances and covariances from (7.66), we obtain the same result as before except for the effect of rounding, namely $s^2(\hat{Y}_h) = .46638$.

Prediction Limits for New Observations

Suppose the Zarthan Company would like to predict sales in two districts. The two districts have the following characteristics:

	District A	*District B*
X_{h1}	220	375
X_{h2}	2,500	3,500

To determine which simultaneous prediction intervals are best here, we shall find S as given in (7.56a) and B as given in (7.57a), assuming the family confidence coefficient is to be .90:

$$S^2 = 2F(.90; 2, 12) = 2(2.81) = 5.62$$

or:

$$S = 2.37$$

and:

$$B = t(1 - .10/2(2); 12) = t(.975; 12) = 2.179$$

Hence, the Bonferroni limits are more efficient here.

For district A, we shall use the results we found when estimating mean sales, since the levels of the independent variables are the same as before. We have:

$$\hat{Y}_A = 135.57$$

Further, we found for district A:

$$\mathbf{X}_h'(\mathbf{X}'\mathbf{X})^{-1}\mathbf{X}_h = .09835$$

Hence:

$$s^2(Y_{\text{A(new)}}) = MSE(1 + .09835) = 4.740(1.09835) = 5.20618$$

or:

$$s(Y_{\text{A(new)}}) = 2.28171$$

In similar fashion, we obtain:

$$\hat{Y}_{\text{B}} = 221.65$$
$$s(Y_{\text{B(new)}}) = 2.34526$$

We found before that $B = 2.179$. Hence, the simultaneous Bonferroni prediction intervals, with family confidence coefficient .90, are by (7.57):

$$135.57 - (2.179)(2.28171) \leq Y_{\text{A(new)}} \leq 135.57 + (2.179)(2.28171)$$

or:

$$130.6 \leq Y_{\text{A(new)}} \leq 140.5$$

$$221.65 - (2.179)(2.34526) \leq Y_{\text{B(new)}} \leq 221.65 + (2.179)(2.34526)$$

or:

$$216.5 \leq Y_{\text{B(new)}} \leq 226.8$$

With family confidence coefficient .90, we predict that sales in the two districts will be within the indicated limits. The Zarthan Company considers these prediction limits reasonably precise, and hence useful.

Computer Printout

Figure 7.8 contains an illustrative computer printout for the Zarthan Company example. Regression analysis printouts differ in format from one computer program to another. The format in Figure 7.8 is typical of a style in which items are identified explicitly (in contrast to the more compressed format shown in earlier chapters). The data input matrix for the computer run was set up as follows (for conciseness rows 2 through 14 are not shown here):

$$\begin{array}{ccc} Y & X_1 & X_2 \end{array}$$
$$\begin{bmatrix} 162 & 274 & 2{,}450 \\ \cdot & \cdot & \cdot \\ \cdot & \cdot & \cdot \\ \cdot & \cdot & \cdot \\ 212 & 370 & 2{,}605 \end{bmatrix}$$

Thus Y, X_1, and X_2 are designated on the printout as variables 1, 2, and 3 respectively.

FIGURE 7.8

Computer Printout for Zarthan Company Example

```
MEANS,VARIANCES AND STANDARD DEVIATIONS

  VAR.                  MEAN                VARIANCE             ST. DEV.
  1.000000000E0      1.506000000E2       3.850114290E3       6.204929000E1
  2.000000000E0      2.417333300E2       1.364920952E4       1.168298300E2
  3.000000000E0      2.961866670E3       5.338296952E5       7.306365000E2

SIMPLE CORRELATION MATRIX
   1                 0.99549             0.6393
   0.99549           1                   0.56856
   0.6393            0.56856             1


INDEPENDENT VARIABLES: 2  3,    RESPONSE VARIABLE: 1

  VAR.               COEFF.              ST. ERROR           T-VALUE
  2.000000000E0      4.960000000E¯1      6.050000000E¯3      8.192000000E1
  3.000000000E0      9.200000000E¯3      9.700000000E¯4      9.500000000E0

INTERCEPT     3.45261

ANOVA [SOURCE '1' IS REGRESSION, SOURCE '2' IS RESIDUAL]:
  SOURCE               SS                   DF                 MS
     1              53844.71643              2               26922.35822
     2                 56.88357             12                   4.7403

F-VALUE                   5679.47
S.E. OF ESTIMATE          2.17722
MULTIPLE R-SQUARED        99.89

COVARIANCE MATRIX

  5.908061820E0      1.009520000E¯3     ¯1.970410000E¯3
  1.009520000E¯3     3.666000000E¯5     ¯3.330000000E¯6
 ¯1.970410000E¯3    ¯3.330000000E¯6      9.400000000E¯7

RESIDUALS
 NO.               OBSERVED             ESTIMATED            RESIDUAL
  1                162                  161.89572             0.10428
  2                120                  122.66732            ¯2.66732
  3                223                  224.42938            ¯1.42938
  4                131                  131.24062            ¯0.24062
  5                 67                   67.69928            ¯0.69928
  6                169                  169.68486            ¯0.68486
  7                 81                   79.73194             1.26806
  8                192                  189.672               2.328
  9                116                  119.83202            ¯3.83202
 10                 55                   53.29052             1.70948
 11                252                  253.71506            ¯1.71506
 12                232                  228.69079             3.30921
 13                144                  144.97934            ¯0.97934
 14                103                  100.53307             2.46693
 15                212                  210.93806             1.06194
```

Note that the printout first gives basic information on the variables, namely means, variances, and standard deviations. These are shown in E format. In this format $E2$ stands for 10^2, $E3$ for 10^3, E^-1 for 10^{-1}, and the like. Thus $1.506E2$, the mean for variable 1, stands for 150.6. Other numbers are interpreted similarly; for instance, $4.9600E^-1$ stands for .49600.

The second block of information contains coefficients of simple correlation, arranged in a matrix where the columns and rows each represent Y, X_1, and X_2 in that order. Thus, the coefficient of simple correlation between Y and X_1 is seen to be .99549, between Y and X_2 it is .63930, and between X_1 and X_2 it is .56856.

Next are blocks of regression analysis results. The first such block involves the estimated regression coefficients b_k. Shown, in turn, are the magnitudes of b_1 and b_2, estimated standard deviations of b_1 and b_2, and t^* values for testing whether a given β_k equals 0 ($k = 1, 2$). The magnitude of b_0 is given in a separate line.

The next block contains ANOVA information: key portions of the ANOVA table, the F^* value for the test of whether or not a regression relation exists, $\sqrt{MSE}$, and R^2.

The final two blocks show respectively the elements of $\mathbf{s}^2(\mathbf{b})$ and the magnitudes of Y_i, $\hat{Y}_i$, and e_i ($i = 1, \ldots, 15$).

It may be noted that the results in Figure 7.8 do not coincide precisely with all corresponding results given earlier, partly because of rounding errors and partly because the results are rounded to different extents. In this connection, it should be noted that different computer regression packages may lead to somewhat different results because final results are rounded to different extents, and even more importantly because rounding errors are not handled equally well by all packages. Particularly when there are a number of independent variables, some of which are highly correlated, rounding errors can be a serious source of difficulty. It is a wise policy to investigate a computer regression package before using it, for instance, by comparing its output for a test problem against results known to be accurate.

Caution about Hidden Extrapolations

Before concluding this illustration of multiple regression analysis, we should caution again about making estimates or predictions outside the scope of the model. The danger, of course, is that the model may not be appropriate when extended outside the region of the observations. In multiple regression, it is particularly easy to lose track of this region since the levels of $X_1, \ldots, X_{p-1}$ *jointly* define the region. Thus, one cannot merely look at the ranges of each independent variable. Consider Figure 7.9, where the shaded region is the region of observations for a multiple regression application with two independent variables. The circled dot is within the ranges of the independent variables X_1 and X_2 individually, yet well outside the joint region of observations.

FIGURE 7.9

Region of Observations on X_1 and X_2 Jointly, Compared with Ranges of X_1 and X_2 Individually

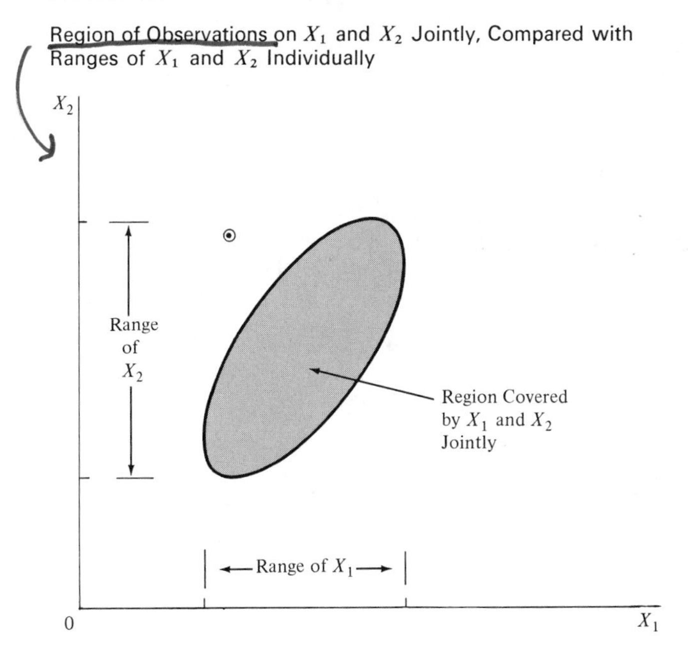

7.9 PROBLEMS IN THE ANALYSIS OF MULTIPLE REGRESSION MODELS

In multiple regression analysis, one is often concerned with the nature and significance of the relations between the independent variables and the dependent variable. Questions that are frequently asked include:

1. What is the relative importance of the effects of the different independent variables?
2. What is the magnitude of the effect of a given independent variable on the dependent variable?
3. Can any independent variable be dropped from the model because it has no effect on the dependent variable?
4. Should any independent variables not yet included in the model be considered for possible inclusion?

If the independent variables included in the model are (1) uncorrelated among themselves, and (2) uncorrelated with any other independent variables that are related to the dependent variable but omitted from the model, relatively simple answers can be given to these questions. Unfortunately, in many regression applications in business and economics, the independent variables are correlated among themselves and with other variables that are not included in the model but are related to the dependent variable. In our

Zarthan Company example, population and per capita income are somewhat correlated, as the scatter plot in Figure 7.10 suggests. As another example, in a regression of family food expenditures on family income, family savings, and age of head of the household, the independent variables will be correlated among themselves. Further, these independent variables will be correlated with other socioeconomic variables not included in the model that do affect family food expenditures, such as family size.

When the independent variables are correlated among themselves, *inter-correlation* or *multicollinearity* among them is said to exist. (Sometimes the latter term is reserved for those instances when the correlation among independent variables is very high or even perfect.) We shall explore now a variety of interrelated problems that are created by multicollinearity among the independent variables.

FIGURE 7.10

Scatter Plot of per Capita Income against Target Population—Zarthan Company Example

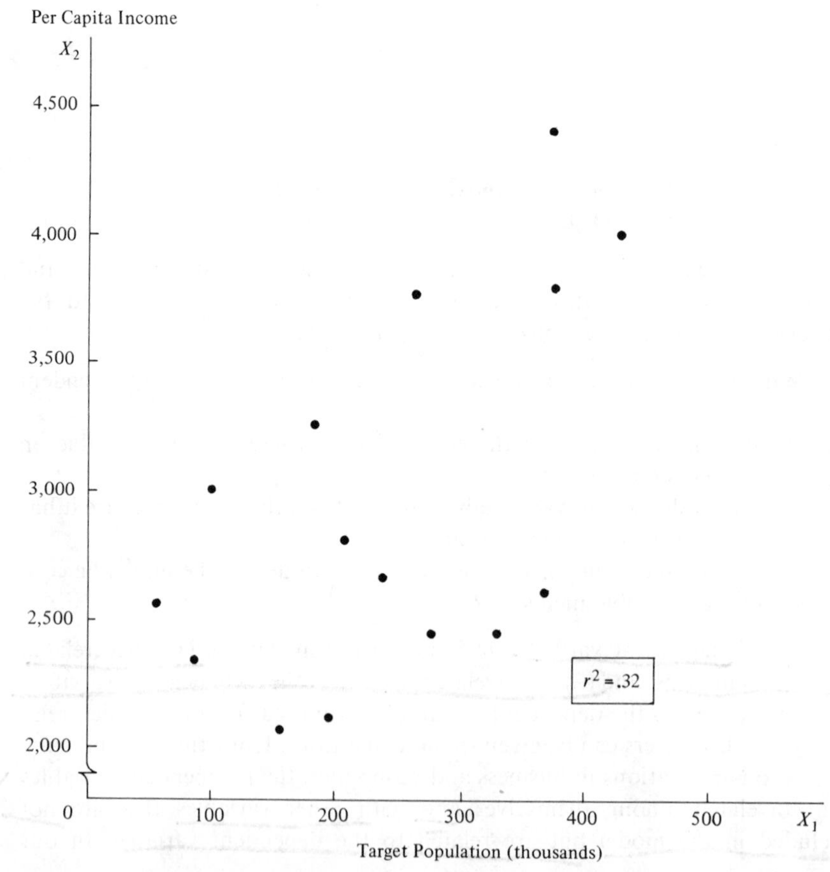

Effect of Multicollinearity on Regression Coefficients

In Table 7.5a, we repeat the ANOVA table for the Zarthan Company example from Table 7.4. This time, we use the notation $SSR(X_1, X_2)$ and $SSE(X_1, X_2)$ to indicate explicitly the independent variables in the model. The reason is that we wish to consider also the regression of sales (Y) on target population (X_1) only. Table 7.5b contains the fitted regression and the analysis of variance table for the regression of Y on X_1 only. These results are obtained in the usual fashion. Only simple regression is involved here, and we do not show any calculations. We use the notation $SSR(X_1)$ and $SSE(X_1)$ to indicate explicitly that these sums of squares refer to the model containing only X_1.

Note first that the regression coefficient for X_1 is not the same in Tables 7.5a and b. Thus, the effect ascribed to X_1 by the fitted response function

TABLE 7.5

ANOVA Tables for Zarthan Company Example

(a) Regression of Y on X_1 and X_2

$\hat{Y} = 3.45 + .496X_1 + .00920X_2$

Source of Variation	SS	df	MS
Regression	$SSR(X_1, X_2) = 53{,}845$	2	$MSR(X_1, X_2) = 26{,}922$
Error	$SSE(X_1, X_2) = 56.9$	12	$MSE(X_1, X_2) = 4.74$
Total	$SSTO = 53{,}902$	14	

(b) Regression of Y on X_1

$\hat{Y} = 22.79 + .529X_1$

Source of Variation	SS	df	MS
Regression	$SSR(X_1) = 53{,}417$	1	$MSR(X_1) = 53{,}417$
Error	$SSE(X_1) = 485$	13	$MSE(X_1) = 37.31$
Total	$SSTO = 53{,}902$	14	

(c) Regression of Y on X_2

$\hat{Y} = -10.21 + .0543X_2$

Source of Variation	SS	df	MS
Regression	$SSR(X_2) = 22{,}030$	1	$MSR(X_2) = 22{,}030$
Error	$SSE(X_2) = 31{,}872$	13	$MSE(X_2) = 2{,}452$
Total	$SSTO = 53{,}902$	14	

varies, depending upon whether only X_1 or both X_1 and X_2 are being considered in the model. The reason for the different regression coefficients is that the independent variables (target population and per capita income) are correlated, as we saw earlier from Figure 7.10. If we were to consider as a third independent variable X_3, the number of drugstores in a district, the regression coefficient for X_1 in the fitted response curve with three independent variables would be different again from the two in Table 7.5, since number of drugstores is highly correlated to target population.

Table 7.5c contains the fitted regression of sales (Y) on per capita income (X_2), and the resulting ANOVA table. Note that the regression coefficient for X_2 here also is different from the one for X_2 in Table 7.5a when both X_1 and X_2 are included in the model.

The important conclusion we must draw is: When independent variables are correlated, the regression coefficient of any independent variable depends on which other independent variables are included in the model. Thus, a regression coefficient does not reflect any inherent effect of the particular independent variable on the dependent variable but only a marginal or partial effect, given whatever other correlated independent variables are included in the model.

Note

To show that the regression coefficient of X_1 is unchanged when X_2 is added to the regression model, provided X_1 and X_2 are uncorrelated, consider the algebraic expression for b_1 in the multiple regression model with two independent variables:

$$(7.68) \qquad b_1 = \frac{\dfrac{\sum (X_{i1} - \bar{X}_1)(Y_i - \bar{Y})}{\sum (X_{i1} - \bar{X}_1)^2} - \left[\dfrac{\sum (Y_i - \bar{Y})^2}{\sum (X_{i1} - \bar{X}_1)^2}\right]^{1/2} r_{Y2} r_{12}}{1 - r_{12}^2}$$

where r_{Y2} denotes the coefficient of simple correlation between Y and X_2, and r_{12} denotes the coefficient of simple correlation between X_1 and X_2.

If X_1 and X_2 are uncorrelated, $r_{12} = 0$, and (7.68) reduces to:

$$(7.68a) \qquad b_1 = \frac{\sum (X_{i1} - \bar{X}_1)(Y_i - \bar{Y})}{\sum (X_{i1} - \bar{X}_1)^2}$$

But (7.68a) is the estimator of the slope for the simple regression of Y on X_1, per (2.10a).

Hence, if X_1 and X_2 are uncorrelated, adding X_2 to the regression model does not change the regression coefficient for X_1. Correspondingly, adding X_1 to the regression model does not change the regression coefficient for X_2 when X_1 and X_2 are uncorrelated.

Effect of Multicollinearity on Regression Sums of Squares

Note from Table 7.5a that the error sum of squares $SSE(X_1, X_2) = 56.9$ when both X_1 and X_2 are included in the model. When only X_2 is included in the model, the error sum of squares is $SSE(X_2) = 31{,}872$ according to

Table 7.5c. Since the variation in Y when X_2 alone is considered is 31,872 but is only 56.9 when both X_1 and X_2 are considered, we may ascribe the difference:

$$SSE(X_2) - SSE(X_1, X_2) = 31{,}872 - 56.9 = 31{,}815$$

to the effect of X_1. We shall denote this difference $SSR(X_1 | X_2)$:

$$SSR(X_1 | X_2) = SSE(X_2) - SSE(X_1, X_2) \tag{7.69}$$

When we fit a regression function containing only X_1, we also obtain a measure of the reduction in variation of Y associated with X_1, namely $SSR(X_1)$. Table 7.5b indicates that for our example $SSR(X_1) = 53{,}417$, which is not the same as $SSR(X_1 | X_2) = 31{,}815$. The reason for the difference is the correlation between X_1 and X_2.

The story is the same for the other independent variable. Let:

$$SSR(X_2 | X_1) = SSE(X_1) - SSE(X_1, X_2) \tag{7.70}$$

For our example, we have:

$$SSR(X_2 | X_1) = 485 - 56.9 = 428$$

This sum of squares is not the same as $SSR(X_2) = 22{,}030$ from Table 7.5c, obtained when sales are regressed only on X_2.

The important conclusion is: When independent variables are correlated, there is no unique sum of squares which can be ascribed to an independent variable as reflecting its effect in reducing the total variation in Y. The reduction in the total variation ascribed to an independent variable must be viewed in the context of the other independent variables included in the model, whenever the independent variables are correlated.

Let us consider the meaning of $SSR(X_1)$ and $SSR(X_1 | X_2)$ further. $SSR(X_1)$ measures the reduction in the variation of Y when X_1 is introduced into the regression and no other independent variable is present. $SSR(X_1 | X_2)$ measures the further reduction in the variation of Y when X_2 is already in the regression model and X_1 is introduced as a second independent variable. Hence the notation $SSR(X_1 | X_2)$ is used to show that the sum of squares measures a reduction in variation of Y associated with X_1, given that X_2 is already included in the model.

The reason why $SSR(X_1 | X_2) = 31{,}815$ is less than $SSR(X_1) = 53{,}417$ in our example should now be apparent. Since X_1 is correlated with X_2, some of its power to reduce the variation in Y is already accounted for by $SSR(X_2)$ when X_2 alone is included in the model. Hence the marginal effect of X_1 in reducing the variation in Y, given that X_2 is in the model, is less than the effect if X_1 were introduced into the model without X_2 being presented. For the same reason, $SSR(X_2 | X_1)$ is less than $SSR(X_2)$.

The terms $SSR(X_2 | X_1)$ and $SSR(X_1 | X_2)$ are frequently called *extra sums of squares*, since they indicate the additional or extra reduction in the error sum of squares achieved by introducing an additional independent variable. Any reduction in the error sum of squares, of course, is always equal to the increase in the regression sum of squares since:

$$SSTO = SSR + SSE$$

Hence, an extra sum of squares can also be thought of as the increase in the regression sum of squares achieved by introducing the new variable. We can write, equivalently to (7.69) and (7.70):

$$SSR(X_1 | X_2) = SSR(X_1, X_2) - SSR(X_2) \tag{7.69a}$$

$$SSR(X_2 | X_1) = SSR(X_1, X_2) - SSR(X_1) \tag{7.70a}$$

Example of Uncorrelated Independent Variables

Table 7.6 contains data for a small-scale experiment on the effect of work crew size (X_1) and level of bonus pay (X_2) on crew productivity score (Y). It is easy to show that X_1 and X_2 are uncorrelated here, that is, $r_{12}^2 = 0$. Table 7.7a contains the fitted regression function and analysis of variance table when both X_1 and X_2 are included in the model. Tables 7.7b and 7.7c contain the same information when only X_1 or X_2 is included in the model.

TABLE 7.6

Work Crew Productivity Example with Uncorrelated Independent Variables

Trial i	*Crew Size* X_{i1}	*Bonus Pay* X_{i2}	*Crew Productivity Score* Y_i
1	4	$2	42
2	4	2	39
3	4	3	48
4	4	3	51
5	6	2	49
6	6	2	53
7	6	3	61
8	6	3	60

Note first that the regression coefficients for X_1 and X_2 are the same, whether only the given independent variable is included in the model or both independent variables are included.

Next, let us find:

$$SSR(X_2 | X_1) = 188.750 - 17.625 = 171.125$$
$$SSR(X_1 | X_2) = 248.750 - 17.625 = 231.125$$

Note that these are the same as $SSR(X_2)$ and $SSR(X_1)$ respectively.

Thus, if the independent variables are uncorrelated, the effects ascribed to them by a first-order model are the same no matter which other independent

TABLE 7.7

ANOVA Tables for Work Crew Productivity Example

(a) Regression of Y on X_1 and X_2

$\hat{Y} = .375 + 5.375X_1 + 9.250X_2$

Source of Variation	*SS*	*df*	*MS*
Regression	$SSR(X_1, X_2) = 402.250$	2	$MSR(X_1, X_2) = 201.125$
Error	$SSE(X_1, X_2) = 17.625$	5	$MSE(X_1, X_2) = 3.525$
Total	$SSTO = 419.875$	7	

(b) Regression of Y on X_1

$\hat{Y} = 23.500 + 5.375X_1$

Source of Variation	*SS*	*df*	*MS*
Regression	$SSR(X_1) = 231.125$	1	$MSR(X_1) = 231.125$
Error	$SSE(X_1) = 188.750$	6	$MSE(X_1) = 31.458$
Total	$SSTO = 419.875$	7	

(c) Regression of Y on X_2

$\hat{Y} = 27.250 + 9.250X_2$

Source of Variation	*SS*	*df*	*MS*
Regression	$SSR(X_2) = 171.125$	1	$MSR(X_2) = 171.125$
Error	$SSE(X_2) = 248.750$	6	$MSE(X_2) = 41.458$
Total	$SSTO = 419.875$	7	

variables are included in the model. This is a strong argument for experimenting whenever possible, since experimental control permits making the independent variables uncorrelated. Unfortunately, much of business and economic data do not lend themselves to experimental control so that one frequently must accept correlated independent variables.

Simultaneous Tests on Regression Coefficients

A not infrequent abuse of multiple regression models is to examine the t^* statistic in (7.42b):

$$t^* = \frac{b_k}{s(b_k)}$$

for each regression coefficient to decide whether or not $\beta_k = 0$ for $k = 1, \ldots, p-1$. Even if a simultaneous inference procedure is used, and often it is not, problems still exist.

Let us consider the first-order regression model with two independent variables:

$$Y_i = \beta_0 + \beta_1 X_{i1} + \beta_2 X_{i2} + \varepsilon_i \qquad \text{Full model} \tag{7.71}$$

If the test on β_1 indicates it is zero, the regression model (7.71) would be:

$$Y_i = \beta_0 + \beta_2 X_{i2} + \varepsilon_i$$

If the test on β_2 indicates it is zero, the regression model (7.71) would be:

$$Y_i = \beta_0 + \beta_1 X_{i1} + \varepsilon_i$$

If the separate tests indicate that $\beta_1 = 0$ and $\beta_2 = 0$, that does not necessarily jointly imply that:

$$Y_i = \beta_0 + \varepsilon_i$$

since each of the tests separately does not consider this alternative.

For an example, consider the data in Table 7.8 for 10 ski resorts in New England during a period of normal snow conditions. The computer output for the regression of visitor days (Y) on miles of intermediate trails (X_1) and lift capacity (X_2) is summarized in Table 7.9a. The proper test for the existence of a regression relation:

TABLE 7.8

Example with Highly Correlated Independent Variables (ski resort example)

Ski Resort i	Miles of Intermediate Trails X_{i1}	Lift Capacity (skiers per hour) X_{i2}	Total Visitor Days during Sample Period Y_i
1	10.5	2,200	19,929
2	2.5	1,000	5,839
3	13.1	3,250	23,696
4	4.0	1,475	9,881
5	14.7	3,800	30,011
6	3.6	1,200	7,241
7	7.1	1,900	11,634
8	22.5	5,575	45,684
9	17.0	4,200	36,476
10	6.4	1,850	12,068

$$C_1: \beta_1 = 0 \text{ and } \beta_2 = 0$$
$$C_2: \text{not both } \beta_1 \text{ and } \beta_2 \text{ equal } 0$$

is the F test of (7.30). The test statistic for our example is (data are in Table 7.9a):

$$F^* = \frac{MSR}{MSE} = \frac{811{,}865{,}088}{2{,}757{,}701} = 294$$

Controlling the level of significance at .05, we require $F(.95; 2, 7) = 4.74$. Since F^* far exceeds this action limit, we conclude C_2, that there is a regression relation between Y and the independent variables X_1 and X_2.

TABLE 7.9

Selected Computer Outputs for Ski Resort Example

(a) Regression of Y on X_1 and X_2

$\hat{Y} = -1{,}806.82 + 1{,}131.03X_1 + 4.00X_2$

Source of Variation	*SS*	*df*	*MS*
Regression	$SSR(X_1, X_2) = 1{,}623{,}730{,}176$	2	$MSR(X_1, X_2) = 811{,}865{,}088$
Error	$SSE(X_1, X_2) = 19{,}303{,}908$	7	$MSE(X_1, X_2) = 2{,}757{,}701$
Total	$SSTO = 1{,}643{,}034{,}084$	9	

Variable	*Estimated Regression Coefficient*	*Estimated Standard Deviation*	t^*
X_1	$b_1 = 1{,}131.03$	$s(b_1) = 615.76$	1.837
X_2	$b_2 = 4.002$	$s(b_2) = 2.71$	1.477

(b) Regression of Y on X_1

$\hat{Y} = -363.98 + 2{,}032.53X_1$

Source of Variation	*SS*	*df*	*MS*
Regression	$SSR(X_1) = 1{,}617{,}707{,}400$	1	$MSR(X_1) = 1{,}617{,}707{,}400$
Error	$SSE(X_1) = 25{,}326{,}684$	8	$MSE(X_1) = 3{,}165{,}836$
Total	$SSTO = 1{,}643{,}034{,}084$	9	

Let us now examine the t^* statistics at a 5 percent family level of significance by the Bonferroni technique. We require $t(.9875; 7) = 2.84$. Since both t^* statistics have absolute values less than this action limit (see Table 7.9a), we would conclude $\beta_1 = 0$ and $\beta_2 = 0$, contrary to the earlier conclusion that not both coefficients equal zero.

To understand this apparently paradoxical result, let us investigate the test of, say:

$$C_1: \beta_2 = 0$$
$$C_2: \beta_2 \neq 0$$

by the general linear test approach. We first obtain the error sum of squares for the full model $SSE(F)$ in (7.71), associated with $n - 3$ degrees of freedom. Next, we formulate the reduced model under C_1:

$$Y_i = \beta_0 + \beta_1 X_{i1} + \varepsilon_i \qquad \text{Reduced model} \tag{7.72}$$

and obtain the error sum of squares $SSE(R)$ for the reduced model. Associated with $SSE(R)$ are $n - 2$ degrees of freedom. The test statistic (3.66) then leads to:

$$F^* = \frac{SSE(R) - SSE(F)}{(n-2) - (n-3)} \div \frac{SSE(F)}{n-3}$$

In our notation recognizing explicitly the independent variables in the model, we have:

$$SSE(F) = SSE(X_1, X_2) = 19{,}303{,}908$$
$$SSE(R) = SSE(X_1) = 25{,}326{,}684$$

so that by (7.70):

$$SSE(R) - SSE(F) = SSE(X_1) - SSE(X_1, X_2) = SSR(X_2 | X_1)$$

Hence:

$$F^* = \frac{SSR(X_2 | X_1)}{1} \div \frac{SSE(X_1, X_2)}{n-3} \tag{7.73}$$

For our example in Table 7.9, we have:

$$SSR(X_2 | X_1) = 25{,}326{,}684 - 19{,}303{,}908 = 6{,}022{,}776$$

and hence:

$$F^* = \frac{6{,}022{,}776}{1} \div \frac{19{,}303{,}908}{7} = 2.18$$

Recall from Table 7.9a that the t^* statistic for testing $\beta_2 = 0$ is $t^* = 1.477$. Hence:

$$(t^*)^2 = (1.477)^2 = 2.18 = F^*$$

Since $F^* = (t^*)^2$, and since there is a corresponding relation between the action limits, the two tests are equivalent. We already knew this holds for simple regression, and now we can see it also holds for multiple regression.

The test statistic F^* in (7.73) indicates clearly that a test on β_2 is a *marginal* test, given that X_1 is in the model. Similarly a test on β_1 is a *marginal* test, given that X_2 is in the model. It is apparent now why the simultaneous tests with the t^* statistics for the two different regression coefficients both led to the conclusion that the regression coefficient equals zero. Table 7.8 shows that X_1 and X_2 are highly correlated; indeed the coefficient of simple determination is $r_{12}^2 = .981$. Hence, if X_1 is already in the model, adding X_2

achieves little more reduction in the variation of Y, and $SSR(X_2|X_1)$ is small. When $SSR(X_2|X_1)$ is small, the test for β_2 leads to a small test statistic F^* and therefore to the conclusion that $\beta_2 = 0$.

Similarly, the explanation why the t test led to the conclusion that $\beta_1 = 0$ is that X_1 is highly correlated with X_2, and X_2 is assumed to be in the model when the t test for β_1 is employed.

Thus, despite the fact that there is a clear relation between the set of independent variables X_1 and X_2 and the dependent variable Y, the separate t tests indicated the respective regression coefficients equal zero because each test considers only the marginal contribution of the independent variable, given that the other is included in the model.

Note

We have just seen that it is possible that a set of independent variables is related to the dependent variable, yet all of the individual tests on the regression coefficients will lead to the conclusion that they equal zero because of the multicollinearity among the independent variables. This apparently paradoxical result is also possible under special circumstances when there is no multicollinearity among the independent variables. The special circumstances are not likely to be found in practice, however.

A Final Comment

The discussion in this section has indicated that the choice of the particular set of independent variables which are to be included in the model is highly important. We shall discuss further the choice of independent variables for the regression model in Chapter 11.

7.10 TESTING HYPOTHESES FOR MULTIPLE REGRESSION MODELS

Extension of Extra Sum of Squares Principle

We defined the extra sum of squares $SSR(X_1|X_2)$ in (7.69):

$$SSR(X_1|X_2) = SSE(X_2) - SSE(X_1, X_2)$$

Likewise, we defined in (7.70):

$$SSR(X_2|X_1) = SSE(X_1) - SSE(X_1, X_2)$$

Extensions for three or more X variables are straightforward. For instance, we define:

$$SSR(X_3|X_1, X_2) = SSE(X_1, X_2) - SSE(X_1, X_2, X_3) \tag{7.74}$$

$SSR(X_3|X_1, X_2)$ measures the reduction in the remaining variation of Y when X_1 and X_2 are already in the model which is achieved by introducing X_3 into the regression model.

We are now in the position to obtain a variety of decompositions for the analysis of variance. Let us consider the case of three X variables. We begin with the simple regression of Y on X_1. For this, the ANOVA identity is:

$$SSTO = SSR(X_1) + SSE(X_1) \tag{7.75}$$

Replacing $SSE(X_1)$ by its equivalent in (7.70), we obtain:

$$SSTO = SSR(X_1) + SSR(X_2|X_1) + SSE(X_1, X_2) \tag{7.75a}$$

Replacing $SSE(X_1, X_2)$ by its equivalent in (7.74), we obtain:

$$SSTO = SSR(X_1) + SSR(X_2|X_1) + SSR(X_3|X_1, X_2) + SSE(X_1, X_2, X_3) \tag{7.75b}$$

Since we have the following identity for multiple regression with three independent variables:

$$SSTO = SSR(X_1, X_2, X_3) + SSE(X_1, X_2, X_3) \tag{7.76}$$

it follows from (7.75b) that:

$$SSR(X_1, X_2, X_3) = SSR(X_1) + SSR(X_2|X_1) + SSR(X_3|X_1, X_2) \tag{7.77}$$

Thus, the regression sum of squares has been decomposed into marginal components, each associated with one degree of freedom. Of course, the order of the independent variables is arbitrary, and other orders can be developed. For instance:

$$SSR(X_1, X_2, X_3) = SSR(X_3) + SSR(X_1|X_3) + SSR(X_2|X_1, X_3) \tag{7.78}$$

Indeed we can define extra sums of squares for two or more independent variables at a time and obtain still other decompositions. For instance, we define:

$$SSR(X_2, X_3|X_1) = SSE(X_1) - SSE(X_1, X_2, X_3) \tag{7.79}$$

Thus, $SSR(X_2, X_3|X_1)$ represents the reduction in the variation of Y gained when X_2 and X_3 are added to the model already containing X_1. There are

TABLE 7.10

Example of ANOVA Table with Decomposition of SSR for Three Independent Variables

Source of Variation	*SS*	*df*	*MS*
Regression	$SSR(X_1, X_2, X_3)$	3	$MSR(X_1, X_2, X_3)$
X_1	$SSR(X_1)$	1	$MSR(X_1)$
X_2, given X_1	$SSR(X_2\|X_1)$	1	$MSR(X_2\|X_1)$
X_3, given X_1 and X_2	$SSR(X_3\|X_1, X_2)$	1	$MSR(X_3\|X_1, X_2)$
Error	$SSE(X_1, X_2, X_3)$	$n-4$	$MSE(X_1, X_2, X_3)$
Total	$SSTO$	$n-1$	

TABLE 7.11

ANOVA Tables with Decomposition of *SSR* for Zarthan Company Example

(a)

Source of Variation	SS	df	MS
Regression	$SSR(X_1, X_2) = 53{,}845$	2	26,922
X_1	$SSR(X_1) = 53{,}417$	1	53,417
X_2, given X_1	$SSR(X_2 \mid X_1) = 428$	1	428
Error	$SSE(X_1, X_2) = 56.9$	12	4.74
Total	$SSTO = 53{,}902$	14	

(b)

Source of Variation	SS	df	MS
Regression	$SSR(X_1, X_2) = 53{,}845$	2	26,922
X_2	$SSR(X_2) = 22{,}030$	1	22,030
X_1, given X_2	$SSR(X_1 \mid X_2) = 31{,}815$	1	31,815
Error	$SSE(X_1, X_2) = 56.9$	12	4.74
Total	$SSTO = 53{,}902$	14	

two degrees of freedom associated with this sum of squares, corresponding to the two X variables being added to the model. We can then make use of the decomposition:

$$SSR(X_1, X_2, X_3) = SSR(X_1) + SSR(X_2, X_3 \mid X_1) \tag{7.80}$$

It is obvious that the number of possible decompositions becomes vast as the number of independent variables increases. Table 7.10 contains the ANOVA table for one possible decomposition for the case of three independent variables, and Table 7.11 contains two possible decompositions for our earlier Zarthan Company example.

Two Basic Types of Tests Considered Earlier

We have already discussed how to conduct two important types of tests in multiple regression analysis:

1. The test whether all $\beta_k = 0$ $(k = 1, \ldots, p - 1)$ was given in (7.30). This test is used to establish whether or not there is any relation between the dependent variable and the set of X variables.

2. The test whether any one $\beta_k = 0$ was given in (7.42). This test is used to determine whether a single X variable can be dropped from the regression model.

Tests That Some Regression Coefficients Equal Zero

We consider now a third class of tests where we wish to determine whether some regression coefficients equal zero. The approach is that of a general linear test, and no new problems arise.

Three Independent Variables. As an illustration, let us consider the multiple regression application in Table 7.12. The data pertain to 14 different

TABLE 7.12

Data for CPU Time Example

Simulation i	*Number of Trials* X_{i1}	*Number of Statements in Program* X_{i2}	X_{i2}^2	*CPU Time (seconds)* Y_i
1	550	458	209,764	445.37
2	600	152	23,104	408.88
3	200	635	403,225	264.61
4	50	128	16,384	73.90
5	350	100	10,000	246.07
6	300	550	302,500	312.00
7	200	577	332,929	250.36
8	175	234	54,756	179.95
9	200	500	250,000	243.76
10	200	491	241,081	243.63
11	175	580	336,400	240.72
12	700	135	18,225	462.60
13	800	162	26,244	531.45
14	650	176	30,976	445.48

computer simulations conducted in designing the layout for a chemical warehouse. The dependent variable is CPU time (the operating time required by the computer's central processing unit to run the simulation), the first independent variable is number of trials in the simulation, and the second independent variable is number of statements in the computer program. Since it is hypothesized that the effect of number of statements (X_2) on CPU time (Y) may be curvilinear, the model being considered is:

$$Y_i = \beta_0 + \beta_1 X_{i1} + \beta_2 X_{i2} + \beta_3 X_{i2}^2 + \varepsilon_i \qquad \text{Full model} \tag{7.81}$$

It is desired that we test whether the number of statements variable can be dropped from the model. Hence, we wish to choose between the alternatives:

$$C_1: \beta_2 = \beta_3 = 0$$

$$C_2: \text{not both } \beta_2 \text{ and } \beta_3 \text{ equal zero}$$

We first fit the full model and obtain (results are shown in Table 7.13a):

$$SSE(F) = SSE(X_1, X_2, X_2^2) = 44.1$$

The model under C_1 is:

$$Y_i = \beta_0 + \beta_1 X_{i1} + \varepsilon_i \qquad \text{Reduced model} \tag{7.82}$$

When we fit the reduced model, we obtain (see Table 7.13b):

$$SSE(R) = SSE(X_1) = 17{,}240.3$$

TABLE 7.13

Computer Results for Computer Simulation Example

(a) Regression of Y on X_1, X_2, and X_2^2

$\hat{Y} = 7.028 + .595X_1 + .323X_2 - .000173X_2^2$

Source of Variation	*SS*	*df*	*MS*
Regression	$SSR(X_1, X_2, X_2^2) = 214{,}691.2$	3	71,563.7
Error	$SSE(X_1, X_2, X_2^2) = 44.1$	10	4.41
Total	$SSTO = 214{,}735.3$	13	

(b) Regression of Y on X_1

$\hat{Y} = 122.71 + .511X_1$

Source of Variation	*SS*	*df*	*MS*
Regression	$SSR(X_1) = 197{,}495.1$	1	197,495.1
Error	$SSE(X_1) = 17{,}240.3$	12	1,436.7
Total	$SSTO = 214{,}735.4$	13	

The general test statistic in (3.66):

$$F^* = \frac{SSE(R) - SSE(F)}{(n-2) - (n-4)} \div \frac{SSE(F)}{n-4}$$

can be simplified because:

$$SSE(R) - SSE(F) = SSE(X_1) - SSE(X_1, X_2, X_2^2) = SSR(X_2, X_2^2 | X_1)$$

Hence, we can write:

$$F^* = \frac{SSR(X_2, X_2^2 | X_1)}{2} \div \frac{SSE(F)}{n-4} \tag{7.83}$$

For our example, we have (see Table 7.13):

$$SSR(X_2, X_2^2 | X_1) = 17{,}240.3 - 44.1 = 17{,}196.2$$

$$F^* = \frac{17{,}196.2}{2} \div \frac{44.1}{10} = 1{,}950$$

Suppose the α risk is to be .01. We require $F(.99; 2, 10) = 7.56$. Since F^* exceeds the action limit, we conclude C_2, that the number of statements variable should not be dropped from the model.

$p - 1$ ***Independent Variables.*** Consider the general multiple regression model:

(7.84) $$Y_i = \beta_0 + \beta_1 X_{i1} + \cdots + \beta_{p-1} X_{i,p-1} + \varepsilon_i \qquad \text{Full model}$$

We wish to test:

(7.85)
$$C_1: \beta_q = \beta_{q+1} = \cdots = \beta_{p-1} = 0$$
$$C_2: \text{not all of the } \beta\text{'s in } C_1 \text{ equal zero}$$

where, for convenience, we arrange the model so that the last $p - q$ coefficients are the ones to be tested. We first fit the full model and obtain $SSE(X_1, \ldots, X_{p-1})$. Then we fit the reduced model:

(7.86) $$Y_i = \beta_0 + \beta_1 X_{i1} + \cdots + \beta_{q-1} X_{i,q-1} + \varepsilon_i \qquad \text{Reduced model}$$

and obtain $SSE(X_1, \ldots, X_{q-1})$. Finally we set up the test statistic:

$$F^* = \frac{SSE(R) - SSE(F)}{(n-q) - (n-p)} \div \frac{SSE(F)}{n-p}$$

or equivalently:

(7.87) $$F^* = \frac{SSR(X_q, \ldots, X_{p-1} | X_1, \ldots, X_{q-1})}{p-q} \div MSE(F)$$

Note that the test statistic (7.87) actually encompasses the first two cases mentioned. If $q = 1$, the test is whether all regression coefficients equal zero. If $q = p - 1$, the test is whether a single regression coefficient equals zero.

Other Tests

Other types of tests are occasionally required. For instance, for the full model:

(7.88) $$Y_i = \beta_0 + \beta_1 X_{i1} + \beta_2 X_{i2} + \beta_3 X_{i3} + \varepsilon_i \qquad \text{Full model}$$

we might wish to test:

(7.89)
$$C_1: \beta_1 = \beta_2 = 5$$
$$C_2: \beta_1 \text{ and } \beta_2 \text{ both do not equal } 5$$

The procedure would be to fit the full model (7.88), then the reduced model:

$$Y_i = \beta_0 + 5(X_{i1} + X_{i2}) + \beta_3 X_{i3} + \varepsilon_i$$

or, since $5(X_{i1} + X_{i2})$ are known constants:

(7.90) $$Y_i - 5(X_{i1} + X_{i2}) = \beta_0 + \beta_3 X_{i3} + \varepsilon_i \qquad \text{Reduced model}$$

and then use the general F^* test statistic (3.66) with 2 and $n - 4$ degrees of freedom.

7.11 COEFFICIENTS OF PARTIAL DETERMINATION

A coefficient of multiple determination R^2, it will be recalled, measures the proportionate reduction in the variation of Y achieved by the introduction of the entire set of X variables considered in the model. A *coefficient of partial determination,* in contrast, measures the marginal contribution of one X variable, when all others are already included in the model.

Two Independent Variables

Let us consider a first-order multiple regression model with two independent variables, as given in (7.1). We define:

(7.91) $$r^2_{Y1.2} = 1 - \frac{SSE(X_1, X_2)}{SSE(X_2)} = \frac{SSE(X_2) - SSE(X_1, X_2)}{SSE(X_2)}$$

or, by (7.69):

(7.91a) $$r^2_{Y1.2} = \frac{SSR(X_1 | X_2)}{SSE(X_2)}$$

Here, $r^2_{Y1.2}$ is the coefficient of partial determination between Y and X_1, given that X_2 is in the model. It measures the proportionate reduction in the variation of Y remaining after X_2 is included in the model which is gained by also including X_1 in the model.

The coefficient of partial determination between Y and X_2, given that X_1 is in the model, is defined:

(7.92) $$r^2_{Y2.1} = \frac{SSR(X_2 | X_1)}{SSE(X_1)}$$

For our earlier Zarthan Company example, these two coefficients of partial determination are (data derived from Table 7.5):

$$r^2_{Y1.2} = \frac{31{,}872 - 56.9}{31{,}872} = .998$$

$$r^2_{Y2.1} = \frac{485 - 56.9}{485} = .882$$

Thus in our example when X_1 is added to the model containing X_2, the error sum of squares falls from 31,872 to 56.9, being reduced by 99.8 percent. Correspondingly when X_2 is added to the model containing X_1, the error sum of squares falls from 485 to 56.9, being reduced by 88.2 percent.

General Case

The generalization of coefficients of partial determination to three or more independent variables in the model is immediate. For instance:

(7.93a)
$$r^2_{Y1.23} = \frac{SSR(X_1|X_2, X_3)}{SSE(X_2, X_3)}$$

(7.93b)
$$r^2_{Y2.13} = \frac{SSR(X_2|X_1, X_3)}{SSE(X_1, X_3)}$$

(7.93c)
$$r^2_{Y3.12} = \frac{SSR(X_3|X_1, X_2)}{SSE(X_1, X_2)}$$

Note that in the subscript to r^2, the entries to the left of the dot show in turn the variable taken as the response and the variable being added. The entries to the right of the dot show the X variables already in the model.

Comments

1. The coefficients of partial determination can take on values between 0 and 1, as the definitions readily indicate.

2. A coefficient of partial determination can be interpreted as a coefficient of simple determination. Consider a multiple regression model with two independent variables. Suppose we regress Y on X_2, and obtain the residuals:

$$Y_i - \hat{Y}_i(X_2)$$

where $\hat{Y}_i(X_2)$ denotes the fitted values of Y when X_2 is in the model.

Suppose we further regress X_1 on X_2 and obtain the residuals:

$$X_{i1} - \hat{X}_{i1}(X_2)$$

where $\hat{X}_{i1}(X_2)$ denotes the fitted values of X_1 in the regression of X_1 on X_2. The coefficient of simple determination r^2 between these two sets of residuals equals the coefficient of partial determination $r^2_{Y1.2}$. Thus, this coefficient measures the relation between Y and X_1 when both of these have been adjusted for their linear relationships to X_2.

3. The coefficients of partial determination can be expressed in terms of simple or other partial correlation coefficients. For example:

(7.94)
$$r^2_{Y2.1} = \frac{(r_{Y2} - r_{12} r_{Y1})^2}{(1 - r^2_{12})(1 - r^2_{Y1})}$$

(7.95)
$$r^2_{Y2.13} = \frac{(r_{Y2.3} - r_{12.3} r_{Y1.3})^2}{(1 - r^2_{12.3})(1 - r^2_{Y1.3})}$$

Extensions are straightforward.

Coefficients of Partial Correlation

The square root of the coefficient of partial determination is called the *coefficient of partial correlation*. This coefficient is frequently used in practice, although it does not have as clear a meaning as the coefficient of partial determination.

For our Zarthan Company example, we have:

$$r_{Y1.2} = \sqrt{.998} = .999$$

$$r_{Y2.1} = \sqrt{.882} = .939$$

Usually the partial correlation coefficient is given the same sign as that of the corresponding regression coefficient in the fitted regression function. For our Zarthan Company example, we had (see Table 7.5a): $b_1 = +.496$ and $b_2 = +.00920$. Since these are both positive, each of the partial correlation coefficients above is taken as positive.

Partial correlation coefficients are frequently used in computer routines for finding the best independent variable to be selected next for inclusion in the regression model. We shall discuss this use in Chapter 11.

7.12 STANDARDIZED REGRESSION COEFFICIENTS

Standardized regression coefficients may be used to facilitate comparisons between regression coefficients. Ordinarily, it is difficult to compare regression coefficients because of differences in the units involved. We cite two examples.

1. Suppose that the following fitted response function has been obtained:

$$\hat{Y} = 200 + 20{,}000X_1 + 0.2X_2$$

On seeing this response function, one may be tempted to conclude that X_1 is the only important independent variable, and that X_2 has little effect on the dependent variable Y. A little reflection should make one wary of this conclusion. The reason is that we do not know the units involved. Suppose the units are:

Y	in dollars
X_1	in thousand dollars
X_2	in cents

In that event, the effect on the mean response of a \$1,000 increase in X_1 when X_2 is constant would be exactly the same as the effect of a \$1,000 increase in X_2 when X_1 is constant, despite the difference in the regression coefficients.

2. In our Zarthan Company example, we cannot make any comparison between b_1 and b_2 because b_1 is in units of one gross of jars per thousand persons while b_2 is in units of one gross of jars per dollar of per capita income.

Standardized regression coefficients, also sometimes called *beta coefficients*, are defined as follows:

$$(7.96) \qquad B_k = b_k \left[\frac{\dfrac{\sum_i (X_{ik} - \bar{X}_k)^2}{n-1}}{\dfrac{\sum_i (Y_i - \bar{Y})^2}{n-1}} \right]^{1/2} = b_k \left[\frac{\sum_i (X_{ik} - \bar{X}_k)^2}{\sum_i (Y_i - \bar{Y})^2} \right]^{1/2}$$

The effect of the term in brackets is to make B_k dimensionless. The coefficient B_k reflects the change in the mean response (in units of standard deviations of Y) per unit change in the independent variable X_k (in units of standard deviations of X_k), when all other independent variables are held constant.

For our Zarthan Company example, we obtain the following standardized regression coefficients:

$$B_1 = .496 \left[\frac{191{,}089}{53{,}902} \right]^{1/2} = .934$$

$$B_2 = .00920 \left[\frac{7{,}473{,}616}{53{,}902} \right]^{1/2} = .108$$

Sometimes, these standardized regression coefficients would be interpreted as showing that target population (X_1) has a greater impact than per capita income (X_2) on sales because B_1 is larger than B_2. The cautions in interpreting regression coefficients mentioned earlier apply to standardized regression coefficients as well: they show the effect of the given independent variable in the context of the other independent variables in the model. Changing the other independent variables in the model will usually change the standardized regression coefficients when the independent variables are correlated among themselves.

Further, the standardized regression coefficients are affected by the spacing of the independent variables, which may be quite arbitrary.

Hence, it is ordinarily not wise to interpret a standardized regression coefficient as reflecting the importance of the independent variable.

PROBLEMS

7.1. Refer to Figure 7.3a. By how much approximately does mean yield increase when rainfall increases from 9 to 11 inches and temperature is held constant? Could you have answered this question if rainfall and temperature interact?

7.2. Sketch a set of contour curves that represents the hill response surface in Figure 7.2. Does the hill response surface in Figure 7.2 reflect interacting or noninteracting variables?

7.3. Sketch a set of contour curves for the response surface:

$$E(Y) = 25 + 3X_1 + 5X_2 + 2X_1X_2$$

Explain the meaning of the regression coefficients.

7.4. For each of the following models, indicate whether it is a linear regression model, an intrinsically linear regression model, or neither of these. In the case of an intrinsically linear model, state how it can be expressed in the form of (7.7) by a suitable transformation:

1. $Y_i = \beta_0 + \beta_1 X_{i1} + \beta_2 \log X_{i2} + \beta_3 X_{i1}^2 + \varepsilon_i$
2. $Y_i = \varepsilon_i \exp(\beta_0 + \beta_1 X_{i1} + \beta_2 X_{i2}^2)$
3. $Y_i = \beta_0 + \log(\beta_1 X_{i1}) + \beta_2 X_{i2} + \varepsilon_i$

7.5. Set up the **X** matrix and **β** vector for each of the following models (assume $i = 1, \ldots, 4$):

1. $Y_i = \beta_0 + \beta_1 X_{i1} + \beta_2 X_{i1} X_{i2} + \varepsilon_i$
2. $Y_i = \beta_1 X_{i1} + \beta_2 X_{i2} + \beta_3 X_{i1}^2 + \varepsilon_i$
3. $\log Y_i = \beta_0 + \beta_1 X_{i1} + \beta_2 X_{i2} + \varepsilon_i$

7.6. Why is it not meaningful to attach a sign to the coefficient of multiple correlation R as we do for the coefficient of simple correlation r?

7.7. In a small-scale study of the relation between degree of brand liking (Y) and moisture content (X_1) and sweetness (X_2) of the product, the following results were obtained (data are coded):

i:	1	2	3	4	5	6
X_{i1}:	4	4	6	6	8	8
X_{i2}:	2	6	2	6	2	6
Y_i:	64	81	72	91	83	96

Assume model (7.1) with normal error terms is appropriate. Using matrix methods, find: (1) the vector of the estimated regression coefficients; (2) the vector of residuals; (3) SSE; (4) SSR; (5) the estimated variance-covariance matrix of **b**; (6) point estimate of $E(Y_h)$ when $X_{h1} = 5$, $X_{h2} = 4$; (7) estimated variance of $\hat{Y}_h$ when $X_{h1} = 5$, $X_{h2} = 4$.

7.8. Refer to Problem 7.7.

a) Calculate the coefficient of multiple determination R^2.
b) Show that the coefficient of simple determination r^2 between Y_i and $\hat{Y}_i$ equals R^2.

7.9. Refer to Problem 7.7.

a) Obtain an interval estimate of $E(Y_h)$ when $X_{h1} = 5$, $X_{h2} = 4$. Use a 95 percent confidence coefficient.
b) Obtain a prediction interval for a new observation $Y_{h(\text{new})}$ when $X_{h1} = 5$, $X_{h2} = 4$. Use a 99 percent confidence coefficient.
c) Test whether X_2 could be dropped from the regression model. Use a level of significance of .01. What is your conclusion?

7.10. The observations below, taken on 10 incoming shipments of chemicals in drums arriving at a warehouse, show number of drums in shipment (X_1), total weight of shipment (X_2, in hundred pounds), and number of man-minutes required to handle the shipment (Y):

i:	1	2	3	4	5	6	7	8	9	10
X_{i1}:	7	18	5	14	11	5	23	9	16	5
X_{i2}:	5.11	16.70	3.20	7.00	11.00	4.00	22.10	7.00	10.60	4.80
Y_i:	58	152	41	93	101	38	203	78	117	44

Assume model (7.1) with normal error terms is appropriate.

a) Obtain the estimated regression equation. How is b_1 here interpreted? How is b_2 here interpreted?

b) Test whether there is a regression relation, using a level of significance of .05. What does your test result imply about β_1 and β_2? Does it imply anything about β_0?

c) Estimate β_1 and β_2 jointly by the Bonferroni procedure, using a 95 percent family confidence coefficient. Interpret your results.

7.11. Refer to Problem 7.10.

a) Management desires simultaneous interval estimates of the mean handling times for five typical shipments specified to be as follows:

	1	2	3	4	5
X_1:	5	6	10	14	20
X_2:	3.20	4.80	7.00	10.00	18.00

Obtain the family of estimates, using a 90 percent family confidence coefficient. Employ the Working-Hotelling type bounds or the Bonferroni procedure, whichever is more efficient.

b) For the observations in Problem 7.10, would you consider a shipment of 20 drums with a weight of 5 hundred pounds to be within the scope of the model? What about a shipment of 20 drums with a weight of 19 hundred pounds? Support your answer by preparing a relevant chart.

7.12. Refer to Problem 7.10. Four separate shipments with the following characteristics will arrive in the next day or two:

	1	2	3	4
X_1:	9	12	15	18
X_2:	7.20	9.00	12.50	16.50

Management desires predictions of the handling times for these shipments so that the actual handling times can be compared with the predicted times to determine whether any actual times are "out of line." Develop the needed predictions, using the most efficient approach and a family confidence coefficient of 90 percent.

7.13. Refer to Problem 7.10.

a) Obtain the residuals and make appropriate residual plots to ascertain whether model (7.1) with normal error terms is apt. Summarize your findings.

b) Can you conduct a formal test for lack of fit here?

7.14. Refer to the Zarthan Company example on page 234. The company's sales manager has suggested that the predictive ability of the model could be greatly improved if promotional expenditures were added to the model, since he knows from experience that these expenditures have a substantial impact on sales. The company allocates its total promotional budget proportionately to the target populations in the districts. Thus a district containing 4.7 percent of the total target population receives 4.7 percent of the total promotional budget. Evaluate the sales manager's suggestion.

7.15. A speaker stated, in a workshop on applied regression analysis: "In business and the social sciences, some degree of multicollinearity in survey data is practically inevitable." Does this statement apply equally to experimental data?

7.16. Refer to Problem 7.10.

a) Does $SSR(X_1)$ equal $SSR(X_1|X_2)$? Does $SSR(X_2)$ equal $SSR(X_2|X_1)$? What does this imply?

b) Does $SSR(X_1)+SSR(X_2|X_1)$ equal $SSR(X_2)+SSR(X_1|X_2)$? Must this always be the case?

7.17. a) Define each of the following extra sums of squares: (1) $SSR(X_3|X_1)$; (2) $SSR(X_3, X_4|X_1)$; (3) $SSR(X_4|X_1, X_2, X_3)$.

b) For a multiple regression model with five X variables, which is the relevant extra sum of squares for testing whether or not $\beta_5=0$? Whether or not $\beta_2=\beta_4=0$?

7.18. An undergraduate, working for a campus apparel shop serving student customers, studied the relation between monthly allowance received by customer (X_1), number of years customer is in college (X_2), and dollar sales to customer to date (Y). The predictive model considered was:

$$Y_i=\beta_0+\beta_1 X_{i1}+\beta_2 X_{i2}+\beta_3 X_{i1}^2+\varepsilon_i$$

State the reduced models for testing whether or not: (1) $\beta_1=\beta_3=0$; (2) $\beta_0=0$; (3) $\beta_3=5$; (4) $\beta_0=10$; (5) $\beta_1=\beta_2$.

7.19. Refer to Problems 7.10 and 7.16. Would you advise dropping X_1 from the model, given that X_2 is to be retained? Would you advise dropping X_2, given that X_1 is to be retained? Use a level of significance of .05 in each case.

7.20. Refer to the computer simulation example on page 262. An observer states that both the number of trials variable (X_1) and the second-order term for the number of statements variable (X_2^2) can be dropped from model (7.81). Conduct the appropriate test and state your findings. Use a level of significance of .01.

7.21. Refer to the work crew productivity example on page 254.

a) Calculate r_{Y1}^2, r_{Y2}^2, $r_{Y1.2}^2$, $r_{Y2.1}^2$, and R^2. Explain what each coefficient measures and interpret your results.

b) Are any of the results obtained in part (a) special because the two independent variables are uncorrelated?

c) Obtain the standardized regression coefficients. How do you interpret these coefficients? Do they have a special meaning here because the independent variables are uncorrelated? (Hint: Obtain r_{Y1}, r_{Y2}.)

7.22. Refer to Problem 7.10.

a) Calculate r_{Y1}^2, r_{Y2}^2, r_{12}^2, $r_{Y1.2}^2$, $r_{Y2.1}^2$, and R^2. Explain what each coefficient measures and interpret your results.

b) Obtain the standardized regression coefficients. How do you interpret these coefficients here?

7.23. (Data set, to be used in accordance with teacher's instructions.) To study several questions of interest, a personnel officer in a governmental agency administered five newly developed aptitude tests to each of 25 applicants for entry-level clerical positions in the agency. For purposes of the study all 25 applicants were accepted for positions irrespective of their test scores. After a probationary period each applicant was rated for proficiency on the job. The scores on the five tests (X_1, X_2, X_3, X_4, X_5) and the job proficiency scores (Y) were as follows:

	Test Score					*Job Proficiency Score*
Subject	X_1	X_2	X_3	X_4	X_5	Y
1	86	110	100	95	87	88
2	62	97	99	99	100	80
3	110	107	103	101	103	96
4	101	117	93	91	95	76
5	100	101	95	102	88	80
6	78	85	95	94	84	73
7	120	77	80	89	74	58
8	105	122	116	112	102	116
9	112	119	106	110	105	104
10	120	89	105	87	97	99
11	87	81	90	90	88	64
12	133	120	113	101	108	126
13	140	121	96	100	89	94
14	84	113	98	85	78	71
15	106	102	109	99	109	111
16	109	129	102	101	108	109
17	104	83	100	93	102	100
18	150	118	107	108	110	127
19	98	125	108	100	95	99
20	120	94	95	96	90	82
21	74	121	91	95	85	67
22	96	114	114	91	103	109
23	104	73	93	80	80	78
24	94	121	115	85	104	115
25	91	129	97	105	83	83

8

Polynomial Regression

IN THIS CHAPTER, we consider one important type of curvilinear response model, namely the polynomial regression model. This is the most frequently used curvilinear response model in practice, because of its ease in handling as a special case of the general linear regression model (7.18). First, we discuss some commonly used polynomial regression models. Then we present two cases to illustrate some of the major problems encountered with polynomial regression.

8.1 POLYNOMIAL REGRESSION MODELS

Polynomial regression models can contain one, two, or more than two independent variables. Further, the independent variable can be present in various powers. We illustrate now some major possibilities.

One Independent Variable—Second Degree

The model:

$$Y_i = \beta_0 + \beta_1 X_i + \beta_2 X_i^2 + \varepsilon_i \tag{8.1}$$

is said to be a *second-order model with one independent variable*, because the independent variable appears in the second degree. The regression coefficients in polynomial regression are frequently written in a slightly different fashion, to reflect the pattern of the exponents:

$$Y_i = \beta_0 + \beta_1 X_i + \beta_{11} X_i^2 + \varepsilon_i \tag{8.1a}$$

We shall employ this latter notation in this chapter.

The response function for model (8.1a) is:

$$E(Y) = \beta_0 + \beta_1 X + \beta_{11} X^2 \tag{8.2}$$

which is a parabola and is frequently called a *quadratic* response function. Figure 8.1 contains two examples of second-order polynomial response functions.

FIGURE 8.1

Examples of Second-Order Polynomial Response Functions

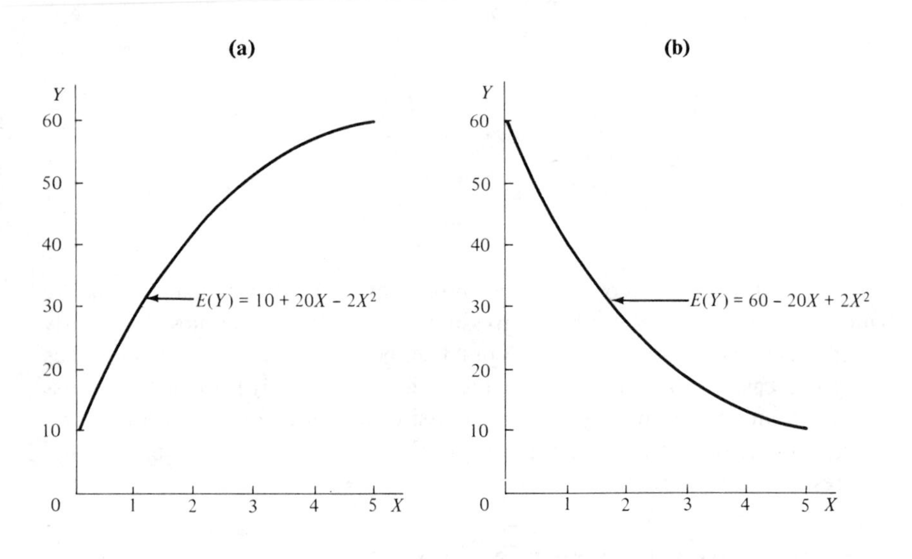

The regression coefficient β_0 represents the mean response of Y when $X = 0$ if the scope of the model includes $X = 0$. Otherwise, β_0 has no separate meaning of its own in the model. The regression coefficient β_1 is often called the *linear effect coefficient* while β_{11} is called the *curvature effect coefficient.*

Uses of Second-Order Model. The second-order polynomial response function (8.2) has two basic types of uses:

1. When the true response function is indeed a second-degree polynomial, containing additive linear and curvature effect components.
2. When the true response function is unknown (or complex), but a second-degree polynomial is a good approximation to the true function.

The second type of use is the more common one, and entails a special danger, that of extrapolation. Consider again Figure 8.1a. This response function may fit the data at hand very well. If, however, information about $E(Y)$ is sought for a larger value of X, extrapolation of this response function leads to the result shown in Figure 8.2, namely a turning down of the response

FIGURE 8.2

Extrapolation of Second-Order Polynomial Response Function in Figure 8.1a

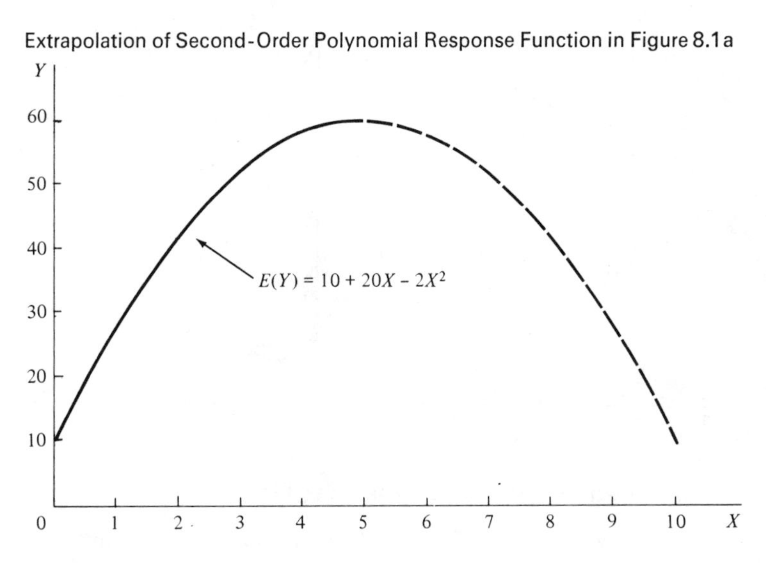

function, which may not be in accord with reality. Polynomial regressions of all types, especially those of higher order, share this danger of extrapolation. They may provide good fits for the data at hand, but may turn in unexpected directions when extrapolated beyond the range of the data.

One Independent Variable—Third Degree

The model:

$$Y_i = \beta_0 + \beta_1 X_i + \beta_{11} X_i^2 + \beta_{111} X_i^3 + \varepsilon_i \tag{8.3}$$

is a *third-order model with one independent variable*. The response function for model (8.3) is:

$$E(Y) = \beta_0 + \beta_1 X + \beta_{11} X^2 + \beta_{111} X^3 \tag{8.4}$$

Figure 8.3 contains two examples of third-order polynomial response functions.

FIGURE 8.3

Examples of Third-Order Polynomial Response Functions

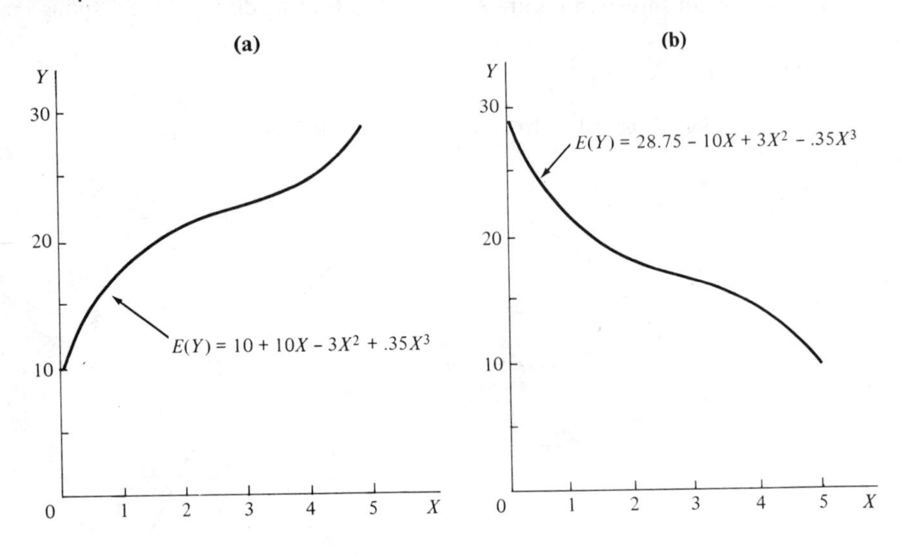

One Independent Variable—Higher Degrees

Polynomial models with the independent variable present in higher powers than the third are not often employed. The interpretation of the coefficients becomes difficult for such models, and they may be highly erratic for even small extrapolations. It must be recognized in this connection that a polynomial model of sufficiently high degree can always be found to fit the data perfectly. For instance, the fitted polynomial regression function for one independent variable of degree $n - 1$ will pass through all n observed Y values. One needs to be wary therefore of using high degree polynomials for the sole purpose of obtaining a good fit. Such regression functions may not show clearly the basic elements of the regression relation between X and Y and may lead to erratic extrapolations.

Two Independent Variables—Second Degree

The model:

$$Y_i = \beta_0 + \beta_1 X_{i1} + \beta_2 X_{i2} + \beta_{11} X_{i1}^2 + \beta_{22} X_{i2}^2 + \beta_{12} X_{i1} X_{i2} + \varepsilon_i \tag{8.5}$$

is a *second-order model with two independent variables*. The response surface is:

$$E(Y) = \beta_0 + \beta_1 X_1 + \beta_2 X_2 + \beta_{11} X_1^2 + \beta_{22} X_2^2 + \beta_{12} X_1 X_2 \tag{8.6}$$

which is the equation of a conic section. Note that model (8.5) contains separate linear and curvature components for each of the two independent variables and a cross-product term. The latter represents the interaction effects between X_1 and X_2, as we noted in Chapter 7. The coefficient β_{12} is often called the *interaction effect coefficient*.

The second-degree response surface for two independent variables, given in (8.6), represents two basic types of surfaces illustrated in Figure 7.2:

1. A hill
2. A saddle

Stationary and rising ridges constitute limiting cases of these two basic types of response surfaces.

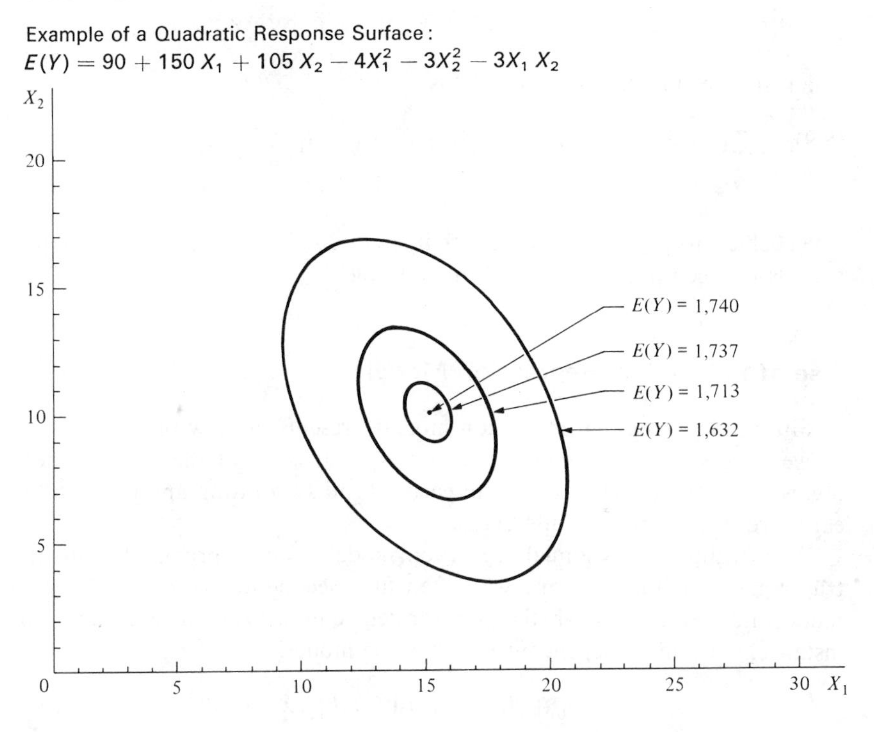

FIGURE 8.4

Example of a Quadratic Response Surface:
$E(Y) = 90 + 150\,X_1 + 105\,X_2 - 4X_1^2 - 3X_2^2 - 3X_1\,X_2$

Usually, it is easiest to portray the second-order response surface (8.6) in terms of contour lines. Figure 8.4 contains a representation, in terms of contour lines, of the response function:

$$(8.7) \qquad E(Y) = 90 + 150X_1 + 105X_2 - 4X_1^2 - 3X_2^2 - 3X_1X_2$$

Note that this response surface is a hill, with a maximum at $X_1 = 15$ and $X_2 = 10$.

Polynomial models in two (or more) independent variables are well adapted to situations where the response function is unknown and a suitable model is to be developed empirically.

Note

The cross-product term $\beta_{12} X_1 X_2$ in (8.6) is considered to be a second-order term, the same as $\beta_{11} X_1^2$ or $\beta_{22} X_2^2$. The reason can be seen readily by writing the latter terms as $\beta_{11} X_1 X_1$ and $\beta_{22} X_2 X_2$ respectively.

Three Independent Variables—Second Degree

The *second-order model with three independent variables* is:

$$Y_i = \beta_0 + \beta_1 X_{i1} + \beta_2 X_{i2} + \beta_3 X_{i3} + \beta_{11} X_{i1}^2 + \beta_{22} X_{i2}^2 + \beta_{33} X_{i3}^2 + \beta_{12} X_{i1} X_{i2} + \beta_{13} X_{i1} X_{i3} + \beta_{23} X_{i2} X_{i3} + \varepsilon_i \tag{8.8}$$

The response surface for this model is:

$$E(Y) = \beta_0 + \beta_1 X_1 + \beta_2 X_2 + \beta_3 X_3 + \beta_{11} X_1^2 + \beta_{22} X_2^2 + \beta_{33} X_3^2 + \beta_{12} X_1 X_2 + \beta_{13} X_1 X_3 + \beta_{23} X_2 X_3 \tag{8.9}$$

The coefficients β_{12}, β_{13}, and β_{23} are interaction effects coefficients for interactions between pairs of independent variables.

Use of Polynomial Regression Models

Fitting of polynomial regression models presents no new problems since, as we have seen in Chapter 7, they are special cases of the general linear regression model (7.18). Hence, all earlier results on fitting apply, as do the earlier results on making inferences.

When using a polynomial regression model as an approximation to the true regression function, one will often fit a second-degree or third-degree model and then explore whether a lower degree model is really adequate. For instance, with one independent variable, the model:

$$Y_i = \beta_0 + \beta_1 X_i + \beta_{11} X_i^2 + \beta_{111} X_i^3 + \varepsilon_i$$

may be fitted with the hope that the cubic term and perhaps even the quadratic term can be dropped. Thus, one would wish to test whether or not $\beta_{111} = 0$, or whether or not both $\beta_{11} = 0$ and $\beta_{111} = 0$. Similar tests would often be conducted with polynomial regression models for two or more independent variables.

8.2 EXAMPLE 1—ONE INDEPENDENT VARIABLE

We illustrate now some of the major types of analyses usually conducted with polynomial regression models with one independent variable.

Setting

A staff analyst for a cafeteria chain wishes to investigate the relation between the number of self-service coffee dispensers in a cafeteria line and sales of coffee. Fourteen cafeterias are chosen for the experiment, because they are similar in such respects as total volume of business, type of clientele, and location. The number of self-service dispensers which are placed in the test cafeterias varies from zero (coffee is dispensed here by a line attendant) to seven and is assigned randomly to each cafeteria.

Table 8.1 contains the results of the experimental study. Sales are measured in hundreds of gallons of coffee sold.

TABLE 8.1

Data for Cafeteria Coffee Sales Example

Cafeteria i	*Number of Dispensers* X_i	*Coffee Sales (hundred gallons)* Y_i
1	0	508.1
2	0	498.4
3	1	568.2
4	1	577.3
5	2	651.7
6	2	657.0
7	4	755.3
8	4	758.9
9	5	787.6
10	5	792.1
11	6	841.4
12	6	831.8
13	7	854.7
14	7	871.4

Fitting of Model

The analyst believes that the relation between sales and number of self-service dispensers is quadratic in the range of observations; sales should increase as the number of dispensers is greater, but if the space is cluttered with dispensers this increase becomes retarded. Hence, he would like to fit the quadratic model:

$$Y_i = \beta_0 + \beta_1 X_i + \beta_{11} X_i^2 + \varepsilon_i \tag{8.10}$$

He further anticipates that the error terms ε_i will be fairly normally distributed, with constant variance.

The **Y** and **X** matrices for this application are given in Table 8.2. Note that the **X** matrix contains a column of 1's, a column of the X observations, and a

TABLE 8.2

Data Matrices for Cafeteria Coffee Sales Example

$$
\mathbf{Y} = \begin{bmatrix} 508.1 \\ 498.4 \\ 568.2 \\ 577.3 \\ 651.7 \\ 657.0 \\ 755.3 \\ 758.9 \\ 787.6 \\ 792.1 \\ 841.4 \\ 831.8 \\ 854.7 \\ 871.4 \end{bmatrix}
\qquad
\mathbf{X} = \begin{bmatrix} 1 & 0 & 0 \\ 1 & 0 & 0 \\ 1 & 1 & 1 \\ 1 & 1 & 1 \\ 1 & 2 & 4 \\ 1 & 2 & 4 \\ 1 & 4 & 16 \\ 1 & 4 & 16 \\ 1 & 5 & 25 \\ 1 & 5 & 25 \\ 1 & 6 & 36 \\ 1 & 6 & 36 \\ 1 & 7 & 49 \\ 1 & 7 & 49 \end{bmatrix}
$$

(columns of $\mathbf{X}$: 1, X, X^2)

TABLE 8.3

Regression Results for Cafeteria Coffee Sales Example

(a) Regression Coefficients

Regression Coefficient	*Estimated Regression Coefficient*	*Estimated Standard Deviation*	t^*
β_0	503.346	4.795	104.97
β_1	78.941	3.455	22.85
β_{11}	−3.969	.482	−8.23

(b) Analysis of Variance

Source of Variation	*SS*	*df*	*MS*
Regression	225,031.8	2	112,515.9
Error	711.7	11	64.7
Total	225,743.5	13	

(c) $(\mathbf{X}'\mathbf{X})^{-1}$ *Matrix*

$$
\begin{bmatrix} .35537190 & -.19318182 & .02169421 \\ -.19318182 & .18452381 & -.02489177 \\ .02169421 & -.02489177 & .00359111 \end{bmatrix}
$$

column of the X^2 values. From this point on, the calculations are routine. We could do the matrix calculations manually, as illustrated in Chapter 7, or with a multiple regression program on a computer. Since no new problems are encountered, we shall simply present the basic computer output, including the $(\mathbf{X'X})^{-1}$ matrix, in Table 8.3.

FIGURE 8.5

Second-Order Polynomial Regression for Cafeteria Coffee Sales Example

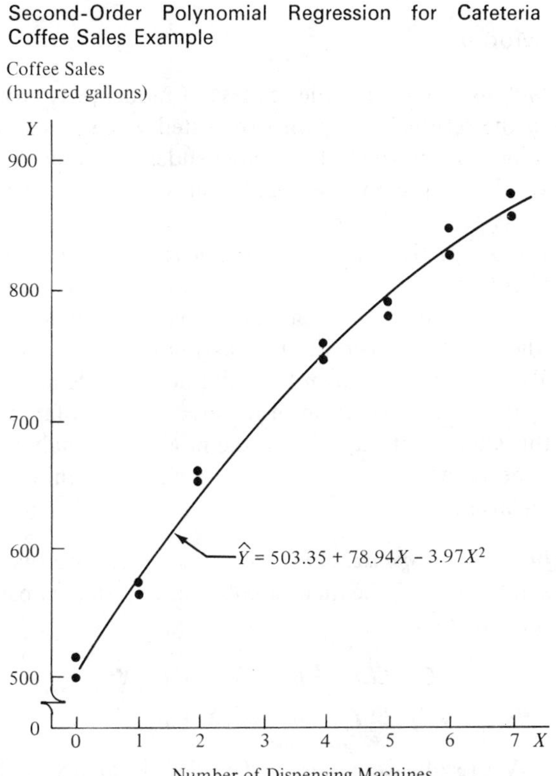

The fitted regression equation is:

$$\hat{Y} = 503.35 + 78.94X - 3.97X^2 \tag{8.11}$$

This response function is plotted in Figure 8.5, together with the original data.

Algebraic Version of Normal Equations. The algebraic version of the least squares normal equations:

$$\mathbf{X'Xb} = \mathbf{X'Y}$$

for the second-order polynomial model (8.10) can be readily obtained from (7.64), by replacing X_{i1} by X_i and X_{i2} by X_i^2. This yields the normal equations:

$$
\begin{aligned}
\sum Y_i &= nb_0 + b_1 \sum X_i + b_{11} \sum X_i^2 \\
\sum X_i Y_i &= b_0 \sum X_i + b_1 \sum X_i^2 + b_{11} \sum X_i^3 \\
\sum X_i^2 Y_i &= b_0 \sum X_i^2 + b_1 \sum X_i^3 + b_{11} \sum X_i^4
\end{aligned} \tag{8.12}
$$

Aptness of Model

Residual Analysis. To study the aptness of model (8.10) for his data, the analyst plotted the residuals e_i against the fitted values, as shown in Figure 8.6a, and also against the level of the independent variable X_i, as shown in Figure 8.6b. We do not present the calculations of the e_i, as these are routine.

There are no systematic departures from 0 evident in the residuals as either $\hat{Y}$ or X increases, suggesting that the quadratic response function is a good fit. Figure 8.5 makes this point also. Further, there is no tendency in Figures 8.6a and 8.6b for the spread in the residuals to vary systematically, so that it appears that the constant error variance assumption is a reasonable one. A normal probability plot, not shown here, did not provide any strong support for concluding that the distribution of the error terms is far from normal.

Based on this study of the aptness of the model, the analyst was willing to conclude that the normal error model (8.10), with constant error variance, is appropriate for his study.

Test for Quadratic Response Function. Since there are two repeat observations for each level of X, the analyst could have tested as part of his study of the aptness of the model:

$$
\begin{aligned}
&C_1: E(Y) = \beta_0 + \beta_1 X + \beta_{11} X^2 \\
&C_2: E(Y) \neq \beta_0 + \beta_1 X + \beta_{11} X^2
\end{aligned} \tag{8.13}
$$

The basic ANOVA results were presented earlier in Table 8.3b. The pure error sum of squares is obtained as follows from the data in Table 8.1:

$$
SSPE = (508.1 - 503.25)^2 + (498.4 - 503.25)^2 + (568.2 - 572.75)^2 + \cdots + (871.4 - 863.05)^2 = 304.6
$$

Note that $\bar{Y}_1 = 503.25$ for $X = 0$, $\bar{Y}_2 = 572.75$ for $X = 1$, and so on. There are $14 - 7 = 7$ degrees of freedom associated with $SSPE$. Hence, we have:

$$
MSPE = \frac{SSPE}{7} = \frac{304.6}{7} = 43.5
$$

Now we are in a position to obtain the lack of fit sum of squares by (4.11):

$$
SSLF = SSE - SSPE = 711.7 - 304.6 = 407.1
$$

FIGURE 8.6

Residual Plots for Cafeteria Coffee Sales Example

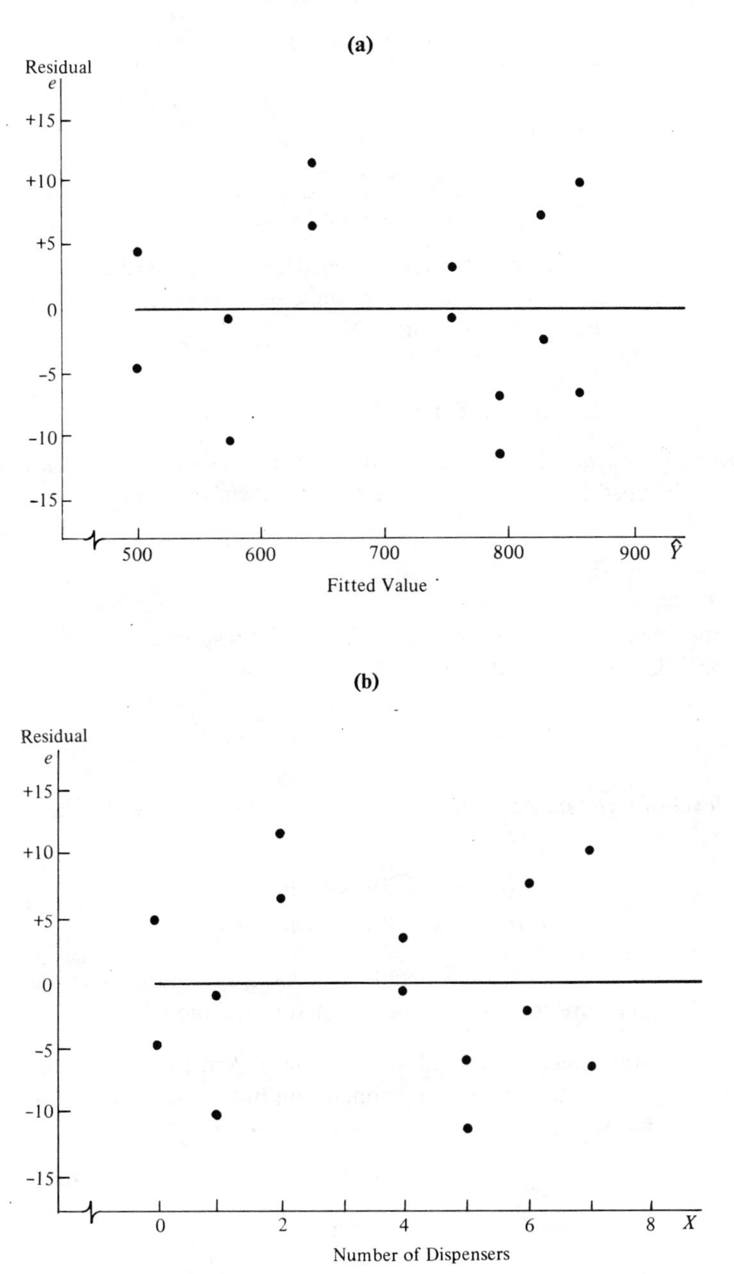

There are $7 - 3 = 4$ degrees of freedom associated with $SSLF$. (Remember that three parameters had to be estimated for the fitted regression equation.) Hence we have:

$$MSLF = \frac{SSLF}{4} = \frac{407.1}{4} = 101.8$$

Thus, test statistic (4.14) here is:

$$F^* = \frac{MSLF}{MSPE} = \frac{101.8}{43.5} = 2.34$$

Assuming the level of significance is to be .05, we require $F(.95; 4, 7) = 4.12$. Since $F^* = 2.34$ is less than this action limit, we conclude C_1, that the quadratic response function is appropriate.

Test Whether β_{11} Equals Zero

t Test Approach. The analyst next studied whether the quadratic term could be dropped from the model. He therefore wished to test:

$$\begin{aligned} C_1&: \beta_{11} = 0 \\ C_2&: \beta_{11} \neq 0 \end{aligned} \tag{8.14}$$

C_1 implies that there is no curvature effect in the response function.

Table 8.3a indicates that:

$$t^* = \frac{b_{11}}{s(b_{11})} = \frac{-3.969}{.482} = -8.23$$

For a level of significance of .05, we require $t(.975; 11) = 2.201$. The decision rule is:

$$\begin{aligned} &\text{If } |t^*| \leq 2.201, \text{ conclude } C_1 \\ &\text{If } |t^*| > 2.201, \text{ conclude } C_2 \end{aligned}$$

Since $|t^*| = 8.23 > 2.201$, we conclude C_2, that a curvature effect does exist, so that the curvature term should be retained in the model.

ANOVA ***Approach.*** The analyst could have used the general linear test approach to choose the appropriate conclusion in (8.14). Under C_2 in (8.14), the reduced model is:

$$Y_i = \beta_0 + \beta_1 X_i + \varepsilon_i \qquad \text{Reduced model} \tag{8.15}$$

When this reduced model is fitted, we obtain:

$$\hat{Y} = 527.33 + 51.43X$$

$$SSE(R) = 5{,}099.4$$

Earlier, we found for the full model (see Table 8.3b):

$$SSE(F) = 711.7$$

Using the explicit notation:

$$SSE(F) = SSE(X, X^2)$$
$$SSE(R) = SSE(X)$$

we obtain by (7.70) the extra sum of squares:

$$SSR(X^2 \mid X) = SSE(X) - SSE(X, X^2) = 5{,}099.4 - 711.7 = 4{,}387.7$$

Table 8.4 contains the expanded analysis of variance, with SSR decomposed into the required extra sum of squares components.

TABLE 8.4

ANOVA for Cafeteria Coffee Sales Example

Source of Variation	*SS*	*df*	*MS*
Regression	$SSR(X, X^2) = 225{,}031.8$	2	112,515.9
Linear	$SSR(X) = 220{,}644.1$	1	220,644.1
Curvature	$SSR(X^2 \mid X) = 4{,}387.7$	1	4,387.7
Error	$SSE = 711.7$	11	64.7
Total	$SSTO = 225{,}743.5$	13	

The test statistic by (7.73) then is:

$$F^* = \frac{SSR(X^2 \mid X)}{1} \div MSE(X, X^2) = 4{,}387.7 \div 64.7 = 67.8$$

For a 5 percent level of significance, we need $F(.95; 1, 11) = 4.84$. Since F^* is greater than this action limit, we are led to conclude C_2, as by the t test.

Note

One may observe here the relation discussed in the previous chapter between the t and F tests as to whether a regression coefficient equals zero. We have for the test statistic:

$$(t^*)^2 = (-8.23)^2 = 67.8 = F^*$$

A corresponding relation holds between the action limits:

$$[t(.975; 11)]^2 = (2.201)^2 = 4.84 = F(.95; 1, 11)$$

Estimation of Regression Coefficients

The analyst next wished to obtain confidence bounds on the two regression coefficients β_1 and β_{11}, with a family confidence coefficient .90. The Bonferroni method is to be used, in view of its simplicity and the ease of interpreting the results.

Here $s = 2$ statements are desired; hence by (7.44a), we have:

$$B = t(1 - .10/2(2); 11) = t(.975; 11) = 2.201$$

From Table 8.3a, we find:

$$b_1 = 78.941 \qquad s(b_1) = 3.455$$
$$b_{11} = -3.969 \qquad s(b_{11}) = .482$$

The Bonferroni confidence intervals therefore are, by (7.44):

$$78.941 - (2.201)(3.455) \leq \beta_1 \leq 78.941 + (2.201)(3.455)$$

or:

$$71.34 \leq \beta_1 \leq 86.55$$

$$-3.969 - (2.201)(.482) \leq \beta_{11} \leq -3.969 + (2.201)(.482)$$

or:

$$-5.03 \leq \beta_{11} \leq -2.91$$

The analyst was satisfied with the precision of these two statements, feeling that the intervals are narrow enough to give him reliable simultaneous information about the comparative magnitudes of the linear and curvature effects.

Coefficient of Multiple Determination

For a descriptive measure of the degree of relation between coffee sales and number of dispensing machines, the analyst calculated the coefficient of multiple determination using the data in Table 8.3b:

$$R^2 = \frac{SSR}{SSTO} = \frac{225{,}031.8}{225{,}743.5} = .997$$

This measure shows that the variation in coffee sales is reduced by 99.7 percent when the quadratic relation to the number of dispensing machines is utilized.

Note that the coefficient of multiple determination is the relevant measure here, not the coefficient of simple determination r^2, since model (8.10) is a multiple regression model even though it contains only one independent variable. Sometimes in curvilinear regression, the coefficient of multiple correlation R is called the *correlation index*.

Estimation of Mean Response

The analyst was particularly interested in the mean response for $X_h = 3$ dispensing machines. He wished to estimate this mean response with a 98 percent confidence coefficient.

The proper interval estimate is given by (7.50). For our example, we have:

$$\mathbf{X}_h = \begin{bmatrix} 1 \\ X_h \\ X_h^2 \end{bmatrix} = \begin{bmatrix} 1 \\ 3 \\ 9 \end{bmatrix}$$

The estimated mean response corresponding to $\mathbf{X}_h$ is, by (7.46):

$$\hat{Y}_h = [1 \quad 3 \quad 9]\begin{bmatrix} 503.346 \\ 78.941 \\ -3.969 \end{bmatrix} = 704.448$$

Next, using the results in Table 8.3 for MSE and $(\mathbf{X}'\mathbf{X})^{-1}$, we obtain when substituting into (7.49):

$$s^2(\hat{Y}_h) = MSE(\mathbf{X}_h'(\mathbf{X}'\mathbf{X})^{-1}\mathbf{X}_h)$$

$$= 64.700[1 \quad 3 \quad 9]\begin{bmatrix} .35537 & -.19318 & .021694 \\ -.19318 & .18452 & -.024892 \\ .021694 & -.024892 & .0035911 \end{bmatrix}\begin{bmatrix} 1 \\ 3 \\ 9 \end{bmatrix}$$

$$= 12.563$$

or:

$$s(\hat{Y}_h) = 3.544$$

We require $t(.99; 11) = 2.718$. Hence, we obtain:

$$704.448 - (2.718)(3.544) \le E(Y_3) \le 704.448 + (2.718)(3.544)$$

or:

$$694.8 \le E(Y_3) \le 714.1$$

With confidence coefficient .98, the analyst can conclude that the mean of the distribution of coffee sales when three dispensing machines are used is somewhere between 694.8 and 714.1 hundred gallons.

8.3 EXAMPLE 2—TWO INDEPENDENT VARIABLES

We shall discuss now another example of polynomial regression, this one involving two independent variables. Rather than carrying this example through all of the various analytical stages as we did the first example, we shall focus here on the analysis of interaction effects and curvature effects.

Setting

Table 8.5 shows, for a sample of 18 physicians in the 35 to 39 age group holding policies with a certain life insurance company, average annual income during the past 5 years (X_1), risk aversion score (X_2), and amount of life insurance carried (Y). Risk aversion was measured by a standard questionnaire administered to each physician in the sample; the higher the score the greater the degree of risk aversion.

TABLE 8.5

Data for Life Insurance Example

Physician i	*Average Annual Income (thousand dollars)* X_{i1}	*Risk Aversion Score* X_{i2}	*Amount of Life Insurance Carried (thousand dollars)* Y_i
1	47.35	7	140
2	29.26	5	45
3	52.14	10	180
4	32.15	6	60
5	40.86	4	90
6	19.18	5	10
7	27.23	4	35
8	25.60	6	35
9	54.14	9	190
10	26.72	5	35
11	38.84	2	75
12	32.99	7	70
13	32.95	4	55
14	21.69	3	10
15	27.90	5	40
16	56.70	1	175
17	37.69	8	95
18	39.94	6	95

A company analyst, studying the relation of average annual income and risk aversion to amount of life insurance carried by physicians in the given age group holding company policies, expected that a quadratic relation would hold between income and amount of life insurance carried. However, he would not have been surprised if aversion to risk showed only linear effects and no curvature effects on amount of life insurance carried, and he was quite uncertain whether or not the two variables interact in their effects on amount of life insurance carried. Hence, he fitted the second-order polynomial regression model:

$$Y_i = \beta_0 + \beta_1 X_{i1} + \beta_2 X_{i2} + \beta_{11} X_{i1}^2 + \beta_{22} X_{i2}^2 + \beta_{12} X_{i1} X_{i2} + \varepsilon_i \tag{8.16}$$

with the intention of first testing for the presence of interaction effects, and then for curvature effects of aversion to risk.

Development of Model

First, the analyst tested for the presence of interaction effects:

$$(8.17) \qquad \begin{aligned} C_1&: \beta_{12} = 0 \\ C_2&: \beta_{12} \neq 0 \end{aligned}$$

Table 8.6 summarizes the basic data for the fit of model (8.16). The analyst could have used the statistic $t^* = -1.40$ from Table 8.6 for testing (8.17). He chose, however, to use the general linear test approach, and fitted the reduced model under C_1 in (8.17):

$$(8.18) \qquad Y_i = \beta_0 + \beta_1 X_{i1} + \beta_2 X_{i2} + \beta_{11} X_{i1}^2 + \beta_{22} X_{i2}^2 + \varepsilon_i$$

Table 8.7 summarizes the basic data for the fit of this reduced model.

Using the general linear test approach, the appropriate test statistic is:

$$F^* = \frac{\dfrac{SSE(X_1, X_2, X_1^2, X_2^2) - SSE(X_1, X_2, X_1^2, X_2^2, X_1X_2)}{13 - 12}}{\dfrac{SSE(X_1, X_2, X_1^2, X_2^2, X_1X_2)}{12}}$$

$$= \frac{\dfrac{21.6 - 18.6}{1}}{\dfrac{18.6}{12}} = 1.94$$

TABLE 8.6

Regression Results for Model (8.16)

(a) Regression Coefficients

Regression Coefficient	*Estimated Regression Coefficient*	*Estimated Standard Deviation*	t^*
β_0	−46.704	4.374	−10.68
β_1	1.017	.228	4.46
β_2	3.727	.963	3.87
β_{11}	.0501	.00307	16.32
β_{22}	.119	.0858	1.39
β_{12}	−.0196	.0140	−1.40

(b) Analysis of Variance

Source of Variation	*SS*	*df*	*MS*
Regression	55,105.0	5	11,021.00
Error	18.6	12	1.55
Total	55,123.6	17	

TABLE 8.7

Regression Results for Model (8.18)

(a) Regression Coefficients

Regression Coefficient	*Estimated Regression Coefficient*	*Estimated Standard Deviation*	t^*
β_0	-43.507	3.867	-11.25
β_1	.930	.227	4.10
β_2	3.181	.913	3.48
β_{11}	.0502	.00318	15.79
β_{22}	.0828	.0849	.98

(b) Analysis of Variance

Source of Variation	*SS*	*df*	*MS*
Regression	55,102.0	4	13,775.50
Error	21.6	13	1.66
Total	55,123.6	17	

Assuming that the level of significance is to be held at .05, we require $F(.95; 1, 12) = 4.75$. Since $F^* = 1.94$ is less than this action limit, we conclude C_1, that no interaction effect exists. This result was welcome to the analyst, as it simplifies the interpretation of the effects of the two independent variables.

At this point, the analyst decided to tentatively adopt the no-interaction model in (8.18), but he still wished to examine whether a curvature effect for aversion to risk exists. The appropriate t^* statistic in Table 8.7 is:

$$t^* = \frac{b_{22}}{s(b_{22})} = \frac{.0828}{.0849} = .98$$

For a 5 percent level of significance, we have $t(.975; 13) = 2.160$. Since the absolute value of t^* is less than this action limit, we conclude C_1, that there is no curvature effect for aversion to risk. Hence the analyst decided to adopt the revised model:

$$Y_i = \beta_0 + \beta_1 X_{i1} + \beta_2 X_{i2} + \beta_{11} X_{i1}^2 + \varepsilon_i \tag{8.19}$$

and obtained the estimated response function:

$$\hat{Y} = -44.535 + .840X_1 + 4.060X_2 + .0519X_1^2$$

He then used this fitted response function for further investigation of the effects of average annual income and aversion to risk on amount of life insurance carried in the population under study.

Note

When multiple tests on the same data are conducted, there exist, as noted earlier, problems with respect to the level of significance for the family of inferences. In the example just cited, the analyst was willing to conduct two tests on the same data, one for interaction effects and one for curvature effects of aversion to risk. The reason was that he knew by the Bonferroni inequality (5.21a) that the family level of significance for the two tests (each conducted at the 5 percent level of significance) could not exceed 10 percent.

In Chapter 11, we shall discuss more fully the empirical determination of an appropriate regression model.

8.4 ESTIMATING THE MAXIMUM OR MINIMUM OF A QUADRATIC REGRESSION FUNCTION

Sometimes in quadratic regression, we wish to estimate the maximum (or minimum) mean response of the regression function, and/or the level of X at which the maximum (or minimum) occurs. Figure 8.2 illustrates a quadratic response function with a maximum mean response.

Given the estimated quadratic response function:

$$\hat{Y} = b_0 + b_1 X + b_{11} X^2 \tag{8.20}$$

the maximum (minimum) occurs at the level X_m:

$$X_m = \frac{-b_1}{2b_{11}} \tag{8.21}$$

and the estimated mean response at X_m is:

$$\hat{Y}_m = b_0 - \frac{b_1^2}{4b_{11}} \tag{8.22}$$

$\hat{Y}_m$ is a maximum if b_{11} is negative, and a minimum if b_{11} is positive.

Example

For our earlier cafeteria coffee sales example, the fitted regression curve was:

$$\hat{Y} = 503.35 + 78.94X - 3.97X^2$$

Assuming that the quadratic regression function is appropriate for larger X values than those in the study, we would estimate that maximum mean coffee sales occur at:

$$X_m = \frac{-78.94}{2(-3.97)} = 10$$

and the estimated mean response there is:

$$\hat{Y}_m = 503.35 - \frac{(78.94)^2}{4(-3.97)} = 896$$

Thus, if the quadratic response function is valid for larger numbers of coffee dispensers than in the study, we would estimate that maximum mean coffee sales occur with 10 dispensers, and that these are 896 hundred gallons.

Comments

1. To derive (8.21), we differentiate $\hat{Y}$ in (8.20) with respect to X, and set this derivative equal to 0:

$$\frac{d\hat{Y}}{dX} = \frac{d}{dX}(b_0 + b_1 X + b_{11} X^2) = b_1 + 2b_{11} X = 0$$

and obtain:

$$X_m = \frac{-b_1}{2b_{11}}$$

Substituting this value into the fitted response function (8.20), we find:

$$\hat{Y}_m = b_0 + b_1\left(\frac{-b_1}{2b_{11}}\right) + b_{11}\left(\frac{-b_1}{2b_{11}}\right)^2$$

$$= b_0 - \frac{b_1^2}{4b_{11}}$$

2. For large samples, the approximate estimated variance of X_m is:

$$s^2(X_m) = \frac{b_1^2}{4b_{11}^2}\left[\frac{s^2(b_1)}{b_1^2} + \frac{s^2(b_{11})}{b_{11}^2} - \frac{2s(b_1, b_{11})}{b_1 b_{11}}\right] \tag{8.23}$$

This approximate estimated variance can be used to construct a confidence interval for the true X level at which the maximum (minimum) occurs. Approximate confidence intervals for $E(Y_m)$ can also be obtained. These are discussed in Reference 8.1.

8.5 SOME FURTHER COMMENTS ON POLYNOMIAL REGRESSION

1. The use of polynomial models in X is not without drawbacks. Such models can be more expensive in degrees of freedom than alternative nonlinear models or linear models with transformed variables. Another potential drawback is that multicollinearity is unavoidable. In general this will not be troublesome. However, if the levels of X are restricted to a narrow range, the degree of multicollinearity in the columns of the **X** matrix can be quite high.

2. Sometimes a quadratic response function is fitted for the purpose of establishing the linearity of the response function when repeat observations are not available for directly testing the linearity of the response function. Fitting the quadratic model:

$$Y_i = \beta_0 + \beta_1 X_i + \beta_{11} X_i^2 + \varepsilon_i \tag{8.24}$$

and testing whether $\beta_{11} = 0$ does not necessarily establish that a linear response function is appropriate. Figure 8.7 provides an example. If sample data were obtained for the response function in Figure 8.7, model (8.24) fitted, and a test on β_{11} made, it likely would lead to the conclusion that $\beta_{11} = 0$.

FIGURE 8.7

Example of Curvilinear Response Function

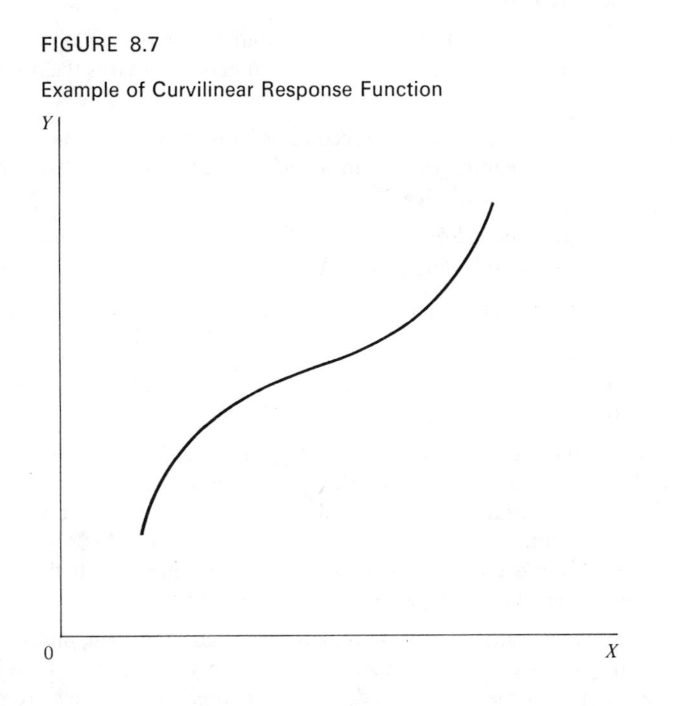

Yet a linear response function clearly would not be appropriate. Examination of residuals would disclose this lack of fit, and should always accompany formal testing of polynomial regression coefficients.

3. When a polynomial regression model with one independent variable is employed, the decomposition of SSR into extra sum of squares components proceeds logically as follows:

$$SSR(X)$$
$$SSR(X^2 | X)$$
$$SSR(X^3 | X, X^2)$$
etc.

PROBLEMS

8.1. A junior investment analyst, in a research seminar on municipal bonds, used a polynomial regression model of relatively high order and obtained an R of .997 in the regression of net interest yield of bond (Y) on industrial diversity index of municipality (X) for seven bond issues. A classmate, unimpressed, said: "You overfitted. Your curve follows the random effects in the data."
   a) Comment on the criticism.
   b) Might R_a defined in (7.33) be more appropriate than R as a descriptive measure here?

8.2. Refer to Figure 7.3b. Indicate a polynomial response function that might be apt here, and explain why you chose this function. What is the order of your polynomial?

8.3. (Calculus needed.) Refer to the second-order response function (8.2). Explain precisely the meaning of the linear effect coefficient β_1 and the curvature effect coefficient β_{11}.

8.4. Consider the second-order model (8.1a) with one independent variable. For each of the following sets of X values:

Set 1:	1.0	1.4	.9	1.3	1.8	.7	1.2	2.3
Set 2:	12	2	123	17	215	71	283	38

find the coefficient of correlation between X and X^2. What generalization does this suggest?

8.5. Refer to the fitted regression equation (8.11) for the coffee dispenser study.
   a) Estimate the mean coffee sales when there are five dispensing machines in a cafeteria. Use an interval estimate, with a 95 percent confidence coefficient.
   b) Predict the coffee sales in a particular cafeteria with five dispensing machines. Use a 95 percent confidence coefficient.

8.6. An operations analyst in a multinational electronics firm, studying factors affecting production in a piecework operation where earnings are based on the number of pieces produced, selected two employees each from various age groups and obtained data on their productivity last year (X–age, in years; Y–productivity data, coded):

i:	1	2	3	4	5	6	7	8	9
X_i:	20	20	25	25	30	30	35	35	40
Y_i:	93	85	99	105	109	114	109	111	100

i:	10	11	12	13	14	15	16	17	18
X_i:	40	45	45	50	50	55	55	60	60
Y_i:	105	94	101	94	98	105	108	109	112

The analyst recognized that the relation between age and productivity is complex, in part because earnings targets (which he could not measure) shift in complex ways with age. However, he believed that for purposes of estimating mean responses, the response function could be approximated suitably by a polynomial of third order, and that perhaps the cubic term and possibly even the quadratic term might be able to be dropped.

a) Fit the third-order polynomial regression model (8.3). Obtain the residuals, and study the aptness of model (8.3) with normal error terms. Summarize your findings. Were you able to make a formal test for lack of fit?

b) Assuming that model (8.3) is apt, determine whether the cubic and/or the quadratic terms can be dropped. Use a level of significance of .025 in any tests. What model do you recommend the analyst employ?

8.7. Refer to Problem 8.6, and employ the model arrived at in part (b) of that problem.

a) Estimate the mean productivity for workers of age 20, 30, 40, and 50 respectively, with a 90 percent family confidence coefficient. Justify the estimation procedure you employed. State your findings.

b) How useful do you think the fitted regression function is for estimating the mean productivity of workers aged 65?

8.8. Students comprising firm A in a computerized marketing game have approached you for assistance in analyzing the relation between promotional expenditures (X) and demand for their firm's product (Y) in the firm's home territory. They believe that the following characteristics hold in this relation: (1) demand in the home territory is affected primarily by promotional expenditures; (2) the relation is either quadratic or linear within the range of X levels of interest to the firm. The team has provided the observations shown below for the 14 periods covered in the game to date (X in thousand dollars, Y in million units), and has stated that these observations span the X levels of interest.

i:	1	2	3	4	5	6	7
X_i:	17	15	25	10	18	15	20
Y_i:	5.51	5.45	5.35	5.25	5.47	5.38	5.50

i:	8	9	10	11	12	13	14
X_i:	25	17	13	20	23	25	15
Y_i:	5.43	5.48	5.33	5.49	5.50	5.39	5.40

Assume that the second-order model (8.1a) with normal error terms applies.

a) Fit this model, and test whether a regression relation exists. Use a level of significance of .05. What do you conclude?

b) Can the quadratic term be dropped from the model? Use a level of significance of .01.

8.9. Refer to Problem 8.8. Obtain the residuals for the second-order regression model (8.1a) and evaluate the aptness of the model. Summarize your findings. Were you able to conduct a formal test for lack of fit?

8.10. Refer to Problem 8.8 and assume that the second-order regression model (8.1a) is appropriate. At what level of promotional expenditures does the maximum in the estimated response function occur? What is the estimated mean response at this level of promotional expenditures?

8.11. Refer to Problem 8.8. Someone who is familiar with this computerized marketing game enters the discussion. He states that in the system of equations on which the game is based, a quadratic relation does hold between promotional expenditures and mean demand in the firm's home territory. He believes that another significant variable related to expected demand in the home territory is the ratio of the firm's selling price to the average competitive selling price; however he does not recall whether this price ratio has both curvature and linear effects or only linear effects. He also does not recall whether price ratio and promotional expenditures interact in affecting demand. The firm's price ratios are ascertained for the 14 periods and are found to be as follows:

i:	1	2	3	4	5	6	7
Ratio:	.94	.98	1.05	.94	1.01	1.06	1.00

i:	8	9	10	11	12	13	14
Ratio:	.96	.99	1.02	1.01	.95	1.00	1.03

Investigate whether the price ratio should be included as a second independent variable in the model and, if so, whether a linear effect is adequate or a quadratic term is required and whether an interaction term is also needed. Explain the method of analysis you employed, and state the regression model you would recommend to the students. Did you use residual plots and/or other analyses to examine the aptness of the model?

CITED REFERENCE

8.1. Williams, E. J. *Regression Analysis*. New York: John Wiley & Sons, Inc., 1959.

9

Indicator Variables

THROUGHOUT the previous chapters on regression analysis, we have utilized quantitative variables in the regression models considered. Quantitative variables take on values on a well-defined scale; examples are income, age, temperature, and amounts of liquid assets.

Many variables of interest in business and economics, however, are not quantitative but rather are qualitative. Examples of qualitative variables are sex (male, female), purchase status (purchase, no purchase), and disability status (not disabled, partly disabled, fully disabled).

Qualitative variables can be used in a multiple regression model just as quantitative variables can, as we shall explain in this chapter. First, we take up the case where some or all of the independent variables are qualitative. Then we turn to the case where the dependent variable is qualitative.

9.1 ONE INDEPENDENT QUALITATIVE VARIABLE

An economist wished to relate the speed with which a particular insurance innovation is adopted (Y) to the size of the insurance firm (X_1) and the type of firm. The dependent variable is measured by the number of months elapsed between the time the first firm adopted the innovation and the time the given firm adopted the innovation. The independent variable X_1 is quantitative, and is measured by the amount of total assets of the firm. The second independent variable, type of firm, is qualitative and is composed of two classes—stock companies and mutual companies. In order that such a qualitative variable can be used in a regression model, quantitative indicators for the classes of the qualitative variable must be found.

Indicator Variables

There are many ways of quantitatively identifying the classes of a qualitative variable. We shall use indicator variables which take on the values 0 and 1. These indicator variables are easy to use, but they are by no means the only way to quantify a qualitative variable.

For our example, where the qualitative variable has two classes, we might define two indicator variables X_2 and X_3 as follows:

(9.1)
$$X_2 = \begin{matrix} 1 \text{ if stock company} \\ 0 \text{ otherwise} \end{matrix}$$
$$X_3 = \begin{matrix} 1 \text{ if mutual company} \\ 0 \text{ otherwise} \end{matrix}$$

Assuming that a first-order model is to be employed, it would be:

(9.2) $$Y_i = \beta_0 X_{i0} + \beta_1 X_{i1} + \beta_2 X_{i2} + \beta_3 X_{i3} + \varepsilon_i \qquad \text{where } X_{i0} \equiv 1$$

The intuitive approach of setting up an indicator variable for each class of the qualitative variable unfortunately leads to computational difficulties. To see why, suppose we have $n = 4$ observations, the first two being stock firms for which $X_2 = 1$ and $X_3 = 0$, and the second two being mutual firms for which $X_2 = 0$ and $X_3 = 1$. The $\mathbf{X}$ matrix would then be:

$$\mathbf{X} = \begin{array}{c} \begin{matrix} X_0 & X_1 & X_2 & X_3 \end{matrix} \\ \begin{bmatrix} 1 & X_{11} & 1 & 0 \\ 1 & X_{21} & 1 & 0 \\ 1 & X_{31} & 0 & 1 \\ 1 & X_{41} & 0 & 1 \end{bmatrix} \end{array}$$

Note that the X_0 column is equal to the sum of the X_2 and X_3 columns, so that the columns are linearly dependent according to definition (6.18). This has a serious effect on the $\mathbf{X'X}$ matrix:

$$\mathbf{X'X} = \begin{bmatrix} 1 & 1 & 1 & 1 \\ X_{11} & X_{21} & X_{31} & X_{41} \\ 1 & 1 & 0 & 0 \\ 0 & 0 & 1 & 1 \end{bmatrix} \begin{bmatrix} 1 & X_{11} & 1 & 0 \\ 1 & X_{21} & 1 & 0 \\ 1 & X_{31} & 0 & 1 \\ 1 & X_{41} & 0 & 1 \end{bmatrix}$$

$$= \begin{bmatrix} 4 & \sum_{i=1}^{4} X_{i1} & 2 & 2 \\ \sum_{i=1}^{4} X_{i1} & \sum_{i=1}^{4} X_{i1}^2 & \sum_{i=1}^{2} X_{i1} & \sum_{i=3}^{4} X_{i1} \\ 2 & \sum_{i=1}^{2} X_{i1} & 2 & 0 \\ 2 & \sum_{i=3}^{4} X_{i1} & 0 & 2 \end{bmatrix}$$

It is quickly apparent that the first column of the $\mathbf{X'X}$ matrix equals the sum of the last two columns, so that the columns are linearly dependent. Hence, the $\mathbf{X'X}$ matrix does not have an inverse and no unique estimators of the regression coefficients can be found.

A simple way out of this difficulty is to drop one of the indicator variables. In our example, for instance, we might drop X_3. While dropping one indicator variable is not the only way out of the difficulty, it leads to simple interpretations of the parameters. In general, therefore, we shall follow the principle:

(9.3) A qualitative variable with c classes will be represented by $c - 1$ indicator variables, each taking on the values 0 and 1.

Note

Indicator variables are frequently also called *dummy variables* or *binary variables*. The latter term has reference to the binary number system containing only 0 and 1.

Interpretation of Regression Parameters

Returning to our example, suppose that we drop the indicator variable X_3 from model (9.2) so that the model becomes:

$$Y_i = \beta_0 + \beta_1 X_{i1} + \beta_2 X_{i2} + \varepsilon_i \tag{9.4}$$

where:

$$X_{i2} = \begin{matrix} 1 \text{ if stock company} \\ 0 \text{ otherwise} \end{matrix}$$

The response function for this model is:

$$E(Y) = \beta_0 + \beta_1 X_1 + \beta_2 X_2 \tag{9.5}$$

To understand the meaning of the parameters of this model, consider first the case of a mutual firm. For such a firm, $X_2 = 0$ and we have:

$$E(Y) = \beta_0 + \beta_1 X_1 + \beta_2(0) = \beta_0 + \beta_1 X_1 \qquad \text{Mutual firm} \tag{9.5a}$$

Thus, the response function for a mutual firm is a straight line, with Y intercept β_0 and slope β_1. This response function is shown in Figure 9.1.

For a stock firm, $X_2 = 1$ and the response function (9.5) is:

$$E(Y) = \beta_0 + \beta_1 X_1 + \beta_2(1) = (\beta_0 + \beta_2) + \beta_1 X_1 \qquad \text{Stock firm} \tag{9.5b}$$

This also is a straight line, with the same slope β_1 but with Y intercept $\beta_0 + \beta_2$. This response function is also shown in Figure 9.1.

The meaning of the parameters in the response function (9.5) is now clear. With reference to our earlier example, the mean time elapsed before the innovation is adopted, $E(Y)$, is a linear function of size of firm (X_1), with the same slope β_1 for both types of firms. β_2 indicates how much higher

FIGURE 9.1

Illustration of Meaning of Regression Parameters for Model (9.4) with Indicator Variable X_2

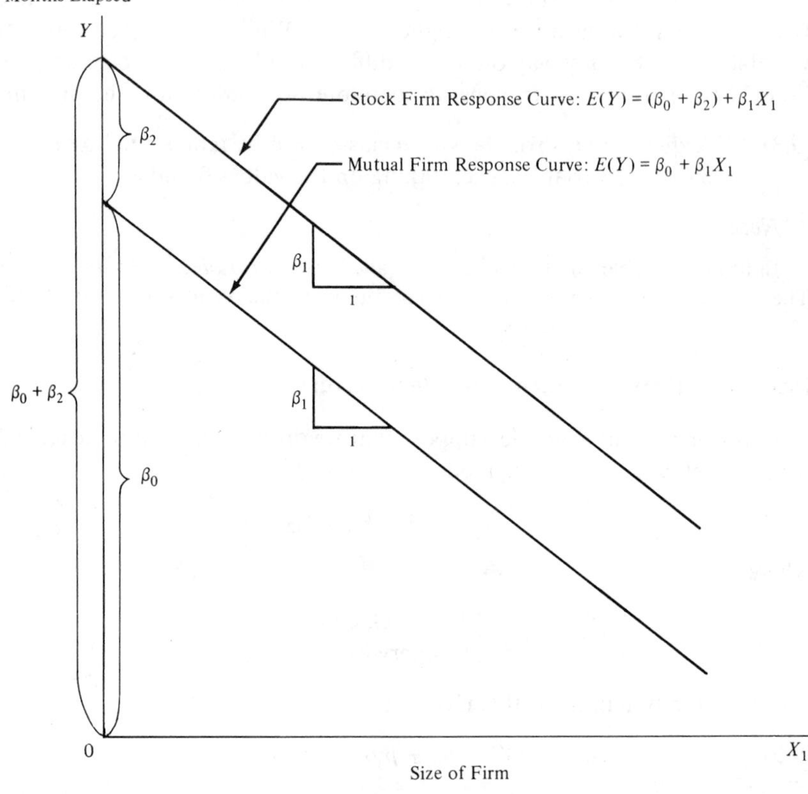

(lower) the response function for stock firms is than the one for mutual firms. Thus, β_2 measures the differential effect of type of firm. In general, β_2 shows how much higher (lower) the mean response line is for the class coded 1 than the line for the class coded 0.

Example

With reference to our earlier illustration, the economist studied 10 mutual firms and 10 stock firms. The data are shown in Table 9.1. The **Y** and **X** data matrices are shown in Table 9.2. Note that $X_2 = 1$ for each stock firm and $X_2 = 0$ for each mutual firm.

Given the **Y** and **X** matrices in Table 9.2, fitting the regression model (9.4)

TABLE 9.1

Data for Insurance Innovation Study

Firm i	Number of Months Elapsed Y_i	Size of Firm (million dollars) X_{i1}	Type of Firm
1	17	151	Mutual
2	26	92	Mutual
3	21	175	Mutual
4	30	31	Mutual
5	22	104	Mutual
6	0	277	Mutual
7	12	210	Mutual
8	19	120	Mutual
9	4	290	Mutual
10	16	238	Mutual
11	28	164	Stock
12	15	272	Stock
13	11	295	Stock
14	38	68	Stock
15	31	85	Stock
16	21	224	Stock
17	20	166	Stock
18	13	305	Stock
19	30	124	Stock
20	14	246	Stock

TABLE 9.2

Data Matrices for Insurance Innovation Study

$$
\mathbf{Y} = \begin{bmatrix} 17 \\ 26 \\ 21 \\ 30 \\ 22 \\ 0 \\ 12 \\ 19 \\ 4 \\ 16 \\ 28 \\ 15 \\ 11 \\ 38 \\ 31 \\ 21 \\ 20 \\ 13 \\ 30 \\ 14 \end{bmatrix}
\qquad
\mathbf{X} = \begin{bmatrix} 1 & 151 & 0 \\ 1 & 92 & 0 \\ 1 & 175 & 0 \\ 1 & 31 & 0 \\ 1 & 104 & 0 \\ 1 & 277 & 0 \\ 1 & 210 & 0 \\ 1 & 120 & 0 \\ 1 & 290 & 0 \\ 1 & 238 & 0 \\ 1 & 164 & 1 \\ 1 & 272 & 1 \\ 1 & 295 & 1 \\ 1 & 68 & 1 \\ 1 & 85 & 1 \\ 1 & 224 & 1 \\ 1 & 166 & 1 \\ 1 & 305 & 1 \\ 1 & 124 & 1 \\ 1 & 246 & 1 \end{bmatrix}
$$

(Columns of $\mathbf{X}$: X_0, X_1, X_2)

TABLE 9.3

Regression Results for Model (9.4) Fit to Insurance Innovation Data

(a) Regression Coefficients

Regression Coefficient	*Estimated Regression Coefficient*	*Estimated Standard Deviation*	t^*
β_0	33.87407	1.81386	18.68
β_1	−.10174	.00889	−11.44
β_2	8.05547	1.45911	5.52

(b) Analysis of Variance

Source of Variation	*SS*	*df*	*MS*
Regression	1,504.41	2	752.20
Error	176.39	17	10.38
Total	1,680.80	19	

is straightforward. Table 9.3 presents the key results from a computer printout. The fitted response function is:

$$\hat{Y} = 33.87407 - .10174X_1 + 8.05547X_2$$

Figure 9.2 contains the fitted response function for each type of firm, together with the actual observations.

The economist was most interested in the effect of type of firm (X_2) on the elapsed time for the innovation to be adopted. He therefore desired to obtain a 95 percent confidence interval for β_2. We require $t(.975; 17) = 2.110$, and obtain from the data in Table 9.3:

$$\begin{aligned} 4.97675 = 8.05547 - (2.110)(1.45911) &\leq \beta_2 \\ &\leq 8.05547 + (2.110)(1.45911) = 11.13419 \end{aligned}$$

Thus, with 95 percent confidence, we would conclude that mutual companies tended to adopt the particular innovation studied somewhere between 5 and 11 months earlier, on the average, than stock companies.

A formal test of:

$$C_1: \beta_2 = 0$$
$$C_2: \beta_2 \neq 0$$

with level of significance .05 would lead to C_2, that type of firm has an effect, since the confidence interval does not cover $\beta_2 = 0$.

The economist also carried out other analyses, some of which will be described shortly.

FIGURE 9.2

Fitted Regression Functions for Model (9.4)

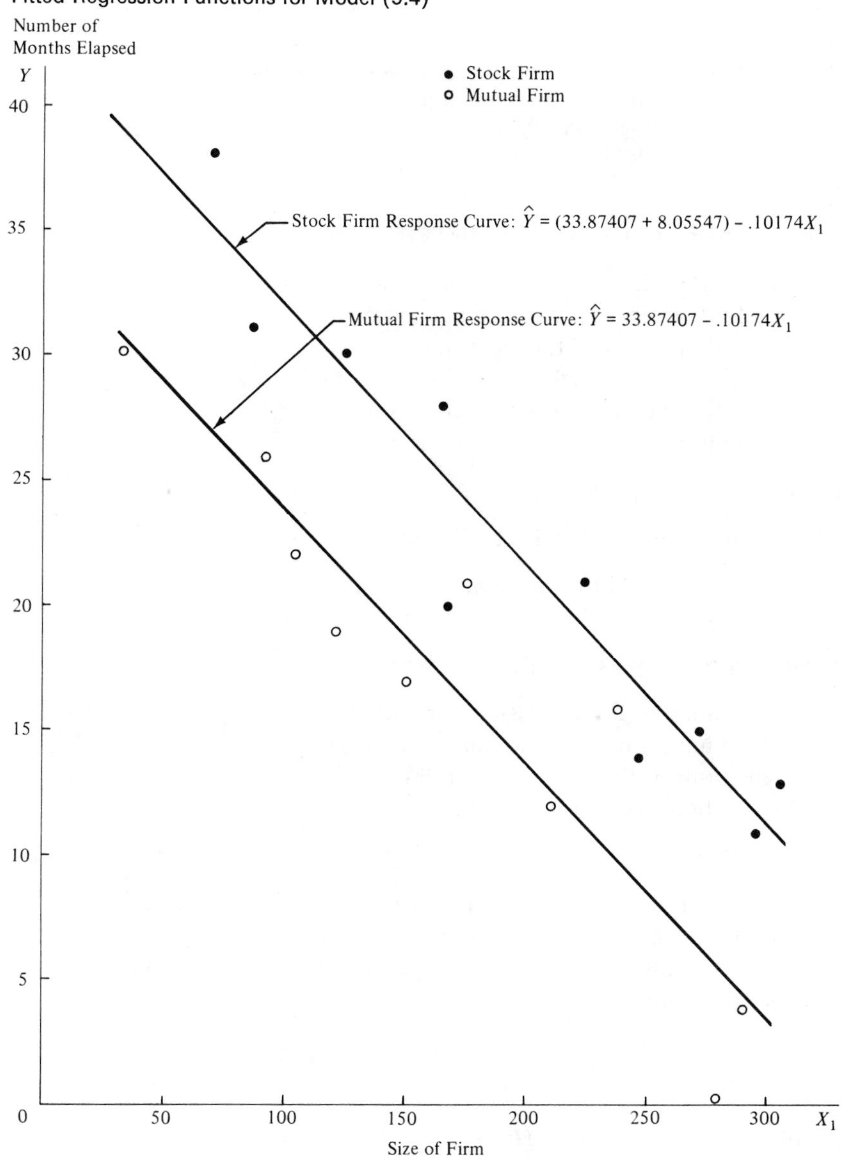

Note

The reader may wonder why we did not simply fit separate regressions for stock firms and mutual firms in our example, and instead adopted the approach of fitting one regression with an indicator variable. There are two reasons for this here. First, the model assumed equal slopes for each type of firm. This common slope β_1 can best be estimated by pooling the two types of firms. Second, other inferences, such as pertaining to β_0 and β_2, can be made more precisely by working with one regression model containing an indicator variable since more degrees of freedom will then be associated with *MSE*.

9.2 MODEL CONTAINING INTERACTION EFFECTS

In our earlier illustration, the economist actually did not begin his analysis with model (9.4) since he expected interaction effects between size and type of firm. Even though one of the independent variables in the regression model is qualitative, interaction effects are introduced into the model in the usual manner, by including cross-product terms. A first-order model with an interaction term for our example would be:

$$Y_i = \beta_0 + \beta_1 X_{i1} + \beta_2 X_{i2} + \beta_3 X_{i1} X_{i2} + \varepsilon_i \tag{9.6}$$

The response function for this model is:

$$E(Y) = \beta_0 + \beta_1 X_1 + \beta_2 X_2 + \beta_3 X_1 X_2 \tag{9.7}$$

Meaning of Regression Parameters

The meaning of the regression parameters in the response function (9.7) can best be understood by examining the nature of the response function for each type of firm. For a mutual firm, $X_2 = 0$ and hence $X_1 X_2 = 0$. The response function for a mutual firm therefore is:

$$E(Y) = \beta_0 + \beta_1 X_1 + \beta_2(0) + \beta_3(0) = \beta_0 + \beta_1 X_1 \qquad \text{Mutual firm} \tag{9.7a}$$

This response function is shown in Figure 9.3. Note that the Y intercept is β_0 and the slope is β_1 for the response function for mutual firms.

For stock firms, $X_2 = 1$ and hence $X_1 X_2 = X_1$. The response function for a stock firm therefore is:

$$E(Y) = \beta_0 + \beta_1 X_1 + \beta_2(1) + \beta_3 X_1$$

or:

$$E(Y) = (\beta_0 + \beta_2) + (\beta_1 + \beta_3) X_1 \qquad \text{Stock firm} \tag{9.7b}$$

This response function is also shown in Figure 9.3. Note that the response function for stock firms has Y intercept $\beta_0 + \beta_2$ and slope $\beta_1 + \beta_3$.

It is thus clear that β_2 indicates how much greater (smaller) is the Y intercept for the class coded 1 than for the class coded 0, and similarly β_3 indicates

FIGURE 9.3

Illustration of Meaning of Regression Parameters for Model (9.6) with Indicator Variable X_2 and Interaction Term

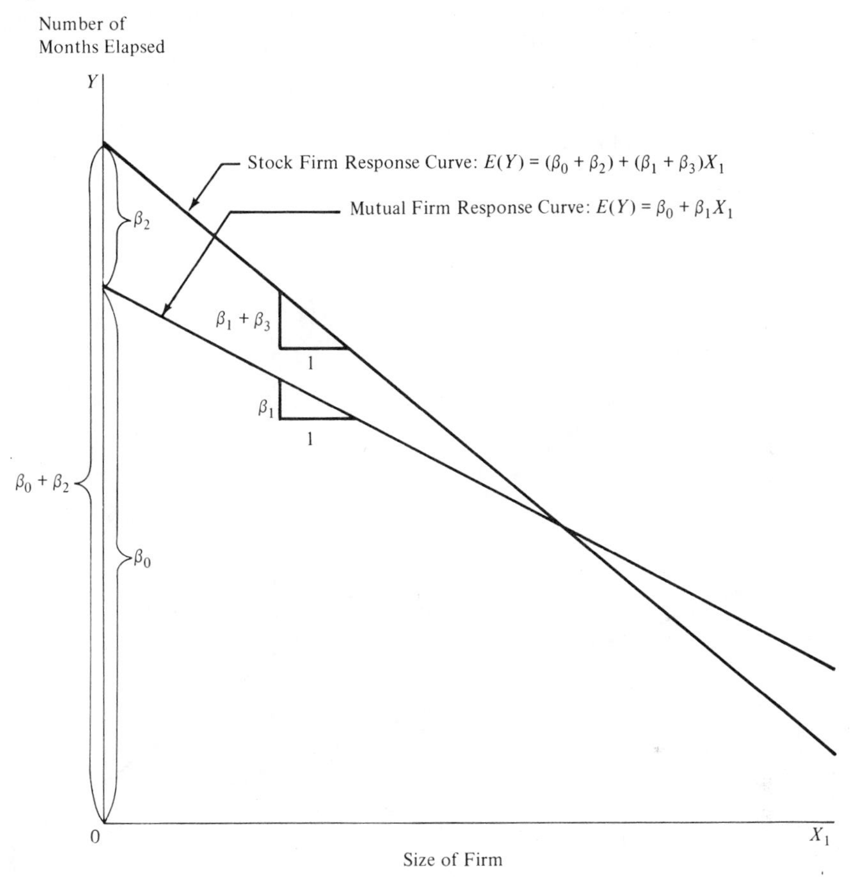

how much greater (smaller) is the slope for the class coded 1 than for the class coded 0. Because both the intercept and the slope differ for the two classes in model (9.6), it is no longer true that β_2 indicates how much higher (lower) one response line is than the other. Figure 9.3 makes it clear that the effect of type of firm with model (9.6) depends on the size of the firm. For smaller firms, according to Figure 9.3, mutual firms tend to innovate more quickly, but for larger firms stock firms tend to innovate more quickly. Thus, when interaction effects are present, the effect of the qualitative variable can only be studied by comparing the regression functions for each class of the qualitative variable. Figure 9.4 illustrates another possible interaction situation. Here, mutual firms tend to introduce the innovation more quickly than stock firms for all sizes of firms in the scope of the model, but the differential effect is much smaller for large firms than for small ones.

FIGURE 9.4

Another Illustration of Model (9.6)

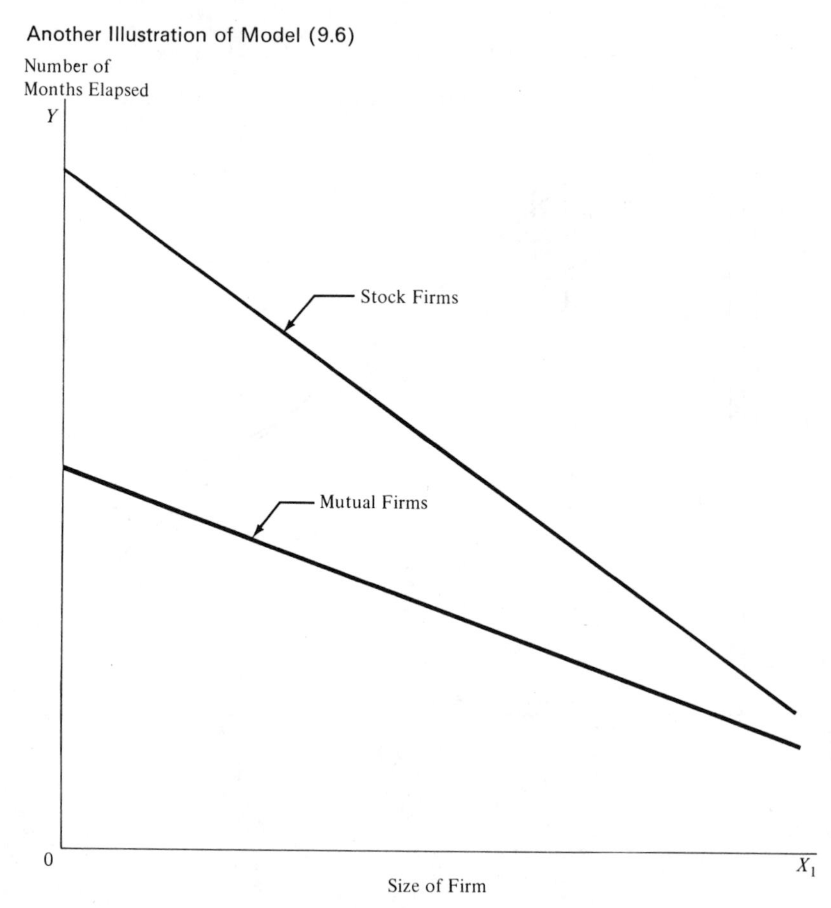

Example

Since the economist anticipated that interaction effects between size and type of firm may be present, he actually first wished to fit model (9.6):

$$Y_i = \beta_0 + \beta_1 X_{i1} + \beta_2 X_{i2} + \beta_3 X_{i1} X_{i2} + \varepsilon_i$$

Table 9.4 shows the **X** matrix for this model. The **Y** matrix is the same as in Table 9.2. Note that the $X_1 X_2$ column in the **X** matrix in Table 9.4 contains 0 for mutual companies and X_{i1} for stock companies.

TABLE 9.4

X Matrix for Fitting Model (9.6) in Insurance Innovation Study

$$
\mathbf{X} = \begin{array}{c} X_0 \quad X_1 \quad X_2 \quad X_1X_2 \\ \begin{bmatrix} 1 & 151 & 0 & 0 \\ 1 & 92 & 0 & 0 \\ 1 & 175 & 0 & 0 \\ 1 & 31 & 0 & 0 \\ 1 & 104 & 0 & 0 \\ 1 & 277 & 0 & 0 \\ 1 & 210 & 0 & 0 \\ 1 & 120 & 0 & 0 \\ 1 & 290 & 0 & 0 \\ 1 & 238 & 0 & 0 \\ 1 & 164 & 1 & 164 \\ 1 & 272 & 1 & 272 \\ 1 & 295 & 1 & 295 \\ 1 & 68 & 1 & 68 \\ 1 & 85 & 1 & 85 \\ 1 & 224 & 1 & 224 \\ 1 & 166 & 1 & 166 \\ 1 & 305 & 1 & 305 \\ 1 & 124 & 1 & 124 \\ 1 & 246 & 1 & 246 \end{bmatrix} \end{array}
$$

Given the **Y** and **X** matrices, the regression fit is routine. Basic results from the computer printout are shown in Table 9.5. To test for the presence of interaction effects:

$$C_1: \beta_3 = 0$$
$$C_2: \beta_3 \neq 0$$

the economist used the t^* statistic from Table 9.5:

$$t^* = \frac{b_3}{s(b_3)} = \frac{-.0004171}{.01833} = -.02$$

For level of significance .05, we require $t(.975; 16) = 2.120$. Since $|t^*| = .02 < 2.120$, we conclude that $\beta_3 = 0$, that is, that no interaction effects are present. It was because of this result that the economist adopted model (9.4) which we discussed earlier.

Note

Fitting model (9.6) yields the same results as fitting separate regressions for stock firms and mutual firms. An advantage of using model (9.6) with an indicator variable is that one regression run on the computer will yield both fitted regressions.

TABLE 9.5

Regression Results for Model (9.6) Fit to Insurance Innovation Data

(a) Regression Coefficients

Regression Coefficient	*Estimated Regression Coefficient*	*Estimated Standard Deviation*	t^*
β_0	33.83837	2.44065	13.86
β_1	−.10153	.01305	−7.78
β_2	8.13125	3.65405	2.23
β_3	−.0004171	.01833	−.02

(b) Analysis of Variance

Source of Variation	*SS*	*df*	*MS*
Regression	1,504.42	3	501.47
Error	176.38	16	11.02
Total	1,680.80	19	

Another advantage is that tests for comparing the regression functions for the different classes of the qualitative variable can be clearly seen to be tests of regression coefficients in a general linear model. For instance, Figure 9.3 makes it clear for our example that the test whether the two regression functions have the same slope involves:

$$C_1: \beta_3 = 0$$
$$C_2: \beta_3 \neq 0$$

Similarly, the test whether the two regression functions in our example are identical would involve:

$$C_1: \beta_2 = \beta_3 = 0$$
$$C_2: \text{not both } \beta_2 = 0 \text{ and } \beta_3 = 0$$

Indeed, this latter test with model (9.6) is the basis for the comparison of two regression lines discussed in Section 5.6. Fitting the full model is the equivalent of fitting separate regressions for each class of the qualitative variable. If $\beta_2 = 0$ and $\beta_3 = 0$, the reduced model consists of a common regression line for both classes of the qualitative variable.

9.3 MORE COMPLEX MODELS

We now briefly consider more complex models involving qualitative independent variables.

Qualitative Variable with More Than Two Classes

If a qualitative independent variable has more than two classes, we require additional indicator variables in the regression model. Consider the regression of tool wear (Y) on tool speed (X_1), where we wish to include also tool model (M1, M2, M3, M4) as an independent variable. Since the qualitative variable (tool model) has four classes, we require three indicator variables. Let us define them as follows:

(9.8)
$$X_2 = \begin{matrix} 1 \text{ if tool model is M1} \\ 0 \text{ otherwise} \end{matrix}$$
$$X_3 = \begin{matrix} 1 \text{ if tool model is M2} \\ 0 \text{ otherwise} \end{matrix}$$
$$X_4 = \begin{matrix} 1 \text{ if tool model is M3} \\ 0 \text{ otherwise} \end{matrix}$$

First-Order Model. A first-order model would be:

(9.9)
$$Y_i = \beta_0 + \beta_1 X_{i1} + \beta_2 X_{i2} + \beta_3 X_{i3} + \beta_4 X_{i4} + \varepsilon_i$$

For this model, the data input for the **X** matrix would be as follows:

Tool Model	X_0	X_1	X_2	X_3	X_4
M1	1	X_{i1}	1	0	0
M2	1	X_{i1}	0	1	0
M3	1	X_{i1}	0	0	1
M4	1	X_{i1}	0	0	0

The response function for model (9.9) is:

(9.10)
$$E(Y) = \beta_0 + \beta_1 X_1 + \beta_2 X_2 + \beta_3 X_3 + \beta_4 X_4$$

To see the meaning of the regression parameters, consider first the response function for tool models M4 for which $X_2 = 0$, $X_3 = 0$, and $X_4 = 0$:

(9.10a)
$$E(Y) = \beta_0 + \beta_1 X_1 \qquad \text{Tool models M4}$$

For tool models M1, $X_2 = 1$, $X_3 = 0$, and $X_4 = 0$ and the response function is:

(9.10b)
$$E(Y) = \beta_0 + \beta_1 X_1 + \beta_2 = (\beta_0 + \beta_2) + \beta_1 X_1 \qquad \text{Tool models M1}$$

Similarly, the response functions for tool models M2 and M3 turn out to be:

(9.10c)
$$E(Y) = (\beta_0 + \beta_3) + \beta_1 X_1 \qquad \text{Tool models M2}$$

(9.10d)
$$E(Y) = (\beta_0 + \beta_4) + \beta_1 X_1 \qquad \text{Tool models M3}$$

It is thus clear that the response function (9.10) implies that the regression of tool wear on tool speed is linear, with the same slope for all types of tool models. The coefficients β_2, β_3, and β_4 indicate respectively how much higher (lower) the response functions for tool models M1, M2, and M3 are than the one for tool models M4. Thus, β_2, β_3, and β_4 measure the differential effects of the qualitative variable classes on the height of the response functions, always compared with the class for which $X_2 = X_3 = X_4 = 0$. Figure 9.5 illustrates a possible arrangement of these response functions.

FIGURE 9.5

Illustration of Model (9.9) for Tool Wear Example

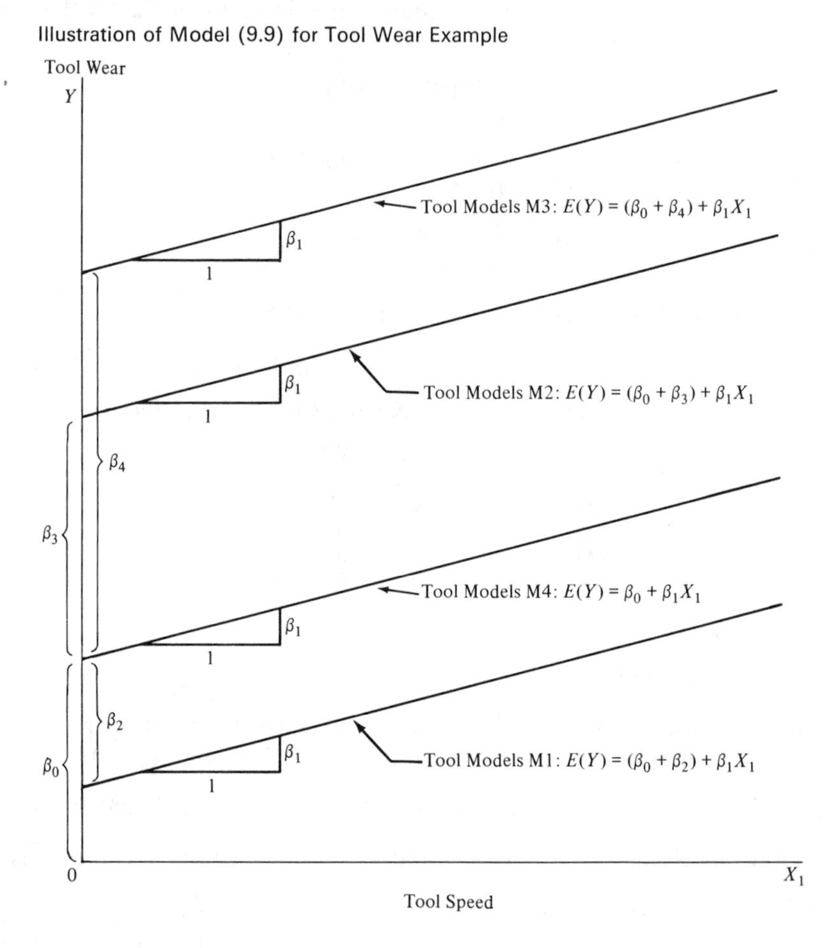

When using model (9.9), one may wish to estimate differential effects other than against tool models M4. For instance, $\beta_4 - \beta_3$ measures how much higher (lower) the response function for tool models M3 is than the response function for tool models M2, as may be seen by comparing (9.10c) and (9.10d). The point estimator of this quantity is, of course, $b_4 - b_3$, and the estimated variance of this estimator is:

$$s^2(b_4 - b_3) = s^2(b_4) + s^2(b_3) - 2s(b_4, b_3) \tag{9.11}$$

The needed variances and covariance can be readily obtained from the estimated variance-covariance matrix of the regression coefficients.

First-Order Model with Interactions Added. If interaction effects between tool speed and tool model are present in our previous illustration, model (9.9) would be modified as follows:

$$Y_i = \beta_0 + \beta_1 X_{i1} + \beta_2 X_{i2} + \beta_3 X_{i3} + \beta_4 X_{i4} + \beta_5 X_{i1} X_{i2} + \beta_6 X_{i1} X_{i3} + \beta_7 X_{i1} X_{i4} + \varepsilon_i \tag{9.12}$$

The response function of this regression model is as follows for tool models M4, for which $X_2 = 0$, $X_3 = 0$, $X_4 = 0$:

$$E(Y) = \beta_0 + \beta_1 X_1 \qquad \text{Tool models M4} \tag{9.13a}$$

Similarly, we find for the other tool models:

$$E(Y) = (\beta_0 + \beta_2) + (\beta_1 + \beta_5)X_1 \qquad \text{Tool models M1} \tag{9.13b}$$

$$E(Y) = (\beta_0 + \beta_3) + (\beta_1 + \beta_6)X_1 \qquad \text{Tool models M2} \tag{9.13c}$$

$$E(Y) = (\beta_0 + \beta_4) + (\beta_1 + \beta_7)X_1 \qquad \text{Tool models M3} \tag{9.13d}$$

Thus, the interaction model (9.12) implies that each tool model has its own regression line, with different intercepts and slopes for the different tool models.

More Than One Qualitative Independent Variable

Models can readily be constructed for cases where two or more of the independent variables are qualitative. Consider the regression of advertising expenditures (Y) on sales (X_1), type of firm (incorporated, not incorporated), and quality of sales management (high, low). We may define:

$$\begin{aligned} X_2 &= \begin{matrix} 1 \text{ if firm incorporated} \\ 0 \text{ otherwise} \end{matrix} \\ X_3 &= \begin{matrix} 1 \text{ if quality of sales management high} \\ 0 \text{ otherwise} \end{matrix} \end{aligned} \tag{9.14}$$

First-Order Model. A first-order model for the above example is:

$$Y_i = \beta_0 + \beta_1 X_{i1} + \beta_2 X_{i2} + \beta_3 X_{i3} + \varepsilon_i \tag{9.15}$$

This model implies that the response function of advertising expenditures on sales is linear, with the same slope for all "type of firm–quality of sales management" combinations, and β_2 and β_3 indicate the additive differential effects of type of firm and quality of sales management on the height of the line.

First-Order Model with Certain Interactions Added. A first-order model with interaction effects between pairs of the independent variables added would be:

$$Y_i = \beta_0 + \beta_1 X_{i1} + \beta_2 X_{i2} + \beta_3 X_{i3} + \beta_4 X_{i1} X_{i2} + \beta_5 X_{i1} X_{i3} + \beta_6 X_{i2} X_{i3} + \varepsilon_i \tag{9.16}$$

Note the implications of this model:

Type of Firm	*Quality of Sales Management*	*Response Function*
Incorp.	High	$E(Y) = (\beta_0 + \beta_2 + \beta_3 + \beta_6) + (\beta_1 + \beta_4 + \beta_5)X_1$
Not incorp.	High	$E(Y) = (\beta_0 + \beta_3) + (\beta_1 + \beta_5)X_1$
Incorp.	Low	$E(Y) = (\beta_0 + \beta_2) + (\beta_1 + \beta_4)X_1$
Not incorp.	Low	$E(Y) = \beta_0 + \beta_1 X_1$

Not only are all response functions different for the various "type of firm–quality of sales management" combinations, but the differential effects of one qualitative variable on the intercept depend on the particular class of the other qualitative variable. For instance, when we move from "not incorporated–low quality" to "incorporated–low quality," the intercept changes by β_2. But if we move from "not incorporated–high quality" to "incorporated–high quality," the intercept changes by $\beta_2 + \beta_6$.

Qualitative Independent Variables Only

Regression models containing only qualitative independent variables can also be constructed. With reference to our previous example, we could regress advertising expenditures only on type of firm and quality of sales management. The first-order model then would be:

$$Y_i = \beta_0 + \beta_2 X_{i2} + \beta_3 X_{i3} + \varepsilon_i \tag{9.17}$$

where X_{i2} and X_{i3} are defined in (9.14).

Comments

1. Models in which all independent variables are qualitative are called *analysis of variance* models. These are discussed fully in Parts III and IV.

2. Models in which there are some quantitative and some qualitative independent variables are called *covariance* models. Chapter 22 discusses covariance models where the chief independent variables of interest are qualitative, and one or more quantitative independent variables are introduced primarily to reduce the variance of the error terms.

9.4 OTHER USES OF INDEPENDENT INDICATOR VARIABLES

Piecewise Linear Regression

Sometimes the regression of Y on X follows a particular linear relation in some range of X, but follows a different linear relation elsewhere. For instance, unit cost (Y) regressed on lot size may follow a certain linear regression up to $X_p = 500$, at which point the slope changes because of some operating efficiencies only possible with lot sizes of more than 500. Figure 9.6 illustrates this situation.

FIGURE 9.6

Illustration of Piecewise Linear Regression

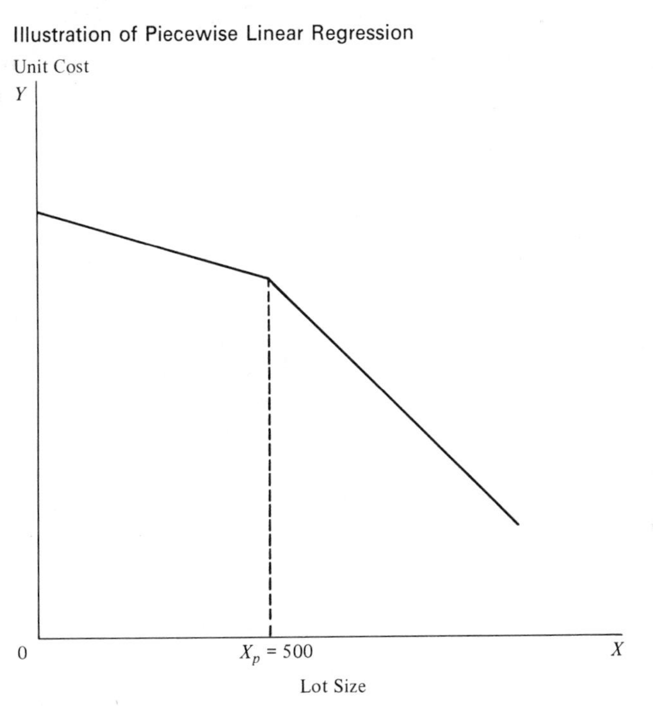

We consider now how indicator variables may be used to fit piecewise linear regressions consisting of two pieces. We take up the case where X_p, the point where the slope changes, is known.

We return to our lot size illustration, for which it is known that the slope changes at $X_p = 500$. The model for our illustration may be expressed as follows:

$$Y_i = \beta_0 + \beta_1 X_{i1} + \beta_2(X_{i1} - 500)X_{i2} + \varepsilon_i \tag{9.18}$$

where X_{i1} is lot size and X_{i2} is an indicator variable defined as follows:

$$X_{i2} = \begin{matrix} 1 \text{ if } X_{i1} > 500 \\ 0 \text{ if } X_{i1} \leq 500 \end{matrix}$$

To check that (9.18) does provide a two-piecewise linear regression, consider the response function of (9.18):

$$E(Y) = \beta_0 + \beta_1 X_1 + \beta_2(X_1 - 500)X_2 \tag{9.19}$$

When $X_1 \leq 500$, $X_2 = 0$ so that (9.19) becomes:

$$E(Y) = \beta_0 + \beta_1 X_1 \qquad X_1 \leq 500 \tag{9.19a}$$

On the other hand, when $X_1 > 500$, $X_2 = 1$ and we obtain:

$$E(Y) = (\beta_0 - 500\beta_2) + (\beta_1 + \beta_2)X_1 \qquad X_1 > 500 \tag{9.19b}$$

Thus, β_1 and $\beta_1 + \beta_2$ are the slopes of the two regression lines, and β_0 and $(\beta_0 - 500\beta_2)$ are the two Y intercepts. These parameters are shown in Figure 9.7.

FIGURE 9.7

Illustration of Parameters of Model (9.18)

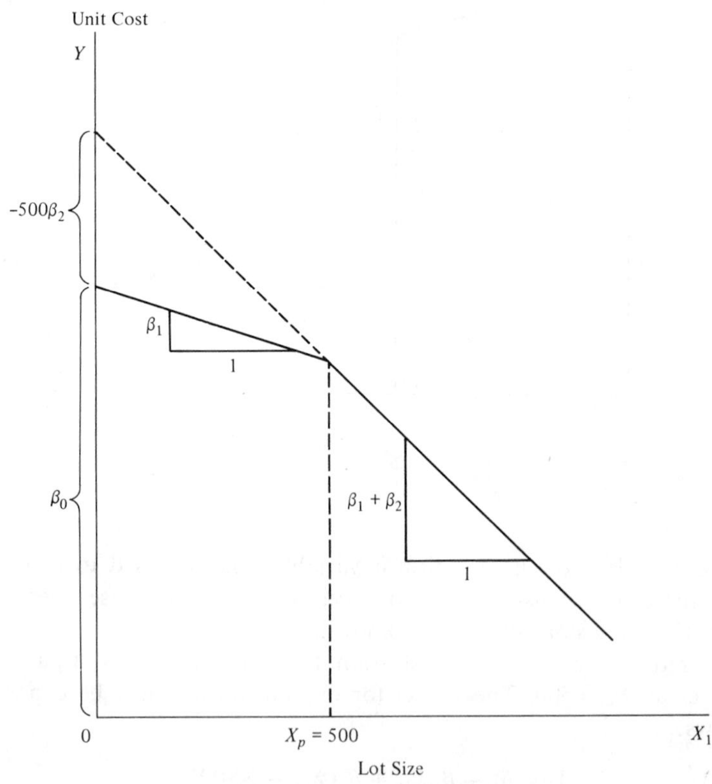

To illustrate the use of model (9.18), consider Table 9.6a which contains eight observations on unit costs for given lot sizes. It is known that the response function slope changes at $X_p = 500$ so that model (9.18) is to be employed. Table 9.6b contains the **X** matrix for our example. The left column of **X** is a column of 1's as usual. The next column contains X_{i1}. The final column on the right contains $(X_{i1} - 500)X_{i2}$, which consists of 0's for all lot sizes up to 500, and of $X_{i1} - 500$ for all lot sizes above 500. The fitting of regression model (9.18) at this point becomes routine. The fitted response function is:

$$\hat{Y} = 5.89545 - .00395X_1 - .00389(X_1 - 500)X_2$$

TABLE 9.6

Data and **X** Matrix for Piecewise Linear Regression Example

(a)			(b)			
Lot i	Unit Cost (dollars) Y_i	Lot Size X_i		X_0	X_1	$(X_1 - 500)X_2$
1	2.57	650		1	650	150
2	4.40	340		1	340	0
3	4.52	400		1	400	0
4	1.39	800	**X** =	1	800	300
5	4.75	300		1	300	0
6	3.55	570		1	570	70
7	2.49	720		1	720	220
8	3.77	480		1	480	0

From this fitted model, expected unit cost is estimated to decline by .00395 for each increase of one in lot size up to 500 and by .00784 thereafter.

Note

The extension of model (9.18) to more than two-piecewise regression lines is straightforward. For instance, if the slope of the regression line in our earlier lot size illustration actually changes at both $X = 500$ and $X = 800$, the model would be:

$$Y_i = \beta_0 + \beta_1 X_{i1} + \beta_2(X_{i1} - 500)X_{i2} + \beta_3(X_{i1} - 800)X_{i3} + \varepsilon_i \tag{9.20}$$

where:

X_{i1} = lot size

$X_{i2} = \begin{cases} 1 \text{ if } X_{i1} > 500 \\ 0 \text{ otherwise} \end{cases}$

$X_{i3} = \begin{cases} 1 \text{ if } X_{i1} > 800 \\ 0 \text{ otherwise} \end{cases}$

Discontinuity in Regression Function

Sometimes the linear regression function may not only change its slope at some value X_p, but may also have a jump point there. Figure 9.8 illustrates this case. Another indicator variable must then be introduced to take care of the jump. Suppose time required to solve a task successfully (Y) is to be regressed on complexity of task (X), when complexity of task is measured on a quantitative scale from 0 to 100. It is known that the slope of the response line changes at $X_p = 40$, and it is believed that the regression relation may be discontinuous there. We therefore set up the model:

$$Y_i = \beta_0 + \beta_1 X_{i1} + \beta_2(X_{i1} - 40)X_{i2} + \beta_3 X_{i3} + \varepsilon_i \tag{9.21}$$

where X_{i1} is complexity of the task and X_{i2} and X_{i3} are indicator variables defined as follows:

$$X_{i2} = \begin{matrix} 1 \text{ if } X_{i1} > 40 \\ 0 \text{ otherwise} \end{matrix}$$

$$X_{i3} = \begin{matrix} 1 \text{ if } X_{i1} > 40 \\ 0 \text{ otherwise} \end{matrix}$$

FIGURE 9.8

Illustration of Model (9.21) for Discontinuous Linear Regression

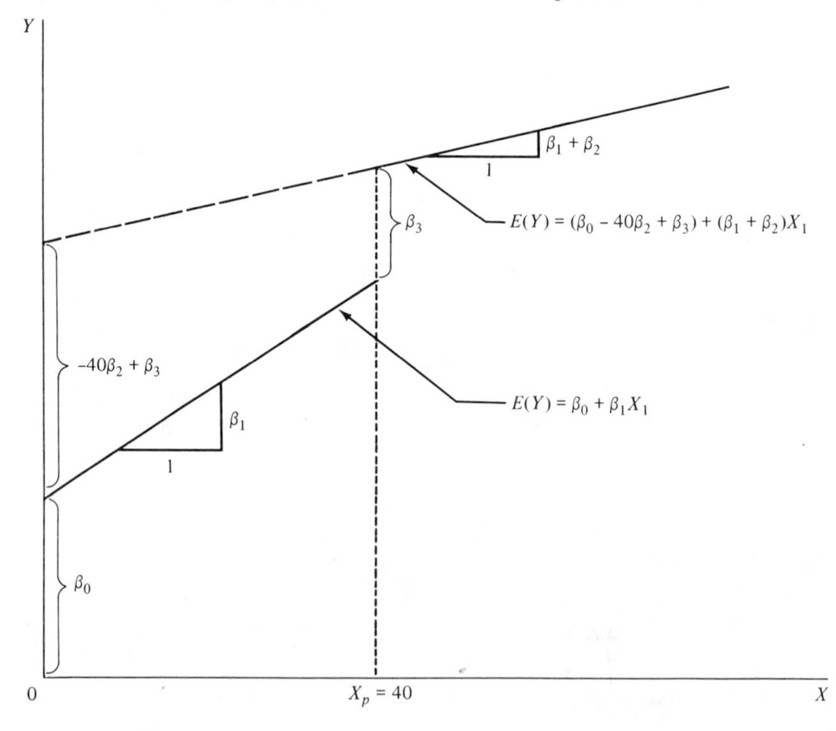

The response function for model (9.21) is:

(9.22) $$E(Y) = \beta_0 + \beta_1 X_1 + \beta_2(X_1 - 40)X_2 + \beta_3 X_3$$

When $X_1 \leq 40$, then $X_2 = 0$ and $X_3 = 0$, so (9.22) becomes:

(9.22a) $$E(Y) = \beta_0 + \beta_1 X_1 \qquad X_1 \leq 40$$

Similarly, when $X_1 > 40$, then $X_2 = 1$ and $X_3 = 1$, so (9.22) becomes:

(9.22b) $$E(Y) = (\beta_0 - 40\beta_2 + \beta_3) + (\beta_1 + \beta_2)X_1 \qquad X_1 > 40$$

These two response functions are shown in Figure 9.8, together with the parameters involved. Note that β_3 represents the difference in the mean responses for the two regression lines at $X_p = 40$ and β_2 represents the difference in the two slopes.

The estimation of the regression coefficients for model (9.21) presents no new problems. One may test whether or not $\beta_3 = 0$ in the usual manner. If it is concluded that $\beta_3 = 0$, the regression function is continuous at X_p so that the earlier piecewise linear regression model applies.

Time Series Applications

Economists and business analysts frequently use time series data in regression analysis. For instance, savings (Y) may be regressed on income (X), where both the savings and income data pertain to a number of years. The model employed might be:

(9.23) $$Y_t = \beta_0 + \beta_1 X_t \qquad t = 1, \ldots, n$$

where Y_t and X_t are savings and income respectively for time period t. Suppose that the period covered includes both peacetime and wartime years, and that this factor should be recognized since it is anticipated that savings in wartime years tend to be higher. The following model might then be appropriate:

(9.24) $$Y_t = \beta_0 + \beta_1 X_{t1} + \beta_2 X_{t2} + \varepsilon_t$$

where X_{t1} represents income and X_{t2} is an indicator variable defined as follows:

$$X_{t2} = \begin{matrix} 1 \text{ if period } t \text{ peacetime} \\ 0 \text{ if period } t \text{ not peacetime} \end{matrix}$$

Note that model (9.24) assumes that the marginal propensity to save (β_1) is constant in both peacetime and wartime years, and that only the height of the response curve is affected by this qualitative variable.

Another use of indicator variables in time series applications occurs when monthly or quarterly data are used. Suppose that quarterly sales (Y) are regressed on quarterly advertising expenditures (X_1) and quarterly disposable personal income (X_2). If seasonal effects also have an influence on quarterly sales, a first-order model incorporating seasonal effects would be:

(9.25) $$Y_t = \beta_0 + \beta_1 X_{t1} + \beta_2 X_{t2} + \beta_3 X_{t3} + \beta_4 X_{t4} + \beta_5 X_{t5} + \varepsilon_t$$

where:

$$X_{t3} = \begin{matrix} 1 \text{ if period } t \text{ is first quarter} \\ 0 \text{ otherwise} \end{matrix}$$

$$X_{t4} = \begin{matrix} 1 \text{ if period } t \text{ is second quarter} \\ 0 \text{ otherwise} \end{matrix}$$

$$X_{t5} = \begin{matrix} 1 \text{ if period } t \text{ is third quarter} \\ 0 \text{ otherwise} \end{matrix}$$

9.5 SOME CONSIDERATIONS IN USING INDEPENDENT INDICATOR VARIABLES

Indicator Variables versus Allocated Codes

An alternative to the use of indicator variables for a qualitative independent variable is to employ *allocated codes*. Consider, for instance, the independent variable "frequency of product use" which has three classes:

a) Frequent user
b) Occasional user
c) Nonuser

Two indicator variables, say X_1 and X_2, could be employed to represent the qualitative variable as follows:

Class	X_1	X_2
Frequent user	1	0
Occasional user	0	1
Nonuser	0	0

Assuming no other independent variables, the model would be:

(9.26) $$Y_i = \beta_0 + \beta_1 X_{i1} + \beta_2 X_{i2} + \varepsilon_i$$

As noted earlier, β_1 in this model measures the differential effect:

$$E(Y \mid \text{frequent user}) - E(Y \mid \text{nonuser})$$

and β_2 measures:

$$E(Y \mid \text{occasional user}) - E(Y \mid \text{nonuser})$$

Thus, β_1 and β_2 measure the differential effects of user status relative to the class of nonusers.

An alternative to indicator variables often employed is to use one independent quantitative variable with allocated codes, say:

Class	X_1
Frequent user	3
Occasional user	2
Nonuser	1

The allocated codes are, of course, arbitrary and could be other sets of numbers. The model with allocated codes for our example would be:

$$Y_i = \beta_0 + \beta_1 X_{i1} + \varepsilon_i \tag{9.27}$$

The basic difficulty with allocated codes is that they define a metric for the classes of the qualitative variable which may or may not be reasonable. To see the nature of this problem concretely, consider the mean responses with model (9.27) for the three classes of the qualitative variable:

Class	$E(Y)$
Frequent user	$E(Y) = \beta_0 + \beta_1(3) = \beta_0 + 3\beta_1$
Occasional user	$E(Y) = \beta_0 + \beta_1(2) = \beta_0 + 2\beta_1$
Nonuser	$E(Y) = \beta_0 + \beta_1(1) = \beta_0 + \beta_1$

Note the key implication:

$$E(Y \mid \text{frequent user}) - E(Y \mid \text{occasional user}) = E(Y \mid \text{occasional user}) - E(Y \mid \text{nonuser}) = \beta_1$$

Thus, the coding 1, 2, 3 implies equal distances between the three user classes, which may or may not be in accord with reality. Other allocated codes may, of course, imply different spacings of the classes of the qualitative variable, but these would ordinarily still be arbitrary.

Indicator variables, in contrast, make no assumptions about the spacing of the classes and rely on the data to show the differential effects that occurred.

Indicator Variables versus Quantitative Variables

If an independent variable is quantitative, such as age, one can nevertheless use indicator variables instead. For instance, the quantitative variable age may be transformed by grouping ages into classes:

Under 21
21–34
35–49
etc.

Indicator variables may then be used for the classes of this new independent variable. At first sight, this may seem to be a foolish approach because information about the actual ages is thrown away. Furthermore, additional parameters are placed into the model, which leads to a reduction of the degrees of freedom associated with MSE. For example, if c age classes are set up, $c - 1$ regression parameters are required for the age variable in the model with the indicator variable approach in lieu of the one parameter if the original quantitative variable age were used in the model.

Nevertheless, there are occasions when replacement of a quantitative variable by indicator variables may be appropriate. Consider, for example, a large-scale survey in which the relation between liquid assets (Y) and age (X) of head of household is to be studied. Two thousand households will be included in the study, so that the loss of 10 or 20 degrees of freedom is immaterial. The analyst is very much in doubt about the shape of the regression function, and hence may prefer the indicator variable approach in order to obtain information about the shape without making any assumptions about the functional form of the regression function.

Another alternative, also utilizing indicator variables, is available to the analyst in doubt about the functional form of the regression function. He could use the quantitative variable age, but employ piecewise linear regression with a number of pieces. Again, this approach loses degrees of freedom for estimating MSE but this is of no concern in large-scale studies. The benefit would be that information about the shape of the regression function is obtained without making strong assumptions about its functional form.

9.6 DEPENDENT INDICATOR VARIABLE

In a variety of applications, the dependent variable of interest has only two possible outcomes, and therefore can be represented by an indicator variable taking on values 0 and 1. Consider, for example, an analysis of whether or not business firms have an industrial relations department, according to size of firm. The dependent variable in this study was defined to have two possible outcomes:

Firm has industrial relations department
Firm does not have industrial relations department

which may be coded 1 and 0 respectively (or vice versa).

As another example, consider a study of labor force participation of wives, as a function of age of wife, number of children, and family income. In this study, the dependent variable Y was defined to have the two possible outcomes:

Wife in labor force
Wife not in labor force

Again, these outcomes may be coded 1 and 0 respectively.

A final example is a study of liability insurance possession, according to age of head of household, amount of liquid assets, and type of occupation of head of household. Here, the dependent variable Y was defined to have the two possible outcomes:

Household has liability insurance policy
Household does not have liability insurance policy

These examples show the wide range of applications in which the dependent variable is dichotomous, and hence may be represented by an indicator variable. A dependent dichotomous variable, taking on the values 0 and 1, is sometimes said to involve *quantal responses* or *binary responses*.

Meaning of Response Function when Dependent Variable Is Binary

Consider the simple linear regression model:

$$Y_i = \beta_0 + \beta_1 X_i + \varepsilon_i \qquad Y_i = 0, 1 \tag{9.28}$$

where the response is binary so that Y_i is an indicator variable. The expected response $E(Y_i)$ has a special meaning in this case. Since $E(\varepsilon_i) = 0$ as usual, we have:

$$E(Y_i) = \beta_0 + \beta_1 X_i \tag{9.29}$$

Consider now Y_i as an ordinary Bernoulli random variable for which we can state the probability distribution as follows:

Y_i	*Probability*
0	$P(Y_i = 0) = 1 - p_i = q_i$
1	$P(Y_i = 1) = p_i$

Thus, p_i is the probability that $Y_i = 1$ and q_i is the probability that $Y_i = 0$. By the ordinary definition of expected value of a random variable in (1.12), we obtain:

$$E(Y_i) = 0(q_i) + 1(p_i) = p_i \tag{9.30}$$

Equating (9.29) and (9.30), we thus find:

$$\beta_0 + \beta_1 X_i = p_i \tag{9.31}$$

The mean response $E(Y_i) = \beta_0 + \beta_1 X_i$ as given by the response function is therefore simply the probability that $Y_i = 1$ when the level of the independent variable is X_i. This interpretation of the mean response applies whether the response function is a simple linear one as here, or a complex multiple regression one. The mean response, when the dependent variable is an indicator variable, always represents the probability that $Y = 1$ for the given levels of the independent variables. Figure 9.9 illustrates a simple linear

FIGURE 9.9

Illustration of Response Function when Dependent Variable Is Binary

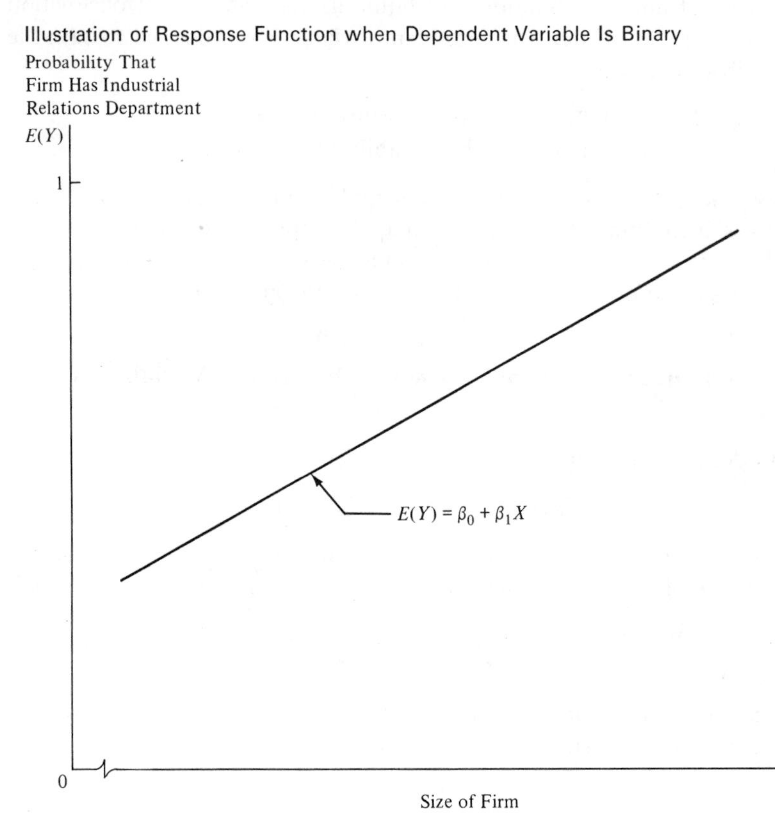

response function for a dependent indicator variable. Here, the indicator variable Y refers to whether or not a firm has an industrial relations department, and the independent variable X is size of firm. The response function in Figure 9.9 shows the probability that firms of given size have an industrial relations department.

Special Problems when Dependent Variable Is Binary

Special problems arise, unfortunately, when the dependent variable is an indicator variable. We shall consider three of these now, using a simple linear regression model as an illustration.

1. ***Nonnormal Error Terms.*** For a binary dependent variable, the error terms $\varepsilon_i = Y_i - (\beta_0 + \beta_1 X_i)$ can take on only two values:

(9.32a) $$\text{When } Y_i = 1: \varepsilon_i = 1 - \beta_0 - \beta_1 X_i$$

(9.32b) $$\text{When } Y_i = 0: \varepsilon_i = -\beta_0 - \beta_1 X_i$$

Clearly, the normal error regression model (3.1), assuming that the ε_i are normally distributed, will no longer be appropriate.

2. ***Nonconstant Error Variance.*** Another problem with the error terms ε_i is that they do not have equal variances when the dependent variable is an indicator variable. To see this, we shall obtain $\sigma^2(Y_i)$ for the simple linear regression model (9.28):

$$\sigma^2(Y_i) = E\{[Y_i - E(Y_i)]^2\} = (1 - p_i)^2 p_i + (0 - p_i)^2(1 - p_i)$$

or:

$$\sigma^2(Y_i) = p_i(1 - p_i) = [E(Y_i)][1 - E(Y_i)] \tag{9.33}$$

The variance of ε_i is, of course, the same as that of Y_i because $\varepsilon_i = Y_i - p_i$ and p_i is a constant:

$$\sigma^2(\varepsilon_i) = p_i(1 - p_i) = [E(Y_i)][1 - E(Y_i)] \tag{9.34}$$

or:

$$\sigma^2(\varepsilon_i) = (\beta_0 + \beta_1 X_i)(1 - \beta_0 - \beta_1 X_i) \tag{9.34a}$$

Note from (9.34a) that $\sigma^2(\varepsilon_i)$ depends on X_i. Hence the error term variances will differ at different levels of X, and ordinary least squares will no longer be optimal.

3. ***Constraints on Response Function.*** Since the response function represents probabilities when the dependent variable is an indicator variable, the mean responses should be constrained as follows:

$$0 \leq E(Y) = p \leq 1 \tag{9.35}$$

Many response functions do not automatically possess this constraint. A linear response function, for instance, may fall outside the constraint limits within the range of the independent variable in the scope of the model.

These three problems create difficulties, but solutions can be found. Problem 2 concerning unequal error variances, for instance, can be handled by using weighted least squares. Problem 3 about the constraints on the response function can be handled by making sure that the mean responses for the model fitted do not fall below 0 or above 1 for levels of X within the scope of the model, or else by using a model which automatically meets the constraints.

Finally, even though the error terms are not normal when the dependent variable is binary, the method of least squares still provides unbiased estimators which, under quite general conditions, are asymptotically normal. Hence, when the sample size is large, inferences concerning the regression coefficients and mean responses can be made in the same fashion as when the error terms are assumed to be normally distributed.

9.7 LINEAR REGRESSION WITH DEPENDENT INDICATOR VARIABLE

We consider now the fitting of linear response functions when the dependent variable is binary. Then we shall explain the fitting of curvilinear response functions.

Illustration

A small-scale investigation was undertaken to study the effect of computer programming experience on ability to complete a complex programming task, including debugging, within a specified time. Twenty-five persons were selected for the study. They had varying amounts of programming experience (measured in months of experience), as shown in Table 9.7, column 1. All persons were given the same programming task, and the results of their success in the task are shown in Table 9.7, column 2. The results are coded in binary fashion: if the task was completed successfully in the allotted time,

TABLE 9.7

Data for Programming Task Example

Person i	(1) *Months of Experience* X_i	(2) *Task Success* Y_i	(3) $\hat{Y}_i$	(4) $\hat{w}_i$
1	14	0	.34920	4.4003
2	29	0	.82212	6.8382
3	6	0	.09697	11.4196
4	25	1	.69601	4.7263
5	18	1	.47531	4.0098
6	4	0	.03392	30.5196
7	18	0	.47531	4.0098
8	12	0	.28614	4.8956
9	22	1	.60142	4.1717
10	6	0	.09697	11.4196
11	30	1	.85365	8.0044
12	11	0	.25461	5.2691
13	30	1	.85365	8.0044
14	5	0	.06544	16.3502
15	20	1	.53837	4.0237
16	13	0	.31767	4.6135
17	9	0	.19156	6.4573
18	32	1	.91671	13.0967
19	24	0	.66448	4.4854
20	13	1	.31767	4.6135
21	19	0	.50684	4.0007
22	4	0	.03392	30.5196
23	28	1	.79059	6.0403
24	22	1	.60142	4.1717
25	8	1	.16003	7.4394

FIGURE 9.10

Scatter Plot of Data for Programming Task Example

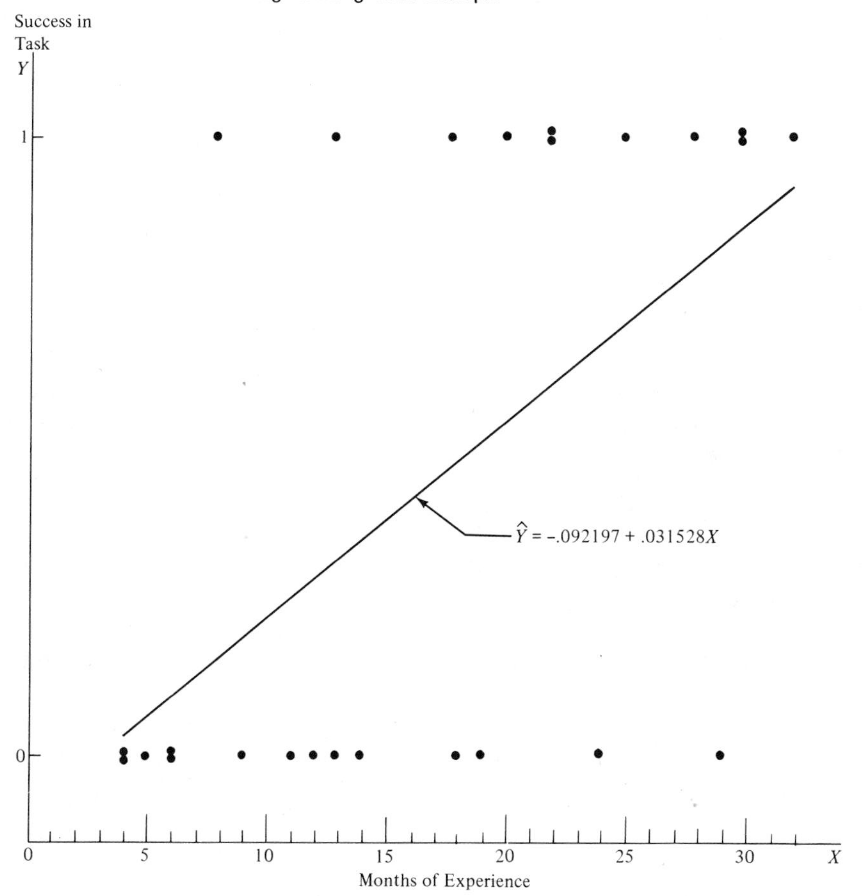

$Y = 1$, and if the task was not completed successfully, $Y = 0$. Figure 9.10 contains a scatter plot of the data. This plot is not too informative because of the nature of the dependent variable, other than to indicate that ability to complete the task successfully appears to increase with amount of experience. At this point, it was decided to fit the linear regression model (9.28).

Ordinary Least Squares Fit

One approach to fitting model (9.28) is to fit it by ordinary least squares despite the unequal error variances. The estimated regression coefficients will still be unbiased, but they will no longer have the minimum variance property among the class of unbiased linear estimators. Thus, the use of ordinary least squares may lead to inefficient estimates, that is, estimates with larger variances than could be obtained with weighted procedures.

The ordinary least squares fit to the data in Table 9.7 leads to the following results:

Coefficient	*Estimated Coefficient*	*Estimated Standard Deviation*
β_0	$b_0 = -.092197$	$s(b_0) = .183272$
β_1	$b_1 = .031528$	$s(b_1) = .009606$

The estimated response function therefore is:

$$\hat{Y} = -.092197 + .031528X \tag{9.36}$$

This response function is shown in Figure 9.10. It may be used in the ordinary manner. For instance, the estimated mean response for persons with $X = 14$ months experience is:

$$\hat{Y}_{14} = -.092197 + .031528(14) = .34920$$

Thus, we estimate that the probability is .349 that a person with 14 months experience will successfully complete the programming task.

As noted earlier from the scatter plot, this probability increases with increasing experience. The coefficient b_1 indicates that the estimated probability increases by .0315 for each additional month of experience. Clearly, the model cannot be extrapolated much beyond the range of the data here or else the estimated probabilities would be negative or exceed 1, meaningless results. For $X = 40$ months experience, for instance, $\hat{Y}_{40} = 1.17$.

Weighted Least Squares

It was pointed out in Chapter 4 that weighted least squares provides efficient estimates when the error variances are unequal. Since we know from (9.34) that:

$$\sigma^2(\varepsilon_i) = p_i(1 - p_i) = [E(Y_i)][1 - E(Y_i)]$$

it would appear that using weighted least squares with weights:

$$w_i = \frac{1}{p_i(1 - p_i)} = \frac{1}{[E(Y_i)][1 - E(Y_i)]} \tag{9.37}$$

is the proper approach to take. There exists a difficulty, however, in carrying this out, namely that the w_i in (9.37) involve the unknown parameters β_0 and β_1 which are to be estimated, since $E(Y_i) = \beta_0 + \beta_1 X_i$.

A way out of this difficulty is to use a two-stage least squares procedure:

1. Stage 1—fit the regression model by ordinary least squares.
2. Stage 2—estimate the weights w_i from the results of stage 1:

$$\hat{w}_i = \frac{1}{\hat{Y}_i(1 - \hat{Y}_i)} \tag{9.38}$$

and then use these estimated weights in obtaining weighted least squares estimates for the regression model.

To apply this approach in our example, we first need to find the estimated weights $\hat{w}_i$ from an ordinary least squares fit. This ordinary least squares fit has been obtained earlier in (9.36). Hence, we are ready to calculate $\hat{Y}_i$ and then $1/(\hat{Y}_i)(1 - \hat{Y}_i)$ for each observation. For the first observation, for instance, $X_1 = 14$ so that $\hat{Y}_1 = .34920$ as found earlier. Hence the estimated weight for the first observation is:

$$\hat{w}_1 = \frac{1}{(.34920)(.65080)} = 4.4003$$

In the same manner, the other weights are calculated. In Table 9.7, columns 3 and 4 contain respectively the fitted values and the estimated weights.

At this point, we could use the normal equations for weighted least squares in (4.51), with w_i replaced by $\hat{w}_i$, to obtain the estimated regression coefficients. Alternatively, we could use matrix calculations. Let us define the matrix **W** as follows:

$$\mathbf{W} = \begin{bmatrix} \hat{w}_1 & 0 & 0 & \cdots & 0 \\ 0 & \hat{w}_2 & 0 & \cdots & 0 \\ \vdots & \vdots & \vdots & & \vdots \\ 0 & 0 & 0 & \cdots & \hat{w}_n \end{bmatrix} \tag{9.39}$$

It can be shown that the weighted least squares estimators of the regression coefficients in matrix notation are:

$$\mathbf{b} = (\mathbf{X'WX})^{-1}\mathbf{X'WY} \tag{9.40}$$

and the estimated variance-covariance matrix of the regression coefficients is:

$$\mathbf{s}^2(\mathbf{b}) = MSE(\mathbf{X'WX})^{-1} \tag{9.41}$$

A computer run for our example, using weighted least squares with the weights as given in Table 9.7, led to the following results:

Regression Coefficient	*Estimated Regression Coefficient*	*Estimated Standard Deviation*
β_0	$b_0 = -.117113$	$s(b_0) = .111841$
β_1	$b_1 = .032672$	$s(b_1) = .006644$

The fitted response function by the two-stage least squares approach therefore is:

(9.42) $$\hat{Y} = -.117113 + .032672X$$

Note that this estimated response function does not differ markedly from the one obtained by unweighted least squares in (9.36), although the estimated standard deviations of the regression coefficients are now somewhat smaller.

Comments

1. If the mean responses range between about .2 and .8 for the scope of the model, there is little to be gained from weighted least squares since the error variances will differ but little. Only if the mean responses range below .2 and/or above .8 will the error variances be sufficiently unequal to make weighted least squares worthwhile.

2. In our example, the fitted response function does not fall below 0 or above 1 within the range of the data (4 months to 32 months experience). If it did, a curvilinear response function would have to be employed. We discuss one important curvilinear model below.

3. The weighted least squares approach could employ additional stages, with the weights refined at each stage. Usually, however, the gain from additional iterations is not large. For our example, for instance, another iteration led to the following results:

$$b_0 = -.125665 \qquad s(b_0) = .084555$$
$$b_1 = .033075 \qquad s(b_1) = .005885$$

Note that there is relatively little decrease from the previous stage in the estimated standard deviations and only small changes in the regression coefficients.

4. If there are repeat observations at the different X levels, the procedure of fitting a linear response function can be simplified. Let $X_1, \ldots, X_c$ denote the X levels, $\bar{p}_j$ the proportion of 1's at X_j ($j = 1, \ldots, c$), and n_j the number of observations at X_j. Fitting the sample proportions $\bar{p}_j$ with weights:

(9.43) $$w_j = n_j$$

leads to the identical estimated response function as ordinary least squares applied to the individual Y observations. If there are hundreds of Y observations located at a limited number of X_j levels, much computational effort can be saved by fitting the sample proportions $\bar{p}_j$.

If weighted least squares estimates are to be obtained, only one stage is required when sample proportions $\bar{p}_j$ are available. The weights would be as follows:

(9.44) $$\hat{w}_j = \frac{n_j}{\bar{p}_j(1 - \bar{p}_j)}$$

where $\hat{w}_j$ denotes the estimated weight.

We discuss fitting a regression model to grouped data more fully below.

5. While we have illustrated only simple linear regression, the extension to multiple regression is straightforward. The matrix formulas (9.40) and (9.41) for weighted least squares, for instance, apply as they stand.

9.8 LOGISTIC RESPONSE FUNCTION

Both theoretical and empirical considerations suggest that when the dependent variable is an indicator variable, the shape of the response function will frequently be curvilinear. Figure 9.11 contains a curvilinear response

FIGURE 9.11

Example of Logistic Response Function

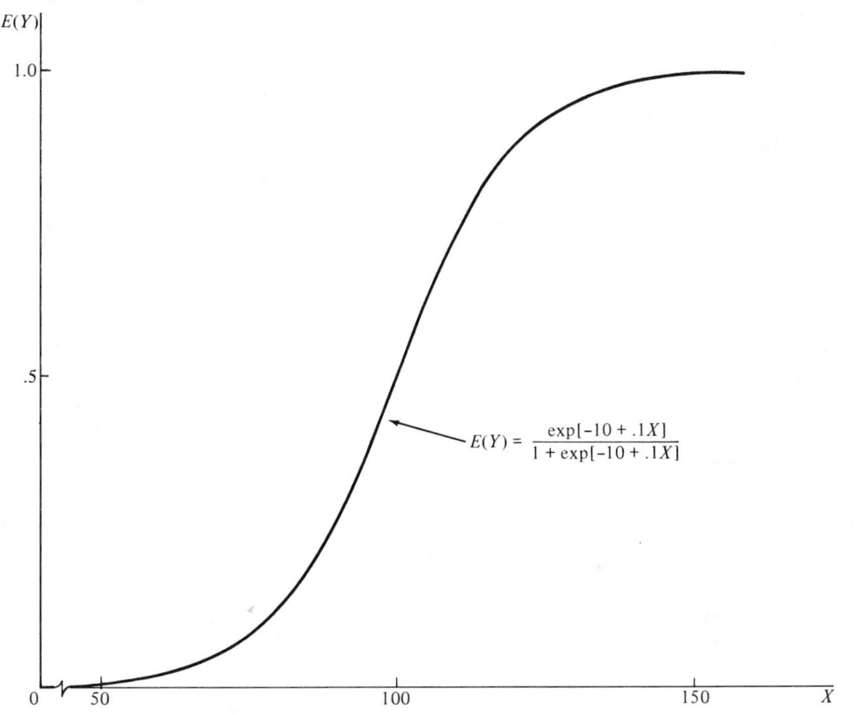

function which has been found appropriate in many instances involving a binary dependent variable. Note that this response function is shaped like a tilted *S*, and that it has asymptotes at 0 and 1. The latter feature assures that the constraints on $E(Y)$ in (9.35) are automatically met. The response function plotted in Figure 9.11 is called the *logistic* function and is given by:

$$E(Y) = \frac{\exp(\beta_0 + \beta_1 X)}{1 + \exp(\beta_0 + \beta_1 X)} \tag{9.45}$$

An interesting property of the logistic function is that it can easily be linearized. Let us denote $E(Y) = p$, since the mean response is a probability when the dependent variable is an indicator variable. Then if we make the transformation:

(9.46)
$$p' = \log_e\left(\frac{p}{1-p}\right)$$

we obtain from (9.45):

(9.47)
$$p' = \beta_0 + \beta_1 X$$

The transformation (9.46) is called the *logistic* or *logit* transformation of the probability p.

Fitting of Logistic Function

The fitting of the transformed logistic response function (9.47) is relatively simple when there are repeat observations at each X value. We now explain the fitting procedure for this case, which arises frequently in practice. We cite two examples:

1. A pricing experiment involves showing a new product to a consumer, providing information about it, and then asking the consumer whether he would buy the product at a given price. Five prices are studied, and n persons are exposed to a given price. The dependent variable is binary (would purchase, would not purchase); the independent variable is price and has five classes.
2. Four hundred heads of households are asked to indicate on a 10-point scale their intent to buy a new car within the next 12 months. A year later, each household is interviewed to determine whether a new car was purchased. The dependent variable is binary (did purchase, did not purchase). The independent variable is the measure of intent to buy and has 10 classes.

We shall denote the X levels at which observations are obtained $X_1, \ldots, X_c$. The number of observations at level X_j will be denoted n_j ($j = 1, \ldots, c$). It can be shown that we need consider only the total number of 1's at each X level, and not the individual Y values. Let R_j be the number of 1's at the level X_j. Hence, the proportion of 1's at the level X_j, denoted by $\bar{p}_j$, is:

(9.48)
$$\bar{p}_j = \frac{R_j}{n_j}$$

Table 9.8 on page 332 contains the X_j, n_j, R_j, and $\bar{p}_j$ values for an example we shall discuss shortly. X_j refers to the price reduction offered by a coupon, n_j is the number of households which received a coupon with price reduction X_j, R_j is the number of these households that redeemed the coupon, and $\bar{p}_j$ is the proportion of households receiving a coupon with price reduction X_j that did redeem the coupon.

We fit the transformed logistic response function (9.47):

$$p' = \beta_0 + \beta_1 X$$

by making the logistic transformation (9.46) on the sample proportion:

$$\bar{p}'_j = \log_e\left(\frac{\bar{p}_j}{1 - \bar{p}_j}\right) \tag{9.49}$$

and using $\bar{p}'_j$ as the dependent variable. If tables of logarithms to the base e are not readily available, one can use a table of logarithms to the base 10 as follows:

$$\bar{p}'_j = 2.302585 \log_{10}\left(\frac{\bar{p}_j}{1 - \bar{p}_j}\right) \tag{9.50}$$

since:

$$\log_e(y) = 2.302585 \log_{10}(y)$$

The logistic transformation, while linearizing the response function, does not eliminate the unequal variances of the error terms. Hence, weighted least squares should be used. It can be shown that, when n_j is reasonably large, the approximate variance of $\bar{p}'_j$ is:

$$\sigma^2(\bar{p}'_j) = \frac{1}{n_j p_j(1 - p_j)} \tag{9.51}$$

which is estimated by:

$$s^2(\bar{p}'_j) = \frac{1}{n_j \bar{p}_j(1 - \bar{p}_j)} \tag{9.52}$$

Hence, the estimated weights to be used in the weighted least squares computations are:

$$\hat{w}_j = n_j \bar{p}_j(1 - \bar{p}_j) \tag{9.53}$$

Note carefully that the use of the estimated weights (9.53) requires the sample sizes n_j to be reasonably large.

The fitting of a linear regression model, with independent variable X and dependent variable $\bar{p}'_j$, using weighted least squares, is straightforward. Once the fitted response function has been obtained:

$$\hat{p}' = b_0 + b_1 X \tag{9.54}$$

it can be transformed back into the original units, if desired:

$$\hat{p} = \frac{\exp(b_0 + b_1 X)}{1 + \exp(b_0 + b_1 X)} \tag{9.55}$$

Example

In a study of the effectiveness of coupons offering a price reduction on a given product, 1,000 homes were selected and a coupon and advertising material for the product were mailed to each. The coupons offered different price reductions (5, 10, 15, 20, and 30 cents), and 200 homes were assigned at random to each of the price reduction categories. The independent variable X in this study is the amount of price reduction, and the dependent variable Y is a binary variable indicating whether or not the coupon was redeemed within a six-month period.

It was expected that the logistic response function would be an appropriate description of the relation between price reduction and probability that the coupon is utilized. Since there were repeat observations at each X_j, and since the number of repeat observations at each X_j was large ($n_j = 200$, for all j), the procedure described earlier could be used for fitting the logistic response function.

Table 9.8 contains the basic data for this experimental study in columns 1 through 4. The transformed proportions $\bar{p}'_j$ are shown in column 5. For instance, for $X_1 = 5$, we have:

$$\bar{p}'_1 = 2.302585 \log_{10}\left(\frac{.160}{1 - .160}\right) = -1.65823$$

TABLE 9.8

Data for Coupon Effectiveness Example

(1) Price Reduction X_j	(2) Number of Households n_j	(3) Number of Coupons Redeemed R_j	(4) Proportion of Coupons Redeemed $\bar{p}_j$	(5) Transformed Proportion $\bar{p}'_j$	(6) Weight $\hat{w}_j$
5	200	32	.160	−1.65823	26.880
10	200	51	.255	−1.07212	37.995
15	200	70	.350	−.61904	45.500
20	200	103	.515	.06002	49.955
30	200	148	.740	1.04597	38.480

Finally, the approximate weights $\hat{w}_j$ are shown in column 6 of Table 9.8. For instance, for $X_1 = 5$, we have:

$$\hat{w}_1 = n_1 \bar{p}_1(1 - \bar{p}_1) = 200(.160)(.840) = 26.880$$

Prior to the fitting of the logistic model, the transformed proportions $\bar{p}'_j$ were plotted against X_j. This plot is shown in Figure 9.12. It appears from there that a linear response function would fit the transformed proportions

FIGURE 9.12

Plot of Data for Coupon Effectiveness Example and Fitted Logistic Response Function

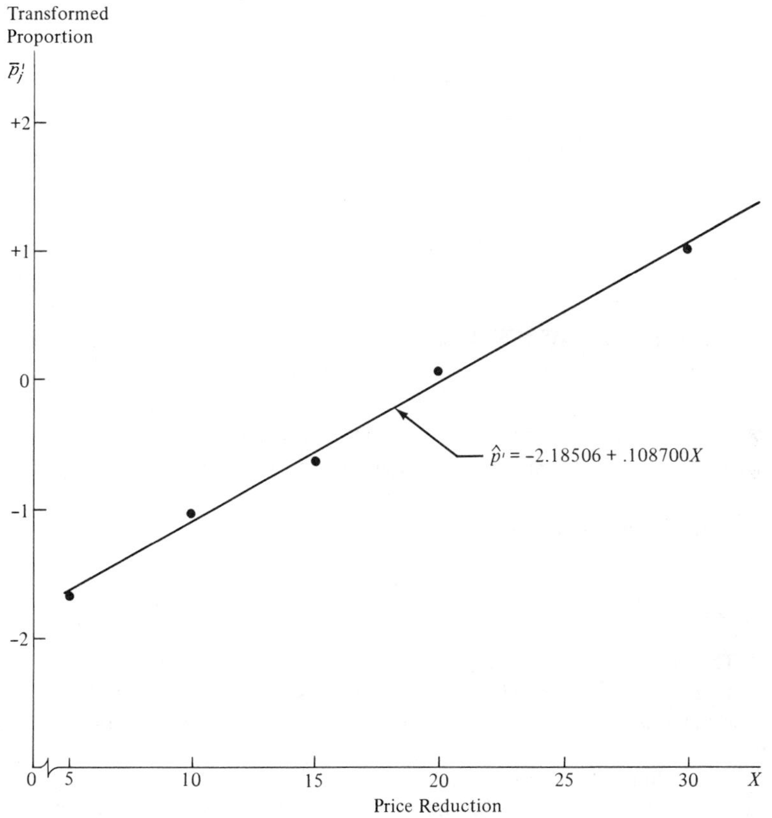

well. Hence, it was decided to proceed with fitting the transformed logistic model (9.47).

The fitting of the transformed logistic model by weighted least squares is straightforward. The normal equations (4.51) could be used, with weights given in column 6 of Table 9.8. Alternatively, matrix calculations could be utilized, with the matrix **W** in (9.39) having on its main diagonal the weights in column 6 of Table 9.8. A computer run, using weighted least squares, led to the following fitted response function:

$$\hat{p}' = -2.18506 + .108700X \tag{9.56}$$

This response function is plotted in Figure 9.12, and it appears to fit the data well.

The fitted response function (9.56) is used in the ordinary manner. To estimate the probability of a coupon redemption if the price reduction is, say, 25 cents, we first substitute in (9.56):

$$\hat{p}' = -2.18506 + .108700(25) = .53244$$

and then transform to the original variable:

$$\text{antilog}_{10}\left(\frac{.53244}{2.302585}\right) = 1.70308 = \frac{\hat{p}}{1 - \hat{p}}$$

so that:

$$\hat{p} = .630$$

when $X = 25$ cents.

The interpretation of the slope $b_1 = .1087$ in the fitted logistic response function (9.56) is not simple, unlike the straightforward interpretation of the slope in a linear regression model. The reason is that the effect of increasing X by a unit varies for the logistic model according to the location of the starting point on the X scale. One interpretation of b_1 is found in the property of the logistic function that the "odds" ratio $\hat{p}/(1 - \hat{p})$ is multiplied by $\exp(b_1)$ for any unit increase in X.

Comments

1. The procedure illustrated assumes that no $\bar{p}_j = 0$ or 1. If this should occur, or if some $\bar{p}_j$ are very close to 0 or 1, modifications in the transformation (9.49) and in the weights (9.53) should be made. Reference 9.1 discusses appropriate modifications.

2. A curvilinear response function of almost the same shape as the logistic function (9.45) is obtained by transforming the $\bar{p}_j$ by means of the cumulative normal distribution. This transformation is called a *probit* transformation. Reference 9.2 describes this transformation. The probit transformation leads to more complex calculations than the logistic transformation, but it does have the desirable property that inferences can be made more readily with it than with the logistic transformation.

3. The transformed logistic model (9.47) can easily be extended into a multiple regression model. For, say, two independent variables, the transformed logistic response function would be:

$$p' = \beta_0 + \beta_1 X_1 + \beta_2 X_2 \tag{9.57}$$

where p' is the logistic transformation (9.46) of p.

4. The logistic model sometimes is motivated by threshold theory. Consider the breaking strength of concrete blocks, measured in pounds per square inch. Each block is assumed to have a threshold T_i, such that it will break if pressure equal to or greater than T_i is applied and it will not break if smaller pressure is applied. A block can be tested for only one pressure level, since some weakening of the block occurs with the first test. Thus, it is not possible to find the actual threshold of each

block, but only whether or not the threshold level is above or below a particular pressure applied to the block. Of course, different pressures can be applied to different blocks in order to obtain information about the breaking strength thresholds of the blocks in the population.

Let $Y_i = 1$ if the block breaks when pressure X_i is applied, and $Y_i = 0$ if the block does not break. The earlier statement of threshold theory then implies:

$$\begin{aligned} Y_i &= 1 \text{ whenever } T_i \leq X_i \\ Y_i &= 0 \text{ whenever } T_i > X_i \end{aligned} \tag{9.58}$$

It follows that for given pressure X_i applied to a block selected at random:

$$p_i = P(Y_i = 1 \mid X_i) = P(T_i \leq X_i) \tag{9.59}$$

Now $P(T \leq X)$ is the cumulative probability distribution of the thresholds of all blocks in the population. If this cumulative probability distribution of thresholds is logistic:

$$P(T \leq X) = \frac{\exp(\beta_0 + \beta_1 X)}{1 + \exp(\beta_0 + \beta_1 X)}$$

we arrive at the logistic response model (9.45).

Incidentally, if the probability distribution of thresholds were normal, the probit response model in comment 2 above would be the relevant one according to threshold theory.

PROBLEMS

9.1. A student, using a model that included indicator variables, was upset when he received only the following output on his multiple regression printout: XTRANSPOSE X SINGULAR. What is the most likely source of the difficulty?

9.2. A marketing research trainee in the national office of a chain of shoe stores used the following response function to study seasonal (winter, spring, summer, fall) effects on sales of a certain line of shoes: $E(Y) = \beta_0 + \beta_1 X_1 + \beta_2 X_2 + \beta_3 X_3$. The X's are indicator variables defined as follows: $X_1 = 1$ if season is winter and 0 otherwise; $X_2 = 1$ if season is spring and 0 otherwise; $X_3 = 1$ if season is fall and 0 otherwise. After fitting the model he tested the regression coefficients β_j $(j = 0, \ldots, 3)$ and came to the following set of conclusions at an .05 family level of significance: $\beta_0 \neq 0$, $\beta_1 = 0$, $\beta_2 \neq 0$, $\beta_3 \neq 0$. In his report he then wrote: "Results of regression analysis show that climatic and other seasonal factors have no influence in determining sales of this shoe line in the winter. Seasonal influences do exist in the other seasons." Do you agree with this interpretation of the test results?

9.3. Refer to model (9.2) for the insurance innovation study. Suppose X_0 were dropped to eliminate the linear dependence in the $\mathbf{X}$ matrix, so that the model is: $Y_i = \beta_1 X_{i1} + \beta_2 X_{i2} + \beta_3 X_{i3} + \varepsilon_i$. What is the meaning here of the regression coefficients β_1, β_2, β_3?

9.4. Refer to model (9.4). Portray graphically the response curves for this model if Y_i is pounds of smoked codfish sold in supermarket i during test period,

X_{i1} is total sales (in hundred dollars) in delicatessen department of supermarket i during test period, $X_{i2} = 1$ if the smoked codfish is prepackaged by supermarket i and 0 otherwise, $\beta_0 = 4.3$, $\beta_1 = 1.0$, $\beta_2 = -2.0$.

9.5. Refer to Problem 3.6. The users of the desk calculators are either training institutions that use a student model or business firms that employ a commercial model. An analyst at Tri-City wishes to use the observations given in Problem 3.6 together with information on type of calculator to estimate the effect of model (S–student, C–commercial) on number of minutes spent on the service call. Records show that the models serviced in the 18 calls were:

i:	1	2	3	4	5	6	7	8	9
Model:	C	S	C	C	C	S	S	C	C

i:	10	11	12	13	14	15	16	17	18
Model:	C	C	S	S	C	S	C	C	C

Assume that the number of machines variable appears in a linear term only in the model and that no interaction variable is required.

a) Specify the model you will employ and explain the meaning of all regression coefficients.
b) Fit your model and make the investigation desired by the analyst. State your conclusions.
c) Why would the analyst wish to include the number of machines variable in the model when his interest is to estimate the effect of type of model on number of minutes spent on the call?

9.6. Refer to Problem 9.5. Make residual plots which will aid in evaluating the assumption that there are no interaction effects. Summarize your findings.

9.7. Refer to model (9.6). Portray graphically the response curves for this model if Y_i is product preference by respondent i, X_{i1} is age of respondent i, $X_{i2} = 1$ if respondent i is female and 0 if male, $\beta_0 = 40$, $\beta_1 = .5$, $\beta_2 = 0$, $\beta_3 = .3$. Describe the nature of the interaction effect.

9.8. Refer to models (9.4) and (9.6). Would the conclusion that $\beta_2 = 0$ have the same implication for each of these models?

9.9. Refer to the tool wear regression model (9.12). Suppose the indicator variables had been defined as follows: $X_{i2} = 1$ if tool model i is M2 and 0 otherwise; $X_{i3} = 1$ if tool model i is M3 and 0 otherwise; $X_{i4} = 1$ if tool model i is M4 and 0 otherwise. Indicate the meaning of each of the following: (1) β_3; (2) $\beta_4 - \beta_3$; (3) β_1; (4) β_7; (5) $\beta_6 - \beta_5$.

9.10. Refer to Problem 5.17.

a) Answer part (b) by fitting a regression model with indicator variables.
b) Answer part (c) by setting up a confidence interval for the appropriate regression coefficient in the regression model with indicator variables.

9.11. Global Electronics periodically imports shipments of a certain large part used as a component in several of its products. The size of the shipment varies depending upon production schedules. For handling and distribution to assembling plants, shipments of size 250 thousand parts or less are sent to warehouse A; larger shipments are sent to warehouse B since this warehouse has specialized equipment that provides greater economies of scale for large shipments. The data below were collected on the 10 most recent shipments; X is the size of the shipment (in thousand parts) and Y is the direct cost of handling the shipment in the warehouse (in thousand dollars).

i:	1	2	3	4	5	6	7	8	9	10
X_i:	225	350	150	200	175	180	325	290	400	125
Y_i:	11.95	14.50	8.83	11.10	10.03	9.97	13.75	13.30	15.00	8.09

a) Specify the model you would use to study the relation between the dependent and independent variables here.
b) Fit your model, and plot the fitted response function with the data on a chart. Is there any indication that greater economies of scale are obtained in handling relatively large shipments than relatively small ones? Could you put this question to a formal test?
c) For relatively small shipments, what is the point estimate of the increase in expected handling cost for each increase of one thousand in the size of the shipment? What is the corresponding estimate for relatively large shipments?

9.12. In time series analysis, the X variable representing time usually is defined to take on values 0, 1, 2, etc. for the successive time periods. Does this represent an allocated code when the time periods are 1970, 1971, 1972, etc.?

9.13. Refer to Figure 9.10. A student stated: "The least squares line cannot be correct as it stands since it is obvious that when Y can take on only values of 0 and 1 the least squares line has to be horizontal." Comment. For the programming task example, what would be the implication if the least squares line *were* horizontal?

9.14. A psychologist made a small-scale study to examine the nature of the relation, if any, between an employee's emotional stability and his ability to perform in a task group. The employee's emotional stability (X) was measured by a lengthy written test, and his ability to perform in a task group ($Y = 1$ if able, $Y = 0$ if unable) was evaluated by his supervisor. The results were:

i:	1	2	3	4	5	6	7	8
X_i:	471	619	584	631	399	481	624	589
Y_i:	0	1	0	1	0	1	1	1

a) Fit a linear response function by the two-stage least squares procedure.
b) Would ordinary least squares have been about equally effective here?
c) Is there any evidence available from this small-scale study whether or not a linear response function is appropriate for the range of X values encountered?

9.15. In an experiment on advertising recall, 1,500 cooperating housewives were divided at random into six groups of 250 each. Those in group 1 received a mail advertisement setting forth claims for a new product. Those in group 2 received two mailings of the advertisement spaced several days apart, and so on through group 6 which received six mailings. Fourteen days after the last mailing for a group, each housewife in that group was interviewed to ascertain whether she accurately recalled the advertising claims. She was scored 1 if she recalled accurately and 0 if she did not. The results are shown below; X_j denotes the number of mailings received by group j and R_j denotes the number of housewives in that group who recalled accurately.

j:	1	2	3	4	5	6
X_j:	1	2	3	4	5	6
R_j:	29	51	93	128	170	197

a) Fit a logistic response function, using transformation (9.49) and weighted least squares.
b) Plot the logistic response function and the original data. Does the fit appear to be a good one?
c) What is the estimated increase in the probability that a housewife will recall accurately if she receives two mailings instead of one? What is the corresponding estimated increase for six mailings instead of five?

CITED REFERENCES

9.1. Cox, D. R. *The Analysis of Binary Data.* London: Methuen & Co. Ltd., 1970.

9.2. Finney, D. J. *Probit Analysis.* 2d ed. Cambridge: Cambridge University Press, 1952.

10

Topics in Regression Analysis—II

IN THIS CHAPTER, we take up three selected topics in regression analysis: multicollinearity, reparameterization to improve computational accuracy, and autocorrelation.

10.1 MULTICOLLINEARITY

When we first discussed multiple regression, in Chapter 7, we noted some key problems that arise when the independent variables which are being considered for the model are correlated among themselves:

1. Adding or deleting an independent variable changes the regression coefficients.
2. The extra sum of squares associated with an independent variable varies, depending upon which independent variables already are included in the model.
3. The estimated regression coefficients individually may not be statistically significant even though a definite statistical relation exists between the dependent variable and the set of independent variables.

We shall now expand on the topic of multicollinearity because high intercorrelations among independent variables are frequently found in business and economics. For example, the independent variables family income and assets or the independent variables store sales and number of employees would tend to be correlated highly.

Nature of Problem

To see the essential nature of the problem of multicollinearity, we shall employ a simple example. The data in Table 10.1 refer to four sample observations on a dependent variable and two independent variables. Mr. A was asked to fit the multiple regression model:

$$E(Y) = \beta_0 + \beta_1 X_1 + \beta_2 X_2 \tag{10.1}$$

He returned in a short time with the fitted model:

$$\hat{Y} = -87 + X_1 + 18X_2 \tag{10.2}$$

He was proud of this model because it fit the data perfectly. The fitted values are shown in Table 10.1.

TABLE 10.1

Example of Perfectly Correlated Independent Variables

Observation i	X_{i1}	X_{i2}	Y_i	Fitted Values: *Model* (10.2)	Fitted Values: *Model* (10.3)
1	2	6	23	23	23
2	8	9	83	83	83
3	6	8	63	63	63
4	10	10	103	103	103

Model (10.2): $\hat{Y} = -87 + X_1 + 18X_2$

Model (10.3): $\hat{Y} = -7 + 9X_1 + 2X_2$

It so happened that Mr. B also was asked to fit model (10.1) to the same data, and he arrived at the fitted model:

$$\hat{Y} = -7 + 9X_1 + 2X_2 \tag{10.3}$$

Again, this model fits perfectly, as shown in Table 10.1.

Indeed, it can be shown that infinitely many models will fit the data in Table 10.1 perfectly. The reason is that the independent variables X_1 and X_2 are perfectly related, according to the relation:

$$X_2 = 5 + .5X_1 \tag{10.4}$$

Note carefully that fitted models (10.2) and (10.3) are entirely different response surfaces. The regression coefficients are different, and the fitted values will differ when X_1 and X_2 do not follow relation (10.4). For example, the fitted value for model (10.2) when $X_1 = 5$ and $X_2 = 5$ is:

$$\hat{Y} = -87 + 5 + 18(5) = 8$$

while the fitted value for model (10.3) is:

$$\hat{Y} = -7 + 9(5) + 2(5) = 48$$

Thus, when X_1 and X_2 are perfectly related and, as in our example, the data do not contain any random error component, many different response functions will lead to the same perfectly fitted values for the observations and to the same fitted values for any other X_1, X_2 combinations following the relation between X_1 and X_2. Yet these response functions are not the same and will lead to different fitted values for X_1, X_2 combinations that do not follow the relation between X_1 and X_2.

Two key implications of this example are:

1. The perfect relation between X_1 and X_2 did not inhibit our ability to obtain a good fit to the data.

2. Since many different models provide the same good fit, one cannot interpret any one set of regression coefficients as reflecting the effects of the different independent variables. Thus in fitted model (10.2), $b_1 = 1$ and $b_2 = 18$ do not imply that X_2 is the key independent variable and X_1 plays little role, because model (10.3) provides an equally good fit and its regression coefficients have opposite magnitudes.

Effects of Multicollinearity

In actual practice, we seldom find independent variables that are perfectly related or data that do not contain some random error component. Nevertheless, the implications just noted for our idealized example still have relevance.

1. The fact that some or all independent variables are correlated among themselves does not, in general, inhibit our ability to obtain a good fit nor does it tend to affect inferences about mean responses or predictions of new observations, provided these inferences are made within the region of observations. (Figure 7.9 on p. 249 provides an illustration of the concept of the region of observations for the case of two independent variables.)

2. The counterpart in real life to the many different regression functions providing equally good fits to the data in our idealized example is that the estimated regression coefficients tend to have large sampling variability when the independent variables are highly correlated. Thus, the estimated regression coefficients tend to vary widely from one sample to the next when the independent variables are highly correlated. As a result, only imprecise information may be available about the individual true regression coefficients and, indeed, each of the estimated regression coefficients individually may be statistically not significant even though a definite statistical relation exists between the dependent variable and the set of independent variables.

Example

To illustrate these basic points, consider the data in Table 10.2 on number of hours worked (Y), number of interviews completed (X_1), and number of miles driven (X_2) by 14 part-time interviewers in a survey to pretest a new product concept.

TABLE 10.2

Example of Highly Correlated Independent Variables

Observation i	*Number of Hours Worked* Y_i	*Number of Interviews Completed* X_{i1}	*Number of Miles Driven* X_{i2}
1	52.1	17	35.7
2	24.6	6	11.4
3	49.2	13	28.6
4	30.0	11	25.8
5	82.2	23	50.6
6	42.4	16	27.2
7	55.7	15	31.3
8	21.1	5	10.0
9	27.7	10	18.9
10	36.3	12	25.2
11	69.1	20	39.9
12	38.8	12	32.5
13	22.8	8	13.6
14	34.7	8	19.0

Suppose that we regress number of hours worked (Y) on number of interviews completed (X_1) only. The results of a least squares fit of the response function:

$$E(Y) = \beta_0 + \beta_1 X_1 \tag{10.5}$$

are shown in Table 10.3a. The coefficient of determination r_{Y1}^2 (the notation shows that the relation between Y and X_1 is being considered) is:

$$r_{Y1}^2 = \frac{SSR}{SSTO} = \frac{3,790.51}{4,244.75} = .8930$$

which indicates that the variability in Y is reduced by 89.3 percent by considering independent variable X_1. Also note that $s(b_1)$ is small relatively:

$$\frac{s(b_1)}{b_1} = \frac{.32727}{3.2749} = .0999$$

Let us now introduce the second independent variable X_2. It is highly correlated with X_1, as a glance at Table 10.2 will confirm. Indeed, the coeffi-

TABLE 10.3

Regression Results for Interviewer Example

(a) Regression of Y on X_1

$\hat{Y} = .73698 + 3.2749X_1$

Source of Variation	*SS*	*df*	*MS*
Regression	3,790.51	1	3,790.51
Error	454.24	12	37.85
Total	4,244.75	13	

Variable	*Estimated Regression Coefficient*	*Estimated Standard Deviation*	t^*
X_1	$b_1 = 3.2749$	$s(b_1) = .32727$	10.01

(b) Regression of Y on X_1 and X_2

$\hat{Y} = .72588 + 1.84395X_1 + .68164X_2$

Source of Variation	*SS*	*df*	*MS*
Regression	3,852.84	2	1,926.42
Error	391.91	11	35.63
Total	4,244.75	13	

Variable	*Estimated Regression Coefficient*	*Estimated Standard Deviation*	t^*
X_1	$b_1 = 1.84395$	$s(b_1) = 1.12749$	1.64
X_2	$b_2 = .68164$	$s(b_2) = .51535$	1.32

cient of determination between the two independent variables X_1 and X_2, denoted r_{12}^2, is $r_{12}^2 = .921$. The results of fitting the response function:

$$E(Y) = \beta_0 + \beta_1 X_1 + \beta_2 X_2 \tag{10.6}$$

are shown in Table 10.3b. Note the following:

1. The fit of model (10.6) has not been made worse in the sense of a higher error sum of squares *SSE*, despite the introduction of a highly correlated independent variable. Indeed, we noted earlier that *SSE* can never increase as the result of introducing another independent variable. (*MSE* can increase if the reduction in *SSE* is not adequate to compensate for the loss of one degree of freedom.) For our example, the coefficient of multiple determination is high, namely:

$$R^2 = \frac{SSR}{SSTO} = \frac{3{,}852.84}{4{,}244.75} = .9077$$

indicating that the variability of Y is reduced by 90.8 percent when both X_1 and X_2 are considered. Further, $MSE = 35.63$ now, as compared with $MSE = 37.85$ when only X_1 is included in the model.

2. The estimated regression coefficients for model (10.6) have large sampling variability. For example, the relative sampling variation of b_1 now is:

$$\frac{s(b_1)}{b_1} = \frac{1.12749}{1.84395} = .6115$$

as compared to .0999 when only X_1 is included in the model. Indeed, here again (as in the example in Chapter 7) separate examinations of b_1 and b_2 would lead to the conclusion, at the level of significance .05, that $\beta_1 = 0$ and $\beta_2 = 0$, whereas a test of the entire regression relation, based on:

$$F^* = \frac{MSR}{MSE} = \frac{1{,}926.42}{35.63} = 54.07$$

would lead to the conclusion that a regression relation does exist.

Comments

1. When the independent variables are highly correlated, not only do the estimated regression coefficients tend to be quite imprecise, but the true regression coefficients tend to lose their meaning. Suppose the two independent variables in a first-order model to predict sales of a product are price level (X_1) and income (X_2). β_1 then indicates the change in $E(Y)$ associated with a unit increase in X_1, when X_2 is held constant. If, in fact, a change in price level is almost always accompanied by a corresponding change in income, the usefulness of the measure β_1 becomes somewhat diminished because it considers the effect of a change in price level with no change in income. The same holds, of course, for β_2 which reflects the effect of a unit increase in income when price level is held constant.

2. Just as high intercorrelations between the independent variables tend to make the estimated regression coefficients imprecise (i.e., erratic from sample to sample), so do the coefficients of partial correlation between the dependent variable and each of the independent variables tend to become erratic from sample to sample when the independent variables are highly correlated.

3. The effect of intercorrelations between the independent variables on the standard deviations of the estimated regression coefficients can be seen readily when the variables in the model are transformed. Consider the model with two independent variables:

$$Y_i = \beta_0 + \beta_1 X_{i1} + \beta_2 X_{i2} + \varepsilon_i \tag{10.7}$$

Let us define the following simple functions of the standardized variables:

$$Y_i' = \frac{1}{\sqrt{n-1}} \left(\frac{Y_i - \bar{Y}}{s_Y} \right) \tag{10.8a}$$

$$X_{i1}' = \frac{1}{\sqrt{n-1}} \left(\frac{X_{i1} - \bar{X}_1}{s_1} \right) \tag{10.8b}$$

(10.8c)
$$X'_{i2} = \frac{1}{\sqrt{n-1}}\left(\frac{X_{i2} - \bar{X}_2}{s_2}\right)$$

where:

(10.9a)
$$s_Y^2 = \frac{\sum (Y_i - \bar{Y})^2}{n-1}$$

(10.9b)
$$s_1^2 = \frac{\sum (X_{i1} - \bar{X}_1)^2}{n-1}$$

(10.9c)
$$s_2^2 = \frac{\sum (X_{i2} - \bar{X}_2)^2}{n-1}$$

It will be shown in the next section that the reparameterized model with the transformed variables, corresponding to the original model (10.7), is:

(10.10)
$$Y'_i = \beta'_1 X'_{i1} + \beta'_2 X'_{i2} + \varepsilon'_i$$

and that the $(\mathbf{X'X})^{-1}$ matrix for the transformed model (10.10) is:

(10.11)
$$(\mathbf{X'X})^{-1} = \frac{1}{1 - r_{12}^2}\begin{bmatrix} 1 & -r_{12} \\ -r_{12} & 1 \end{bmatrix}$$

where r_{12} is the coefficient of correlation between X_1 and X_2. Hence, the variance-covariance matrix of the estimated regression coefficients is:

(10.12)
$$\boldsymbol{\sigma}^2(\mathbf{b}') = (\sigma')^2 \frac{1}{1 - r_{12}^2}\begin{bmatrix} 1 & -r_{12} \\ -r_{12} & 1 \end{bmatrix}$$

where $(\sigma')^2$ is the error term variance for the transformed model (10.10) and $\mathbf{b}'$ denotes the vector of the estimated regression coefficients for the transformed model.

Thus, the estimated regression coefficients b'_1 and b'_2 have the same variance:

(10.13)
$$\sigma^2(b'_1) = \sigma^2(b'_2) = \frac{(\sigma')^2}{1 - r_{12}^2}$$

which becomes larger as the correlation between X_1 and X_2 increases. Indeed, as X_1 and X_2 approach perfect correlation (i.e., as r_{12}^2 approaches 1), the variances of b'_1 and b'_2 become larger without limit.

4. We have noted that high multicollinearity is usually not a problem when the purpose of the regression analysis is to make inferences on the response function or predictions of new observations, provided these inferences are made within the range of observations. For instance, the estimated mean number of hours worked in our interviewer example when the only independent variable included in the model is number of interviews (X_1), together with its estimated standard deviation, are as follows for $X_{h1} = 10$ (calculations not shown):

$$\hat{Y}_h = 33.49 \qquad s(\hat{Y}_h) = 1.85$$

When the highly correlated independent variable number of miles driven (X_2) is also included in the model, the estimated mean number of hours worked, together with its estimated standard deviation, are as follows for $X_{h1} = 10$ and $X_{h2} = 20$:

$$\hat{Y}_h = 32.80 \qquad s(\hat{Y}_h) = 1.87$$

Thus, the precision of the estimated mean response is equally good as before, despite the addition of the second independent variable which is highly correlated with the first one. This stability in the precision of the estimated mean response occurred despite the fact that the estimated standard deviation of b_1 became substantially larger when X_2 was added to the model (see Table 10.3). The essential reason for the stability is that the covariance between b_1 and b_2 is negative, and plays a strong counteracting influence to the increase in $s^2(b_1)$ in determining the value of $s^2(\hat{Y}_h)$ as given in (7.67).

Remedial Measures

Various remedial measures have been proposed for the difficulties caused by multicollinearity when the main purpose of the regression analysis is to assess the effects of each of the independent variables. One proposal is to drop one or several independent variables from the model in order to lessen the multicollinearity and thereby reduce the standard errors of the estimated regression coefficients of the independent variables remaining in the model. Yet this action will not help to assess the effects of the independent variables for two reasons. First, no information is obtained about the dropped independent variables. Second, the magnitudes of the regression coefficients for the independent variables remaining in the model are affected by the correlated independent variables not included in the model.

Sometimes it is possible to add some observations which break the pattern of multicollinearity. Often, however, this option is not available. In business and economics, for instance, many independent variables cannot be controlled, so that new observations will tend to show the same intercorrelation patterns as the earlier observations.

In some economic studies, the regression coefficients for different independent variables have been estimated from different sets of data to avoid the problems of multicollinearity. Demand studies, for instance, have used both cross-section and time series data to this end. Suppose the independent variables in a demand study are price and income, and the relation to be estimated is:

$$Y_i = \beta_0 + \beta_1 X_{i1} + \beta_2 X_{i2} + \varepsilon_i \tag{10.14}$$

where Y is demand, X_1 is income, and X_2 is price. The income coefficient β_1 is then estimated from cross-section data. The demand variable Y is thereupon adjusted:

$$Y_i' = Y_i - b_1 X_{i1} \tag{10.15}$$

Finally, the price coefficient β_2 is estimated by regressing the adjusted demand variable Y' on X_2.

Some Further Comments

1. When severe multicollinearity is suspected because of excessively wide confidence intervals for some or all β_k, the correlation matrix (see next section) should be obtained to indicate which pairs of X variables are highly correlated.

2. It will be noted in the next section that a near-zero determinant of $\mathbf{X'X}$ is a potential source of serious roundoff errors in least squares results. Severe multicollinearity has the effect of making this determinant come close to zero. Thus under severe multicollinearity the regression coefficients may be subject to large roundoff errors as well as large sampling variances. For this reason, it may be desirable to drop one or several independent variables in the presence of severe multicollinearity even when concern is with inferences about the mean response or predictions of new observations, so as to increase the accuracy of the calculations.

3. When a large number of intercorrelated independent variables are candidates for inclusion in the regression model, it may be desirable to drop some of them not only for reasons of improving computational accuracy but also to reduce the computational labor. In Chapter 11, we discuss the selection of independent variables to be included in the model.

10.2 REPARAMETERIZATION TO IMPROVE COMPUTATIONAL ACCURACY

Roundoff Errors in Least Squares Calculations

Least squares results tend to be sensitive to rounding of data in intermediate stages of calculations. When the number of independent variables is small—say three or less—roundoff effects can be controlled by carrying a sufficient number of digits in intermediate calculations. (Some computer programs have an option whereby twice the usual number of digits can be carried, such as 16 instead of 8.) But this expedient becomes increasingly inefficient as the number of independent variables becomes larger.

Roundoff errors tend to enter into least squares calculations primarily when the inverse of $\mathbf{X'X}$ is taken. Of course, any errors in $(\mathbf{X'X})^{-1}$ may be magnified when calculating $\mathbf{b}$ or making other subsequent calculations. The danger of serious roundoff errors in $(\mathbf{X'X})^{-1}$ is particularly great when: (1) $\mathbf{X'X}$ has a determinant which is close to zero, and/or (2) the elements of $\mathbf{X'X}$ differ substantially in order of magnitude. The first condition arises when some or all of the independent variables are highly intercorrelated. A remedial measure, as noted in Section 10.1, is to discard one or several of the intercorrelated independent variables in an effort to shift the determinant away from near zero so that roundoff errors will not have as severe an effect.

The second condition arises when the variables have substantially different magnitudes, so that the entries in the $\mathbf{X'X}$ matrix cover a wide range, say from 15 to 49,000,000. A solution for this condition is to transform the variables and thereby reparameterize the regression model. We now address ourselves to this second problem. The transformation which we will take up is called the *standardized variable transformation*. It makes all entries in the $\mathbf{X'X}$ matrix for the transformed variables fall between -1 and $+1$, so that the calculation of the inverse matrix becomes much less subject to roundoff errors due to dissimilar orders of magnitudes than with the original variables. Many computer regression packages automatically use this transformation to obtain the basic regression results and then retransform to the original variables. In any case, users of computer programs are well advised to check that the regression package to be employed makes appropriate provisions to prevent roundoff errors from getting out of hand.

Standardized Variable Transformation

We shall illustrate the standardized variable transformation for the case of two independent variables. The basic regression model which we shall assume is the usual first-order one:

$$Y_i = \beta_0 + \beta_1 X_{i1} + \beta_2 X_{i2} + \varepsilon_i \tag{10.16}$$

If we wish to use the deviations $(X_{i1} - \bar{X}_1)$ and $(X_{i2} - \bar{X}_2)$, (10.16) must be modified by adding and subtracting the same terms:

$$Y_i = (\beta_0 + \beta_1 \bar{X}_1 + \beta_2 \bar{X}_2) + \beta_1(X_{i1} - \bar{X}_1) + \beta_2(X_{i2} - \bar{X}_2) + \varepsilon_i$$

or:

$$Y_i = \beta_0' + \beta_1(X_{i1} - \bar{X}_1) + \beta_2(X_{i2} - \bar{X}_2) + \varepsilon_i \tag{10.17}$$

where:

$$\beta_0' = \beta_0 + \beta_1 \bar{X}_1 + \beta_2 \bar{X}_2 \tag{10.18}$$

It can be shown that the least squares estimator of β_0' is always $\bar{Y}$. Hence, we can rewrite (10.17) as follows:

$$Y_i - \bar{Y} = \beta_1(X_{i1} - \bar{X}_1) + \beta_2(X_{i2} - \bar{X}_2) + \varepsilon_i \tag{10.19}$$

While it might appear that by eliminating one parameter from the model we have been able to increase the degrees of freedom available for MSE by one, this is not so since the dependent variable observations $Y_i - \bar{Y}$ are now subject to the restriction $\sum (Y_i - \bar{Y}) = 0$.

The standardized variable transformation goes one step beyond model (10.19) by first expressing each deviation variable in units of its standard deviation:

$$\frac{Y_i - \bar{Y}}{s_Y} \qquad \frac{X_{i1} - \bar{X}_1}{s_1} \qquad \frac{X_{i2} - \bar{X}_2}{s_2} \tag{10.20}$$

where s_Y, s_1, and s_2 are the respective standard deviations of Y, X_1, and X_2 as defined in (10.9). We then consider the functions:

$$Y_i' = \frac{1}{\sqrt{n-1}}\left(\frac{Y_i - \bar{Y}}{s_Y}\right) \tag{10.21a}$$

$$X_{i1}' = \frac{1}{\sqrt{n-1}}\left(\frac{X_{i1} - \bar{X}_1}{s_1}\right) \tag{10.21b}$$

$$X_{i2}' = \frac{1}{\sqrt{n-1}}\left(\frac{X_{i2} - \bar{X}_2}{s_2}\right) \tag{10.21c}$$

Thus, the transformed variables Y', X_1', and X_2' are simple functions of the standardized variables in (10.20).

Reparameterized Model

The regression model with the transformed variables Y', X_1', X_2' as defined in (10.21) is a simple extension of model (10.19):

$$Y_i' = \beta_1' X_{i1}' + \beta_2' X_{i2}' + \varepsilon_i' \tag{10.22}$$

It is easy to show that the new parameters β_1' and β_2' in (10.22) and the original parameters β_0, β_1, and β_2 in (10.16) are related as follows:

$$\beta_1 = \left(\frac{s_Y}{s_1}\right)\beta_1' \tag{10.23a}$$

$$\beta_2 = \left(\frac{s_Y}{s_2}\right)\beta_2' \tag{10.23b}$$

$$\beta_0 = \bar{Y} - \beta_1\bar{X}_1 - \beta_2\bar{X}_2 \tag{10.23c}$$

Thus, the new regression coefficients β_1' and β_2' and the original regression coefficients β_1 and β_2 are related by simple scaling factors involving ratios of standard deviations.

X′X Matrix for Transformed Variables

The **X** matrix for the transformed variables in model (10.22) is:

$$\mathbf{X} = \begin{bmatrix} X_{11}' & X_{12}' \\ X_{21}' & X_{22}' \\ \cdot & \cdot \\ \cdot & \cdot \\ \cdot & \cdot \\ X_{n1}' & X_{n2}' \end{bmatrix}$$

Remember that model (10.22) does not contain an intercept term; hence there is no column of 1's in the **X** matrix. The **X′X** matrix then is:

$$\mathbf{X'X} = \begin{bmatrix} X_{11}' & X_{21}' & \cdots & X_{n1}' \\ X_{12}' & X_{22}' & \cdots & X_{n2}' \end{bmatrix} \begin{bmatrix} X_{11}' & X_{12}' \\ X_{21}' & X_{22}' \\ \cdot & \cdot \\ \cdot & \cdot \\ \cdot & \cdot \\ X_{n1}' & X_{n2}' \end{bmatrix} \tag{10.24}$$

$$= \begin{bmatrix} \sum (X_{i1}')^2 & \sum X_{i1}' X_{i2}' \\ \sum X_{i2}' X_{i1}' & \sum (X_{i2}')^2 \end{bmatrix}$$

Let us now consider the elements of this **X′X** matrix. First, we have:

$$\sum (X'_{i1})^2 = \sum \left(\frac{X_{i1} - \bar{X}_1}{\sqrt{n-1}\, s_1} \right)^2 = \frac{\sum (X_{i1} - \bar{X}_1)^2}{n-1} \div s_1^2 = 1$$

Similarly:

$$\sum (X'_{i2})^2 = 1$$

Finally:

$$\sum X'_{i1} X'_{i2} = \sum \left(\frac{X_{i1} - \bar{X}_1}{\sqrt{n-1}\, s_1} \right) \left(\frac{X_{i2} - \bar{X}_2}{\sqrt{n-1}\, s_2} \right)$$

$$= \frac{1}{n-1} \frac{\sum (X_{i1} - \bar{X}_1)(X_{i2} - \bar{X}_2)}{s_1 s_2}$$

$$= \frac{\sum (X_{i1} - \bar{X}_1)(X_{i2} - \bar{X}_2)}{[\sum (X_{i1} - \bar{X}_1)^2]^{1/2} [\sum (X_{i2} - \bar{X}_2)^2]^{1/2}}$$

But this equals r_{12}, the coefficient of correlation between X_1 and X_2, by (3.71). Since $\sum X'_{i1} X'_{i2} = \sum X'_{i2} X'_{i1}$, we find that the **X′X** matrix for the transformed variables is:

(10.25)
$$\mathbf{X'X} = \begin{bmatrix} 1 & r_{12} \\ r_{12} & 1 \end{bmatrix}$$

By (3.70), r_{12} must fall between -1 and $+1$. Hence it follows that the elements of the **X′X** matrix for the transformed variables in (10.21) must have values between -1 and $+1$. While our example dealt with the case of two independent variables, the same result follows for any number of independent variables transformed according to the principle of (10.21).

The matrix:

$$\begin{bmatrix} 1 & r_{12} \\ r_{12} & 1 \end{bmatrix}$$

is called the *correlation matrix of the independent variables.*

Regression Calculations

The least squares estimators b'_1 and b'_2 for the transformed model (10.22) are obtained in the usual fashion. The inverse of the **X′X** matrix is:

(10.26)
$$(\mathbf{X'X})^{-1} = \frac{1}{1 - r_{12}^2} \begin{bmatrix} 1 & -r_{12} \\ -r_{12} & 1 \end{bmatrix}$$

It is easy to show that the **X′Y** matrix for the transformed variables is:

(10.27)
$$\mathbf{X'Y} = \begin{bmatrix} r_{Y1} \\ r_{Y2} \end{bmatrix}$$

where r_{Y1} and r_{Y2} are the coefficients of correlation between Y and X_1 and between Y and X_2 respectively. Hence, the estimated regression coefficients for the reparameterized model (10.22) are, by (7.21):

$$\mathbf{b}' = \frac{1}{1 - r_{12}^2} \begin{bmatrix} 1 & -r_{12} \\ -r_{12} & 1 \end{bmatrix} \begin{bmatrix} r_{Y1} \\ r_{Y2} \end{bmatrix} = \frac{1}{1 - r_{12}^2} \begin{bmatrix} r_{Y1} - r_{12} r_{Y2} \\ r_{Y2} - r_{12} r_{Y1} \end{bmatrix} \tag{10.28}$$

The return to the estimated regression coefficients for the original model is accomplished by employing the relations in (10.23):

$$b_1 = \left(\frac{s_Y}{s_1}\right) b_1' \tag{10.29a}$$

$$b_2 = \left(\frac{s_Y}{s_2}\right) b_2' \tag{10.29b}$$

$$b_0 = \bar{Y} - b_1 \bar{X}_1 - b_2 \bar{X}_2 \tag{10.29c}$$

Comments

1. Some computer packages present both the regression coefficients b_k for the original model as well as the coefficients b_k' for the transformed model. The latter are sometimes labeled *beta coefficients* in printouts.

2. The regression coefficients b_k' for the reparameterized model are the same as the standardized regression coefficients B_k discussed in Section 7.12, as a comparison of (7.96) with (10.29) makes clear. Thus, use of the transformed variables in (10.21) automatically leads to the standardized regression coefficients.

3. Some computer printouts show the magnitude of the determinant of the correlation matrix of the independent variables. A near-zero value for this determinant implies both a high degree of multicollinearity among the independent variables and a high potential for roundoff errors. In the simple linear regression case this determinant is seen to be $1 - r_{12}^2$, which approaches 0 as r_{12}^2 approaches 1.

4. When the correlation matrix of the independent variables is augmented by a row and column for Y, it is called the *correlation matrix*. A correlation matrix shows the coefficients of correlation for all pairs of dependent and independent variables. This information is useful in a variety of tasks—for instance, in selecting the final independent variables to be included in the model. Many computer programs display the correlation matrix in the printout.

For the case of two independent variables, the correlation matrix is as follows:

$$\begin{bmatrix} 1 & r_{Y1} & r_{Y2} \\ r_{Y1} & 1 & r_{12} \\ r_{Y2} & r_{12} & 1 \end{bmatrix}$$

Since the correlation matrix is symmetric, the lower (or upper) triangular block of elements is frequently omitted in computer printouts.

10.3 AUTOCORRELATION

The basic regression models considered so far have assumed that the random error terms ε_i are either uncorrelated random variables or independent normal random variables. In business and economics, many regression applications involve time series data. For such data, the assumption of uncorrelated or independent error terms is often not appropriate; rather the error terms are frequently correlated positively over time. Error terms correlated over time are said to be *autocorrelated* or *serially correlated.*

A major cause of positively autocorrelated error terms in business and economic regression applications involving time series data is the omission of one or several key variables from the model. When time-ordered effects of such "missing" key variables are positively correlated, the error terms in the regression model will tend to be positively autocorrelated since the error terms include effects of missing variables. Suppose, for example, that annual sales of a product are regressed against average yearly price over a period of 30 years. If population size has an important effect on sales, its omission from the model may lead to the error terms being positively autocorrelated because the effect of population size on sales likely is positively correlated over time.

Another common cause of positively autocorrelated error terms in economic data is systematic coverage errors in the dependent variable time series, which errors often tend to be positively correlated over time.

Problems of Autocorrelation

If the error terms in the regression model are positively autocorrelated, the use of the regular least squares procedures has a number of important consequences. We summarize these first, and then discuss them in more detail:

1. The regular least squares regression coefficients are still unbiased, but no longer have the minimum variance property and may be quite inefficient.
2. *MSE* may seriously underestimate the variance of the error terms.
3. $s(b_k)$ calculated according to the regular least squares procedure may seriously underestimate the true standard deviation of the estimated regression coefficient with that procedure.
4. The confidence intervals and tests using the t and F distributions, discussed earlier, are no longer strictly applicable.

To illustrate these problems intuitively, we shall consider the simple regression model with time series data:

$$Y_t = \beta_0 + \beta_1 X_t + \varepsilon_t$$

Here, Y_t and X_t are observations for period t. Let us assume that the error terms ε_t are positively autocorrelated as follows:

$$\varepsilon_t = \varepsilon_{t-1} + u_t$$

The u_t, called *disturbances*, are independent normal random variables. Thus, any error term ε_t is the sum of the previous error term ε_{t-1} and a new disturbance term u_t. We shall assume here that the u_t have mean 0 and variance 1.

In Table 10.4, column 1, we show 10 random observations on the normal

TABLE 10.4

Example of Positively Autocorrelated Error Terms

t	(1) u_t	(2) $\varepsilon_{t-1} + u_t = \varepsilon_t$
0	—	3.0
1	+.5	3.0 + .5 = 3.5
2	−.7	3.5 − .7 = 2.8
3	+.3	2.8 + .3 = 3.1
4	0	3.1 + 0 = 3.1
5	−2.3	3.1 − 2.3 = .8
6	−1.9	.8 − 1.9 = −1.1
7	+.2	−1.1 + .2 = − .9
8	−.3	− .9 − .3 = −1.2
9	+.2	−1.2 + .2 = −1.0
10	−.1	−1.0 − .1 = −1.1

variable u_t with mean 0 and variance 1, selected from a table of random normal deviates. Suppose now that $\varepsilon_0 = 3.0$; we obtain then:

$$\varepsilon_1 = \varepsilon_0 + u_1 = 3.0 + .5 = 3.5$$

$$\varepsilon_2 = \varepsilon_1 + u_2 = 3.5 - .7 = 2.8$$

etc.

These error terms are shown in Table 10.4, column 2, and they are plotted in Figure 10.1. Note the systematic pattern in these error terms. Their positive relation over time is shown by the fact that adjacent error terms tend to be of the same magnitude.

Suppose that X_t in the regression model represents time, such that $X_1 = 1$, $X_2 = 2$ etc. Further, suppose we know that $\beta_0 = 2$ and $\beta_1 = .5$. Figure 10.2a contains the true regression line and the observed Y values based on the error terms in Figure 10.1. Figure 10.2b contains the estimated regression line fitted by ordinary least squares methods and the observed Y values. Notice that the fitted regression line differs sharply from the true regression line

FIGURE 10.1

Example of Positively Autocorrelated Error Terms

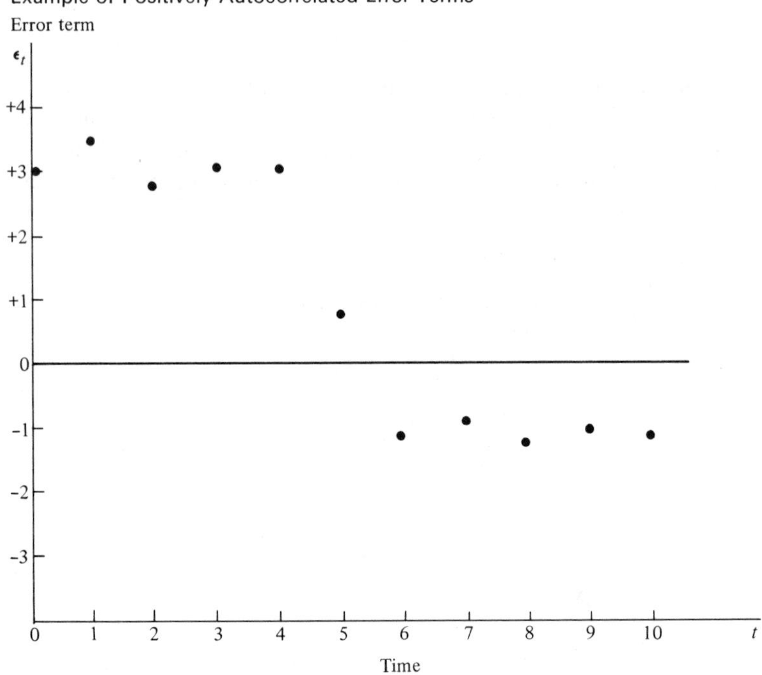

because the initial ε_0 value was large and the succeeding positively autocorrelated error terms tended to be large for some time. This persistency pattern in the positively autocorrelated error terms leads to a fitted regression line far from the true one. Had the initial ε_0 value been small, say $\varepsilon_0 = -.2$, and the disturbances different, a sharply different fitted regression line might have been obtained because of the persistency pattern, as shown in Figure 10.2c. This variation from sample to sample in the fitted regression lines because of the positively autocorrelated error terms may be so substantial as to lead to large variances of the estimated regression coefficients when regular least squares methods are used.

Another key problem with applying regular least squares methods when the error terms are positively autocorrelated, as mentioned before, is that MSE may seriously underestimate the variance of the ε_t. Figure 10.2 makes this clear. Note that the variability of the Y's around the fitted regression line in Figure 10.2b is substantially smaller than the variability of the Y's around the true regression line in Figure 10.2a. This is one of the factors leading to an indication of greater precision of the regression coefficients than is actually the case when regular least squares methods are used in the presence of positively autocorrelated errors.

FIGURE 10.2

Regression with Positively Autocorrelated Error Terms

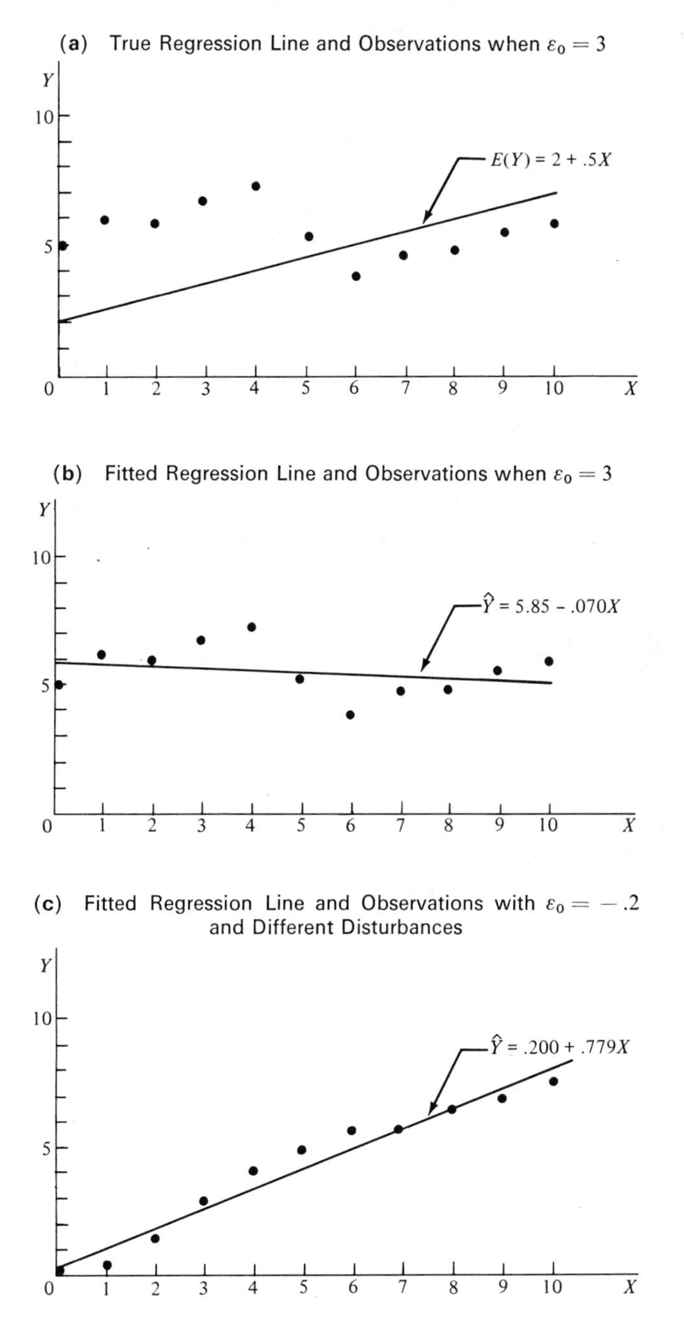

The chief remedy for these problems is to develop a model which recognizes the autocorrelations among the error terms in an appropriate fashion. We turn now to one such autocorrelation model. It is a simple model, yet experience suggests that it is widely applicable in business and economics when the error terms are serially correlated.

First-Order Autoregressive Error Model

The simple linear regression model for one independent variable, with the random error terms following a first-order autoregressive process, is:

$$(10.30) \qquad \begin{aligned} Y_t &= \beta_0 + \beta_1 X_t + \varepsilon_t \\ \varepsilon_t &= \rho\varepsilon_{t-1} + u_t \end{aligned}$$

where:

ρ is a parameter such that $|\rho| < 1$
u_t are independent $N(0, \sigma^2)$

Note that (10.30) is identical to the simple linear regression model (3.1) except for the structure of the error terms. Each error term in model (10.30) consists of a fraction of the previous error term (when $\rho > 0$) plus a new disturbance term u_t. The parameter ρ is called the *autocorrelation parameter*.

Properties of Error Terms

It is instructive to expand the expression for ε_t:

$$\varepsilon_t = \rho\varepsilon_{t-1} + u_t$$

Since $\varepsilon_{t-1} = \rho\varepsilon_{t-2} + u_{t-1}$, we obtain:

$$\varepsilon_t = \rho(\rho\varepsilon_{t-2} + u_{t-1}) + u_t = \rho^2\varepsilon_{t-2} + \rho u_{t-1} + u_t$$

Replacing now ε_{t-2} by $\rho\varepsilon_{t-3} + u_{t-2}$, we obtain:

$$\varepsilon_t = \rho^3\varepsilon_{t-3} + \rho^2 u_{t-2} + \rho u_{t-1} + u_t$$

Continuing in this fashion, we find:

$$(10.31) \qquad \varepsilon_t = \sum_{s=0}^{\infty} \rho^s u_{t-s}$$

Thus, the error term ε_t in period t is a linear combination of the current and preceding disturbance terms. When $0 < \rho < 1$, (10.31) indicates that the further the period is in the past, the smaller is the weight of that disturbance term in determining ε_t.

Mean. Since $E(u_t) = 0$ according to model (10.30) for all t, it follows from (10.31) that:

$$(10.32) \qquad E(\varepsilon_t) = 0$$

Thus, ε_t has expectation zero, just as for the simple regression model with uncorrelated error terms.

Variance. Since the u_t are independent according to model (10.30) with variance σ^2, it follows from (10.31) that:

$$\sigma^2(\varepsilon_t) = \sum_{s=0}^{\infty} \rho^{2s}\sigma^2(u_{t-s}) = \sigma^2 \sum_{s=0}^{\infty} \rho^{2s}$$

Now for $|\rho| < 1$, it is known that:

$$\sum_{s=0}^{\infty} \rho^{2s} = \frac{1}{1-\rho^2}$$

Hence, we have:

$$\sigma^2(\varepsilon_t) = \frac{\sigma^2}{1-\rho^2} \tag{10.33}$$

We thus see that the error terms have constant variance, just as for the simple regression model with uncorrelated error terms.

Covariance. To find the covariance of ε_t and ε_{t-1}, we need to recognize that:

$$\sigma^2(\varepsilon_t) = E(\varepsilon_t^2)$$
$$\sigma(\varepsilon_t, \varepsilon_{t-1}) = E(\varepsilon_t \varepsilon_{t-1})$$

These results follow from theorems (1.14a) and (1.19a) respectively, since $E(\varepsilon_t) = 0$ by (10.32).

By (10.31), we have:

$$E(\varepsilon_t \varepsilon_{t-1}) = E[(u_t + \rho u_{t-1} + \rho^2 u_{t-2} + \cdots)(u_{t-1} + \rho u_{t-2} + \rho^2 u_{t-3} + \cdots)]$$

which can be rewritten:

$$\begin{aligned} E(\varepsilon_t \varepsilon_{t-1}) &= E\{[u_t + \rho(u_{t-1} + \rho u_{t-2} + \cdots)][u_{t-1} + \rho u_{t-2} + \rho^2 u_{t-3} + \cdots]\} \\ &= E[u_t(u_{t-1} + \rho u_{t-2} + \rho^2 u_{t-3} + \cdots)] \\ &\qquad + E[\rho(u_{t-1} + \rho u_{t-2} + \rho^2 u_{t-3} + \cdots)^2] \end{aligned}$$

Since $E(u_t u_{t-s}) = 0$ for all $s \neq 0$ by the assumed independence of the u_t, the first term drops out and we obtain:

$$E(\varepsilon_t \varepsilon_{t-1}) = \rho E(\varepsilon_{t-1}^2) = \rho\sigma^2(\varepsilon_{t-1})$$

Hence by (10.33), which holds for all t, we have:

$$\sigma(\varepsilon_t, \varepsilon_{t-1}) = \rho\left(\frac{\sigma^2}{1-\rho^2}\right) \tag{10.34}$$

In general, it can be shown that:

$$\sigma(\varepsilon_t, \varepsilon_{t-s}) = \rho^s\left(\frac{\sigma^2}{1-\rho^2}\right) \qquad s \neq 0 \tag{10.35}$$

Thus, the error terms for model (10.30) are autocorrelated unless the autocorrelation parameter ρ equals zero.

Durbin-Watson Test for Autocorrelation

When regression analysis is based upon time series data, it is often desirable to test whether or not the error terms in the regression model are uncorrelated. One test widely used is the Durbin-Watson test. This test assumes the first-order autoregressive error model (10.30) with the values of the independent variable fixed, or a corresponding multiple regression model. The test consists of determining whether or not the autocorrelation parameter ρ is zero. Note from model (10.30) that if $\rho = 0$, $\varepsilon_t = u_t$. Hence the error terms ε_t are then independent since the u_t are independent.

In view of the fact that correlated error terms in business and economic applications tend to show positive serial correlation, the usual test alternatives considered are:

(10.36)
$$\begin{aligned} C_1&: \rho = 0 \\ C_2&: \rho > 0 \end{aligned}$$

The test statistic D is obtained by first fitting the ordinary least squares regression line and calculating the residuals:

(10.37)
$$e_t = Y_t - \hat{Y}_t$$

and then calculating the statistic:

(10.38)
$$D = \frac{\sum_{t=2}^{n} (e_t - e_{t-1})^2}{\sum_{t=1}^{n} e_t^2}$$

where n is the number of observations.

An exact test procedure is not available, but Durbin and Watson have obtained lower and upper bounds d_L and d_U such that a value of D outside these bounds leads to a definite decision. The decision rule for testing between the alternatives in (10.36) is:

(10.39)
$$\begin{aligned} &\text{If } D > d_U, \text{ conclude } C_1 \\ &\text{If } D < d_L, \text{ conclude } C_2 \\ &\text{If } d_L \leq D \leq d_U, \text{ the test is inconclusive} \end{aligned}$$

Small values of D lead to the conclusion that $\rho > 0$ because the adjacent error terms ε_t and ε_{t-1} tend to be of the same magnitude when they are positively autocorrelated. Hence the differences in the residuals, $e_t - e_{t-1}$, would tend to be small when $\rho > 0$, leading to a small numerator in D and hence to a small test statistic D.

Table A–6 contains the bounds d_L and d_U for various sample sizes (n), for two levels of significance (.05 and .01), and for various numbers of X variables $(p - 1)$ in the regression model.

Example. The Blaisdell Company wished to predict its sales by using as a predictor variable disposable personal income. In Table 10.5, columns 1 and 2 contain seasonally adjusted quarterly data on company sales and disposable personal income respectively for the period 1967–71. A scatter plot (not shown) suggested that a linear regression model is appropriate. The market research analyst was, however, concerned whether or not the error terms could be assumed to be uncorrelated. He therefore used the Durbin-Watson test with the alternatives:

$$C_1: \rho = 0$$
$$C_2: \rho > 0$$

He fitted an ordinary least squares regression line to the data in Table 10.5. The results are shown in Table 10.6a. He then calculated the residuals e_t, which are shown in column 3 of Table 10.5. Note how the residuals consistently are above or below the fitted values for extended periods. Autocorrelation

TABLE 10.5

Durbin-Watson Test Calculations for Blaisdell Company Example (sales and disposable personal income data are seasonally adjusted)

Year and Quarter	t	(1) *Company Sales (million dollars)* Y_t	(2) *Disposable Personal Income (billion dollars)* X_t	(3) *Residuals* e_t	(4) $e_t - e_{t-1}$	(5) $(e_t - e_{t-1})^2$	(6) e_t^2
1967: 1	1	34.97	133.6	−.99655	———	———	.99311190
2	2	35.35	135.4	−.98090	+.01565	.00024492	.96216481
3	3	35.92	137.6	−.85622	+.12468	.01554510	.73311269
4	4	36.64	140.0	−.62203	+.23419	.05484496	.38692132
1968: 1	5	37.65	143.8	−.38122	+.24081	.05798946	.14532869
2	6	38.57	147.1	−.12920	+.25202	.06351408	.01669264
3	7	39.13	148.8	.08669	+.21589	.04660849	.00751516
4	8	39.89	151.4	.32040	+.23371	.05462036	.10265616
1969: 1	9	40.42	153.3	.46580	+.14540	.02114116	.21696964
2	10	41.40	156.5	.79806	+.33226	.11039671	.63689976
3	11	42.34	160.8	.86766	+.06960	.00484416	.75283388
4	12	43.05	163.6	1.01089	+.14323	.02051483	1.02189859
1970: 1	13	44.04	166.9	1.33291	+.32202	.10369688	1.77664907
2	14	45.18	171.4	1.56202	+.22911	.05249139	2.43990648
3	15	45.70	174.0	1.55573	−.00629	.00003956	2.42029583
4	16	46.09	175.4	1.66235	+.10662	.01136782	2.76340752
1971: 1	17	46.29	180.5	.83001	−.83234	.69278988	.68891660
2	18	44.34	184.9	−2.01063	−2.84064	8.06923561	4.04263300
3	19	44.50	187.1	−2.29595	−.28532	.08140750	5.27138640
4	20	44.90	188.7	−2.21982	+.07613	.00579578	4.92760083
Total						9.46708865	30.30690097

TABLE 10.6

Regression Results for Blaisdell Company Example

(a) Original Variables Y_t and X_t

Regression Coefficient	Estimated Regression Coefficient	Estimated Standard Deviation
β_0	$b_0 = 8.92339$	$s(b_0) = 2.67208$
β_1	$b_1 = .20242$	$s(b_1) = .01660$

$$\hat{Y} = 8.92339 + .20242X$$

(b) Transformed Variables $Y_t' = Y_t - rY_{t-1}$ and $X_t' = X_t - rX_{t-1}$

Regression Coefficient	Estimated Regression Coefficient	Estimated Standard Deviation
β_0'	$b_0' = 4.02457$	$s(b_0') = 1.31639$
$\beta_1' = \beta_1$	$b_1' = .04862$	$s(b_1') = .06483$

$$b_0 = \frac{b_0'}{1-r} = \frac{4.02457}{1-.89100} = 36.92266$$

$$s(b_0) = \frac{s(b_0')}{1-r} = \frac{1.31639}{1-.89100} = 12.07697$$

(c) First Differences $Y_t' = Y_t - Y_{t-1}$ and $X_t' = X_t - X_{t-1}$

Regression Coefficient	Estimated Regression Coefficient	Estimated Standard Deviation
$\beta_1' = \beta_1$	$b_1 = .15068$	$s(b_1) = .05241$

in the error terms is suggested when such a pattern is obtained despite the fact that an appropriate regression function has been employed.

Columns 4, 5, and 6 of Table 10.5 contain the necessary calculations for the test statistic D. The analyst then obtained:

$$D = \frac{\sum_{t=2}^{20}(e_t - e_{t-1})^2}{\sum_{t=1}^{20} e_t^2} = \frac{9.4671}{30.3069} = .312$$

Using a level of significance of .01, he found in Table A–6 for $n = 20$ and $p - 1 = 1$:

$$d_L = .95$$

$$d_U = 1.15$$

Since $D = .312$ falls below $d_L = .95$, decision rule (10.39) indicates that the appropriate conclusion is C_2, namely that the error terms are positively autocorrelated.

Comments

1. If a test for negative autocorrelation is required, the test statistic to be used is $4 - D$, where D is defined as above. The test is then conducted in the same manner described for testing for positive autocorrelation. That is, if the quantity $4 - D$ falls below d_L, we conclude $\rho < 0$, that negative autocorrelation exists, and so on.

2. A two-sided test for C_1: $\rho = 0$ versus C_2: $\rho \neq 0$ can be made by employing both one-sided tests separately. The Type I risk with the two-sided test is 2α, where α is the Type I risk with each one-sided test.

3. When the Durbin-Watson test employing the bounds d_L and d_U gives indeterminate results, in principle more observations are required. Of course, with time series data it may be impossible to obtain more observations, or additional observations may lie in the future and be obtainable only with great delay. Durbin and Watson (Ref. 10.1) do give an approximate test which may be used when the bounds test is indeterminate, but the degrees of freedom should be larger than about 40 before this approximate test will give more than a rough indication of whether autocorrelation exists.

4. While the Durbin-Watson test is widely used, other tests for autocorrelation are available. One such alternative test, due to Theil and Nagar, is found in Reference 10.2.

Estimation of Regression Parameters

When the autocorrelation parameter ρ in model (10.30) is not zero, it is desirable to recognize the autocorrelated structure of the error terms for estimating the regression parameters. Two suggested methods of doing so will now be discussed, and our earlier Blaisdell Company example will be used to illustrate each.

Iterative Approach. The iterative approach is motivated by an interesting property of model (10.30). Consider the transformed dependent variable:

$$Y_t' = Y_t - \rho Y_{t-1}$$

Substituting in this expression for Y_t and Y_{t-1} according to model (10.30), we obtain:

$$\begin{aligned} Y_t' &= (\beta_0 + \beta_1 X_t + \varepsilon_t) - \rho(\beta_0 + \beta_1 X_{t-1} + \varepsilon_{t-1}) \\ &= \beta_0(1 - \rho) + \beta_1(X_t - \rho X_{t-1}) + (\varepsilon_t - \rho\varepsilon_{t-1}) \end{aligned}$$

But by (10.30), $\varepsilon_t - \rho\varepsilon_{t-1} = u_t$. Hence:

$$Y_t' = \beta_0(1 - \rho) + \beta_1(X_t - \rho X_{t-1}) + u_t$$

where the u_t are independent error terms. Thus, when we use the transformed variables:

(10.40a) $$Y_t' = Y_t - \rho Y_{t-1}$$

(10.40b) $$X_t' = X_t - \rho X_{t-1}$$

the reparameterized regression model:

(10.41) $$Y_t' = \beta_0' + \beta_1' X_t' + u_t$$

has independent error terms. This means that ordinary least squares methods have their usual optimum properties with model (10.41).

The parameters in the original model (10.30) are related to the parameters in the transformed model (10.41) as follows:

(10.42a) $$\beta_0 = \frac{\beta_0'}{1 - \rho}$$

(10.42b) $$\beta_1 = \beta_1'$$

The transformed model (10.41) unfortunately cannot be used directly because the autocorrelation parameter ρ needed to obtain the transformed variables in (10.40) is unknown. We can, however, estimate ρ. Note that the autoregressive error process assumed in model (10.30) can be viewed as a regression through the origin:

$$\varepsilon_t = \rho\varepsilon_{t-1} + u_t$$

where ε_t is the dependent variable, ε_{t-1} the independent variable, u_t the error term, and ρ the slope of the line through the origin. Since the ε_t and ε_{t-1} are unknown we use the residuals e_t and e_{t-1}, obtained by ordinary least squares methods, as the dependent and independent variables respectively, and estimate ρ by fitting a straight line through the origin. From our previous discussion of regression through the origin, we know by (5.32) that the estimate of the slope ρ, denoted by r, is:

(10.43) $$r = \frac{\sum_{t=2}^{n} e_{t-1}e_t}{\sum_{t=2}^{n} e_{t-1}^2}$$

We now obtain the transformed variables:

(10.44a) $$Y_t' = Y_t - rY_{t-1}$$

(10.44b) $$X_t' = X_t - rX_{t-1}$$

and use ordinary least squares with these transformed variables.

The Durbin-Watson test is then employed to test whether the error terms for the transformed model are uncorrelated. If the test indicates that they are uncorrelated, the procedure terminates. Otherwise, the parameter ρ is re-estimated from the new residuals for the regression model with the original variables, using the regression coefficients derived from the fit of the regression model with the transformed variables. A new set of transformed variables is then obtained with the new r. This process may be continued for several iterations until the Durbin-Watson test suggests that the error terms in the transformed model are uncorrelated.

This iterative approach does not always work properly. A major reason is that the estimate r in (10.43), when the error terms are positively autocorrelated, tends to underestimate the autocorrelation parameter ρ. This bias, when serious, can significantly reduce the effectiveness of the iterative approach.

Example. We demonstrate the iterative approach for our Blaisdell Company example, although this is one of the times when the iterative approach does not appear to work effectively. The necessary calculations for estimating the autocorrelation parameter ρ, based on the residuals obtained with ordinary least squares applied to the original variables, appear in columns 3 and 4 of Table 10.7. Hence we estimate:

$$r = \frac{22.61300}{25.37930} = .89100$$

We now obtain the transformed variables Y_t' and X_t' in (10.44a) and (10.44b):

$$Y_t' = Y_t - .89100Y_{t-1}$$
$$X_t' = X_t - .89100X_{t-1}$$

These are shown in Table 10.8. Ordinary least squares fitting of linear regression is now used with these transformed variables. The results are shown in Table 10.6b.

TABLE 10.7

Calculations for Estimating ρ for Blaisdell Company Example

t	(1) Dependent Variable e_t	(2) Independent Variable e_{t-1}	(3) $e_t e_{t-1}$	(4) e_{t-1}^2
1	−.99655	———	———	———
2	−.98090	−.99655	.97751590	.99311190
3	−.85622	−.98090	.83986620	.96216481
4	−.62203	−.85622	.53259453	.73311269
5	−.38122	−.62203	.23713028	.38692132
6	−.12920	−.38122	.04925362	.14532869
7	.08669	−.12920	−.01120035	.01669264
8	.32040	.08669	.02777548	.00751516
9	.46580	.32040	.14924232	.10265616
10	.79806	.46580	.37173635	.21696964
11	.86766	.79806	.69244474	.63689976
12	1.01089	.86766	.87710882	.75283388
13	1.33291	1.01089	1.34742539	1.02189859
14	1.56202	1.33291	2.08203208	1.77664907
15	1.55573	1.56202	2.43008137	2.43990648
16	1.66235	1.55573	2.58616777	2.42029583
17	.83001	1.66235	1.37976712	2.76340752
18	−2.01063	.83001	−1.66884301	.68891660
19	−2.29595	−2.01063	4.61630595	4.04263300
20	−2.21982	−2.29595	5.09659573	5.27138640
Total			22.61300029	25.37930014

TABLE 10.8

Transformed Variables for First Iteration for Blaisdell Company Example

t	(1) Y_t	(2) X_t	(3) $Y_t' = Y_t - .89100 Y_{t-1}$	(4) $X_t' = X_t - .89100 X_{t-1}$
1	34.97	133.6	—	—
2	35.35	135.4	4.1917	16.362
3	35.92	137.6	4.4232	16.959
4	36.64	140.0	4.6353	17.398
5	37.65	143.8	5.0038	19.060
6	38.57	147.1	5.0238	18.974
7	39.13	148.8	4.7641	17.734
8	39.89	151.4	5.0252	18.819
9	40.42	153.3	4.8780	18.403
10	41.40	156.5	5.3858	19.910
11	42.34	160.8	5.4526	21.358
12	43.05	163.6	5.3251	20.327
13	44.04	166.9	5.6824	21.132
14	45.18	171.4	5.9404	22.692
15	45.70	174.0	5.4446	21.283
16	46.09	175.4	5.3713	20.366
17	46.29	180.5	5.2238	24.219
18	44.34	184.9	3.0956	24.074
19	44.50	187.1	4.9931	22.354
20	44.90	188.7	5.2505	21.994

Based on the fitted regression for the transformed variables in Table 10.6b, residuals were obtained and the Durbin-Watson statistic calculated. The result was (calculations not shown) $D = 1.37$. From Table A–6, we find for $\alpha = .01$, $p - 1 = 1$, and $n = 19$:

$$d_L = .93 \qquad d_U = 1.13$$

Since $D = 1.37 > d_U = 1.13$, we conclude that the autocorrelation coefficient for the error terms in the model with the transformed variables is zero. Hence, the estimated regression coefficients for the model with the original variables are (see Table 10.6b):

$$b_0 = 36.92266 \qquad s(b_0) = 12.07697$$
$$b_1 = .04862 \qquad s(b_1) = .06483$$

The estimated regression coefficient b_1 differs sharply from that obtained with ordinary least squares (see Table 10.6a), and is indeed not statistically significant. It also differs substantially from the regression coefficient obtained with the first differences approach, to be discussed next. It would therefore appear that the iterative approach did not work well in this example.

First Differences Approach. Some economists and statisticians have suggested that instead of iterative estimation of ρ, which is not always successful, the autocorrelation parameter be assumed to equal 1. If $\rho = 1$, the transformed model (10.41) becomes:

$$Y_t' = \beta_1' X_t' + u_t \tag{10.45}$$

since $\beta_0' = \beta_0(1 - \rho)$. Thus the regression coefficient $\beta_1' = \beta_1$ can be directly estimated by regular least squares methods for regression through the origin with the transformed variables:

$$Y_t' = Y_t - Y_{t-1} \tag{10.46a}$$

$$X_t' = X_t - X_{t-1} \tag{10.46b}$$

Note that these transformed variables are ordinary first differences. It has been found that this first differences approach is effective in a variety of applications in reducing the autocorrelations of the error terms, and of course it is much simpler than the iterative approach.

Example. Table 10.9 contains the transformed variables Y_t' and X_t', based on the first differences transformation in (10.46) for our Blaisdell Company example. Application of the ordinary least squares method of estimating a linear regression through the origin led to the results shown in Table 10.6c. Note that the estimated regression coefficient $b_1 = .15068$ is similar to that obtained with ordinary least squares applied to the original variables ($b_1 = .20242$), but has an appreciably higher standard error. We

TABLE 10.9

First Differences for Blaisdell Company Data

t	(1) Y_t	(2) X_t	(3) $Y_t' = Y_t - Y_{t-1}$	(4) $X_t' = X_t - X_{t-1}$
1	34.97	133.6	—	—
2	35.35	135.4	.38	1.8
3	35.92	137.6	.57	2.2
4	36.64	140.0	.72	2.4
5	37.65	143.8	1.01	3.8
6	38.57	147.1	.92	3.3
7	39.13	148.8	.56	1.7
8	39.89	151.4	.76	2.6
9	40.42	153.3	.53	1.9
10	41.40	156.5	.98	3.2
11	42.34	160.8	.94	4.3
12	43.05	163.6	.71	2.8
13	44.04	166.9	.99	3.3
14	45.18	171.4	1.14	4.5
15	45.70	174.0	.52	2.6
16	46.09	175.4	.39	1.4
17	46.29	180.5	.20	5.1
18	44.34	184.9	−1.95	4.4
19	44.50	187.1	.16	2.2
20	44.90	188.7	.40	1.6

reiterate that the standard error $s(b_1) = .01660$ for ordinary least squares applied to the original variables is likely to be a serious understatement of the true standard error with that method.

Note

Sometimes the first differences approach can overcorrect, leading to negative autocorrelations in the error terms. Hence, it may be appropriate to use a two-sided Durbin-Watson test when testing for autocorrelation with first differences data. One complication arises here. The first differences model (10.45) has no intercept term, but the Durbin-Watson test requires a fitted regression with an intercept term. A valid test for autocorrelation in a no-intercept model can be carried out by fitting for this purpose a regression function with an intercept term. Of course, the fitted no-intercept model is still the model of basic interest.

Comments

1. The first-order autoregressive error process in model (10.30) is the simplest kind. A second-order process would be:

(10.47) $$\varepsilon_t = \rho_1 \varepsilon_{t-1} + \rho_2 \varepsilon_{t-2} + u_t$$

Still higher-order processes could be postulated. In practice, the first-order error process in model (10.30) appears to be satisfactory for many applications.

2. The autocorrelated error structure can also be used to advantage in situations where predictions of the dependent variable are to be made. Johnston (Ref. 10.3) discusses this problem.

PROBLEMS

10.1. Refer to the example of perfectly correlated independent variables in Table 10.1.

a) Develop another model, like models (10.2) and (10.3), that fits the data perfectly.

b) What is the intersection of the infinitely many response surfaces that fit the data perfectly?

10.2. The progress report of a research analyst to his supervisor stated: "All the estimated regression coefficients in our model with three independent variables to predict sales are statistically significant. Our new preliminary model with seven independent variables, which includes the three variables of our smaller model, is less satisfactory because only two of the seven regression coefficients are statistically significant. Yet in some initial trials the expanded model is giving more precise sales predictions than the smaller model. The reasons for this disturbing contradiction are now being investigated." Comment.

10.3. An assistant in the district sales office of a national cosmetics firm obtained data, shown below, on advertising expenditures and sales last year in the district's 10 territories. X_1 denotes expenditures for point-of-sale displays in beauty salons and department stores (in thousand dollars), while X_2 and X_3 represent the corresponding expenditures for local media advertising and prorated share of national media advertising respectively. Y denotes sales (in thousand cases). The assistant was instructed to estimate the increase in expected sales when X_1 is increased by one thousand dollars and X_2 and X_3 are held constant, and was told to use the ordinary multiple regression model with linear terms for the independent variables and normal error terms.

i:	1	2	3	4	5	6	7	8	9	10
X_{i1}:	4.2	6.5	3.0	2.1	2.9	7.2	4.8	4.3	2.6	3.1
X_{i2}:	4.0	6.5	3.5	2.0	3.0	7.0	5.0	4.0	2.5	3.0
X_{i3}:	3.0	5.0	4.0	3.0	4.0	3.0	4.5	5.0	5.0	4.0
Y_i:	8.26	14.70	9.73	5.62	7.84	12.18	8.56	10.77	7.56	8.90

a) State the model and fit it to the data.

b) Test whether there is a regression relation between sales and the three independent variables. Use a level of significance of .05. Also test for each of the regression coefficients β_k $(k = 1, 2, 3)$ individually whether or not $\beta_k = 0$. Use a level of significance of .05 each time.

c) Obtain the correlation matrix.

d) What do the results in parts (b) and (c) suggest about the suitability of the data for the research objective?

e) The assistant eventually decided to drop X_2 and X_3 from the model "to clear up the picture." Fit the assistant's revised model. Is he now in a better position to achieve his research objective?

f) Why would an experiment here be more effective in providing suitable data to meet the research objective? How would you design such an experiment? What model would you employ?

10.4. Two authors wrote as follows: "Our research utilized a multiple regression model. Two of the independent variables important in our theory turned out to be highly correlated in our observations. This made it difficult to assess the individual effects of each of these variables separately. We retained both variables in our model, however, because the coefficient of multiple determination was very high which made this difficulty unimportant." Comment.

10.5. Refer to the work crew productivity example data in Table 7.6.

a) For the variables transformed according to (10.21), obtain: (1) $\mathbf{X'X}$; (2) $\mathbf{X'Y}$; (3) $\mathbf{b'}$; (4) $\mathbf{s^2(b')}$.

b) Show that the regression coefficients obtained in part (a3) are the same as the standardized regression coefficients according to (7.96).

10.6. Derive the relations between the β_k and β_k' $(k = 1, 2)$ shown in (10.23a) and (10.23b).

10.7. Derive the expression for $\mathbf{X'Y}$ in (10.27) for the transformed variables.

10.8. Refer to the first-order autoregressive error model (10.30). Suppose Y_t is company's percent share of the market, X_t is company's selling price as a percent of average competitive selling price, $\beta_0 = 50$, $\beta_1 = -.30$, $\rho = .6$, $\sigma^2 = 1$, and $\varepsilon_0 = 2.403$. Let X_t and u_t be as follows for $t = 1, \ldots, 10$:

t:	1	2	3	4	5	6	7	8	9	10
X_t:	100	115	120	90	85	75	70	95	105	110
u_t:	.764	.509	−.242	−1.808	−.485	.501	−.539	.434	−.299	.030

a) Plot the true regression line. Generate the observations Y_t $(t = 1, \ldots, 10)$, and plot these. Fit a least squares line to the observations and plot it also. How does your fitted regression line relate to the true line?

b) Repeat the steps in part (a) but this time let $\rho = 0$. In which of the two cases does the fitted regression line come closer to the true line? Is this the expected outcome?

10.9. Refer to Problem 10.8. Suppose $\rho = -.7$. Obtain the error terms ε_t and plot them in the format of Figure 10.1.

10.10. Refer to Problems 10.8 and 10.9.

a) For each of the cases $\rho = .6$, $\rho = 0$, and $\rho = -.7$, obtain the successive error term differences $\varepsilon_t - \varepsilon_{t-1}$ $(t = 2, \ldots, 10)$.

b) Compare the magnitudes of the successive differences $\varepsilon_t - \varepsilon_{t-1}$ for the three cases. What generalization does this suggest?

c) For which of the three cases is $\sum (\varepsilon_t - \varepsilon_{t-1})^2$ smallest? For which is it largest?

10.11. A speaker, in discussing problems in using regression analysis for cost control, stated: "The recent floods south of here increased the operating costs of some of the outlets of a fast-food chain for which we provide management services. Since cost data for the outlets are put into our data matrix in geographic order, the cost data for the outlets in the flood area are in adjoining rows. We know in advance that the residuals for these rows will all be on the plus side—that is, they won't be in a random order. I'm not sure, but I think this calls for an autocorrelated model." Discuss.

10.12. For each of the following tests concerning the autocorrelation parameter ρ in model (10.30), state the appropriate decision rule based on the Durbin-Watson statistic for a sample of size 20 and three independent variables: (1) $C_1: \rho = 0$, $C_2: \rho \neq 0$, $\alpha = .02$; (2) $C_1: \rho = 0$, $C_2: \rho < 0$, $\alpha = .05$; (3) $C_1: \rho = 0$, $C_2: \rho > 0$, $\alpha = .01$.

10.13. A staff analyst for a manufacturer of electronic components has compiled the data shown below on value of annual production of certain types of electronic equipment (X, in billions of constant dollars) and value of the firm's components used in this production (Y, in millions of constant dollars). The analyst assumes that model (10.30) is appropriate and wishes to decide whether $\rho = 0$ or $\rho > 0$, using a level of significance of .05. Make the test. What conclusion is indicated?

t:	1	2	3	4	5	6	7	8
X_t:	2.052	2.026	2.002	1.949	1.942	1.887	1.968	2.053
Y_t:	103.4	101.9	100.8	98.0	97.3	93.5	97.5	102.2

t:	9	10	11	12	13	14	15	16
X_t:	2.102	2.113	2.058	2.060	2.035	2.080	2.102	2.150
Y_t:	105.0	107.2	105.1	103.9	103.0	104.8	105.0	107.2

10.14. Refer to Problem 10.13. The staff analyst wishes to use the iterative method to estimate the regression coefficients in model (10.30).

a) Use one iteration to obtain estimates of the regression coefficients β_0 and β_1. Also obtain the estimated standard deviations of these estimates.

b) Test whether any positive autocorrelation remains after the first iteration; use $\alpha = .05$. What do you conclude?
c) Does the iterative approach appear to have been effective here?

10.15. Refer to Problem 10.13. The staff analyst wishes to compare the results obtained with the iterative approach with those for the first differences approach.
a) Estimate the regression coefficient β_1 by the first differences approach, and obtain the estimated standard deviation of this estimate.
b) Compare the results obtained in part (a) with those in Problem 10.14a. Summarize your findings.
c) Using the results of part (a), obtain an interval estimate for β_1 by the method of first differences with a 95 percent confidence coefficient.

10.16. Refer to Problem 10.15. Test whether or not the error terms with the first differences approach are autocorrelated, using a two-sided test and a level of significance of .02. What do you conclude? Why is a two-sided test meaningful here?

10.17. The data below show seasonally adjusted quarterly sales for the McGill Company (Y, in million dollars) and for the entire industry (X, in million dollars), for the most recent 20 quarters.

t:	1	2	3	4	5	6	7
X_t:	127.3	130.0	132.7	129.4	135.0	137.1	141.2
Y_t:	20.96	21.40	21.96	21.52	22.39	22.76	23.48

t:	8	9	10	11	12	13	14
X_t:	142.8	145.5	145.3	148.3	146.4	150.2	153.1
Y_t:	23.66	24.10	24.01	24.54	24.30	25.00	25.64

t:	15	16	17	18	19	20
X_t:	157.3	160.7	164.2	165.6	168.7	171.7
Y_t:	26.36	26.98	27.52	27.78	28.24	28.78

The first-order autoregressive error model (10.30) is to be employed.
a) Would you expect the autocorrelation parameter ρ to be positive, negative, or zero here?
b) Fit the linear regression model by ordinary least squares, obtain the residuals, and plot them against time. What do you find?
c) Conduct a formal test for positive autocorrelation, using $\alpha = .01$. What conclusion is indicated?

10.18. Refer to Problem 10.17.
a) Use one iteration with the iterative method to estimate the parameters β_0 and β_1 in model (10.30). Also obtain the estimated standard deviations of these estimates.
b) Test whether any positive autocorrelation remains after the first iteration; use $\alpha = .01$. What do you conclude?
c) Does the iterative approach appear to have been effective here?

10.19. Refer to Problem 10.17.
a) Estimate the regression coefficient β_1 in model (10.30) by the first

differences approach; also obtain the estimated standard deviation of this estimate.

b) Compare your results in part (a) with those in Problem 10.18a. State your findings.

c) Using the results of part (a), obtain an interval estimate for β_1 by the method of first differences with a 99 percent confidence coefficient.

10.20. Refer to Problem 10.19. Test whether or not the error terms with the first differences approach are autocorrelated, using a two-sided test and $\alpha = .02$. What do you conclude? What would be the likely implication if negative autocorrelation were found to be present?

10.21. Suppose the autoregressive error process for the model $Y_t = \beta_0 + \beta_1 X_t + \varepsilon_t$ is that given by (10.47).

a) What would be the transformed variables Y_t' and X_t' to be used with the iterative method?

b) How would you estimate the parameters ρ_1 and ρ_2 for use with the iterative method?

CITED REFERENCES

10.1. Durbin, J., and Watson, G. S. "Testing for Serial Correlation in Least Squares Regression. II." *Biometrika*, Vol. 38 (1951), pp. 159–78.

10.2. Theil, H., and Nagar, A. L. "Testing the Independence of Regression Disturbances." *Journal of the American Statistical Association*, Vol. 56 (1961), pp. 793–806.

10.3. Johnston, J. *Econometric Methods*. New York: McGraw-Hill Book Co., 1963.

11

Search for "Best" Set of Independent Variables

ONE OF THE MOST difficult problems in regression analysis often is the selection of the set of independent variables to be included in the model. In this chapter, we take up several search methods for helping to find a "best" set of independent variables. These search methods are heavily computer-oriented.

11.1 NATURE OF PROBLEM

As we have seen in previous chapters, regression analysis has three major uses: (1) description, (2) control, and (3) prediction. For each of these uses, the investigator must specify the set of independent variables to be employed for describing, controlling, and/or predicting the dependent variable.

In some fields, theory can aid in selecting the independent variables to be employed and in specifying the functional form of the regression relation. Often in these fields, controlled experiments can be undertaken to furnish data, on the basis of which the regression parameters can be estimated and the theoretical form of the regression function tested.

In many other subject matter fields, however, including the social and behavioral sciences and management, serviceable theoretical models are relatively rare. To complicate matters further, the available theoretical models may involve independent variables that are not directly measurable, such as a family's future earnings over the next 10 years. Under these conditions, investigators are often forced to prospect for independent variables that could conceivably be related to the dependent variable under study. Obviously, such a set of independent variables will be large. For example, a company's sales of portable dishwashers in a district may be affected by population size, per capita income, percent of population in urban areas, percent of population under 50 years old, percent of families with children at home, etc. etc.!

After such a lengthy list has been compiled, some of the independent variables can be screened out. An independent variable (1) may not be fundamental to the problem; (2) may be subject to large measurement errors; and (3) may effectively duplicate another independent variable in the list. Other independent variables which cannot be measured either may be deleted or replaced by proxy variables which are highly correlated with them.

Typically, the number of independent variables that remain after this initial screening is still large. Further, many of these variables will be highly intercorrelated. Hence, the investigator usually will wish to reduce the number of independent variables to be used in the final model. There are several reasons for this. A regression model with a large number of independent variables is expensive to maintain. Further, regression models with a limited number of independent variables are easier to analyze and understand. Finally, the presence of many highly intercorrelated independent variables may add little to the predictive power of the model, while detracting from its descriptive abilities and increasing the problem of roundoff errors (as we have seen in Chapter 10).

The problem then is how to shorten the list of independent variables so as to obtain, in some sense, a "best" selection of independent variables. This "best" set needs to be small enough so that maintenance costs are manageable and analysis is facilitated, yet it must be large enough so that adequate description, control, or prediction is possible. The search procedures for finding a "best" set of independent variables to be described in this chapter all presuppose that the investigator has established the functional form of the regression relation, whether given variables are to appear in linear form, quadratic form, etc.; whether the independent variables are first transformed, such as by a logarithmic transformation; and whether any interaction terms are to be included. At this point, the search procedures are employed to reduce the number of X variables.

None of the search procedures can be proven to yield invariably *the* "best" set of independent variables. Indeed, often there is no unique "best" set. The entire variable-selection process is pragmatic, with large doses of subjective judgment. The principal advantage of automatic search procedures is that they enable the investigator to focus his judgments on the pivotal areas of the problem.

Note

All too often, unwary investigators will screen the set of independent variables by fitting the regression model containing the entire set of potential independent variables and then simply dropping all those for which the t^* statistic:

(11.1) $$t_k^* = \frac{b_k}{s(b_k)}$$

has a small absolute value. As we know from Chapter 10, this procedure can lead to serious difficulties. Suppose X_3 and X_7 are highly correlated and each is closely related to the dependent variable. Because of the collinearity between X_3 and X_7, both t_3^* and t_7^* may be small absolutely, leading to the dropping of both variables with this procedure.

Clearly, a good search procedure must be able to handle important intercorrelated independent variables in such a way that not all of them will be dropped.

11.2 EXAMPLE

In order to illustrate the search procedures to be discussed in the following sections, we shall use a relatively simple example which has four potential independent variables. By limiting the number of potential independent variables, we shall be able to explain the search procedures without overwhelming the reader with masses of computer printouts.

A forest manager was interested in predicting the volume of wood five years hence in a given forest area containing mixed softwood trees. To study the feasibility of predicting wood volume by regression methods, an investigation was undertaken in which the forest area was divided into plots and 55 plots were selected at random. For each of these sample plots, the following measurements were made:

X_1 volume of wood in plot (cubic feet)
X_2 number of trees in plot
X_3 average age of trees in plot (years)
X_4 average volume of wood per tree in plot ($X_1 \div X_2$)

These constitute the potential independent variables for a predictive regression model of the wood volume five years hence.

Five years later, the volume of wood in each sample plot was again measured:

Y volume of wood in plot five years later (cubic feet)

This is the dependent variable. The data on the potential independent variables and on the dependent variable are presented in Table 11.1.

The forest manager wished to obtain the "best" set of independent variables for predicting Y. He assigned the task to an analyst, who first obtained the correlation matrix for all the variables from a computer run. This matrix provides valuable basic information on the nature of the problems to be encountered. Table 11.2 contains the correlation matrix but does not show the duplicate terms below the main diagonal.

Table 11.2 indicates clearly that all of the independent variables are related to Y, with X_1 showing a very high degree of relationship and X_4 a high degree of relationship. The correlation matrix further shows quickly that the potential independent variables are intercorrelated. In particular, the correlations

TABLE 11.1

Data for Forest Inventory Example

Plot	Wood Volume 1st Measurement X_1	Number of Trees X_2	Average Age of Trees X_3	Average Volume per Tree $(X_1 \div X_2)$ X_4	Wood Volume 2d Measurement Y
1	170	47	50	3.617	200
2	310	30	70	10.333	358
3	131	26	50	5.038	167
4	83	19	50	4.368	111
5	231	38	70	6.079	290
6	21	9	50	2.333	36
7	80	27	50	2.963	133
8	91	23	50	3.957	135
9	369	67	90	5.507	399
10	256	25	90	10.240	279
11	321	42	90	7.643	280
12	104	33	30	3.152	146
13	88	34	30	2.588	123
14	86	30	30	2.867	122
15	84	29	30	2.897	103
16	239	52	70	4.596	264
17	323	51	90	6.333	360
18	45	18	30	2.500	55
19	389	50	70	7.780	421
20	61	31	50	1.968	89
21	58	6	30	9.667	98
22	239	45	90	5.311	262
23	126	17	80	7.412	147
24	273	29	70	9.414	287
25	83	17	80	4.882	73
26	63	23	50	2.739	86
27	60	31	50	1.935	104
28	310	65	50	4.769	330
29	67	21	70	3.190	82
30	75	24	50	3.125	112
31	11	7	50	1.571	24
32	165	31	80	5.323	185
33	279	26	80	10.731	301
34	306	26	90	11.769	284
35	477	30	90	15.900	467
36	368	23	90	16.000	410
37	224	34	30	6.588	279
38	151	29	50	5.207	182
39	109	25	50	4.360	126
40	148	41	50	3.610	160
41	112	18	70	6.222	141
42	21	13	30	1.615	31
43	56	33	30	1.697	78
44	29	18	40	1.611	45
45	72	37	50	1.946	93
46	201	47	60	4.277	220
47	170	33	50	5.152	207
48	140	27	50	5.185	170
49	69	33	40	2.091	92
50	79	9	50	8.778	99
51	67	20	40	3.350	85
52	125	15	80	8.333	136
53	583	28	60	20.821	653
54	89	26	70	3.423	111
55	18	4	40	4.500	22

TABLE 11.2

Correlation Matrix for Forest Inventory Example

	Y	X_1	X_2	X_3	X_4
Y	1.000	.989	.544	.572	.797
X_1		1.000	.537	.626	.805
X_2			1.000	.263	−.0164
X_3				1.000	.547
X_4					1.000

between X_1 and respectively X_3 and X_4 are relatively high. Also X_3 and X_4 are moderately highly correlated. It is precisely the presence of many intercorrelations among the independent variables that led the forest manager to want to screen the variables to obtain the "best" set.

11.3 ALL POSSIBLE REGRESSIONS

The all possible regressions search procedure calls for an examination of all possible regression equations involving the potential independent variables and selection of the "best" equation according to some criterion. When this search procedure is used with our forest inventory example, for instance, 16 different regression equations are required, as shown in Table 11.3 on page 376. First, there is the regression equation with no independent variables. Then there are the regression equations with one independent variable (X_1, X_2, X_3, X_4), with two independent variables (X_1 and X_2, X_1 and X_3, X_1 and X_4, X_2 and X_3, X_2 and X_4, X_3 and X_4), and so on.

Different criteria for comparing the fitted models may be used with the all possible regressions search procedure. We shall discuss three—R_p^2, MSE_p, and C_p. Before doing so, we need to develop some notation. Let us denote the number of potential independent variables in the set to be considered by $P - 1$. We assume throughout this chapter that all regression equations contain an intercept term β_0. Hence, the regression equation containing all potential independent variables contains P parameters, and the equation with no independent variables contains one parameter (β_0).

The number of independent variables actually used in a regression equation will be denoted by $p - 1$, as always, so that there are p parameters in this regression equation. Thus, we have:

$$1 \leq p \leq P \tag{11.2}$$

The all possible regressions approach assumes that the number of observations n exceeds the maximum number of potential parameters:

$$n > P \tag{11.3}$$

and, indeed, it is highly desirable that n be substantially larger than P so that sound results can be obtained.

R_p^2 Criterion

The R_p^2 criterion calls for an examination of the coefficient of multiple determination R^2, defined in (7.31), in order to select the best set of independent variables. We show the number of parameters in the regression equation as a subscript of R^2. Thus, R_p^2 indicates that there are p parameters, or $p - 1$ independent variables, in the regression equation on which R_p^2 is based.

Since R_p^2 is a ratio of sums of squares:

$$R_p^2 = \frac{SSR_p}{SSTO} = 1 - \frac{SSE_p}{SSTO} \tag{11.4}$$

and the denominator is constant for all possible regressions, R_p^2 varies inversely with the error sums of squares SSE_p. But we know that SSE_p can never increase as additional independent variables are included in the model. Thus, R_p^2 will be a maximum when all $P - 1$ potential independent variables are included in the regression model. The reason for using the R_p^2 criterion with all possible regressions therefore cannot be to maximize R_p^2. Rather, the intent is to find the point where adding more independent variables is not worthwhile because it leads to a very small increase in R_p^2. Often, this point is reached when only a limited number of independent variables are included in the regression model. Clearly, the determination of where diminishing returns set in is a judgmental one.

Example. Table 11.3 contains the R_p^2 values for all possible regressions for our forest inventory example. The data were obtained from a series of

TABLE 11.3

R_p^2, MSE_p, and C_p Values for All Possible Regressions for Forest Inventory Example

Independent Variables in Model	p	*df*	SSE_p	R_p^2	MSE_p	C_p
None	1	54	902,773	0	16,718	2,996.9
X_1	2	53	18,894	.979	356	12.8
X_2	2	53	635,810	.296	11,996	2,097.0
X_3	2	53	607,248	.327	11,458	2,000.5
X_4	2	53	329,851	.635	6,224	1,063.4
X_1, X_2	3	52	18,711	.979	360	14.2
X_1, X_3	3	52	15,523	.983	299	3.4
X_1, X_4	3	52	18,894	.979	363	14.8
X_2, X_3	3	52	457,049	.494	8,789	1,495.1
X_2, X_4	3	52	49,816	.945	958	119.3
X_3, X_4	3	52	305,899	.661	5,883	984.4
X_1, X_2, X_3	4	51	15,474	.983	303	5.3
X_1, X_2, X_4	4	51	17,955	.980	352	13.7
X_1, X_3, X_4	4	51	15,495	.983	304	5.3
X_2, X_3, X_4	4	51	49,496	.945	971	120.2
X_1, X_2, X_3, X_4	5	50	14,780	.984	296	4.9

computer runs. For instance, when X_1 is the only independent variable in the regression model, we obtain:

$$R_2^2 = 1 - \frac{SSE(X_1)}{SSTO} = 1 - \frac{18{,}894}{902{,}773} = .979$$

Note that $SSTO = SSE_1 = 902{,}773$.

The R_p^2 values are plotted in Figure 11.1. The maximum R_p^2 value for each p, denoted $\max(R_p^2)$, appears at the top of the graph. These points are

FIGURE 11.1

R_p^2 Plot for Forest Inventory Example

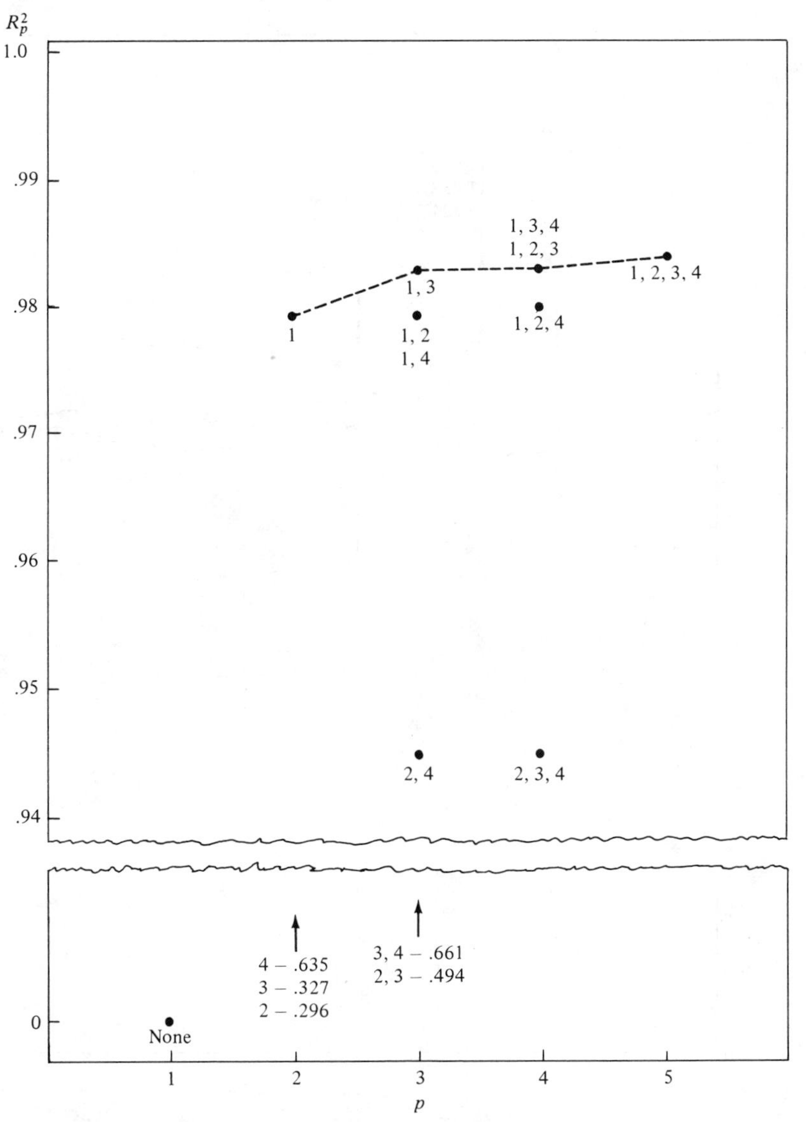

connected by straight lines to show the impact of adding additional independent variables. Figure 11.1 makes it clear that little increase in $\max(R_p^2)$ takes place after two independent variables are included in the model. Note that X_1 is included in the model for each $\max(R_p^2)$ when $p \geq 2$, and that X_3 is included in each such model for $p \geq 3$. Hence, Figure 11.1 suggests that X_1 and X_3 should be included in the regression model, and the forest manager might well decide to stop there in view of the small gains achieved by adding further independent variables.

FIGURE 11.2

MSE_p Plot for Forest Inventory Example

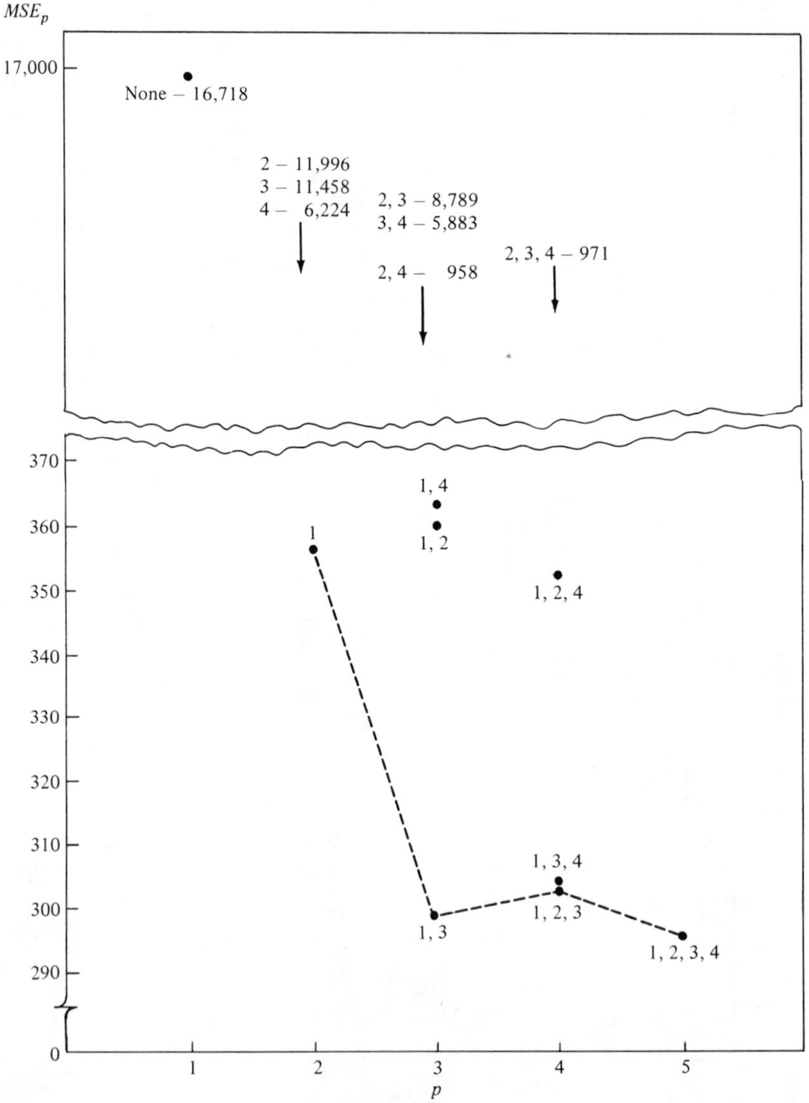

MSE_p Criterion

Since R_p^2 does not take account of the number of parameters in the model, and since $\max(R_p^2)$ can never decrease as p increases, the use of MSE_p as a criterion has been suggested. MSE_p does take the number of parameters in the model into account through the degrees of freedom. Indeed, $\min(MSE_p)$ can increase as p increases, if the reduction in SSE_p becomes so small that it is not sufficient to offset the loss of an additional degree of freedom. Users of the MSE_p criterion either seek to find the set of independent variables which minimizes MSE_p, or a set for which MSE_p is so close to the minimum that adding more independent variables is not worthwhile.

Example. The MSE_p values for all possible regressions for our forest inventory example are shown in Table 11.3. For instance, if the regression model contains only X_1, we have:

$$MSE_2 = \frac{SSE(X_1)}{n-2} = \frac{18{,}894}{53} = 356$$

Figure 11.2 contains the MSE_p plot for our example. We have connected the $\min(MSE_p)$ values for each p by straight lines. The story which Figure 11.2 tells is very similar to that told by Figure 11.1. Independent variables X_1 and X_3 appear to be the leading candidates for inclusion. Gains in the reduction of $\min(MSE_p)$ after X_1 and X_3 are included are so small that the forest manager again might decide to stop with these two independent variables if he were using the MSE_p criterion.

C_p Criterion

This criterion is concerned with the *total squared error* of the n fitted observations for any given regression model. The total squared error has a bias component and a random error component. The proposed measure Γ_p is a standardized function of the total squared error:

$$\Gamma_p = \frac{1}{\sigma^2}\Bigg[\underbrace{\sum_{i=1}^{n}(v_i - \eta_i)^2}_{\text{Bias component}} + \underbrace{\sum_{i=1}^{n}\sigma^2(\hat{Y}_i)}_{\text{Random error component}}\Bigg] \tag{11.5}$$

where:

$$\begin{aligned} v_i &= E(Y_i) \text{ according to true regression relation} \\ \eta_i &= E(Y_i) \text{ according to fitted model} \\ \sigma^2(\hat{Y}_i) &= \text{variance of fitted value } \hat{Y}_i \\ \sigma^2 &= \text{true error variance} \end{aligned}$$

The model which includes all $P-1$ potential independent variables is assumed to have been carefully chosen so as to yield an unbiased estimate of σ^2. We shall denote this estimate $\hat{\sigma}^2$:

$$\hat{\sigma}^2 = MSE(X_1, \ldots, X_{P-1}) \tag{11.6}$$

It can then be shown that an estimator of Γ_p is C_p:

$$C_p = \frac{SSE_p}{\hat{\sigma}^2} - (n - 2p) \tag{11.7}$$

When there is no bias in a regression equation with $p-1$ independent variables, C_p has an expected value of p:

$$E(C_p \mid v_i \equiv \eta_i) = p \tag{11.8}$$

Thus, when the C_p values for all possible regressions are plotted against p, those regressions with little bias will tend to fall near the line $C_p = p$. Those regressions with substantial bias will tend to fall above this line.

When using the C_p criterion, one seeks to identify the set of independent variables that leads to the smallest C_p value. If that C_p value should contain a substantial bias component, one may prefer a set of independent variables that leads to a slightly larger C_p value which does not contain a substantial bias component. Reference 11.1 contains extended discussions of applications of the C_p criterion.

Example. Table 11.3 contains the C_p values for all possible regressions for our forest inventory example. For instance, when X_1 is the only independent variable in the regression model, we have:

$$C_2 = \frac{SSE(X_1)}{MSE(X_1, X_2, X_3, X_4)} - [n - (2)(2)]$$

$$= \frac{18{,}894}{296} - (55 - 4) = 12.8$$

The C_p values for all possible regressions are plotted in Figure 11.3. We find again that use of X_1 and X_3 in the model is suggested. The C_p value for that set of independent variables is the minimum one, and appears to contain only a small bias component. While the use of the entire set X_1, X_2, X_3, X_4 may lead to a slightly lower bias, the total squared error for this set is larger than the one for the set X_1, X_3.

Note

To see why C_p as defined in (11.7) is an estimator of Γ_p, we need to utilize two results that we shall simply state. First, it can be shown that:

$$\sum_{i=1}^{n} \sigma^2(\hat{Y}_i) = p\sigma^2 \tag{11.9}$$

Thus, the total random error of the n fitted values $\hat{Y}_i$ increases as the number of variables in the regression model increases.

FIGURE 11.3

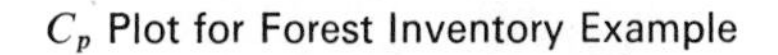

C_p Plot for Forest Inventory Example

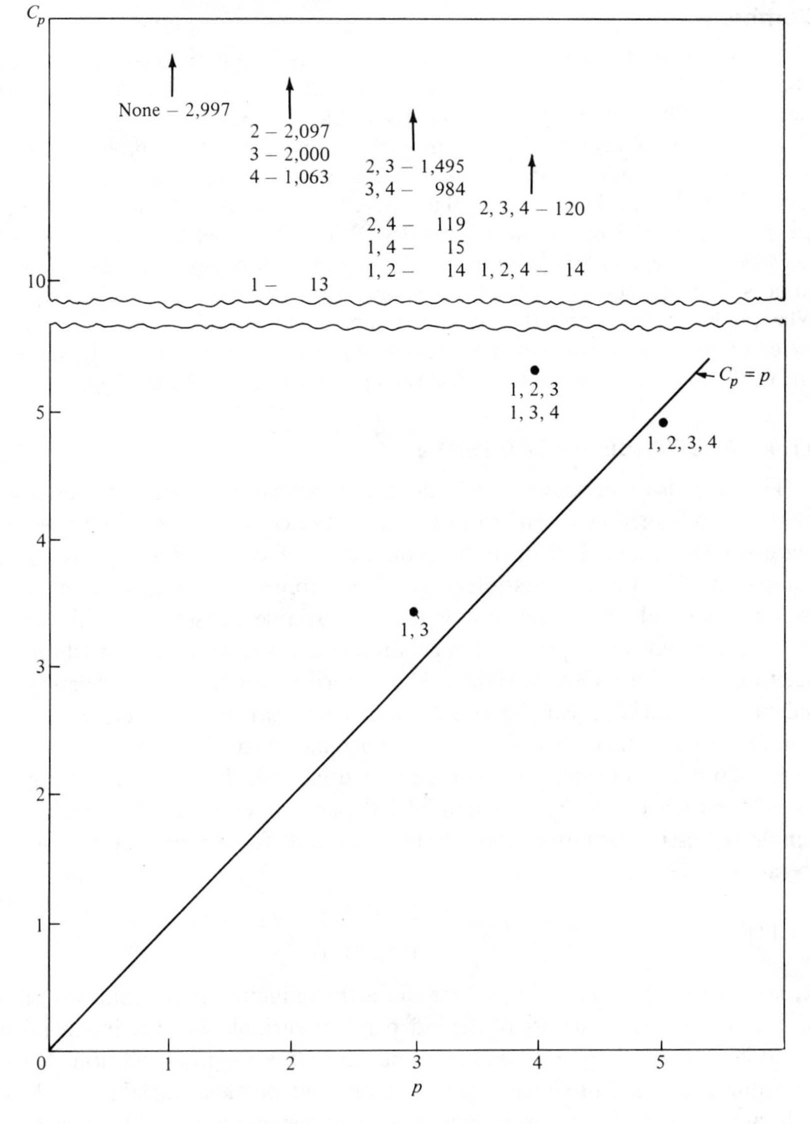

Further, it can be shown that:

$$E(SSE_p) = \sum (\nu_i - \eta_i)^2 + (n - p)\sigma^2 \tag{11.10}$$

Hence, Γ_p in (11.5) can be expressed as follows:

$$\Gamma_p = \frac{1}{\sigma^2}\,[E(SSE_p) - (n-p)\sigma^2 + p\sigma^2]$$

$$= \frac{E(SSE_p)}{\sigma^2} - (n - 2p)$$

Replacing $E(SSE_p)$ by the estimator SSE_p and using $\hat{\sigma}^2$ as defined in (11.6) as an estimator of σ^2, yields C_p in (11.7).

Comments

1. In the forest inventory example, all three criteria appear to lead to the same "best" set of independent variables. This is not always the case. Different criteria may suggest different sets of independent variables as "best."

2. A major disadvantage of the all possible regressions search procedure is the amount of computational effort required. Since each potential independent variable either can be included or excluded, there are $2^{(P-1)}$ possible regressions when there are $P-1$ potential independent variables. When $P-1=10$, there are already 1,024 possible regressions. With the availability of large computers today, running all possible regressions for 10 or 12 potential variables is not very time consuming. When, however, the potential candidates number 30 to 50 (which is not infrequently the case), the all possible regressions search approach is simply too costly. Instead, one of the other approaches to be discussed will then need to be employed.

11.4 STEPWISE REGRESSION

The stepwise regression search method is probably the most widely used of the search methods that do not require the computation of all possible regressions. It was developed to economize on computational efforts, as compared with the all possible regressions approach, while arriving at a reasonably good "best" set of independent variables. Essentially, this search method computes a sequence of regression equations, at each step adding or deleting an independent variable. The criterion for adding or deleting an independent variable can be stated equivalently in terms of error sum of squares reduction, coefficient of partial correlation, or F^* statistic.

1. To begin, the stepwise regression routine calculates all simple regressions, for each of the $P-1$ potential independent variables. For each such simple regression equation, the F^* statistic (3.58) for testing whether or not the slope is zero is obtained:

$$F_k^* = \frac{MSR(X_k)}{MSE(X_k)} \tag{11.11}$$

Recall that $MSR(X_k) = SSR(X_k)$ measures the reduction in the total variation of Y associated with the use of the independent variable X_k. The independent variable with the largest F^* value is the candidate for first addition. If this F^* value exceeds a predetermined level, the independent variable is added. Otherwise, the program terminates with no independent variable considered sufficiently helpful to enter the regression model.

2. Assume X_7 is the independent variable entered at step 1. The stepwise regression routine now calculates all regressions with two independent variables, where X_7 is one of the pair. For each such regression, the F^* statistic:

$$F_k^* = \frac{MSR(X_k | X_7)}{MSE(X_7, X_k)} = \left[\frac{b_k}{s(b_k)}\right]^2 \tag{11.12}$$

is obtained. This is the statistic for testing whether or not $\beta_k = 0$ when X_7 and X_k are the independent variables in the model. The independent variable with the largest F^* value is the candidate for addition at the second stage. If this F^* value exceeds a predetermined level, the second independent variable is added. Otherwise the program terminates.

3. Suppose X_3 is added at the second stage. Now the stepwise regression routine examines whether any of the other independent variables already in the model should be dropped. For our illustration, there is at this stage only one other independent variable in the model, X_7, so that only one F^* statistic is obtained:

$$F_7^* = \frac{MSR(X_7 | X_3)}{MSE(X_3, X_7)} \tag{11.13}$$

At later stages, there would be a number of these F^* statistics, for each of the variables in the model besides the one last added. The variable for which this F^* value is smallest is the candidate for deletion. If this F^* value falls below a predetermined limit, the independent variable is dropped from the model; otherwise, it is retained.

4. Suppose X_7 is retained, so that both X_3 and X_7 are now in the model. The stepwise regression routine now examines which independent variable is the next candidate for addition, then examines whether any of the variables already in the model should now be dropped, and so on until no further independent variables can either be added or deleted, at which point the search terminates.

It should be noted that the stepwise regression routine allows an independent variable, brought into the model at an earlier stage, to be dropped subsequently if it is no longer helpful in conjunction with variables added at later stages.

Example. Figure 11.4 shows the computer printout obtained when a particular stepwise regression routine (the U.C.L.A. BMD computer routine) was applied to our forest inventory example. The F limits for adding or deleting a variable were both specified to be 4.0, as shown at the top left of Figure 11.4. Since the degrees of freedom associated with MSE vary, depending on the number of independent variables in the model, and since repeated tests on the same data are undertaken, fixed F limits for adding or deleting a variable have no precise probabilistic meaning. Note, however, that $F(.95; 1, 60) = 4.00$, so that the specified F limits of 4.0 would correspond roughly to a level of significance of .05 for any single test based on near 55 degrees of freedom. We shall now follow through the steps.

1. Adding the first independent variable with the largest F^* value (11.11) is equivalent to adding the independent variable for which the coefficient of correlation:

$$r_{Yk} = \left[\frac{SSR(X_k)}{SSTO}\right]^{1/2} \tag{11.14}$$

FIGURE 11.4

Stepwise Regression for Forest Inventory Example

```
SUB-PROBLM      1
 DEPENDENT VARIABLE              5
 MAXIMUM NUMBER OF STEPS        10
 F-LEVEL FOR INCLUSION     4.000000
 F-LEVEL FOR DELETION      4.000000
 TOLERANCE LEVEL            .001000


 STEP NUMBER     1
 VARIABLE ENTERED      1

 MULTIPLE R                  .9895
 STD. ERROR OF EST.        18.8808

 ANALYSIS OF VARIANCE
                      DF     SUM OF SQUARES     MEAN SQUARE      F RATIO
        REGRESSION     1       883879.788        883879.788     2479.445
        RESIDUAL      53        18893.594           356.483

                  VARIABLES IN EQUATION                .                     VARIABLES NOT IN EQUATION
                                                       .
  VARIABLE      COEFFICIENT   STD. ERROR  F TO REMOVE  .   VARIABLE    PARTIAL CORR.      TOLERANCE      F TO ENTER
                                                       .
   (CONSTANT     22.50483 )                            .
   1              1.01238       .02033    2479.4451    .      2            .09843          .7112          .5088
                                                       .      3          -.42236           .6077        11.2902
                                                       .      4          -.00043           .3517          .0000

 STEP NUMBER     2
 VARIABLE ENTERED      3

 MULTIPLE R                  .9914
 STD. ERROR OF EST.        17.2778

 ANALYSIS OF VARIANCE
                      DF     SUM OF SQUARES     MEAN SQUARE      F RATIO
        REGRESSION     2       887250.163        443625.082     1486.065
        RESIDUAL      52        15523.219           298.523

                  VARIABLES IN EQUATION                .                     VARIABLES NOT IN EQUATION
                                                       .
  VARIABLE      COEFFICIENT   STD. ERROR  F TO REMOVE  .   VARIABLE    PARTIAL CORR.      TOLERANCE      F TO ENTER
                                                       .
   (CONSTANT     43.44943 )                            .
   1              1.06261       .02387    1982.1715    .      2            .05644          .7021          .1630
   3              -.50983       .15173      11.2902    .      4            .04278          .3487          .0935

F-LEVEL INSUFFICIENT FOR FURTHER COMPUTATION
```

is largest absolutely. We see from Table 11.2 that X_1 is most highly correlated with Y. From Figure 11.4, we find that the F^* statistic for X_1 is:

$$F_1^* = \frac{MSR(X_1)}{MSE(X_1)} = \frac{883{,}879.788}{356.483} = 2{,}479.445$$

Since this value exceeds 4.0, X_1 is added to the model. The printout in Figure 11.4 does not contain the data underlying the selection of X_1. It only shows, under step 1, some characteristics of the regression model containing X_1.

2. Next, all regression equations containing X_1 and another independent variable are obtained, and the F^* statistics calculated. They are:

$$F_k^* = \frac{MSR(X_k \mid X_1)}{MSE(X_1, X_k)}$$

These are shown in step 1 at the right, under the heading "F to enter." X_3 has the highest F^* value, which exceeds 4.0, so that X_3 now enters the model.

3. Step 2 in Figure 11.4 summarizes the situation at this point. X_1 and X_3 are now in the model. Next, a test is undertaken whether X_1 should be dropped. The F^* statistic is shown under the heading "F to remove" in step 2:

$$F_1^* = \frac{MSR(X_1 | X_3)}{MSE(X_1, X_3)} = 1{,}982.1715$$

Since this F^* value exceeds 4.0, X_1 is not dropped.

4. Next, all regressions containing X_1, X_3, and one of the remaining potential independent variables are obtained. The appropriate F^* statistics are:

$$F_k^* = \frac{MSR(X_k | X_1, X_3)}{MSE(X_1, X_3, X_k)}$$

These are shown at the right under the heading "F to enter" in step 2. Since none of these exceeds 4.0, no new independent variable is added and the routine terminates.

Once again we find that X_1 and X_3 are suggested as the "best" set of independent variables.

Comments

1. The routine employed conducts a test to screen out any independent variable that is too highly correlated with independent variables already in the model. The tolerance values in the printout pertain to this screening.

2. The stepwise regression routine employed prints out the partial correlation coefficients at each stage. These could be used equivalently to the F^* values for screening the independent variables, and indeed some routines actually use the partial correlation coefficients for screening.

3. The F limits for adding and deleting a variable need not be the same, as in our example. Frequently the F limit for deleting an independent variable is specified to be smaller than the F limit for adding a variable.

4. The F limits for adding and deleting a variable need not be selected in terms of approximate significance levels, but may be determined descriptively in terms of error reduction. For instance, an F limit of 1.0 for adding a variable may be specified with the thought that the marginal error reduction associated with the added variable should be at least as great as the remaining error mean square once that variable has been added.

5. A limitation of the stepwise regression search approach is that it presumes there is a single "best" set of independent variables and seeks to identify it. As noted earlier, there is often no unique "best" set. Hence, some statisticians suggest that all possible regressions with the same number of independent variables as in the stepwise regression solution be run subsequently to study whether some other sets of independent variables might be better.

Another limitation of the stepwise regression routine is that it sometimes arrives at an unreasonable "best" set when the independent variables are very highly correlated. In our forest inventory example, for instance, if $X_5 = X_1^2$, $X_6 = X_3^2$, and $X_7 = X_1 X_3$ had been included in the list of variables to be considered for inclusion in the model, the stepwise regression procedure would have identified X_1 and

$X_7 = X_1X_3$ as the "best" set. The variable X_3 (average age of trees), a meaningful variable from a descriptive point of view, would have been excluded because of the pattern of intercorrelations in the screened variables.

11.5 OTHER SEARCH PROCEDURES

There are many other search procedures which have been proposed to find a "best" set of independent variables. We mention three of these.

Forward Selection

This search procedure is a simplified version of stepwise regression, omitting the test whether a variable once entered into the model should be dropped.

Backward Elimination

This search procedure is the opposite of forward selection. It begins with the model containing all potential independent variables and identifies the one with the smallest F^* value. For instance, the F^* value for X_1 is:

$$F_1^* = \frac{MSR(X_1 \mid X_2, \ldots, X_{P-1})}{MSE(X_1, \ldots, X_{P-1})} \tag{11.15}$$

If the minimum F^* value is less than a predetermined limit, that independent variable is dropped. The model with the remaining $P-2$ independent variables is then fitted, and the next candidate for dropping is identified. This process continues until no further independent variables can be dropped.

The backward elimination procedure requires more computations than the forward selection method, since it starts with the biggest possible model. However, it does have the advantage of showing analysts the implications of models with many variables.

t-Directed Search

This search procedure first fits the model with all $P-1$ potential independent variables. All variables for which the absolute value of the t^* statistic:

$$t_k^* = \frac{b_k}{s(b_k)} \tag{11.16}$$

exceeds a predetermined level are then automatically retained. The subset of all possible regressions consisting of those equations which contain the automatically retained variables (among others) is then obtained for screening by a specified criterion. If, for example, four variables are being considered and it turns out that X_1 and X_4 are to be automatically retained, the regressions (X_1, X_4), (X_1, X_4, X_2), (X_1, X_4, X_3), and (X_1, X_4, X_2, X_3) would be studied with this approach.

11.6 IMPLEMENTATION OF SEARCH PROCEDURES

Options and Refinements

Our discussion of the major search procedures for identifying a "best" set of independent variables has focused on the main conceptual issues and not on options, variations, and refinements available with particular computer packages. It is essential that the specific features of the package to be employed are fully understood so that intelligent use of the package can be made. In some packages, there is an option for regression equations through the origin. Some packages permit variables to be brought into the model and tested in pairs or other groupings instead of singly, to save computing time or for other reasons. Some packages, once a "best" regression equation is identified, will calculate all the possible regression equations in the same number of variables and will develop information for each equation so that a final choice can be made by the user. Some stepwise programs have options for forcing variables into the regression equation; such variables are not removed even if their F^* values become too low.

The diversity of these options and special features serves to emphasize a point made earlier: there is no unique way of searching for a "best" set of independent variables, and subjective elements must play an important role in the search process.

Completion of Model Building Process

The screening of variables by a computerized selection process is only one step in the building of a regression model. Once the set of independent variables has been identified, the resulting model needs to be studied for its aptness by the methods of Chapter 4. If repeat observations are available, a formal test for lack of fit can be made. In any case, a variety of residual plots can be employed to identify the nature of lack of fit, outliers, and other deficiencies.

Frequently, the original set of $P - 1$ potential independent variables may exclude cross-product terms and powers of the independent variables to keep the selection problem within reasonable bounds. Residual plots against such "missing" variables, or augmenting the model of "best" independent variables by adding cross-product and/or power terms, can be useful in identifying ways in which the model fit can be improved further.

Thus, the automatic selection process needs to be augmented by a variety of interactive analyses in order to arrive at a final regression model which serves as a reasonable basis for description, control, and/or prediction.

Cautions in Use of Final Model

The model-building process, as we have just noted, requires repeated analyses on the same set of data in order to arrive at a model which fits the data well. A consequence is that the model may be subject to *prediction bias*, that is, the indicated predictive ability of the model for the data on which the model is based may be greater than the model's predictive ability for new data. The prediction bias arises because the choice of the final model is so uniquely related to the observations at hand. The prediction bias may be particularly large when the effects of independent variables are small.

It is good statistical practice to measure the prediction bias by observing the predictive power of the model on a new set of data. If necessary, some of the original data can be kept aside for this calibration of predictive power and the model derived only from the remaining data.

Often, a predictive model is desired for values of the independent variables which cover only a portion of the entire observation space. In that case, it is good practice to test the stability of the regression model by fitting it to that portion of the observations which fall in the space of future interest and comparing the regression results with those for the model based on all observations. Similarly, if the data are time series, it is often desirable to study the stability of the regression model over time by fitting the model also to the most recent data alone, and comparing results.

In this connection, it is worthwhile repeating an earlier caution. When the independent variables are highly intercorrelated, use of the model for prediction for values of the independent variables that do not follow the past pattern of multicollinearity becomes highly suspect.

PROBLEMS

11.1. A speaker stated: "In a well-designed experiment involving quantitative independent variables, a procedure for screening the independent variables after the observations are obtained is not necessary." Discuss.

11.2. An educational researcher wishes to predict the grade point averages in graduate work for applicants to the Graduate School. List a dozen variables that might be useful independent variables here.

11.3. A student noted in a seminar: "Since the R_p^2 criterion for all possible regressions can never decrease with additional independent variables, why don't we use the adjusted coefficient of multiple determination R_a^2 as defined in (7.33)? This measure can decrease, and we can then seek to maximize this criterion." Comment. Is the proposed criterion related to the MSE_p criterion?

11.4. Using the result in (3.28b), find $\sum_{i=1}^{n} \sigma^2(\hat{Y}_i)$ and confirm that it is the same as indicated by (11.9) for $p=2$.

11.5. In stepwise regression, what advantage is there in using a relatively large F limit for adding variables? What advantage is there in using a smaller F limit for adding variables?

11.6. In stepwise regression, why should the F limit for deleting variables never exceed the F limit for adding variables?

11.7. Draw a flowchart of each of the following selection methods: (1) stepwise regression; (2) forward selection; (3) backward elimination.

11.8. An economic consultant, retained by a large employment agency in a metropolitan area to develop a regression equation for predicting monthly agency revenues (Y), decided that three economic indicators for the area were potentially useful as independent variables, namely average weekly overtime hours of production workers in manufacturing (X_1), number of job vacancies in manufacturing (X_2), and index of help wanted advertising in newspapers (X_3). Monthly observations on agency revenues and the three independent variables (all seasonally adjusted) were obtained for the past 25 months. The ANOVA table for the model $Y_i = \beta_0 + \beta_1 X_{i1} + \beta_2 X_{i2} + \beta_3 X_{i3} + \varepsilon_i$ was found to be as follows:

Source of Variation	*SS*	*df*	*MS*
Regression	5,409.89	3	1,803.30
Error	16.35	21	.78
Total	5,426.24	24	

a) Test to determine whether a regression relation exists between Y and the set of independent variables. Use a level of significance of .05. What do you find?

b) Does the conclusion that a regression relation exists imply that the consultant need not screen the independent variables and can just use them all? If a regression relation had not existed, what would this imply about screening the independent variables?

11.9. Refer to Problem 11.8. The consultant decided to screen the independent variables to determine the best set for predicting agency revenues. The regression sums of squares for all possible regressions were found to be as follows:

Independent Variables in Model	*SSR*	*Independent Variables in Model*	*SSR*
X_1	2,970.64	X_1, X_2	5,123.80
X_2	3,654.85	X_1, X_3	5,409.59
X_3	3,584.54	X_2, X_3	3,741.30
		X_1, X_2, X_3	5,409.89

a) For each of the following criteria, indicate which set of independent variables you would recommend as the best set for predicting Y: (1) R_p^2; (2) MSE_p; (3) C_p. Support your recommendations with graphs.

b) Did the three criteria in part (a) identify the same best set? Does this always happen?

c) Would stepwise regression have any advantages here as a screening procedure over all possible regressions?

d) An observer states: "There are only three variables, so why screen? You might as well use all three." Discuss.

11.10. Refer to Problems 11.8 and 11.9. The consultant was interested to learn how the stepwise selection procedure and some of its variations would perform in this application.

a) Determine the set of variables that is selected as best by the stepwise regression procedure, using F limits of 4.30 to add or delete a variable as the case may be. Show your steps.

b) Determine the set of variables that is selected as best by the forward selection procedure, using an F limit of 4.30 to add a variable. Show your steps.

c) Determine the set of variables that is selected as best by the backward elimination procedure, using an F limit of 4.30 to delete a variable. Show your steps.

d) Compare the results of the three selection procedures. How consistent are these results? How do the results compare with those for all possible regressions in Problem 11.9? State your findings.

11.11. Refer to Problem 11.10a. To what level of significance in any individual test do the F limits of 4.30 roughly correspond?

11.12. Refer to Problems 11.8 and 11.9. What would be the best set of independent variables according to the t-directed search procedure if a t limit of 2.080 is employed and the C_p criterion is utilized? How efficient was this procedure here in obtaining the best set of independent variables, as compared with all possible regressions?

11.13. An analyst, wishing to screen 10 potential independent variables to predict operating costs in the 26 branch offices of a bank, fitted a model containing all 10 independent variables, calculated $F_k^* = MSR(X_k | X_1, \ldots, X_{k-1}, X_{k+1}, \ldots, X_{10}) \div MSE(X_1, \ldots, X_{10})$ in turn for $k = 1, \ldots, 10$, and selected those independent variables for which F_k^* exceeded 4.54. What procedure did the analyst employ? Is this procedure an effective one for highly intercorrelated independent variables?

11.14. Refer to Problem 11.10a. Suppose the consultant "forced" X_2 into the best set for administrative reasons by arbitrarily entering it into the set first and not removing it even if its F^* value becomes too low. Which set of variables (including X_2) is now selected as best by the stepwise regression procedure if F limits of 4.30 are used to add or delete a variable? Did the forced inclusion of X_2 affect the selection of the other variables included in the best set? Will this always happen?

11.15. Two researchers investigated the factors affecting summer attendance at privately operated beaches on Lake Ontario, and collected information on attendance and on 11 independent variables for 42 beaches. Two summers were studied, of relatively hot and relatively cool weather respectively. Stepwise regression was to be used to screen the potential independent variables.

a) Should the screening be done for both summers combined or should it be done separately for each summer? Explain the problems involved and how you might handle them.

b) Will the stepwise regression screening procedure select those independent variables that are most important in a causal sense for determining beach attendance?

11.16. Data on sales last year (Y, in thousand squares) in 15 sales districts are given below for a maker of asphalt roofing shingles. Shown also are promotional expenditures (X_1, in thousand dollars), number of active accounts (X_2), number of competing brands (X_3), and district potential (X_4, coded) for each of the districts.

District i	X_{i1}	X_{i2}	X_{i3}	X_{i4}	Y_i
1	5.5	31	10	8	79.3
2	2.5	55	8	6	200.1
3	8.0	67	12	9	163.2
4	3.0	50	7	16	200.1
5	3.0	38	8	15	146.0
6	2.9	71	12	17	177.7
7	8.0	30	12	8	30.9
8	9.0	56	5	10	291.9
9	4.0	42	8	4	160.0
10	6.5	73	5	16	339.4
11	5.5	60	11	7	159.6
12	5.0	44	12	12	86.3
13	6.0	50	6	6	237.5
14	5.0	39	10	4	107.2
15	3.5	55	10	4	155.0

It is believed that the model need contain only linear effects, and that no interaction terms are required.

a) Fit all possible regressions.

b) For each of the following criteria, indicate which set of independent variables you would recommend as the best set for predicting Y: (1) R_p^2; (2) MSE_p; (3) C_p. Support your recommendations with graphs.

c) Based on your analysis in part (b), which set of independent variables would you recommend as best? Justify your recommendation.

d) For the set of independent variables recommended in part (c), study the aptness of the model containing only linear effects and no interaction terms. Summarize your findings. Should you modify the model? If so, how would you proceed?

11.17. Refer to Problem 11.16.

a) Using stepwise regression, find the best set of independent variables to predict sales. Explain your choice of F limits for adding and deleting variables.

b) Were substantially fewer model fittings required with stepwise regression here than with all possible regressions?

c) Examine all possible regression models containing the number of independent variables in your best set in part (a). Are there grounds for changing your best set? Which set of independent variables do you recommend as best?

d) For the set of independent variables recommended in part (c), examine the aptness of the model containing only linear effects and no interaction terms. Summarize your findings. Should you modify the model? If so, how would you proceed?

11.18. Refer to Problem 7.23. The personnel officer wishes to identify the best set of tests for predicting job proficiency score. He believes that only linear effects are present, and that there are no interaction effects.

a) Fit all possible regressions.

b) For each of the following criteria, indicate which set of independent variables you would recommend as the best set for predicting Y: (1) R_p^2; (2) MSE_p; (3) C_p. Support your recommendations with graphs.

c) Based on your analysis in part (b), which set of independent variables would you recommend as best? Justify your recommendation.

d) For the set of independent variables recommended in part (c), study the aptness of the model containing only linear effects and no interaction terms. Summarize your findings. Should you modify the model? If so, how would you proceed?

11.19. Refer to Problem 11.18.

a) Using stepwise regression, find the best set of independent variables to predict job proficiency score. Explain your choice of F limits for adding and deleting variables.

b) Were substantially fewer model fittings required with stepwise regression here than with all possible regressions?

c) Examine all possible regression models containing the number of independent variables in your best set in part (a). Are there grounds for changing your best set? Which set of independent variables do you recommend as best?

d) For the set of independent variables recommended in part (c), examine the aptness of the model containing only linear effects and no interaction terms. Summarize your findings. Should you modify the model? If so, how would you proceed?

CITED REFERENCE

11.1. Daniel, Cuthbert, and Wood, Fred S. *Fitting Equations to Data*. New York: Wiley-Interscience, 1971.

12

Normal Correlation Models

THE PURPOSE of this chapter is to indicate the relation between regression analysis, discussed in Chapters 2–11, and normal correlation models. We first take up bivariate normal correlation models, and then consider multivariate normal ones.

12.1 DISTINCTION BETWEEN REGRESSION AND CORRELATION MODELS

As we know, the basic regression models taken up in this book assume that the independent variables $X_1, \ldots, X_{p-1}$ are fixed constants, and primary interest exists in making inferences about the dependent variable Y on the basis of the independent variables.

We saw in Chapter 3, for the case of a single independent variable, that the regression analysis for a normal error regression model is applicable even when X is a random variable, provided that the conditional distributions of Y follow certain specifications and the marginal distribution of X does not involve the regression model parameters β_0, β_1, and σ^2. Thus, in the case where X is a random variable, only the conditional distributions of Y were specified, and a restriction was placed on the marginal distribution of X. We did not, however, seek to completely specify the joint distribution of X and Y. While the discussion in Chapter 3 dealt with only a single independent variable, all of the points apply when the regression model contains a number of independent variables which are random.

Correlation models, like regression models with random independent variables, consist of variables all of which are random. Correlation models differ from regression models by specifying the joint distribution of the variables completely. Furthermore, the variables in a correlation model play a symmetrical role, with no one variable automatically designated as the dependent variable. Correlation models are employed to study the nature of the relations between the variables, and also may be used for making inferences about any one of the variables on the basis of the others.

Thus, an analyst may use a correlation model for the two variables "height of person" and "weight of person" in a study of a sample of persons, each variable being taken as random. He might wish to study the relation between the two variables. He also might be interested in making inferences about weight of a person on the basis of his height, in making inferences about height on the basis of weight, or in both.

Other examples where a correlation model may be appropriate are:

1. To study the relation between service station sales of gasoline and sales of auxiliary products.
2. To study the relation between company net income determined by generally accepted accounting principles and net income according to tax regulations.

The correlation model most widely employed is the normal correlation model. We discuss it now for the case of two variables.

12.2 BIVARIATE NORMAL DISTRIBUTION

The normal correlation model for the case of two variables is based on the *bivariate normal distribution*. Let us denote the two variables as Y_1 and Y_2. (We do not use the notation X and Y in this chapter because both variables play a symmetrical role in correlation analysis.) We say that Y_1 and Y_2 are *jointly normally distributed* if their joint probability distribution is the bivariate normal distribution.

Density Function

The density function for the bivariate normal distribution is as follows:

$$f(Y_1, Y_2) = \frac{1}{2\pi\sigma_1\sigma_2\sqrt{1-\rho_{12}^2}} \exp\left\{-\frac{1}{2(1-\rho_{12}^2)}\left[\left(\frac{Y_1-\mu_1}{\sigma_1}\right)^2 - 2\rho_{12}\left(\frac{Y_1-\mu_1}{\sigma_1}\right)\left(\frac{Y_2-\mu_2}{\sigma_2}\right) + \left(\frac{Y_2-\mu_2}{\sigma_2}\right)^2\right]\right\} \tag{12.1}$$

Note that this density function involves five parameters: $\mu_1, \mu_2, \sigma_1, \sigma_2, \rho_{12}$. We shall explain the meaning of these parameters shortly. First, let us consider a graphic representation of the bivariate normal distribution.

Graphic Representation

Figure 12.1 contains a graphic representation of a bivariate normal distribution. It is a surface in three-dimensional space. For every pair of Y_1, Y_2 values, there is a density $f(Y_1, Y_2)$ represented by the height of the surface at that point. The surface is continuous, and probability corresponds to volume under the surface.

FIGURE 12.1

Example of Bivariate Normal Distribution

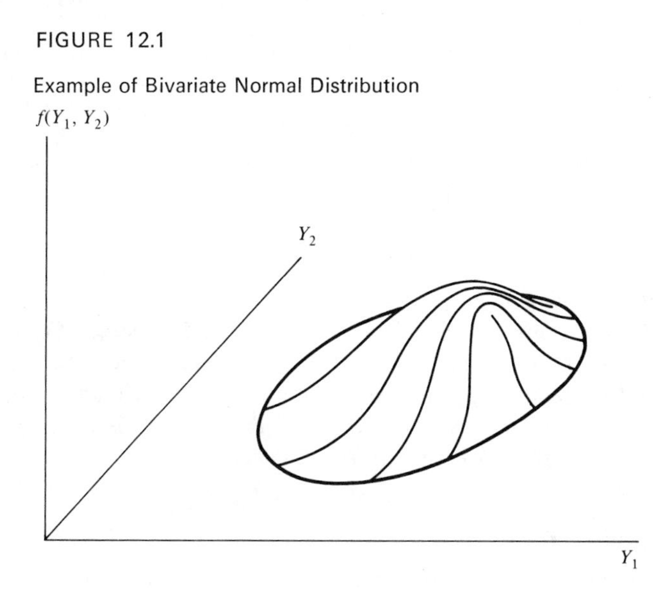

Marginal Distributions

If Y_1 and Y_2 are jointly normally distributed, it can be shown that their marginal distributions have the following characteristics:

(12.2a) The marginal distribution of Y_1 is normal with mean μ_1 and standard deviation σ_1:

$$f_1(Y_1) = \frac{1}{\sqrt{2\pi}\,\sigma_1} \exp\left[-\frac{1}{2}\left(\frac{Y_1 - \mu_1}{\sigma_1}\right)^2\right]$$

(12.2b) The marginal distribution of Y_2 is normal with mean μ_2 and standard deviation σ_2:

$$f_2(Y_2) = \frac{1}{\sqrt{2\pi}\,\sigma_2} \exp\left[-\frac{1}{2}\left(\frac{Y_2 - \mu_2}{\sigma_2}\right)^2\right]$$

Thus, when Y_1 and Y_2 are jointly normally distributed, each of the two variables by itself is normally distributed. It is not generally true, however, that if Y_1 and Y_2 are each normally distributed, they must be jointly normally distributed in accord with (12.1).

Meaning of Parameters

The five parameters of the bivariate normal density function (12.1) have the following meaning:

1. μ_1 and σ_1 are, respectively, the mean and standard deviation of the marginal distribution of Y_1.

2. μ_2 and σ_2 are, respectively, the mean and standard deviation of the marginal distribution of Y_2.

3. ρ_{12} is the *coefficient of correlation* between Y_1 and Y_2. It is defined as follows:

$$\rho_{12} = \frac{\sigma_{12}}{\sigma_1 \sigma_2} \tag{12.3}$$

where σ_{12} is the covariance between Y_1 and Y_2:

$$\sigma_{12} = E[(Y_1 - \mu_1)(Y_2 - \mu_2)] \tag{12.4}$$

If Y_1 and Y_2 are independent, $\sigma_{12} = 0$ according to (1.23) so that $\rho_{12} = 0$ then. If Y_1 and Y_2 are positively related, σ_{12} is positive and so is ρ_{12}. On the other hand, if Y_1 and Y_2 are negatively related, σ_{12} is negative and so is ρ_{12}. The coefficient of correlation ρ_{12} is a pure number, and can take on any value between -1 and $+1$. It assumes $+1$ if Y_1 and Y_2 are perfectly positively related, and -1 if the perfect relation is a negative one.

Contour Representation

Bivariate normal distributions frequently are portrayed in terms of a contour diagram. A contour curve on such a diagram is composed of all the points on the surface that are equidistant from the $Y_1 Y_2$ plane. To put this another way, a contour curve is composed of all Y_1, Y_2 outcomes which have constant density $f(Y_1, Y_2)$. Thus we can picture a contour as the cross section obtained by slicing a bivariate normal surface horizontally at a fixed distance above the $Y_1 Y_2$ plane, as in Figure 12.2.

Figure 12.3 presents a contour diagram for the bivariate normal surface of Figure 12.1. It is a property of the bivariate normal distribution that all contour curves are ellipses. Note that the ellipses have a common center, at (μ_1, μ_2), and common major and minor axes. Also note that the higher the horizontal cross section of the surface is above the $Y_1 Y_2$ plane, the smaller is the corresponding contour ellipse.

Figure 12.4 illustrates the effects of different parameter values on the location and shape of the bivariate normal surface. Note that when Y_1 and Y_2 are positively related so that $\rho_{12} > 0$, the principal axis has positive slope, implying that the surface tends to run along a line with positive slope. When Y_1 and Y_2 are negatively related so that $\rho_{12} < 0$, the principal axis has a negative slope, implying that the surface tends to run along a line with negative slope.

FIGURE 12.2

Contour Ellipse for Bivariate Normal Surface

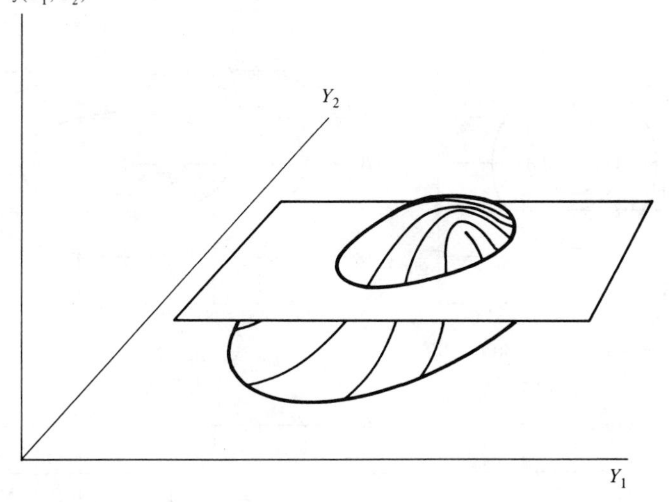

FIGURE 12.3

Contour Diagram for Bivariate Normal Surface in Figure 12.1

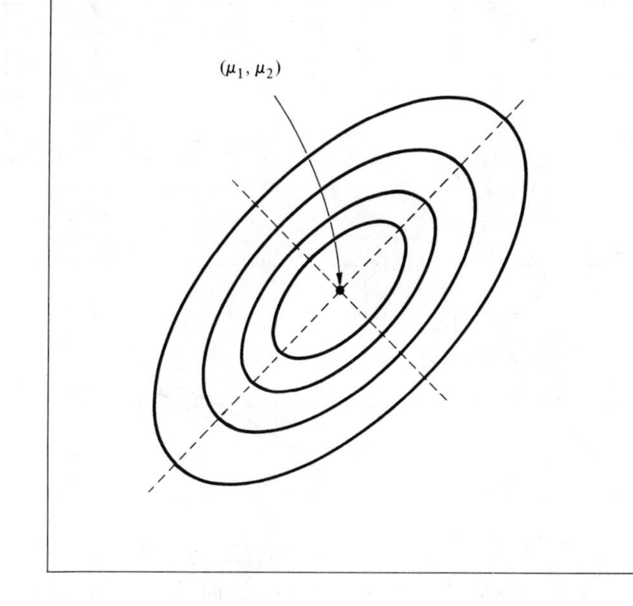

FIGURE 12.4

Effects of Parameter Values on Location and Shape of Bivariate Normal Distribution

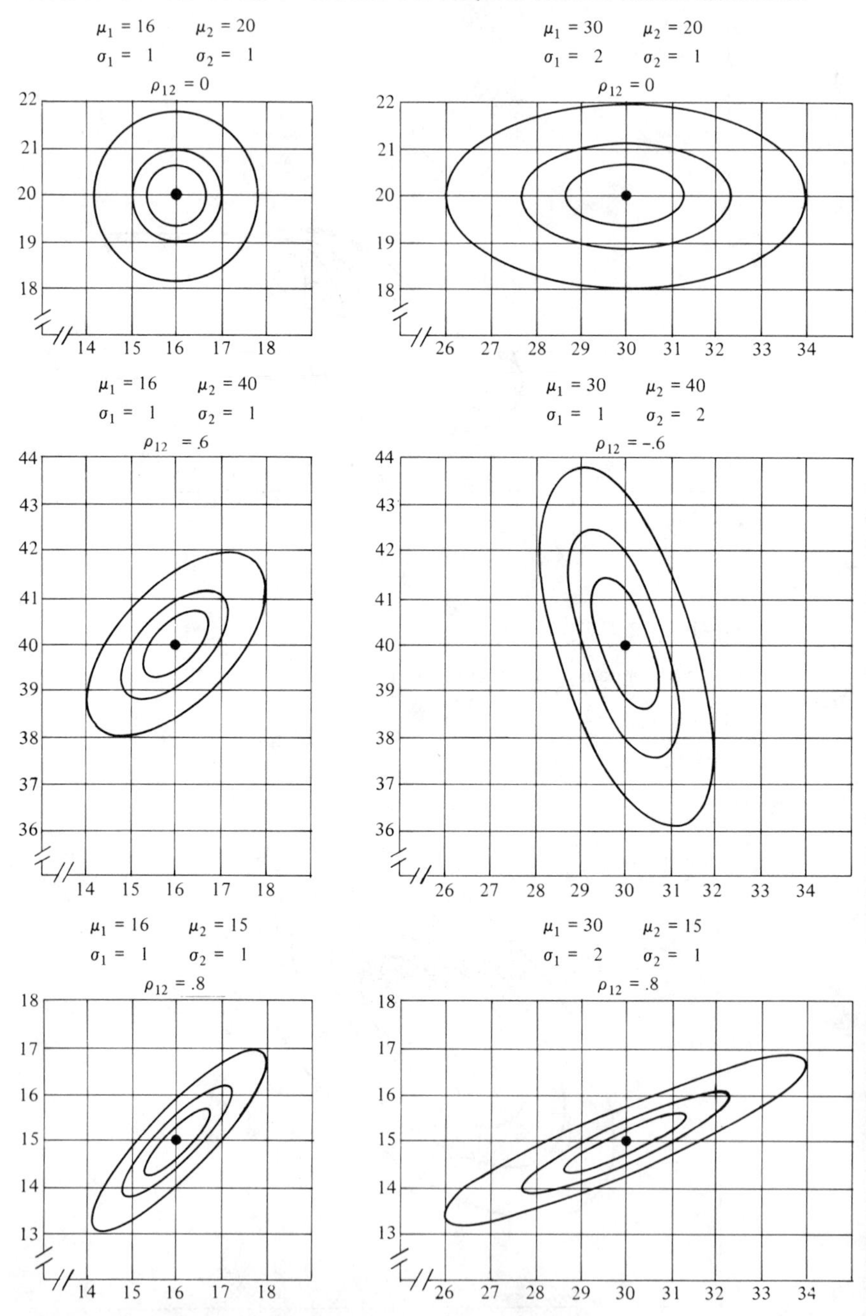

Source: Adapted, with permission, from Helen M. Walker and Joseph Lev, *Statistical Inference* (New York: Holt, Rinehart & Winston, Inc., 1953), p. 250.

Figure 12.4 also demonstrates how the mean values μ_1 and μ_2 affect the location of the surface, and how the standard deviations σ_1 and σ_2, together with the correlation coefficient ρ_{12}, affect the shape of the surface.

12.3 CONDITIONAL INFERENCES

As noted, one principal use of bivariate correlation models is to make conditional inferences regarding one variable, given the other variable. Suppose Y_1 represents a service station's gasoline sales and Y_2 its sales of auxiliary products. We may then wish to predict a service station's auxiliary sales Y_2, given that its gasoline sales are $Y_1 = \$5{,}500$.

Such conditional inferences require the use of conditional probability distributions, which we discuss next.

Conditional Probability Distributions of Y_1

The conditional density function of Y_1, for any given value of Y_2, is denoted $f(Y_1 | Y_2)$, and is defined as follows:

(12.5)
$$f(Y_1 | Y_2) = \frac{f(Y_1, Y_2)}{f_2(Y_2)}$$

where $f(Y_1, Y_2)$ is the joint density function of Y_1 and Y_2, and $f_2(Y_2)$ is the marginal density function of Y_2. When Y_1 and Y_2 are jointly normally distributed according to (12.1), it can be shown that:

(12.6) The conditional probability distribution of Y_1, for any given value of Y_2, is normal with mean $\alpha_{1.2} + \beta_{12} Y_2$ and standard deviation $\sigma_{1.2}$:

$$f(Y_1 | Y_2) = \frac{1}{\sqrt{2\pi}\,\sigma_{1.2}} \exp\left[-\frac{1}{2}\left(\frac{Y_1 - \alpha_{1.2} - \beta_{12} Y_2}{\sigma_{1.2}} \right)^2 \right]$$

The parameters $\alpha_{1.2}$, β_{12}, and $\sigma_{1.2}$ of the conditional probability distributions of Y_1 are functions of the parameters of the joint probability distribution (12.1), as follows:

(12.7a)
$$\alpha_{1.2} = \mu_1 - \mu_2 \rho_{12} \frac{\sigma_1}{\sigma_2}$$

(12.7b)
$$\beta_{12} = \rho_{12} \frac{\sigma_1}{\sigma_2}$$

(12.7c)
$$\sigma_{1.2}^2 = \sigma_1^2(1 - \rho_{12}^2)$$

Important Characteristics of Conditional Distributions. Three important characteristics of the conditional probability distributions of Y_1 are normality, linear regression, and constant variance. We take up each of these in turn.

1. The conditional probability distribution of Y_1, for any given value of Y_2, is normal. Imagine that we slice a bivariate normal distribution vertically at a given value of Y_2, say at Y_{h2}. That is, we slice it parallel to the Y_1 axis. This slicing is shown in Figure 12.5. The exposed cross section has the shape of a normal distribution and, after being scaled so that its area is one, portrays the conditional probability distribution of Y_1, given that $Y_2 = Y_{h2}$.

FIGURE 12.5

Cross Section of Bivariate Normal Distribution at Y_{h2}

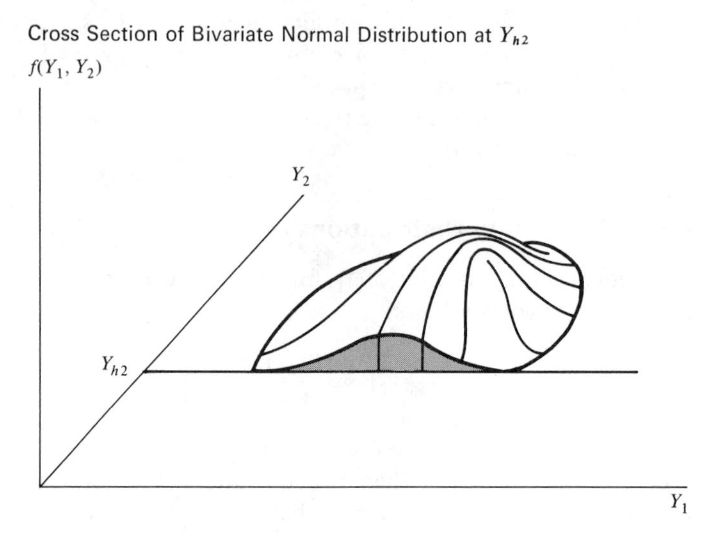

This property of normality holds no matter what the value Y_{h2} is. Thus, whenever we slice the bivariate normal distribution parallel to the Y_1 axis, we obtain (after proper scaling) a normal conditional probability distribution.

2. The means of the conditional probability distributions of Y_1 fall on a straight line, and hence are a linear function of Y_2:

$$E(Y_1 | Y_2) = \alpha_{1.2} + \beta_{12} Y_2 \tag{12.8}$$

Here $\alpha_{1.2}$ is the intercept parameter and β_{12} the slope parameter. Thus, the relation between the conditional means and Y_2 is given by a linear regression function.

3. All conditional probability distributions of Y_1 have the same standard deviation $\sigma_{1.2}$. Thus, no matter where we slice the bivariate normal distribution parallel to the Y_1 axis, the resulting conditional probability distribution (after scaling to have an area of one) has the same standard deviation. Hence, constant variances characterize the conditional probability distributions of Y_1.

Equivalence to Normal Error Regression Model. Suppose that we select a random sample of observations (Y_1, Y_2) from a bivariate normal population and wish to make conditional inferences about Y_1, given Y_2. The preceding discussion makes it clear that the normal error regression model (2.25) is entirely applicable because:

1. The Y_1 observations are independent.
2. The Y_1 observations, when Y_2 is considered given or fixed, are normally distributed with mean $E(Y_1 | Y_2) = \alpha_{1.2} + \beta_{12} Y_2$ and constant variance $\sigma_{1.2}^2$.

Conditional Probability Distributions of Y_2

The random variables Y_1 and Y_2 play symmetrical roles in the bivariate normal probability distribution (12.1). Hence, it follows:

(12.9) The conditional probability distribution of Y_2, for any given value of Y_1, is normal with mean $\alpha_{2.1} + \beta_{21} Y_1$ and standard deviation $\sigma_{2.1}$:

$$f(Y_2 | Y_1) = \frac{1}{\sqrt{2\pi}\,\sigma_{2.1}} \exp\left[-\frac{1}{2}\left(\frac{Y_2 - \alpha_{2.1} - \beta_{21} Y_1}{\sigma_{2.1}}\right)^2 \right]$$

The parameters $\alpha_{2.1}$, β_{21}, and $\sigma_{2.1}$ of the conditional probability distributions of Y_2 are functions of the parameters of the joint probability distribution (12.1), as follows:

$$\alpha_{2.1} = \mu_2 - \mu_1 \rho_{12} \frac{\sigma_2}{\sigma_1} \tag{12.10a}$$

$$\beta_{21} = \rho_{12} \frac{\sigma_2}{\sigma_1} \tag{12.10b}$$

$$\sigma_{2.1}^2 = \sigma_2^2(1 - \rho_{12}^2) \tag{12.10c}$$

The parameter $\alpha_{2.1}$ is the intercept of the line of regression of Y_2 on Y_1, and the parameter β_{21} is the slope of this line.

Again, we find that the conditional correlation model of Y_2, for given Y_1, is the equivalent of the normal error regression model (2.25).

Comments

1. The notation for the parameters of the conditional correlation models departs somewhat from our previous notation for regression models. The symbol α is now used to denote the regression intercept. The subscript 1.2 to α indicates that Y_1 is regressed on Y_2. Similarly, the subscript 2.1 to α indicates that Y_2 is regressed on Y_1. The symbol β_{12} indicates that it is the slope term in the regression of Y_1 on Y_2, while β_{21} is the slope term in the regression of Y_2 on Y_1. Finally, $\sigma_{2.1}$ is the standard deviation of the conditional probability distributions of Y_2, for any given Y_1, while $\sigma_{1.2}$ is the standard deviation of the conditional probability distributions of Y_1, for any given Y_2. This notation can be extended straightforwardly for multivariate correlation models.

2. Two distinct regressions are involved in a bivariate normal model, that of Y_1 on Y_2 when Y_2 is fixed, and that of Y_2 on Y_1 when Y_1 is fixed. In general, the two regression lines are not the same. For instance, the two slopes β_{12} and β_{21} are the same only if $\sigma_1 = \sigma_2$, as can be seen from (12.7b) and (12.10b).

FIGURE 12.6

Illustration of Relation between Lines of Regression and Contour Ellipses

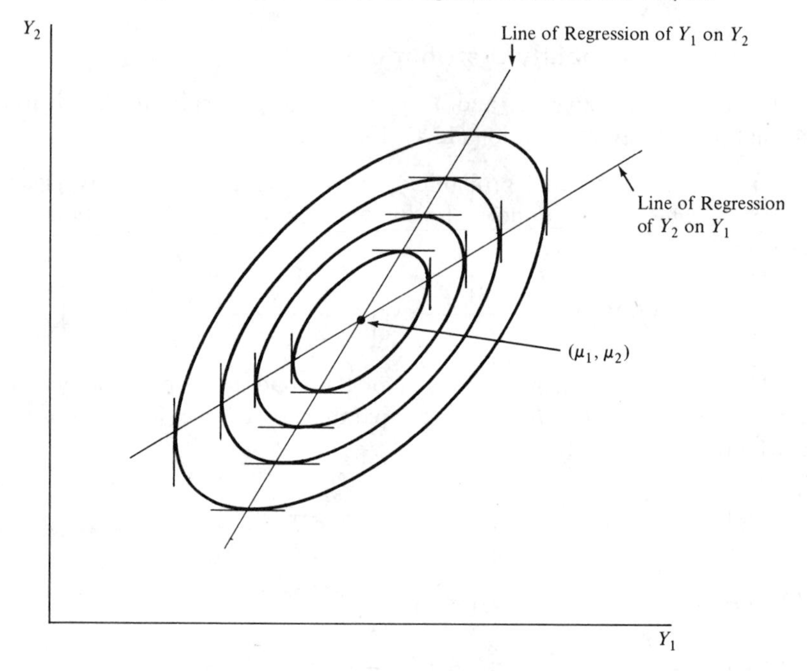

3. Figure 12.6 illustrates the relation of the two regression lines to the contour ellipses. Note that both regression lines go through the point (μ_1, μ_2). If $\rho_{12} = 0$, the two regression lines intersect at right angles. The larger absolutely is ρ_{12}, the more the two regression lines come together.

Use of Regression Analysis

In view of the equivalence of each of the conditional bivariate normal correlation models (12.6) and (12.9) with the normal error regression model (2.25), all conditional inferences with these correlation models can be made by means of the usual regression methods. Thus, if a researcher has data which can be appropriately described as having been generated from a bivariate normal distribution and he wishes to make inferences about Y_2, given a particular value of Y_1, he would simply use ordinary regression techniques. Thus, he would estimate the regression equation of Y_2 on Y_1 by means of (2.12), he would estimate the slope of the regression equation by means of the interval estimate (3.15), he would predict a new observation Y_2, given the value of Y_1, by means of (3.36), and so on. Computer regression packages can be used in the usual manner. To avoid notational problems, the researcher indeed may be well advised to relabel his variables according to the regression usage: $Y = Y_2$, $X = Y_1$. Of course, if the researcher wished to make conditional inferences on Y_1, for given values of Y_2, the notation correspondences would be: $Y = Y_1$, $X = Y_2$.

Note

When obtaining interval estimates for the conditional correlation models, the confidence coefficient refers to repeated samples where pairs of observations (Y_1, Y_2) are obtained from the bivariate normal population. We noted a similar point for regression models where the independent variable X is a random variable.

12.4 INFERENCES ON ρ_{12}

A principal use of correlation models is to study the relationships between the variables. In a bivariate normal model, the parameter ρ_{12} and its square, ρ_{12}^2, provide information about the degree of relationship between the two variables Y_1 and Y_2. Of the two measures, ρ_{12}^2 is the more meaningful one.

Coefficient of Determination

The square of the coefficient of correlation ρ_{12} is called the *coefficient of determination*. We noted earlier in (12.7c) and (12.10c) that:

(12.11a) $$\sigma_{1.2}^2 = \sigma_1^2(1 - \rho_{12}^2)$$

(12.11b) $$\sigma_{2.1}^2 = \sigma_2^2(1 - \rho_{12}^2)$$

We can rewrite these expressions as follows:

(12.12a) $$\rho_{12}^2 = \frac{\sigma_1^2 - \sigma_{1.2}^2}{\sigma_1^2}$$

(12.12b) $$\rho_{12}^2 = \frac{\sigma_2^2 - \sigma_{2.1}^2}{\sigma_2^2}$$

The meaning of ρ_{12}^2 is now clear. Consider first (12.12a). ρ_{12}^2 measures how much smaller relatively is the variability in any conditional distribution of Y_1, for a given level of Y_2, than is the variability in the marginal distribution of Y_1. Thus, ρ_{12}^2 measures the relative reduction in the variability of Y_1 associated with the use of the variable Y_2. Correspondingly, (12.12b) shows that ρ_{12}^2 also measures the relative reduction in the variability of Y_2 associated with the use of the variable Y_1.

It can be shown that:

(12.13) $$0 \leq \rho_{12}^2 \leq 1$$

$\rho_{12}^2 = 0$ if Y_1 and Y_2 are independent, so that the variances of each variable in the conditional probability distributions are then no smaller than the variance in the marginal distribution. $\rho_{12}^2 = 1$ if there is no variability in the conditional probability distributions for each variable, so that perfect predictions of either variable can be made from the other.

Note

The interpretation of ρ_{12}^2 as measuring the relative reduction in the conditional variances as compared with the marginal variance is valid for the case of a bivariate normal population, but not for many other bivariate populations. Of course, the interpretation implies nothing in a causal sense.

Point Estimators of ρ_{12} and ρ_{12}^2

The maximum likelihood estimator of ρ_{12} is denoted r_{12}, and is given by:

(12.14)
$$r_{12} = \frac{\sum (Y_{i1} - \bar{Y}_1)(Y_{i2} - \bar{Y}_2)}{[\sum (Y_{i1} - \bar{Y}_1)^2]^{1/2}[\sum (Y_{i2} - \bar{Y}_2)^2]^{1/2}}$$

The estimator is biased (unless $\rho = 0$ or 1), but the bias is small if n is large.

The coefficient of determination ρ_{12}^2 is estimated by the square of the sample coefficient of correlation; r_{12}^2.

Note

The maximum likelihood estimator (12.14) of the population correlation coefficient is the same as the descriptive coefficient of correlation in (3.71) for the regression model. With a correlation model, r_{12} is an estimator of a parameter. With a regression model, in contrast, r_{12} is only a descriptive measure which reflects the proportion of the total sum of squares that is partitioned into the regression sum of squares. Another difference is that in the standard regression case where the X_i are fixed, the magnitude of r_{12} can be arbitrarily affected by the spacing pattern chosen for the X_i. For the bivariate normal model, on the other hand, the magnitude of r_{12} cannot be so affected since neither variable is under the control of the investigator.

Test Whether $\rho_{12} = 0$

When the population is a bivariate normal one, it is frequently desired to test between:

(12.15)
$$C_1: \rho_{12} = 0$$
$$C_2: \rho_{12} \neq 0$$

The reason for interest in this test is, as we have seen, that in the case where Y_1 and Y_2 are jointly normally distributed, $\rho_{12} = 0$ implies that Y_1 and Y_2 are independent.

We can use regression procedures for the test since (12.7b) implies that the following alternatives are equivalent to those in (12.15):

(12.15a)
$$\begin{aligned} &C_1: \beta_{12} = 0 \\ &C_2: \beta_{12} \neq 0 \end{aligned}$$

and (12.10b) implies that the following alternatives are also equivalent to the ones in (12.15):

(12.15b)
$$\begin{aligned} &C_1: \beta_{21} = 0 \\ &C_2: \beta_{21} \neq 0 \end{aligned}$$

It can be shown that the statistics for testing either (12.15a) or (12.15b) can be expressed directly in terms of r_{12}:

(12.16)
$$t^* = \frac{r_{12}\sqrt{n-2}}{\sqrt{1-r_{12}^2}}$$

If C_1 holds, t^* follows the $t(n-2)$ distribution. The appropriate decision rule to control the Type I error at α is:

(12.17)
$$\begin{aligned} &\text{If } |t^*| \leq t(1-\alpha/2; n-2), \text{ conclude } C_1 \\ &\text{If } |t^*| > t(1-\alpha/2; n-2), \text{ conclude } C_2 \end{aligned}$$

Interval Estimation of ρ_{12}

Because the sampling distribution of r_{12} is complicated when $\rho_{12} \neq 0$, interval estimation of ρ_{12} is usually done by means of a transformation.

z′ Transformation. This transformation, due to R. A. Fisher, is as follows:

(12.18)
$$z' = \frac{1}{2}\log_e\left(\frac{1+r_{12}}{1-r_{12}}\right)$$

When n is large (25 or more is a useful rule of thumb according to a number of statisticians), the distribution of z' approximately:

1. Is normal
2. With mean:

(12.19)
$$\zeta = \frac{1}{2}\log_e\left(\frac{1+\rho_{12}}{1-\rho_{12}}\right)$$

3. And variance:

(12.20)
$$\sigma^2(z') = \frac{1}{n-3}$$

Note that the transformations from r_{12} to z' in (12.18) and from ρ_{12} to ζ in (12.19) are identical. Also note that the variance of z' is a known constant, depending only on the sample size n.

Table A–7 gives paired values for the left and right sides of (12.18) and (12.19), thus eliminating tedious calculations. For instance, if r_{12} or ρ_{12} equals .25, Table A–7 indicates that z' or ζ equals .2554, and vice versa. The values on the two sides of the transformation always have the same sign. Thus, if r_{12} or ρ_{12} is negative, a minus sign is attached to the value in Table A–7. For instance, if $r_{12} = -.25$, $z' = -.2554$.

Interval Estimate. Since z' is approximately normally distributed for large n, it follows that the standardized statistic:

$$\frac{z' - \zeta}{\sigma(z')} \tag{12.21}$$

is approximately a standard normal variable when n is large. Therefore, a $1 - \alpha$ confidence interval for ζ is:

$$z' - z(1 - \alpha/2)\sigma(z') \leq \zeta \leq z' + z(1 - \alpha/2)\sigma(z') \tag{12.22}$$

where $z(1 - \alpha/2)$ is the $(1 - \alpha/2)100$ percentile of the standard normal distribution. These percentiles are given in Table A–1. The $1 - \alpha$ confidence limits for ρ_{12} are then obtained by transforming the limits on ζ by means of (12.19).

Comments

1. As usual, the confidence interval for ρ_{12} can be employed to test whether or not ρ_{12} has a specified value—say, .5—by noting whether or not the specified value falls within the confidence limits.

2. Confidence limits for ρ_{12}^2 can be obtained by squaring the respective confidence limits for ρ_{12}.

Example

An economist investigated food purchasing patterns by households in a midwestern city. He selected 200 households with family incomes between \$7,500 and \$12,500, and ascertained from each household, among other things, the proportions of the food budget expended for beef and poultry, respectively. He expected these to be negatively related, and wished to estimate the coefficient of correlation, with a 95 percent interval estimate. The economist had some supporting evidence which suggested that the joint distribution of the two variables did not depart markedly from a bivariate normal one.

The point estimate of ρ_{12} turned out to be $r_{12} = -.61$ (data and calculations not shown). To obtain a 95 percent confidence interval estimate, we require:

$$z' = -.7089 \text{ when } r_{12} = -.61 \qquad \text{(from Table A–7)}$$

$$\sigma(z') = \frac{1}{\sqrt{200 - 3}} = .07125$$

$$z(.975) = 1.96 \qquad \text{(from Table A–1)}$$

Hence, the confidence interval for ζ, by (12.22), is:

$$-.849 = -.7089 - (1.96)(.07125) \leq \zeta \leq -.7089 + (1.96)(.07125) = -.569$$

Using Table A–7 to transform back to ρ_{12}, we obtain:

$$-.69 \leq \rho_{12} \leq -.51$$

This confidence interval was sufficiently precise to be useful to the economist, confirming the negative relation and indicating that the degree of relation is moderately high.

Caution

Correlation models, like regression models, do not express any causal relations. Earlier cautions about drawing conclusions as to causality from regression findings apply equally to correlation studies. Correlation findings can be useful in analyzing causal relationships, but they do not by themselves establish causal patterns.

12.5 MULTIVARIATE NORMAL DISTRIBUTION

The normal correlation model for the case of p variables $Y_1, \ldots, Y_p$ is based on the *multivariate normal distribution*. This distribution is an extension of the bivariate normal distribution, and has corresponding properties. In particular, if $Y_1, \ldots, Y_p$ are jointly normally distributed (i.e., they follow the multivariate normal distribution), the marginal probability distribution of each variable Y_k is normal, with mean μ_k and standard deviation σ_k.

Conditional Inferences

One major use of multivariate correlation models is to make conditional inferences on one variable when the other variables have given values. It can be shown in the case of the multivariate normal distribution that:

(12.23) The conditional probability distributions of Y_k, for any given set of values for the other variables, are normal with mean given by a linear regression function and constant variance.

Suppose Y_1, Y_2, Y_3, and Y_4 are jointly normally distributed. The conditional probability distribution of, say, Y_3, when Y_1, Y_2, and Y_4 are fixed at any specified levels, has the characteristics:

1. It is normal.
2. It has a mean given by:

$$E(Y_3 \mid Y_1, Y_2, Y_4) = \alpha_{3.124} + \beta_{31.24}Y_1 + \beta_{32.14}Y_2 + \beta_{34.12}Y_4$$

3. It has constant variance $\sigma_{3.124}^2$.

The notation is a straightforward extension of that for the bivariate correlation case. Thus, $\beta_{31.24}$ denotes the regression coefficient of Y_1 when Y_3 is regressed on Y_1, Y_2, and Y_4.

It is clear from the above that the conditional multivariate correlation models are equivalent to the normal error multiple regression model (7.7). Hence, in the case where the variables are jointly normally distributed, all inferences on one variable conditional on the other variables being fixed are carried out by the usual multiple regression techniques. For instance, an interval estimate of a conditional mean would be obtained by (7.50), or a prediction of a new observation on the variable of interest, given the values of the other variables, would be obtained by (7.54). To facilitate use of the regression formulas, it may be helpful to relabel the variable of interest Y and the other variables X's. Computer multiple regression packages can be used in ordinary fashion.

Coefficients of Multiple Correlation and Determination

A major use of multivariate correlation models is to study the relationships between the variables. One set of measures useful to this end consists of the *coefficients of multiple determination* and the *coefficients of multiple correlation.*

Meaning of Coefficients. There is a coefficient of multiple determination associated with each variable. Suppose that Y_1, Y_2, Y_3, and Y_4 are included in the correlation model. The coefficient of multiple determination associated with, say, Y_1 is denoted $\rho^2_{1.234}$, and is defined as follows:

$$\rho^2_{1.234} = \frac{\sigma_1^2 - \sigma^2_{1.234}}{\sigma_1^2} \tag{12.24}$$

where $\sigma^2_{1.234}$ is the variance of the conditional distributions of Y_1, when the other variables are fixed. Thus, $\rho^2_{1.234}$ measures how much smaller, relatively, is the variability in the conditional distributions of Y_1, when the other variables are fixed at given values, than is the variability in the marginal distribution of Y_1. The other coefficients of multiple determination are defined and interpreted in similar fashion.

It can be shown that coefficients of multiple determination take on values between 0 and 1. Thus:

$$0 \leq \rho^2_{1.234} \leq 1 \tag{12.25}$$

Let us consider the significance of the limiting values. If $\rho^2_{1.234} = 0$, no reduction in the variability of Y_1 takes place by considering the other variables. Equivalently, $\rho^2_{1.234} = 0$ in the case of the normal correlation model implies that Y_1 is independent of Y_2, Y_3, and Y_4, so that $\rho_{12} = \rho_{13} = \rho_{14} = 0$.

At the other extreme, if $\rho^2_{1.234} = 1$, the conditional distributions of Y_1, given Y_2, Y_3, and Y_4, have no variability so that perfect predictions can be made of Y_1 from knowledge of the other variables.

The positive square root of a coefficient of multiple determination is the coefficient of multiple correlation. Thus, for Y_1 we have in our example:

(12.26) $$\rho_{1.234} = \sqrt{\rho_{1.234}^2}$$

$\rho_{1.234}$ can be viewed as a simple correlation coefficient, namely between (1) Y_1, and (2) $\alpha_{1.234} + \beta_{12.34}Y_2 + \beta_{13.24}Y_3 + \beta_{14.23}Y_4$.

Estimation of Coefficients. Coefficients of multiple determination and correlation are estimated by (7.31) and (7.34) respectively, the descriptive coefficients in the regression case. Thus, to estimate $\rho_{1.234}^2$, we need the total sum of squares for Y_1, denoted $SSTO(Y_1)$. Then we require the regression sum of squares when Y_1 is regressed on Y_2, Y_3, and Y_4, denoted $SSR(Y_2, Y_3, Y_4)$. The estimator, denoted $R_{1.234}^2$, is:

(12.27) $$R_{1.234}^2 = \frac{SSR(Y_2, Y_3, Y_4)}{SSTO(Y_1)}$$

The positive square root of $R_{1.234}^2$, denoted $R_{1.234}$, is the estimated coefficient of multiple correlation for Y_1.

Testing of Coefficients. To test, say:

(12.28) $$\begin{aligned} &C_1: \rho_{1.234} = 0 \\ &C_2: \rho_{1.234} \neq 0 \end{aligned}$$

we can utilize the equivalent test on the regression coefficients:

(12.28a) $$\begin{aligned} &C_1: \beta_{12.34} = \beta_{13.24} = \beta_{14.23} = 0 \\ &C_2: \text{not all regression coefficients are zero} \end{aligned}$$

It turns out that the F^* statistic for this test, given in (7.30b), can be expressed directly in terms of $R_{1.234}^2$, as follows:

(12.29) $$F^* = \frac{R_{1.234}^2}{1 - R_{1.234}^2} \frac{n-4}{3}$$

In general, the factor on the right is $(n - q - 1)/q$ when there are q predictor variables.

If C_1 holds, F^* follows the $F(q, n - q - 1)$ distribution, so that the decision rule to control the Type I error risk at α is set up in the usual fashion:

(12.30) $$\begin{aligned} &\text{If } F^* \leq F(1 - \alpha; q, n - q - 1), \text{ conclude } C_1 \\ &\text{If } F^* > F(1 - \alpha; q, n - q - 1), \text{ conclude } C_2 \end{aligned}$$

Coefficients of Partial Correlation and Determination

Meaning of Coefficients. Suppose again that four variables Y_1, Y_2, Y_3, and Y_4 are included in a multivariate normal correlation model. Consider now the correlation between Y_1 and Y_2 in the conditional joint distribution when each of the variables Y_3 and Y_4 is fixed at a given level. When the variables are jointly normally distributed, this correlation does not depend on the levels where Y_3 and Y_4 are fixed, and is given by:

$$\rho_{12.34} = \frac{\sigma_{12.34}}{\sigma_{1.34}\sigma_{2.34}} \tag{12.31}$$

$\rho_{12.34}$ is called the *coefficient of partial correlation* between Y_1 and Y_2, when Y_3 and Y_4 are fixed.

The square of $\rho_{12.34}$ is called the *coefficient of partial determination*, and is denoted by $\rho_{12.34}^2$. The following relation holds:

$$\rho_{12.34}^2 = \frac{\sigma_{1.34}^2 - \sigma_{1.234}^2}{\sigma_{1.34}^2} \tag{12.32}$$

Thus, $\rho_{12.34}^2$ measures how much smaller, relatively, is the variability in the conditional distributions of Y_1, given Y_2, Y_3, and Y_4, than it is in the conditional distributions of Y_1, given Y_3 and Y_4 only.

The other partial correlation coefficients are defined and interpreted in similar fashion. For instance, $\rho_{12.3}$ measures the correlation between Y_1 and Y_2, when only Y_3 is given.

A coefficient of partial correlation $\rho_{12.3}$ is called a *first-order* coefficient, a coefficient $\rho_{12.34}$ a *second-order* coefficient, and so on. All partial correlation coefficients measure the correlation between two variables; the order of the coefficient simply indicates how many other variables are fixed in the joint conditional bivariate distribution.

Point Estimation of Coefficients. Point estimators of the coefficients of partial determination and correlation are the descriptive regression measures encountered earlier; see, for instance, (7.91). Thus, to estimate $\rho_{12.3}^2$, we regress Y_1 on Y_3 and obtain the error sum of squares $SSE(Y_3)$. Next, we regress Y_1 on Y_2 and Y_3 and obtain the error sum of squares $SSE(Y_2, Y_3)$. The estimator of $\rho_{12.3}^2$ then is:

$$r_{12.3}^2 = \frac{SSE(Y_3) - SSE(Y_2, Y_3)}{SSE(Y_3)} = \frac{SSR(Y_2 | Y_3)}{SSE(Y_3)} \tag{12.33}$$

Tests concerning Coefficients. Researchers often wish to test whether two variables are correlated in the conditional probability distributions when other variables are given. Thus, when four variables Y_1, Y_2, Y_3, and Y_4 are being considered and the correlation between Y_1 and Y_2, when Y_3 and Y_4 are given, is of interest, the following test may be desired to see if Y_1 and Y_2 are correlated in the conditional joint distributions:

$$\begin{aligned} &C_1: \rho_{12.34} = 0 \\ &C_2: \rho_{12.34} \neq 0 \end{aligned} \tag{12.34}$$

Such tests concerning partial correlation coefficients can be carried out via the test statistic (12.16) for the bivariate case, with r_{12} replaced by the estimated partial correlation coefficient and n replaced by $n - q$, where q is the number of variables that are held fixed.

Interval Estimation of Coefficients. Interval estimates of partial correlation coefficients are obtained via the z' transformation (12.18) in identical fashion to that for simple correlation coefficients. Only the standard deviation $\sigma(z')$ need be modified. It is, for q variables being held fixed:

$$\sigma(z') = \frac{1}{\sqrt{n - q - 3}} \tag{12.35}$$

Example

An operations analyst, wishing to study the relations between three types of test scores made by applicants for entry-level clerical positions in a large insurance company, drew a sample of 250 such applicants from recent records and ascertained their scores. The variables were:

Y_1 verbal aptitude test score
Y_2 reading aptitude test score
Y_3 personal interview score

The multivariate normal model was considered to be applicable for the study.

The analyst's initial interest was in the partial correlation coefficient $\rho_{23.1}$. Since his computer program did not have an option for calculating partial correlation coefficients by a single command (many programs have such an option), the analyst ran two separate regressions, Y_2 on Y_1 and Y_2 on Y_1 and Y_3. He obtained in turn: $SSE(Y_1) = 6{,}340$ and $SSE(Y_1, Y_3) = 6{,}086$. The point estimate of $\rho^2_{23.1}$ then was calculated corresponding to (12.33):

$$r^2_{23.1} = \frac{SSE(Y_1) - SSE(Y_1, Y_3)}{SSE(Y_1)} = \frac{6{,}340 - 6{,}086}{6{,}340} = .040$$

Hence $r_{23.1} = .20$. (The sign of $r_{23.1}$ here is positive because the regression coefficient for Y_3 when Y_2 is regressed on Y_1 and Y_3 is positive.) Desiring a confidence interval with a 95 percent confidence coefficient, the analyst required:

$$z' = .2027 \qquad \text{when } r_{23.1} = .20$$

$$\sigma(z') = \frac{1}{\sqrt{246}} = .06376$$

$$z(.975) = 1.96$$

The confidence interval for ζ was, by (12.22):

$$.0777 = .2027 - (1.96)(.06376) \le \zeta \le .2027 + (1.96)(.06376) = .3277$$

In transforming from ζ to ρ, the analyst obtained:

$$.08 \leq \rho_{23.1} \leq .32$$

Thus the coefficient of partial correlation between reading aptitude score and interview score, with verbal aptitude score fixed, was at best relatively low and could, indeed, be close to zero.

Next the analyst ascertained, from a printout of the regression of Y_2 on Y_3, that $r_{23} = .83$. For a 95 percent confidence interval he then obtained:

$$.79 \leq \rho_{23} \leq .86$$

Thus ρ_{23} turned out to be substantially higher than $\rho_{23.1}$.

The comparative magnitudes of $\rho_{23.1}$ and ρ_{23} suggested to the analyst (and further investigation verified) that in the company's interviewing procedure the interview scores (Y_3) depended in good part on verbal skills. Also, the reading aptitude scores (Y_2) with the test used by the company tended to be heavily influenced by verbal aptitude. Thus when verbal aptitude is not considered, the degree of relation between Y_2 and Y_3 was relatively high since applicants with relatively good (poor) verbal aptitude tended to score well (poorly) in both the reading aptitude test and the personal interview. However, among applicants at any given level of verbal aptitude score the degree of relationship between reading score and interview score tended to be low.

PROBLEMS

12.1. A management trainee in a production department, wishing to study the relation between weight of rough casting and machining time to produce the finished block, selected castings so that the weights would be spaced equally apart in the sample and observed the corresponding machining times. Would you recommend that he use a regression or a correlation model?

12.2. A social scientist stated: "The conditions for the bivariate and multivariate normal distributions are so rarely met in my experience that I feel much safer using a regression model." Comment.

12.3. Refer to Figures 12.1 and 12.3. Where in the $Y_1 Y_2$ plane is the height of the bivariate normal surface greatest? How can this point be ascertained from inspection of the density function (12.1)?

12.4. Plot a contour diagram for the bivariate normal distribution with parameters $\mu_1 = 100$, $\mu_2 = 50$, $\sigma_1 = 3$, $\sigma_2 = 4$, $\rho_{12} = .80$.

12.5. Refer to Problem 12.4.

a) State the characteristics of the marginal distribution of Y_1.

b) State the characteristics of the conditional distribution of Y_2, when $Y_1 = 105$.

c) State the characteristics of the conditional distribution of Y_1, when $Y_2 = 55$.

12.6. Refer to Figure 12.6. For any specified value Y_{h2}, can you identify where the conditional density $f(Y_1 | Y_{h2})$ is maximized? Can this result be ascertained also from inspection of the conditional density function (12.6)?

12.7. The random variables Y_1 and Y_2 are jointly normally distributed. Show that if $\rho_{12} = 0$, Y_1 and Y_2 are independent random variables.

12.8. (Calculus required.)

a) Obtain the maximum likelihood estimators of the parameters of the bivariate normal distribution (12.1).

b) Using the results in part (a), obtain the maximum likelihood estimators of the parameters of the conditional probability distribution of Y_1 for any value of Y_2, as given in (12.7).

c) Show that the maximum likelihood estimators of $\alpha_{1.2}$ and β_{12} are the same as the least squares estimators (2.10) for the regression coefficients in the regression model.

12.9. Refer to Problem 12.4.

a) Give the two regression lines for this model.

b) Why are there two regression lines for a bivariate normal distribution and not just one? What is the meaning of each?

c) Must β_{12} and β_{21} have the same sign?

12.10. Refer to Problem 3.2. Explain whether each of the following would be affected if the bivariate normal model (12.1) were employed instead of the normal error regression model (3.1): (1) the point estimates of the regression coefficients; (2) the confidence limits; (3) the interpretation of the confidence coefficient 95 percent.

12.11. The observations below show assessed value for property tax purposes (Y_1, in thousand dollars) and sales price (Y_2, in thousand dollars) for a sample of 10 parcels of residential property sold recently in "arm's length" transactions in a tax district. Assume the bivariate normal model (12.1) applies.

i:	1	2	3	4	5	6	7	8	9	10
Y_{i1}:	13.9	16.0	10.3	11.8	16.7	12.5	10.0	11.4	13.9	12.2
Y_{i2}:	28.6	34.7	21.0	25.5	39.8	24.0	19.4	22.5	28.3	25.0

a) What does ρ_{12} measure here? Is this measure affected by which variable is denoted Y_1 and which is denoted Y_2?

b) Would $\rho_{12} = 1$ imply here that assessed value and sales price are identical for each parcel in the population?

12.12. Refer to Problem 12.11. The analyst wishes to determine whether or not Y_1 and Y_2 are statistically independent.

a) State the alternative conclusions and conduct the appropriate test, using a level of significance of .01. What do you find?

b) To test $\rho_{12} = .7$ versus $\rho_{12} \neq .7$, could you have used test statistic (12.16)?

12.13. Show that test statistics (3.18) and (12.16) are equivalent.

12.14. An engineer, desiring to estimate the coefficient of correlation ρ_{12} between rate of water flow at point A in a stream (Y_1) and concurrent rate of flow at point B (Y_2), obtained $r_{12} = .87$ in a sample of 150 observations. Assume the bivariate normal model (12.1) is appropriate.

a) Obtain a confidence interval for ρ_{12}, using a confidence coefficient of 95 percent. How do you interpret your confidence coefficient here?

b) Refer to your confidence interval in part (a). Give the corresponding confidence interval for ρ_{12}^2. Interpret the meaning of your confidence interval.

12.15. A building construction consultant wishes to estimate the degree of relationship between cost of bid preparation (Y_1) and amount of bid (Y_2) for his clients. In a sample of 84 bids prepared by clients, r_{12} is found to be .41. Assume the bivariate normal model (12.1) applies.

a) Obtain the desired estimate, using an interval estimate with a 90 percent confidence coefficient.

b) Can you conclude from your confidence interval in part (a) whether or not the two variables are independent?

12.16. Show that test statistic (7.30b) for $p = 4$ is equivalent to test statistic (12.29).

12.17. A marketing research analyst in a pharmaceutical firm wishes to study the relation between the following variables in the target population:

Y_1 socioeconomic status score of housewife
Y_2 magazine media exposure score of housewife
Y_3 awareness score of housewife for firm's new product

He believes it is reasonable to assume that the variables follow a multivariate normal distribution.

a) Explain what is measured by each of the following: (1) $\sigma_{3.12}^2$; (2) $\sigma_{31.2}$; (3) $\sigma_{32.1}$; (4) $\sigma_{1.3}^2$.

b) Explain the meaning of each of the following: (1) $\rho_{2.13}^2$; (2) ρ_{23}^2; (3) $\rho_{12.3}^2$; (4) $\rho_{23.1}^2$.

12.18. Refer to Problem 12.17. The following observations were obtained in a small pilot study involving a random sample of 40 housewives:

Housewife i	Y_{i1}	Y_{i2}	Y_{i3}	Housewife i	Y_{i1}	Y_{i2}	Y_{i3}
1	76	81	53	21	81	86	66
2	92	89	78	22	57	51	32
3	85	84	66	23	49	52	28
4	63	70	46	24	71	72	54
5	50	48	38	25	70	66	55
6	89	93	68	26	91	86	67
7	35	29	20	27	95	91	78
8	67	68	41	28	81	79	60
9	94	98	77	29	40	38	15
10	83	79	61	30	41	43	18
11	24	28	12	31	88	91	72
12	31	37	14	32	63	63	40
13	42	36	17	33	59	54	34
14	71	71	53	34	24	26	11
15	87	89	71	35	96	91	79
16	28	23	3	36	74	80	52
17	36	36	12	37	30	38	14
18	61	58	35	38	81	85	65
19	74	70	47	39	96	92	78
20	66	60	40	40	38	40	16

a) The analyst first wished to test $\rho_{3.12} = 0$ versus $\rho_{3.12} \neq 0$. Perform the test, using a level of significance of .01. What conclusion is indicated? What is the implication for the analyst if $\rho_{3.12} = 0$?

b) The analyst next wished to estimate $\rho_{31.2}$ and $\rho_{32.1}$. Obtain the appropriate interval estimates, using a confidence coefficient of 95 percent for each.

c) Estimate all other coefficients of partial correlation and all coefficients of simple correlation. Interpret your results, and summarize your findings.

part III

Basic Analysis of Variance

13

Single-Factor Analysis of Variance

ANALYSIS of variance is a versatile statistical tool for studying the relation between a dependent variable and one or more independent variables. It does not require making assumptions about the nature of the statistical relation, nor does it require that the independent variables are quantitative.

In the present chapter we consider first the relation between analysis of variance and regression. We then take up the basic elements of single-factor analysis of variance, which is appropriate when one independent variable is being studied. In the remainder of Part III we continue our discussion of single-factor analysis of variance. In Part IV we shall take up multifactor analysis of variance, where two or more independent variables are under investigation.

13.1 RELATION BETWEEN REGRESSION AND ANALYSIS OF VARIANCE

Regression analysis, as we have seen, is concerned with the statistical relation between one or more independent variables and a dependent variable. Both the independent and dependent variables in ordinary regression models are quantitative. (We exclude from this discussion the use of indicator variables in regression models, described in Chapter 9.) The regression function describes the nature of the statistical relation between the mean response and the level(s) of the independent variable(s).

We encountered the analysis of variance in our consideration of regression. It was used there for a variety of tests concerning the regression coefficients, the fit of the regression model, and the like. The analysis of variance is actually much more general than its use with regression models indicated. Analysis of variance models are a basic type of statistical model. They are concerned, like regression models, with the statistical relation between one or more independent variables and a dependent variable. Like regression, the analysis of variance is appropriate for both survey data and data based on formal experiments. Further like ordinary regression, the dependent variable for analysis of variance is a quantitative variable. Analysis of variance models differ from ordinary regression models in two key respects:

1. The independent variables in analysis of variance models may be qualitative (sex, geographic location, plant shift, etc.).
2. If the independent variables are quantitative, no assumption is made in analysis of variance models about the nature of the statistical relation between them and the dependent variable. Thus, the problem of specifying the type of regression function, encountered in ordinary regression analysis, does not arise in analysis of variance models.

Illustration

Figure 13.1 illustrates the essential differences between regression and analysis of variance models for the case where the independent variable is quantitative. Shown in Figure 13.1a is the regression model for a pricing study involving three different price levels, X = \$50, \$60, \$70. Note that the XY plane has been rotated from its usual position so that the Y axis faces the viewer. For each level of the independent variable, there is a probability distribution of sales volumes. The means of these probability distributions fall on the regression curve, which describes the statistical relation between price level and mean sales volume.

The analysis of variance model for the same study is shown in Figure 13.1b. The three price levels are treated as separate populations, each leading to a probability distribution of sales volumes. The quantitative differences in the three price levels, and their statistical relation to expected sales volume, are not considered by the analysis of variance model.

Figure 13.2 illustrates the analysis of variance model for a study of the effects of four different types of incentive pay systems on employee productivity. Here, each type of incentive pay system corresponds to a different population, and there is associated with each a probability distribution of employee productivities. Since type of incentive pay system is a qualitative variable, Figure 13.2 does not contain a corresponding regression model representation.

Choice between Two Types of Models

When the independent variables are qualitative, there is no fundamental choice available between regression and analysis of variance models. The situation differs, however, when the independent variables are quantitative.

FIGURE 13.1

Relation between Regression and Analysis of Variance Models

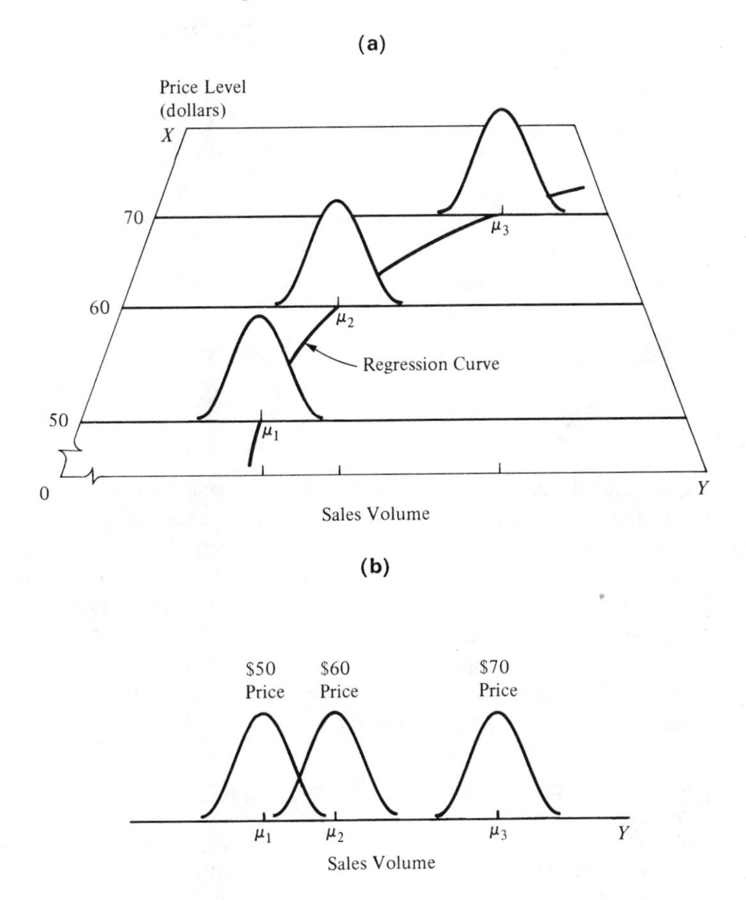

Here the choice involves on one side an analysis not requiring a specification of the nature of the statistical relation (analysis of variance) and on the other an analysis requiring this specification (ordinary regression analysis). If there is substantial doubt about the nature of the statistical relation, a strategy sometimes followed is to first employ an analysis of variance model to study the effects of the independent variables on the dependent variable without restrictive assumptions on the nature of the statistical relation, and then to turn to regression analysis to exploit the quantitative character of the independent variables.

FIGURE 13.2

Analysis of Variance Model Representation for Incentive Pay Example

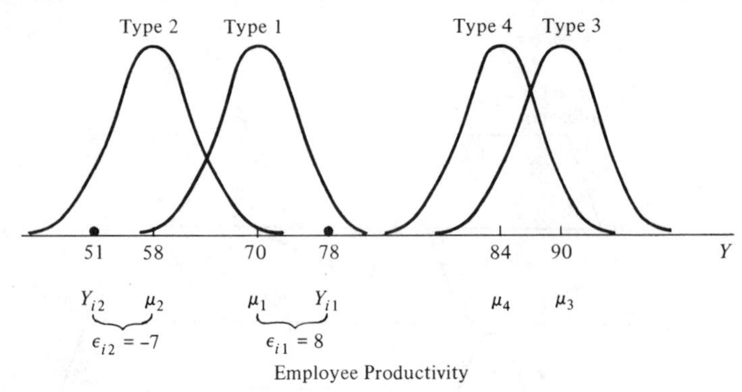

Note

If you have studied Chapter 9, you will know that regression analysis by means of indicator variables can handle qualitative independent variables, and it can also handle by the same means quantitative independent variables without making assumptions about the nature of the statistical relationship. When indicator variables are so used with regression models, the same results will be obtained as with analysis of variance models. The reason why analysis of variance exists as a distinct statistical methodology is that the structure of the independent indicator variables permits computational simplifications which are explicitly recognized in the statistical procedures for the analysis of variance.

13.2 FACTORS, FACTOR LEVELS, AND TREATMENTS

The analysis of variance has its own terminology. We shall introduce some of it now. Note that many of the concepts have already been encountered in regression analysis.

Factor

A *factor* is an independent variable to be studied in an investigation. For instance, in an investigation of the effect of price on sales of a luxury item, the factor being studied is price. Similarly, in a study comparing the appeal of four different television programs, the factor under investigation is type of television program.

Factor Level

A *level* of a factor is a particular form of that factor. In the pricing study mentioned earlier, three prices were used, namely, \$50, \$60, and \$70. Each of these prices is a level of the factor under study, and we say that the price factor has three levels in this study. As another example, in a study of the effect of color of the questionnaire paper on response rate in a mail survey, color of paper is the factor under study and each different color used is a level of that factor.

Single-Factor and Multifactor Studies

Investigations differ as to the number of factors studied. Some are *single-factor* studies, where only one factor is of concern. For instance, the study of the appeal of four different television programs mentioned earlier is an example of a single-factor study. In *multifactor* investigations, two or more factors are investigated simultaneously. An example of a multifactor investigation is a study of the effects of temperature and concentration of the solvent upon the yield of a certain chemical process. Here, two factors, temperature and concentration, are studied simultaneously to obtain information about their effects. Multifactor investigations will be discussed in Part IV; now, we shall focus on single-factor studies.

Experimental and Classification Factors

A factor may be categorized as to whether it is an *experimental* factor or a *classification* factor. In any investigation based on survey data, the factors under study are classification factors. A *classification* factor pertains to the characteristic of the units under study and is not under the control of the investigator. For instance, in a study of the effects of education and experience of salesmen on their sales volumes, where the study is based on a sample of salesmen presently employed by the company, education and experience would be classification factors. The reason is that these refer to the characteristics of the salesmen in the study and were not manipulated experimentally. On the other hand, an *experimental* factor is one where the level of the factor is assigned at random to the experimental unit. An example is the factor "temperature" in an experiment on the effect of five cooking temperatures on the fluffiness of omelets prepared from a mix. The experimental units here are different packages of the mix, and the choice of which particular temperature is to be used with a given package of mix is made at random.

We have pointed out before that experimental data provide firmer foundations for conclusions than do survey data. If experimental factors were only to appear in experimental studies and classification factors only in survey studies, there would be no need to distinguish between these two types of factors. However, classification factors can be found in experimental studies and therefore it is important to recognize them as such, since the inferences pertaining to these factors will not be as clear as for experimental factors. An example may clarify this point. An appliance manufacturer operates three training centers in the United States for training mechanics to service the company's products. Suppose that two different training programs are to be studied, and that at each of the three centers trainees will be assigned at random to one of the two training programs. One may view this as a two-factor study, the factors being "training program" and "training center." If it should turn out that the same training program is superior to the other in all three centers, the evidence would be quite clear as to the effect of the training programs since at each center the trainees were assigned at random to the two programs.

On the other hand, differences between the three training centers cannot be interpreted as clearly because "training center" is a classification factor. One center may excel for any number of reasons, such as because its staff is doing a better training job, it has better facilities, or because trainees assigned to it come from a geographic region in which better education is provided.

The analysis of classification factors can be most useful. However, one needs to be much more careful in the analysis and support it with external evidence. In our training center example, one would need external evidence as to whether or not the education of trainees at the three centers is the same, whether or not the facilities are equal, and the like.

Qualitative and Quantitative Factors

Finally, we need to distinguish between *qualitative* and *quantitative* factors. A *qualitative* factor is one where the levels differ by some qualitative attribute. Examples are type of advertisement or brand of rust inhibitor. On the other hand, a *quantitative* factor is one where each level is described by a numerical quantity on a scale. Examples are temperature in degrees centigrade, age in years, or price in dollars.

Treatments

In single-factor studies, a treatment corresponds to a factor level. Thus, in a study of five advertisements, each advertisement is a treatment. In multi-factor studies, a treatment corresponds to a combination of factor levels. Thus, in a study of the effects on sales volume of price ($.25, $.29) and package color (red, blue), each combination, such as "red package color–$.25 price," is a treatment. This particular study contains four treatments since there are four "price–package color" combinations.

13.3 USES OF ANALYSIS OF VARIANCE MODELS

Analysis of variance models basically are used to analyze the effects of the independent variable or variables under study on the dependent variable. The purpose of the study affects the resulting use of the analysis. We illustrate these points with three examples:

Example 1. A company was using a particular type of advertisement, the results of which were satisfactory to the company. In a study, the company investigated the present type of advertisement as well as three other types.

The analysis of variance model was then utilized to determine whether any of the three alternate types of advertisements is superior to the present type and, if so, which one is best.

Example 2. Four machines in a plant were studied with respect to the diameters of ball bearings they were turning out. The purpose of the analysis of variance model study was to determine whether substantial differences between the machines existed. If so, the machines would need to be calibrated.

Example 3. In a sample study of employees of an organization, the employees' absentee rates were studied according to length of service, sex, marital status, and type of job of the employee. The analysis of variance model for this multifactor investigation was studied to determine whether the effects of these independent variables interact in important ways and which of the independent variables shows the greatest statistical relation to the dependent variable.

These three examples demonstrate that analysis of variance models for single-factor studies are utilized to compare different factor level effects, to ascertain the "best" factor level, and the like. In multifactor studies, the analysis of variance model is employed to determine whether the different factors interact, which factors are the key ones, which factor combinations are "best," and so on.

13.4 MODEL I—FIXED EFFECTS

Distinction between Models I and II

We shall consider two alternate analysis of variance models. Model I, to be taken up here, applies to such cases as a comparison of five different advertisements or a comparison of four different rust inhibitors, where the conclusions will pertain to just those factor levels included in the study. Model II, to be discussed in Chapter 16, applies to a different type of situation, namely where the conclusions will extend to a population of factor levels of which the levels in the study are a sample. Consider, for instance, a company which owns several hundred retail stores throughout the country. Seven of these stores are selected at random, and a sample of employees from each store is then chosen and asked in a confidential interview for an evaluation of the management of the store. The seven stores in the study constitute the seven levels of the factor under study, namely retail stores. In this case, however, management is not just interested in the seven stores included in the study but wishes to generalize the study results to all of the retail stores it owns. Another example when model II is applicable is when three machines out of 75 in a plant are selected at random and their output studied for a period of 10 days. The three machines constitute the three factor levels in this study, but interest is not just in the three machines in the study but in all machines in the plant.

Thus, the essential difference between situations where model I and model II are applicable is that model I is the relevant one when the factor levels are chosen because of intrinsic interest in them (e.g., five different advertisements) and they are not considered as a sample from a larger population. Model II is appropriate when the factor levels constitute a sample from a larger population (e.g., three machines out of 75) and interest is in this larger population.

Basic Ideas

The basic elements of analysis of variance model I for a single-factor study are quite simple. Corresponding to each factor level, there is a probability distribution of responses. For example, in a study of the effects of four types of incentive pay on employee productivity, there is a probability distribution of employee productivities for each type of incentive pay. The analysis of variance model I assumes that:

1. Each of the probability distributions is normal.
2. Each probability distribution has the same variance (standard deviation).
3. The observations for each factor level are random observations from the corresponding probability distribution and are independent of the observations for any other factor level.

Figure 13.2 illustrates these conditions. Note the normality of the probability distributions and the constant variability. The probability distributions differ only with respect to their means. Differences in the means hence reflect the essential factor level effects, and it is for this reason that the analysis of variance focuses on the mean responses for the different factor levels. The analysis of the sample data from each of the factor level probability distributions therefore usually proceeds in two steps:

1. Determine whether or not the factor level means are the same.
2. If the factor level means are not the same, examine how they differ and what the implications of the differences are.

In this chapter, we consider step 1, the testing procedure for determining whether or not the factor level means are equal. In the next chapter, we take up the analysis of the factor level effects when the means are not equal.

Formulation of Model I

There are r levels of the factor under study (e.g., four types of incentive pay), and we shall denote any one of these by the index j ($j = 1, \ldots, r$). The number of observations for the jth factor level is denoted by n_j, and the total number of observations in the study is denoted by n_T, where:

$$n_T = \sum_{j=1}^{r} n_j \tag{13.1}$$

We shall use the index i to represent any one observation for the jth factor level. Hence, $i = 1, \ldots, n_j$. The ith observation for the jth factor level is represented by Y_{ij}; for instance, Y_{ij} is the productivity of the ith employee in the jth plant, or the sales volume of the ith store featuring the jth type of shelf display.

Model I can now be stated as follows:

$$Y_{ij} = \mu_j + \varepsilon_{ij} \tag{13.2}$$

where:

Y_{ij} is the value of the response variable in the ith trial for the jth factor level or treatment
μ_j are parameters
ε_{ij} are independent $N(0, \sigma^2)$
$i = 1, \ldots, n_j; j = 1, \ldots, r$

Important Features of Model

1. The observed value of Y in the ith trial for the jth factor level or treatment is the sum of two components: (1) a constant term μ_j, and (2) a random error term ε_{ij}.

2. Since $E(\varepsilon_{ij}) = 0$, it follows that:

$$E(Y_{ij}) = \mu_j \tag{13.3}$$

Thus, all observations for the jth factor level have the same expectation μ_j.

3. Since μ_j is a constant, it follows from (1.15a) that:

$$\sigma^2(Y_{ij}) = \sigma^2(\varepsilon_{ij}) = \sigma^2 \tag{13.4}$$

Thus, all observations have the same variance, regardless of factor level.

4. Since each ε_{ij} is normally distributed, so is each Y_{ij}. This follows from (1.31) because Y_{ij} is a linear function of ε_{ij}.

5. The error terms are assumed to be independent. Hence the error term outcome on any one trial has no effect on the error term for the outcome of any other trial for the same factor level or for a different factor level. Since the ε_{ij} are independent, so are the observations Y_{ij}.

6. In view of these features, model (13.2) can be restated as follows:

$$Y_{ij} \text{ are independent } N(\mu_j, \sigma^2) \tag{13.5}$$

Example

Suppose that analysis of variance model (13.2) is applicable to our earlier incentive pay study illustration and that the parameters are as follows:

$$\mu_1 = 70 \qquad \mu_2 = 58 \qquad \mu_3 = 90 \qquad \mu_4 = 84$$
$$\sigma = 4$$

Figure 13.2 contains a representation of this model. Note that employee productivities for incentive pay type 1 according to this model are normally distributed with mean $\mu_1 = 70$ and standard deviation $\sigma = 4$.

Suppose that in the ith trial of incentive pay type 1, the observed productivity is $Y_{i1} = 78$. In that case, the error term value is $\varepsilon_{i1} = 8$, for we have:

$$\varepsilon_{i1} = Y_{i1} - \mu_1 = 78 - 70 = 8$$

Figure 13.2 shows this observation Y_{i1}. Note that the deviation of Y_{i1} from the mean μ_1 represents the error term ε_{i1}. This figure also shows the observation $Y_{i2} = 51$, for which the error term value is $\varepsilon_{i2} = -7$.

Interpretation of Factor Level Means

Survey Data. In a survey situation, the factor level means μ_j correspond to the means for the different factor level populations. For instance, in a study of the productivity of employees in each of three shifts operated in a plant, the populations consist of the employee productivities for each of the three shifts. The population mean μ_1 is the mean productivity for employees in shift 1; μ_2 and μ_3 are interpreted similarly. The variance σ^2 refers to the variability of employee productivities within a shift.

Experimental Data. In an experimental study, the factor level means μ_j stand for the mean response that would be obtained if the jth treatment were applied to all units in the population of experimental units about which inferences are to be drawn. Similarly, the variance σ^2 refers to the variability of responses if any experimental treatment were applied to the entire population of experimental units. For instance, in a study of three different training programs in which 90 employees participate, a third of these employees is assigned at random to each of the three programs. The mean μ_1 here denotes the mean response—for instance, the mean gain in productivity—if training program 1 were given to each employee in the population of experimental units; the means μ_2 and μ_3 are interpreted correspondingly. The variance σ^2 denotes the variability in productivity gains if any training program were given to each employee in the population of experimental units.

Comments

1. Model I, like any other model, is not likely to be met exactly by any real-world situation. However, it will be met approximately in a number of cases. As we shall note later, the statistical procedures based on model I are quite robust so that even if the actual conditions of the situation under study differ substantially from those of model I, the statistical analysis may still be appropriate to a high degree of approximation.

2. At times, all treatments under study are given to each of the units in the study. For instance, a person may be asked to use toothpaste A for a week and then give it a rating, which is followed by a week's use for toothpastes B and C each.

In this type of case, model I is usually not appropriate since the several responses by the same person for the different treatments under study are likely to be correlated. For instance, if a person prefers tooth powder to toothpaste, all of his ratings for the different toothpastes are likely to be low. In Chapter 23, we take up models for this situation.

Alternate Formulation of Model I

For single-factor investigations, the parameterization of model I in (13.2) is completely adequate. For multifactor experiments, on the other hand, a different form of parameterization is more useful. We develop this alternate parameterization now, since it is especially simple to see the relationship between the two in the case of single-factor studies.

The alternate parameterization for model I is:

(13.6) $$Y_{ij} = \mu_{.} + \tau_j + \varepsilon_{ij}$$

where:

- $\mu_{.}$ is a constant component common to all observations
- τ_j is the effect of the jth factor level (a constant for each factor level)
- ε_{ij} are independent $N(0, \sigma^2)$
- $i = 1, \ldots, n_j; j = 1, \ldots, r$

Note by comparing (13.2) and (13.6) that the only difference between the two models is that the mean response μ_j in (13.2) is broken into two parts in (13.6):

(13.7) $$\underset{(13.2)}{\mu_j} = \underset{(13.6)}{\mu_{.} + \tau_j}$$

Thus, τ_j, which is called the *effect of the jth factor level*, is defined as follows:

(13.8) $$\tau_j = \mu_j - \mu_{.}$$

Definition of $\mu_{.}$. The splitting up of the factor level mean μ_j into two components, the overall constant $\mu_{.}$ and the specific factor level effect τ_j, can be done in many ways. For example, $\mu_{.}$ often is defined as the unweighted mean of all factor level means:

(13.9) $$\mu_{.} = \frac{\sum_{j=1}^{r} \mu_j}{r}$$ or weighted $\mu_{.} = \frac{\sum_{j=1}^{r} n_j \mu_j}{n_T}$

This definition implies:

(13.10) $$\sum_{j=1}^{r} \tau_j = 0$$

because by (13.8):

$$\sum \tau_j = \sum (\mu_j - \mu_{.}) = \sum \mu_j - r\mu_{.}$$

and by (13.9):

$$\sum \mu_j = r\mu_.$$

Thus, the definition of the overall constant $\mu_.$ in (13.9) implies a restriction on the τ_j, in this case that their sum must be zero.

For our incentive pay example, where $\mu_1 = 70$, $\mu_2 = 58$, $\mu_3 = 90$, and $\mu_4 = 84$, we obtain when $\mu_.$ is defined according to (13.9):

$$\mu_. = \frac{70 + 58 + 90 + 84}{4} = 75.5$$

Hence:

$$\tau_1 = 70 - 75.5 = -\ \ 5.5$$
$$\tau_2 = 58 - 75.5 = -17.5$$
$$\tau_3 = 90 - 75.5 = \ \ 14.5$$
$$\tau_4 = 84 - 75.5 = \ \ \ 8.5$$

The first factor level effect $\tau_1 = -5.5$ indicates that the mean employee productivity for incentive pay type 1 is 5.5 less than the average productivity for all four types of incentive pay.

Alternate Definition of $\mu_.$**.** The constant $\mu_.$ can also be defined as some weighted average of the factor level means μ_j:

(13.11) $$\mu_. = \sum_{j=1}^{r} w_j \mu_j$$

where the w_j are weights defined so that $\sum w_j = 1$. The restriction on the τ_j then is:

(13.12) $$\sum_{j=1}^{r} w_j \tau_j = 0$$

This follows in the same fashion as (13.10).

Frequently, the weights used are the relative sample sizes n_j/n_T since with these weights simplifications in computations arise. In that case, the overall constant $\mu_.$ is:

(13.13) $$\mu_. = \frac{\sum_{j=1}^{r} n_j \mu_j}{n_T}$$

and the restriction on the τ_j is:

(13.14) $$\sum_{j=1}^{r} n_j \tau_j = 0$$

When the sample sizes for all factor levels are the same, a desirable situation frequently as we shall see, it does not matter whether the constant component $\mu_.$ is defined as an unweighted or weighted average of the μ_j; the result is the same in either case.

The choice of the definition of $\mu_.$ should depend on the meaningfulness of the resulting measures of factor level effects τ_j. However, to simplify the discussion, we shall utilize, unless otherwise noted, the definition of $\mu_.$ in (13.13).

Note

As mentioned earlier, we are usually interested initially in whether or not the factor level means μ_j are equal, for instance, whether or not five different advertisements have equal mean sales effectiveness. With our first parameterization model (13.2), the two alternate conclusions for the decision problem would be stated as follows:

(13.15)
$$C_1: \mu_1 = \mu_2 = \cdots = \mu_r$$
$$C_2: \text{not all } \mu_j \text{ are equal}$$

The same two alternatives would be stated as follows with the second parameterization model (13.6):

(13.16)
$$C_1: \tau_1 = \tau_2 = \cdots = \tau_r = 0$$
$$C_2: \text{not all } \tau_j \text{ equal } 0$$

The equivalence of the two forms can be readily established. The equality of the factor level means $\mu_1 = \mu_2 = \cdots = \mu_r$ in (13.15) implies that all τ_j are equal. This follows from (13.7) since the common term $\mu_.$ drops out for each pair of factor levels. Further, the equality of the factor level means implies that all $\tau_j = 0$, whether the restriction on the τ_j is of the form in (13.10) or (13.12). In either case, the restriction can be satisfied in only one way given the equality of the τ_j, namely that $\tau_j \equiv 0$. Thus, it is equivalent to state that all factor level means μ_j are equal or that all factor level effects τ_j are zero.

13.5 ESTIMATION OF PARAMETERS

The parameters of the analysis of variance model are ordinarily unknown and must be estimated from sample data, obtained either by survey or experiment. As with regression, we use the method of least squares to provide estimators of the model parameters. Before turning to these estimators, we shall describe an example to be used throughout the remainder of this chapter and shall develop needed notation.

Example

The Kenton Food Company wishes to test four different package designs for a new breakfast cereal. Ten stores, with approximately equal sales volumes, were selected as the experimental units. Each store was randomly assigned one of the package designs, with two of the package designs assigned to three stores each, and the other two designs to two stores each. Other relevant conditions besides package design, such as price, amount and location of shelf space, and special promotional efforts, were kept the same for all stores in the experiment. Sales, in number of cases, were observed for the study period, and the results are recorded in Table 13.1a.

TABLE 13.1

Number of Cases Sold by Stores, for Each of Four Package Designs—Kenton Food Company Example

(a) Sample Data

	Package Design				
Store	1	2	3	4	*Total*
1	12	14	19	24	
2	18	12	17	30	
3		13	21		
Total	30	39	57	54	180
Mean	15	13	19	27	18
Number of stores	2	3	3	2	10

(b) Symbolic Notation

	Factor Level (j)				
Sample Unit (i)	1	2	3	4	*Total*
1	Y_{11}	Y_{12}	Y_{13}	Y_{14}	
2	Y_{21}	Y_{22}	Y_{23}	Y_{24}	
3		Y_{32}	Y_{33}		
Total	$Y_{.1}$	$Y_{.2}$	$Y_{.3}$	$Y_{.4}$	$Y_{..}$
Mean	$\bar{Y}_{.1}$	$\bar{Y}_{.2}$	$\bar{Y}_{.3}$	$\bar{Y}_{.4}$	$\bar{Y}_{..}$
Number of sample units	n_1	n_2	n_3	n_4	n_T

Note

In this type of situation it would be more reasonable to assign an equal number of stores to each package design if there is equal interest in each of the four designs. We are utilizing unequal sample sizes to demonstrate the analytical procedure in full generality. Also, the small sample sizes in this example are not indicative of sample sizes needed for this type of study. Here, we simply wish to keep computations to a minimum.

Notation

Table 13.1b shows the symbolic notation for the Kenton Food Company data in Table 13.1a. Y_{ij}, as explained earlier, represents the observation on the ith unit for the jth factor level; here, Y_{ij} is the number of cases sold by the ith store assigned to the jth package design. For example, Y_{11} represents the sales of the first store assigned package design 1. For our example, $Y_{11} =$ 12 cases. Similarly, sales of the second store assigned package design 3 are $Y_{23} = 17$ cases.

The total of the observations for the jth factor level is denoted $Y_{.j}$:

$$Y_{.j} = \sum_{i=1}^{n_j} Y_{ij} \tag{13.17}$$

Thus, the dot in $Y_{.j}$ indicates an aggregation over the i index; in our example, the aggregation is over all stores assigned to the jth package design. For instance, the total sales for all stores assigned to package design 1 are, according to Table 13.1, $Y_{.1} = 30$ cases. Similarly, total sales for all stores assigned to package design 4 are $Y_{.4} = 54$ cases.

The sample mean for the jth factor level is denoted $\overline{Y}_{.j}$:

$$\overline{Y}_{.j} = \frac{\sum_i Y_{ij}}{n_j} = \frac{Y_{.j}}{n_j} \tag{13.18}$$

In our illustration, the mean number of cases sold by stores assigned package design 1 is $\overline{Y}_{.1} = 15$. Thus, the dot in the subscript indicates that the averaging was done over i (stores).

The total of all observations in the study is denoted $Y_{..}$:

$$Y_{..} = \sum_{j=1}^{r} \sum_{i=1}^{n_j} Y_{ij} \tag{13.19}$$

where the two dots indicate aggregation over both the i and j indexes (in our example, over all stores for any one package design and then over all package designs).

Finally, the mean for all observations is denoted $\overline{Y}_{..}$:

$$\overline{Y}_{..} = \frac{\sum_j \sum_i Y_{ij}}{n_T} = \frac{Y_{..}}{n_T} \tag{13.20}$$

The dots here indicate that the averaging was done over both i and j. For our example, we have from Table 13.1:

$$\overline{Y}_{..} = \frac{180}{10} = 18$$

Least Squares Estimators

Model (13.2). According to the least squares criterion, the sum of the squared deviations of the observations around their expected values must be minimized with respect to the parameters. For model (13.2), we have by (13.3) that:

$$E(Y_{ij}) = \mu_j$$

Hence, the quantity to be minimized is:

(13.21) $$Q = \sum_j \sum_i (Y_{ij} - \mu_j)^2$$

Now (13.21) can be written as follows:

(13.21a) $$Q = \sum_i (Y_{i1} - \mu_1)^2 + \sum_i (Y_{i2} - \mu_2)^2 + \cdots + \sum_i (Y_{ir} - \mu_r)^2$$

Note that each of the parameters appears in only one of the component sums in (13.21a). Hence, Q can be minimized by minimizing each of the component sums. It is well known that the sample mean minimizes a sum of squared deviations. Hence, the least squares estimator of μ_j, denoted $\hat{\mu}_j$, is:

(13.22) $$\hat{\mu}_j = \bar{Y}_{.j}$$

Example. For our Kenton Food Company illustration, the least squares estimates of the model parameters are as follows according to Table 13.1:

Parameter	*Least Squares Estimate*
μ_1	$\bar{Y}_{.1} = 15$
μ_2	$\bar{Y}_{.2} = 13$
μ_3	$\bar{Y}_{.3} = 19$
μ_4	$\bar{Y}_{.4} = 27$

Thus, we estimate that the mean sales per store with package design 1 are 15 cases for the population of stores under study, for package design 2 the mean sales are 13 cases, and so on.

Model (13.6). With the alternate parameterization model (13.6), the quantity Q to be minimized with respect to the parameters is:

(13.23) $$Q = \sum_j \sum_i (Y_{ij} - \mu_. - \tau_j)^2$$

If the common component $\mu_.$ is defined as:

(13.24) $$\mu_. = \frac{\sum n_j \mu_j}{n_T}$$

so that the restriction on the τ_j is:

(13.25) $$\sum n_j \tau_j = 0$$

the least squares estimator for $\mu_.$, denoted by $\hat{\mu}_.$, is:

(13.26) $$\hat{\mu}_. = \bar{Y}_{..}$$

and the least squares estimator of τ_j, denoted by $\hat{\tau}_j$, is:

(13.27) $$\hat{\tau}_j = \bar{Y}_{.j} - \bar{Y}_{..}$$

Example. For our Kenton Food Company illustration, the least squares estimate of the common component $\mu_.$ is, according to Table 13.1a:

$$\hat{\mu}_. = \bar{Y}_{..} = 18$$

The least squares estimates of the package design effects τ_j are:

Parameter	*Estimate*
τ_1	$\bar{Y}_{.1} - \bar{Y}_{..} = 15 - 18 = -3$
τ_2	$\bar{Y}_{.2} - \bar{Y}_{..} = 13 - 18 = -5$
τ_3	$\bar{Y}_{.3} - \bar{Y}_{..} = 19 - 18 = 1$
τ_4	$\bar{Y}_{.4} - \bar{Y}_{..} = 27 - 18 = 9$

Thus, the experiment indicates that mean sales per store for package design 1 are 3 cases lower than the mean sales for all four package designs, that the mean sales for package design 4 are 9 cases higher, and so on.

Comments

1. The least squares estimators given earlier are also maximum likelihood estimators for the normal error models (13.2) and (13.6) respectively. Hence, they have all of the desirable properties mentioned in Chapter 2 for the regression estimators. For example, they are minimum variance unbiased estimators.

2. To derive the least squares estimator of μ_j, we need to minimize, with respect to μ_j, the jth component sum of squares in (13.21a):

$$Q_j = \sum_i (Y_{ij} - \mu_j)^2 \tag{13.28}$$

Differentiating with respect to μ_j, we obtain:

$$\frac{dQ_j}{d\mu_j} = \sum_i -2(Y_{ij} - \mu_j) \tag{13.29}$$

When we set (13.29) equal to zero, we find the result in (13.22):

$$-2\sum_{i=1}^{n_j}(Y_{ij} - \hat{\mu}_j) = 0$$

$$\sum_i Y_{ij} = n_j \hat{\mu}_j$$

$$\hat{\mu}_j = \bar{Y}_{.j}$$

Residuals

Residuals are highly useful for examining the aptness of the analysis of variance model in a given application. The residual e_{ij} is defined as follows:

$$e_{ij} = Y_{ij} - \hat{\mu}_j = Y_{ij} - \bar{Y}_{.j} \tag{13.30}$$

Thus, a residual represents the deviation of an observation from the respective factor level sample mean.

This definition of a residual also holds for the alternate parameterization model (13.6), since:

$$\hat{\mu}_. + \hat{\tau}_j = \bar{Y}_{..} + (\bar{Y}_{.j} - \bar{Y}_{..}) = \bar{Y}_{.j} \tag{13.31}$$

Example. Table 13.2 contains the residuals for the Kenton Food Company illustration. For instance, from Table 13.1, we find:

$$e_{11} = Y_{11} - \bar{Y}_{.1} = 12 - 15 = -3$$
$$e_{12} = Y_{12} - \bar{Y}_{.2} = 14 - 13 = +1$$

TABLE 13.2

Residuals for Kenton Food Company Example

	Package Design				
Store	1	2	3	4	*Total*
1	−3	+1	0	−3	
2	+3	−1	−2	+3	
3		0	+2		
Total	0	0	0	0	0

Note from Table 13.2 that the residuals sum to zero for each factor level. This illustrates that the residuals for analysis of variance model (13.2), or for the alternate model (13.6), are subject to the following r constraints:

$$\sum_i e_{ij} = 0 \qquad \text{for } j = 1, \ldots, r \tag{13.32}$$

The use of residuals for examining the aptness of an analysis of variance model will be discussed in Chapter 15.

13.6 ANALYSIS OF VARIANCE

Just as the analysis of variance for a regression model partitions the total sum of squares into the regression sum of squares and the error sum of squares, so a corresponding partitioning exists for the analysis of variance model (13.2).

Partitioning of *SSTO*

The total variability of the Y's, not using any information about factor levels, is measured in terms of the deviations of the observations Y_{ij} around the overall mean $\bar{Y}_{..}$:

$$Y_{ij} - \bar{Y}_{..} \tag{13.33}$$

The conventional measure of the total variability is the sum of the squares of these deviations, denoted as for regression by *SSTO* for *total sum of squares*:

$$SSTO = \sum_j \sum_i (Y_{ij} - \bar{Y}_{..})^2 \tag{13.34}$$

When we utilize information about the factor levels, the deviations reflecting the uncertainty remaining in the data are those of each observation Y_{ij} around its respective factor level mean $\overline{Y}_{.j}$:

(13.35) $$Y_{ij} - \overline{Y}_{.j}$$

The difference between the deviations (13.33) and (13.35) reflects the difference between the factor level mean and the overall mean:

(13.36) $$(Y_{ij} - \overline{Y}_{..}) - (Y_{ij} - \overline{Y}_{.j}) = \overline{Y}_{.j} - \overline{Y}_{..}$$

Note how we can now decompose the total deviation $Y_{ij} - \overline{Y}_{..}$ into two components:

(13.37) $$\underbrace{Y_{ij} - \overline{Y}_{..}}_{\substack{\text{Total}\\ \text{deviation}}} = \underbrace{\overline{Y}_{.j} - \overline{Y}_{..}}_{\substack{\text{Deviation of}\\ \text{factor level}\\ \text{mean around}\\ \text{overall mean}}} + \underbrace{Y_{ij} - \overline{Y}_{.j}}_{\substack{\text{Deviation}\\ \text{around}\\ \text{factor}\\ \text{level mean}}}$$

Thus, the total deviation $Y_{ij} - \overline{Y}_{..}$ can be viewed as the sum of two components:

1. The deviation of the factor level mean around the overall mean.
2. The deviation of Y_{ij} around its factor level mean.

Figure 13.3 illustrates this decomposition for the Kenton Food Company data.

When we sum and then square (13.37), the cross products on the right drop out and we obtain:

(13.38) $$\sum_j \sum_i (Y_{ij} - \overline{Y}_{..})^2 = \sum_j n_j(\overline{Y}_{.j} - \overline{Y}_{..})^2 + \sum_j \sum_i (Y_{ij} - \overline{Y}_{.j})^2$$

The term on the left is *SSTO*, as defined in (13.34). The first term on the right will be denoted *SSTR*, standing for *treatment sum of squares*:

(13.39a) $$SSTR = \sum_j n_j(\overline{Y}_{.j} - \overline{Y}_{..})^2$$

The second term on the right in (13.38) is denoted *SSE*, standing for *error sum of squares*:

(13.39b) $$SSE = \sum_j \sum_i (Y_{ij} - \overline{Y}_{.j})^2$$

Thus, (13.38) can be written equivalently:

(13.40) $$SSTO = SSTR + SSE$$

The correspondence to the regression decomposition in (3.47a) is readily apparent.

The total sum of squares for the analysis of variance model is therefore made up of these two components:

1. *SSE*: A measure of the random variation of the observations around the respective factor level means. The less variation among the observations for each factor level, the smaller is *SSE*. If $SSE = 0$, all observations for a factor level are the same, and this holds for all factor levels. The more the observations for a factor level differ among themselves, the larger *SSE* will be.

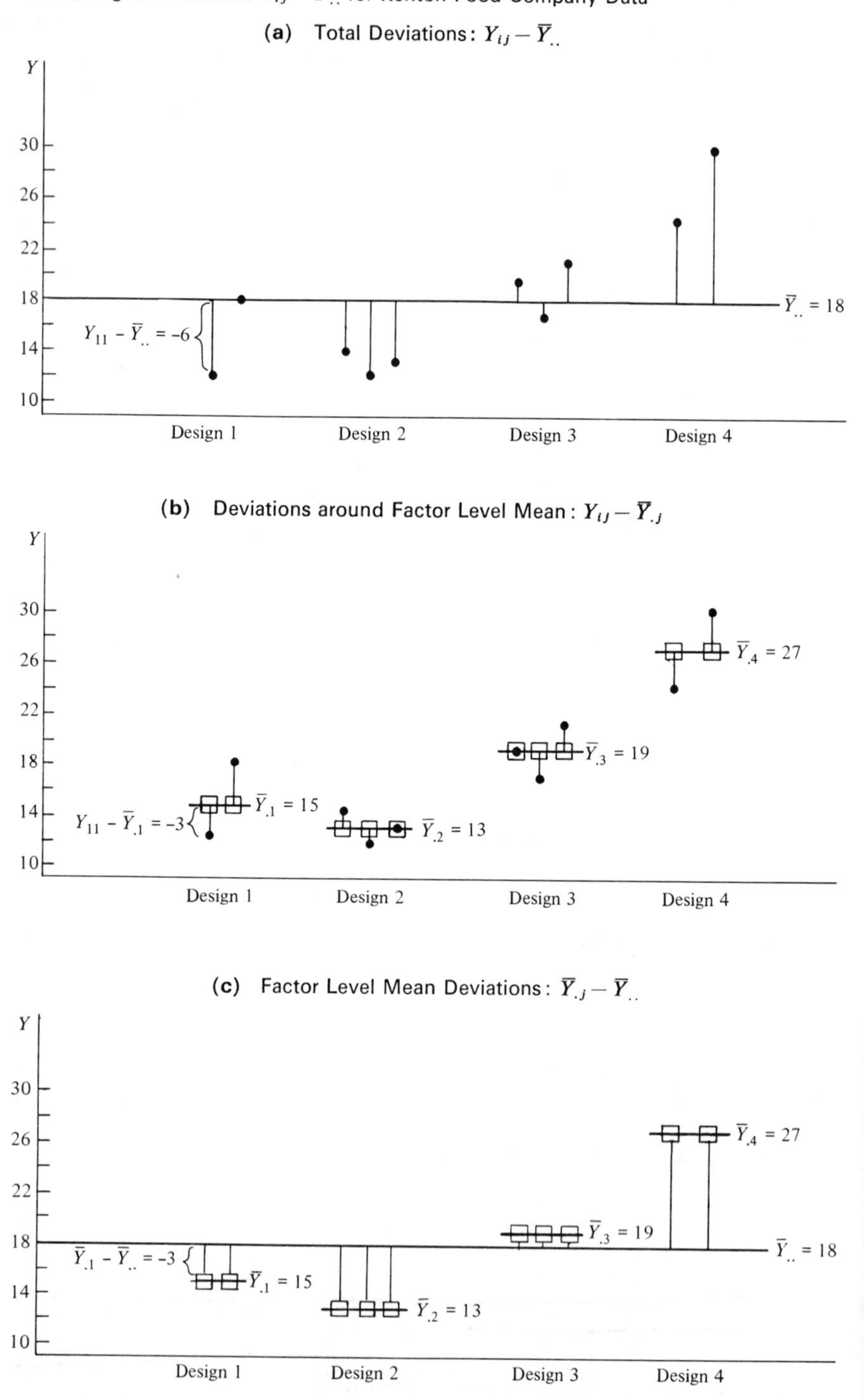

FIGURE 13.3

Partitioning of Deviations $Y_{ij} - \bar{Y}_{..}$ for Kenton Food Company Data

2. *SSTR*: A measure of the extent of differences between factor level means, based on the deviations of the factor level sample means $\bar{Y}_{.j}$ around the overall mean $\bar{Y}_{..}$. If all factor level sample means $\bar{Y}_{.j}$ are the same, $SSTR = 0$. The more the factor level means differ, the larger will $SSTR$ be.

Comments

1. To prove (13.38), we begin by considering (13.37):

$$(Y_{ij} - \bar{Y}_{..}) = (\bar{Y}_{.j} - \bar{Y}_{..}) + (Y_{ij} - \bar{Y}_{.j})$$

Squaring both sides we obtain:

$$(Y_{ij} - \bar{Y}_{..})^2 = (\bar{Y}_{.j} - \bar{Y}_{..})^2 + (Y_{ij} - \bar{Y}_{.j})^2 + 2(\bar{Y}_{.j} - \bar{Y}_{..})(Y_{ij} - \bar{Y}_{.j})$$

If we sum over all sample observations in the study (i.e., over both i and j), we obtain:

(13.41) $$\sum_j \sum_i (Y_{ij} - \bar{Y}_{..})^2 = \sum_j \sum_i (\bar{Y}_{.j} - \bar{Y}_{..})^2 + \sum_j \sum_i (Y_{ij} - \bar{Y}_{.j})^2 + \sum_j \sum_i 2(\bar{Y}_{.j} - \bar{Y}_{..})(Y_{ij} - \bar{Y}_{.j})$$

The first term on the right in (13.41) equals:

(13.42) $$\sum_j \sum_i (\bar{Y}_{.j} - \bar{Y}_{..})^2 = \sum_j n_j(\bar{Y}_{.j} - \bar{Y}_{..})^2$$

since $(\bar{Y}_{.j} - \bar{Y}_{..})^2$ is constant when summed over i; hence n_j such terms are picked up for the summation over i.

The third term on the right in (13.41) is zero:

(13.43) $$\sum_j \sum_i 2(\bar{Y}_{.j} - \bar{Y}_{..})(Y_{ij} - \bar{Y}_{.j}) = 2 \sum_j (\bar{Y}_{.j} - \bar{Y}_{..}) \sum_i (Y_{ij} - \bar{Y}_{.j}) = 0$$

This follows because $(\bar{Y}_{.j} - \bar{Y}_{..})$ is constant for the summation over i; hence it can be brought in front of the summation sign over i. Further, $\sum_i (Y_{ij} - \bar{Y}_{.j}) = 0$, since the sum of the deviations around the arithmetic mean is zero.

Thus, (13.41) reduces to (13.38).

2. The squared factor level mean deviations $(\bar{Y}_{.j} - \bar{Y}_{..})^2$ in $SSTR$ in (13.39a) are weighted by the number of observations n_j at that factor level. The reason is that for each of the observations at factor level j, the deviation component $\bar{Y}_{.j} - \bar{Y}_{..}$ is the same, as Figure 13.3c makes clear.

Computational Formulas. For hand computing, the definitional formulas for *SSTO*, *SSTR*, and *SSE* given earlier are usually not too convenient. Useful computational formulas for hand computing, which are algebraically identical to the definitional formulas, are:

(13.44a) $$SSTO = \sum_j \sum_i Y_{ij}^2 - \frac{Y_{..}^2}{n_T}$$

(13.44b) $$SSTR = \sum_j \frac{Y_{.j}^2}{n_j} - \frac{Y_{..}^2}{n_T}$$

(13.44c) $$SSE = \sum_j \sum_i Y_{ij}^2 - \sum_j \frac{Y_{.j}^2}{n_j}$$

Example. The analysis of variance breakdown of the total sum of squares for our Kenton Food Company data in Table 13.1a is obtained as follows, using the computational formulas in (13.44):

$$SSTO = (12)^2 + (18)^2 + (14)^2 + \cdots + (30)^2 - \frac{(180)^2}{10}$$

$$= 3{,}544 - 3{,}240$$

$$= 304$$

$$SSTR = \frac{(30)^2}{2} + \frac{(39)^2}{3} + \frac{(57)^2}{3} + \frac{(54)^2}{2} - \frac{(180)^2}{10}$$

$$= 3{,}498 - 3{,}240$$

$$= 258$$

$$SSE = 3{,}544 - 3{,}498$$

$$= 46$$

Thus, the decomposition of $SSTO$ for our example is:

$$304 = 258 + 46$$

$$SSTO = SSTR + SSE$$

Note that much of the total variation in the observations is associated with variation in the factor level means.

Breakdown of Degrees of Freedom

Corresponding to the decomposition of the total sum of squares, we can also obtain a breakdown of the associated degrees of freedom.

$SSTO$ has $n_T - 1$ degrees of freedom associated with it. There are altogether n_T observations, but there is one constraint on the deviations $(Y_{ij} - \bar{Y}_{..})$, namely, $\sum \sum (Y_{ij} - \bar{Y}_{..}) = 0$.

$SSTR$ has $r - 1$ degrees of freedom associated with it. There are r factor level means $\bar{Y}_{.j}$, but there is one constraint on the deviations $(\bar{Y}_{.j} - \bar{Y}_{..})$, namely $\sum n_j(\bar{Y}_{.j} - \bar{Y}_{..}) = 0$.

SSE has $n_T - r$ degrees of freedom associated with it. This can be readily seen by considering the component of SSE for the jth factor level:

$$\sum_{i=1}^{n_j} (Y_{ij} - \bar{Y}_{.j})^2 \tag{13.45}$$

The expression in (13.45) is the equivalent of a total sum of squares considering only the jth factor level. Hence, there are $n_j - 1$ degrees of freedom associated with this sum of squares. Since SSE is a sum of squares consisting of component sums of squares such as the one in (13.45), the degrees of freedom associated with SSE is the sum of the component degrees of freedom:

$$(n_1 - 1) + (n_2 - 1) + \cdots + (n_r - 1) = n_T - r \tag{13.46}$$

Example. For the Kenton Food Company illustration, for which $n_T = 10$ and $r = 4$, the degrees of freedom associated with the three sums of squares are as follows:

SS	*df*
SSTO	$10 - 1 = 9$
SSTR	$4 - 1 = 3$
SSE	$10 - 4 = 6$

Note that degrees of freedom, like sums of squares, are additive:

$$9 = 3 + 6$$

Mean Squares

The mean squares, as usual, are obtained by dividing each sum of squares by its associated degrees of freedom. We therefore have:

$$MSTR = \frac{SSTR}{r - 1} \tag{13.47a}$$

$$MSE = \frac{SSE}{n_T - r} \tag{13.47b}$$

Here, $MSTR$ stands for *treatment mean square* and MSE, as before, stands for *error mean square*.

Example. For our Kenton Food Company example, we obtain from earlier results:

$$MSTR = \frac{258}{3} = 86$$

$$MSE = \frac{46}{6} = 7.67$$

Note that the two mean squares do not add to $SSTO/(n_T - 1) = 304/9 = 33.8$. Thus, mean squares here, as in regression, are not additive.

Analysis of Variance Table

The breakdowns of the total sum of squares and degrees of freedom, together with the resulting mean squares, are presented in an ANOVA table, such as Table 13.3. The ANOVA table for our Kenton Food Company example is presented in Table 13.4.

TABLE 13.3

ANOVA Table for Single-Factor Study

Source of Variation	*SS*	*df*	*MS*	*E(MS)*
Between treatments	$SSTR = \sum n_j(\overline{Y}_{.j} - \overline{Y}_{..})^2$	$r - 1$	$MSTR = \frac{SSTR}{r-1}$	$\sigma^2 + \frac{1}{r-1} \sum n_j(\mu_j - \mu_.)^2$
Error (within treatments)	$SSE = \sum \sum (Y_{ij} - \overline{Y}_{.j})^2$	$n_T - r$	$MSE = \frac{SSE}{n_T - r}$	σ^2
Total	$SSTO = \sum \sum (Y_{ij} - \overline{Y}_{..})^2$	$n_T - 1$		

TABLE 13.4

ANOVA Table for Package Design Study

Source of Variation	*SS*	*df*	*MS*
Between designs	258	3	86
Error	46	6	7.67
Total	304	9	

Expected Mean Squares

The expected values of MSE and $MSTR$ can be shown to be as follows:

(13.48a) $$E(MSE) = \sigma^2$$

(13.48b) $$E(MSTR) = \sigma^2 + \frac{\sum n_j(\mu_j - \mu_.)^2}{r - 1}$$

or:

(13.48c) $$E(MSTR) = \sigma^2 + \frac{\sum n_j \tau_j^2}{r - 1}$$

where $\mu_.$ is defined as in (13.13). These expected values are shown in the $E(MS)$ column of Table 13.3.

Two important features of the results in (13.48) deserve attention:

1. MSE is an unbiased estimator of the variance of the error terms ε_{ij}, whether or not the factor level means μ_j are equal. This is intuitively reasonable since the variability of the observations within each factor level is not affected by the magnitudes of the factor level means.

2. If all factor level means μ_j are equal and hence equal to $\mu_.$ (or equivalently if all factor level effects τ_j are zero), $E(MSTR) = \sigma^2$ since the second term on the right in (13.48b) is then zero. Hence, $MSTR$ and MSE both estimate the error variance σ^2 if all factor level means μ_j are equal. If, however, the factor level means are not equal, $MSTR$ tends, on the average, to be larger than MSE, since the second term in (13.48b) will then be positive. This is intuitively reasonable, as shown in Figure 13.4. The situation portrayed there assumes that $n_j \equiv n$. If all μ_j are equal, then all $\bar{Y}_{.j}$ follow the same sampling distribution, with mean $\mu_.$ and variance σ^2/n; this is portrayed in Figure 13.4a. If the μ_j are not equal, on the other hand, the $\bar{Y}_{.j}$ follow different sampling distributions, each with the same variability σ^2/n but centered on different means μ_j. One such possibility is shown in Figure 13.4b. Hence, the $\bar{Y}_{.j}$ will tend to differ more from each other if the μ_j differ than if the μ_j are equal, and consequently $MSTR$ will tend to be larger when the factor level means are not the same than when they are equal. This property of $MSTR$ is utilized in constructing the statistical test discussed in the next section to determine

FIGURE 13.4

Sampling Distributions of $\overline{Y}_{.j}$

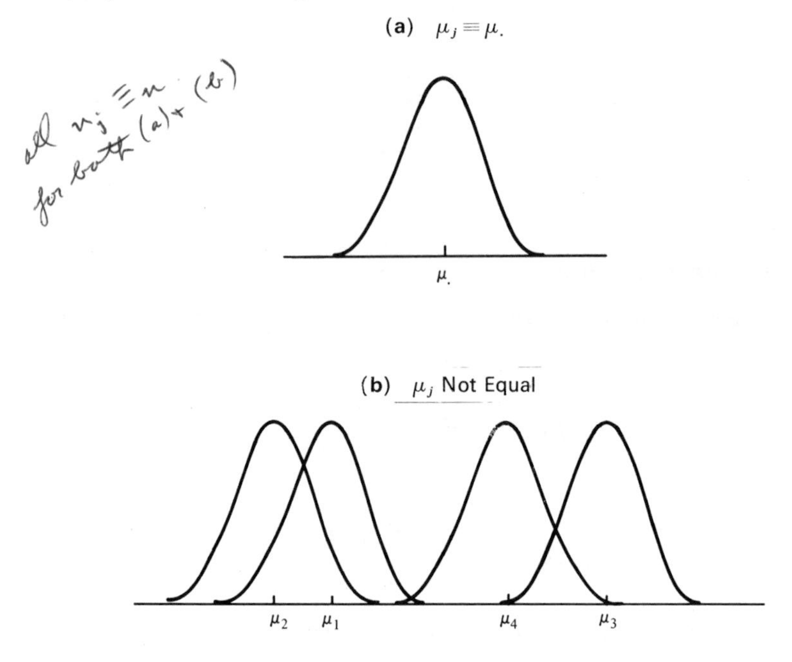

whether or not the factor level means μ_j are the same. If $MSTR$ and MSE are of the same order of magnitude, this is taken to suggest that the factor level means μ_j are equal. If $MSTR$ is substantially larger than MSE, this is taken to suggest that the μ_j are not equal.

Derivation of $E(MSE)$**.** To find the expected value of MSE, we first note that MSE can be expressed as follows:

$$MSE = \frac{1}{n_T - r} \sum_j \sum_i (Y_{ij} - \overline{Y}_{.j})^2 \tag{13.49}$$

$$= \frac{1}{n_T - r} \sum_j \left[(n_j - 1) \frac{\sum_i (Y_{ij} - \overline{Y}_{.j})^2}{n_j - 1} \right]$$

Now let us denote the usual sample variance of the observations for the jth factor level as s_j^2:

$$s_j^2 = \frac{\sum_i (Y_{ij} - \overline{Y}_{.j})^2}{n_j - 1} \tag{13.50}$$

Hence, (13.49) can be expressed as follows:

$$(13.51) \qquad MSE = \frac{1}{n_T - r} \sum_j (n_j - 1)s_j^2$$

Since it is well known that the sample variance (13.50) is an unbiased estimator of the population variance, which in our case is σ^2 for all factor levels, we obtain:

$$E(MSE) = \frac{1}{n_T - r} \sum_j (n_j - 1)E(s_j^2)$$

$$= \frac{1}{n_T - r} \sum_j (n_j - 1)\sigma^2$$

or:

$$E(MSE) = \sigma^2$$

Derivation of $E(MSTR)$**.** To simplify the proof, we shall assume that all sample sizes n_j are the same, namely $n_j \equiv n$. The general result in (13.48b) becomes for this special case:

$$(13.52) \qquad E(MSTR) = \sigma^2 + \frac{n \sum (\mu_j - \mu_.)^2}{r - 1} \qquad \text{when } n_j \equiv n$$

Further, when all factor level sample sizes are n, $MSTR$ as defined in (13.39a) and (13.47a) becomes:

$$(13.53) \qquad MSTR = \frac{n \sum (\bar{Y}_{.j} - \bar{Y}_{..})^2}{r - 1} \qquad \text{when } n_j \equiv n$$

To derive (13.52), consider the model formulation for Y_{ij} in (13.2):

$$Y_{ij} = \mu_j + \varepsilon_{ij}$$

Averaging the Y_{ij} for the jth factor level, we obtain:

$$(13.54) \qquad \bar{Y}_{.j} = \mu_j + \bar{\varepsilon}_{.j}$$

where $\bar{\varepsilon}_{.j}$ is the average of the ε_{ij} for the jth factor level:

$$(13.55) \qquad \bar{\varepsilon}_{.j} = \frac{\sum_i \varepsilon_{ij}}{n}$$

Averaging the Y_{ij} over all factor levels, we obtain:

$$(13.56) \qquad \bar{Y}_{..} = \mu_. + \bar{\varepsilon}_{..}$$

where $\mu_.$ is as defined in (13.9) and $\bar{\varepsilon}_{..}$ is the average of all ε_{ij}:

$$(13.57) \qquad \bar{\varepsilon}_{..} = \frac{\sum \sum \varepsilon_{ij}}{n_T}$$

Since the sample sizes are equal, we also have:

(13.58) $$\bar{Y}_{..} = \frac{\sum \bar{Y}_{.j}}{r} \qquad \bar{\varepsilon}_{..} = \frac{\sum \bar{\varepsilon}_{.j}}{r}$$

Using (13.54) and (13.56), we obtain:

(13.59) $$\bar{Y}_{.j} - \bar{Y}_{..} = (\mu_j + \bar{\varepsilon}_{.j}) - (\mu_. + \bar{\varepsilon}_{..}) = (\mu_j - \mu_.) + (\bar{\varepsilon}_{.j} - \bar{\varepsilon}_{..})$$

If we square $(\bar{Y}_{.j} - \bar{Y}_{..})$ and sum over the factor levels, we therefore obtain:

(13.60) $$\sum (\bar{Y}_{.j} - \bar{Y}_{..})^2 = \sum (\mu_j - \mu_.)^2 + \sum (\bar{\varepsilon}_{.j} - \bar{\varepsilon}_{..})^2 + 2 \sum (\mu_j - \mu_.)(\bar{\varepsilon}_{.j} - \bar{\varepsilon}_{..})$$

We now wish to find $E[\sum (\bar{Y}_{.j} - \bar{Y}_{..})^2]$, and therefore need to find the expected value of each term on the right in (13.60):

1. Since $\sum (\mu_j - \mu_.)^2$ is a constant, its expectation is:

(13.61) $$E[\sum (\mu_j - \mu_.)^2] = \sum (\mu_j - \mu_.)^2$$

2. Before finding the expectation of the second term on the right, consider first the expression:

$$\frac{\sum (\bar{\varepsilon}_{.j} - \bar{\varepsilon}_{..})^2}{r - 1}$$

This is an ordinary sample variance, since $\bar{\varepsilon}_{..}$ is the sample mean of the r terms $\bar{\varepsilon}_{.j}$ per (13.58). We further know that the sample variance is an unbiased estimator of the variance of the variable, in this case $\bar{\varepsilon}_{.j}$. But $\bar{\varepsilon}_{.j}$ is just the mean of n independent error terms ε_{ij} by (13.55). Hence:

$$\sigma^2(\bar{\varepsilon}_{.j}) = \frac{\sigma^2(\varepsilon_{ij})}{n} = \frac{\sigma^2}{n}$$

Therefore:

$$E\left[\frac{\sum (\bar{\varepsilon}_{.j} - \bar{\varepsilon}_{..})^2}{r - 1}\right] = \frac{\sigma^2}{n}$$

so that:

(13.62) $$E[\sum (\bar{\varepsilon}_{.j} - \bar{\varepsilon}_{..})^2] = \frac{(r-1)\sigma^2}{n}$$

3. Since both $\bar{\varepsilon}_{.j}$ and $\bar{\varepsilon}_{..}$ are means of ε_{ij} terms, all of which have expectation 0, it follows that:

$$E(\bar{\varepsilon}_{.j}) = 0 \qquad E(\bar{\varepsilon}_{..}) = 0$$

Hence:

(13.63) $$E[2 \sum (\mu_j - \mu_.)(\bar{\varepsilon}_{.j} - \bar{\varepsilon}_{..})] = 2 \sum (\mu_j - \mu_.)E(\bar{\varepsilon}_{.j} - \bar{\varepsilon}_{..}) = 0$$

We have thus shown, by (13.61), (13.62), and (13.63), that:

$$E[\sum(\bar{Y}_{.j} - \bar{Y}_{..})^2] = \sum(\mu_j - \mu_.)^2 + \frac{(r-1)\sigma^2}{n}$$

But then (13.52) follows at once:

$$E(MSTR) = E\left[\frac{n\sum(\bar{Y}_{.j} - \bar{Y}_{..})^2}{r-1}\right] = \frac{n}{r-1}\left[\sum(\mu_j - \mu_.)^2 + \frac{(r-1)\sigma^2}{n}\right]$$

$$= \sigma^2 + \frac{n\sum(\mu_j - \mu_.)^2}{r-1}$$

13.7 *F* TEST FOR EQUALITY OF FACTOR LEVEL MEANS

It is customary to begin the analysis of a single-factor study by determining whether or not the factor level means μ_j are equal. If, for instance, the four package designs in our Kenton Food Company example lead to the same mean sales volumes, there is no need for further analysis, such as to determine which design is best or how two particular designs compare in stimulating sales.

Thus, the alternative conclusions we wish to consider are:

(13.64)
$$C_1: \mu_1 = \mu_2 = \cdots = \mu_r$$
$$C_2: \text{not all } \mu_j \text{ are equal}$$

Alternatively, we could express these conclusions as follows:

(13.64a)
$$C_1: \tau_1 = \tau_2 = \cdots = \tau_r = 0$$
$$C_2: \text{not all } \tau_j = 0$$

Test Statistic

The test statistic to be used for choosing between the alternatives in (13.64) is:

(13.65)
$$F^* = \frac{MSTR}{MSE}$$

Note that *MSTR* here plays the role corresponding to *MSR* for a regression model.

Large values of F^* support C_2, since *MSTR* will tend to exceed *MSE* when C_2 holds, as we saw from (13.48). Values of F^* near 1 support C_1, since both *MSTR* and *MSE* have the same expected value when C_1 holds. Hence, the appropriate test is an upper-tail one.

Distribution of F^*

When all treatment means μ_j are equal, each observation Y_{ij} has the same expected value, namely $E(Y_{ij}) \equiv \mu_.$. In view of the additivity of sums of squares and degrees of freedom, Cochran's theorem (3.59) then implies:

$$\text{When } C_1 \text{ holds, } \frac{SSE}{\sigma^2} \text{ and } \frac{SSTR}{\sigma^2} \text{ are independent } \chi^2 \text{ variables}$$

It follows in the same fashion as for regression:

$$\text{When } C_1 \text{ holds, } F^* \text{ is distributed as } F(r-1, n_T - r)$$

If C_2 holds, that is, if the μ_j are not all equal, F^* does *not* follow the F distribution. Rather, it follows a complex distribution called the *noncentral F distribution.* We shall make use of the noncentral F distribution when we discuss the power of the F test.

Note

$SSTR$ and SSE are independent even if all μ_j are not equal. Intuitively, this can be seen because SSE is the variability within the factor level samples, and this within-sample variability is not affected by the magnitudes of the factor level means. $SSTR$, on the other hand, is solely based on the factor level means $\bar{Y}_{.j}$.

Construction of Decision Rule

Usually, the risk of making a Type I error is controlled in constructing the decision rule. This provides protection against making further, more detailed, analyses of the factor effects when in fact there are no differences in the factor level means. The Type II error can also be controlled, as we shall see later, through sample size determination.

Since we know that F^* is distributed as $F(r-1, n_T - r)$ if C_1 holds, and that large values of F^* lead to conclusion C_2, the appropriate decision rule to control the level of significance at α is:

$$\begin{aligned} &\text{If } F^* \leq F(1-\alpha; r-1, n_T - r), \text{ conclude } C_1 \\ &\text{If } F^* > F(1-\alpha; r-1, n_T - r), \text{ conclude } C_2 \end{aligned} \tag{13.66}$$

where $F(1-\alpha; r-1, n_T - r)$ is the $(1-\alpha)100$ percentile of the appropriate F distribution.

Example

For our Kenton Food Company illustration, we wish to test whether or not mean sales are the same for the four package designs:

$$\begin{aligned} &C_1: \mu_1 = \mu_2 = \mu_3 = \mu_4 \\ &C_2: \text{not all } \mu_j \text{ are equal} \end{aligned}$$

Suppose that management wishes to control the risk of making a Type I error at $\alpha = .05$. We therefore require $F(.95; 3, 6)$, where the degrees of freedom are those shown in Table 13.4. From Table A–4 in the Appendix, we find $F(.95; 3, 6) = 4.76$. Hence the decision rule is:

$$\text{If } F^* \leq 4.76, \text{ conclude } C_1$$
$$\text{If } F^* > 4.76, \text{ conclude } C_2$$

The test statistic, using the data from Table 13.4, is:

$$F^* = \frac{MSTR}{MSE} = \frac{86}{7.67} = 11.2$$

Since $F^* = 11.2 > 4.76$, we conclude C_2, that the factor level means μ_j are not equal, or that the four different package designs do not lead to the same mean sales volume. Thus we conclude that there is a relation between package design and sales volume.

This conclusion did not surprise the sales manager of the Kenton Food Company. He conducted the study in the first place because he expected the four package designs to have different effects on sales volume and was interested in finding out the nature of these differences. In the next chapter we discuss the second stage of the analysis, namely how to study the nature of the factor level effects when differences exist.

General Linear Test Approach

The F test just explained is a test of a linear statistical model. Hence it can be obtained by the general method explained in Chapter 3:

1. The full model is model (13.2):

(13.67) $$Y_{ij} = \mu_j + \varepsilon_{ij} \qquad \text{Full model}$$

Fitting the full model leads to the least squares estimators $\hat{\mu}_j = \bar{Y}_{.j}$ per (13.22), and the resulting error sum of squares:

$$SSE(F) = \sum \sum (Y_{ij} - \bar{Y}_{.j})^2$$

2. The reduced model under C_1 is:

(13.68) $$Y_{ij} = \mu_. + \varepsilon_{ij} \qquad \text{Reduced model}$$

where $\mu_.$ is the common mean for all factor levels. Fitting the reduced model leads to the least squares estimator $\hat{\mu}_. = \bar{Y}_{..}$ and the resulting error sum of squares:

$$SSE(R) = \sum \sum (Y_{ij} - \bar{Y}_{..})^2$$

3. The general test statistic (3.66) then becomes:

$$F^* = \frac{SSE(R) - SSE(F)}{(n_T - 1) - (n_T - r)} \div \frac{SSE(F)}{n_T - r}$$

Note that $SSE(R)$ has $n_T - 1$ degrees of freedom associated with it because one parameter ($\mu_.$) had to be estimated, and $SSE(F)$ has $n_T - r$ degrees of freedom because r parameters (the r μ_j's) had to be estimated.

Since we have, according to (13.34) and (13.39b) respectively:

$$SSE(R) = SSTO$$
$$SSE(F) = SSE$$

and since by (13.40) $SSTO - SSE = SSTR$, we obtain the test statistic (13.65):

$$F^* = \frac{SSTR}{r-1} \div \frac{SSE}{n_T - r} = \frac{MSTR}{MSE}$$

Note

If there are only two factor levels so that $r = 2$, it can easily be shown that the test statistic F^* in (13.65) is the equivalent of the two-population two-sided t test in Table 1.2a. The F test here has $(1, n_T - 2)$ degrees of freedom and the t test has $n_1 + n_2 - 2$ or $n_T - 2$ degrees of freedom; thus both tests lead to equivalent critical regions. For comparing two population means, the t test generally is to be preferred since it can be used both to conduct two-sided and one-sided tests (see Table 1.2); the F test can be used only for a two-sided test.

13.8 COMPUTER INPUT AND OUTPUT

For analysis of variance as for regression analysis, packaged computer programs are readily available for handling calculations that would have been highly tedious or simply not feasible in earlier days. We will assume, as in our regression analysis discussions, that computers are used in handling analysis of variance calculations for all but the simplest data sets.

For regression analysis, a single "multiple regression" packaged program frequently suffices for a variety of applications such as simple linear regression, polynomial regression, and so on. Packaged analysis of variance programs, on the other hand, are usually keyed to specific designs. Thus a library may contain a specific program for single-factor analysis of variance and other specific programs for additional, widely used designs that we will discuss in later chapters. Formats for input of observations and output of results often differ somewhat among the different analysis of variance programs in a library, and may differ substantially from one library to another. We will illustrate typical formats as we take up various designs, beginning now with formats for single-factor analysis of variance.

Input

One form of input for the Kenton Food Company data in Table 13.1 requires two columns:

Treatment	*Observation*
1	12
1	18
2	14
2	12
2	13
.	.
.	.
.	.
4	30

Here the treatment identifications are entered in one column and the observations are entered in another.

In a different input format, the observations are entered in rows. An example would be, for the Kenton data:

(2)	12	18	
(3)	14	12	13
(3)	19	17	21
(2)	24	30	

The numbers in parentheses are the sample sizes for the respective treatments. In particular formats these numbers may be in some other location than is shown here. The sample size numbers are redundant in our Kenton Food Company illustration, but they are required in the format because in many cases the observations for a treatment will run beyond a single row.

Still another input format for the Kenton data is shown at the top of the printout in Figure 13.5. Here the observations are entered in columns, with each column containing the observations for a given treatment. The program calls for dummy zeros to be entered where there are no "legitimate" observations, but these entries do not affect the results—for instance, they do not enter into the calculated treatment means $\bar{Y}_{.j}$. If any legitimate zero outcomes occur in the sample observations Y_{ij}, *all* the observed Y_{ij} are "coded" or transformed by adding a suitable constant c such that $Y'_{ij} = Y_{ij} + c \neq 0$ for all observed Y_{ij}, where Y'_{ij} is the transformed observation. This coding does not affect sums of squares, degrees of freedom, and mean squares, and hence the ANOVA table is the same as for the original data. The coding does change the factor level means by a constant amount, so that the least squares estimator of μ_j is $\bar{Y}'_{.j} - c$.

Output

Figure 13.5 illustrates, for the Kenton Food Company example, a rectangular output format that corresponds to the simple linear regression format presented in Figure 2.12. The format statement explains the items in the output matrix. (We have added some annotations to facilitate understanding of

FIGURE 13.5

Example of Computer Printout for Kenton Food Company Case

```
ONE-WAY ANOVA


DATA

12  14  19  24
18  12  17  30
 0  13  21   0


DATA IS AN M×N MATRIX, WHERE M IS THE NUMBER OF OBSERVATIONS
FOR THE TREATMENT WITH THE MAXIMUM NUMBER OF OBSERVATIONS
AND N IS THE NUMBER OF TREATMENTS. THE MATRIX BELOW IS
AN (N+3)×4 MATRIX MADE UP AS FOLLOWS:

ROW I, I=1,...N:
          I, NO. OF OBSERVATIONS FOR TREATMENT I,
          MEAN OF TREATMENT I, 0
ROW N+1:  DF, SS, MS, AND F-RATIO FOR TREATMENTS
ROW N+2:  DF, SS, AND MS FOR ERROR, 0
ROW N+3:  DF, AND SS FOR TOTAL, SQUARE ROOT OF ERROR MS, 0
```

	j	n_j	$\bar{Y}_{.j}$	
	1	2	15	0
	2	3	13	0
	3	3	19	0
	4	2	27	0
Source	**df**	**SS**	**MS**	
Treatments	3	258	86	F^* → 11.2173913
Error	6	46	7.666666667	0
Total	9	304	$\sqrt{MSE}$ → 2.768874621	0

the output.) Note that the first four rows give the sample sizes and treatment sample means shown earlier in Table 13.1. The remainder of the output matrix gives the analysis of variance results shown in Table 13.4, together with the square root of MSE and the test statistic for the F test for equality of factor level means.

The results in Figure 13.5 coincide with those obtained by hand calculations in Tables 13.1 and 13.4. This is because roundoff errors are no problem in calculations involving the simple data set of our Kenton example. With more complex data sets, roundoff effects can arise, so that somewhat differing results may be obtained with alternate procedures or programs.

We noted, in discussing roundoff errors in regression analysis, that least squares results tend to be highly sensitive to roundoff effects and that different packaged regression programs do not control such effects equally well.

Similarly, different analysis of variance packages differ in quality, and it is a good idea to check a library program before using it for the first time. One procedure is to test the program on a complex data set for which accurate results are already known.

13.9 POWER OF F TEST

The power of the F test is of importance for assessing the discriminating ability of the decision rule employed, and also for determining needed sample sizes. By the power of the F test, we refer to the probability that the decision rule will lead to conclusion C_2 when in fact C_2 holds. Specifically, the power is given by the following expression:

$$\text{Power} = P\{F^* > F(1-\alpha; r-1, n_T - r) \,|\, \phi\} \tag{13.69}$$

where ϕ is the *noncentrality parameter*, that is, a measure of how unequal the μ_j are:

$$\phi = \frac{1}{\sigma}\sqrt{\frac{\sum n_j(\mu_j - \mu_.)^2}{r}} = \frac{1}{\sigma}\sqrt{\frac{\sum n_j \tau_j^2}{r}} \tag{13.70}$$

where $\mu_.$ is as defined in (13.13). When all factor level samples are of equal size n, the parameter ϕ becomes:

$$\phi = \frac{1}{\sigma}\sqrt{\frac{n}{r}\sum (\mu_j - \mu_.)^2} = \frac{1}{\sigma}\sqrt{\frac{n}{r}\sum \tau_j^2} \qquad \text{when } n_j \equiv n \tag{13.70a}$$

To determine power probabilities, we need to utilize the noncentral F distribution since this is the sampling distribution of F^* when C_2 holds. The resulting calculations are quite complex. Fortunately, charts have been prepared which make the determination of power probabilities quite simple. Table A–8 contains the Pearson-Hartley charts of the power of the F test. The proper curve to utilize depends on the number of factor levels, the sample sizes, and the level of significance employed in the decision rule. The Pearson-Hartley charts are used as follows:

1. Each page refers to a different ν_1, the number of degrees of freedom for the numerator of F^*. For our situation, $\nu_1 = r - 1$, or the number of factor levels minus one. Table A–8 contains charts for $\nu_1 = 2, 3, 4, 5$, and 6 as shown in the upper left-hand corner of each chart.
2. Two levels of significance, denoted by α, are used in the charts, namely $\alpha = .05$ and $\alpha = .01$. There are two X scales, depending on which level of significance is employed. Further, the left set of curves on each chart refers to $\alpha = .05$, the right set to $\alpha = .01$.
3. There are separate curves for different values of ν_2, the degrees of freedom for the denominator of F^*. In our case, $\nu_2 = n_T - r$. The curves are indexed according to the value of ν_2 at the top of the chart. Since only selected values of ν_2 are used in the charts, one needs to interpolate for intermediate values of ν_2.

4. The X scale represents ϕ, the noncentrality parameter defined in (13.70).

5. Finally, the Y scale gives the power $1 - \beta$, where β is the risk of making a Type II error.

Examples

1. Consider the case where $\nu_1 = 2$, $\nu_2 = 10$, $\phi = 3$, and $\alpha = .05$. We then find from page 819 that the power is $1 - \beta = .983$ approximately.

2. Suppose that for our Kenton Food Company example, the analyst wishes to know the power of the decision rule on page 449 when there are substantial differences between the factor level means. Specifically, he wishes to consider the case when $\tau_1 = -3.5$, $\tau_2 = -3$, $\tau_3 = +2$, $\tau_4 = +5$. Recall that, say, $\tau_4 = 5$ means that μ_4 is 5 cases greater than $\mu_.$, the average for all four package designs. Thus, the specified value of ϕ is:

$$\phi = \frac{1}{\sigma}\left[\frac{2(-3.5)^2 + 3(-3)^2 + 3(2)^2 + 2(5)^2}{4}\right]^{1/2}$$

$$= \frac{1}{\sigma}(5.33)$$

Note that we still need to know σ, the variability of the error terms ε_{ij} in the model. Suppose that from past experience it is known that $\sigma = 2.5$ cases approximately. Then we have:

$$\phi = \frac{1}{2.5}(5.33) = 2.13$$

Further, we have for this example:

$$\nu_1 = r - 1 = 3$$

$$\nu_2 = n_T - r = 6$$

$$\alpha = .05$$

Hence, from the chart on page 820 we find that the power is $1 - \beta = .72$ approximately. In other words, there are about 72 chances in 100 that the decision rule will lead to the detection of differences in the mean sales volumes for the four package designs when the differences are the ones specified earlier.

Comments

1. If the decision rule in our package design example had been set up with a level of significance $\alpha = .01$, the power would have been $1 - \beta = .36$ approximately. Thus, the smaller the specified α risk, the smaller is the power for any given ϕ, and hence the larger the risk of a Type II error.

2. Any given value of ϕ encompasses many different combinations of factor level means μ_j or factor level effects τ_j. Thus, in our example, $\tau_1 = -3.5$, $\tau_2 = -3$, $\tau_3 = +2$, $\tau_4 = +5$ and $\tau_1 = +5$, $\tau_2 = +2$, $\tau_3 = -3$, $\tau_4 = -3.5$ lead to the same value of $\phi = 2.13$ and hence to the same power.

3. The effect of misspecifying σ in determining the power unfortunately can be great. In our Kenton Food Company example, if $\sigma = 3.0$ instead of 2.5, ϕ would be 1.78 and the power (for $\alpha = .05$) would be .56 instead of .72.

4. The larger is ϕ, that is, the larger are the differences between the factor level means, the higher is the power and hence the smaller is the probability of making a Type II error, for a given risk α of making a Type I error.

5. Since many single-factor studies are undertaken because of the expectation that the factor level means differ and it is desired to investigate these differences, the α risk used in constructing the decision rule for determining whether or not the factor level means are equal is often set relatively high (e.g., .05 or .10 instead of .01) so as to increase the power of the test.

6. The Pearson-Hartley chart for $\nu_1 = 1$ is not reproduced in Table A–8 since this case corresponds to the comparison of two population means. As was noted earlier, the F test is the equivalent of the two-sided t test for this case, and the power charts for the two-sided t test presented in Table A–5 can then be used, with non-centrality parameter:

$$\delta = \frac{|\mu_1 - \mu_2|}{\sigma\sqrt{\frac{1}{n_1} + \frac{1}{n_2}}} \tag{13.71}$$

and degrees of freedom $n_1 + n_2 - 2$.

PROBLEMS

13.1. A company wishes to study job satisfaction of employees according to length of service. It plans to classify employees into 3 length of service groups (less than 2 years, 2–5 years, more than 5 years), and select 24 employees at random from each group for intensive interviewing. Is length of service here a classification or an experimental factor? What are the implications of your answer?

13.2. Refer to Problem 13.1. Suppose $\mu_1 = 60$, $\mu_2 = 75$, $\mu_3 = 90$, $\sigma = 3$, and that model (13.2) is applicable. Draw a representation of this model for the present example.

13.3. Derive the restriction in (13.14) when the overall constant $\mu_.$ is defined according to (13.13).

13.4. (Calculus needed.) Show that the least squares estimators in (13.22) are also maximum likelihood estimators for model (13.2).

13.5. Show that the result in (13.44b) is the algebraic equivalent of (13.39a).

13.6. Refer to Problem 13.1. Suppose that the parameters are as given in Problem 13.2. Find $E(MSTR)$ and $E(MSE)$ for this case. What is the implication of $E(MSTR)$ being larger than $E(MSE)$ here?

13.7. Show that the test statistic F^* in (13.65) is the equivalent of the two-population two-sided t test in Table 1.2a.

13.8. A student calculated the sums of squares for a single-factor analysis of variance and obtained the following results:

$$SSTR = 1{,}214.2$$
$$SSE = -10.8$$
$$SSTO = 1{,}203.4$$

Can these results be correct? Discuss.

13.9. Refer to Problem 13.1. The study was conducted as planned, and the results below were obtained from a computer run. The job satisfaction scores are indexes ranging from 0 (highly dissatisfied) to 100 (highly satisfied).

$$n_1 = 24 \qquad n_2 = 24 \qquad n_3 = 24$$
$$\bar{Y}_{.1} = 62.4 \qquad \bar{Y}_{.2} = 78.3 \qquad \bar{Y}_{.3} = 84.3$$

$$SSTR = 6{,}147.36$$
$$SSE = 1{,}724.64$$
$$SSTO = 7{,}872.00$$

a) Set up the appropriate analysis of variance table.
b) Test whether or not the mean job satisfaction scores for the three groups are equal. Use a level of significance of .01.
c) What should be the next step in the analysis?

13.10. A breakfast cereal manufacturer used a different agent for handling premium distributions for each of its five products. It was desired to study the timeliness with which the premiums are distributed, and at random 20 transactions for each agent were selected. The delay (in days) for handling the transaction was determined for each sample case, and the results were as follows:

	Agent				
Transaction	C	F	H	L	X
1	13	23	28	20	10
2	18	29	31	18	8
3	12	24	33	23	12
4	10	20	27	24	14
5	17	21	30	29	11
6	15	20	31	22	12
7	13	25	28	22	18
8	17	27	29	28	13
9	16	28	34	26	7
10	16	27	21	17	12
11	14	27	36	21	11
12	17	23	29	20	14
13	14	21	31	20	11
14	13	28	32	18	14
15	12	23	34	24	16
16	11	24	31	25	8
17	18	21	33	21	7
18	19	24	34	17	10
19	16	24	36	20	12
20	15	22	33	24	11

a) Set up the appropriate analysis of variance table.
b) Test whether or not the mean delay time varies for the five agents. Use a level of significance of .01.
c) What should be the next step in the analysis?
d) Is "agent" a classification or experimental factor here? What is the implication of this?

13.11. A consumer organization wished to study whether or not the cash offer for a used car varies with the characteristics of the owner of the car. It selected a one-year-old medium-price car for which cash offers were solicited from 24 used-car dealers. Four "owners" were used in the study (denoted A, B, C, D) and each was sent to six dealers selected at random from the 24 in the study. The cash offers (in hundreds of dollars) were as follows:

	Owner			
Dealer	A	B	C	D
1	25	21	26	20
2	24	25	29	23
3	26	23	27	24
4	27	22	28	26
5	23	20	29	21
6	28	22	28	23

a) Set up the appropriate analysis of variance table.
b) Test whether or not the mean cash offers vary for the four different owners. Use a level of significance of .05.
c) What should be the next step in the analysis?
d) Is "owner" a classification or experimental factor here? What is the implication of this?

13.12. Refer to Problem 13.1.

a) If the F test for determining whether or not the treatment means are equal will be conducted at a level of significance of .05, what is the power of the test when the treatment means in fact are as shown in Problem 13.2? Assume that $\sigma = 3$.
b) What would be the power if the treatment sample sizes were $n = 10$, everything else remaining unchanged?

14

Analysis of Factor Effects

THE F TEST for determining whether or not the factor level means μ_j differ, discussed in the previous chapter, is a preliminary test to establish whether detailed analysis of the factor level effects is warranted. If the F test leads to the conclusion that the factor level means μ_j are equal, the implication is that there is no relation between the factor and the dependent variable. On the other hand, if the F test leads to the conclusion that the factor level means μ_j differ, the implication is that there is a relation between the factor and the dependent variable. In that case, a thorough analysis of the nature of the factor level effects is usually undertaken. This is done in two principal ways:

1. A decomposition of the treatment sum of squares, and testing of hypotheses of interest.
2. A direct comparison of factor level effects, using estimation techniques.

We shall illustrate each of these two avenues of approach in turn, but will concentrate on the estimation approach in view of its greater usefulness. Throughout this chapter, we assume the fixed effects analysis of variance model (13.2):

(14.1a) $$Y_{ij} = \mu_j + \varepsilon_{ij}$$

where:

μ_j are parameters
ε_{ij} are independent $N(0, \sigma^2)$

or the equivalent model (13.6):

(14.1b) $$Y_{ij} = \mu_. + \tau_j + \varepsilon_{ij}$$

where:

$\mu_.$ is an overall constant
τ_j are parameters subject to the constraint $\sum n_j \tau_j = 0$
ε_{ij} are independent $N(0, \sigma^2)$

14.1 DECOMPOSITION OF *SSTR*

Decomposition of a sum of squares to further the analysis has been encountered in regression. There we begin the analysis with the basic decomposition:

Source of Variation	*SS*
Regression	*SSR*
Error	*SSE*
Total	*SSTO*

When we wish to study whether, say, the quadratic term is needed in the regression model (8.1a):

$$Y_i = \beta_0 + \beta_1 X_i + \beta_{11} X_i^2 + \varepsilon_i$$

we undertake the further decomposition:

Source of Variation	*SS*
Regression	*SSR*
Linear	$SSR(X)$
Curvature	$SSR(X^2 \mid X)$
Error	*SSE*
Total	*SSTO*

Similar to the decomposition of the regression sum of squares *SSR*, in analysis of variance we can decompose the treatment sum of squares *SSTR* into components of interest to the analyst. Many such decompositions can be made. We shall illustrate one type that is frequently useful.

Group Effect Decomposition

Suppose, for our Kenton Food Company example of the previous chapter, that two of the four package designs used 3-color printing and the other two used 5-color printing. We might then be interested in whether the mean sales for the 3-color and 5-color design groups differ, and whether within each color group the two designs differ.

This analysis can be carried out by means of a *group effect decomposition*. Such a decomposition is appropriate when the r factor levels or treatments can be classified into relevant groups and interest is in: (1) whether or not the different group means are the same, and (2) whether or not the several factor level means within each group are the same.

We need to develop some further notation to handle the group effect decomposition. The number of classes or groups will be denoted by c, and the number of factor levels in the gth group ($g = 1, \ldots, c$) will be denoted r_g. Thus, we have:

$$r = \sum_{g=1}^{c} r_g \tag{14.2}$$

where r is the total number of factor levels in the study.

We denote the sum of the observations for the gth group as Y_g:

$$Y_g = \sum_{g} Y_{.j} \tag{14.3}$$

where $Y_{.j}$ is the sum of the observations for the jth factor level as defined in (13.17), and the summation is over all factor levels in the gth group. It follows that:

$$Y_{..} = \sum_{g=1}^{c} Y_g \tag{14.4}$$

where $Y_{..}$ is the sum of all observations as defined in (13.19).

Finally, the total number of observations for the gth group will be denoted N_g:

$$N_g = \sum_{g} n_j \tag{14.5}$$

where n_j is the number of observations for the jth factor level, and the summation is over all factor levels in the gth group. It follows that:

$$n_T = \sum_{g=1}^{c} N_g \tag{14.6}$$

where n_T is the total number of observations in the study, as defined in (13.1).

The group effect decomposition of $SSTR$ can now be stated as follows:

$$SSTR = SS(\text{between groups}) + SS(\text{within 1st group}) + \cdots + SS(\text{within } c\text{th group}) \tag{14.7}$$

where:

$$SS(\text{between groups}) = \sum_{g=1}^{c} \frac{Y_g^2}{N_g} - \frac{Y_{..}^2}{n_T} \tag{14.8a}$$

$$SS(\text{within } g\text{th group}) = \sum_{g} \frac{Y_{.j}^2}{n_j} - \frac{Y_g^2}{N_g} \tag{14.8b}$$

Note that there is a different within-group sum of squares for each of the c groups.

There are $c - 1$ degrees of freedom associated with SS(between groups) and $r_g - 1$ degrees of freedom with SS(within gth group). The resulting mean squares are:

(14.9a) $$MS(\text{between groups}) = \frac{SS(\text{between groups})}{c-1}$$

(14.9b) $$MS(\text{within } g\text{th group}) = \frac{SS(\text{within } g\text{th group})}{r_g - 1}$$

The group effect decomposition of $SSTR$, along with the degrees of freedom and mean squares, are shown in Table 14.1.

TABLE 14.1

Group Effect Decomposition of $SSTR$

Source of Variation	*SS*	*df*	*MS*
Treatments	$SSTR$	$r-1$	$MSTR$
Between groups	SS(between groups)	$c-1$	MS(between groups)
Within 1st group	SS(within 1st group)	$r_1 - 1$	MS(within 1st group)
⋮	⋮	⋮	⋮
Within cth group	SS(within cth group)	$r_c - 1$	MS(within cth group)
Error	SSE	$n_T - r$	MSE
Total	$SSTO$	$n_T - 1$	

The test statistic for determining whether or not the group means differ is:

(14.10a) $$F^* = \frac{MS(\text{between groups})}{MSE}$$

and the test statistic for determining whether or not the factor level means within the gth group differ is:

(14.10b) $$F^* = \frac{MS(\text{within } g\text{th group})}{MSE}$$

In view of the additivity of sums of squares and degrees of freedom, Cochran's theorem (3.59) applies when all factor level means are the same, so that the F^* statistics in (14.10a) and (14.10b) follow the F distribution if the means are equal. Hence, the decision rules to control the risk of making a Type I error are set up in the usual way. We shall illustrate these various points now.

Example

We return to the Kenton Food Company illustration of the previous chapter and note that package designs 1 and 2 are the 3-color designs and

TABLE 14.2

Results for Kenton Food Company Example Obtained in Chapter 13

	Package Design				
	1	2	3	4	*Total*
n_j	2	3	3	2	10
$Y_{.j}$	30	39	57	54	180
$\bar{Y}_{.j}$	15	13	19	27	18

Source of Variation	*SS*	*df*	*MS*
Between designs	258	3	86
Error	46	6	7.67
Total	304	9	

Package Design	*Characteristics*
1	3-color, with cartoons
2	3-color, without cartoons
3	5-color, with cartoons
4	5-color, without cartoons

package designs 3 and 4 are the 5-color designs. We have then from Table 14.2, which summarizes the results from the previous chapter:

3-Color Group	*5-Color Group*	*Total*
$r_1 = 2$	$r_2 = 2$	$r = 4$
$Y_1 = 30 + 39 = 69$	$Y_2 = 57 + 54 = 111$	$Y_{..} = 69 + 111 = 180$
$N_1 = 2 + 3 = 5$	$N_2 = 3 + 2 = 5$	$n_T = 5 + 5 = 10$

Substituting into (14.8a), we obtain:

$$SS(\text{between groups}) = \frac{(69)^2}{5} + \frac{(111)^2}{5} - \frac{(180)^2}{10} = 176.4$$

Using next (14.8b) for each of the two groups, we find:

$$SS(\text{within 3-color group}) = \frac{(30)^2}{2} + \frac{(39)^2}{3} - \frac{(69)^2}{5} = 4.8$$

$$SS(\text{within 5-color group}) = \frac{(57)^2}{3} + \frac{(54)^2}{2} - \frac{(111)^2}{5} = 76.8$$

The resulting ANOVA table is shown in Table 14.3. Note that the degrees of freedom associated with the between-groups sum of squares is $2 - 1 = 1$, and the degrees of freedom associated with each within-group sum of squares is $2 - 1 = 1$.

To test for the various effects of interest, we shall use a level of significance of $\alpha = .05$ in each case. The needed mean squares are found in Table 14.3.

TABLE 14.3

ANOVA Table for Package Design Study

Source of Variation	*SS*	*df*	*MS*
Between designs	258	3	86
Between 3- and 5-color design groups	176.4	1	176.4
Within 3-color design group	4.8	1	4.8
Within 5-color design group	76.8	1	76.8
Error	46	6	7.67
Total	304	9	

1. *Test whether mean sales for the two 3-color designs are the same* ($C_1: \mu_1 = \mu_2$; $C_2: \mu_1 \neq \mu_2$).

Test statistic:

$$F^* = \frac{4.8}{7.67} = .63$$

Decision rule:

If $F^* \leq F(.95; 1, 6) = 5.99$, conclude that mean sales are equal
If $F^* > 5.99$, conclude that mean sales differ

Conclusion: No difference between mean sales for the two 3-color designs.

2. *Test whether mean sales for the two 5-color designs are the same* ($C_1: \mu_3 = \mu_4$; $C_2: \mu_3 \neq \mu_4$).

Test statistic:

$$F^* = \frac{76.8}{7.67} = 10.0$$

Decision rule:

If $F^* \leq F(.95; 1, 6) = 5.99$, conclude that mean sales are equal
If $F^* > 5.99$, conclude that mean sales differ

Conclusion: Mean sales for the two 5-color designs differ.

3. *Test whether mean sales for 3-color and 5-color design groups are the same* (the precise nature of C_1 and C_2 for this case is discussed on p. 471).

Test statistic:

$$F^* = \frac{176.4}{7.67} = 23.0$$

Decision rule:

If $F^* \leq F(.95; 1, 6) = 5.99$, conclude that mean sales are equal
If $F^* > 5.99$, conclude that mean sales differ

Conclusion: Mean sales for 3-color design group differs from mean sales for 5-color design group.

Thus, the additional analysis suggests that the principal reasons why we found in Chapter 13 that the four package designs differ in sales response are that mean sales differ between 3- and 5-color designs and further that the two 5-color designs differ in sales effectiveness.

Comments

1. The decomposition of $SSTR$ in Table 14.1 is called an *orthogonal decomposition.* An orthogonal decomposition is one where the component sums of squares add to the total sum of squares ($SSTR$ in our case), and likewise for the degrees of freedom. Thus, the decomposition of $SSTO$ into $SSTR$ and SSE in Table 14.1 is also an orthogonal decomposition.

It is a property of an orthogonal decomposition that the component sums of squares are independently distributed for the analysis of variance model (14.1). For instance, the variation between the mean sales for the 3-color and 5-color design groups, which is reflected by SS(between groups), is not affected by the variation between mean sales of the designs within either group.

2. Often, it is not possible to study all questions of interest with a single orthogonal decomposition. For example, package designs 2 and 4 in our illustration did not utilize cartoons, whereas 1 and 3 did. The analyst might therefore also wish to study: (1) whether or not the use of cartoons affects mean sales, (2) whether the two designs with cartoons differ in sales effectiveness, and (3) whether the two designs without cartoons differ in sales effectiveness. The decomposition of $SSTR$ for these questions would follow the same principles of group decomposition just explained. However, the component sums of squares for the cartoon decomposition would not be independent of the component sums of squares for the color decomposition.

3. A decomposition of $SSTR$ is usually not adequate analysis by itself, and should be followed by estimation of the magnitudes of the effects found to exist. In our illustration, for instance, one would wish to estimate by how much mean sales for 5-color designs and 3-color designs differ, and by how much mean sales for the two 5-color designs differ. Thus, one is eventually led to the estimation approach, to be discussed next.

4. A difficulty with decomposition of $SSTR$ and subsequent separate testing of components is that the level of significance and power, insofar as the *set* of tests is concerned, is affected. The use of multiple tests on the same data is the counterpart of multiple comparisons for estimation purposes discussed in Chapter 5. Suppose the three F^* statistics obtained from Table 14.3 were independent (they are really dependent). If there are no differences between the means, $P\{F^* \leq F(.95; 1, 6) = 5.99\} = .95$. Hence the probability that all three F^* statistics are less than 5.99,

assuming independence, is $(.95)^3 = .857$ if no differences between the factor level means exist. Thus, the level of significance that at least one of the three tests leads to conclusion C_2 (means differ), when there are no differences in the means, would be $1 - .86 = .14$, and not .05. We see then that the level of significance and power for a *set* or *family* of tests is not the same as that for an *individual* test. Actually, the F^* statistics are dependent since they all have the same denominator MSE. It often therefore becomes difficult to determine the actual level of significance and power for a family of tests obtained as a result of a decomposition of $SSTR$, and it would be still more difficult to determine these if the tests resulted from a number of decompositions. Multiple comparisons with a family confidence coefficient, on the other hand, can be used readily to make inferences with specified protection for the entire family of estimates. These will be discussed shortly.

14.2 ESTIMATION OF FACTOR EFFECTS

If the F test indicates that the factor level means μ_j differ, one may proceed directly to estimate the factor effects of interest. As noted above, one also usually arrives at this point after making one or more decompositions of $SSTR$. Estimates of factor effects usually employed include:

1. Estimation of a factor level mean μ_j.
2. Estimation of the difference between two factor level means.
3. Estimation of a contrast among factor level means.

We shall discuss each of these three types of estimation problems in turn.

Estimation of Factor Level Mean

An unbiased point estimator of the factor level mean μ_j was obtained in (13.22):

(14.11)
$$\hat{\mu}_j = \overline{Y}_{.j}$$

This estimator has mean and variance:

(14.12a)
$$E(\overline{Y}_{.j}) = \mu_j$$

(14.12b)
$$\sigma^2(\overline{Y}_{.j}) = \frac{\sigma^2}{n_j}$$

The latter result follows because (13.54) indicates that $\overline{Y}_{.j} = \mu_j + \bar{\varepsilon}_{.j}$, the sum of a constant plus a mean of n_j independent ε_{ij} terms, each of which has variance σ^2. Further, $\overline{Y}_{.j}$ is normally distributed because the error terms ε_{ij} are normally distributed.

The estimated variance of $\overline{Y}_{.j}$ is denoted $s^2(\overline{Y}_{.j})$, and is obtained as usual by replacing σ^2 in (14.12b) by the unbiased point estimator MSE:

(14.13)
$$s^2(\overline{Y}_{.j}) = \frac{MSE}{n_j}$$

The estimated standard deviation $s(\overline{Y}_{.j})$ is the positive square root of (14.13).

It can be shown that:

(14.14) $$\frac{\bar{Y}_{.j} - \mu_j}{s(\bar{Y}_{.j})} \text{ is distributed as } t(n_T - r) \text{ for model (14.1)}$$

where the degrees of freedom are those associated with MSE. The result (14.14) follows from the definition of t in (1.39) since: (1) $\bar{Y}_{.j}$ is normally distributed, and (2) MSE/σ^2 is distributed independently of $\bar{Y}_{.j}$ as $\chi^2(n_T - r)/(n_T - r)$ according to the following theorem:

(14.15) For model (14.1), SSE/σ^2 is distributed as χ^2, with $n_T - r$ degrees of freedom, and is independent of $\bar{Y}_{.1}, \ldots, \bar{Y}_{.r}$.

It follows directly from (14.14) that the confidence interval for μ_j, with confidence coefficient $1 - \alpha$, is:

(14.16) $$\bar{Y}_{.j} - t(1 - \alpha/2; n_T - r)s(\bar{Y}_{.j}) \leq \mu_j \leq \bar{Y}_{.j} + t(1 - \alpha/2; n_T - r)s(\bar{Y}_{.j})$$

Example. In the Kenton Food Company illustration, the sales manager wished to estimate mean sales for package design 1 with a 95 percent confidence coefficient.

Using the results from Table 14.2, we have:

$$\bar{Y}_{.1} = 15 \qquad n_1 = 2 \qquad MSE = 7.67$$

We require $t(.975; 6)$. From Table A–2 in the Appendix, we obtain $t(.975; 6) = 2.447$. Finally, we need $s(\bar{Y}_{.1})$:

$$s(\bar{Y}_{.1}) = \left(\frac{MSE}{n_1}\right)^{1/2} = \left(\frac{7.67}{2}\right)^{1/2} = 1.958$$

Hence we obtain the confidence interval:

$$15 - (2.447)(1.958) \leq \mu_1 \leq 15 + (2.447)(1.958)$$

or:

$$10.2 \leq \mu_1 \leq 19.8$$

Thus, the mean sales per store for package design 1 are estimated to be between 10.2 and 19.8 cases (95 percent confidence coefficient).

Estimation of Difference between Two Factor Level Means

Frequently two treatments or factor levels are to be compared by estimating the difference between the two factor level means, say, μ_j and $\mu_{j'}$:

(14.17) $$\mu_j - \mu_{j'}$$

Such a difference between two factor level means will be called a *pairwise comparison*. A point estimator of (14.17), denoted as D, is:

(14.18) $$D = \bar{Y}_{.j} - \bar{Y}_{.j'}$$

This point estimator is unbiased:

(14.19) $$E(D) = \mu_j - \mu_{j'}$$

Since $\overline{Y}_{.j}$ and $\overline{Y}_{.j'}$ are independent, the variance of D follows from (1.26b):

(14.20) $$\sigma^2(D) = \sigma^2(\overline{Y}_{.j}) + \sigma^2(\overline{Y}_{.j'}) = \sigma^2\left[\frac{1}{n_j} + \frac{1}{n_{j'}}\right]$$

The estimated variance of D, denoted $s^2(D)$, is given by:

(14.21) $$s^2(D) = MSE\left[\frac{1}{n_j} + \frac{1}{n_{j'}}\right]$$

Finally, D is normally distributed by (1.35), because D is a linear combination of independent normal variables.

It follows from these characteristics, theorem (14.15), and the definition of t in (1.39) that:

(14.22) $$\frac{D - (\mu_j - \mu_{j'})}{s(D)} \text{ is distributed as } t(n_T - r) \text{ for model (14.1)}$$

Hence, the $1 - \alpha$ confidence interval for $\mu_j - \mu_{j'}$ is:

(14.23) $$D - t(1 - \alpha/2; n_T - r)s(D) \leq \mu_j - \mu_{j'} \leq D + t(1 - \alpha/2; n_T - r)s(D)$$

Example. We noted earlier that the two 5-color designs differed in terms of sales effectiveness. We wish now to estimate the difference in mean sales for these two designs, using a 95 percent confidence coefficient. From Table 14.2 we have:

$$\overline{Y}_{.3} = 19 \qquad n_3 = 3 \qquad MSE = 7.67$$
$$\overline{Y}_{.4} = 27 \qquad n_4 = 2$$

Hence:

$$D = \overline{Y}_{.3} - \overline{Y}_{.4} = 19 - 27 = -8$$

The estimated variance of D is:

$$s^2(D) = MSE\left[\frac{1}{n_3} + \frac{1}{n_4}\right] = 7.67\left[\frac{1}{3} + \frac{1}{2}\right] = 6.392$$

so that the estimated standard deviation of D is:

$$s(D) = \sqrt{6.392} = 2.528$$

We require $t(.975; 6) = 2.447$. The desired confidence interval therefore is:

$$-8 - (2.447)(2.528) \leq \mu_3 - \mu_4 \leq -8 + (2.447)(2.528)$$

or:

$$-14.2 \leq \mu_3 - \mu_4 \leq -1.8$$

Thus we estimate, with confidence coefficient .95, that the mean sales for package design 3 fall short of those for package design 4 by somewhere between 1.8 and 14.2 cases per store.

Estimation of Contrast

A *contrast* is a comparison involving two or more factor level means. It is the most general type of comparison we shall consider, and includes the previous case of a pairwise difference between two factor level means in (14.17). A contrast will be denoted by L, and is defined as follows:

$$L = \sum_{j=1}^{r} c_j \mu_j \tag{14.24}$$

where the c_j are suitable coefficients subject to the restriction:

$$\sum_{j=1}^{r} c_j = 0 \tag{14.24a}$$

Illustrations of Contrasts. We illustrate some contrasts by reference to our Kenton Food Company example:

1. To compare the mean sales for the two 3-color designs:

$$L = \mu_1 - \mu_2$$

Here, $c_1 = 1$, $c_2 = -1$, $c_3 = 0$, $c_4 = 0$, and $\sum c_j = 0$.

2. To compare the mean sales for the 3-color and 5-color design groups:

$$L = \frac{\mu_1 + \mu_2}{2} - \frac{\mu_3 + \mu_4}{2}$$

Here, $c_1 = 1/2$, $c_2 = 1/2$, $c_3 = -1/2$, $c_4 = -1/2$, and $\sum c_j = 0$.

3. To compare the mean sales for designs with and without cartoons:

$$L = \frac{\mu_1 + \mu_3}{2} - \frac{\mu_2 + \mu_4}{2}$$

Here, $c_1 = 1/2$, $c_2 = -1/2$, $c_3 = 1/2$, $c_4 = -1/2$, and $\sum c_j = 0$.

Note that the first contrast is simply a pairwise comparison. In the second and third contrasts, we compare averages of several factor level means. The averages used here are unweighted averages of the means μ_j; these are ordinarily the ones of interest. Only in special cases would one be interested in weighted averages of the μ_j to describe the mean response for a group of several factor levels.

Confidence Interval for L. An unbiased estimator of L is:

$$\hat{L} = \sum_{j=1}^{r} c_j \bar{Y}_{.j} \tag{14.25}$$

Since the $\bar{Y}_{.j}$ are independent, the variance of $\hat{L}$ according to (1.26) is:

$$\sigma^2(\hat{L}) = \sum_{j=1}^{r} c_j^2 \sigma^2(\bar{Y}_{.j}) = \sigma^2 \sum_{j=1}^{r} \frac{c_j^2}{n_j} \tag{14.26}$$

An unbiased estimator of this variance is:

$$s^2(\hat{L}) = MSE \sum_{j=1}^{r} \frac{c_j^2}{n_j} \tag{14.27}$$

$\hat{L}$ is normally distributed by (1.35), because it is a linear combination of independent normal random variables. It can be shown by theorem (14.15), the characteristics of $\hat{L}$ just mentioned, and the definition of t, that:

$$\frac{\hat{L} - L}{s(\hat{L})} \text{ is distributed as } t(n_T - r) \text{ for model (14.1)} \tag{14.28}$$

Consequently, the $1 - \alpha$ confidence interval for L is:

$$\hat{L} - t(1 - \alpha/2; n_T - r)s(\hat{L}) \leq L \leq \hat{L} + t(1 - \alpha/2; n_T - r)s(\hat{L}) \tag{14.29}$$

Example. In the Kenton Food Company illustration, the mean sales for the 3-color design group differed from the mean sales for the 5-color design group. Let us estimate this effect. We wish to estimate:

$$L = \frac{\mu_1 + \mu_2}{2} - \frac{\mu_3 + \mu_4}{2}$$

The point estimate is (see data in Table 14.2):

$$\hat{L} = \frac{\bar{Y}_{.1} + \bar{Y}_{.2}}{2} - \frac{\bar{Y}_{.3} + \bar{Y}_{.4}}{2} = \frac{15 + 13}{2} - \frac{19 + 27}{2} = -9$$

Since $c_1 = 1/2$, $c_2 = 1/2$, $c_3 = -1/2$, $c_4 = -1/2$, we obtain:

$$\sum \frac{c_j^2}{n_j} = \frac{(\frac{1}{2})^2}{2} + \frac{(\frac{1}{2})^2}{3} + \frac{(-\frac{1}{2})^2}{3} + \frac{(-\frac{1}{2})^2}{2} = \frac{5}{12} = .4167$$

and:

$$s^2(\hat{L}) = MSE \sum \frac{c_j^2}{n_j} = 7.67(.4167) = 3.196$$

or:

$$s(\hat{L}) = 1.79$$

For a confidence coefficient of 95 percent, we require $t(.975; 6) = 2.447$, and the confidence interval for L is:

$$\hat{L} - t(.975; 6)s(\hat{L}) \leq L \leq \hat{L} + t(.975; 6)s(\hat{L})$$
$$-9 - (2.447)(1.79) \leq L \leq -9 + (2.447)(1.79)$$
$$-13.4 \leq L \leq -4.6$$

Therefore, we conclude that mean sales for the 3-color design group fall below those for the 5-color design group by somewhere between 4.6 and 13.4 cases per store (95 percent confidence coefficient).

Note

Theorem (14.28) enables us to test any hypothesis concerning a contrast L by means of the t test. Alternatively, we can use a confidence interval for testing indirectly a hypothesis concerning L. For instance, if in our earlier illustration we wish to decide whether or not the mean sales of the two color groups differ:

$$C_1: \frac{\mu_1 + \mu_2}{2} = \frac{\mu_3 + \mu_4}{2}$$

$$C_2: \frac{\mu_1 + \mu_2}{2} \neq \frac{\mu_3 + \mu_4}{2}$$

we can use the two-sided confidence interval in (14.29). C_1 implies that $L = (\mu_1 + \mu_2)/2 - (\mu_3 + \mu_4)/2 = 0$, while C_2 implies that $L \neq 0$. For our example, the 95 percent confidence interval does not include 0. Hence, a test using a level of significance of .05 would lead to conclusion C_2, that the two groups do not have the same mean sales. Of course, the confidence interval we obtained provides additional information, namely how much greater are the mean sales for the 5-color group.

Orthogonal Contrasts. Two estimated contrasts $\hat{L}_1$ and $\hat{L}_2$:

(14.30) $$\hat{L}_1 = \sum_{j=1}^{r} c_{1j} \bar{Y}_{.j} \qquad \hat{L}_2 = \sum_{j=1}^{r} c_{2j} \bar{Y}_{.j}$$

where:

$$\sum_{j=1}^{r} c_{1j} = 0 \qquad \sum_{j=1}^{r} c_{2j} = 0$$

are said to be *orthogonal* if:

(14.31) $$\sum_{j=1}^{r} \frac{c_{1j} c_{2j}}{n_j} = 0$$

In the event that all sample sizes n_j are equal, that is, $n_j \equiv n$, condition (14.31) reduces to:

(14.31a) $$\sum_{j=1}^{r} c_{1j} c_{2j} = 0 \qquad \text{when } n_j \equiv n$$

Suppose we consider $\hat{L}_1 = \bar{Y}_{.1} - \bar{Y}_{.2}$ and $\hat{L}_2 = \bar{Y}_{.3} - \bar{Y}_{.4}$ for our Kenton Food Company illustration. $\hat{L}_1$ compares the two 3-color designs, and $\hat{L}_2$ compares the two 5-color designs. We have for these two contrasts:

	Package Design			
	1	2	3	4
c_{1j}	1	−1	0	0
c_{2j}	0	0	1	−1
n_j	2	3	3	2

so that:

$$\sum \frac{c_{1j}c_{2j}}{n_j} = \frac{(1)(0)}{2} + \frac{(-1)(0)}{3} + \frac{(0)(1)}{3} + \frac{(0)(-1)}{2} = 0$$

Thus, $\hat{L}_1$ and $\hat{L}_2$ are orthogonal contrasts.

There exists a definite relationship between orthogonal contrasts and orthogonal decomposition. It can be shown that the sum of squares associated with any contrast $\hat{L}$ is:

$$SS(\hat{L}) = \frac{(\hat{L})^2}{\sum_{j=1}^{r} \frac{c_j^2}{n_j}} \tag{14.32}$$

Thus, for contrast $\hat{L}_1 = \overline{Y}_{.1} - \overline{Y}_{.2}$ we have, using the results in Table 14.2:

$$\hat{L}_1 = 15 - 13 = 2$$

$$\sum \frac{c_{1j}^2}{n_j} = \frac{(1)^2}{2} + \frac{(-1)^2}{3} + \frac{(0)^2}{3} + \frac{(0)^2}{2} = \frac{5}{6}$$

$$SS(\hat{L}_1) = \frac{(2)^2}{\frac{5}{6}} = 4.8$$

Note that this sum of squares is equal to SS(within 3-color group) in Table 14.3.

Similarly, for contrast $\hat{L}_2 = \overline{Y}_{.3} - \overline{Y}_{.4}$ we have:

$$SS(\hat{L}_2) = \frac{(19 - 27)^2}{\frac{(0)^2}{2} + \frac{(0)^2}{3} + \frac{(1)^2}{3} + \frac{(-1)^2}{2}} = 76.8$$

which is the same as SS(within 5-color group) in Table 14.3.

Finally, consider the following contrast:

$$\hat{L}_3 = \frac{2\overline{Y}_{.1} + 3\overline{Y}_{.2}}{5} - \frac{3\overline{Y}_{.3} + 2\overline{Y}_{.4}}{5}$$

This compares the 3-color designs with the 5-color designs. It is unlike our earlier comparison between the two color groups in that $\hat{L}_3$ involves a weighted average for the two 3-color designs, and similarly for the two 5-color designs. The weights are proportional to the sample sizes. (Often, comparisons which give equal weight to the designs in each group would be more reasonable, since weighting by sample size may not reflect properly the importance of each mean.)

It is easy to show that $\hat{L}_3$ is orthogonal to $\hat{L}_1$ and $\hat{L}_2$, and that $SS(\hat{L}_3) =$ 176.4, the same as SS(between 3- and 5-color groups) in Table 14.3. Thus, the three orthogonal contrasts $\hat{L}_1$, $\hat{L}_2$, and $\hat{L}_3$ correspond to the orthogonal decomposition of $SSTR$ in Table 14.3.

Need for Multiple Comparison Procedures

The procedures for analyzing the factor level effects discussed up to this point have two important limitations:

1. The confidence coefficient $1 - \alpha$ (or level of significance α) applies only to that particular estimate (test), and not to a series of estimates (tests).
2. The confidence coefficient $1 - \alpha$ (or level of significance α) is appropriate only if the estimate (test) was not suggested by the data.

The first limitation is familiar from regression analysis. It is particularly serious for analysis of variance models because frequently many different comparisons are of interest there, and one needs to piece the different findings together. Consider the very simple case where three different advertisements are being compared for their effectiveness in stimulating sales. The following estimates of their comparative effectiveness have been obtained, each with a 95 percent statement confidence coefficient:

$$59 \leq \mu_2 - \mu_1 \leq 62$$
$$-2 \leq \mu_3 - \mu_1 \leq 3$$
$$58 \leq \mu_2 - \mu_3 \leq 64$$

It would be natural here to piece the different comparisons together and conclude that advertisement 2 leads to highest mean sales, while advertisements 1 and 3 are substantially less effective and do not differ much among themselves. One would therefore like a family confidence coefficient for this family of statements, providing known assurance that all statements in the family are correct.

The second limitation of the procedures for analyzing factor level effects, namely that the estimate (test) must not be suggested by the data, is an important one in exploratory investigations where many new questions may be suggested once the data are being analyzed. Sometimes, the process of studying effects suggested by the data is called *data snooping*. Analysts often have the tendency to investigate comparisons where the effect appears to be large from the sample data. Now, effects may appear large because in fact they are, or because a random occurrence made them appear large even though they are not. Consequently, investigating only comparisons for which the effect appears to be large implies a smaller confidence coefficient (or a larger level of significance) than the specified one if in fact the effect is small or nonexistent. Thus, it can be shown that if six factor levels are being studied and the analyst will always compare the smallest and largest factor level means by using confidence interval (14.23), with a 95 percent confidence coefficient, the interval estimate will suggest a real effect 40 percent of the time when indeed there is no difference between any of the factor level means (Ref. 14.1). With a greater number of factor levels, the likelihood of an erroneous indication of a real effect would be even greater.

One solution to this problem of making comparisons suggested by initial analysis of the data is to use a multiple comparison procedure where the family of statements includes all the possible statements one anticipates might be made after the data are looked at. For instance, if one is investigating five factor level means and decides in advance that he is principally interested in three pairwise comparisons but may wish to study some others that will appear interesting, one can use the family of all pairwise comparisons as the basis for obtaining an appropriate confidence coefficient for the comparisons suggested by the data.

In the next three sections, we shall discuss three multiple comparison procedures for analysis of variance models which permit the family confidence coefficient to be controlled, two of which allow data snooping to be undertaken naturally without affecting the confidence coefficient. Two of these methods, the Scheffé and Bonferroni methods, have been encountered before. The third, the Tukey method, is new and will be discussed first.

14.3 TUKEY METHOD OF MULTIPLE COMPARISONS

The Tukey method of multiple comparisons which we will consider here applies when:

1. All factor level sample sizes are equal, in other words, $n_j \equiv n$.
2. The family of interest is the set of all pairwise comparisons of factor level means, in other words, the family consists of estimates of all pairs $\mu_j - \mu_{j'}$.

Studentized Range Distribution

The Tukey method utilizes the *studentized range distribution.* Suppose that we have r independent observations $Y_1, \ldots, Y_r$ from a normal distribution with mean μ and variance σ^2. Let w be the range for this set of observations; thus:

(14.33) $$w = \max(Y_j) - \min(Y_j)$$

Suppose further that we have an estimate s^2 of the variance σ^2 which is based on ν degrees of freedom and is independent of the Y's. Then, the ratio w/s is called the *studentized range.* It is denoted by:

(14.34) $$q(r, \nu) = \frac{w}{s}$$

where the arguments in parentheses remind us that the distribution of q depends on r and ν. The distribution of q has been tabulated, and certain percentiles are presented in Table A–9.

This table is simple to use. Suppose that $r = 5$, $v = 10$. The 95th percentile is then $q(.95; 5, 10) = 4.65$, which means:

$$P\left\{\frac{w}{s} = q(5, 10) \leq 4.65\right\} = .95$$

Thus, with five normal Y observations, the probability is .95 that their range is not more than 4.65 times as great as an independent sample standard deviation based on 10 degrees of freedom.

Multiple Comparison Confidence Intervals

The Tukey multiple comparison confidence intervals for all pairwise comparisons $\mu_j - \mu_{j'}$, with a family confidence coefficient $1 - \alpha$, are as follows:

(14.35) $$D - Ts(D) \leq \mu_j - \mu_{j'} \leq D + Ts(D)$$

where:

(14.36a) $$D = \bar{Y}_{.j} - \bar{Y}_{.j'}$$

(14.36b) $$s^2(D) = s^2(\bar{Y}_{.j}) + s^2(\bar{Y}_{.j'}) = \frac{MSE}{n} + \frac{MSE}{n} = \frac{2MSE}{n}$$

(14.36c) $$T = \frac{1}{\sqrt{2}} q(1 - \alpha; r, n_T - r)$$

Remember that it is assumed that $n_j \equiv n$, so that $n_T = rn$.

The family confidence coefficient $1 - \alpha$ here refers to the proportion of correct families of pairwise comparisons when repeated sets of samples are selected and all pairwise confidence intervals are calculated each time. A family of pairwise comparisons is considered to be correct if every pairwise comparison in the family is correct. Thus, when the family confidence coefficient is $1 - \alpha$, all pairwise comparisons in the family will be correct in $(1 - \alpha)100$ percent of the families.

Example

In a study of the effectiveness of different rust inhibitors, four brands (A, B, C, D) were tested, each on a different set of five units. The basic results were as follows:

j	*Inhibitor*	n_j	$\bar{Y}_{.j}$
1	A	5	43
2	B	5	89
3	C	5	67
4	D	5	40

$MSE = 4.50$

The higher the mean, the more effective is the rust inhibitor.

The analysis of variance is shown in Table 14.4. Using a level of significance of .05 for testing whether or not the four rust inhibitors differ in effectiveness, we require $F(.95; 3, 16) = 3.24$. The test statistic, using the mean squares from Table 14.4, is:

$$F^* = \frac{MSTR}{MSE} = \frac{2{,}631.25}{4.50} = 584.72$$

Since $F^* = 584.72 > 3.24$, we conclude that the four rust inhibitors differ in effectiveness.

TABLE 14.4

ANOVA Table for Rust Inhibitor Example

Source of Variation	*SS*	*df*	*MS*
Between rust inhibitors	7,893.75	3	2,631.25
Error	72.00	16	4.50
Total	7,965.75	19	

To examine the nature of the differences, it was desired to estimate all pairwise comparisons by means of the Tukey procedure, using a family confidence coefficient of 95 percent. Since $r = 4$ and $n_T - r = 16$, the required percentile of the studentized range distribution is $q(.95; 4, 16)$. From Table A–9, we find $q(.95; 4, 16) = 4.05$. Hence by (14.36c):

$$T = \frac{1}{\sqrt{2}}(4.05) = 2.86$$

Further, we need $s(D)$. Using (14.36b), we find:

$$s(D) = \left(\frac{2MSE}{n}\right)^{1/2} = \left(\frac{2(4.5)}{5}\right)^{1/2} = 1.34$$

Hence we obtain:

$$Ts(D) = 2.86(1.34) = 3.8$$

The pairwise confidence intervals, with a 95 percent family confidence coefficient, therefore are:

$$42.2 = (89 - 43) - 3.8 \le \mu_2 - \mu_1 \le (89 - 43) + 3.8 = 49.8$$
$$20.2 = (67 - 43) - 3.8 \le \mu_3 - \mu_1 \le (67 - 43) + 3.8 = 27.8$$
$$-.8 = (43 - 40) - 3.8 \le \mu_1 - \mu_4 \le (43 - 40) + 3.8 = 6.8$$
$$18.2 = (89 - 67) - 3.8 \le \mu_2 - \mu_3 \le (89 - 67) + 3.8 = 25.8$$
$$45.2 = (89 - 40) - 3.8 \le \mu_2 - \mu_4 \le (89 - 40) + 3.8 = 52.8$$
$$23.2 = (67 - 40) - 3.8 \le \mu_3 - \mu_4 \le (67 - 40) + 3.8 = 30.8$$

To study the composite information, we shall first order the rust inhibitors from poorest to best according to the mean response $\overline{Y}_{.j}$:

Brand D A C B

The pairwise comparisons indicate that all but one of the differences (D and A) are statistically significant (confidence interval does not cover zero). We shall show this as follows:

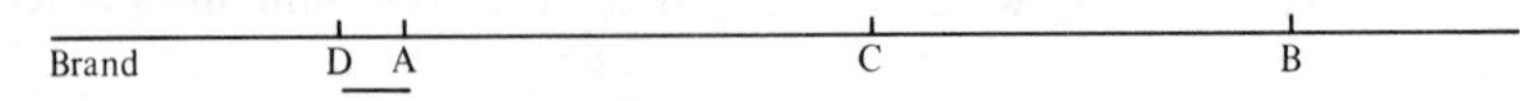

The line between D and A indicates that there is no clear evidence whether D or A is the better rust inhibitor. The absence of a line signifies that a difference in performance has been found. Thus, the multiple comparison procedure permits us to infer, with a 95 percent family confidence coefficient for the chain of conclusions, that B is the best inhibitor (better by somewhere between 18 and 26 units than the second best), C is second best, and A and D follow behind with little or no difference between them.

Comments

1. The Tukey method, as noted earlier, is only applicable if all factor level sample sizes are the same. If in an experiment, a few observations are lost (e.g., due to sickness of subject, store closed because of strike) but the balance between sample sizes is fairly well maintained, it may still be satisfactory to employ the Tukey method, using for n the average sample size.

2. If not all pairwise comparisons are of interest, the confidence coefficient for the family of comparisons being considered will be greater than the specification $1 - \alpha$ used in setting up the Tukey intervals. Thus, the confidence coefficient $1 - \alpha$ with the Tukey method serves as a guaranteed minimum level when not all pairwise comparisons are of interest.

3. The Tukey method can be used for data snooping, as long as the effects to be studied on the basis of preliminary data analysis are pairwise comparisons.

4. The Tukey method can be modified to handle general contrasts between factor level means. We do not discuss this modification, since the Scheffé method (to be discussed next) is to be preferred for this situation.

Derivation of Tukey Confidence Intervals

Consider the deviations:

$$(\overline{Y}_{.1} - \mu_1), \ldots, (\overline{Y}_{.r} - \mu_r) \tag{14.37}$$

where each of the $\overline{Y}_{.j}$ is based on n observations and model (14.1) is assumed. These deviations are independent variables (because the error terms are independent), they are normally distributed (because the error terms are independent normal variables), they have the same expectation zero (because μ_j is subtracted from $\overline{Y}_{.j}$), and they have the same variance σ^2/n. Further, MSE/n is an estimate of σ^2/n which is independent of the $(\overline{Y}_{.j} - \mu_j)$ per theorem (14.15). Thus, it follows from the definition of the studentized range q in (14.34) that:

$$\frac{\max(\overline{Y}_{.j} - \mu_j) - \min(\overline{Y}_{.j} - \mu_j)}{\sqrt{\frac{MSE}{n}}} = q(r, n_T - r) \tag{14.38}$$

where $n_T - r$ is the number of degrees of freedom associated with SSE, $\max(\overline{Y}_{.j} - \mu_j)$ is the largest deviation, and $\min(\overline{Y}_{.j} - \mu_j)$ is the smallest deviation.

In view of (14.38), we can write the following probability statement:

$$P\left\{\frac{\max(\overline{Y}_{.j} - \mu_j) - \min(\overline{Y}_{.j} - \mu_j)}{\sqrt{MSE/n}} \leq q(1-\alpha; r, n_T - r)\right\} = 1 - \alpha \tag{14.39}$$

Note now that the following inequality holds for *all* pairs of factor levels j and j':

$$|(\overline{Y}_{.j} - \mu_j) - (\overline{Y}_{.j'} - \mu_{j'})| \leq \max(\overline{Y}_{.j} - \mu_j) - \min(\overline{Y}_{.j} - \mu_j) \tag{14.40}$$

The absolute sign is needed at the left since the factor levels j and j' are not ordered so that we may be subtracting the larger deviation from the smaller. To put this another way, we are merely concerned here with the difference between the two factor level deviations regardless of direction.

Since the inequality (14.40) holds for all pairs of factor levels j and j', it follows from (14.39) that:

$$P\left\{\left|\frac{(\overline{Y}_{.j} - \mu_j) - (\overline{Y}_{.j'} - \mu_{j'})}{\sqrt{MSE/n}}\right| \leq q(1-\alpha; r, n_T - r)\right\} = 1 - \alpha \tag{14.41}$$

for all $(r)(r-1)/2$ pairwise comparisons among the r factor levels. By rearranging the inequality in (14.41) and using the definitions of $s(D)$ in (14.36b) and T in (14.36c), we obtain the Tukey multiple comparison confidence intervals in (14.35).

14.4 SCHEFFÉ METHOD OF MULTIPLE COMPARISONS

The Scheffé method of multiple comparisons was encountered previously for regression models. It is also applicable for analysis of variance models. It applies for analysis of variance models:

1. Regardless whether or not the factor level sample sizes are equal.
2. When the family of statements is the set of estimates of all possible contrasts:

$$L = \sum c_j \mu_j \qquad \text{where } \sum c_j = 0 \tag{14.42}$$

Thus, infinitely many statements belong to this family.

We noted earlier that an unbiased estimator of L is:

(14.43) $$\hat{L} = \sum c_j \bar{Y}_{.j}$$

for which the estimated variance is:

(14.44) $$s^2(\hat{L}) = MSE \sum \frac{c_j^2}{n_j}$$

Scheffé showed the probability is $1 - \alpha$ that *all* statements of the type:

(14.45) $$\hat{L} - Ss(\hat{L}) \leq L \leq \hat{L} + Ss(\hat{L})$$

are correct simultaneously, where $\hat{L}$ and $s(\hat{L})$ are given by (14.43) and (14.44) respectively, and S is given by:

(14.45a) $$S^2 = (r-1)F(1-\alpha; r-1, n_T - r)$$

Thus, if we were to calculate the confidence intervals for all conceivable contrasts by use of (14.45), then in $(1-\alpha)100$ percent of repetitions of the experiment, the entire set of confidence intervals in the family would be correct.

Example

Suppose that in our earlier Kenton Food Company example dealing with the four cereal package designs, we wish to estimate the following contrasts with a family confidence coefficient of 90 percent:

Comparison of 3-color and 5-color designs:

$$L_1 = \frac{\mu_1 + \mu_2}{2} - \frac{\mu_3 + \mu_4}{2}$$

Comparison of designs with and without cartoons:

$$L_2 = \frac{\mu_1 + \mu_3}{2} - \frac{\mu_2 + \mu_4}{2}$$

Comparison of the two 3-color designs:

$$L_3 = \mu_1 - \mu_2$$

Comparison of the two 5-color designs:

$$L_4 = \mu_3 - \mu_4$$

Consider the estimation of L_1. Earlier, we found:

$$\hat{L}_1 = -9$$

$$s(\hat{L}_1) = 1.79$$

Since $r = 4$ and $n_T - r = 6$ (see Table 14.2), we have:

$$S^2 = (r-1)F(1-\alpha; r-1, n_T - r) = 3F(.90; 3, 6) = 3(3.29) = 9.87$$

or:

$$S = 3.14$$

Hence, the confidence interval for L_1 by the Scheffé multiple comparison method is:

$$-9 - (3.14)(1.79) \le L_1 \le -9 + (3.14)(1.79)$$
$$-14.6 \le L_1 \le -3.4$$

In similar fashion, we obtain the other desired confidence intervals, and the entire set is:

$$-14.6 \le L_1 \le -3.4$$
$$-8.6 \le L_2 \le 2.6$$
$$-5.9 \le L_3 \le 9.9$$
$$-15.9 \le L_4 \le -.1$$

This set of confidence intervals has a family confidence coefficient of 90 percent, so that any chain of conclusions derived from the intervals has associated with it this confidence coefficient. The principal conclusions drawn by the sales manager from the above set of estimates were as follows: Five-color designs lead to higher mean sales than 3-color designs, the increase being somewhere between 3 and 15 cases. No overall effect of cartoons in the package design was indicated, although the 5-color design with a cartoon led to mean sales which were lower than those for the 5-color design without a cartoon.

Comments

1. If in our Kenton Food Company example we wished to estimate a single contrast with statement confidence coefficient .90, the required t value would be $t(.95; 6) = 1.943$. This t value is smaller than the Scheffé multiple $S = 3.14$, so that the single confidence interval would be somewhat narrower. The increased width of the interval with the Scheffé method is the price paid for a known confidence coefficient for a family of statements and a chain of conclusions drawn from them, and for the possibility of making comparisons not specified in advance of the data analysis.

2. Since applications of the Scheffé method never involve all conceivable contrasts, the confidence coefficient for the finite family of statements actually considered will be greater than $1 - \alpha$. Thus, when we state the confidence coefficient is $1 - \alpha$ with the Scheffé method, we really mean it is guaranteed to be at least $1 - \alpha$. For this reason, it has been suggested that the confidence coefficient $1 - \alpha$ used with the Scheffé method be below the level ordinarily used, since $1 - \alpha$ is a lower bound and

the actual confidence coefficient will be greater. Confidence coefficients of 90 percent and 95 percent with the Scheffé method are frequently mentioned.

3. The Scheffé method can be used for a wide variety of data snooping, since the family of statements contains all possible contrasts.

Comparison of Scheffé Method with Tukey Method

1. The Tukey method can only be used with equal sample sizes for all factor levels; the Scheffé method is applicable whether or not the sample sizes are equal.

2. If only pairwise comparisons are to be made and all factor levels have equal sample sizes, the Tukey method gives narrower confidence limits, and is therefore the preferred method.

3. In the case of general contrasts, the Scheffé method tends to give narrower confidence limits, and is therefore the preferred method.

4. The Scheffé method has the property that if the test based on F^* indicates that the factor level means μ_j are not equal, the corresponding Scheffé multiple comparison procedure will find at least one contrast (out of all possible contrasts) that differs significantly from zero (the confidence interval does not cover zero). It may be, though, that this contrast is not one of those estimated by the analyst.

14.5 BONFERRONI MULTIPLE COMPARISON METHOD

The Bonferroni method of multiple comparisons was encountered earlier for regression models. It is also applicable for analysis of variance models:

1. Regardless whether or not the factor level sample sizes are equal.
2. When the family of interest is the particular set of estimated contrasts specified by the user.

If the family consists of s statements, the Bonferroni inequality (5.23) implies that the confidence coefficient is at least $1-\alpha$ that all of the following confidence intervals are correct:

(14.46) $$\hat{L}_i - Bs(\hat{L}_i) \le L_i \le \hat{L}_i + Bs(\hat{L}_i) \qquad i = 1, \ldots, s$$

where:

(14.46a) $$B = t(1-\alpha/2s; n_T - r)$$

Example

Suppose the sales manager of the Kenton Food Company is basically interested in estimating the following two contrasts, with a family confidence coefficient of .975:

Comparison of 3-color and 5-color designs:

$$L_1 = \frac{\mu_1 + \mu_2}{2} - \frac{\mu_3 + \mu_4}{2}$$

Comparison of designs with and without cartoons:

$$L_2 = \frac{\mu_1 + \mu_3}{2} - \frac{\mu_2 + \mu_4}{2}$$

Earlier we found:

$$\hat{L}_1 = -9 \qquad s(\hat{L}_1) = 1.79$$
$$\hat{L}_2 = -3 \qquad s(\hat{L}_2) = 1.79$$

For a 97.5 percent family confidence coefficient with the Bonferroni method, we require:

$$B = t(1 - .025/(2)(2); 6) = t(.99375; 6) = 3.57$$

(We shall explain in comment 3 below how to obtain percentiles for the t distribution that are not tabled.)

We can now complete the confidence intervals for the two contrasts. For L_1, we have:

$$-9 - (3.57)(1.79) \leq L_1 \leq -9 + (3.57)(1.79)$$

or:

$$-15.4 \leq L_1 \leq -2.6$$

Similarly, we obtain the other confidence interval:

$$-9.4 \leq L_2 \leq 3.4$$

These confidence intervals have a guaranteed family confidence coefficient of 97.5 percent, which means that in at least 97.5 percent of repetitions of the experiment, both intervals will be correct.

Again, we would conclude from this family of estimates that mean sales for 5-color designs are higher than those for 3-color designs (by somewhere between 3 and 15 cases per store), and that no overall effect of cartoons in the package design is indicated.

The Scheffé multiple for a 97.5 percent family confidence coefficient in this case would have been:

$$S^2 = 3F(.975; 3, 6) = 3(6.60) = 19.8$$

or:

$$S = 4.45$$

as compared to the Bonferroni multiple $B = 3.57$. Thus the Scheffé method here would have led to wider confidence intervals than the Bonferroni method.

Comments

1. The Bonferroni method can be used whether the factor level sample sizes are equal or unequal, and for pairwise comparisons as well as for general contrasts.

2. It is not necessary that all contrasts be estimated with statement confidence coefficients $1 - (\alpha/s)$ for the Bonferroni family confidence coefficient to be $1 - \alpha$. Different statement confidence coefficients $1 - \alpha_i$ may be used, depending upon the importance of each statement, provided that $\alpha_1 + \alpha_2 + \cdots + \alpha_s = \alpha$.

3. To obtain an untabled percentile of the t distribution, a linear interpolation in Table A–2 may ordinarily be used to give a reasonably correct result as long as the number of degrees of freedom is not minimal. In our illustration of the Bonferroni method, we required $t(.99375; 6)$. From Table A–2, we know that:

$$t(.99; 6) = 3.143$$
$$t(.995; 6) = 3.707$$

Linear interpolation therefore gives us:

$$t(.99375; 6) = 3.143 + \frac{.99375 - .99}{.995 - .99}(3.707 - 3.143) = 3.57$$

Comparison of Bonferroni Method with Scheffé and Tukey Methods

1. If all pairwise comparisons are of interest (and the sample sizes are equal), the Tukey method is superior to the Bonferroni method in the sense of leading to narrower confidence intervals. If not all pairwise comparisons are to be considered, however, the Bonferroni method may be the better.

2. The Bonferroni method will be better than the Scheffé method if the number of contrasts to be estimated is about the same as the number of factor levels, or less. Indeed, the number of statements to be made must exceed the number of factor levels by a considerable amount before the Scheffé method becomes better.

3. In any given problem, one may compute the Bonferroni confidence intervals as well as the Scheffé confidence intervals and, when appropriate, the Tukey confidence intervals, and select the set exhibiting the greatest precision. This choice is proper, since it does not depend on the observed data.

4. The Bonferroni multiple comparison method does not lend itself to data snooping unless one can specify in advance the family of statements one may be interested in, and provided this family is not large. On the other hand, the Tukey and Scheffé methods involve families of statements that lend themselves naturally to data snooping.

5. There are still other methods of making multiple comparisons. Many of these are designed for special cases, such as comparing experimental treatments with a control treatment. A good reference book on multiple comparisons has been written by Miller (Ref. 14.2).

14.6 ANALYSIS OF FACTOR EFFECTS WHEN FACTOR QUANTITATIVE

When the factor under investigation is quantitative, the analysis of factor effects can be carried beyond the point of multiple comparisons to include a study of the nature of the response function. Consider an experimental study undertaken to investigate the effect of the price of a product on sales. Five different price levels are investigated (28 cents, 29 cents, 30 cents, 31 cents, and 32 cents), and the experimental unit is a store. After a preliminary test whether or not mean sales differ for the price levels studied, the analyst might use multiple comparisons to examine whether "odd pricing" at 29 cents actually leads to higher sales than "even pricing" at 28 cents, as well as other questions of interest to him. In addition, the analyst may wish to study whether mean sales are a specified function of price, in the range of prices studied in the experiment. Further, once the relation has been established the analyst may wish to use it for estimating sales volumes at various price levels not studied.

The methods of regression analysis discussed in Parts I and II are, of course, appropriate for the analysis of the response function. It should only be noted that the single-factor studies discussed in this chapter almost always involve replications at the different factor levels, so that the lack of fit of a specified response function can be tested. For this purpose, it should be remembered that the analysis of variance error sum of squares in (13.39b) is identical to the pure error sum of squares of Chapter 4 in (4.8). We illustrate this relation in the following example.

Example

In a study to reduce raw material costs in a glass working firm, an operations analyst collected the experimental data in Table 14.5 on the number of ac-

TABLE 14.5

Data for Piecework Trainees Example

	Treatment (hours of training)			
i	1 (6 hours)	2 (8 hours)	3 (10 hours)	4 (12 hours)
1	40	53	53	63
2	39	48	58	62
3	39	49	56	59
4	36	50	59	61
5	42	51	53	62
6	43	50	59	62
7	41	48	58	61

ceptable units produced from equal amounts of raw material by 28 entry-level piecework employees who had received special training as part of the experiment. Four training levels were used (6, 8, 10, and 12 hours), with 7 of the employees being assigned at random to each level. The higher the number of acceptable pieces, the more efficient is the employee in utilizing the raw material.

Preliminary Analysis. The analyst first tested whether or not the mean number of acceptable pieces is the same for the four training levels. Model (14.1a) was employed:

$$Y_{ij} = \mu_j + \varepsilon_{ij} \tag{14.47}$$

The alternate conclusions and appropriate test statistic were:

$$C_1: \mu_1 = \mu_2 = \mu_3 = \mu_4$$
$$C_2: \text{not all } \mu_j \text{ are equal}$$

$$F^* = \frac{MSTR}{MSE}$$

A computer program for single-factor ANOVA gave the output shown in Figure 14.1. The format differs from that of Figure 13.5 by entries being labeled instead of being keyed to a legend. Note from the MS column of the ANOVA table that $MSTR = 602.89286$ and $MSE = 4.2619$.

Residual analysis (to be discussed in Chapter 15) showed model (14.1a) to be apt. Therefore, the analyst proceeded in his test, using $\alpha = .05$ since analysis of power probabilities showed that this significance level would give an acceptable balance between risks of Type I and Type II errors. The decision rule therefore was:

$$\text{If } F^* \leq F(.95; 3, 24) = 3.01, \text{ conclude } C_1$$
$$\text{If } F^* > 3.01, \text{ conclude } C_2$$

From the printout in Figure 14.1, we have:

$$F^* = \frac{MSTR}{MSE} = 141.5$$

Since $F^* = 141.5 > 3.01$, the analyst concluded C_2, that training level effects differed and that further analysis of them is warranted.

Investigation of Treatment Effects. The analyst's interest next centered on multiple comparisons of all pairs of treatment means. All treatment sample sizes being equal, a Tukey multiple comparison option in the analyst's computer package was used. It gave the output shown in the lower portion of Figure 14.1. For instance, the confidence interval comparing 6 and 8 hours of training is:

$$6.81 \leq \mu_2 - \mu_1 \leq 12.90$$

FIGURE 14.1

Computer ANOVA Output for Piecework Trainees Example

```
SINGLE FACTOR ANOVA

TREATMENT        N          MEAN              STD. DEV.
   1             7          40                 2.3094
   2             7          49.8571            1.7728
   3             7          56.5714            2.6367
   4             7          61.4286            1.2724
   0            28          51.9643            8.4129

TREATMENT 0 IS ENTIRE DATA SET


ANOVA

  SOURCE           SS              DF             MS
   BETW.       1808.67857          3          602.89286←MSTR
   ERROR        102.28571         24            4.2619 ←MSE

F* STAT  = 141.461← MSTR/MSE

S. E. OF EST.  = 2.0644←√MSE



TUKEY INTERVALS FOR PAIRWISE COMPARISONS

LIST OF TREATMENTS:    1,2,3,4

FAMILY CONFIDENCE COEFFICIENT:   .95

LIMITS ARE FOR INTERVAL ESTIMATE OF TRUE MEAN OF TREATMENT J
MINUS TRUE MEAN OF TREATMENT J' IN THIS FORMAT:

          J      J'         LOWER LIMIT        UPPER LIMIT
          4      3             1.81                7.90
          4      2             8.53               14.62
          4      1            18.38               24.47
          3      2             3.67                9.76
          3      1            13.53               19.62
          2      1             6.81               12.90
```

Two points are to be noted in particular from the confidence intervals in Figure 14.1: (1) None of the intervals covers zero, thus all estimated differences are statistically significant. (2) There is some indication that differences between the means for adjoining treatments diminish as the number of hours of training increases; that is, diminishing returns appear to be obtained as the length of training is increased.

Estimation of Response Function. These findings were in accord with the analyst's expectations. He had surmised that the treatment means μ_j would most likely follow a quadratic response function with respect to training level. He now wished to investigate this point further by fitting a quadratic regression model. The model to be fitted and tested was:

$$Y_{ij} = \beta_0 + \beta_1 X_j + \beta_{11} X_j^2 + \varepsilon_{ij} \tag{14.48}$$

where Y_{ij} and ε_{ij} are defined as above, the β's are regression coefficients, and X_j denotes the number of hours of training in the jth training level. The **X** and **Y** matrices for the regression analysis are given in Table 14.6. A computer run of a multiple regression package yielded the estimated regression function:

$$\hat{Y} = -3.73571 + 9.17500X - .31250X^2 \tag{14.49}$$

TABLE 14.6

X and **Y** Matrices for Piecework Trainees Example

Y	X (col 1)	X (col 2)	X (col 3)
40	1	6	36
39	1	6	36
39	1	6	36
36	1	6	36
42	1	6	36
43	1	6	36
41	1	6	36
53	1	8	64
48	1	8	64
49	1	8	64
50	1	8	64
51	1	8	64
50	1	8	64
48	1	8	64
53	1	10	100
58	1	10	100
56	1	10	100
59	1	10	100
53	1	10	100
59	1	10	100
58	1	10	100
63	1	12	144
62	1	12	144
59	1	12	144
61	1	12	144
62	1	12	144
62	1	12	144
61	1	12	144

TABLE 14.7

Analyses of Variance for Piecework Trainees Example

(a) Regression Model

Source of Variation	*SS*	*df*	*MS*
Regression	1,808.10	2	904.05
Error	102.87	25	4.11
Total	1,910.97	27	

(b) Analysis of Variance Model

Source of Variation	*SS*	*df*	*MS*
Treatments	1,808.68	3	602.89
Error	102.29	24	4.26
Total	1,910.97	27	

(c) ANOVA for Lack of Fit Test

Source of Variation	*SS*	*df*	*MS*
Regression	1,808.10	2	904.05
Error	102.87	25	4.11
Lack of fit	.58	1	.58
Pure error	102.29	24	4.26
Total	1,910.97	27	

and the analysis of variance for the regression model (14.48) shown in Table 14.7a. For completeness, we repeat in Table 14.7b the analysis of variance for the ANOVA model (14.47).

Since the data contained replicates, the analyst could test the regression model (14.48) for lack of fit. We noted earlier that the ANOVA error sum of squares in (13.39b) is identical to the regression pure error sum of squares in (4.8). Both measure variation around the group means. Hence the lack of fit sum of squares can be readily obtained from previous results:

$$\text{(14.50)} \qquad SSLF = \underset{\text{(Table 14.7a)}}{SSE} - \underset{\text{(Table 14.7b)}}{SSPE} = 102.87 - 102.29 = .58$$

When .58 is divided by $c - 3 = 4 - 3 = 1$ degree of freedom, we obtain $MSLF$. Table 14.7c contains the analysis of variance for the regression model, with the error sum of squares and degrees of freedom broken down into lack of fit and pure error components.

The alternate conclusions for the test of lack of fit are:

$$C_1: E(Y) = \beta_0 + \beta_1 X + \beta_{11} X^2$$
$$C_2: E(Y) \neq \beta_0 + \beta_1 X + \beta_{11} X^2$$

and the appropriate test statistic is:

$$F^* = \frac{MSLF}{MSPE}$$

For $\alpha = .05$, the decision rule is:

$$\text{If } F^* \leq F(.95; 1, 24) = 4.26, \text{ conclude } C_1$$
$$\text{If } F^* > 4.26, \text{ conclude } C_2$$

Since:

$$F^* = \frac{.58}{4.26} = .136$$

the analyst concluded that the quadratic response function is a good fit. Consequently he used the fitted function in (14.49) in further evaluation of the relation between mean number of acceptable pieces produced and level of training.

Note

The lack of fit sum of squares (14.50) can also be obtained as follows:

$$(14.51) \qquad SSLF = \underset{\text{(Table 14.7b)}}{SSTR} - \underset{\text{(Table 14.7a)}}{SSR} = 1{,}808.68 - 1{,}808.10 = .58$$

Thus $SSLF$ can be viewed as an estimate of the extent to which treatment differentials reflected in $SSTR$ are not accounted for by the response function. In our present example $SSLF$ is comparatively negligible. The quadratic regression function accounts for $(1{,}808.10 \div 1{,}808.68)100 = 99.97$ percent of the treatment differentials reflected in $SSTR$.

PROBLEMS

14.1. Refer to Problem 13.11. Owners A and D were male, while owners B and C were female.

a) Decompose the treatment sum of squares so that you can examine differences in offers for male and female owners, and also differences within each sex group.

b) Make appropriate tests, at a level of significance of .01 each, and summarize the results. Are the several tests independent of each other?

c) Consider the following two contrasts:

$$\hat{L}_1 = \bar{Y}_{.1} + \bar{Y}_{.4} - \bar{Y}_{.2} - \bar{Y}_{.3}$$

$$\hat{L}_2 = \frac{\bar{Y}_{.1} + \bar{Y}_{.4}}{2} - \frac{\bar{Y}_{.2} + \bar{Y}_{.3}}{2}$$

where $\bar{Y}_{.1}$ is the sample mean for owner A, etc. Calculate the sum of squares associated with each contrast. From the point of view of testing, does it matter which contrast is considered? From the point of view of estimation, is one contrast more meaningful than the other?

14.2. Refer to the rust inhibitor example on page 474. Brands B and D are nationally advertised; brands A and C are private brands.

a) Decompose the treatment sum of squares so that you can examine differences between nationally advertised and private brands, and also differences within each brand group.

b) Make appropriate tests, at a level of significance of .05 each, and summarize your results. Are the several tests independent of each other?

c) Brands B and C are higher priced; brands A and D are cheaper. Decompose the treatment sum of squares in appropriate fashion and test for differences in effectiveness between and within price groups. Use a level of significance of .05 for each test and summarize your results.

d) Are the sums of squares in part (c) independent of those in part (a)?

14.3. Refer to the rust inhibitor example on page 474. Consider the following three contrasts:

$$\hat{L}_1 = \frac{\bar{Y}_{.1} + \bar{Y}_{.3}}{2} - \frac{\bar{Y}_{.2} + \bar{Y}_{.4}}{2}$$

$$\hat{L}_2 = \frac{\bar{Y}_{.2} + \bar{Y}_{.3}}{2} - \frac{\bar{Y}_{.1} + \bar{Y}_{.4}}{2}$$

$$\hat{L}_3 = \bar{Y}_{.1} - \bar{Y}_{.4}$$

a) Are $\hat{L}_1$ and $\hat{L}_3$ orthogonal contrasts?

b) Are $\hat{L}_2$ and $\hat{L}_3$ orthogonal contrasts?

c) The covariance of $\hat{L}_2$ and $\hat{L}_3$ is zero. What generalization does this suggest? What is the implication of this generalization?

14.4.* Refer to the rust inhibitor example on page 474. Brands B and D are nationally advertised; brands A and C are private brands. Estimate the contrast:

$$L = \frac{\mu_2 + \mu_4}{2} - \frac{\mu_1 + \mu_3}{2}$$

with a 95 percent confidence interval. What does your confidence interval indicate about the comparative effectiveness of the two groups of brands?

14.5. Refer to Problem 13.10.

a) Obtain confidence intervals for all pairwise comparisons with the Tukey procedure, using a family confidence coefficient of .90.

b) Analyze your results.

c) If you had employed the Bonferroni or Scheffé procedures, what would have been the B and S multiples respectively? Which procedure is most efficient here?

Need MSE.

14.6.* Refer to Problem 13.11.

a) Obtain confidence intervals for all pairwise comparisons with the Tukey procedure, using a family confidence coefficient of .90.

b) Analyze your results.

c) If you had employed the Bonferroni or Scheffé procedures, what would have been the B and S multiples respectively? Which procedure is most efficient here?

14.7. Refer to Problem 13.11. The analyst wished to study differences in mean cash offers for male owners (A and D) and female owners (B and C), and for older owners (A and C) and younger owners (B and D).

a) What are the relevant contrasts to be estimated?

b) Estimate these contrasts, using a family confidence coefficient of .95.

c) Analyze your results and summarize them.

d) Justify the use of the multiple comparison procedure you employed in part (b). Is it the most efficient procedure in this instance?

14.8. A single-factor study consisted of six treatments, with the sample size for each treatment being $n = 10$.

a) Assuming pairwise comparisons are to be made with a 90 percent family confidence coefficient, find the B, S, and T multiples if the number of comparisons in the family is 2; 3; 10. What generalization is suggested by your results?

b) Assuming a variety of contrasts are to be estimated with a 90 percent family confidence coefficient, find the B and S multiples if the number of comparisons in the family is 2; 10. What generalization is suggested by your results?

14.9. Refer to Figure 14.1. If we wish to estimate the treatment mean μ_1 by means of an interval estimate, would it be valid to use for the estimated variance of $\overline{Y}_{.1}$:

$$s^2(\overline{Y}_{.1}) = \frac{s_1^2}{n_1} = \frac{(2.3094)^2}{7}$$

instead of (14.13):

$$s^2(\overline{Y}_{.1}) = \frac{MSE}{n_1} = \frac{4.2619}{7}$$

What is the advantage of using (14.13)? What is a disadvantage?

14.10. Refer to the bank example on page 113. Suppose the analyst had wished to employ analysis of variance initially to determine whether the value of the gift (directly proportional to minimum deposit) had any effect on the number of new accounts opened.

a) State the analysis of variance model.

b) Obtain the appropriate analysis of variance table.

c) Test whether the value of the gift had any effect on the number of new accounts opened. Use a level of significance of .05.

d) Make pairwise comparisons with a 90 percent family confidence coefficient to obtain information on the shape of the response function. Analyze your results.

e) Show that $SSLF = 14{,}320.6$ in Table 4.4 can be obtained also by means of (14.51).

14.11. Refer to Problem 13.10. The five agents had the following number of employees:

Agent	*Number of Employees*	*Agent*	*Number of Employees*
C	12	L	8
F	7	X	15
H	6		

a) To study whether the regression of delay time on size of agency (measured by number of employees) is linear, fit a linear regression model.

b) Test whether the linear regression is a good fit. Use a level of significance of .05. Discuss your findings.

CITED REFERENCES

14.1. Cochran, William G., and Cox, Gertrude M. *Experimental Designs*, p. 74. 2d ed. New York: John Wiley & Sons, Inc., 1957.

14.2. Miller, Rupert G., Jr. *Simultaneous Statistical Inference*. New York: McGraw-Hill Book Co., 1966.

15

Implementation of ANOVA Model

IMPLEMENTATION of an analysis of variance model requires determination of appropriate sample sizes and, once the data are at hand, an examination of the aptness of the model. In this chapter, we first consider the planning of sample sizes. Then we take up a number of topics related to examining the aptness of the analysis of variance model.

15.1 PLANNING OF SAMPLE SIZES WITH POWER APPROACH

For analysis of variance problems, as for other statistical problems, it is important to plan the sample sizes so that needed protection against both Type I and Type II errors can be obtained, or so that the estimates of interest have sufficient precision to be useful.

We shall generally assume in this discussion that all factor levels are to have equal sample sizes, reflecting that they are about equally important. Indeed, if major interest lies in pairwise comparisons of all factor level means, it can be shown that equal sample sizes maximize the precision of the various comparisons. Another reason for equal sample sizes is that certain departures from the assumed ANOVA model are not troublesome, as will be noted in Section 15.7, if all factor levels have the same sample size. There will be times, however, when unequal sample sizes are appropriate. For instance, when four experimental treatments are each to be compared to a control, it may be reasonable to make the sample size for the control larger. We shall comment later on the planning of sample sizes for such a case.

Planning of sample sizes can be approached in terms of controlling the risks of making Type I and II errors, in terms of controlling the widths of desired confidence intervals, or in terms of a combination of these two. In this section we consider planning of sample sizes with the power approach, which permits controlling the risks of making Type I and II errors. One procedure for implementing the power approach would be to use the power charts for the F test presented in Table A–8. A trial and error process, however, would be required with these charts. Fortunately, other charts are available which furnish the appropriate sample sizes directly. These charts are presented in Table A–10. They are applicable when all factor levels are to have equal sample sizes, that is, when $n_j \equiv n$.

Use of Table A–10 Charts

The planning of sample sizes by means of the charts in Table A–10 is done in terms of the testing framework for deciding whether or not the factor level means differ:

(15.1)
$$C_1: \mu_1 = \mu_2 = \cdots = \mu_r$$
$$C_2: \text{not all } \mu_j \text{ are equal}$$

The charts use a slightly different noncentrality parameter than the one defined in (13.70a), namely:

(15.2)
$$\phi' = \frac{1}{\sigma}\sqrt{\frac{\sum(\mu_j - \mu_.)^2}{r}} = \frac{1}{\sigma}\sqrt{\frac{\sum \tau_j^2}{r}}$$

The sole difference is that ϕ' does not involve n, the common sample size for each factor level, because we now wish to determine this.

The following specifications need be made in using the charts in Table A–10:

1. The level α at which the risk of making a Type I error is to be controlled.
2. The value of ϕ' at which the risk of making a Type II error is to be controlled.
3. The level β at which the risk of making a Type II error is to be controlled when ϕ' has the specified value.

The most difficult specification to make among these three no doubt is the second, which involves the determination of how much the factor level means μ_j must differ so that it becomes important to recognize the differences and conclude C_2. We shall discuss below at some length how the specification of ϕ' might be undertaken.

When using the sample size charts in Table A–10, two α levels are available at which the risk of a Type I error can be controlled ($\alpha = .01, .05$). The Type II error risk can be controlled at five β levels ($\beta = .5, .3, .2, .1, .05$) through the specification on the power P as defined in (13.69). The relation between β and P is, of course:

(15.3)
$$P = 1 - \beta$$

The X axis of the charts contains the noncentrality parameter ϕ', and the Y axis the needed sample size n for each treatment. Charts are available for $r = 2, 3, 4, 5$ treatments.

Example. Suppose that we are to determine whether or not four different brands of snow tires have the same mean tread life (in thousands of miles) and that we have the following specifications:

$$r = 4 \qquad \alpha = .05$$
$$\phi' = .70 \qquad \beta = .10 \quad \text{or} \quad P = .90$$

We turn to the chart for $r = 4$, find $\phi' = .70$ on the X scale corresponding to $\alpha = .05$, locate the curve for $P = .90$ in the left set of curves corresponding to $\alpha = .05$, and read the ordinate of this curve at $\phi' = .70$. We find this to be approximately 8. Hence, eight snow tires of each brand should be tested to provide the specified protection.

Specification of ϕ'

It is usually not an easy matter to arrive at a meaningful specification of ϕ' for which it is desired to control the risk of a Type II error. We shall discuss three possible approaches to specifying ϕ'.

Specifying Each Factor Level Mean. One way of arriving at a specification of ϕ' is to determine particular factor level means for which it would be desired to conclude C_2 with high probability. For instance, such a set of means in our snow tire example might be as follows:

$$\mu_1 = 25 \qquad \mu_2 = 24 \qquad \mu_3 = 30 \qquad \mu_4 = 29 \text{ (thousand miles)}$$

Since ϕ' involves the deviations $\mu_j - \mu_.$, the above specification could equally well be made in terms of the deviations around the overall mean $\mu_. = 27$, namely:

$$\tau_1 = -2 \qquad \tau_2 = -3 \qquad \tau_3 = 3 \qquad \tau_4 = 2$$

Suppose it is known from past experience that the standard deviation of the tread life of tires is $\sigma = 2$ (thousand miles) approximately. Then using (15.2), we have:

$$\phi' = \frac{1}{2}\sqrt{\frac{(-2)^2 + (-3)^2 + (3)^2 + (2)^2}{4}} = 1.27$$

With this approach, it is usually difficult to single out one set of factor level means on which the control of the Type II error risks should be based. For instance, it might also be important to have a small risk of making a Type II error for the set:

$$\tau_1 = 2 \qquad \tau_2 = 1 \qquad \tau_3 = -2 \qquad \tau_4 = -1$$

For this set, we have:

$$\phi' = \frac{1}{2}\sqrt{\frac{(2)^2 + (1)^2 + (-2)^2 + (-1)^2}{4}} = .79$$

Note that this value of ϕ' differs substantially from that obtained before.

Since power varies directly with ϕ', one can calculate ϕ' for several sets of factor level means, for each of which it is important to control the risk of a Type II error, and use the *smallest* ϕ' for finding the sample sizes. If this is done, the risks of making Type II errors for sets of factor level means which are associated with larger values of ϕ' will be below the controlled level.

Specifying the Maximum Difference. A second method of specifying the noncentrality parameter ϕ' for which it is important to conclude C_2 is in terms of the difference between the two extreme factor level means. For instance, in our snow tire example it might be specified that it is important to conclude that the four brands of snow tires have different mean tread lives when the difference between any two brands of tires is 3.0 (thousand miles) or greater.

Many patterns of factor level means fit this specification, and usually lead to different values of ϕ'. For instance, we might have:

$$\mu_1 = 24 \qquad \mu_2 = 27 \qquad \mu_3 = 25 \qquad \mu_4 = 26$$

or:

$$\mu_1 = 25 \qquad \mu_2 = 25 \qquad \mu_3 = 28 \qquad \mu_4 = 28$$

For each of these patterns, the maximum difference between any pair of means is 3, yet their associated ϕ' values differ. In the spirit of the first approach, we shall consider the pattern which leads to the smallest value of ϕ' and determine the sample sizes using this value. Then, all other patterns with the same maximum difference between pairs will have the same or a smaller risk of making a Type II error.

It can be shown that ϕ' is minimized, given the specification on the maximum difference between any two factor level means, when all the other factor level means are equal to the overall mean $\mu_{.}$. Thus, a pattern of factor level means which minimizes ϕ' for our example above is:

$$\mu_1 = 25 \qquad \mu_2 = 28 \qquad \mu_3 = 26.5 \qquad \mu_4 = 26.5$$

or in terms of the τ_j:

$$\tau_1 = -1.5 \qquad \tau_2 = +1.5 \qquad \tau_3 = 0 \qquad \tau_4 = 0$$

and ϕ' for this set is:

$$\phi' = \frac{1}{2}\sqrt{\frac{(-1.5)^2 + (1.5)^2 + (0)^2 + (0)^2}{4}} = .53$$

Note that it does not matter which factor levels are taken as the two extreme ones; the value of ϕ' will always be the same.

In general, if the specified maximum difference is d, the minimum value of ϕ' is:

$$\phi' = \frac{d}{\sigma}\sqrt{\frac{1}{2r}} \tag{15.4}$$

Thus for our snow tire example where $\sigma = 2$ and $r = 4$, if the maximum difference between pairs of brands for which it is important to recognize differences in mean tread life is $d = 3$ (thousand miles), we obtain:

$$\phi' = \frac{3}{2}\sqrt{\frac{1}{(2)(4)}} = .53$$

Assuming that the other specifications are:

$$\alpha = .05 \qquad \beta = .10 \quad \text{or} \quad P = .90$$

we find from Table A–10 that the needed number of tires for each brand is approximately 13 tires.

Specifying a Standard Deviation Multiple. This method of specifying ϕ' can sometimes be employed. To illustrate its use, suppose that a plant has r coil winding machines, each of which turns out about the same amount of production. A certain characteristic of the wound coils must fall within specified limits for the coil to be acceptable. The inherent variability of this characteristic for each machine, as measured by the standard deviation, is σ. If all machines are calibrated at the same mean level of the characteristic, then the variability of coils selected at random from the plant's output is σ. However, if the machines are not calibrated at the same level, the variability of coils selected at random from the plant's output is larger, namely:

$$\left[\sigma^2 + \frac{\sum(\mu_j - \mu_.)^2}{r}\right]^{1/2} = \left[\sigma^2 + \frac{\sum \tau_j^2}{r}\right]^{1/2} \tag{15.5}$$

Suppose that some increase in the variability of the coils, due to the coil winding machines being out of calibration, could be tolerated without jeopardizing the acceptability of the coils, but after a certain point it would be important to correct the lack of calibration. Let us express this point relative to the inherent variability σ of each machine, and denote it as G:

$$G = \frac{\left[\sigma^2 + \frac{\sum \tau_j^2}{r}\right]^{1/2}}{\sigma} = \left[1 + \frac{1}{\sigma^2}\frac{\sum \tau_j^2}{r}\right]^{1/2} \tag{15.6}$$

We now square each side of (15.6), and using (15.2) obtain:

$$G^2 = 1 + \frac{1}{\sigma^2}\frac{\sum \tau_j^2}{r} = 1 + (\phi')^2$$

or:

$$\phi' = \sqrt{G^2 - 1} \tag{15.7}$$

Hence, we can obtain the specified value of ϕ' directly from G.

To illustrate this approach, assume that in our coil winding example an increase in the variability of 20 percent or more due to lack of calibration would cause substantial difficulties in meeting acceptance specifications, and this lack of calibration should then be found and corrected. Thus, the specification is $G = 1.2$. Hence, by (15.7) we have:

$$\phi' = \sqrt{G^2 - 1} = \sqrt{(1.2)^2 - 1} = .66$$

Suppose that the other specifications are:

$$r = 5 \qquad \alpha = .05 \qquad \beta = .05 \quad \text{or} \quad P = .95$$

From Table A–10, we then find $n = 10$ approximately. Thus, a sample of 10 coils from each machine provides the specified degree of protection against failing to detect an increase of 20 percent in the variability of the coils due to lack of calibration. If the increase is more than 20 percent, the protection will be greater than the specified level.

It should be noted that this method of specifying the noncentrality parameter ϕ' does not require explicitly an advance estimate of the standard deviation σ. This is not as much of an advantage as it might seem, however, because a meaningful specification of the ratio G frequently will require knowledge, or an advance estimate, of the standard deviation σ.

Comments

1. The exact specification of the noncentrality parameter ϕ' has great effect on the sample size when ϕ' is small, but much less effect when ϕ' is larger. For instance, when $r = 3$, $\alpha = .05$, $\beta = .10$, we have from Table A–10:

ϕ'	n
.25	70
.30	47
.60	13
.65	11

Thus, unless ϕ' is quite small, one need not be too concerned about some imprecision in specifying ϕ'.

2. Decreasing either α or β or both increases the required sample size. For instance, when $r = 4$, $\alpha = .05$, $\phi' = .5$, we have:

$\beta = 1 - P$	n
.5	7
.1	15
.05	17

3. An error in the advance estimate of σ can cause a substantial miscalculation of needed sample size, although the needed sample size often will still be "in the same ball park." For instance, when $r = 5$, $\alpha = .05$, $\beta = .10$, $\sigma\phi' = 1.6$, we have:

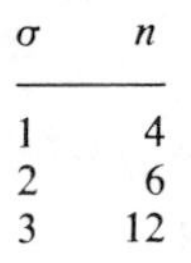

σ	n
1	4
2	6
3	12

Notice that the sample size is quite sensitive to a change in σ from 2 to 3, but not so much when σ changes from 2 to 1. In view of the usual approximate nature of the advance estimate of σ, it is generally desirable to investigate the needed sample size for a range of likely values of σ before deciding on the sample size to be employed.

15.2 PLANNING OF SAMPLE SIZES WITH ESTIMATION APPROACH

The estimation approach to planning sample sizes may be used either in conjunction with the control of Type I and II errors or by itself. The essence of the approach is to specify the major comparisons of interest and to determine the expected widths of the confidence intervals for various sample sizes, given an advance judgment of the standard deviation. The approach is iterative, starting with an initial judgment of needed sample sizes. This initial judgment may be based on the needed sample sizes to control the risks of Type I and II errors, if these have been previously obtained. If the anticipated widths of the confidence intervals, based on the initial sample sizes, are satisfactory, the iteration process is terminated. If one or more widths are too great, larger sample sizes should be tried next. If the widths are unnecessarily tight, smaller sample sizes should be tried next. This process is continued until those sample sizes are found which yield satisfactory anticipated widths.

Example

Suppose that we are to plan sample sizes for our earlier snow tire example by means of the estimation approach. Suppose further that the sample sizes for each tire brand are to be equal, that is, $n_j \equiv n$.

Management has indicated it wishes three types of estimates:

1. A comparison of the mean tread life for each pair of brands:

$$\mu_j - \mu_{j'}$$

2. A comparison of the mean tread life for the two high priced brands and the two low priced brands:

$$\frac{\mu_1 + \mu_4}{2} - \frac{\mu_2 + \mu_3}{2}$$

3. A comparison of the mean tread life for the national brands and the local brand:

$$\frac{\mu_1 + \mu_2 + \mu_4}{3} - \mu_3$$

Management further indicated that it wishes a family confidence coefficient of .95 for the entire set of comparisons.

We first need an advance estimate of the standard deviation of the tread lives of tires. Suppose that from past experience we judge it to be approximately $\sigma = 2$ (thousand miles). Next we require an initial judgment of needed sample size, and shall consider $n = 10$ as a starting point.

We know from (14.26) that the variance of an estimated contrast $\hat{L}$ when $n_j \equiv n$ is:

$$\sigma^2(\hat{L}) = \frac{\sigma^2}{n} \sum c_j^2 \qquad \text{when } n_j \equiv n$$

Hence, our anticipations of the standard deviations of the required estimators, given $\sigma = 2$ and $n = 10$, are:

Contrast	*Anticipated Variance*	*Anticipated Standard Deviation*
Pairwise comparisons	$\frac{2^2}{10}[1^2 + (-1)^2] = .80$	.89
High and low priced brands	$\frac{2^2}{10}\left[\left(\frac{1}{2}\right)^2 + \left(\frac{1}{2}\right)^2 + \left(-\frac{1}{2}\right)^2 + \left(-\frac{1}{2}\right)^2\right] = .40$	.63
National and local brands	$\frac{2^2}{10}\left[\left(\frac{1}{3}\right)^2 + \left(\frac{1}{3}\right)^2 + \left(\frac{1}{3}\right)^2 + (-1)^2\right] = .53$	.73

We shall employ the Scheffé multiple comparison procedure, and therefore require the Scheffé multiple S in (14.45a) for $r = 4$, $n_T = nr = 10(4) = 40$, and $1 - \alpha = .95$:

$$S^2 = (r-1)F(1-\alpha; r-1, n_T - r) = 3F(.95; 3, 36) = 3(2.87) = 8.61$$

or:

$$S = 2.93$$

Hence, the anticipated widths of the confidence intervals are:

Contrast	*Anticipated Width of Confidence Interval* $= \pm S\sigma(\hat{L})$
Pairwise comparisons	$\pm(2.93)(.89) = \pm 2.61$ (thousand miles)
High and low priced brands	$\pm(2.93)(.63) = \pm 1.85$ (thousand miles)
National and local brands	$\pm(2.93)(.73) = \pm 2.14$ (thousand miles)

Management was satisfied with these anticipated widths. However, it was decided to increase the sample sizes from 10 to 15 in case the actual standard deviation of the tread lives of tires is somewhat greater than the anticipated value $\sigma = 2$ (thousand miles).

Comments

1. Since one cannot usually be certain that the advance judgment of the standard deviation is correct, it is advisable to study a range of values for the standard deviation before making a decision on sample size.

2. If the sample sizes for the factor levels are to be unequal, one can still use the iterative procedure with the estimation approach just described. For instance, suppose that four new flavors of a fruit syrup are each to be compared with a control, the present flavor. One may therefore wish to make the control sample size larger in order to increase the precision of these key comparisons. Suppose the control sample size is to be twice as large as the other factor level sample sizes. Then we can represent the control sample size as $2n$ and the other sample sizes as n. We then proceed as before with an initial specification for n, but now utilize formula (14.26) for the variance of an estimated contrast in its general form since the sample sizes are unequal.

15.3 PLANNING OF SAMPLE SIZES TO FIND "BEST" TREATMENT

There are occasions when the chief purpose of the study is to ascertain the factor level or treatment with the highest or lowest mean. In our snow tire example, for instance, it may be desired to determine which of the four brands has the longest mean tread life.

Bechhofer has developed a table, shown in Table A–11, which enables us to determine the necessary sample size so that with probability $1 - \alpha$ the highest (lowest) sample factor level mean is from the factor level with the highest (lowest) population mean. We need to specify the probability $1 - \alpha$, the standard deviation σ, and the smallest difference λ between the highest (lowest) and second highest (second lowest) factor level means which it is important to recognize. Table A–11 assumes that equal sample sizes are to be used for all factor levels, and that the analysis of variance model (14.1) is appropriate.

Example

Suppose that in our snow tire example, the chief objective is to identify the brand with the longest mean tread life. There are $r = 4$ brands. We anticipate, as before, that $\sigma = 2$ (thousand miles). Further, assume we are informed that a difference $\lambda = 1$ (thousand miles) between the highest and second highest brand means is important to recognize, and that the probability is to be $1 - \alpha = .90$ or greater that we identify correctly the brand with the highest mean tread life when $\lambda \geq 1$.

The entry in Table A–11 is $\lambda\sqrt{n}/\sigma$. For $r = 4$ and probability $1 - \alpha = .90$, we find from Table A–11 that $\lambda\sqrt{n}/\sigma = 2.4516$. Hence, since the λ specification is $\lambda = 1$:

$$\frac{(1)\sqrt{n}}{2} = 2.4516$$

$$\sqrt{n} = 4.9032$$

$$n = 24$$

Thus, sample sizes of 24 tires for each brand provide an assurance of at least .90 that the brand with the highest sample mean is the brand with the highest population mean, when the mean tread life for the best brand exceeds that of the second best by at least 1 (thousand miles) and when $\sigma = 2$ (thousand miles).

Note

If the anticipated standard deviation is not accurate, the probability of identifying the population with the highest (lowest) mean correctly is, of course, affected. This is no different from the other approaches, where a misjudgment of the standard deviation affects the risks of making a Type II error or the widths of the confidence intervals actually obtained.

15.4 RESIDUAL ANALYSIS

When discussing regression analysis, we emphasized the importance of examining the aptness of the regression model and pointed out the effectiveness of residual plots for spotting major departures from the assumed model. Examination of aptness is no less important with an analysis of variance model, and residual plots again are most helpful for spotting major departures from the assumed model.

Since residual plots for ANOVA models are used in a similar manner as those for regression analysis, we shall give only a brief discussion of them here. Before doing so, we should note again the proper sequence of using any statistical model:

1. First, one should examine whether the proposed model is appropriate for the set of data at hand.
2. If the proposed model is not appropriate, corrective measures such as transformations of the data may have to be taken, or the model may need to be modified.
3. Only after this review of the aptness of the model has been made and any necessary corrective measures completed and their effectiveness ascertained, should the analysis of the data be undertaken.

It is not necessary, nor is it usually possible, that the ANOVA model fits perfectly. As will be noted later, the ANOVA model is reasonably robust against certain types of departures from the model, such as the error terms not being exactly normally distributed. The major purpose of the examination of the aptness of the model is therefore to detect serious departures from the conditions assumed by the model.

Residuals

The residuals e_{ij} for the ANOVA model (14.1) were defined in (13.30):

$$e_{ij} = Y_{ij} - \overline{Y}_{.j} \tag{15.8}$$

As in regression, standardized residuals:

$$\frac{e_{ij}}{\sqrt{MSE}} \tag{15.9}$$

are sometimes helpful. When the sample sizes for each factor level are not small, one can use the standardized residuals:

$$\frac{e_{ij}}{s_j} \tag{15.10}$$

where s_j is the sample standard deviation for the observations from the jth factor level, as defined in (13.50).

Residual Plots

Residual plots for an analysis of variance model are shown in Figure 15.1, which contains the residuals of Table 13.2 for the Kenton study, plotted against observation number. It is helpful to plot the residuals for each factor level or treatment separately, but side by side. The plotting of residuals within a factor level may be by time order when appropriate, or in arbitrary sequence otherwise.

We consider now briefly how residual plots reveal various possible departures from analysis of variance model (14.1). None of these possible departures is strongly suggested in Figure 15.1 for the Kenton Food Company data.

FIGURE 15.1

Residual Plots for Kenton Food Company Example

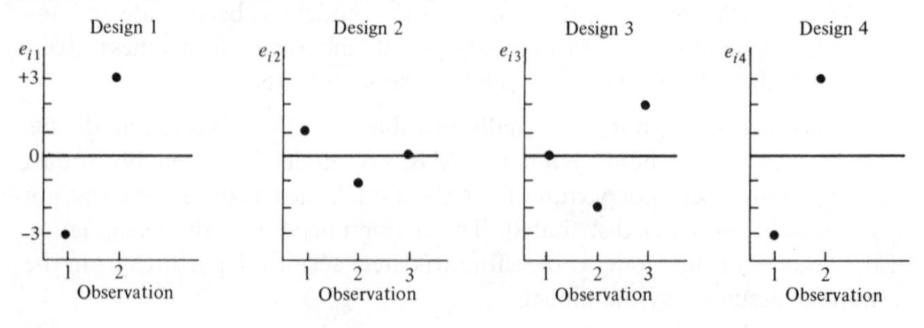

Nonconstancy of Error Variance

ANOVA model (14.1) requires that the error terms ε_{ij} have constant variance for all factor levels. When this condition is met, the residual plot should show about the same extent of scatter of the residuals around 0 for each factor level. Figure 15.2 illustrates residual plots which suggest that

FIGURE 15.2

Residual Plots Suggesting Unequal Factor Level Variances

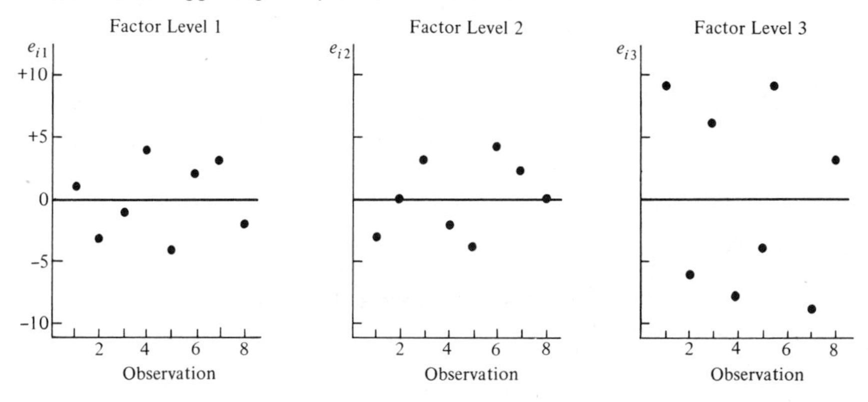

the error variances are not constant. It would appear that the error terms for factor level 3 have a larger variance than those for the other two factor levels, because the extent of scatter is substantially greater for factor level 3 than for the other factor levels.

A number of statistical tests have been developed for formally examining the equality of r variances; two of these tests will be discussed in Section 15.6.

Nonindependence of Error Terms

Whenever data are obtained in a time sequence, the residuals should be plotted in time order to examine if they are serially correlated. Figure 15.3 contains the residuals for an experiment on group interactions. Three different treatments were applied, and the group interactions recorded on tapes. Seven replications were made for each treatment. Afterwards, the experimenter measured the number of interactions by listening to the tapes in randomized order. Figure 15.3 strongly suggests that the experimenter discerned a larger number of interactions as he gained experience in listening to the tapes, as a result of which the residuals in Figure 15.3 appear to be serially correlated. In this instance, an inclusion in the model of a linear term for the time effect might be sufficient to assure independence of the error terms in the revised model.

FIGURE 15.3

Residual Plots for Group Interaction Study Illustrating Time-Related Effect

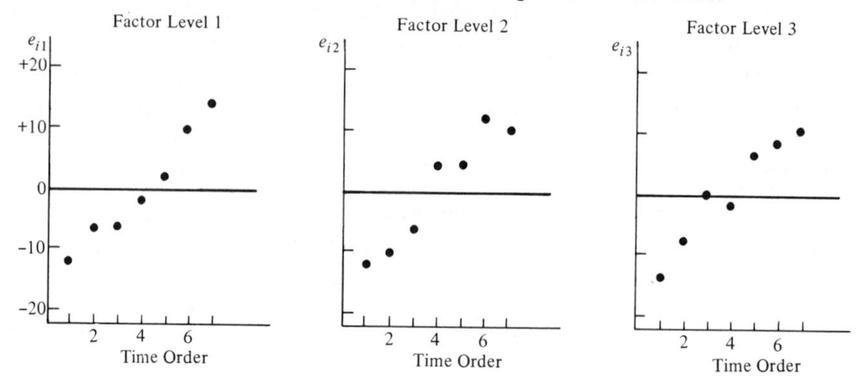

FIGURE 15.4

Residual Plots Illustrating Decreasing Error Variance Over Time

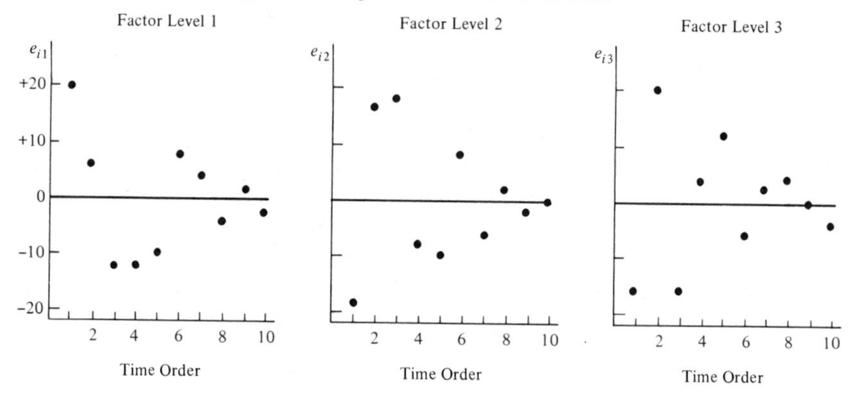

Time-related effects may also lead to increases or decreases in the error variance over time. For instance, an experimenter may make more precise measurements over time. Figure 15.4 portrays residual plots where the error variance decreases over time.

Outliers

Residual plots easily show outlier observations, that is, observations which differ from the factor level mean $\overline{Y}_{.j}$ by far more than do other observations. As noted in Chapter 4, it would appear to be wise practice to reject outlier observations only if they can be identified as being due to such specific causes as instrumentation malfunctioning, observer measurement blunder, or recording error.

Omission of Important Independent Variables

Residual analysis may also be used to study whether or not the single-factor ANOVA model is too coarse a model. Suppose that in a learning experiment involving three motivational treatments, the residuals shown in Figure 15.5 are obtained. The residual plots in Figure 15.5 show no unusual patterns. The experimenter wondered, however, whether the treatment effects differ according to the sex of the subject. The residuals for male subjects are circled in Figure 15.5, those for females are not. The results in Figure 15.5

FIGURE 15.5

Residual Plots Illustrating Omission of Important Independent Variable

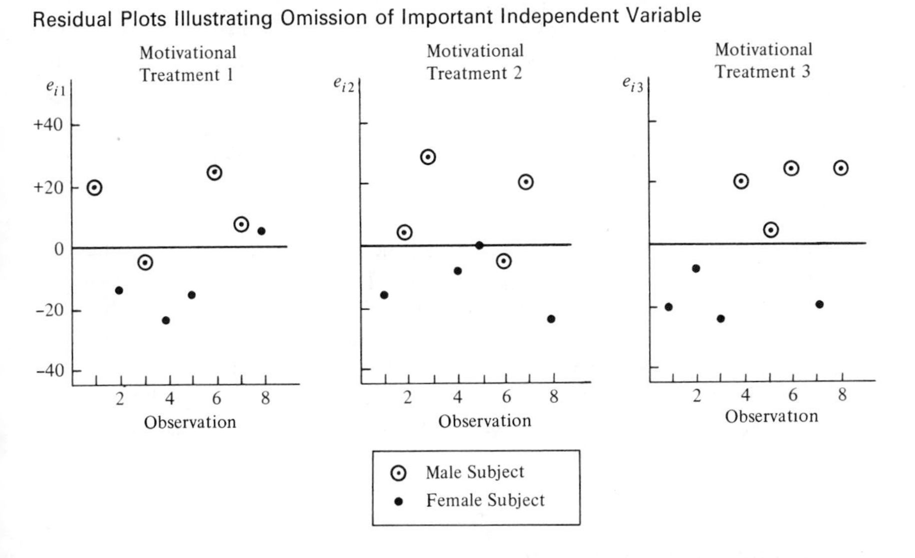

suggest strongly that for each of the motivational treatments studied, the treatment effects do differ according to sex. Hence, a multifactor model recognizing both motivational treatment and sex of subject as independent variables might be more useful.

It should be noted that residual analysis here does not invalidate the original single-factor model. Rather, the residual analysis points out that the original model overlooks differences in treatment effects which may be important to recognize. Since there are usually many independent variables which have some effect on the dependent variable, the analyst needs to identify those for residual analysis which are most likely to have an important effect on the dependent variable.

Nonnormality of Error Terms

Nonnormality of the error terms can be studied by plotting the residuals e_{ij} for each factor level in the form of a frequency distribution to see whether or not these distributions differ markedly from a normal distribution. Another possibility is to plot the cumulative distribution of the residuals on normal probability paper (see Chapter 4) and see whether or not the points fall approximately on a straight line. Chi-square goodness of fit tests or related tests may also be employed. If the factor level sample sizes are small, one can combine the e_{ij} for all treatments into one group.

Alternatively, one can work with the standardized residuals (15.10). If normality exists and n_j is not small, one would expect roughly 95 percent of the standardized residuals for any factor level to fall between -2 and $+2$, and so on. If the n_j are small, one can work with the standardized residuals (15.9) and analyze them for all factor levels combined.

It should be noted for all of these analyses that one reason why nonnormality may be indicated is failure of the error terms ε_{ij} to have equal variances.

Note

As we pointed out for regression models, the residuals e_{ij} are not independent random variables. For the ANOVA model (14.1), they are subject to restrictions stated earlier in (13.32). Consequently, statistical tests which require independent observations are not exactly appropriate for residuals. If, however, the number of residuals for each factor level is large, the effect of the correlations will be small. It has been noted that graphic plots of residuals are less subject to the effects of correlation than are statistical tests because graphic plots contain the individual residuals and not simply functions of them.

15.5 TRANSFORMATIONS

If residual or other analysis indicates that the ANOVA model is not appropriate for the data at hand, a choice of possible corrective measures is available. One is to modify the model. This approach may have the disadvantage at times of leading to fairly complex analysis. Another approach is to use transformations on the data. A third approach, useful when lack of normality is the basic difficulty, is to employ nonparametric tests such as the median test or the Kruskal-Wallis test based on ranks (discussed in Chapter 16).

The use of transformations is the principal topic of this section. Since transformations were discussed earlier in connection with regression analysis, our discussion here will be brief.

Transformations to Stabilize Variances

There are several types of situations in which the variances of the error terms are not constant, each of which requires a different type of transformation to stabilize the variance.

Variance Proportional to μ_j. When the variance of the error terms for each factor level (denoted by σ_j^2) varies proportionally to the factor level mean μ_j, the sample statistics $s_j^2/\bar{Y}_{.j}$ would tend to be constant. Here s_j^2 is the sample variance for the jth factor level observations, as defined in (13.50). This type of situation is often found when the observed variable Y is a count, such as the number of attempts by a subject before the correct solution is found. For this case, a square root transformation is helpful for stabilizing the variances:

(15.11) If σ_j^2 proportional to μ_j:

$$Y' = \sqrt{Y} \quad \text{or} \quad Y' = \sqrt{Y} + \sqrt{Y+1}$$

Standard Deviation Proportional to μ_j. When the standard deviation of the error terms for each factor level is proportional to the mean, $s_j/\bar{Y}_{.j}$ tends to be constant for the different factor levels. A helpful transformation in this case for stabilizing the variances is the logarithmic transformation:

(15.12) If σ_j proportional to μ_j: $\quad Y' = \log Y$

Standard Deviation Proportional to μ_j^2. When the error term standard deviation is proportional to the square of the factor level mean, $s_j/\bar{Y}_{.j}^2$ tends to be constant. An appropriate transformation here is the reciprocal transformation:

(15.13) If σ_j proportional to μ_j^2: $\quad Y' = \dfrac{1}{Y}$

Dependent Variable Is a Proportion. At times, the observed variable Y_{ij} is a proportion $\bar{p}_{ij}$. For instance, the treatments may be different training procedures, the unit of observation is a company training class, and the observed variable Y_{ij} is the proportion of employees in the ith class for the jth training procedure who benefited substantially by the training. Note that n_j here refers to the number of classes receiving the jth training procedure, not to the number of students.

It is well known that in the binomial case the variance of the sample proportion depends on the true proportion:

(15.14)
$$\sigma^2(\bar{p}_{ij}) = \frac{p_j(1-p_j)}{m}$$

Here p_j is the population proportion for the jth treatment and m is the number of cases on which each sample proportion is based. Since $\sigma^2(\bar{p}_{ij})$ depends on the treatment proportion p_j, the variances of the error terms will not be stable if the treatment proportions p_j differ. An appropriate transformation for this case is the arc sine transformation:

(15.15) If observation is a proportion: $\bar{p}' = 2 \arcsin \sqrt{\bar{p}}$

Tables to facilitate this transformation have been prepared, such as the one in Reference 15.1 which incorporates a slight refinement over transformation (15.15) to improve the variance stabilization.

When the proportions $\bar{p}_{ij}$ are based on different numbers of cases (for instance, in our earlier illustration there may be different numbers of employees in each training class), transformation (15.15) should be employed together with a weighted least squares analysis as described in Chapter 4.

Transformations to Correct Lack of Normality

If the error terms are normally distributed but with unequal variances, a transformation of the observations to stabilize the variances would destroy the normality. Fortunately, in practice, lack of normality and unequal variances tend to go hand in hand. Further, the transformation which helps to correct the lack of constancy of error variances usually also is effective in making the distribution of the error terms more normal. It is wise policy, however, as mentioned in Chapter 4 to check the residuals after a transformation has been applied to make sure that the transformation is effective in both stabilizing the variances and making the distribution of the error terms reasonably normal.

Comments

1. If a transformation of the observations is required, one can work completely with the transformed data for testing the equality of factor level means. On the other hand, it is usually desirable when making estimates to change the confidence intervals based on the transformed variable back to intervals in the original variable since it is ordinarily easier then to understand the significance of the results. For instance, in a pricing experiment most persons can understand better that the increase in mean sales with a reduction in price is $\mu_2 - \mu_1$, rather than a corresponding statement about $\log \mu_2 - \log \mu_1$.

2. It is generally advisable to plot s_j against $\bar{Y}_{.j}$ so that one can study the nature of the relationship between them, for instance, whether it is linear or quadratic. The choice of an appropriate transformation is thereby facilitated.

3. The transformations suggested earlier are obtained by the following argument. Suppose that Y is a random variable with mean μ and variance σ_Y^2 which is a function of μ, say $\sigma_Y^2 = g(\mu)$. For a transformation of Y, say $Y' = h(Y)$, it can be shown using a Taylor series expansion that:

(15.16) $$\sigma_{Y'}^2 \simeq [h'(\mu)]^2 g(\mu)$$

where $h'(\mu)$ is the first derivative of $h(Y)$ at μ. We seek that transformation h for which $\sigma_{Y'}^2$ is a constant, for convenience say 1. Hence, we wish:

$$[h'(\mu)]^2 g(\mu) = 1$$

or:

$$h'(\mu) = \frac{1}{\sqrt{g(\mu)}} \tag{15.17}$$

Formula (15.17) is a differential equation whose solution is (neglecting the arbitrary constant):

$$h(\mu) = \int \frac{d\mu}{\sqrt{g(\mu)}} \tag{15.18}$$

To illustrate the use of (15.18), suppose σ_Y^2 is proportional to μ, say $\sigma_Y^2 = k\mu = g(\mu)$. We obtain from (15.18):

$$h(\mu) = \int \frac{d\mu}{\sqrt{k\mu}} = \frac{2}{\sqrt{k}} \sqrt{\mu}$$

Thus, a square root transformation $Y' = \sqrt{Y}$ will yield a constant variance of Y', and $Y' = (2/\sqrt{k})\sqrt{Y}$ will yield a constant variance equal to 1.

Formula (15.16) makes it clear that the transformations obtained by this procedure only approximately stabilize the variances. It is therefore important to inspect the residuals for the transformed variable to see how effectively the variances have actually been stabilized.

15.6 TESTS FOR EQUALITY OF VARIANCES

Several formal tests are available for studying whether or not r populations have equal variances, as required by the ANOVA model. We shall consider two of these, the Bartlett test and the Hartley test. Both assume that each of the r populations is normal. Also, it is assumed by both tests that independent random samples are obtained from each population, in other words, that the error terms ε_{ij} are independent. The Bartlett test is an all-purpose test, and can be used for equal and unequal sample sizes. The Hartley test is applicable only if the sample sizes are equal, and it is designed to be sensitive against substantial differences between the largest and the smallest population variances.

Bartlett Test

The basic idea underlying the Bartlett test is simple. If $s_1^2, \ldots, s_r^2$ are the sample variances as defined in (13.50) from r normal populations, then it was shown in (13.51) that the weighted arithmetic average of these, using the associated degrees of freedom as weights, is the mean square error:

$$MSE = \frac{1}{n_T - r} \sum (n_j - 1)s_j^2 \tag{15.19}$$

Similarly, the weighted geometric average of the s_j^2, denoted $GMSE$, is:

$$GMSE = [(s_1^2)^{n_1 - 1}(s_2^2)^{n_2 - 1} \cdots (s_r^2)^{n_r - 1}]^{1/(n_T - r)} \tag{15.20}$$

It can be shown that for any given set of s_j^2 values, the following relation holds between these two averages:

$$GMSE \leq MSE \tag{15.21}$$

The two averages are equal if all s_j^2 are equal; the greater the variation between the s_j^2, the further apart the two averages will be. Hence, if the ratio $MSE/GMSE$ is close to 1, we have evidence that the population variances are equal. If $MSE/GMSE$ is large, it would indicate that the population variances are unequal. The same conclusions follow if we consider $\log(MSE/GMSE) = \log MSE - \log GMSE$.

Bartlett has shown that a function of $\log MSE - \log GMSE$ follows, for large sample sizes, approximately the χ^2 distribution with $r - 1$ degrees of freedom when the population variances are equal. This test statistic is:

$$B = \frac{2.302585}{C}(n_T - r)(\log_{10} MSE - \log_{10} GMSE) \tag{15.22}$$

which reduces to:

$$B = \frac{2.302585}{C}\left[(n_T - r)\log_{10} MSE - \sum_{j=1}^{r}(n_j - 1)\log_{10} s_j^2\right] \tag{15.22a}$$

where the term 2.302585 is a factor to convert logarithms on the base 10 to logarithms on the base e, and C is:

$$C = 1 + \frac{1}{3(r-1)}\left[\sum_{j=1}^{r}\frac{1}{n_j - 1} - \frac{1}{n_T - r}\right] \tag{15.22b}$$

The term C is always greater than 1.

Thus, for deciding between:

$$\begin{aligned} &C_1: \sigma_1^2 = \sigma_2^2 = \cdots = \sigma_r^2 \\ &C_2: \text{not all } \sigma_j^2 \text{ are equal} \end{aligned} \tag{15.23}$$

we calculate the statistic B. Since B is approximately distributed as χ^2 with $r - 1$ degrees of freedom when C_1 holds, and since we saw that large values of B lead to conclusion C_2, the appropriate decision rule for controlling the risk of a Type I error at α is:

$$\begin{aligned} &\text{If } B \leq \chi^2(1 - \alpha; r - 1), \text{ conclude } C_1 \\ &\text{If } B > \chi^2(1 - \alpha; r - 1), \text{ conclude } C_2 \end{aligned} \tag{15.24}$$

where $\chi^2(1 - \alpha; r - 1)$ is the $(1 - \alpha)100$ percentile of the χ^2 distribution with $r - 1$ degrees of freedom. Percentiles of the χ^2 distribution are given in Table A–3.

Example. Table 15.1 contains data from a study on the time required to complete a certain production operation for each of the three plant shifts. Note that 20 cycles of the operation were performed by shift one, 17 by shift

TABLE 15.1

Bartlett Test for Equality of Three Population Variances

Population	s_j^2	$n_j - 1$	$(n_j - 1)s_j^2$	$\log_{10} s_j^2$	$(n_j - 1)\log_{10} s_j^2$
1	415	19	7,885	2.61805	49.74295
2	698	16	11,168	2.84386	45.50176
3	384	20	7,680	2.58433	51.68660
Total		55	26,733		146.93131

$$MSE = 26{,}733 \div 55 = 486.05$$
$$\log_{10} MSE = 2.68668$$

two, and 21 by shift three. We now wish to determine by the Bartlett test whether the shift variances σ_j^2 are the same for all three shifts. The needed calculations for the test are shown in Table 15.1. Substituting into (15.22b) and then (15.22a), we obtain:

$$C = 1 + \frac{1}{3(3-1)}\left[\left(\frac{1}{19} + \frac{1}{16} + \frac{1}{20}\right) - \frac{1}{55}\right] = 1.02449$$

and:

$$B = \frac{2.302585}{1.02449}[55(2.68668) - 146.93131] = \frac{1.92517}{1.02449} = 1.88$$

Assume that we are to control the risk of making a Type I error at .05. We therefore require $\chi^2(.95; 3-1)$. From Table A–3, we find that $\chi^2(.95; 2) = 5.99$. Hence the decision rule is:

$$\text{If } B \leq 5.99, \text{ conclude } C_1$$
$$\text{If } B > 5.99, \text{ conclude } C_2$$

Since $B = 1.88$, we would conclude C_1, that the three population variances are equal.

Comments

1. Note in the above example that we could have avoided calculating the denominator C. Even before dividing by C, we can see that the numerator of the test statistic falls below the action limit 5.99. Since $C > 1$ always, the effect of dividing by C is to make the test statistic B still smaller. Thus, one may compute the numerator of B first, and only calculate the denominator C if it can affect the outcome.

2. If C_1 holds, the test statistic B follows approximately the χ^2 distribution when the sample sizes n_j are reasonably large. It has been suggested that $n_j \geq 5$ is sufficient for the χ^2 approximation to be a good one.

3. The Bartlett test is quite sensitive to departures from normality. That is, if the populations in fact are not normal, the actual level of significance may differ substantially from the specified one.

4. As will be noted in the next section, the F test is not much affected by unequal variances if the factor level sample sizes are equal, as long as the differences in the variances are not unusually large. Hence, a fairly low α level in conducting the Bartlett test may be justified when the equality of variances is tested for determining the aptness of the ANOVA model and the sample sizes are equal, since only large differences between variances need be detected.

Hartley Test

If the sample sizes n_j are all equal ($n_j \equiv n$), a simple test for deciding between:

$$\begin{aligned} &C_1: \sigma_1^2 = \sigma_2^2 = \cdots = \sigma_r^2 \\ &C_2: \text{not all } \sigma_j^2 \text{ are equal} \end{aligned} \tag{15.25}$$

is due to Hartley. It is based solely on the largest sample variance, denoted $\max(s_j^2)$, and the smallest sample variance, denoted $\min(s_j^2)$. The test statistic is:

$$H = \frac{\max(s_j^2)}{\min(s_j^2)} \tag{15.26}$$

Clearly, values of H near 1 support C_1, and large values of H support C_2. The distribution of H when C_1 holds has been tabulated, and selected percentiles are presented in Table A–12. The distribution of H depends on the number of factor levels r and the sample sizes n. As we noted earlier, the Hartley test, like the Bartlett test, assumes normal populations.

The appropriate decision rule for controlling the risk of making a Type I error at α is:

$$\begin{aligned} &\text{If } H \le H(1-\alpha; r, n), \text{ conclude } C_1 \\ &\text{If } H > H(1-\alpha; r, n), \text{ conclude } C_2 \end{aligned} \tag{15.27}$$

where $H(1-\alpha; r, n)$ is the $(1-\alpha)100$ percentile of the distribution of H when C_1 holds, for r factor levels and sample sizes of n.

Example. In a study of the appeal of four different television commercials, 10 observations were made for each commercial. The sample variances were as follows:

$$s_1^2 = 193 \qquad s_2^2 = 146 \qquad s_3^2 = 215 \qquad s_4^2 = 128$$

Before proceeding with the analysis of variance, it was desired to use the Hartley test to determine whether or not the four treatment variances are equal:

$$\begin{aligned} &C_1: \sigma_1^2 = \sigma_2^2 = \sigma_3^2 = \sigma_4^2 \\ &C_2: \text{not all } \sigma_j^2 \text{ are equal} \end{aligned}$$

The level of significance is to be controlled at .05. For $r = 4$ and $n = 10$, we require from Table A–12 $H(.95; 4, 10) = 6.31$. Hence, the appropriate decision rule is:

$$\text{If } H \leq 6.31, \text{ conclude } C_1$$
$$\text{If } H > 6.31, \text{ conclude } C_2$$

We have $\max(s_j^2) = 215$ and $\min(s_j^2) = 128$. Hence:

$$H = \frac{215}{128} = 1.68$$

We therefore conclude C_1, that all four treatment variances are equal.

Comments

1. The Hartley test strictly requires equal sample sizes. If the sample sizes are unequal but do not differ greatly, the Hartley test may still be used as an approximate test. For this purpose, the average sample size would be used for entering Table A–12.
2. The Hartley test, like the Bartlett test, is quite sensitive to departures from the assumption of normal populations.
3. For the same reasons noted for the Bartlett test, a low α level may be justified when the Hartley test is used for determining the aptness of the ANOVA model with respect to equal treatment variances.

15.7 EFFECTS OF DEPARTURES FROM MODEL

In preceding sections, we considered how residual analysis and other statistical techniques can be helpful in assessing the aptness of the ANOVA model for the data at hand. We also discussed the use of transformations, chiefly for stabilizing variances but also for obtaining as a by-product error distributions more nearly normal. The question now arises what are the effects of any remaining departures from the model on the inferences made. A thorough review of the many studies investigating these effects has been made by Scheffé (Ref. 15.2). Here, we shall summarize the findings.

Nonnormality

For the fixed effects model, lack of normality is not an important matter, provided the departure from normality is not of extreme form. It may be noted in this connection that *kurtosis* of the error distribution (either more or less peaked than a normal distribution) is more important than skewness of the distribution in terms of the effects on inferences.

The point estimators of factor level means and contrasts are unbiased whether or not the populations are normal. The F test for the equality of factor level means is but little affected by lack of normality, either in terms of the level of significance or power of the test. Hence, the F test is a *robust* test against departures from normality. For instance, the specified level of significance might be .05 whereas the actual level for a nonnormal error distribution might be .04 or .065. Typically, the achieved level of significance in the presence of nonnormality is slightly higher than the specified one, and the achieved power of the test is slightly less than the calculated one. Single interval estimates of factor level means and contrasts and the Scheffé multiple comparison method also are not much affected by lack of normality provided the sample sizes are not extremely small.

For the random effects model (to be discussed in the next chapter), lack of normality has more serious implications. The estimators of the variance components are still unbiased, but the actual confidence coefficient for interval estimates may be substantially different from the specified one.

Unequal Error Variances

If the error variances are unequal, the F test for the equality of means with the fixed effects model is only slightly affected if all factor level sample sizes are equal. Specifically, unequal error variances then raise the actual level of significance only slightly higher than the specified level. Similarly, the Scheffé multiple comparison procedure based on the F distribution is not affected to any substantial extent by unequal variances if the sample sizes are equal. Thus, the F test and related analyses are robust against unequal variances if the sample sizes are equal. Single comparisons between factor level means, on the other hand, can be substantially affected by unequal variances, so that the actual and specified confidence coefficients may differ markedly in these cases.

The use of equal sample sizes for all factor levels not only tends to minimize the effects of unequal variances on inferences with the F distribution but also simplifies calculational procedures. Thus here at least, simplicity and robustness go hand in hand.

For the random effects model, unequal error variances can have pronounced effects on inferences about the variance components, even with equal sample sizes.

Nonindependence of Error Terms

Lack of independence of the error terms can have serious effects on inferences in the analysis of variance, for both fixed and random effects models. Since this defect is often difficult to correct, it is important to prevent it in the first place whenever feasible. The use of randomization in those stages of a study which are likely to lead to correlated error terms can be a most important insurance policy. In the case of survey data, however, randomization may not be possible. Here, in the presence of correlated error terms, one may be able to modify the model. For instance, in the earlier discussion based on Figure 15.3, it was noted that inclusion in the model of a linear term for the learning effect of the analyst might remove the correlation of the error terms.

Modification of the model because of correlated error terms may also be necessary in experimental studies. In one case, the experimenter asked each of 10 subjects to give ratings to four new flavors of a fruit syrup and to the standard flavor, on a scale from 0 to 100. He applied the single-factor analysis of variance model but found high degrees of correlation in the residuals for each subject. He thereupon modified his model to a randomized complete block design model (see Chapter 23), which is often appropriate when the same subject is given each of the different treatments and differences between subjects are expected.

PROBLEMS

15.1. A market researcher stated in a seminar: "The power approach to determining sample size for analysis of variance problems is not meaningful, and only the estimation approach should be used. We never conduct a study where all treatment means are expected to be equal; hence we are always interested in a variety of estimates." Discuss.

15.2. Refer to Problem 13.1. Suppose the company had not yet decided on sample size but did wish to select the same number of employees from each length of service group.

a) What would be the required sample size if a maximum difference of 6 points between any two treatment means is to be detected with probability of at least .90, and the level of significance for the test is to be .05? Assume that $\sigma = 3$ points approximately.

b) Suppose pairwise comparisons are of primary importance. What would be the needed sample size if the precision of all pairwise comparisons is to be ± 2 points, with a 90 percent family confidence coefficient? Assume again that $\sigma = 3$ points approximately.

c) If the sample sizes in part (b) were employed, what would be the minimum power of the test (using $\alpha = .05$) when the maximum difference between any two groups is 6 points?

15.3. Refer to Problem 13.11. Suppose the sample sizes had not yet been determined but it had been decided to use the same number of used-car dealers for each owner.

a) What would be the required sample sizes if a maximum difference of 2 hundred dollars between any two owners is to be recognized with probability of at least .80, and the level of significance for the test is to be .01? Assume that $\sigma = 1.2$ hundred dollars approximately.

b) Suppose primary interest were in two comparisons: (1) between male owners (A and D) and female owners (B and C), and (2) between older owners (A and C) and younger owners (B and D). These two comparisons are to be estimated with a family confidence coefficient of 90 percent by the Bonferroni method, and the precision of each estimate is to be ± 1 hundred dollars. What are the needed sample sizes? Assume again that $\sigma = 1.2$ hundred dollars approximately.

15.4. Refer to the coil winding example on page 496. Suppose the test for equality of means for the five coil winding machines is to be conducted with $\alpha = .05$, and an increase in the standard deviation of 10 percent or more due to lack of calibration is to be detected with probability .90. What are the required sample sizes in this case?

15.5. Refer to Problem 13.10. Suppose the sample sizes had not yet been determined but it had been specified that the same number of transactions should be sampled for each agent. The objective of the study is to identify the agent with the smallest mean delay time.

a) What sample sizes would you suggest if it is important to identify the correct best agent when there is a difference of .5 days between the best and second best agents, and the probability of a correct identification in this case is to be .95? Assume that $\sigma = 1.2$ days.

b) Why do you think the approach utilized in part (a) does not consider the risk of an incorrect identification when the treatment means are the same or practically the same?

15.6. Refer to Figures 15.3 and 15.4. Both show effects that are related to time. What feature of the residual plots enables you to diagnose that in one case the error variance changes over time whereas in the other case the effect is of a different nature?

15.7. Refer to Figure 15.3. How would you modify the basic analysis of variance model (14.1) to include a linear term for the time effect?

15.8. For a single-factor study involving three treatments, demonstrate how the residual plots would appear if the error variance increased over time. Develop another set of residual plots which show the effect of error terms negatively correlated over time. Do your plotted residuals sum to zero for each treatment? Must they?

15.9. Refer to Problem 13.11.

a) Develop appropriate residual plots to investigate the aptness of the single-factor ANOVA model (14.1) for the data at hand.

b) Analyze your results.

c) Would a logarithmic transformation improve the aptness of model (14.1)? Develop the transformed data and obtain the appropriate residual plots.

15.10. Refer to Problem 13.11. Test whether or not the error variances for the four treatments are equal, using the Bartlett test with a level of significance of .01. What conclusion do you reach?

15.11. Refer to Problem 13.11. The analyst for the consumer organization was concerned whether any important independent variables were being omitted. In particular, he wondered whether type of used-car dealer (independent; affiliated with new-car dealer) affected the amount of cash offer. The relevant information is as follows (*I*–independent; *A*–affiliated):

	Owner			
Dealer	A	B	C	D
1	*A*	*I*	*A*	*I*
2	*I*	*A*	*A*	*A*
3	*A*	*A*	*I*	*A*
4	*I*	*A*	*I*	*A*
5	*I*	*I*	*A*	*I*
6	*A*	*I*	*I*	*I*

a) Develop residual plots to investigate whether type of used-car dealer is an important independent variable.
b) Analyze your results.

15.12. Refer to Problem 13.10.
a) Develop appropriate residual plots to investigate the aptness of the single-factor ANOVA model (14.1) for the data at hand.
b) Analyze your results.
c) Would a square root transformation improve the aptness of model (14.1)? Develop the transformed data and obtain the appropriate residual plots.

15.13. Refer to Problem 13.10. Test whether or not the error variances for the five treatments are equal, using the Hartley test with a level of significance of .01. What conclusion do you reach?

15.14. In a single-factor study with five treatments, the following results were obtained:

Treatment	n_j	$\bar{Y}_{.j}$	s_j^2
1	21	412	914
2	21	395	1,123
3	21	286	417
4	21	502	850
5	21	489	695

$MSE = 799.8$

a) Use the Bartlett test, with a level of significance of .01, to determine whether or not the treatment error variances are equal. What conclusion do you reach?
b) Would you reach the same conclusion as in part (a) with the Hartley test?
c) Would these tests be appropriate if the distribution of the error terms were far from normal?

CITED REFERENCES

15.1. Owen, Donald B. *Handbook of Statistical Tables*. Reading, Mass.: Addison-Wesley Publishing Co., Inc., 1962.

15.2. Scheffé, Henry. *The Analysis of Variance*. New York: John Wiley & Sons, Inc., 1959.

16

Topics in Analysis of Variance—I

IN THIS CHAPTER, we discuss three selected topics in analysis of variance: some alternative tests to the F test for deciding whether or not several population means are equal; model II for single-factor analysis of variance, appropriate when the treatment effects are random; and use of the regression approach to single-factor analysis of variance.

16.1 SOME ALTERNATIVES TO THE F TEST

There are two chief reasons for considering alternatives to the F test in (13.65) for studying whether or not r population means are equal: (1) greater simplicity of the alternative test, and (2) less restrictive model required by the alternative test.

Studentized Range Test

This test, which utilizes the studentized range distribution discussed in Section 14.3, has the merit of simplicity. It is only applicable when all factor level sample sizes are equal, that is, when $n_j \equiv n$. Analysis of variance model (14.1) is assumed, as before, to be appropriate.

For choosing between the two alternatives:

$$\begin{aligned} &C_1: \mu_1 = \mu_2 = \cdots = \mu_r \\ &C_2: \text{not all } \mu_j \text{ are equal} \end{aligned} \tag{16.1}$$

the studentized range test uses the test statistic:

$$q^* = \frac{\max(\overline{Y}_{.j}) - \min(\overline{Y}_{.j})}{\sqrt{\dfrac{MSE}{n}}} \tag{16.2}$$

where $\max(\bar{Y}_{.j})$ is the largest factor level sample mean $\bar{Y}_{.j}$ and $\min(\bar{Y}_{.j})$ is the smallest.

When C_1 holds, q^* is distributed as $q(r, n_T - r)$ as defined in (14.34). As is intuitively clear, large values of q^* lead to conclusion C_2, that the population means are unequal, and smaller values of q^* lead to conclusion C_1, that the population means are equal. Thus, the decision rule for controlling the risk of a Type I error at α is:

$$\text{If } q^* \leq q(1-\alpha; r, n_T - r), \text{ conclude } C_1$$
$$\text{If } q^* > q(1-\alpha; r, n_T - r), \text{ conclude } C_2 \tag{16.3}$$

where $q(1-\alpha; r, n_T - r)$ is the $(1-\alpha)100$ percentile of the appropriate q distribution.

Example. We consider again the rust inhibitor example of page 474. The sample results were:

$$\bar{Y}_{.1} = 43 \qquad MSE = 4.5$$
$$\bar{Y}_{.2} = 89 \qquad n = 5$$
$$\bar{Y}_{.3} = 67 \qquad n_T = 20$$
$$\bar{Y}_{.4} = 40$$

We wish to test:

$$C_1: \mu_1 = \mu_2 = \mu_3 = \mu_4$$
$$C_2: \text{not all } \mu_j \text{ are equal}$$

by means of the studentized range test. The test statistic for our example is:

$$q^* = \frac{89 - 40}{\sqrt{\dfrac{4.5}{5}}} = \frac{49}{\sqrt{.9}} = 52$$

Suppose a level of significance of .05 is specified. We then require:

$$q(.95; 4, 20 - 4)$$

From Table A-9, we find that $q(.95; 4, 16) = 4.05$. Hence the appropriate decision rule is:

$$\text{If } q^* \leq 4.05, \text{ conclude } C_1$$
$$\text{If } q^* > 4.05, \text{ conclude } C_2$$

Since $q^* = 52 > 4.05$ we conclude C_2, that the four rust inhibitors have different means.

Comments

1. The studentized range test has the merit of being somewhat simpler to perform than the F test.

2. If one is interested in testing C_1 primarily against an alternative C_2 based only on the maximum difference between any two population means, then the studentized range test is more efficient than the F test.

Kruskal-Wallis Rank Test

If the populations are far from normal—for instance, if they are highly skewed—the F test may be quite inappropriate. One possibility then is to use the Kruskal-Wallis test based on the ranks of the observations to test whether the population means are equal.

For this test, the only assumption required about the population distributions is that they are continuous and of the same shape. Thus, the population distributions must have the same variability, skewness, etc., but may differ as to the location of the mean. It is also assumed that the samples from the different populations are independent random ones.

First, all n_T observations are ranked from 1 to n_T. Let $\bar{R}_{.j}$ be the mean of the ranks for the jth factor level and $\bar{R}_{..}$ the overall mean rank. The test statistic is then simply:

$$K = \frac{\sum_{j=1}^{r} n_j(\bar{R}_{.j} - \bar{R}_{..})^2}{\frac{n_T(n_T+1)}{12}} \tag{16.4}$$

The numerator is the usual treatment sum of squares, but with the data expressed in ranks, and the denominator is the variance of the ranks 1, 2, 3, $\ldots, n_T$. The test statistic K can be expressed equivalently as follows:

$$K = \frac{12}{n_T(n_T+1)} \sum_{j=1}^{r} n_j \bar{R}_{.j}^2 - 3(n_T+1) \tag{16.4a}$$

If the n_j are reasonably large (5 or more is the usual advice), K is approximately a χ^2 random variable with $r - 1$ degrees of freedom when C_1 (all μ_j are equal) holds. Large values of K, as expected, lead to C_2 (not all μ_j are equal). Thus, in choosing between:

$$\begin{aligned} &C_1: \mu_1 = \mu_2 = \cdots = \mu_r \\ &C_2: \text{not all } \mu_j \text{ are equal} \end{aligned} \tag{16.5}$$

the appropriate decision rule for controlling the risk of making a Type I error at α is:

$$\begin{aligned} &\text{If } K \leq \chi^2(1-\alpha; r-1), \text{ conclude } C_1 \\ &\text{If } K > \chi^2(1-\alpha; r-1), \text{ conclude } C_2 \end{aligned} \tag{16.6}$$

Example. Servo-Data, Inc., operates three electronic computers at three different locations. The computers are identical as to make and model, but are subject to different degrees of voltage fluctuation in the power lines serving the respective installations. Management wishes to test whether or not the mean lengths of operating time between failures are the same for the three computers. The alternative conclusions are:

$$\begin{aligned} &C_1: \mu_1 = \mu_2 = \mu_3 \\ &C_2: \text{not all } \mu_j \text{ are equal} \end{aligned}$$

TABLE 16.1

Times between Failure for Three Computers (in hours)

	Computer A		*Computer* B		*Computer* C	
Observation	*Time*	*Rank*	*Time*	*Rank*	*Time*	*Rank*
1	105	11	56	8	183	13
2	3	2	43	7	144	12
3	90	10	1	1	219	15
4	217	14	37	5	86	9
5	22	4	14	3	39	6
Mean		8.2		4.8		11.0

Table 16.1 contains the lengths of time between failures for the three different computers, for five failure intervals each. Even though the sample sizes are small, they suggest highly skewed distributions. In the same table, the data are ranked from 1 to 15 and the mean ranks are shown.

Using the computational formula (16.4a), we obtain for the test statistic:

$$K = \frac{12}{(15)(16)} 5[(8.2)^2 + (4.8)^2 + (11.0)^2] - 3(16) = 4.8$$

A level of significance of .10 has been specified. Since $r = 3$, we require $\chi^2(.90; 2)$. From Table A–3, we find that $\chi^2(.90; 2) = 4.61$. The decision rule for choosing between C_1 and C_2 is:

$$\text{If } K \leq 4.61, \text{ conclude } C_1$$
$$\text{If } K > 4.61, \text{ conclude } C_2$$

Since we have $K = 4.8 > 4.61$, we conclude C_2, that the mean time between failures differs for the three computers.

Comments

1. The Kruskal-Wallis test, like the F test, does not require equal sample sizes.

2. If the n_j are small so that the χ^2 approximation is not appropriate, special tables should be used for conducting the Kruskal-Wallis test; see, for instance, the tables by Owen in Reference 16.1.

3. In case of ties among some observations, each of the tied observations is given the mean of the ranks involved. Thus, if two observations are tied for what would otherwise have been the third and fourth ranked positions, each would be given the mean value 3.5. If a large number of ties exist, the test statistic in (16.4) needs to be modified.

4. The Kruskal-Wallis test can also be used to choose among the alternatives:

(16.7)
$$C_1\text{: all populations are identical}$$
$$C_2\text{: all populations are not identical}$$

This statement of the decision problem avoids the earlier assumption that all populations are identical except for the location of the means. If conclusion C_2 in (16.7) is reached, however, one cannot identify the reason for the difference. For example, the means might differ, or the variances, or the nature of the skewness, or some combination of these.

Median Test

The median test is another test which may be utilized when populations are far from normally distributed. With this test, one would be interested in choosing between the two alternatives:

$$\begin{aligned} &C_1\text{: all population medians are equal} \\ &C_2\text{: not all population medians are equal} \end{aligned} \tag{16.8}$$

The median test assumes only that all populations are of the same shape, but they may differ as to the location of the median. Also it is assumed that the samples from the different populations are independent random ones.

All sample data are combined to determine the median value for the combined sample. For each factor level, the number of observations above this median value and the number not above it are then ascertained. Finally, a test for homogeneity is conducted using the test statistic:

$$X^2 = \sum_i^2 \sum_j^r \frac{(f_{ij} - F_{ij})^2}{F_{ij}} \tag{16.9}$$

where f_{ij} $(i = 1, 2; j = 1, \ldots, r)$ is the observed frequency in a cell and F_{ij} is the expected frequency when all population medians are equal.

When the sample sizes are reasonably large, the test statistic X^2 is distributed approximately as χ^2 with $r - 1$ degrees of freedom when C_1 holds. Large values of X^2 lead to conclusion C_2. Hence, the appropriate decision rule for controlling the risk of making a Type I error at α is:

$$\begin{aligned} &\text{If } X^2 \leq \chi^2(1 - \alpha; r - 1), \text{ conclude } C_1 \\ &\text{If } X^2 > \chi^2(1 - \alpha; r - 1), \text{ conclude } C_2 \end{aligned} \tag{16.10}$$

Example. We consider again the Servo-Data example dealing with the time between failures for three different computers. Fifteen additional observations on time between failures were made for each computer to provide more precise information. The median number of hours between failures for the combined sample of 60 observations was 64 hours. Table 16.2 summarizes the results for the three samples (the original data are not shown). The expected frequencies when all population medians are equal are shown in parentheses in Table 16.2. These are obtained by allocating the total frequencies in each row to the three computers in proportion to the total number of observations for each computer. In our example, 20 observations were made on each computer; hence the 30 frequencies above the median are allocated

TABLE 16.2

Times between Failure for Three Computers

	Computer			
	A	B	C	*Total*
Number of observations above median (64 hours)	13 (10)	3 (10)	14 (10)	30
Number of observations not above	7 (10)	17 (10)	6 (10)	30
Total	20	20	20	60

equally to each computer. Similarly, the 30 frequencies not above the median are allocated equally to each computer. These then are the frequencies which are expected if the three population medians are equal.

The test statistic is calculated using (16.9):

$$X^2 = \frac{(13-10)^2}{10} + \frac{(7-10)^2}{10} + \cdots + \frac{(6-10)^2}{10} = 14.8$$

Suppose it has been specified that the level of significance is to be .05. We therefore require $\chi^2(.95; 2) = 5.99$. The decision rule is:

$$\text{If } X^2 \leq 5.99, \text{ conclude } C_1$$
$$\text{If } X^2 > 5.99, \text{ conclude } C_2$$

Since $X^2 = 14.8 > 5.99$, we conclude C_2, that the median number of hours between failure is not the same for the three computers.

Note

Like the Kruskal-Wallis test, the median test does not require equal sample sizes.

16.2 ANOVA MODEL II—RANDOM EFFECTS

As we noted earlier, there are occasions when the employed factor levels are not of intrinsic interest in themselves, but constitute a sample from a larger population of factor levels. Model II is designed for this type of situation. Consider, for instance, Apex Enterprises, a company that builds roadside restaurants carrying one of several promoted trade names, leases franchises to individuals to operate the restaurants, and provides management services. This company employs a large number of personnel officers who interview applicants for jobs in the restaurants. At the end of the interview the personnel officer assigns a subjective rating between 0 and 100 to indicate the applicant's potential value on the job. Suppose now that five personnel officers are selected at random and each is assigned four candidates at random. In this case, the company would not wish to make inferences concerning the five personnel officers who happened to be selected but rather about the population of all personnel officers. Questions of interest might be: How great is the variation in ratings between all personnel officers? Or: What is the mean rating by all personnel officers?

The distinction between this situation, for which model II is designed, and one where model I is appropriate can be seen readily by modifying our example slightly. If a smaller company had only five personnel officers who were all included in the study and interest was limited to these five officers, model I would be relevant since the factor levels (the five personnel officers) are then not considered a sample from a larger population. A repetition of the experiment for the smaller company would involve the same five personnel officers, but in the case of the large company, a repetition would involve a new sample of five personnel officers which likely would consist of different officers.

Model

Model II for single-factor analysis of variance is as follows:

(16.11) $$Y_{ij} = \mu_j + \varepsilon_{ij}$$

where:

μ_j are independent $N(\mu_., \sigma_\tau^2)$
ε_{ij} are independent $N(0, \sigma^2)$
μ_j and ε_{ij} are independent random variables
$i = 1, \ldots, n_j; j = 1, \ldots, r$

Model (16.11) is similar in appearance to fixed effects model (13.2). The main distinction is that the factor level means μ_j are constants for model I but are random variables for model II. Hence, model II is often called a *random effects* model.

Meaning of Model Terms. We shall explain the meaning of the model terms with reference to our earlier personnel officers example. The term μ_j corresponds in our example to the mean of all ratings by the jth personnel officer if he interviewed all prospective employees. The expected value of μ_j is $\mu_.$. Thus, $\mu_.$ represents in our example the mean rating for all prospective employees by all personnel officers. The variability of the μ_j is measured by the variance σ_τ^2. The more the different personnel officers vary in their mean ratings (for instance, some may rate consistently higher than others), the greater will be σ_τ^2. On the other hand, if all personnel officers rate at the same mean level, all μ_j will be equal to $\mu_.$ and then $\sigma_\tau^2 = 0$.

The term ε_{ij} represents in our example the variation associated with the differing potential values of different prospective employees. Note that model (16.11) assumes that all ε_{ij} have the same variance σ^2. This means that the distributions of ratings for prospective employees by the different personnel officers are assumed to have the same variability. The distributions for the different personnel officers may differ, however, with respect to the mean levels of the distributions.

Figure 16.1 illustrates the ANOVA model II. On the top is shown the distribution of the μ_j, which is normal. A number of μ_j (two in the illustration) are selected at random from this distribution. Each in turn leads to a distribution of $Y_{ij} = \mu_j + \varepsilon_{ij}$, which are all normal distributions. A number of Y_{ij} (two each in the illustration) are then selected from each of these distributions

FIGURE 16.1

Representation of ANOVA Model II

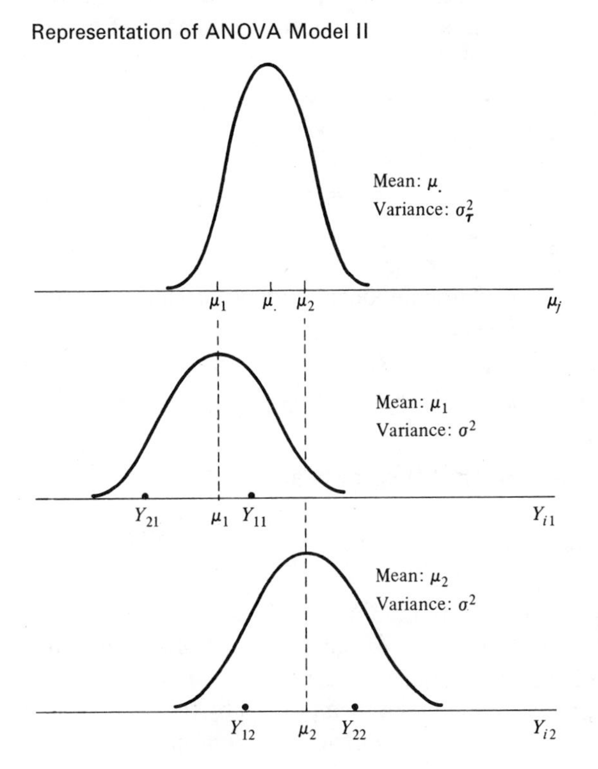

Important Features of Model. 1. The expected value of an observation Y_{ij} is:

$$E(Y_{ij}) = \mu_. \tag{16.12}$$

because we have by (16.11):

$$\begin{aligned} E(Y_{ij}) &= E(\mu_j) + E(\varepsilon_{ij}) \\ &= \mu_. + 0 \\ &= \mu_. \end{aligned}$$

Note that this expectation averages over both the selection of μ_j and that of ε_{ij}.

2. The variance of Y_{ij} is:

$$\sigma^2(Y_{ij}) = \sigma_\tau^2 + \sigma^2 \tag{16.13}$$

This result follows because model II assumes that μ_j and ε_{ij} are independent random variables, and $\sigma^2(\mu_j) = \sigma_\tau^2$ and $\sigma^2(\varepsilon_{ij}) = \sigma^2$ according to model (16.11). Because the variance of Y in this model is the sum of two components, σ_τ^2 and σ^2, this model is sometimes called the *components of variance* model.

3. Finally, the Y_{ij} are normally distributed because they are linear combinations of the independent normal variables μ_j and ε_{ij}. Note, however, that the Y_{ij} are not independent because a number of different Y_{ij} have the same component μ_j.

Alternate Formulation of Model II. We can express model (16.11) in the following alternate form:

$$Y_{ij} = \mu_. + \tau_j + \varepsilon_{ij} \tag{16.14}$$

where:

$\mu_.$ is a constant component common to all observations
τ_j are independent $N(0, \sigma_\tau^2)$
ε_{ij} are independent $N(0, \sigma^2)$
τ_j and ε_{ij} are independent
$i = 1, \ldots, n_j; j = 1, \ldots, r$

Model formulation (16.14) corresponds to the fixed effects model (13.6) in that:

$$\tau_j = \mu_j - \mu_. \tag{16.15}$$

Here, however, the τ_j are random variables whereas they are fixed quantities for model I. With reference to our personnel officers example, τ_j represents the effect of the jth personnel officer. That is, τ_j measures by how much the mean evaluation of all potential employees by the jth personnel officer differs from the mean evaluation by all personnel officers.

Note

At times, the population of μ_j's is relatively small and should be treated as a finite population. This can be done, but we do not discuss this case here. If the population of the μ_j is finite but large, little is lost in treating it as an infinite population. We did this, in fact, in our illustration of the personnel officers. The number of officers is finite, but since there are many, we treated the population of μ_j's as an infinite one. Thus, there are two basic situations when the population of μ_j's is treated as infinite—when the population is finite but large, and when interest centers in the underlying *process* generating the μ_j.

Questions of Interest

When model II is appropriate, one is usually not interested in inferences about the particular μ_j included in the study, such as which is the highest or lowest, but rather in inferences about the entire population of the μ_j. Specifically, interest often centers on the mean of the μ_j, $\mu_.$, and on the variability of the μ_j, measured by σ_τ^2. In our personnel officers example, for instance, management would not ordinarily be as interested in the mean ratings of the five personnel officers who happened to be included in the study as in the mean rating by all personnel officers and in the effect of variability among all personnel officers.

While σ_τ^2 is a direct measure of the variability of the μ_j, the effect of this variability is often measured more meaningfully by the ratio:

$$\frac{\sigma_\tau^2}{\sigma_\tau^2 + \sigma^2} \tag{16.16}$$

Note the following characteristics of this ratio:

1. The ratio takes on values between 0 (when $\sigma^2 = \infty$) and 1 (when $\sigma^2 = 0$).
2. The denominator is $\sigma^2(Y_{ij})$ according to (16.13).
3. In view of property 2, the ratio measures the proportion of the total variability of the Y_{ij} which is accounted for by the variability of the μ_j.

With reference to our personnel officers example, the denominator of the ratio measures the variability of ratings for all candidates by all personnel officers, and the numerator measures the variability of the mean ratings by all personnel officers. The ratio then measures the proportion of the total variability of ratings which is accounted for by differences among the personnel officers. If the ratio is near zero, the effect of differences among personnel officers on the total variability is relatively insignificant. On the other hand, if the ratio is large, say .5 or more, then much of the total variability is accounted for by differences between personnel officers, and management may wish to study the advisability of giving the personnel officers more training to improve uniformity of ratings between officers.

Test Whether $\sigma_\tau^2 = 0$

We first consider how to decide between:

$$\begin{aligned} &C_1: \sigma_\tau^2 = 0 \\ &C_2: \sigma_\tau^2 \neq 0 \end{aligned} \tag{16.17}$$

C_1 implies that all μ_j are equal, that is, $\mu_j \equiv \mu_.$. C_2 implies that the μ_j differ. For our personnel officers example, C_1 implies that the mean ratings for all personnel officers are equal, while C_2 implies that they differ.

Despite the fact that model II differs from model I, the analysis of variance for a single-factor analysis is conducted in identical fashion. (This is not always the case in more complex situations.) The difference between the two models appears in the expected mean squares. It can be shown in a similar manner to that employed in our derivation for model I that, for model II:

$$E(MSE) = \sigma^2 \tag{16.18}$$

$$E(MSTR) = \sigma^2 + n'\sigma_\tau^2 \tag{16.19}$$

where:

$$n' = \frac{1}{r-1}\left[\sum n_j - \frac{\sum n_j^2}{\sum n_j}\right] \tag{16.19a}$$

We may view n' as an average n_j. If all $n_j = n$, $n' = n$.

It is clear from (16.18) and (16.19) that if $\sigma_\tau^2 = 0$, MSE and $MSTR$ have the same expection σ^2. Otherwise, $E(MSTR) > E(MSE)$ since $n' > 0$ always. Hence, large values of the test statistic:

$$F^* = \frac{MSTR}{MSE} \tag{16.20}$$

will lead to conclusion C_2 in (16.17). Since F^* again follows the F distribution when C_1 holds, the decision rule for controlling the risk of making a Type I error at α is the same one as for model I:

$$\begin{aligned} &\text{If } F^* \leq F(1-\alpha; r-1, n_T - r), \text{ conclude } C_1 \\ &\text{If } F^* > F(1-\alpha; r-1, n_T - r), \text{ conclude } C_2 \end{aligned} \tag{16.21}$$

Example. Table 16.3 contains the results of a study by Apex Enterprises on the evaluation ratings of potential employees by its personnel officers. Five personnel officers were selected at random, and each was assigned at random 4 prospective employees. The ANOVA calculations are routine and were done by a computer program package. The results are shown in Table 16.4, which also shows the expected mean squares in general and for this particular example.

TABLE 16.3

Ratings by Five Personnel Officers

	Officer (j)					
Candidate (i)	A	B	C	D	E	*Mean*
1	76	58	49	74	66	
2	64	75	63	71	74	
3	85	81	62	85	81	
4	75	66	46	90	79	
Mean	$\bar{Y}_{.1} = 75$	$\bar{Y}_{.2} = 70$	$\bar{Y}_{.3} = 55$	$\bar{Y}_{.4} = 80$	$\bar{Y}_{.5} = 75$	$\bar{Y}_{..} = 71$

TABLE 16.4

ANOVA Table for Single-Factor Model II Personnel Officers Study

Source of Variation	*SS*	*df*	*MS*	*E(MS) General*	*E(MS) Example*
Between personnel officers	$SSTR = 1{,}480$	4	$MSTR = 370$	$\sigma^2 + n'\sigma_\tau^2$	$\sigma^2 + 4\sigma_\tau^2$
Error (within personnel officers)	$SSE = 1{,}134$	15	$MSE = 75.6$	σ^2	σ^2
Total	$SSTO = 2{,}614$	19			

$$n' = \frac{1}{r-1}\left[\sum n_j - \frac{\sum n_j^2}{\sum n_j}\right]$$

$$(n' = n \text{ if all } n_j = n)$$

The appropriate test statistic, using the data from Table 16.4, is:

$$F^* = \frac{370}{75.6} = 4.89$$

Assuming that we are to control the risk of making a Type I error at .05, we require $F(.95; 4, 15) = 3.06$. Hence, the decision rule is:

$$\text{If } F^* \leq 3.06, \text{ conclude } C_1$$
$$\text{If } F^* > 3.06, \text{ conclude } C_2$$

Since $F^* = 4.89 > 3.06$, we conclude C_2, that $\sigma_\tau^2 \neq 0$ or that the mean ratings of the personnel officers differ.

Note

We shall illustrate the derivation of an expected mean square for model II by sketching the development for deriving $E(MSTR)$ in (16.19) when $n_j \equiv n$. The proof parallels that for model I. By model (16.14), we can write:

$$\bar{Y}_{.j} = \mu. + \tau_j + \bar{\varepsilon}_{.j}$$
$$\bar{Y}_{..} = \mu. + \bar{\tau}. + \bar{\varepsilon}_{..}$$

where $\bar{\varepsilon}_{.j}$ and $\bar{\varepsilon}_{..}$ are defined in (13.55) and (13.57) respectively, and:

$$\bar{\tau}. = \frac{\sum \tau_j}{r}$$

Hence, corresponding to (13.59), we obtain:

$$\bar{Y}_{.j} - \bar{Y}_{..} = (\tau_j - \bar{\tau}.) + (\bar{\varepsilon}_{.j} - \bar{\varepsilon}_{..})$$

so that:

$$\sum(\bar{Y}_{.j} - \bar{Y}_{..})^2 = \sum(\tau_j - \bar{\tau}.)^2 + \sum(\bar{\varepsilon}_{.j} - \bar{\varepsilon}_{..})^2 + 2\sum(\tau_j - \bar{\tau}.)(\bar{\varepsilon}_{.j} - \bar{\varepsilon}_{..})$$

When we take the expectation, the cross-product term drops out because of the independence of the τ_j and ε_{ij} and because these variables have expectation zero. From (13.62) we know that:

$$E[\sum(\bar{\varepsilon}_{.j} - \bar{\varepsilon}_{..})^2] = \frac{(r-1)\sigma^2}{n}$$

Lastly, since $\sum(\tau_j - \bar{\tau}_.)^2$ is the numerator of an ordinary sample variance for r independent τ_j observations, it follows from the unbiasedness of the sample variance that:

$$E[\sum(\tau_j - \bar{\tau}_.)^2] = (r-1)\sigma_\tau^2$$

Hence we obtain:

$$E\left[\frac{n}{r-1}\sum(\bar{Y}_{.j} - \bar{Y}_{..})^2\right] = \frac{n}{r-1}\left[(r-1)\sigma_\tau^2 + \frac{r-1}{n}\sigma^2\right] = n\sigma_\tau^2 + \sigma^2$$

which is the result in (16.19) for the case $n_j \equiv n$.

Estimation of $\mu_.$

When ANOVA model II is applicable, one is frequently interested in estimating the overall mean $\mu_.$. We shall assume in developing an interval estimate for $\mu_.$ that *all factor level sample sizes are equal*, that is, that $n_j \equiv n$. We know from (16.12) that:

$$E(Y_{ij}) = \mu_.$$

Hence, an unbiased estimator of $\mu_.$ is:

(16.22) $$\hat{\mu}_. = \bar{Y}_{..}$$

It can be shown that the variance of this estimator is:

(16.23) $$\sigma^2(\bar{Y}_{..}) = \frac{\sigma_\tau^2}{r} + \frac{\sigma^2}{n_T} = \frac{n\sigma_\tau^2 + \sigma^2}{n_T}$$

(Remember that $n_T = rn$ here.)

Formula (16.23) shows that the variance of $\bar{Y}_{..}$ is made up of two components. The first corresponds to the variance of a sample mean based on r observations when sampling from the population of μ_j's, and reflects the contribution due to sampling the factor levels. The second component corresponds to the variance of a sample mean based on n_T observations when sampling from the populations of Y_{ij}'s, given the μ_j, and reflects the contribution due to variation within factor levels.

An unbiased estimator of $\sigma^2(\bar{Y}_{..})$ is:

(16.24) $$s^2(\bar{Y}_{..}) = \frac{MSTR}{n_T}$$

This estimator is unbiased because we know from (16.19) that when $n_j \equiv n$:

(16.25) $$E(MSTR) = n\sigma_\tau^2 + \sigma^2 \qquad \text{when } n_j \equiv n$$

(Remember that $n' = n$ when $n_j \equiv n$.)

It can be shown that:

(16.26) $$\frac{\bar{Y}_{..} - \mu_.}{s(\bar{Y}_{..})} \text{ is distributed as } t(r-1) \text{ for model (16.11)}$$

Hence we obtain in usual fashion the confidence interval:

(16.27) $$\bar{Y}_{..} - t(1-\alpha/2; r-1)s(\bar{Y}_{..}) \le \mu_. \le \bar{Y}_{..} + t(1-\alpha/2; r-1)s(\bar{Y}_{..})$$

Example. Management of Apex Enterprises wishes to estimate the mean rating for all prospective employees by all personnel officers, with a 90 percent confidence coefficient. We have from Tables 16.3 and 16.4:

$$\bar{Y}_{..} = 71 \qquad MSTR = 370 \qquad n_T = 20$$

We require $t(.95; 4) = 2.132$ and:

$$s(\bar{Y}_{..}) = \left(\frac{370}{20}\right)^{1/2} = 4.301$$

Hence the desired 90 percent confidence interval is:

$$71 - (2.132)(4.301) \le \mu_. \le 71 + (2.132)(4.301)$$

or:

$$62 \le \mu_. \le 80$$

Thus, we conclude that the mean rating assigned by all personnel officers to all prospective employees is between 62 and 80 (90 percent confidence coefficient). The interval estimate is not too precise because of the relatively small sample sizes of personnel officers and potential employees.

Note

The variance of $\bar{Y}_{..}$ in (16.23) can be derived readily. First, we consider:

$$\bar{Y}_{.j} = \mu_. + \tau_j + \bar{\varepsilon}_{.j}$$

where $\bar{\varepsilon}_{.j}$ is defined in (13.55). Because of the independence of τ_j and the ε_{ij}, we have:

$$\sigma^2(\bar{Y}_{.j}) = \sigma_\tau^2 + \frac{\sigma^2}{n}$$

Remember that $\bar{\varepsilon}_{.j}$ is just an ordinary mean of n independent ε_{ij} observations.

For the case $n_j \equiv n$ which we are considering here, we have:

$$\bar{Y}_{..} = \frac{\sum_{j=1}^{r} \bar{Y}_{.j}}{r}$$

In view of the independence of the τ_j and the ε_{ij} among themselves and between each other, it follows that the $\bar{Y}_{.j}$ are independent so that:

$$\sigma^2(\bar{Y}_{..}) = \frac{\sigma^2(\bar{Y}_{.j})}{r} = \frac{\sigma_\tau^2}{r} + \frac{\sigma^2}{rn} = \frac{n\sigma_\tau^2 + \sigma^2}{n_T}$$

Note

Estimation of $\sigma_\tau^2/(\sigma_\tau^2 + \sigma^2)$

As noted earlier, the ratio $\sigma_\tau^2/(\sigma_\tau^2 + \sigma^2)$ reveals meaningfully the effect of the extent of variation between the μ_j. To develop an interval estimate for this ratio, we shall assume that *all factor level sample sizes are equal*, that is, that $n_j \equiv n$.

We begin by obtaining confidence limits for the ratio σ_τ^2/σ^2. First, we need to note that $MSTR$ and MSE are independent random variables for model II, just as for model I. It can be shown further, when $n_j \equiv n$, that:

$$\frac{MSTR}{n\sigma_\tau^2 + \sigma^2} \div \frac{MSE}{\sigma^2} = F(r-1, n_T - r) \tag{16.28}$$

Hence, we can write the probability statement:

$$P\left\{F(\alpha/2; r-1, n_T - r) \le \frac{MSTR}{n\sigma_\tau^2 + \sigma^2} \div \frac{MSE}{\sigma^2} \le F(1-\alpha/2; r-1, n_T - r)\right\} = 1 - \alpha \tag{16.29}$$

Rearranging the inequalities, we obtain the following confidence limits for σ_τ^2/σ^2:

$$L_L = \frac{1}{n}\left[\frac{MSTR}{MSE}\frac{1}{F(1-\alpha/2; r-1, n_T - r)} - 1\right] \tag{16.30a}$$

$$L_U = \frac{1}{n}\left[\frac{MSTR}{MSE}\frac{1}{F(\alpha/2; r-1, n_T - r)} - 1\right] \tag{16.30b}$$

where L_L is the lower confidence limit and L_U the upper.

The confidence interval for $\sigma_\tau^2/(\sigma_\tau^2 + \sigma^2)$ can now be readily obtained and is as follows:

$$\frac{L_L}{1 + L_L} \le \frac{\sigma_\tau^2}{\sigma_\tau^2 + \sigma^2} \le \frac{L_U}{1 + L_U} \tag{16.31}$$

Example. Management of Apex Enterprises wishes a 90 percent confidence interval for $\sigma_\tau^2/(\sigma_\tau^2 + \sigma^2)$. From previous work, we have:

$$MSTR = 370 \qquad MSE = 75.6 \qquad n = 4 \qquad r = 5 \qquad n_T = 20$$

For a 90 percent confidence coefficient, we require:

$$F(.05; 4, 15) = .170 \qquad F(.95; 4, 15) = 3.06$$

Hence, the 90 percent confidence interval for σ_τ^2/σ^2 is by (16.30):

$$\frac{1}{4}\left(\frac{370}{75.6}\frac{1}{3.06} - 1\right) \leq \frac{\sigma_\tau^2}{\sigma^2} \leq \frac{1}{4}\left(\frac{370}{75.6}\frac{1}{.170} - 1\right)$$

$$.15 \leq \frac{\sigma_\tau^2}{\sigma^2} \leq 6.9$$

Finally, the confidence interval for $\sigma_\tau^2/(\sigma_\tau^2 + \sigma^2)$ is by (16.31):

$$.13 = \frac{.15}{1.15} \leq \frac{\sigma_\tau^2}{\sigma_\tau^2 + \sigma^2} \leq \frac{6.9}{7.9} = .87$$

Hence, we conclude that the variability of the mean ratings for the different personnel officers accounts for somewhere between 13 and 87 percent of the total variance of the ratings (90 percent confidence coefficient). Note that this interval estimate is not very precise. The reason is the relatively small sample sizes. The confidence interval does indicate, though, that the variability of personnel officers accounts for at least 13 percent of the total variability and is therefore not negligible.

Comments

1. It may happen occasionally that the lower limit of the confidence interval for σ_τ^2/σ^2 is negative. Since this ratio cannot be negative, the usual practice is to consider the lower limit L_L in (16.30a) to be zero in that case.

2. If one-sided or two-sided tests concerning the relative magnitudes of σ_τ^2 and σ^2 are desired, such as the following (where c is a specified constant):

$$C_1: \sigma_\tau^2 \leq c\sigma^2 \qquad C_1: \sigma_\tau^2 = c\sigma^2$$
$$C_2: \sigma_\tau^2 > c\sigma^2 \qquad C_2: \sigma_\tau^2 \neq c\sigma^2$$

the decision rule can be constructed by utilizing (16.28). Alternatively, one-sided or two-sided confidence intervals can be set up from which the appropriate conclusion can be drawn. For instance, suppose that for our personnel officers example we are considering:

$$C_1: \sigma_\tau^2 = \frac{1}{2}\sigma^2$$

$$C_2: \sigma_\tau^2 \neq \frac{1}{2}\sigma^2$$

Since the 90 percent confidence interval for σ_τ^2/σ^2 above (corresponding to a level of significance of .10) includes .5, the conclusion to be reached is C_1.

3. The ratio σ_τ^2/σ^2 is of relevance in planning investigations. In our example dealing with the personnel officers, suppose that the mean rating $\mu_.$ is to be estimated, and that the costs of including in the study a personnel officer and a candidate are c_1 and c_2 respectively. For a given total budget C, the ratio σ_τ^2/σ^2 is the determining variable for finding the optimum balance between number of personnel officers and candidates to include in the study so as to minimize the variance of the estimator. If the populations are not large, the model will need to take account of their finite nature.

Estimation of σ^2 and σ_τ^2

At times, interest exists in estimating σ^2 and σ_τ^2 separately. An unbiased estimator of σ^2, according to (16.18), is:

$$\hat{\sigma}^2 = MSE \tag{16.32}$$

A confidence interval for σ^2 can be obtained in the usual fashion by means of (1.64); here, the degrees of freedom will be $n_T - r$.

An unbiased point estimator of σ_τ^2 is also available. Since by (16.18) and (16.19), we have:

$$E(MSE) = \sigma^2$$
$$E(MSTR) = \sigma^2 + n'\sigma_\tau^2$$

it follows that:

$$\hat{\sigma}_\tau^2 = \frac{MSTR - MSE}{n'} \tag{16.33}$$

is an unbiased estimator of σ_τ^2. Occasionally, this point estimator will turn out to be negative. Since a variance cannot be negative, the usual practice is to consider the point estimator to be zero in that event. Only approximate confidence intervals for σ_τ^2 are available. These are discussed by Scheffé in Reference 16.2.

Example. For our personnel officers example, a 90 percent confidence interval for σ^2 requires:

$$MSE = 75.6 \qquad \chi^2(.05; 15) = 7.26 \qquad \chi^2(.95; 15) = 25.0$$

Using (1.64), we find:

$$45.4 = \frac{(15)(75.6)}{25.0} \le \sigma^2 \le \frac{(15)(75.6)}{7.26} = 156.2$$

An unbiased point estimate of σ_τ^2 requires:

$$MSE = 75.6 \qquad MSTR = 370 \qquad n' = n = 4$$

Hence, by (16.33) we find:

$$\hat{\sigma}_\tau^2 = \frac{370 - 75.6}{4} = 73.6$$

16.3 REGRESSION APPROACH TO SINGLE-FACTOR ANALYSIS OF VARIANCE

(This section presupposes familiarity with Chapter 9. If the reader has not yet studied Chapter 9, he should read Sections 9.1 through 9.3 before proceeding further.)

Correspondence between Regression and ANOVA Models

We noted in Chapter 9 that regression analysis can handle qualitative independent variables by means of *indicator variables* which take on the values 0 and 1. We also formulated principle (9.3), which stated that a qualitative variable with r classes will be represented by $r - 1$ such indicator variables.

To illustrate the regression approach to single-factor analysis of variance with fixed effects, consider a study with $r = 3$ treatments and $n = 2$ observations for each treatment. The analysis of variance model for this case is:

(16.34)
$$Y_{ij} = \mu_. + \tau_j + \varepsilon_{ij} \qquad i = 1, 2; j = 1, 2, 3$$

Since the independent variable has three classes (factor levels), we shall use two indicator variables in the regression model. Let us define them as follows:

(16.35)
$$X_1 = \begin{matrix} 1 \text{ if observation from factor level 1} \\ 0 \text{ otherwise} \end{matrix}$$
$$X_2 = \begin{matrix} 1 \text{ if observation from factor level 2} \\ 0 \text{ otherwise} \end{matrix}$$

With these indicator variables to represent the three classes of the independent variable, the regression model for our illustration is as follows:

(16.36)
$$Y_{ij} = \beta_0 + \beta_1 X_{ij1} + \beta_2 X_{ij2} + \varepsilon_{ij} \qquad i = 1, 2; j = 1, 2, 3$$

where X_{ij1} is the value of X_1 for the ith observation from the jth factor level, and X_{ij2} is the value of X_2 for the ith observation from the jth factor level.

We can readily find the correspondence between the ANOVA parameters $\mu_.$ and τ_j and the regression parameters β_0, β_1, β_2. Consider the ith observation from factor level 3. The ANOVA model (16.34) gives for this case:

(16.37)
$$Y_{i3} = \mu_. + \tau_3 + \varepsilon_{i3}$$

With the regression model, the indicator variables for the ith observation from factor level 3 are, according to (16.35):

$$X_{i31} = 0 \qquad X_{i32} = 0$$

Hence, regression model (16.36) becomes for this case:

(16.38)
$$Y_{i3} = \beta_0 + \beta_1(0) + \beta_2(0) + \varepsilon_{i3} = \beta_0 + \varepsilon_{i3}$$

Equating (16.37) and (16.38), we find the correspondence:

(16.39a)
$$\beta_0 = \mu_. + \tau_3 = \mu_3$$

Similarly, the other two correspondences can be derived:

(16.39b) $$\beta_1 = \tau_1 - \tau_3 = \mu_1 - \mu_3$$

(16.39c) $$\beta_2 = \tau_2 - \tau_3 = \mu_2 - \mu_3$$

To generalize, we shall follow the convention that the last factor level (r) is the one for which there is no indicator variable equal to 1. The regression model for single-factor analysis of variance is then:

(16.40) $$Y_{ij} = \beta_0 + \beta_1 X_{ij1} + \beta_2 X_{ij2} + \cdots + \beta_{r-1} X_{ij,r-1} + \varepsilon_{ij}$$

where:

$$X_{ij1} = \begin{matrix} 1 \text{ if observation from factor level 1} \\ 0 \text{ otherwise} \end{matrix}$$

$$\vdots \qquad \qquad \vdots$$

$$X_{ij,r-1} = \begin{matrix} 1 \text{ if observation from factor level } r-1 \\ 0 \text{ otherwise} \end{matrix}$$

The correspondences between the ANOVA model parameters and the regression model parameters then are:

(16.41a) $$\beta_0 = \mu_. + \tau_r = \mu_r$$

(16.41b) $$\beta_j = \tau_j - \tau_r = \mu_j - \mu_r \qquad j = 1, \ldots, r-1$$

Thus, the regression parameter β_0 equals the mean response for the rth factor level, and β_j measures the difference between the mean responses for the jth factor level and the rth factor level.

Note

The regression approach has not generally been utilized for ordinary analysis of variance problems. The reason is that the **X** matrix for analysis of variance problems usually contains the 0's and 1's for the indicator variables in a certain structure which permits computational simplifications that are explicitly recognized in the statistical procedures for analysis of variance. We take up the regression approach to analysis of variance here, and in later chapters, for two principal reasons. First, we see that analysis of variance models are encompassed by the general linear statistical model (7.18) considered in Chapter 7. Second, the availability of high-speed and large-size computers today permits the use of the regression approach to analysis of variance whereas earlier the regression calculations would have been impractical.

Example

To illustrate the use of the regression approach to analysis of variance, we return to the Kenton Food Company example of Chapter 13, dealing with the four cereal package designs. The data are presented in Table 13.1a. The regression model to be employed is based on the response function:

(16.42) $$E(Y) = \beta_0 + \beta_1 X_1 + \beta_2 X_2 + \beta_3 X_3 \qquad \text{Full model}$$

where:

$$X_j = \begin{matrix} 1 \text{ if observation from } j\text{th factor level} \\ 0 \text{ otherwise} \end{matrix} \qquad j = 1, 2, 3$$

$$\beta_0 = \mu_. + \tau_4 = \mu_4$$
$$\beta_1 = \tau_1 - \tau_4 = \mu_1 - \mu_4$$
$$\beta_2 = \tau_2 - \tau_4 = \mu_2 - \mu_4$$
$$\beta_3 = \tau_3 - \tau_4 = \mu_3 - \mu_4$$

The observation vector **Y** and the matrix **X** for the data in Table 13.1a are shown in Table 16.5. For observation $Y_{11} = 12$, for instance, note that $X_1 = 1$, $X_2 = 0$, $X_3 = 0$, so that we have:

$$E(Y_{11}) = \beta_0 + \beta_1$$

TABLE 16.5

Data Matrices for Model (16.42) Based on Data in Table 13.1a

$$\mathbf{Y} = \begin{bmatrix} 12 \\ 18 \\ 14 \\ 12 \\ 13 \\ 19 \\ 17 \\ 21 \\ 24 \\ 30 \end{bmatrix} \qquad \mathbf{X} = \begin{bmatrix} 1 & 1 & 0 & 0 \\ 1 & 1 & 0 & 0 \\ 1 & 0 & 1 & 0 \\ 1 & 0 & 1 & 0 \\ 1 & 0 & 1 & 0 \\ 1 & 0 & 0 & 1 \\ 1 & 0 & 0 & 1 \\ 1 & 0 & 0 & 1 \\ 1 & 0 & 0 & 0 \\ 1 & 0 & 0 & 0 \end{bmatrix}$$

(Columns 2–4 of **X** are headed X_1, X_2, X_3.)

Since $\beta_0 = \mu_4$ and $\beta_1 = \mu_1 - \mu_4$, we find:

$$E(Y_{11}) = \mu_4 + (\mu_1 - \mu_4) = \mu_1$$

which is precisely the expected value of Y_{11} according to the analysis of variance model.

Similarly, for $Y_{24} = 30$, we have $X_1 = 0$, $X_2 = 0$, $X_3 = 0$, so that:

$$E(Y_{24}) = \beta_0 = \mu_4$$

which again is in accordance with the analysis of variance model.

Fitting of Regression Model

The fitting of the regression model for analysis of variance is straightforward. We assume that a multiple regression computer package is available, and hence need only specify the **Y** observation vector and the **X** matrix. For our example in Table 16.5, a computer run of a multiple regression package led to the results summarized in Table 16.6.

TABLE 16.6

Computer Output for Kenton Food Company Multiple Regression Example

(a) Regression Coefficients

$b_0 = 27.0$ $b_2 = -14.0$
$b_1 = -12.0$ $b_3 = -8.0$

(b) Analysis of Variance

Source of Variation	SS	df	MS
Regression	$SSR = 258$	3	$MSR = 86$
Error	$SSE = 46$	6	$MSE = 7.67$
Total	$SSTO = 304$	9	

(c) Estimated Variance-Covariance Matrix for Regression Coefficients

	b_0	b_1	b_2	b_3
b_0	3.8333			
b_1	−3.8333	7.6667		
b_2	−3.8333	3.8333	6.3889	
b_3	−3.8333	3.8333	3.8333	6.3889

Note

The least squares estimators b_0, b_1, b_2, and b_3 of the regression model parameters for our example turn out to be as expected:

(16.43a) $$b_0 = \bar{Y}_{.4}$$

(16.43b) $$b_1 = \bar{Y}_{.1} - \bar{Y}_{.4}$$

(16.43c) $$b_2 = \bar{Y}_{.2} - \bar{Y}_{.4}$$

(16.43d) $$b_3 = \bar{Y}_{.3} - \bar{Y}_{.4}$$

Test for Factor Level Effects

To test for factor level effects in our example via the regression approach, we follow the general linear test procedure described in Chapter 3. The alternative conclusions with reference to the analysis of variance model for our Kenton Food Company example are:

(16.44)
$$C_1: \tau_1 = \tau_2 = \tau_3 = \tau_4 = 0$$
$$C_2: \text{not all } \tau_j \text{ are zero}$$

We know from Table 16.6 that the error sum of squares when the full model is fitted is:

$$SSE(F) = 46$$

If C_1 holds, the β's in the regression model (16.42) become:

$$\beta_0 = \mu_. \qquad \beta_2 = 0$$
$$\beta_1 = 0 \qquad \beta_3 = 0 \tag{16.45}$$

Hence, the response function for the reduced regression model under C_1 does not include any indicator variables for factor level effects:

$$E(Y) = \beta_0 \qquad \text{Reduced model} \tag{16.46}$$

where:

$$\beta_0 = \mu_.$$

Indeed, model (16.46) simply postulates that all observations have the same mean $\mu_{..}$.

The data matrices for our example for the reduced model (16.46) under C_1 are shown in Table 16.7. As noted in Chapter 3, the least squares estimator of β_0 in model (16.46) is $\bar{Y}_{..}$. Hence the estimated response function is:

$$\hat{Y} = \bar{Y}_{..}$$

and the error sum of squares for the reduced model is:

$$SSE(R) = \sum\sum (Y_{ij} - \bar{Y}_{..})^2 = SSTO$$

We therefore know from Table 16.6 that:

$$SSE(R) = 304$$

TABLE 16.7

Data Matrices for Model (16.46) Based on Data in Table 13.1a

$$\mathbf{Y} = \begin{bmatrix} 12 \\ 18 \\ 14 \\ 12 \\ 13 \\ 19 \\ 17 \\ 21 \\ 24 \\ 30 \end{bmatrix} \qquad \mathbf{X} = \begin{bmatrix} 1 \\ 1 \\ 1 \\ 1 \\ 1 \\ 1 \\ 1 \\ 1 \\ 1 \\ 1 \end{bmatrix}$$

The appropriate test statistic, as discussed in Chapter 3, compares the difference $SSE(R) - SSE(F)$ with $SSE(F)$, each divided by its degrees of freedom. Our test statistic for factor level effects therefore is:

$$F^* = \frac{SSE(R) - SSE(F)}{(n_T - 1) - (n_T - r)} \div \frac{SSE(F)}{n_T - r}$$

or:

$$F^* = \frac{\dfrac{SSE(R) - SSE(F)}{r - 1}}{\dfrac{SSE(F)}{n_T - r}} \tag{16.47}$$

When C_1 holds, F^* follows the $F(r - 1, n_T - r)$ distribution.

For our Kenton Food Company example, we have:

$$SSE(R) = 304 \qquad SSE(F) = 46$$
$$n_T - 1 = 9 \qquad n_T - r = 6$$

Hence:

$$F^* = \frac{304 - 46}{9 - 6} \div \frac{46}{6} = \frac{86}{7.67} = 11.2$$

which is identical to the test statistic obtained with the analysis of variance on page 449. From this point on, the test procedure parallels the one described in Chapter 13.

Note

The regression approach can be reconciled quite easily with the analysis of variance approach. If the full regression model holds, the error sum of squares $SSE(F)$ equals the error sum of squares SSE of the analysis of variance. To see this for our example, first remember that for any observation from factor levels 1, 2, and 3, all X_j's are zero except the indicator variable corresponding to the factor level of the observation. For factor level 4 observations, all indicator variables are zero. Hence:

$$SSE(F) = \sum (Y - \hat{Y})^2 = \sum_{j=1}^{3} \sum_{i} (Y_{ij} - b_0 - b_j)^2 + \sum_{i} (Y_{i4} - b_0)^2 \tag{16.48}$$

Second, recall from (16.43) that the least squares estimators for our example are:

$$b_0 = \bar{Y}_{.4}$$
$$b_j = \bar{Y}_{.j} - \bar{Y}_{.4} \qquad j = 1, 2, 3$$

Hence:

$$SSE(F) = \sum_{j=1}^{4} \sum_{i} (Y_{ij} - \bar{Y}_{.j})^2 = SSE \tag{16.49}$$

Turning now to $SSE(R)$, when we fit the regression model (16.46) under C_1, the least squares estimator of β_0 is $\bar{Y}_{..}$ so that:

$$SSE(R) = \sum (Y - \hat{Y})^2 = \sum \sum (Y_{ij} - \bar{Y}_{..})^2 = SSTO \tag{16.50}$$

Hence, $SSE(R) - SSE(F) = SSTO - SSE$, which by (13.40) equals $SSTR$. Thus, the regression approach compares $SSTR$ with SSE, each divided by its degrees of freedom, which of course is identical to the analysis of variance test.

Estimation of Factor Level Means

With the regression approach, estimation of factor level means is based on the regression coefficients b_0 and b_j. Suppose we wish to estimate $\mu_1 = \mu_. + \tau_1$ for our Kenton Food Company example. From the definitions of the β's in (16.42), it is clear that:

$$\mu_1 = \mu_. + \tau_1 = \beta_0 + \beta_1$$

Hence, an unbiased estimator of μ_1 is:

$$b_0 + b_1 \tag{16.51}$$

Further, we know by (1.25b) that the variance of this estimator is:

$$\sigma^2(b_0 + b_1) = \sigma^2(b_0) + \sigma^2(b_1) + 2\sigma(b_0, b_1) \tag{16.52}$$

Hence, the estimated variance is:

$$s^2(b_0 + b_1) = s^2(b_0) + s^2(b_1) + 2s(b_0, b_1) \tag{16.53}$$

For our Kenton Food Company example, the needed data are in Table 16.6. We obtain:

$$b_0 + b_1 = 27.0 + (-12.0) = 15.0$$

$$s^2(b_0 + b_1) = 3.8333 + 7.6667 + 2(-3.8333) = 3.8334$$

or:

$$s(b_0 + b_1) = 1.96$$

These are the same results we obtained earlier with the analysis of variance on page 466.

As a second illustration, suppose we wish to estimate $\mu_3 - \mu_4$ for our example. We see from the definition of the β's in (16.42) that:

$$\mu_3 - \mu_4 = \tau_3 - \tau_4 = \beta_3$$

Hence, an unbiased estimator is:

$$b_3 \tag{16.54}$$

Its estimated variance is:

$$s^2(b_3) \tag{16.55}$$

For our Kenton Food Company example, we obtain from Table 16.6:

$$b_3 = -8.0$$

and:

$$s^2(b_3) = 6.3889$$

or:

$$s(b_3) = 2.53$$

These are exactly the same results, except for slight rounding effects, that we found earlier on page 467.

Thus, estimation of factor level effects with the regression approach involves no new principles. The factor level effects are simply functions of the regression coefficients, and can be estimated directly from b_0 and the b_j. The estimated variance of an estimated effect can therefore be obtained from the estimated variance-covariance matrix for the regression coefficients in the computer printout. Confidence intervals can be constructed and multiple comparison procedures employed in exactly the same fashion already explained.

Note

It is apparent from the above illustrations that it is usually easier to estimate factor level means, and make comparisons between them, by working directly with the means $\bar{Y}_{.j}$ than with the regression coefficients. The main purpose of the discussion is to show that the regression approach encompasses the analysis of variance, both for testing the equality of factor level means and for estimating the factor level means.

PROBLEMS

16.1. Refer to Problem 13.10. It was suggested to the analyst that the studentized range test is simpler than the regular F test.

a) Conduct the studentized range test, with a level of significance of .01. What conclusion do you reach?
b) Does the conclusion in part (a) differ from the one in Problem 13.10b?
c) What precisely were the simplifications in using the studentized range test rather than the F test?
d) Does the studentized range test avoid any of the assumptions needed for the F test?

16.2. A management consultant was engaged to assist a company in improving the efficiency of communications. As part of this study, he selected 10 executives at random from each of the sales, production, and research and development divisions and studied their communications during the most recent three-month period in great detail. Among other data, he obtained the following information on total costs of long-distance telephone calls (in dollars):

Executive	*Sales*	*Production*	*Research and Development*
1	1,293	640	83
2	823	383	41
3	149	915	512
4	350	219	194
5	764	193	79
6	243	507	84
7	246	321	281
8	102	142	147
9	319	531	641
10	433	482	301

The consultant, after examining the data, decided to employ a nonparametric approach to test whether or not the mean telephone expenses per executive are the same for the three divisions.

a) What feature of the data probably occasioned this decision?
b) Conduct the Kruskal-Wallis rank test, with a level of significance of .05. What conclusion do you reach?
c) What assumptions required for the F test are avoided with the Kruskal-Wallis test?

16.3. Refer to Problem 16.2.

a) Conduct the median test, with a level of significance of .025. What conclusion do you reach?
b) Is there a basic distinction between the assumptions for the median test and those for the Kruskal-Wallis test?

16.4. In each of the following cases, indicate whether ANOVA model I or II is more appropriate, and state your reasons:

1. The treatments are five employees in a large plant.
2. The treatments are three levels of motivation.
3. The treatments are four types of group organizations.
4. The treatments are the two sexes of hospital patients.

16.5. Refer to the Apex enterprises personnel officers example on page 528.

a) Explain with reference to this example over what the expectation in (16.12) is taken.
b) Explain with reference to this example over what the variance in (16.13) is taken.

16.6. Refer to the coil winding example on page 496. Suppose the company owns a large number of machines, and r are selected at random for the study. With reference to this example, what does the ratio in (16.16) measure?

16.7. Refer to formula (16.19a). Show that $n' = n$ when $n_j \equiv n$.

16.8. A trading stamp company operates a large number of redemption centers. A marketing specialist was interested in the number of months elapsed between the time a customer first began saving trading stamps for a redemption and the time of redemption. Five redemption centers were selected at random, and 20 current redemptions at each selected center were analyzed. The results were as follows:

$n_1 = 20$	$n_2 = 20$	$n_3 = 20$	$n_4 = 20$	$n_5 = 20$
$\bar{Y}_{.1} = 12.6$	$\bar{Y}_{.2} = 8.3$	$\bar{Y}_{.3} = 15.2$	$\bar{Y}_{.4} = 9.7$	$\bar{Y}_{.5} = 6.9$

$$SSTR = 898.64$$
$$SSE = 2{,}378.00$$
$$SSTO = 3{,}276.64$$

a) Test whether or not the mean elapsed time differs for the various redemption centers the company operates. Use a level of significance of .10. State the model you employed. What conclusion do you reach?
b) Estimate the mean elapsed time for all redemption centers, using a 95 percent confidence interval.

16.9. Refer to Problem 16.8.
a) Estimate the measure (16.16), using a 95 percent confidence interval. What does this measure refer to in the present example?
b) Obtain a point estimate of σ_τ^2.
c) Estimate σ^2, using a 95 percent confidence interval.

16.10. Refer to the coil winding example on page 496. From the large number of coil winding machines which the company owns, four were selected at random, and six coils were then selected from the day's output of each chosen machine. The results were as follows:

	Machine			
Coil	1	2	3	4
1	201	198	211	206
2	198	196	214	204
3	209	201	207	205
4	197	200	209	208
5	203	204	208	203
6	204	199	210	209

a) Test whether or not the mean value of the coil characteristic differs for the various coil winding machines which the company owns. Use a level of significance of .10. State the model you assumed. What conclusion do you reach?
b) Estimate σ_τ^2/σ^2, using a 90 percent confidence interval. What does the ratio refer to in this example?

16.11. Refer to Problem 16.8. Suppose the primary intent of the study is to estimate the mean elapsed time for all redemption centers ($\mu_.$), and that equal sample sizes are to be selected for each sample redemption center. What are the values of r and n which will minimize the variance (16.23), given that n_T is fixed? Ignore any cost considerations.

16.12. Refer to Problem 13.10.
a) Set up the regression model counterpart to the ANOVA model.
b) State the correspondences between the regression model parameters and the ANOVA model parameters.
c) Test the equality of treatment means by the regression approach, using a level of significance of .01. Is your test statistic the same as that obtained in Problem 13.10b?

d) Obtain an interval estimate by the regression approach for the contrast:

$$L = \frac{\mu_1 + \mu_2}{2} - \frac{\mu_3 + \mu_4}{2}$$

where μ_1 is the mean delay for agent C, etc. Use a confidence coefficient of 95 percent.

16.13. Refer to Problem 13.11.

a) Set up the regression model counterpart to the ANOVA model.

b) State the correspondences between the regression model parameters and the ANOVA model parameters.

c) Test the equality of treatment means by the regression approach, using a level of significance of .05. Is your test statistic the same as that obtained in Problem 13.11b?

d) Obtain an interval estimate by the regression approach for the contrast:

$$L = \frac{\mu_1 + \mu_4}{2} - \frac{\mu_2 + \mu_3}{2}$$

where μ_1 is the mean cash offer for owner A, etc. Use a confidence coefficient of 99 percent.

CITED REFERENCES

16.1. Owen, Donald B. *Handbook of Statistical Tables.* Reading, Mass.: Addison-Wesley Publishing Co., Inc., 1962.

16.2. Scheffé, Henry. *The Analysis of Variance.* New York: John Wiley & Sons, Inc., 1959.

part IV

Multifactor Analysis of Variance

17

Two-Factor Analysis of Variance

IN PART III we considered studies in which the effect of one factor is investigated. Now we are concerned with investigations of the simultaneous effects of two or more factors. In this chapter, we take up the analysis of variance for two-factor studies. In Chapters 18 and 19, we continue the discussion of two-factor studies by taking up the analysis of factor effects, the planning of sample sizes, and a number of other topics. Finally in Chapter 20, we take up the analysis of variance for studies in which three or more factors are being investigated.

17.1 MULTIFACTOR STUDIES

Before focusing specifically on two-factor studies, we shall first make some general remarks about multifactor studies, which encompass investigations of two or more factors.

Examples of Two-Factor Studies

Example 1. A company investigated the effect of selling price and type of promotional campaign on sales of one of its products. Three selling prices (59 cents, 60 cents, 64 cents) were studied, as were two types of promotional campaigns (radio advertising, newspaper advertising). Let us consider selling price to be factor A and promotional campaign to be factor B. Factor A here was studied at three price levels; in general, we use the symbol a to denote the number of levels of A investigated. Factor B was here studied at two levels; we shall use the symbol b to denote the number of levels of B investigated. Each combination of price and promotion campaign was studied, as follows:

Treatment	*Description*
1	59¢ price, radio advertising
2	60¢ price, radio advertising
3	64¢ price, radio advertising
4	59¢ price, newspaper advertising
5	60¢ price, newspaper advertising
6	64¢ price, newspaper advertising

Each of the combinations of a factor level of A and a factor level of B is a *treatment*. Thus, there are $3 \times 2 = 6$ treatments here altogether. In general, the total number of possible treatments in a two-factor study is ab.

Twelve communities throughout the United States, of approximately equal size and similar socioeconomic characteristics, were selected and assigned at random to the treatments such that each treatment was given to two experimental units. As before, we shall use the symbol n for the number of units receiving a given treatment. In the two communities assigned to treatment 1, for instance, the product price was fixed at 59 cents and radio advertising employed, and so on for the other communities in the study.

Example 2. A steel company studied the effect of carbon content and tempering temperature on the strength of steel. Carbon content was investigated at a high level and a low level (the precise definitions of these levels is not important here). Tempering temperature was also studied at a high level and at a low level. Altogether, $2 \times 2 = 4$ treatments were defined for this study:

Treatment	*Description*
1	High carbon level, high tempering temperature
2	High carbon level, low tempering temperature
3	Low carbon level, high tempering temperature
4	Low carbon level, low tempering temperature

These four treatments were then each assigned to three production batches, in randomized fashion.

Example 3. Multifactor studies also can be made with survey data. An analyst wished to study the effects of income (under \$5,000, \$5,000–\$9,999, \$10,000–\$19,999, \$20,000 and over) and stage in the life cycle of the household (stages 1, 2, 3, 4) on appliance purchases. Here, $4 \times 4 = 16$ treatments are defined. These are, in part:

Treatment	*Description*
1	Under \$5,000 income, stage 1
2	Under \$5,000 income, stage 2
⋮	⋮
16	\$20,000 and over income, stage 4

The analyst then selected 20 households with the required income and life-cycle characteristics for each of the "treatment" classes for his study.

Note

When we considered single-factor studies, we did not place any restrictions on the nature of the r factor levels under study. Formally, the ab treatments in a two-factor investigation could be considered as the r factor levels in a single-factor investigation and analyzed according to the methods discussed in Part III. The reason why new methods of analysis are required is that we wish to analyze the ab treatments in special ways that recognize two factors are involved and enable us to obtain information about the effects of each of the two factors as well as about any special joint effects.

Complete and Fractional Factorial Studies

The three examples just cited are *complete factorial studies* because all possible combinations of factor levels for the different factors were included. At times, it is not feasible or desirable to include all possible combinations of factor levels for the different factors; such incomplete factorial studies are called *fractional factorial studies.* For instance, suppose the steel company mentioned in Example 2 wished to study six temperatures, five levels of carbon content, and four methods of cooling the steel. A complete factorial study would then involve $6 \times 5 \times 4 = 120$ treatments. Such a study might be extremely costly and time-consuming. Under these conditions, it is possible to design a fractional factorial study containing only some of the 120 factor level combinations, which will still provide information about the effects of each of the three factors as well as about any important special joint effects of these factors.

The discussion of multifactor investigations in Part IV deals solely with complete factorial studies.

Advantages of Multifactor Studies

Efficiency. Multifactor studies are more efficient than the traditional experimental approach of manipulating only one factor at a time and keeping all other conditions constant. With reference to Example 1, the traditional approach to studying the effect of promotion campaign would have been to keep price constant at a given level and vary only the promotional campaign. An important problem with this approach is the choice of the price level to be held constant. This choice is especially difficult when one is not sure whether the promotional effect is the same at different price levels. Even though the traditional approach devotes all resources to studying the effect of only one factor, it does not yield any more precise information about the effects of that factor than a multifactor experiment of the same size. With reference to Example 1 again, suppose that 12 communities were to be utilized in a traditional study, six assigned to radio advertising and the other six to newspaper advertising, and that the price would be kept constant at 59 cents. For this traditional study, the comparison between the two types of promotional campaigns would be based on two samples of six communities each. The same is true for the two-factor study in Example 1, since each promotional campaign occurs there in three treatments and each treatment has two communities assigned to it.

Amount of Information. The traditional study provides less information than the two-factor study. Specifically in our previous illustration, it does not provide any information about the effect of price, nor about any special joint effects of price and promotional campaign. Information about price effects would require an additional traditional experiment for which promotional campaign would be kept constant at a given level and price varied. Thus, the traditional approach would require a larger sample to provide information about both price and promotion campaign effects, and unless the traditional study were yet further enlarged, it would still not provide full information about any special joint effects of the two factors. Such special joint effects are called *interactions.* Interaction effects were encountered in regression models and will be discussed in the context of analysis of variance models fully below. Here it suffices to point out that interaction effects may be very important. For instance, it might be that the price effect is not large when the promotional campaign is in newspapers, but it is large with radio advertising. Such interaction effects can be readily investigated from factorial studies.

Validity of Findings. In addition to being more efficient and readily providing information about interaction effects, multifactor studies also can strengthen the validity of the findings. Suppose that in Example 1, management was principally interested in investigating the effect of price on sales. If the promotional campaign used in the price study had been newspaper advertising, doubts would exist whether or not the price effect differs for other promotional vehicles. By including type of promotional campaign as another factor in the study, management can get information about the persistence of the price effect with different promotional vehicles, without increasing the number of experimental units in the study. Thus, multifactor studies can include some factors of secondary importance to permit inferences about the primary factors with a greater range of validity.

Comments

1. In studies based on survey data, as for ones utilizing experimental data, multifactor analysis of the data permits a ready evaluation of interaction effects and economizes on the number of cases required for analysis.

2. The advantages of multifactor experiments just described should not lead one to think that the more factors are included in the study, the better. Experiments involving many factors, each at numerous levels, become complex, costly, and time-consuming. It is often better research strategy to begin with a few factors, investigate the effects of these, and extend the investigation in accordance with the results obtained to date. In this way, resources can be devoted principally to the most promising avenues of investigation and a better understanding of the working of the factors can be obtained.

17.2 MEANING OF MODEL ELEMENTS

Before presenting a formal statement of the analysis of variance model for two-factor studies, we shall develop the model elements and discuss their meaning. This will not only be helpful in understanding the ANOVA model but will also provide insights into how the analysis of two-factor studies should proceed. *Throughout this section, we assume that all population means are known.*

Example

We shall consider a simple two-factor study in which the effects of sex and age on learning of a task are of interest. Table 17.1 contains the mean learning time (in minutes) for each treatment (sex–age combination). Note that for simplicity, the age factor has been defined in terms of only three factor levels.

The numbers in Table 17.1 represent the true means, and will be denoted by μ_{ij}, where i refers to the level of factor A ($i = 1, \ldots, a$) and j refers to the level of factor B ($j = 1, \ldots, b$). Thus $\mu_{22} = 11$ indicates that the mean learning time for middle-aged females is 11 minutes.

TABLE 17.1

Age Effect but No Sex Effect, with No Interactions

(a) Mean Learning Times (in minutes)

	Factor B—Age			
Factor A—Sex	$j=1$ *Young*	$j=2$ *Middle*	$j=3$ *Old*	*Row Average*
$i=1$ Male	9 (μ_{11})	11 (μ_{12})	16 (μ_{13})	12 ($\mu_{1.}$)
$i=2$ Female	9 (μ_{21})	11 (μ_{22})	16 (μ_{23})	12 ($\mu_{2.}$)
Column average	9 ($\mu_{.1}$)	11 ($\mu_{.2}$)	16 ($\mu_{.3}$)	12 ($\mu_{..}$)

(b) Specific Age Effects (in minutes)

	$j=1$	$j=2$	$j=3$
$i=1$	-3 ($\beta_{1(1)}$)	-1 ($\beta_{2(1)}$)	$+4$ ($\beta_{3(1)}$)
$i=2$	-3 ($\beta_{1(2)}$)	-1 ($\beta_{2(2)}$)	$+4$ ($\beta_{3(2)}$)
Column average	-3 (β_1)	-1 (β_2)	$+4$ (β_3)

(c) Specific Sex Effects (in minutes)

	$j=1$	$j=2$	$j=3$	*Row Average*
$i=1$	0 ($\alpha_{1(1)}$)	0 ($\alpha_{1(2)}$)	0 ($\alpha_{1(3)}$)	0 (α_1)
$i=2$	0 ($\alpha_{2(1)}$)	0 ($\alpha_{2(2)}$)	0 ($\alpha_{2(3)}$)	0 (α_2)

Factor Level Means

The treatment means in Table 17.1a indicate that the mean learning times for men and women are the same, for each age group. On the other hand, the mean learning time increases with age, for each sex. Thus, sex has no effect on learning time, but age does. This can also be seen quickly from the row averages and column averages shown in Table 17.1a, which in this case tell the complete story. The row averages are the sex *factor level means*, and the column averages the age factor level means. We denote the column average for the first column as $\mu_{.1}$, which is the average of μ_{11} and μ_{21}. In general, the column average for the jth column is:

$$\mu_{.j} = \frac{\sum_{i=1}^{a} \mu_{ij}}{a} \tag{17.1}$$

and the row average for the ith row is:

$$\mu_{i.} = \frac{\sum_{j=1}^{b} \mu_{ij}}{b} \tag{17.2}$$

The overall mean learning time for all ages and both sexes is denoted by $\mu_{..}$, and is defined in the following equivalent fashions:

$$\mu_{..} = \frac{\sum_i \sum_j \mu_{ij}}{ab} \tag{17.3a}$$

$$\mu_{..} = \frac{\sum_i \mu_{i.}}{a} \tag{17.3b}$$

$$\mu_{..} = \frac{\sum_j \mu_{.j}}{b} \tag{17.3c}$$

Specific Effects

When we compare the mean learning time for young males with the row average, we measure the *specific effect* of young age for men. This specific effect, denoted $\beta_{1(1)}$, is:

$$\beta_{1(1)} = 9 - 12 = -3 \tag{17.4}$$

This specific effect implies that the mean learning time for young men is three minutes less than the average learning time for all men. The notation $\beta_{1(1)}$ is used to indicate that we are measuring the specific effect of the first factor level of B, when factor A is at the first level.

As another example, $\beta_{1(2)}$ is the specific effect of young age for women. This effect is:

(17.5) $$\beta_{1(2)} = 9 - 12 = -3$$

the same as for men.

In general, we define $\beta_{j(i)}$ as follows:

(17.6) $$\beta_{j(i)} = \mu_{ij} - \mu_{i.}$$

Table 17.1b contains each of the specific age effects for our learning illustration.

Similarly, there are specific effects of sex. For instance, the specific effect of male sex for young persons, denoted $\alpha_{1(1)}$, is:

(17.7) $$\alpha_{1(1)} = \mu_{11} - \mu_{.1} = 9 - 9 = 0$$

Since this specific effect is 0, male persons who are young have the same mean learning time as all young persons.

In general, we define the specific effect of the ith level of factor A, for the jth level of factor B, as follows:

(17.8) $$\alpha_{i(j)} = \mu_{ij} - \mu_{.j}$$

Table 17.1c contains the specific effects of male and female sex for our learning illustration.

Main Effects

Main Age Effects. To summarize the specific age effects, we shall average them. These averages are shown in Table 17.1b. In this instance, of course, the mean for each age group is the same as its components since the specific age effects are the same for men and women. For young persons, the mean specific age effect is denoted by β_1; it is:

(17.9) $$\beta_1 = \frac{(-3) + (-3)}{2} = -3$$

β_1 is called the *main effect* for factor B at the first level.

It can be shown that β_1 is the equivalent of the difference between the mean learning time for young persons and the mean learning time for all persons:

(17.9a) $$\beta_1 = \mu_{.1} - \mu_{..} = 9 - 12 = -3$$

To show this, we begin with the definition of β_1 as the mean specific age effect for young persons:

$$\beta_1 = \frac{\beta_{1(1)} + \beta_{1(2)}}{2}$$

Using (17.6), we obtain:

$$\beta_1 = \frac{(\mu_{11} - \mu_{1.}) + (\mu_{21} - \mu_{2.})}{2}$$

Rearranging terms and employing (17.1) and (17.3b), we find:

$$\beta_1 = \frac{\mu_{11} + \mu_{21}}{2} - \frac{\mu_{1.} + \mu_{2.}}{2} = \mu_{.1} - \mu_{..}$$

Main Sex Effects. The means of the specific sex effects are shown in Table 17.1c and will be denoted α_i. Note that the main sex effects in Table 17.1c are all zero, indicating that sex does not affect mean learning time.

General Definitions. In general, we define the main effect of factor B at the jth level as follows:

$$\beta_j = \frac{\sum_i \beta_{j(i)}}{a} = \mu_{.j} - \mu_{..} \tag{17.10}$$

Similarly, the main effect of the ith level of factor A is defined:

$$\alpha_i = \frac{\sum_j \alpha_{i(j)}}{b} = \mu_{i.} - \mu_{..} \tag{17.11}$$

It follows from (17.3b) and (17.3c) that:

$$\sum_i \alpha_i = 0 \qquad \sum_j \beta_j = 0 \tag{17.12}$$

Thus, the sum of the main effects for each factor is zero.

Additive Factor Effects

The factor effects in Table 17.1a have an interesting property. Each mean response μ_{ij} can be obtained by adding the respective sex and age main effects to the overall mean $\mu_{..}$. For instance, we have:

$$\mu_{11} = \mu_{..} + \alpha_1 + \beta_1 = 12 + 0 + (-3) = 9$$
$$\mu_{23} = \mu_{..} + \alpha_2 + \beta_3 = 12 + 0 + 4 = 16$$

In general we have for Table 17.1a:

$$\mu_{ij} = \mu_{..} + \alpha_i + \beta_j \qquad \text{Additive factor effects} \tag{17.13}$$

which can be also expressed, using the definitions of α_i in (17.11) and of β_j in (17.10), as:

$$\mu_{ij} = \mu_{i.} + \mu_{.j} - \mu_{..} \qquad \text{Additive factor effects} \tag{17.13a}$$

When all treatment means can be decomposed in the form of (17.13) or (17.13a), we say that the *factors do not interact*, or that *no factor interactions are present*, or that the *factor effects are additive*. The significance of no factor interactions is that the effects of the two factors can be described separately merely by analyzing the factor level means or the factor main effects. Thus, in our illustration in Table 17.1a, the two sex means signify that sex has no influence regardless of age, and the three age means portray the influence of age regardless of sex. The analysis of factor effects is therefore quite simple when there are no factor interactions.

Graphic Presentation

Figure 17.1 presents the mean learning times of Table 17.1a in graphic form. The X axis contains the sex factor levels (denoted by A_1 and A_2), and the Y axis contains learning time. Separate curves are drawn for each of the age factor levels (denoted by B_1, B_2, and B_3). The zero slope of each curve indicates that sex has no effect. The differences in the heights of the three curves show the age effects on learning time.

The points on each curve are conventionally connected by straight lines even though the variable on the X axis (sex, in our example) is not a continuous variable. When the variable on the X axis is qualitative, the slopes of

FIGURE 17.1

Age Effect but No Sex Effect, with No Interactions

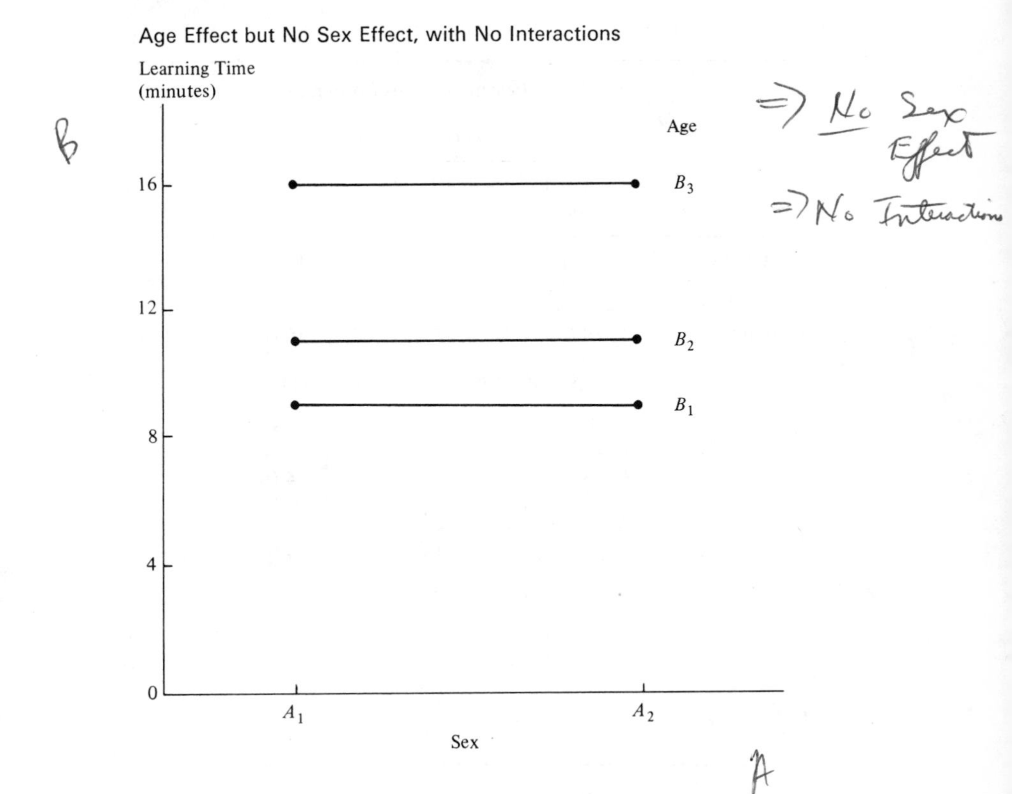

the curves have no meaning, except when the slope is zero, which implies there are no factor level effects. If one of the two factors is a quantitative variable, it is ordinarily advisable to place that factor on the X scale.

A Second Example with Additive Factor Effects

Table 17.2a contains another illustration of factor effects which do not interact, for the same sex-age learning setting as before. The situation here differs from that of Table 17.1a in that not only age but also sex affects the learning time. This is evident from the fact that the mean learning times for men and women are not the same for any age group. Alternatively, this can be seen from Table 17.2c, which shows that the specific sex effects are not zero.

In Table 17.2a, as in Table 17.1a, every mean response can be decomposed according to (17.13):

$$\mu_{ij} = \mu_{..} + \alpha_i + \beta_j$$

For instance:

$$\mu_{11} = \mu_{..} + \alpha_1 + \beta_1 = 12 + 2 + (-3) = 11$$

TABLE 17.2

Age and Sex Effects, with No Interactions

(a) Mean Learning Times (in minutes)

	Factor B—Age			
Factor A—Sex	$j=1$ *Young*	$j=2$ *Middle*	$j=3$ *Old*	*Row Average*
$i=1$ Male	11 (μ_{11})	13 (μ_{12})	18 (μ_{13})	14 ($\mu_{1.}$)
$i=2$ Female	7 (μ_{21})	9 (μ_{22})	14 (μ_{23})	10 ($\mu_{2.}$)
Column average	9 ($\mu_{.1}$)	11 ($\mu_{.2}$)	16 ($\mu_{.3}$)	12 ($\mu_{..}$)

(b) Specific Age Effects (in minutes)

	$j=1$	$j=2$	$j=3$
$i=1$	-3 ($\beta_{1(1)}$)	-1 ($\beta_{2(1)}$)	4 ($\beta_{3(1)}$)
$i=2$	-3 ($\beta_{1(2)}$)	-1 ($\beta_{2(2)}$)	4 ($\beta_{3(2)}$)
Column average	-3 (β_1)	-1 (β_2)	4 (β_3)

(c) Specific Sex Effects (in minutes)

	$j=1$	$j=2$	$j=3$	*Row Average*
$i=1$	2 ($\alpha_{1(1)}$)	2 ($\alpha_{1(2)}$)	2 ($\alpha_{1(3)}$)	2 (α_1)
$i=2$	-2 ($\alpha_{2(1)}$)	-2 ($\alpha_{2(2)}$)	-2 ($\alpha_{2(3)}$)	-2 (α_2)

Hence the two factors do not interact and the factor effects can be analyzed separately by examining the factor level means $\mu_{i.}$ and $\mu_{.j}$ respectively.

Figure 17.2 presents the data from Table 17.2a in graphic form. This time, we have placed age on the X axis and used different curves for each sex. Note that the difference in the heights of the two curves reflects the sex difference and the departure from horizontal for each of the curves reflects the age effect.

FIGURE 17.2

Age and Sex Effects, with No Interactions

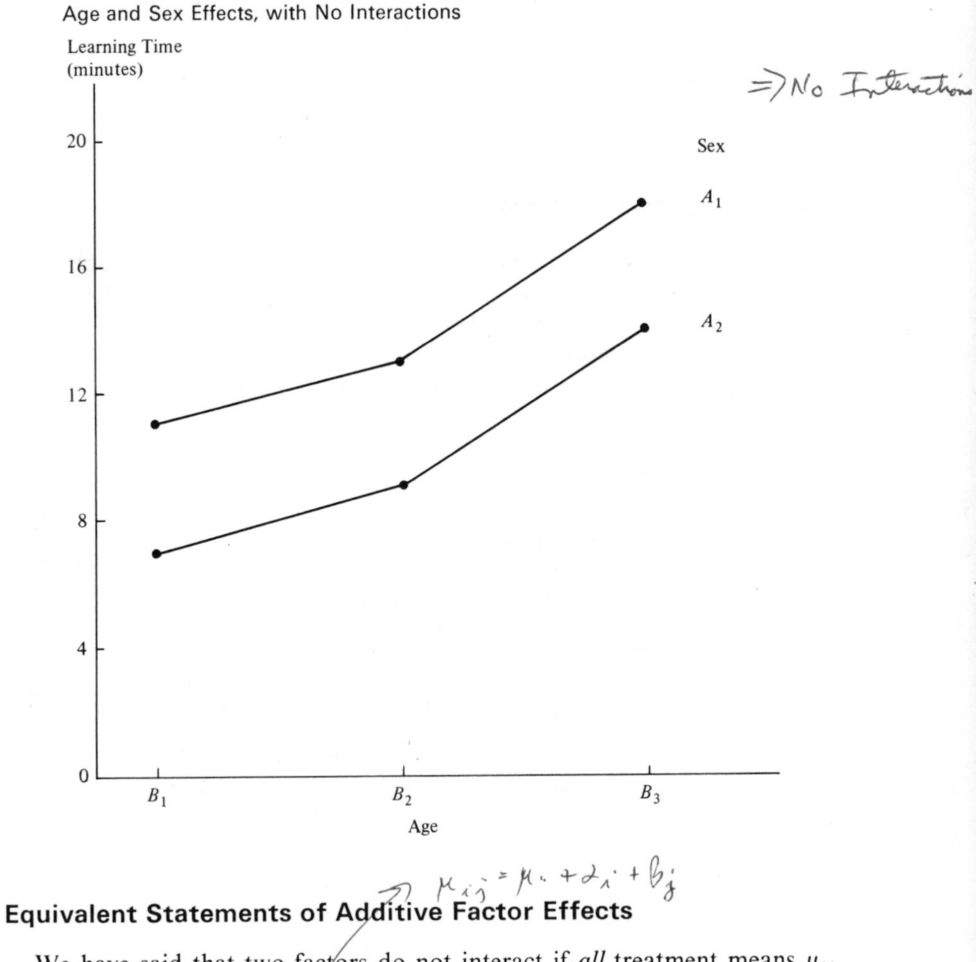

Equivalent Statements of Additive Factor Effects

We have said that two factors do not interact if *all* treatment means μ_{ij} can be expressed according to (17.13). There are a number of other, equivalent, methods of recognizing when two factors do not interact. These are:

1. The difference between the mean responses for any two levels of factor B is the same for all levels of factor A. (Thus, in Table 17.2a, going

from young to middle age leads to an increase of 2 minutes for both males and females, and going from middle age to old leads to an increase of 5 minutes for both males and females.) Note that it is *not* required that the changes, say, between levels 1 and 2 and between levels 2 and 3 of factor B are the same. These, of course, may differ depending upon the nature of the factor B effect.

2. The difference between the mean responses for any two levels of factor A is the same for all levels of factor B. (Thus, in Table 17.2a, going from male to female leads to a decrease of 4 minutes for all three age groups.)

3. The curves for the different levels of a factor are all parallel (such as in Figure 17.2).

All of these conditions are equivalent, implying that the two factors do not interact.

Interacting Factor Effects

Table 17.3a contains an illustration for our sex-age learning setting where the factor effects do interact. The mean learning times for the different sex-age combinations in Table 17.3a indicate that sex has no effect on learning time for young persons, but has a substantial effect for old persons. This differential influence of sex, depending on the age of the person, implies that the age and sex factors interact.

TABLE 17.3

Age and Sex Effects, with Interactions

(a) Mean Learning Times (in minutes)

	Factor B—Age				
Factor A—Sex	$j=1$ *Young*	$j=2$ *Middle*	$j=3$ *Old*	*Row Average*	*Main Sex Effect*
$i=1$ Male	9 (μ_{11})	12 (μ_{12})	18 (μ_{13})	13 ($\mu_{1.}$)	1 (α_1)
$i=2$ Female	9 (μ_{21})	10 (μ_{22})	14 (μ_{23})	11 ($\mu_{2.}$)	-1 (α_2)
Column average	9 ($\mu_{.1}$)	11 ($\mu_{.2}$)	16 ($\mu_{.3}$)	12 ($\mu_{..}$)	—
Main age effect	-3 (β_1)	-1 (β_2)	4 (β_3)	—	—

(b) Interactions (in minutes)

	$j=1$	$j=2$	$j=3$	*Row Average*
$i=1$	-1	0	1	0
$i=2$	1	0	-1	0
Column average	0	0	0	0

Definition of Interaction. We can study the existence of interacting factor effects formally by examining whether or not all treatment means μ_{ij} can be expressed according to (17.13):

$$\mu_{ij} = \mu_{..} + \alpha_i + \beta_j$$

If they can, the factor effects are additive; otherwise, the factor effects are interacting.

For our example in Table 17.3a, the main factor effects α_i and β_j are shown in the margins of the table. It is clear that the factors interact. For instance:

$$\mu_{..} + \alpha_1 + \beta_1 = 12 + 1 + (-3) = 10$$

whereas $\mu_{11} = 9$. If the two factors were additive, μ_{11} would have to be 10.

The difference between the treatment mean μ_{ij} and the value $\mu_{..} + \alpha_i + \beta_j$ which would be expected if the two factors were additive is called the *interaction* of the ith level of factor A with the jth level of factor B, and is denoted by $(\alpha\beta)_{ij}$. Thus, we define $(\alpha\beta)_{ij}$ as follows:

$$(\alpha\beta)_{ij} = \mu_{ij} - (\mu_{..} + \alpha_i + \beta_j) \tag{17.14}$$

Replacing α_i and β_j by their definitions in terms of treatment means in (17.11) and (17.10) respectively, we obtain the alternate definition:

$$(\alpha\beta)_{ij} = \mu_{ij} - \mu_{i.} - \mu_{.j} + \mu_{..} \tag{17.14a}$$

To repeat, the interaction of the ith level of A with the jth level of B, denoted by $(\alpha\beta)_{ij}$, is simply the difference between μ_{ij} and the value which would be expected if the factors were additive. If in fact the two factors are additive, all interactions $(\alpha\beta)_{ij} = 0$.

The interactions for our illustration in Table 17.3a are shown in Table 17.3b. We have, for instance:

$$\begin{aligned}(\alpha\beta)_{13} &= \mu_{13} - (\mu_{..} + \alpha_1 + \beta_3)\\ &= 18 - (12 + 1 + 4)\\ &= 1\end{aligned}$$

Recognition of Interactions. We may recognize whether or not interactions are present in one of the following equivalent fashions:

1. By examining whether all μ_{ij} can be expressed as the sums $\mu_{..} + \alpha_i + \beta_j$.
2. By examining whether the difference between the mean responses for any two levels of factor B is the same for all levels of factor A. (Note in Table 17.3a that the mean learning time increases when going from young to middle-aged persons by 3 minutes for men but only by 1 minute for women.)
3. By examining whether the difference between the mean responses for any two levels of factor A is the same for all levels of factor B. (Note in Table 17.3a that there is no difference between sexes for young persons, but a 4-minute difference for old persons.)

FIGURE 17.3

Age and Sex Effects, with Interactions

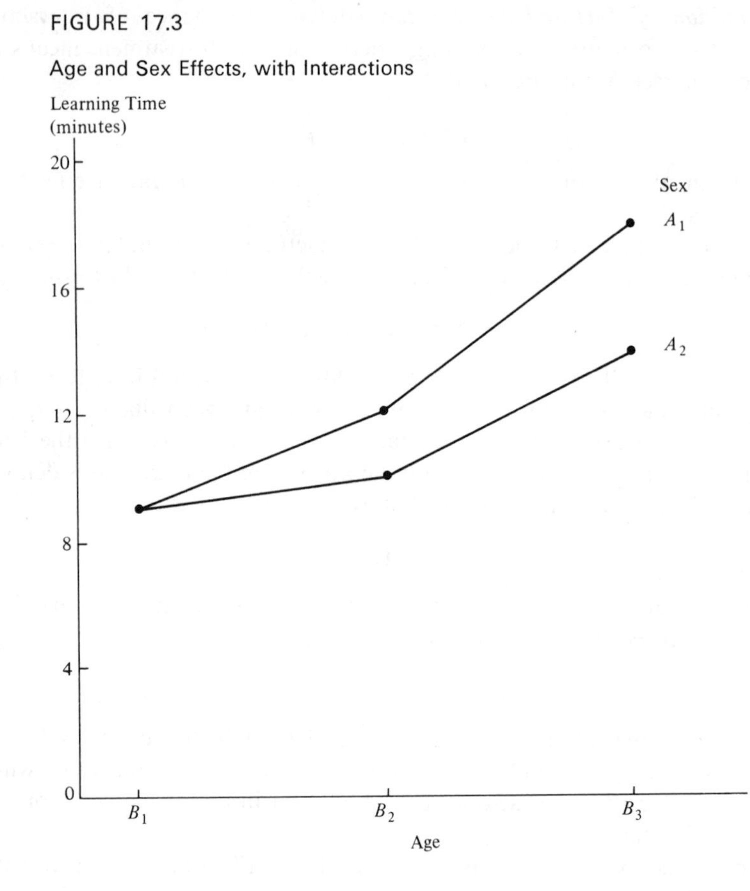

4. By examining whether the factor level curves in a graph are parallel. (Figure 17.3 presents the data in Table 17.3a, with age on the X axis. Note that the sex factor level curves are not parallel.)

Comments

1. Note from Table 17.3b that some interactions are zero even though the two factors are interacting. *All* interactions must be zero in order for the two factors to be additive.

2. The interaction $(\alpha\beta)_{ij}$ may also be interpreted as the difference between the specific effect of factor A at the ith level when factor B is at the jth level $(\alpha_{i(j)})$ and the mean specific effect (main effect) of factor A at the ith level (α_i):

$$(\alpha\beta)_{ij} = \alpha_{i(j)} - \alpha_i \tag{17.15}$$

This may be rewritten, using (17.8) and (17.11):

$$\begin{aligned}(\alpha\beta)_{ij} &= (\mu_{ij} - \mu_{.j}) - (\mu_{i.} - \mu_{..})\\ &= \mu_{ij} - \mu_{i.} - \mu_{.j} + \mu_{..}\end{aligned}$$

Alternatively, $(\alpha\beta)_{ij}$ may be interpreted as the difference between the specific effect of factor B at the jth level when factor A is at the ith level and the main effect of factor B at the jth level:

$$(\alpha\beta)_{ij} = \beta_{j(i)} - \beta_j \tag{17.16}$$

3. Table 17.3b illustrates that interactions sum to zero when added over either rows or columns:

$$\sum_i (\alpha\beta)_{ij} = 0 \tag{17.17a}$$

$$\sum_j (\alpha\beta)_{ij} = 0 \tag{17.17b}$$

Consequently, the sum of all interactions is also zero:

$$\sum_i \sum_j (\alpha\beta)_{ij} = 0 \tag{17.17c}$$

We show this for (17.17a):

$$\begin{aligned}\sum_i (\alpha\beta)_{ij} &= \sum_{i=1}^{a} (\mu_{ij} - \mu_{..} - \alpha_i - \beta_j) \\ &= \sum_i \mu_{ij} - a\mu_{..} - \sum_i \alpha_i - a\beta_j\end{aligned}$$

Now $\sum_i \mu_{ij} = a\mu_{.j}$ by (17.1) and $\sum \alpha_i = 0$ by (17.12). Finally $\beta_j = \mu_{.j} - \mu_{..}$ by (17.10). Hence, we obtain:

$$\sum_i (\alpha\beta)_{ij} = a\mu_{.j} - a\mu_{..} - a(\mu_{.j} - \mu_{..}) = 0$$

Important and Unimportant Interactions

When two factors interact, the question arises whether the factor level means, which are averages of specific treatment means, are meaningful measures. In Table 17.3a, for instance, it may well be argued that the sex factor level means 13 and 11 are misleading measures. They indicate that some difference exists in learning time for men and women, but that this difference is not too great. These factor level means hide the fact that there is no difference in learning time between sexes for young persons, but there is a relatively large difference for old persons. The interactions in Table 17.3a would therefore be considered important ones, implying that one should not discuss the effects of each factor separately in terms of the factor level means. A chart, such as Figure 17.3, presents effectively a description of the interacting effects of the two factors.

There are other times when two factors interact where the interactions are unimportant. Table 17.4 and Figure 17.4 present such a case. Note from Figure 17.4 that the curves are *almost* parallel. Perfectly parallel curves, we know, would indicate there are no interactions. For practical purposes, one may say that the mean learning time for women is 2 minutes less than that

TABLE 17.4

Age and Sex Effects, with Unimportant Interactions

	Factor B—Age			
Factor A—Sex	$j=1$ *Young*	$j=2$ *Middle*	$j=3$ *Old*	*Row Average*
$i=1$ Male	9.75	12.00	17.25	13.0
$i=2$ Female	8.25	10.00	14.75	11.0
Column average	9.0	11.0	16.0	12.0

FIGURE 17.4

Age and Sex Effects, with Unimportant Interactions (curves almost parallel)

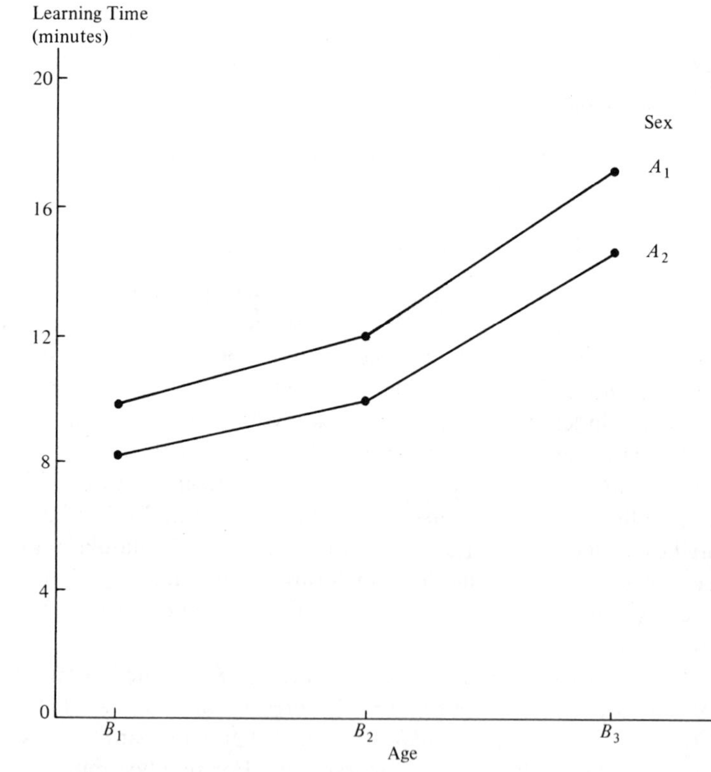

for men, and this statement is approximately true for all age groups. Alternatively, statements based on average learning time for different age groups will hold approximately for both sexes.

Thus, in the case of unimportant interactions, the analysis of factor effects can proceed as for the case of no interactions. Each factor can be studied separately, based on the factor level means $\mu_{i.}$ and $\mu_{.j}$ respectively. This separate analysis of factor effects is, of course, much simpler than a joint analysis for the two factors, based on the treatment means μ_{ij}, which is required when the interactions are important.

Comments

1. The advantage of unimportant (or no) interactions, namely, that one is then able to analyze the factor effects separately, is especially great when the study contains more than two factors.

2. Occasionally, important interactions exist which can be made unimportant by a simple transformation of scale, such as a square, a square root, or a logarithmic transformation. Table 17.5a contains an example of important interactions. If a square root transformation is applied to these data, the resulting treatment means in Table 17.5b show no interacting effects. Essentially, cases where important interactions can be made unimportant by a meaningful transformation may be said to represent instances where the original scale of measurement is not the most appropriate one. Ordinarily, of course, one cannot hope that a simple transformation of scale removes all interactions as in Table 17.5, but only that interactions become unimportant after the transformation.

TABLE 17.5

Illustration of a Removable Interaction

(a) Treatment Means—Original Scale

	Factor B	
Factor A	$j=1$	$j=2$
$i=1$	16	64
$i=2$	49	121
$i=3$	64	144

(b) Treatment Means after Square Root Transformation

	Factor B	
Factor A	$j=1$	$j=2$
$i=1$	4	8
$i=2$	7	11
$i=3$	8	12

Analysis of Interactions

The analysis of interactions can be quite difficult when the interacting effects are complex. In that event, it may not be possible to generalize about factor effects but only to make a series of comparisons of treatment means. There are many occasions, however, when the interactions are of simple structure, such as in Table 17.3a, so that the joint factor effects can be described in a straightforward manner. Table 17.6 provides several additional illustrations of this type.

In Table 17.6a, we have a situation where either raising the pay or increasing the authority of low-paid executives with small authority leads to increased productivity. However, combining both higher pay and greater authority does not lead to further improvements in productivity than increasing either one alone. Table 17.6b represents a case where both higher pay and greater authority are required before any substantial increase in productivity takes place. Table 17.6c portrays a situation where size of crew and personality of foreman do not interact on the productivity per man if the smallest crew size is excluded.

TABLE 17.6

Examples of Different Types of Interactions

(a) Productivity of Executives

	Factor B—Authority	
Factor A—Pay	*Small*	*Great*
Low	50	76
High	74	75

(b) Productivity of Executives

	Factor B—Authority	
Factor A—Pay	*Small*	*Great*
Low	50	52
High	53	75

(c) Productivity per Person in Crew

	Factor B—Personality of Foreman	
Factor A—Crew Size	*Extrovert*	*Introvert*
4 Persons	28	20
6 Persons	22	20
8 Persons	20	18
10 Persons	17	15

Note

It is possible that the two factors interact, yet the main effects for one (or both) factors are zero. This would be the result of interactions in opposite directions that balance out over one (or both) factors. Thus, there would be definite factor effects, but these would not be disclosed by the factor level means. The case of interacting factors with no main effects for one (or both) factors fortunately is unusual. Typically, the interaction effects are smaller than the main effects.

17.3 MODEL I (FIXED EFFECTS) FOR TWO-FACTOR STUDIES

Having explained the model elements, we are now ready to develop the analysis of variance fixed effects model for two-factor studies.

The basic situation is as follows: Factor A is studied at a levels, and these are of intrinsic interest in themselves; in other words, the a levels are not considered a sample from a larger population of factor A levels. Similarly, factor B is studied at b levels, which are of intrinsic interest in themselves. All ab factor level combinations are included in the study. The number of observations for each of the ab treatments is the same, denoted by n, and it is required that $n > 1$. Thus, the total number of observations for the study is:

$$n_T = abn \tag{17.18}$$

The kth observation ($k = 1, \ldots, n$) for the treatment where A is at the ith level and B at the jth level is denoted by Y_{ijk} ($i = 1, \ldots, a$; $j = 1, \ldots, b$). Table 17.7 on page 570 illustrates this notation for an example where A is at three levels, B is at two levels, and two observations are made for each treatment.

Model Not Incorporating Factorial Structure

If we were to regard the ab treatments without considering the factorial structure of the study, we would write the analysis of variance model as follows:

$$Y_{ijk} = \mu_{ij} + \varepsilon_{ijk} \tag{17.19}$$

where:

μ_{ij} are parameters
ε_{ijk} are independent $N(0, \sigma^2)$
$i = 1, \ldots, a; j = 1, \ldots, b; k = 1, \ldots, n$

The parameter μ_{ij} is the mean response for the treatment in which A is at the ith level and B at the jth level.

Note that this formulation is similar to the single-factor model (13.2), except for the two subscripts now needed to identify the treatment.

Meaning of Model Parameters. In a survey situation, the treatment mean μ_{ij} corresponds to the population mean for the elements having the characteristics of the ith level of factor A and the jth level of factor B. For instance, in a study of the productivity of employees in each of three shifts (factor A) and in each of five plants (factor B), the population mean μ_{ij} is the mean productivity per employee in the ith shift at the jth plant.

In an experimental study, the treatment mean μ_{ij} stands for the mean response that would be obtained if the treatment consisting of the ith level of factor A and the jth level of factor B were applied to all units in the population of experimental units about which inferences are to be drawn. For instance, in a study where factor A is type of training program (structured, partially structured, unstructured) and factor B is time of training (during work, after work), $6n$ employees are selected and n are assigned at random to each of the 6 treatments. The mean μ_{ij} here represents the mean response, say, mean gain in productivity, if the ith training program administered during the jth time were given to all employees in the population of experimental units.

Model Incorporating Factorial Structure

Model (17.19) is ordinarily not adequate because it does not explicitly recognize the factorial structure of a two-factor study. We incorporate the factorial structure into the model by noting that the definition of an interaction in (17.14) implies:

$$\mu_{ij} = \mu_{..} + \alpha_i + \beta_j + (\alpha\beta)_{ij} \tag{17.20}$$

This formulation indicates that the mean μ_{ij} for any treatment can be viewed as the sum of four component effects. Specifically, (17.20) states that the mean response for the treatment where factor A is at the ith level and factor B at the jth level is the sum of:

1. An overall constant $\mu_{..}$;
2. The main effect α_i for factor A at the ith level;
3. The main effect β_j for factor B at the jth level;
4. The interaction effect $(\alpha\beta)_{ij}$ when factor A is at the ith level and factor B at the jth level.

Replacing μ_{ij} in model (17.19) by the equivalent expression in (17.20), we obtain the fixed effects analysis of variance model for two-factor studies:

$$Y_{ijk} = \mu_{..} + \alpha_i + \beta_j + (\alpha\beta)_{ij} + \varepsilon_{ijk} \tag{17.21}$$

where:

$\mu_{..}$ is a constant
α_i are constants subject to the restriction $\sum \alpha_i = 0$
β_j are constants subject to the restriction $\sum \beta_j = 0$

$(\alpha\beta)_{ij}$ are constants subject to the restrictions

$$\sum_i (\alpha\beta)_{ij} = 0 \qquad \sum_j (\alpha\beta)_{ij} = 0$$

ε_{ijk} are independent $N(0, \sigma^2)$
$i = 1, \ldots, a; j = 1, \ldots, b; k = 1, \ldots, n$

Important Features of Model. Model (17.21) for two-factor studies differs from model (13.2) for single-factor studies only in that the treatment mean is now expressed in terms of a factor A main effect, a factor B main effect, and an interaction effect. Normality, independent error terms, and constant variances for the error terms are properties of the models for both single-factor and two-factor studies.

Note that for model (17.21), we have:

$$E(Y_{ijk}) = \mu_{..} + \alpha_i + \beta_j + (\alpha\beta)_{ij} \tag{17.22}$$

because $E(\varepsilon_{ijk}) = 0$, and:

$$\sigma^2(Y_{ijk}) = \sigma^2 \tag{17.23}$$

because the error term ε_{ijk} is the only random term on the right-hand side and $\sigma^2(\varepsilon_{ijk}) = \sigma^2$. Finally, the Y's are independent normal random variables because the error terms are independent normal random variables. Hence, we can also state model (17.21) as follows:

$$Y_{ijk} \text{ are independent } N\left(\mu_{..} + \alpha_i + \beta_j + (\alpha\beta)_{ij}, \sigma^2\right) \tag{17.24}$$

17.4 ANALYSIS OF VARIANCE

Illustration

Table 17.7 contains an illustration which we shall employ both in this chapter and the next. The Castle Bakery Company supplies sliced wrapped Italian bread to a large number of supermarkets in a metropolitan area. An experimental study was made of the effect of height of the shelf display (bottom, middle, top) and the width of the shelf display (regular, wide) on sales of this bakery's bread (measured in cases) during the experimental period. Twelve supermarkets, similar in terms of sales volumes and clientele, were utilized in the study. Two stores were assigned at random to each of the six treatments, and the display of the bread in each store followed the treatment specifications for that store. Sales of the bread were recorded, and these results are presented in Table 17.7.

Notation

Table 17.7 illustrates the notation we shall use for two-factor studies. It is a straightforward extension of the notation for single-factor studies. A dot in the subscript indicates aggregation or averaging over the variable represented

TABLE 17.7

Sample Data and Notation for Two-Factor Castle Bakery Example (sales in cases)

Factor A (display height) (i)	*Factor B (display width) (j)* B_1 *(regular)*	B_2 *(wide)*	*Row Total*	*Display Height Average*
A_1 (Bottom)	47 (Y_{111})	46 (Y_{121})		
	43 (Y_{112})	40 (Y_{122})		
Total	90 ($Y_{11.}$)	86 ($Y_{12.}$)	176 ($Y_{1..}$)	
Average	45 ($\bar{Y}_{11.}$)	43 ($\bar{Y}_{12.}$)		44 ($\bar{Y}_{1..}$)
A_2 (Middle)	62 (Y_{211})	67 (Y_{221})		
	68 (Y_{212})	71 (Y_{222})		
Total	130 ($Y_{21.}$)	138 ($Y_{22.}$)	268 ($Y_{2..}$)	
Average	65 ($\bar{Y}_{21.}$)	69 ($\bar{Y}_{22.}$)		67 ($\bar{Y}_{2..}$)
A_3 (Top)	41 (Y_{311})	42 (Y_{321})		
	39 (Y_{312})	46 (Y_{322})		
Total	80 ($Y_{31.}$)	88 ($Y_{32.}$)	168 ($Y_{3..}$)	
Average	40 ($\bar{Y}_{31.}$)	44 ($\bar{Y}_{32.}$)		42 ($\bar{Y}_{3..}$)
Column total	300 ($Y_{.1.}$)	312 ($Y_{.2.}$)	612 ($Y_{...}$)	
Display width average	50 ($\bar{Y}_{.1.}$)	52 ($\bar{Y}_{.2.}$)		51 ($\bar{Y}_{...}$)

by the index. For instance, the sum of the observations for the treatment corresponding to the ith level of factor A and the jth level of factor B is:

(17.25a)
$$Y_{ij.} = \sum_{k=1}^{n} Y_{ijk}$$

The corresponding mean is:

(17.25b)
$$\bar{Y}_{ij.} = \frac{Y_{ij.}}{n}$$

The total of all observations for the ith factor level of A is:

(17.25c)
$$Y_{i..} = \sum_{j}^{b} \sum_{k}^{n} Y_{ijk}$$

and the corresponding mean is:

(17.25d)
$$\bar{Y}_{i..} = \frac{Y_{i..}}{bn}$$

Similarly, for the jth factor level of B the sum of all observations and their mean are denoted by:

(17.25e)
$$Y_{.j.} = \sum_i^a \sum_k^n Y_{ijk}$$

(17.25f)
$$\bar{Y}_{.j.} = \frac{Y_{.j.}}{an}$$

Finally, the sum of all observations in the study is:

(17.25g)
$$Y_{...} = \sum_i^a \sum_j^b \sum_k^n Y_{ijk}$$

and the overall mean is:

(17.25h)
$$\bar{Y}_{...} = \frac{Y_{...}}{nab}$$

Point Estimators of Parameters

Least squares estimators of the parameters for model (17.21) are obtained by minimizing:

(17.26)
$$Q = \sum_i \sum_j \sum_k [Y_{ijk} - \mu_{..} - \alpha_i - \beta_j - (\alpha\beta)_{ij}]^2$$

subject to the restrictions:

$$\sum_i \alpha_i = 0 \qquad \sum_j \beta_j = 0 \qquad \sum_i (\alpha\beta)_{ij} = 0 \qquad \sum_j (\alpha\beta)_{ij} = 0$$

When we perform this minimization, we obtain the following least squares estimators of the parameters:

	Parameter	*Estimator*
(17.27a)	$\mu_{..}$	$\bar{Y}_{...}$
(17.27b)	α_i	$\bar{Y}_{i..} - \bar{Y}_{...}$
(17.27c)	β_j	$\bar{Y}_{.j.} - \bar{Y}_{...}$
(17.27d)	$(\alpha\beta)_{ij}$	$\bar{Y}_{ij.} - \bar{Y}_{i..} - \bar{Y}_{.j.} + \bar{Y}_{...}$
(17.27e)	$\mu_{ij} = \mu_{..} + \alpha_i + \beta_j + (\alpha\beta)_{ij}$	$\bar{Y}_{ij.}$

The correspondence of the least squares estimators to the definitions of the parameters is readily apparent.

Note

The least squares estimators in (17.27) are the same estimators as those obtained by the method of maximum likelihood.

Residuals

The residual e_{ijk} is, as usual, the deviation between the observation and its estimated expected value:

$$e_{ijk} = Y_{ijk} - \bar{Y}_{ij.} \tag{17.28}$$

Residuals are highly useful for assessing the aptness of the two-factor model (17.21), as they also are for the models considered earlier.

Breakdown of Total Sum of Squares

Breakdown of Total Deviation. We shall break down the deviation of an observation Y_{ijk} from the overall mean $\bar{Y}_{...}$ in two stages. First we shall obtain a decomposition of the total deviation $Y_{ijk} - \bar{Y}_{...}$ by viewing the study as consisting of ab treatments:

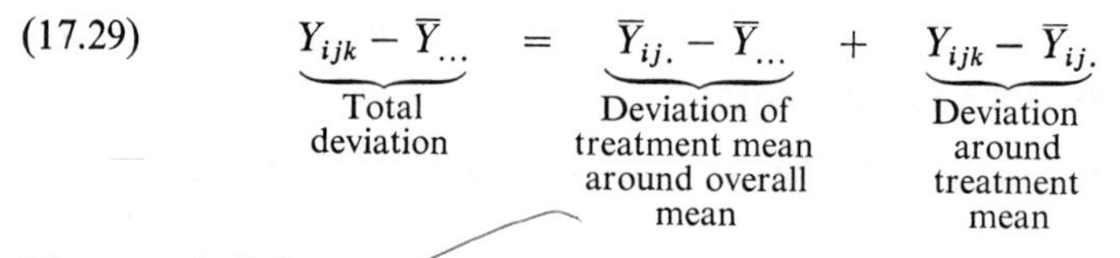

$$\underbrace{Y_{ijk} - \bar{Y}_{...}}_{\substack{\text{Total}\\ \text{deviation}}} = \underbrace{\bar{Y}_{ij.} - \bar{Y}_{...}}_{\substack{\text{Deviation of}\\ \text{treatment mean}\\ \text{around overall}\\ \text{mean}}} + \underbrace{Y_{ijk} - \bar{Y}_{ij.}}_{\substack{\text{Deviation}\\ \text{around}\\ \text{treatment}\\ \text{mean}}} \tag{17.29}$$

Then we shall decompose the treatment mean deviation $\bar{Y}_{ij.} - \bar{Y}_{...}$ in terms of components reflecting the factor A main effect, the factor B main effect, and the AB interaction:

$$\underbrace{\bar{Y}_{ij.} - \bar{Y}_{...}}_{\substack{\text{Deviation of}\\ \text{treatment mean}\\ \text{around overall}\\ \text{mean}}} = \underbrace{\bar{Y}_{i..} - \bar{Y}_{...}}_{\substack{A\text{ main}\\ \text{effect}}} + \underbrace{\bar{Y}_{.j.} - \bar{Y}_{...}}_{\substack{B\text{ main}\\ \text{effect}}} + \underbrace{\bar{Y}_{ij.} - \bar{Y}_{i..} - \bar{Y}_{.j.} + \bar{Y}_{...}}_{AB\text{ interaction effect}} \tag{17.30}$$

Treatment and Error Sums of Squares. When we square (17.29) and sum over all observations, the cross-product term drops out and we obtain:

$$SSTO = SSTR + SSE \tag{17.31}$$

where:

$$SSTO = \sum_i \sum_j \sum_k (Y_{ijk} - \bar{Y}_{...})^2 \tag{17.31a}$$

$$SSTR = n \sum_i \sum_j (\bar{Y}_{ij.} - \bar{Y}_{...})^2 \tag{17.31b}$$

$$SSE = \sum_i \sum_j \sum_k (Y_{ijk} - \bar{Y}_{ij.})^2 \tag{17.31c}$$

$SSTR$ reflects the variability between the ab treatment means and is the ordinary *treatment sum of squares*, and SSE reflects the variability within treatments and is the usual *error sum of squares*. The only difference between these formulas and those for the single-factor case is the use of the two subscripts i and j to designate a treatment.

TABLE 17.8

ANOVA Table Neglecting Factorial Structure for Castle Bakery Example

Source of Variation	*SS*	*df*	*MS*
Between treatments	1,580	5	316
Error	62	6	10.3
Total	1,642	11	

Example. For our Castle Bakery example, Table 17.8 contains the decomposition of the total sum of squares in (17.31). This is the ordinary ANOVA table treating the study as a single-factor one with $ab = r = 6$ treatments. The sums of squares are obtained as follows:

$$SSTO = (47 - 51)^2 + (43 - 51)^2 + (46 - 51)^2 + \cdots + (46 - 51)^2 = 1{,}642$$
$$SSTR = 2[(45 - 51)^2 + (43 - 51)^2 + (65 - 51)^2 + \cdots + (44 - 51)^2] = 1{,}580$$
$$SSE = (47 - 45)^2 + (43 - 45)^2 + (46 - 43)^2 + \cdots + (46 - 44)^2 = 62$$

One could test at this point, by means of (13.65), whether or not the six treatment means are equal. If they are, neither of the two factors has any effect. Ordinarily, however, testing of factor effects is not made until a further decomposition of the treatment sum of squares has been carried out to reflect the factorial nature of the study.

Breakdown of Treatment Sum of Squares. When we square (17.30) and sum over all treatments and over the n observations associated with each treatment mean $\bar{Y}_{ij.}$, all cross-product terms drop out and we obtain:

(17.32) $$SSTR = SSA + SSB + SSAB$$

where:

(17.32a) $$SSA = nb \sum_i (\bar{Y}_{i..} - \bar{Y}_{...})^2$$

(17.32b) $$SSB = na \sum_j (\bar{Y}_{.j.} - \bar{Y}_{...})^2$$

(17.32c) $$SSAB = n \sum_i \sum_j (\bar{Y}_{ij.} - \bar{Y}_{i..} - \bar{Y}_{.j.} + \bar{Y}_{...})^2$$

SSA, called the *factor A sum of squares*, measures the variability of the estimated A factor level means ($\bar{Y}_{i..}$). The more variable they are, the bigger will be SSA. Similarly SSB, called the *factor B sum of squares*, measures the variability of the estimated B factor level means ($\bar{Y}_{.j.}$). Finally $SSAB$, called the *AB interaction sum of squares*, measures the variability of the estimated interactions ($\bar{Y}_{ij.} - \bar{Y}_{i..} - \bar{Y}_{.j.} + \bar{Y}_{...}$) for the ab treatments. Remember that the mean of all interactions is zero, so that the deviations of the estimated interactions around their mean is not explicitly shown as it was in SSA and SSB.

The larger are the estimated interactions (sign disregarded), the larger will be $SSAB$.

The breakdown of $SSTR$ into the components SSA, SSB, and $SSAB$ is an orthogonal decomposition. While many such decompositions are possible, this one is of interest because the three components provide information about the factor A main effects, the factor B main effects, and the AB interactions respectively, as will be seen shortly.

Computational Forms. For hand computing purposes, we ordinarily use the following formulas which are algebraically identical to the definitional formulas previously given:

(17.33a)
$$SSTO = \sum_i \sum_j \sum_k Y_{ijk}^2 - \frac{Y_{...}^2}{nab}$$

(17.33b)
$$SSE = \sum_i \sum_j \sum_k Y_{ijk}^2 - \frac{\sum_i \sum_j Y_{ij.}^2}{n}$$

(17.33c)
$$SSA = \frac{\sum_i Y_{i..}^2}{nb} - \frac{Y_{...}^2}{nab}$$

(17.33d)
$$SSB = \frac{\sum_j Y_{.j.}^2}{na} - \frac{Y_{...}^2}{nab}$$

Ordinarily, the interaction sum of squares is obtained as a remainder:

(17.33e)
$$SSAB = SSTO - SSE - SSA - SSB$$

or from:

(17.33f)
$$SSAB = SSTR - SSA - SSB$$

where:

(17.33g)
$$SSTR = \frac{\sum_i \sum_j Y_{ij.}^2}{n} - \frac{Y_{...}^2}{nab}$$

Example. For our Castle Bakery example, we obtain the following decomposition of $SSTR$, using the data in Table 17.7 and the computational formulas in (17.33):

$$SSA = \frac{(176)^2 + (268)^2 + (168)^2}{(2)(2)} - \frac{(612)^2}{(2)(3)(2)} = 1{,}544$$

$$SSB = \frac{(300)^2 + (312)^2}{(2)(3)} - \frac{(612)^2}{(2)(3)(2)} = 12$$

$$SSAB = 1{,}580 - 1{,}544 - 12 = 24$$

Hence, we have:

$$1{,}580 = 1{,}544 + 12 + 24$$
$$SSTR = SSA + SSB + SSAB$$

Note

Since the decomposition of *SSTR* is orthogonal, we know from Chapter 14 that the component sums of squares can be expressed in terms of orthogonal contrasts of the treatment means. For instance, in our Castle Bakery example, *SSB* is based on the contrast:

$$\hat{L} = \frac{\bar{Y}_{11.} + \bar{Y}_{21.} + \bar{Y}_{31.}}{3} - \frac{\bar{Y}_{12.} + \bar{Y}_{22.} + \bar{Y}_{32.}}{3}$$

which compares the mean sales for the regular display width with that for wide displays. From Table 17.7, we have:

$$\hat{L} = \frac{45 + 65 + 40}{3} - \frac{43 + 69 + 44}{3} = -2$$

The associated sum of squares for this contrast is by (14.32):

$$SS(\hat{L}) = \frac{(\hat{L})^2}{\frac{1}{n}\sum c_j^2} = \frac{(-2)^2}{\frac{1}{2}\left[\left(\frac{1}{3}\right)^2 + \left(\frac{1}{3}\right)^2 + \left(\frac{1}{3}\right)^2 + \left(-\frac{1}{3}\right)^2 + \left(-\frac{1}{3}\right)^2 + \left(-\frac{1}{3}\right)^2\right]} = 12$$

which is equal to *SSB* as obtained earlier.

Recapitulation. Combining the decompositions in (17.31) and (17.32), we have established that:

$$SSTO = SSA + SSB + SSAB + SSE \tag{17.34}$$

where the component sums of squares are defined in (17.33).

For our Castle Bakery example, we have found:

$$1{,}642 = 1{,}544 + 12 + 24 + 62$$
$$SSTO = SSA + SSB + SSAB + SSE$$

Thus, much of the total variability in this instance is associated with the factor *A* (display height) effects.

Breakdown of Degrees of Freedom

We are familiar from single-factor analysis of variance as to how the degrees of freedom are divided between the treatment and error components. Corresponding to the further breakdown of the treatment sum of squares, we can also obtain a breakdown of the associated $ab - 1$ degrees of freedom. *SSA* has a $a - 1$ degrees of freedom associated with it. There are a factor level means $\bar{Y}_{i..}$, and these are subject to one restriction, $\sum(\bar{Y}_{i..} - \bar{Y}_{...}) = 0$.

Similarly, SSB has $b - 1$ degrees of freedom associated with it. The degrees of freedom associated with $SSAB$, the interaction sum of squares, is the remainder:

$$(ab - 1) - (a - 1) - (b - 1) = (a - 1)(b - 1)$$

The degrees of freedom associated with $SSAB$ may be understood as follows: There are ab interaction terms. These are subject to b restrictions since $\sum_i (\overline{Y}_{ij.} - \overline{Y}_{i..} - \overline{Y}_{.j.} + \overline{Y}_{...}) = 0$ for all j. There are a additional restrictions since $\sum_j (\overline{Y}_{ij.} - \overline{Y}_{i..} - \overline{Y}_{.j.} + \overline{Y}_{...}) = 0$ for all i, but only $a - 1$ of these are independent since the last one is implied by the previous b restrictions. Altogether, therefore, there are $b + (a - 1)$ independent restrictions. Hence, the degrees of freedom are:

$$ab - (b + a - 1) = (a - 1)(b - 1)$$

Example. For our Castle Bakery example, SSA has $3 - 1 = 2$ degrees of freedom associated with it, SSB has $2 - 1 = 1$ degrees of freedom, and $SSAB$ has $(3 - 1)(2 - 1) = 2$ degrees of freedom.

Mean Squares

Mean squares are obtained in the usual way by dividing the sums of squares by their associated degrees of freedom. We thus obtain:

(17.35a)
$$MSA = \frac{SSA}{a - 1}$$

(17.35b)
$$MSB = \frac{SSB}{b - 1}$$

(17.35c)
$$MSAB = \frac{SSAB}{(a - 1)(b - 1)}$$

Example. For our Castle Bakery example, these mean squares are:

$$MSA = \frac{1{,}544}{2} = 772$$

$$MSB = \frac{12}{1} = 12$$

$$MSAB = \frac{24}{2} = 12$$

Note that these mean squares do not add to the treatment mean square $MSTR = 1{,}580/5 = 316$. Here again, we see that mean squares are not additive.

TABLE 17.9

ANOVA Table for Two-Factor Fixed Effects Study

Source of Variation	SS	df	MS	E(MS)
Between treatments	$SSTR = n \sum \sum (\bar{Y}_{ij.} - \bar{Y}_{...})^2$	$ab - 1$	$MSTR = \frac{SSTR}{ab - 1}$	$\sigma^2 + \frac{n}{ab - 1} \sum \sum (\mu_{ij} - \mu_{..})^2$
Factor A	$SSA = nb \sum (\bar{Y}_{i..} - \bar{Y}_{...})^2$	$a - 1$	$MSA = \frac{SSA}{a - 1}$	$\sigma^2 + \frac{bn}{a - 1} \sum \alpha_i^2$
Factor B	$SSB = na \sum (\bar{Y}_{.j.} - \bar{Y}_{...})^2$	$b - 1$	$MSB = \frac{SSB}{b - 1}$	$\sigma^2 + \frac{an}{b - 1} \sum \beta_j^2$
Interactions	$SSAB = n \sum \sum (\bar{Y}_{ij.} - \bar{Y}_{i..} - \bar{Y}_{.j.} + \bar{Y}_{...})^2$	$(a - 1)(b - 1)$	$MSAB = \frac{SSAB}{(a - 1)(b - 1)}$	$\sigma^2 + \frac{n}{(a - 1)(b - 1)} \sum \sum (\alpha\beta)_{ij}^2$
Error	$SSE = \sum \sum \sum (Y_{ijk} - \bar{Y}_{ij.})^2$	$ab(n - 1)$	$MSE = \frac{SSE}{ab(n - 1)}$	σ^2
Total	$SSTO = \sum \sum \sum (Y_{ijk} - \bar{Y}_{...})^2$	$nab - 1$		

Expected Mean Squares

It can be shown, along the same lines used for the single-factor case, that the mean squares for the two-factor model (17.21) have the following expectations:

(17.36a) $$E(MSE) = \sigma^2$$

(17.36b) $$E(MSA) = \sigma^2 + nb\frac{\sum \alpha_i^2}{a-1}$$

(17.36c) $$E(MSB) = \sigma^2 + na\frac{\sum \beta_j^2}{b-1}$$

(17.36d) $$E(MSAB) = \sigma^2 + n\frac{\sum\sum (\alpha\beta)_{ij}^2}{(a-1)(b-1)}$$

These expectations show that if there are no factor A main effects (i.e., if all $\mu_{i.}$ are equal, or all $\alpha_i = 0$), MSA and MSE have the same expectation; otherwise MSA tends to be larger than MSE. Similarly, if there are no factor B main effects, MSB and MSE have the same expectation; otherwise MSB tends to be larger than MSE. Finally, if there are no interactions (i.e., if all $(\alpha\beta)_{ij} = 0$) so that the factor effects are additive, $MSAB$ has the same expectation as MSE; otherwise, $MSAB$ tends to be larger than MSE. This suggests that the ratios MSA/MSE, MSB/MSE, and $MSAB/MSE$ will provide information about the main effects and interactions of the two factors respectively, with large values of the ratios indicating the presence of factor effects. We shall see shortly that tests based on these statistics are regular F tests.

Analysis of Variance Table

The breakdown of the total sum of squares into the treatment and error components, and the further breakdown of the treatment sum of squares into the several factor components, is shown in Table 17.9. Also shown there are

TABLE 17.10

ANOVA Table for Two-Factor Castle Bakery Example

Source of Variation	*SS*	*df*	*MS*
Between treatments	1,580	5	316
Factor A (display height)	1,544	2	772
Factor B (display width)	12	1	12
Interactions	24	2	12
Error	62	6	10.3
Total	1,642	11	

the associated degrees of freedom, the mean squares, and the expected mean squares. Table 17.10 contains the two-factor analysis of variance for our Castle Bakery example.

17.5 *F* TESTS

In view of the additivity of sums of squares and degrees of freedom, Cochran's theorem (3.59) applies when no factor effects are present. Hence the test statistics based on the appropriate mean squares then follow the F distribution, leading to the usual type of F test for factor effects.

Test for Interactions

Ordinarily, the analysis of a two-factor study begins with a test to determine whether or not the two factors interact:

(17.37)
$$\begin{aligned} &C_1: \mu_{ij} = \mu_{..} + \alpha_i + \beta_j && \text{for all } i, j \\ &C_2: \mu_{ij} \neq \mu_{..} + \alpha_i + \beta_j && \text{for some } i, j \end{aligned}$$

or equivalently:

(17.37a)
$$\begin{aligned} &C_1: \text{all } (\alpha\beta)_{ij} = 0 \\ &C_2: \text{not all } (\alpha\beta)_{ij} \text{ are zero} \end{aligned}$$

As we noted from an examination of the expected mean squares in Table 17.9, the appropriate test statistic is:

(17.38)
$$F^* = \frac{MSAB}{MSE}$$

Large values of F^* indicate the existence of interactions. When C_1 holds, F^* is distributed as $F((a-1)(b-1), (n-1)ab)$. Hence the appropriate decision rule to control the Type I error at α is:

(17.39)
$$\begin{aligned} &\text{If } F^* \leq F(1-\alpha; (a-1)(b-1), (n-1)ab), \text{ conclude } C_1 \\ &\text{If } F^* > F(1-\alpha; (a-1)(b-1), (n-1)ab), \text{ conclude } C_2 \end{aligned}$$

where $F(1-\alpha; (a-1)(b-1), (n-1)ab)$ is the $(1-\alpha)100$ percentile of the appropriate F distribution.

Test for Factor *A* Effects

Tests for factor A main effects and for factor B main effects ordinarily follow the test for interactions, though we noted earlier that the A and B main effects may not have useful meanings if strong interactions exist. To test whether or not A main effects are present:

(17.40)
$$\begin{aligned} &C_1: \mu_{1.} = \mu_{2.} = \cdots = \mu_{a.} \\ &C_2: \text{not all } \mu_{i.} \text{ are equal} \end{aligned}$$

or equivalently:

$$\begin{aligned}&C_1: \alpha_1 = \alpha_2 = \cdots = \alpha_a = 0\\&C_2: \text{not all } \alpha_i \text{ are zero}\end{aligned} \tag{17.40a}$$

we use the test statistic:

$$F^* = \frac{MSA}{MSE} \tag{17.41}$$

Again, large values of F^* indicate the existence of factor A main effects. Since F^* is distributed as $F(a-1, (n-1)ab)$ when C_1 holds, the appropriate decision rule for controlling the risk of making a Type I error at α is:

$$\begin{aligned}&\text{If } F^* \le F(1-\alpha; a-1, (n-1)ab), \text{ conclude } C_1\\&\text{If } F^* > F(1-\alpha; a-1, (n-1)ab), \text{ conclude } C_2\end{aligned} \tag{17.42}$$

Test for Factor *B* Effects

This test is similar to the one for factor A effects. The alternatives are:

$$\begin{aligned}&C_1: \mu_{.1} = \mu_{.2} = \cdots = \mu_{.b}\\&C_2: \text{not all } \mu_{.j} \text{ are equal}\end{aligned} \tag{17.43}$$

or equivalently:

$$\begin{aligned}&C_1: \beta_1 = \beta_2 = \cdots = \beta_b = 0\\&C_2: \text{not all } \beta_j \text{ are zero}\end{aligned} \tag{17.43a}$$

The test statistic is:

$$F^* = \frac{MSB}{MSE} \tag{17.44}$$

and the appropriate decision rule for controlling the risk of a Type I error at α is:

$$\begin{aligned}&\text{If } F^* \le F(1-\alpha; b-1, (n-1)ab), \text{ conclude } C_1\\&\text{If } F^* > F(1-\alpha; b-1, (n-1)ab), \text{ conclude } C_2\end{aligned} \tag{17.45}$$

Example

We shall investigate in our Castle Bakery example the presence of display height and display width effects, using a level of significance of .05 for each test. First, we begin by testing whether or not interaction effects are present:

$$\begin{aligned}&C_1: \text{all } (\alpha\beta)_{ij} = 0\\&C_2: \text{not all } (\alpha\beta)_{ij} \text{ are zero}\end{aligned}$$

Using the data from Table 17.10 in test statistic (17.38), we obtain:

$$F^* = \frac{12}{10.3} = 1.17$$

Since we are to control the risk of making a Type I error at .05, we require $F(.95; 2, 6) = 5.14$, so that the decision rule is:

$$\text{If } F^* \leq 5.14, \text{ conclude } C_1$$
$$\text{If } F^* > 5.14, \text{ conclude } C_2$$

Since $F^* = 1.17 < 5.14$, we conclude C_1, that display height and display width do not interact in their effects on sales.

Next we turn to testing for display height (factor A) main effects; the alternative conclusions are given in (17.40). The test statistic (17.41) for our example becomes:

$$F^* = \frac{772}{10.3} = 75.0$$

For $\alpha = .05$, we require $F(.95; 2, 6) = 5.14$. Since $F^* = 75.0 > 5.14$, we conclude C_2, that not all factor A level means $\mu_{i.}$ are equal, or that some definite effects associated with height of display level exist.

Finally, we test for display width (factor B) main effects; the alternative conclusions are given in (17.43). The test statistic (17.44) becomes for our example:

$$F^* = \frac{12}{10.3} = 1.17$$

For $\alpha = .05$, we require $F(.95; 1, 6) = 5.99$. Since $F^* = 1.17 < 5.99$, we conclude C_1, that all $\mu_{.j}$ are equal, or that display width has no effect on sales.

Thus, the analysis of variance tests suggest that only display height has an effect on sales for the treatments studied. At this point, it would clearly be desirable to conduct further analyses of the nature of the display height effects. We shall discuss such analyses of the factor effects in Chapter 18.

Comments

1. If the test for interactions is conducted with a level of significance of α_1, that for factor A effects with a level of significance of α_2, and that for factor B effects with a level of significance of α_3, the level of significance α for the *family* of three tests is greater than the individual levels of significance. From the Bonferroni inequality in (5.23), we can derive the inequality:

(17.46) $$\alpha \leq \alpha_1 + \alpha_2 + \alpha_3$$

Bonferroni

For the case considered here, a somewhat tighter inequality can be used, the *Kimball inequality*, which utilizes the fact that the numerators of the three test statistics are independent and the denominator is the same in each case. This inequality states:

$$\alpha \leq 1 - (1 - \alpha_1)(1 - \alpha_2)(1 - \alpha_3) \tag{17.47}$$

For our Castle Bakery example, where $\alpha_1 = \alpha_2 = \alpha_3 = .05$, the Bonferroni inequality yields as the bound for the family level of significance:

$$\alpha \leq .05 + .05 + .05 = .15$$

while the Kimball inequality yields the bound:

$$\alpha \leq 1 - (.95)(.95)(.95) = .143$$

This illustration makes it clear that the level of significance for the family of three tests may be substantially higher than the levels of significance for the individual tests.

2. When the test on interactions leads to the conclusion that the two factors do not interact, it is sometimes suggested that the interaction sum of squares and degrees of freedom be pooled with the error sum of squares and degrees of freedom for purposes of testing factor A and factor B effects. The reason advanced is that when no interactions exist, $E(MSAB) = \sigma^2$, the same expectation as for MSE, so that the new estimator of σ^2 would have a larger number of degrees of freedom associated with it. The pooled mean square would have $SSAB + SSE$ in the numerator (for our example in Table 17.10, $24 + 62 = 86$) and $(a - 1)(b - 1) + (n - 1)ab$ degrees of freedom in the denominator (for our example, $2 + 6 = 8$).

This pooling procedure affects both the level of significance and the power of the tests for factor A and factor B effects, in ways not yet fully understood. It has been suggested therefore by some statisticians that pooling should not be considered unless: (1) the degrees of freedom associated with MSE are small, perhaps 5 or less, and (2) the test statistic $MSAB/MSE$ falls substantially below the action limit of the decision rule, perhaps when $MSAB/MSE < 2$ for $\alpha = .05$. Part (1) of this rule is designed to limit pooling to cases where the gains may be substantial, while part (2) is designed to give reasonable assurance that there are indeed no interactions.

17.6 Computer Input and Output

Computer input and output formats for two-factor ANOVA vary considerably from one package to another. Figure 17.5 shows one example of computer printout, for the Castle Bakery case. Note that the input data are entered in columns, where each column corresponds to a treatment. A specified order for treatments must be adhered to in entering the observations. The first a columns are for the treatments at the successive levels of factor A and the first level of factor B, the next a columns are for the successive levels of factor A and the second level of factor B, and so on. For the Castle Bakery data, the input format thus is:

$$\begin{matrix} Y_{111} & Y_{211} & Y_{311} & Y_{121} & Y_{221} & Y_{321} \\ Y_{112} & Y_{212} & Y_{312} & Y_{122} & Y_{222} & Y_{322} \end{matrix}$$

FIGURE 17.5

Illustration of Computer Printout for Two-Factor ANOVA

```
TWO FACTOR ANOVA


      DATA


 47  62  41  46  67  42
 43  68  39  40  71  46

 NO. OF LEVELS OF 1ST (A) FACTOR: 3
 NO. OF LEVELS OF 2ND (B) FACTOR: 2


ANOVA:
  SOURCE              SS                  DF                 MS
    0               1580                   5                316
    1               1544                   2                772
   10                 12                   1                 12
   11                 24                   2                 12
    2                 62                   6                 10.33333333

TOTAL SS = 1642       TOTAL DF = 11
SOURCES: 0=TRTMT COMBS. 1=1ST FACTOR. 10=2ND FACTOR.
        11=INTERACTION. 2=ERROR

F-STATISTICS:
  TRTMNT COMB.   : 30.58064516
      FACTOR A   : 74.70967742
      FACTOR B   : 1.161290323
    INTERACTION: 1.161290323

      POINT ESTIMATES

 PARAMETER          PT. EST.
OVERALL MEAN        51
    ALPHA'S         ¯7  16  ¯9
    BETA'S          ¯1  1
    GAMMA'S
     2                  ¯2
    ¯1                   1
    ¯1                   1
    SIGMA           3.21455

      MATRIX OF MEANS

 45  43  44
 65  69  67
 40  44  42
 50  52  51

RIGHTMOST COLUMN CONTAINS MEANS OF 1ST FACTOR LEVELS. BOTTOM
     ROW CONTAINS MEANS OF 2ND FACTOR LEVELS. GRAND MEAN IS IN
     LOWER RIGHT-HAND CORNER. ALL OTHER ENTRIES ARE MEANS OF FACTOR
     COMBINATIONS.
```

The only control information to be entered is that on the number of levels of factor A. The program itself then makes the appropriate identifications between columns of treatments.

The remainder of the printout involves output. The first output block shows ANOVA results that we have presented in Table 17.10, while the second block gives the F^* statistics for testing, respectively, for equality of all treatment means, factor A effects, factor B effects, and interaction effects. The next block shows the values of the point estimators of $\mu_{..}$, α_i $(i = 1, 2, 3)$, β_j $(j = 1, 2)$ and interactions $(\alpha\beta)_{ij}$ $(i = 1, 2, 3; j = 1, 2)$.

The final block presents sample means in the following format:

$$\begin{array}{ccc} \bar{Y}_{11.} & \bar{Y}_{12.} & \bar{Y}_{1..} \\ \bar{Y}_{21.} & \bar{Y}_{22.} & \bar{Y}_{2..} \\ \bar{Y}_{31.} & \bar{Y}_{32.} & \bar{Y}_{3..} \\ \bar{Y}_{.1.} & \bar{Y}_{.2.} & \bar{Y}_{...} \end{array}$$

17.7 POWER OF F TESTS

The power of the F tests for interactions, factor A effects, and factor B effects can be ascertained in similar fashion to the single-factor case, using the Pearson-Hartley charts in Table A-8. The noncentrality parameter ϕ and the appropriate degrees of freedom for each of these cases are as follows:

(17.48a) Test for interactions:

$$\phi = \frac{1}{\sigma}\sqrt{\frac{n \sum \sum (\alpha\beta)_{ij}^2}{(a-1)(b-1)+1}}$$

$$\nu_1 = (a-1)(b-1) \qquad \nu_2 = ab(n-1)$$

(17.48b) Test for A effects:

$$\phi = \frac{1}{\sigma}\sqrt{\frac{nb \sum \alpha_i^2}{a}}$$

$$\nu_1 = a-1 \qquad \nu_2 = ab(n-1)$$

(17.48c) Test for B effects:

$$\phi = \frac{1}{\sigma}\sqrt{\frac{na \sum \beta_j^2}{b}}$$

$$\nu_1 = b-1 \qquad \nu_2 = ab(n-1)$$

Example

Consider the test for A main effects (display height) in our Castle Bakery example. We wish to find the power of the test when $\mu_{1.} = 50$, $\mu_{2.} = 55$, $\mu_{3.} = 45$, or equivalently when $\alpha_1 = 0$, $\alpha_2 = +5$, $\alpha_3 = -5$. Assume that we know from past experience that $\sigma = 3$ cases. We have for this test from before:

$$n = 2 \qquad a = 3 \qquad b = 2 \qquad \alpha = .05$$

so that:

$$\phi = \frac{1}{3}\sqrt{\frac{(2)(2)[(0)^2 + (5)^2 + (-5)^2]}{3}} = 2.7$$

For $\nu_1 = 2$, $\nu_2 = 6$, $\alpha = .05$, and $\phi = 2.7$, we find from Table A–8 that the power is about .89. Thus, if $\mu_{1.} = 50$, $\mu_{2.} = 55$, and $\mu_{3.} = 45$ (and $\sigma = 3$), the probability is about .89 that the F test will detect the differences in the display height means.

PROBLEMS

17.1. A student in class discussion stated: "A treatment is a treatment, whether the study involves a single factor or two factors. The number of factors only affects the analysis of the results." Discuss.

17.2. In a two-factor study, the mean responses μ_{ij} for the different factor combinations are as follows:

	Factor B		
Factor A	B_1	B_2	B_3
A_1	23	34	36
A_2	28	39	41

a) What are the factor level means for factor A?
b) What is the specific effect of B_1 when factor A is at level A_2? When factor A is at level A_1?
c) What are the main effects of factor B?
d) Does the fact that $\mu_{12} - \mu_{11} = 11$ while $\mu_{13} - \mu_{12} = 2$ imply that factors A and B interact?
e) Plot the mean responses on a chart so that you can analyze whether the two factors interact. What do you find?

17.3. Verify the specific age effects and the specific sex effects in Table 17.2.

17.4. Verify the interactions in Table 17.3b.

17.5. In a two-factor study, the mean responses μ_{ij} for the different factor combinations are as follows:

	Factor B			
Factor A	B_1	B_2	B_3	B_4
A_1	253	268	271	272
A_2	291	276	273	272

a) What are the main factor B effects? What do these imply about factor B? (Hint: Does factor B have any specific effects?)
b) Plot the mean responses on a chart so that you can analyze whether the two factors interact. How do you recognize that the two factors interact? Describe the nature of the interactions.
c) Would a logarithmic transformation be helpful here in making the interactions relatively unimportant?

17.6. A financial analyst stated in a continuing education class: "I feel uncomfortable about deciding in a given problem whether the interactions are important or unimportant. I would much rather have the statistician make the decision." Discuss.

17.7. (Calculus needed.) Derive the least squares estimators of the parameters in model (17.21).

17.8. Derive the breakdown of $SSTR$ in (17.32).

17.9. In a study of the effect of the experimenter on the responses obtained, 10 male and 10 female personnel officials were shown a neutral photograph of a human face and asked to rate it on a scale from -10 (extreme failure) to $+10$ (extreme success). For half of the officials of each sex, selected at random, the experimenter looked at the official as often as possible while delivering his instructions. For the other half of the officials, the experimenter did not look at the official. The results were as follows:

	High Eye Contact		*Low Eye Contact*	
Official	*Male*	*Female*	*Male*	*Female*
1	+1	+6	−3	0
2	−3	+2	+2	+3
3	+2	+4	−4	−1
4	−4	+1	−1	+6
5	+1	+5	0	+4

a) State the model you will use to analyze the effects of experimenter's eye contact and sex of official on the response.
b) Test whether the factors have any main effects and whether they interact. Use a level of significance of .01 for each test. Summarize your results.
c) Give an upper bound for the family level of significance. What does the family level of significance refer to in this example?
d) What should be the next step in the analysis of the study results?

17.10. A computer software organization was encountering serious problems in predicting the man-day requirements for implementing large-scale programming projects. As part of a study on how to improve predictions, the company asked 24 programmers to predict the number of man-days for a large project about to be undertaken. The programmers were given full details about the specifications for the project. The programmers were classified by type of experience and amount of experience, and after the project was completed the forecasting errors were obtained. The forecasting errors (man-days predicted minus actual man-days) were:

Type of Experience	*Amount of Experience*		
	Under 2 Years	*2–5 Years*	*More than 5 Years*
Small-scale systems only:			
	−278	−109	−46
	−196	−118	−92
	−241	−88	−89
	−188	−96	−58
Large-scale systems:			
	−83	−47	−38
	−44	−52	−33
	−68	−31	−42
	−57	−49	−31

a) State the model you will use to analyze the effects of amount and type of experience on the forecasting error.

b) Test whether the factors have any main effects and whether they interact. Use a level of significance of .025 for each test. Summarize your results.

c) Give an upper bound for the family level of significance. What does the family level of significance refer to in this example?

d) What should be the next step in the analysis of the study results?

17.11. In a two-factor study, factor A had 3 levels, factor B two levels, and $n = 10$.

a) What is the power of the test for factor A main effects when $\alpha_1 = -10$, $\alpha_2 = 8$, $\alpha_3 = 2$? Assume that $\sigma = 20$, and that the level of significance is .05.

b) What is the power of the test for factor B main effects when the difference between the two factor B level means is 15? Again assume that $\sigma = 20$, and that the level of significance is .05.

18

Analysis of Two-Factor Studies

WE DISCUSS in this chapter how to analyze the nature of factor effects in two-factor studies, once the analysis of variance has indicated the presence of such effects. We also take up briefly two major problems in implementing the analysis of variance model for two-factor studies. These are planning of sample sizes and determining the aptness of the model. We continue to consider the two-factor analysis of variance model (17.21), where there are n observations for each treatment.

18.1 ANALYSIS WHEN BOTH FACTORS QUALITATIVE

Strategy for Analysis

Our consideration in Chapter 17 of the meaning of the model elements suggests the following basic strategy for analyzing factor effects in two-factor studies:

1. Examine whether the two factors interact.
2. If they do not, examine the factor effects separately in terms of the factor level means $\mu_{i.}$ and $\mu_{.j}$ respectively.
3. If the factors do interact, examine if the interactions are important or unimportant.
4. If they are unimportant, proceed as in step 2.
5. If they are important, determine whether the interactions can be made unimportant by a meaningful transformation of scale. If so, make the transformation and proceed as in step 2.

6. For important interactions that cannot be made unimportant by a meaningful transformation, analyze the two factor effects jointly in terms of the treatment means μ_{ij}. In some special cases, there may also be interest in the factor level means $\mu_{i.}$ and $\mu_{.j}$.

Step 1 of this strategy, testing for interaction effects, was discussed in Chapter 17. Also discussed there was step 5, the possible diminution of important interactions by a meaningful transformation, as well as how to test for the presence of factor main effects. Now we turn to steps 2 and 6 of the strategy for analysis, namely how to compare factor level means $\mu_{i.}$ and $\mu_{.j}$ when there are no interactions or only unimportant ones, and how to compare treatment means μ_{ij} when there are important interactions.

Analysis of Factor Level Means

As just noted, analysis of factor effects need only involve the factor level means $\mu_{i.}$ and $\mu_{.j}$ when the two factors do not interact, or interact only in unimportant fashion.

Estimation of Factor Level Mean. Unbiased point estimators of $\mu_{i.}$ and $\mu_{.j}$ are:

(18.1a) $$\hat{\mu}_{i.} = \bar{Y}_{i..}$$

(18.1b) $$\hat{\mu}_{.j} = \bar{Y}_{.j.}$$

where $\bar{Y}_{i..}$ and $\bar{Y}_{.j.}$ are defined in (17.25d) and (17.25f). The variance of $\bar{Y}_{i..}$ is:

(18.2a) $$\sigma^2(\bar{Y}_{i..}) = \frac{\sigma^2}{bn}$$

since $\bar{Y}_{i..}$ contains bn independent observations, each with variance σ^2. Similarly, we have:

(18.2b) $$\sigma^2(\bar{Y}_{.j.}) = \frac{\sigma^2}{an}$$

Unbiased estimators of these variances are obtained by replacing σ^2 with MSE:

(18.3a) $$s^2(\bar{Y}_{i..}) = \frac{MSE}{bn}$$

(18.3b) $$s^2(\bar{Y}_{.j.}) = \frac{MSE}{an}$$

Confidence intervals for $\mu_{i.}$ and $\mu_{.j}$ utilize, as usual, the t distribution:

(18.4a) $$\bar{Y}_{i..} - t(1-\alpha/2; (n-1)ab)s(\bar{Y}_{i..}) \leq \mu_{i.} \leq \bar{Y}_{i..} + t(1-\alpha/2; (n-1)ab)s(\bar{Y}_{i..})$$

(18.4b) $$\bar{Y}_{.j.} - t(1-\alpha/2; (n-1)ab)s(\bar{Y}_{.j.}) \leq \mu_{.j} \leq \bar{Y}_{.j.} + t(1-\alpha/2; (n-1)ab)s(\bar{Y}_{.j.})$$

The degress of freedom $(n-1)ab$ are those associated with MSE.

Estimation of Contrast. A contrast among the factor level means $\mu_{i.}$:

(18.5) $$L = \sum c_i \mu_{i.} \qquad \text{where } \sum c_i = 0$$

is estimated unbiasedly by:

(18.6) $$\hat{L} = \sum c_i \bar{Y}_{i..}$$

The variance of this estimator is, because of the independence of the $\bar{Y}_{i..}$:

(18.7) $$\sigma^2(\hat{L}) = \sum c_i^2 \sigma^2(\bar{Y}_{i..}) = \frac{\sigma^2}{bn} \sum c_i^2$$

An unbiased estimator of this variance is:

(18.8) $$s^2(\hat{L}) = \frac{MSE}{bn} \sum c_i^2$$

Finally, the appropriate $1-\alpha$ confidence interval for L is:

(18.9) $$\hat{L} - t(1-\alpha/2; (n-1)ab)s(\hat{L}) \leq L \leq \hat{L} + t(1-\alpha/2; (n-1)ab)s(\hat{L})$$

To estimate a contrast among the factor level means $\mu_{.j}$:

(18.10) $$L = \sum c_j \mu_{.j} \qquad \text{where } \sum c_j = 0$$

we use the estimator:

(18.11) $$\hat{L} = \sum c_j \bar{Y}_{.j.}$$

whose estimated variance is:

(18.12) $$s^2(\hat{L}) = \frac{MSE}{an} \sum c_j^2$$

The $1-\alpha$ confidence interval for L in (18.9) is still appropriate, with $\hat{L}$ and $s(\hat{L})$ now defined in (18.11) and (18.12) respectively.

Multiple Pairwise Comparisons. Usually, more than one comparison is of interest, and the multiple comparison procedures discussed in Chapter 14 can be employed with only minor modifications. If all or a large number of pairwise comparisons among the factor level means $\mu_{i.}$ are to be made, the Tukey procedure of (14.35) is appropriate, with:

(18.13a) $$D = \bar{Y}_{i..} - \bar{Y}_{i'..}$$

(18.13b) $$s^2(D) = \frac{2MSE}{bn}$$

(18.13c) $$T = \frac{1}{\sqrt{2}} q(1 - \alpha; a, (n-1)ab)$$

leading to the confidence intervals:

(18.14) $$D - Ts(D) \leq \mu_{i.} - \mu_{i'.} \leq D + Ts(D)$$

The probability then is $1 - \alpha$ that all statements in the family are correct.

For pairwise comparisons of the factor level means $\mu_{.j}$, the only changes are:

(18.15a) $$D = \bar{Y}_{.j.} - \bar{Y}_{.j'.}$$

(18.15b) $$s^2(D) = \frac{2MSE}{an}$$

(18.15c) $$T = \frac{1}{\sqrt{2}} q(1 - \alpha; b, (n-1)ab)$$

If only a few pairwise comparisons are to be made, the Bonferroni method may be best. All of the above formulas still apply, but the Tukey multiple T is replaced by the Bonferroni multiple B:

(18.16) $$B = t(1 - \alpha/2s; (n-1)ab)$$

where s is the number of statements in the family.

If it is desired to have a family confidence coefficient $1 - \alpha$ for the joint set of pairwise comparisons involving *both* factor A and B means, the Bonferroni method can be used either directly or in conjunction with the Tukey method. To illustrate the latter, suppose the pairwise comparisons for factor A are made with the Tukey procedure with a family confidence coefficient of .95, and likewise for the pairwise comparisons for factor B. The Bonferroni inequality then assures us that the family confidence coefficient is at least .90 that the joint set of comparisons for both factors is entirely correct.

Multiple Contrasts. When a large number of contrasts among the factor level means $\mu_{i.}$ or $\mu_{.j}$ are of interest, the Scheffé method should be used. Suppose the contrasts involve the $\mu_{i.}$, as in (18.5). The unbiased estimator is given in (18.6), and its estimated variance in (18.8). For this case, the Scheffé multiple is defined by:

(18.17) $$S^2 = (a-1)F(1 - \alpha; a-1, (n-1)ab)$$

leading to the confidence intervals:

$$\hat{L} - Ss(\hat{L}) \leq L \leq \hat{L} + Ss(\hat{L}) \tag{18.18}$$

The probability is then $1 - \alpha$ that every confidence interval (18.18) in the family of all possible contrasts is correct.

If contrasts among the factor level means $\mu_{.j}$ are desired, the unbiased point estimator is given in (18.11), its estimated variance is given in (18.12), and the Scheffé confidence intervals (18.18) are appropriate with:

$$S^2 = (b - 1)F(1 - \alpha; b - 1, (n - 1)ab) \tag{18.19}$$

When the number of contrasts of interest is small, the Bonferroni method may be best. Confidence intervals (18.18) would need to be modified by replacing the Scheffé multiple S with the Bonferroni multiple B:

$$B = t(1 - \alpha/2s; (n - 1)ab) \tag{18.20}$$

where s is the number of statements in the family.

When it is desired to obtain a family confidence coefficient for the joint set of contrasts for both factors, several possibilities exist:

1. The Bonferroni method may be used directly, with s representing the total number of statements in the joint set.
2. The Bonferroni method can be used to join the two sets of Scheffé multiple comparison families, in the same way explained earlier for joining two Tukey sets.
3. The Scheffé method can be modified to use the S multiple defined by:

$$S^2 = (a + b - 2)F(1 - \alpha; a + b - 2, (n - 1)ab) \tag{18.21}$$

When this S multiple is used in both sets of multiple comparisons, the probability is $1 - \alpha$ that all statements in the combined family are correct.

One may try each of these three approaches and see which leads to the narrowest confidence intervals, without affecting the validity of the procedure.

Example. We now illustrate these methods for our Castle Bakery example of Chapter 17. It will be recalled that no interactions between display height and display width were found. Hence we shall analyze only the factor level means. Suppose that management wishes to make pairwise comparisons between the factor level means for each factor, with a combined confidence coefficient of 90 percent for all comparisons. The Tukey multiple comparison procedure will be used, with the comparisons for display height assigned a family confidence coefficient of 95 percent, and similarly for the display width comparison. The Bonferroni inequality then guarantees a family confidence coefficient of at least 90 percent for the joint set of comparisons.

For the comparison of display height means ($i = 1$—bottom, 2—middle, 3—top), we have from Tables 17.7 and 17.8:

$$\bar{Y}_{2..} - \bar{Y}_{1..} = 67 - 44 = 23 \qquad MSE = 10.3$$

$$a = 3$$

$$\bar{Y}_{1..} - \bar{Y}_{3..} = 44 - 42 = 2 \qquad b = 2$$

$$n = 2$$

$$\bar{Y}_{2..} - \bar{Y}_{3..} = 67 - 42 = 25 \qquad (n-1)ab = 6$$

Hence:

$$s^2(D) = \frac{(2)(10.3)}{(2)(2)} = 5.15$$

$$q(.95; 3, 6) = 4.34$$

$$T = \frac{4.34}{\sqrt{2}} = 3.07$$

$$Ts(D) = 3.07\sqrt{5.15} = 7.0$$

Similarly, for the comparison of display widths ($j = 1$—regular, 2—wide), we have:

$$\bar{Y}_{.2.} - \bar{Y}_{.1.} = 52 - 50 = 2$$

$$s^2(D) = \frac{(2)(10.3)}{(3)(2)} = 3.43$$

$$q(.95; 2, 6) = 3.46$$

$$T = \frac{3.46}{\sqrt{2}} = 2.45$$

$$Ts(D) = 2.45\sqrt{3.43} = 4.5$$

We obtain therefore the following confidence intervals for all pairwise comparisons of factor level means (90 percent family confidence coefficient for joint set):

$$16 = 23 - 7.0 \le \mu_{2.} - \mu_{1.} \le 23 + 7.0 = 30$$

$$-5 = 2 - 7.0 \le \mu_{1.} - \mu_{3.} \le 2 + 7.0 = 9$$

$$18 = 25 - 7.0 \le \mu_{2.} - \mu_{3.} \le 25 + 7.0 = 32$$

$$-2.5 = 2 - 4.5 \le \mu_{.2} - \mu_{.1} \le 2 + 4.5 = 6.5$$

It was concluded from these confidence intervals that for the product studied and the types of stores in the experiment, the middle shelf height is far better than either the bottom or the top heights, the latter two do not differ significantly in sales effectiveness, and widening the display has no significant effect on sales. All these conclusions are covered by the family confidence coefficient of .90.

Analysis of Treatment Means

When important interactions exist, the analysis of factor effects generally must be based on the treatment means μ_{ij}. Typically, this analysis will involve multiple comparisons.

Multiple Pairwise Comparisons. If pairs of treatment means μ_{ij} are to be compared, either the Tukey or the Bonferroni multiple comparison procedure may be used, depending on which is more advantageous. In effect, the analysis is equivalent to the single-factor case, with the total number of treatments here equal to $r = ab$, the degrees of freedom associated with MSE here equal to $n_T - r = (n-1)ab$, and the estimated treatment mean denoted by $\bar{Y}_{ij.}$. Formula (14.35) for the Tukey multiple comparison procedure then becomes:

(18.22) $$D - Ts(D) \leq \mu_{ij} - \mu_{i'j'} \leq D + Ts(D) \qquad i, j \neq i', j'$$

where:

(18.22a) $$D = \bar{Y}_{ij.} - \bar{Y}_{i'j'.}$$

(18.22b) $$s^2(D) = \frac{2MSE}{n}$$

(18.22c) $$T = \frac{1}{\sqrt{2}} q(1 - \alpha; ab, (n-1)ab)$$

If the Bonferroni method is employed, the multiple in the confidence interval is:

(18.23) $$B = t(1 - \alpha/2s; (n-1)ab)$$

where s is the number of statements in the family.

Multiple Contrasts. The Scheffé multiple comparison procedure for single-factor studies is directly applicable to the estimation of contrasts involving the treatment means μ_{ij}. We denote these contrasts as follows:

(18.24) $$L = \sum\sum c_{ij}\mu_{ij} \qquad \text{where } \sum\sum c_{ij} = 0$$

The point estimator of L is:

(18.25) $$\hat{L} = \sum\sum c_{ij}\bar{Y}_{ij.}$$

and the estimated variance of this estimator is:

(18.26) $$s^2(\hat{L}) = \frac{MSE}{n}\sum\sum c_{ij}^2$$

The Scheffé multiple S is given by:

(18.27) $$S^2 = (ab - 1)F(1 - \alpha; ab - 1, (n-1)ab)$$

and the joint confidence intervals are as usual:

$$\hat{L} - Ss(\hat{L}) \leq L \leq \hat{L} + Ss(\hat{L}) \tag{18.28}$$

If the number of contrasts is small, the Bonferroni procedure may be preferable. The confidence intervals (18.28) would simply be modified by replacing S with B as defined in (18.23).

Example (Adapted from Reference 18.1). A company studied the effect of two factors, brazing material and honeycomb structure material, on the bonding of the material. The observed variable was the number of cells (out of 512 cells in a panel) that are not completely bonded to the core. Twenty-four panels for each of the treatment combinations were studied. The results are summarized in Table 18.1. A glance at these results suggests that the two factors interact. The F test on interactions, not shown here, confirms that brazing material and structure material do interact.

TABLE 18.1

Results of Study on Effects of Brazing Material and Honeycomb Structure Material

(a) Mean Number of Defective Cells per Panel ($n = 24$)

	Brazing Material			
Structure Material	*GE81*	*AG–MN*	*Coast Metal 53*	*Electrolytic Copper*
17–7 PH	88 ($\bar{Y}_{11.}$)	108 ($\bar{Y}_{12.}$)	13 ($\bar{Y}_{13.}$)	23 ($\bar{Y}_{14.}$)
Titanium	63 ($\bar{Y}_{21.}$)	46 ($\bar{Y}_{22.}$)	40 ($\bar{Y}_{23.}$)	29 ($\bar{Y}_{24.}$)

(b) Mean Square Error and Degrees of Freedom

$MSE = 108 \qquad df = (n-1)ab = (24-1)(2)(4) = 184$

Source: Adapted, with permission, from F. C. Leone, "Experimental Design and Analysis of Variance—III Additional Designs and Analyses," in *1959 National Convention Transactions*, American Society for Quality Control, pp. 541–54.

Suppose that management wishes to estimate the following contrasts (among others), some of which were suggested by the data:

$$L_1 = \mu_{14} - \mu_{13}$$

$$L_2 = \mu_{24} - \mu_{13}$$

$$L_3 = \frac{\mu_{23} + \mu_{24}}{2} - \frac{\mu_{13} + \mu_{14}}{2}$$

$$L_4 = \frac{\mu_{11} + \mu_{12}}{2} - \frac{\mu_{21} + \mu_{22}}{2}$$

Suppose further that the family confidence coefficient is to be .90. The Scheffé multiple comparison procedure will be utilized because contrasts other than pairwise comparisons are involved and some of the contrasts were suggested by the data. (If only four contrasts designated in advance were to be estimated, the Bonferroni method would yield more precise estimates here.)

The point estimators of the contrasts are, using the data in Table 18.1:

$$\hat{L}_1 = 23 - 13 = 10$$

$$\hat{L}_2 = 29 - 13 = 16$$

$$\hat{L}_3 = \frac{40 + 29}{2} - \frac{13 + 23}{2} = 16.5$$

$$\hat{L}_4 = \frac{88 + 108}{2} - \frac{63 + 46}{2} = 43.5$$

The estimated variances of these estimates are obtained by (18.26), with $n = 24$:

$$s^2(\hat{L}_1) = s^2(\hat{L}_2) = \frac{108}{24}[(1)^2 + (-1)^2] = 9$$

$$s^2(\hat{L}_3) = s^2(\hat{L}_4) = \frac{108}{24}\left[\left(\frac{1}{2}\right)^2 + \left(\frac{1}{2}\right)^2 + \left(-\frac{1}{2}\right)^2 + \left(-\frac{1}{2}\right)^2\right] = 4.5$$

Finally, we have by (18.27):

$$S^2 = (8 - 1)F(.90; 7, 184) = 7(1.75) = 12.25$$

or:

$$S = 3.50$$

The 90 percent confidence intervals for the family of contrasts then are by (18.28):

$$-.5 = 10 - 3.50\sqrt{9} \le L_1 \le 10 + 3.50\sqrt{9} = 20.5$$

$$5.5 = 16 - 3.50\sqrt{9} \le L_2 \le 16 + 3.50\sqrt{9} = 26.5$$

$$9.1 = 16.5 - 3.50\sqrt{4.5} \le L_3 \le 16.5 + 3.50\sqrt{4.5} = 23.9$$

$$36.1 = 43.5 - 3.50\sqrt{4.5} \le L_4 \le 43.5 + 3.50\sqrt{4.5} = 50.9$$

From these confidence intervals, the following conclusions may be drawn with family confidence coefficient .90:

1. The use of 17–7 PH structure material with Coast Metal 53 brazing material is not significantly better than the same structure material with electrolytic copper as the brazing material, but it is significantly better than any other factor combination. The latter conclusion follows because all other factor combinations perform more poorly than the titanium–electrolytic copper combination, which does perform significantly poorer than the combination of 17–7 PH and Coast Metal 53.

2. The 17–7 PH structure material performs significantly better than titanium when Coast Metal 53 and electrolytic copper are the brazing materials.

3. On the other hand, the 17–7 PH structure material performs significantly poorer than titanium when used with the other two brazing materials.

Additional contrasts, not presented here, would permit still further analyses of the factor effects. Note that the confidence intervals indicate not only which effects are statistically significant but also show the magnitudes of the effects.

18.2 ANALYSIS WHEN ONE OR BOTH FACTORS QUANTITATIVE

When one or both of the factors in a two-factor study are quantitative, the analysis of factor effects can be carried beyond the point of multiple comparisons to include a study of the nature of the response function. Since the familiar methods of regression analysis, discussed earlier, then come into use, we shall only briefly discuss this extension of the analysis. The study of factor effects by multiple comparison techniques can be very helpful in choosing an appropriate functional form for the regression relation.

Analysis of Response Function When One Factor Quantitative

(This section assumes previous study of Chapter 9.)

Consider an experiment in which the effect of type of cake mix (factor A) and temperature (factor B) on the lightness of the cake texture, suitably measured, is to be investigated. Two kinds of cake mixes (G, H) and four temperatures (300°, 315°, 330°, 345°) are studied. The analyst in this case might be interested in extending his investigation of factor effects into the nature of the response function relating cake texture to baking temperature. Since factor A is qualitative, indicator variables will be used to represent it in the response function. If kind of cake mix and baking temperature do not interact, a first-order model might be appropriate:

$$Y_{ijk} = \beta_0 + \beta_1 X_{ijk1} + \beta_2 X_{ijk2} + \varepsilon_{ijk} \tag{18.29}$$

where:

$$X_{ijk1} = \begin{matrix} 1 \text{ if observation from first level of factor } A \\ 0 \text{ otherwise} \end{matrix}$$

X_{ijk2} is the baking temperature for the observation

The β's denote regression coefficients here.

We know from Chapter 9 that model (18.29) implies a linear relation between cake texture and temperature which has constant slope for both kinds of cake mixes but different heights. Figure 9.1 provides an illustration of this model.

If cake mix and baking temperature interact, an appropriate model might be:

$$Y_{ijk} = \beta_0 + \beta_1 X_{ijk1} + \beta_2 X_{ijk2} + \beta_3 X_{ijk1} X_{ijk2} + \varepsilon_{ijk} \tag{18.30}$$

From Chapter 9, we know that this model implies a linear relation between cake texture and temperature with different slopes and intercepts for the two kinds of cake mixes. Figure 9.3 provides an illustration of this model.

If the regression relation is quadratic and no interactions between the two factors exist, a suitable model would be:

$$Y_{ijk} = \beta_0 + \beta_1 X_{ijk1} + \beta_2 X_{ijk2} + \beta_3 X_{ijk2}^2 + \varepsilon_{ijk} \tag{18.31}$$

Analysis of Response Function When Both Factors Quantitative

When both factors are quantitative, the analysis of the nature of the response function involves ordinary multiple regression. We shall denote the two factor variables as X_1 and X_2. A first-order model would then be:

$$Y_{ijk} = \beta_0 + \beta_1 X_{ijk1} + \beta_2 X_{ijk2} + \varepsilon_{ijk} \tag{18.32}$$

where X_{ijk1} and X_{ijk2} are the respective values of X_1 and X_2 for the kth observation from the ith level of factor A and the jth level of factor B. A second-order model, including interactions between the two factors, would be:

$$\begin{aligned} Y_{ijk} = \beta_0 + \beta_1 X_{ijk1} + \beta_2 X_{ijk2} + \beta_3 X_{ijk1}^2 \\ + \beta_4 X_{ijk2}^2 + \beta_5 X_{ijk1} X_{ijk2} + \varepsilon_{ijk} \end{aligned} \tag{18.33}$$

The discussion of response surfaces in Chapter 7 is completely applicable here.

Example (Adapted from Reference 18.2). A study was conducted to improve the efficiency of a burr remover in a wool textile carding machine. The rollers in the burr remover are adjustable as to speed and spacing. Four spacings and three speeds, as shown in Table 18.2, were used in the study. Four replications were made for each treatment. For purposes of condensation, only the treatment means are presented in Table 18.2. The observed variable is a measure of the efficiency of the carding.

An analysis of variance was conducted by a computer package for this two-factor study. The results are summarized in Table 18.3a. The test for interactions utilizes the test statistic:

$$F^* = \frac{MSAB}{MSE} = \frac{4.92}{1.32} = 3.73$$

TABLE 18.2

Observed Treatment Means in Two-Factor Burr Remover Study with Both Factors Quantitative ($n = 4$)

Spacing	Speed 300 rpm	400 rpm	500 rpm
1.0 unit	21.6	22.3	22.9
1.2 units	18.7	19.1	21.6
1.4 units	15.8	17.9	19.4
1.6 units	13.2	16.7	19.5

Source: Reprinted, with permission, from D. R. Cox, *Planning of Experiments* (New York: John Wiley & Sons, Inc., 1958), p. 124.

TABLE 18.3

Analysis of Variance Tables for Burr Remover Study

(a) Analysis of Variance Model

Source of Variation	*SS*	*df*	*MS*
Spacing (A)	232.86	3	77.62
Speed (B)	99.49	2	49.74
Interactions (AB)	29.53	6	4.92
Error	47.61	36	1.32
Total	409.49	47	

(b) Regression Model

$$\hat{Y} = 45.09333 - 25.45000X_1 - .03340X_2 + .03925X_1X_2$$

Source of Variation	*SS*	*df*	*MS*
Regression	352.20	3	117.40
Error	57.29	44	1.30
Total	409.49	47	

(c) ANOVA for Lack of Fit Test

Source of Variation	*SS*	*df*	*MS*
Regression	352.20	3	117.40
Error	57.29	44	1.30
Lack of fit	9.68	8	1.21
Pure error	47.61	36	1.32
Total	409.49	47	

For level of significance .05, we require $F(.95; 6, 36) = 2.36$. Since $F^* = 3.73 > 2.36$, it was concluded that spacing and speed interact. Further analysis suggested that a first-order model with interaction effects added might be appropriate. The response function for this model is:

$$E(Y) = \beta_0 + \beta_1 X_1 + \beta_2 X_2 + \beta_3 X_1 X_2 \tag{18.34}$$

where X_1 represents spacing and X_2 speed. This model was fitted by a multiple regression computer package. The results are summarized in Table 18.3b.

Since replications were used in the study, a test of the appropriateness of the response function can be conducted. The lack of fit sum of squares is readily obtainable because *SSE* for the ANOVA model is the same as *SSPE* for the regression model. Hence, we find:

$$SSLF = \underset{\text{(Table 18.3b)}}{SSE} - \underset{\text{(Table 18.3a)}}{SSPE} = 57.29 - 47.61 = 9.68$$

Table 18.3c contains the decomposition needed for testing lack of fit. The appropriate test statistic is:

$$F^* = \frac{MSLF}{MSE} = \frac{1.21}{1.32} = .92$$

Assuming a level of significance of .05 is to be used, we require $F(.95; 8, 36) = 2.21$. Since $F^* = .92 < 2.21$, we conclude that the response function (18.34) is appropriate.

Let us take a closer look at the estimated response function:

$$\hat{Y} = 45.09333 - 25.45000X_1 - .03340X_2 + .03925X_1X_2$$

Note that b_3 is positive. It implies here, as a glance at Table 18.2 will confirm, that efficiency declines with increased spacing for all speeds (e.g., if $X_2 = 300$, $\hat{Y} = 35.073 - 13.675X_1$), but the decline is smaller for high speeds. Correspondingly, the efficiency increases with speed for all spacings (e.g., if $X_1 = 1.0$, $\hat{Y} = 19.643 + .00585X_2$), but the increase is larger for large spacings.

Comments

1. If a principal purpose of a two-factor study is to estimate the response surface, it is helpful for calculational simplifications to space the factor levels equally, as in the burr remover example of Table 18.2. One can then code these levels in simple fashion. For instance, if three levels are employed, the codes for the levels may be $-1, 0, +1$; if four levels are employed, the codes can be $-3, -1, +1, +3$. These codes are then used in the regression calculations.

2. Special designs have been developed to estimate response surfaces efficiently. An example of such a design for estimating a first-order model is shown in Table 18.4. Each of the two factors is studied at three equally spaced levels, but not all factor combinations are included. Only one replication is made for four of the treatments, while the center treatment has four replications. The center treatment is often called the *center of the design*. This design permits evaluation of quadratic effects if information about these is desired, and the replications at the center of the design provide

TABLE 18.4

Example of a First-Order Design

		Factor B		
		-1	0	$+1$
Factor A	-1	1	—	1
	0	—	4	—
	$+1$	1	—	1

Numbers for each treatment show replication size

an estimate of experimental (pure) error for testing the goodness of fit of the model. Many designs for estimating response surfaces are available, and these are discussed by Cochran and Cox in Reference 18.3.

3. A sequence of small-scale factorial experiments is frequently used in investigations seeking to find the best factor combination, such as the combination of temperature and pressure which maximizes the yield of a chemical process. After each experiment, a response surface is fitted and an evaluation is made of the factor combinations having high yields. The next experiment is then conducted to investigate these factor combinations in more detail. This process is continued until the factor combination having the maximum yield is established with sufficient precision. Various strategies may be employed in determining the sequence of experiments, such as the method of steepest ascent and the single-factor strategy. These are discussed in books on experimental designs such as the one by Cochran and Cox (Ref. 18.3).

18.3 IMPLEMENTATION OF TWO-FACTOR ANOVA MODEL

The two basic problems in implementing a two-factor analysis of variance model—planning sample sizes and evaluating the aptness of the model—are handled in essentially the same fashion as discussed in Chapter 15 for single-factor studies. Hence, we make only a few brief comments.

Planning Sample Sizes

No essentially new problems arise in planning sample size for two-factor studies. In most cases, equal replications are desired for each treatment. One can then plan the sample sizes using either the power approach or the estimation approach discussed in Chapter 15.

With the power approach, one would be concerned typically with both the power of detecting factor A effects and the power of detecting factor B effects. One can first specify values for the set of parameters for which it is important to detect factor A effects, and obtain the needed sample sizes from Table A–10, with $r = a$. The resulting sample size is bn, from which n can be readily obtained. The use of Table A–10 for this purpose is appropriate provided the resulting sample size is not small, specifically provided $a(bn - 1) \geq 20$. If this condition is not met, one should use the Pearson-Hartley power charts in Table A–8. These charts, as we noted earlier, require an iterative approach for determining needed sample sizes.

In the same way, one can then specify values for the set of factor B effects which are important to be detected, and find the needed sample sizes. If the sample sizes obtained from the factor A and factor B power specifications differ substantially, a judgment will need to be made as to the final sample sizes.

Alternatively, or in conjunction with the power approach, one can specify the important contrasts to be estimated and then find the sample sizes which are expected to provide the needed precisions for the desired confidence coefficient. Frequently this approach is more useful than the power approach, although both of these approaches can be used jointly for arriving at a determination of needed sample sizes.

If the purpose of the factorial study is to identify the best of the ab factor combinations, Table A–11 can be used for finding the needed sample sizes, as described in Section 15.3. For this purpose, $r = ab$.

Evaluation of Aptness of Model

No new problems arise in examining the aptness of the analysis of variance two-factor model. The residuals:

$$e_{ijk} = Y_{ijk} - \bar{Y}_{ij.} \tag{18.35}$$

may be examined for normality, constancy of error variance, and independence of error terms in the same fashion as for a single-factor study.

Transformations may be employed to stabilize the error variance and to make the error distributions more normal. Our earlier discussion of this topic in Chapter 15 for the single-factor case applies completely to the two-factor case.

Finally, the earlier discussion on effects of departures from the model applies fully to the two-factor case. In particular, the employment of equal sample sizes for each treatment minimizes the effect of unequal variances.

PROBLEMS

18.1. Why do many analysts first test for interaction effects between the two factors rather than begin the analysis with tests for the main factor effects?

18.2. In a two-factor study, factor A had four levels and factor B had three levels; $n = 12$ for each factor combination. No interactions between factors A and B were noted, and the analyst wishes to estimate three contrasts among the factor level means for A and two constrasts among the factor level means for B. A family confidence coefficient of 90 percent for the joint set of interval estimates is desired. Which of the three procedures described on page 592 will be most efficient here?

18.3. Refer to the Castle Bakery example on page 592, where multiple pairwise comparisons among display height means and among display width means were made by means of the Tukey procedure and combined by the Bonferroni method. How efficient would it have been in this case to use the Bonferroni method entirely, with the family consisting of four statements? How efficient would the Scheffé method of (18.21) have been here?

18.4. Refer to the Castle Bakery example on page 592. Explain why the family confidence coefficient for the joint set of comparisons is at least 90 percent when the family confidence coefficient for each separate set is 95 percent.

18.5. Refer to Problem 17.9. Analyze the results of the study of the effects of eye contact by the experimenter and sex of the personnel official on the response. Explain the reasons why you chose your particular analytical approach and summarize your findings.

18.6. Refer to Problem 17.10. Analyze the results of the study of the effects of type and amount of experience on the forecasting error. Explain the reasons why you chose your particular analytical approach and summarize your findings.

18.7. Refer to Problem 17.10. The mean amounts of experience in the three experience classes were as follows:

Under 2 years:	1.0 years
2–5 years:	3.5 years
More than 5 years:	7.5 years

Assume that the amount of experience for each programmer is close to the mean amount of experience in his class, so that the mean can be used to represent the amount of experience for each programmer in that class. Fit a linear regression of the logarithm of minus the forecasting error on the amount of experience and type of experience, using an indicator variable for the latter. Study the aptness of the regression model and compare the regression functions for the two types of experience classes. Summarize your findings.

18.8. A market research manager was planning a study of the effects of duration of advertising (factor A) and price level (factor B) on sales. No strong interactions were expected, and the primary analysis was to consist of pairwise comparisons of factor level means for each factor. Factor A had three levels and factor B two levels. Equal sample sizes were to be used for each factor combination. What sample sizes would you recommend if the precision of the pairwise comparisons is not to exceed ± 4 thousand dollars, with a family confidence coefficient of 80 percent for the joint set of comparisons? Assume that $\sigma = 8$ thousand dollars, approximately.

18.9. Refer to Problem 17.9. Obtain the residuals and analyze them to determine the appropriateness of the model you employed in Problem 17.9a. Summarize your findings.

18.10. Refer to Problem 17.10. Obtain the residuals and analyze them to determine the appropriateness of the model you employed in Problem 17.10a. Summarize your findings.

CITED REFERENCES

18.1. Leone, Fred C. "Experimental Design and Analysis of Variance—III Additional Designs and Analyses," in *1959 National Convention Transactions*, American Society for Quality Control, pp. 541–54.

18.2. Cox, D. R. *Planning of Experiments.* New York: John Wiley & Sons, Inc., 1958.

18.3. Cochran, William G., and Cox, Gertrude M. *Experimental Designs.* 2d ed. New York: John Wiley & Sons, Inc., 1957.

19

Topics in Analysis of Variance—II

IN THIS CHAPTER, we discuss a number of selected topics in the analysis of variance for two-factor studies. Up to this point, we have assumed there are the same number of observations for each treatment, and this constant sample size is 2 or more. We consider now a variety of situations where one or the other of these conditions is not met. In addition, we take up two other types of models for two-factor studies which are appropriate when the effects of one or both factors may be viewed as random effects. Finally, we shall consider the regression approach to the analysis of variance for two-factor studies.

19.1 ONE OBSERVATION PER TREATMENT

First, we take up the case where there is only one observation for each treatment. We can no longer work with the two-factor fixed effects ANOVA model (17.21) because no estimate of the error variance σ^2 will be available. Recall from (17.31c) that SSE is a sum of squares made up of components measuring the variability within each treatment $\sum_k (Y_{ijk} - \bar{Y}_{ij.})^2$. With only one observation per treatment, there is no variability within a treatment, and SSE will then always be zero.

The only way out of this difficulty is to change the model. A glance at Table 17.9 indicates that if the two factors do not interact, the interaction mean square $MSAB$ has expectation σ^2. Thus, if it is possible to assume that the two factors do not interact, we may use $MSAB$ as the estimator of the error variance σ^2, and proceed with the analysis of factor effects as usual. If it is unreasonable to assume that the two factors do not interact, transformations may be tried to remove the interaction effects.

No-Interaction Model

The two-factor fixed effects ANOVA model with no interactions is:

$$Y_{ij} = \mu_{..} + \alpha_i + \beta_j + \varepsilon_{ij} \tag{19.1}$$

where the model terms are as in (17.21). Note that the third subscript has been dropped from the Y and ε terms because there is now only one observation per treatment.

SSA and SSB are calculated as before from (17.32a) and (17.32b) respectively, with $n = 1$. The interaction sum of squares in (17.32c), with $n = 1$, now serves as the error sum of squares:

$$SSE = \sum_i \sum_j (Y_{ij} - \bar{Y}_{i.} - \bar{Y}_{.j} + \bar{Y}_{..})^2 \tag{19.2}$$

Note that SSE in (19.2) is identical to $SSAB$ in (17.32c), with $n = 1$, the third subscript dropped because there is only one observation per treatment, and the mean $\bar{Y}_{ij.}$ replaced by the observation Y_{ij} for the same reason. The number of degrees of freedom associated with SSE in (19.2) is the same as the number associated with $SSAB$ in (17.32c) for $n = 1$, namely $(a - 1)(b - 1)$.

The analysis of variance table for the case $n = 1$ for the no-interaction model (19.1) is shown in Table 19.1. No new problems arise in the tests for factor A and factor B effects, nor in estimating these effects.

Example

Table 19.2a shows amount of six-month premium charged by a certain automobile insurance firm for specific types and amounts of coverage in a given risk category, classified by size of city (factor A) and geographic region (factor B). Note there is only one observation per cell, namely the amount of the premium for the given factor level combination. This firm adjusts its premiums periodically to reflect comparative loss experience in different localities. An insurance analyst wished to evaluate the effects of the city sizes and geographic regions shown in Table 19.2 on the amount of the premium. He conjectured, from experience in other cases, that interaction effects would not be present in the premiums under analysis. A test for interactions (to be discussed below) did indeed lead to the conclusion that interaction effects are not present, so the analyst adopted model (19.1).

The required sums of squares are obtained as follows (using the definitional formulas):

$$SSA = 2[(120 - 175)^2 + (195 - 175)^2 + (210 - 175)^2] = 9{,}300$$
$$SSB = 3[(190 - 175)^2 + (160 - 175)^2] = 1{,}350$$
$$SSE = (140 - 120 - 190 + 175)^2 + \cdots + (200 - 210 - 160 + 175)^2 = 100$$
$$SSTO = (140 - 175)^2 + \cdots + (200 - 175)^2 = 10{,}750$$

TABLE 19.1

ANOVA Table for No-Interaction Two-Factor Fixed Effects Model, $n = 1$

Source of Variation	*SS*	*df*	*MS*	*E(MS)*
Factor *A*	$SSA = b\sum(\bar{Y}_{i.} - \bar{Y}_{..})^2$	$a - 1$	$MSA = \frac{SSA}{a-1}$	$\sigma^2 + \frac{b}{a-1}\sum\alpha_i^2$
Factor *B*	$SSB = a\sum(\bar{Y}_{.j} - \bar{Y}_{..})^2$	$b - 1$	$MSB = \frac{SSB}{b-1}$	$\sigma^2 + \frac{a}{b-1}\sum\beta_j^2$
Error	$SSE = \sum\sum(Y_{ij} - \bar{Y}_{i.} - \bar{Y}_{.j} + \bar{Y}_{..})^2$	$(a-1)(b-1)$	$MSE = \frac{SSE}{(a-1)(b-1)}$	σ^2
Total	$SSTO = \sum\sum(Y_{ij} - \bar{Y}_{..})^2$	$ab - 1$		

TABLE 19.2

Two-Factor Insurance Premium Study with $n = 1$

(a) Premiums for Automobile Insurance Policy (in dollars)

Size of City (factor A)	Region (factor B) East ($j = 1$)	West ($j = 2$)	Average
Small ($i = 1$)	140	100	120
Medium ($i = 2$)	210	180	195
Large ($i = 3$)	220	200	210
Average	190	160	175

(b) ANOVA Table

Source of Variation	SS	df	MS
Size of city	9,300	2	4,650
Region	1,350	1	1,350
Error	100	2	50
Total	10,750	5	

The ANOVA table is given in Table 19.2b. In the analyst's test of city size (factor A) effects, the alternative conclusions are:

$$C_1: \alpha_1 = \alpha_2 = \alpha_3 = 0$$
$$C_2: \text{not all } \alpha_i \text{ are zero}$$

The F^* statistic takes the usual form:

$$F^* = \frac{MSA}{MSE}$$

and the decision rule is (remember that the denominator of F^* here involves $(a - 1)(b - 1)$ degrees of freedom):

$$\text{If } F^* \leq F(1 - \alpha; a - 1, (a - 1)(b - 1)), \text{ conclude } C_1$$
$$\text{If } F^* > F(1 - \alpha; a - 1, (a - 1)(b - 1)), \text{ conclude } C_2$$

Let $\alpha = .05$. Hence we need $F(.95; 2, 2) = 19.0$. Table 19.2b provides the data for the test statistic:

$$F^* = \frac{4{,}650}{50} = 93$$

Since $F^* = 93 > 19.0$, the conclusion is C_2, that city size effects are present.

The test for geographic region (factor B) effects proceeds similarly, the alternative conclusions being:

$$C_1: \beta_1 = \beta_2 = 0$$
$$C_2: \text{not all } \beta_j \text{ are zero}$$

For $\alpha = .05$ the decision rule is:

$$\text{If } F^* \leq F(.95; 1, 2) = 18.5, \text{ conclude } C_1$$
$$\text{If } F^* > F(.95; 1, 2) = 18.5, \text{ conclude } C_2$$

Table 19.2b shows that:

$$F^* = \frac{MSB}{MSE} = \frac{1,350}{50} = 27$$

Hence the conclusion is C_2, that geographic region effects are present.

Given the findings up to this point—namely, no interactions, factor A effects present, factor B effects present—the analyst then examined the factor level means $\mu_{i.}$ and $\mu_{.j}$ by methods discussed in Section 18.1.

Comments

1. The analysis of two-factor studies with $n = 1$ just outlined depends on the assumption that the two factors do not interact. If one utilizes this analysis when in fact interactions are present, the result is that the actual level of significance for testing factor A and B effects is below the specified one and the actual power of the tests is lower than the expected power. Correspondingly, confidence intervals for contrasts based on the main effects will tend to be too wide. This means that when interactions are present, the analysis is more likely to fail to disclose real effects than anticipated. However, when the analysis based on the no-interactions model does indicate the presence of factor A main effects or of factor B main effects, they may be taken as real effects even though interactions are actually present.

2. Sometimes, the case $n = 1$ is encountered when the observations Y_{ij} are proportions. For instance, the data may consist of the proportion of employees in a plant absent during the past week, with the plants classified by size and geographic area. As noted earlier, the arc sine transformation can be used for such data to stabilize the variances. The transformed data then can be analyzed through the analysis of variance model (19.1), provided that each proportion is based on roughly the same number of cases. If the number of cases differ greatly, a weighted least squares approach should be utilized.

Tukey Test for Additivity

Usually, we would like to have some evidence about the correctness of the needed assumption of no interactions between the two factors when $n = 1$. We describe now a test devised by Tukey which may be used for examining whether or not the two factors are interacting, when $n = 1$.

As we noted earlier, if $n = 1$ we cannot include the interaction terms $(\alpha\beta)_{ij}$ in our model because the error sum of squares SSE in (17.31c) would always equal zero. It is possible, however, to impose some further restrictions on the $(\alpha\beta)_{ij}$ and include the more restricted interaction effects in the model. Suppose we assume that:

$$(\alpha\beta)_{ij} = D\alpha_i\beta_j \tag{19.3}$$

where D is some constant. One motivation for this assumption is that if $(\alpha\beta)_{ij}$ is any second-degree polynomial function of α_i and β_j, then it must be of the form (19.3) because of the restrictions on the α_i, β_j, and $(\alpha\beta)_{ij}$ that the sums over each subscript be 0.

Using (19.3) in a regular two-factor model with interactions, we obtain:

$$Y_{ij} = \mu_{..} + \alpha_i + \beta_j + D\alpha_i\beta_j + \varepsilon_{ij} \tag{19.4}$$

where each of the terms has the usual meaning. Remember there is no third subscript here because $n = 1$. The interaction sum of squares $\sum\sum D^2\alpha_i^2\beta_j^2$ now needs to be obtained. The least squares estimator of D, assuming the other parameters are known, turns out to be:

$$\hat{D} = \frac{\sum_i\sum_j \alpha_i\beta_j Y_{ij}}{\sum_i \alpha_i^2 \sum_j \beta_j^2} \tag{19.5}$$

The usual estimator of α_i is $(\bar{Y}_{i.} - \bar{Y}_{..})$, and that of β_j is $(\bar{Y}_{.j} - \bar{Y}_{..})$. Replacing the parameters in $\hat{D}$ by these estimators, we obtain:

$$\hat{D} = \frac{\sum_i\sum_j (\bar{Y}_{i.} - \bar{Y}_{..})(\bar{Y}_{.j} - \bar{Y}_{..})Y_{ij}}{\sum_i (\bar{Y}_{i.} - \bar{Y}_{..})^2 \sum_j (\bar{Y}_{.j} - \bar{Y}_{..})^2} \tag{19.5a}$$

The sample counterpart of the interaction sum of squares $\sum\sum D^2\alpha_i^2\beta_j^2$ therefore is:

$$SSAB = \frac{\left[\sum_i\sum_j (\bar{Y}_{i.} - \bar{Y}_{..})(\bar{Y}_{.j} - \bar{Y}_{..})Y_{ij}\right]^2}{\sum_i (\bar{Y}_{i.} - \bar{Y}_{..})^2 \sum_j (\bar{Y}_{.j} - \bar{Y}_{..})^2} \tag{19.6}$$

This expression can be expanded out for computational simplifications.

We stated earlier that SSE in (19.2):

$$SSE = \sum\sum (Y_{ij} - \bar{Y}_{i.} - \bar{Y}_{.j} + \bar{Y}_{..})^2$$

reflects only the random variability of the error terms if indeed there are no interactions. If interaction effects exist, however, SSE reflects not only the random error variation but also interaction effects. Hence, a pure error sum of squares may be obtained as follows:

$$SSPE = SSE - SSAB \tag{19.7}$$

where SSE is defined in (19.2) and $SSAB$ in (19.6).

It can then be shown that if $D = 0$, that is, if no interactions exist, $SSPE$ and $SSAB$ are independently distributed, and the test statistic:

$$F^* = \frac{SSAB}{1} \div \frac{SSPE}{ab - a - b} \tag{19.8}$$

is distributed as $F(1, ab - a - b)$. Note that one degree of freedom is associated with $SSAB$, and $(a - 1)(b - 1) - 1 = ab - a - b$ degrees of freedom are associated with $SSPE$.

Thus, for testing:

$$\begin{aligned} &C_1\text{: } D = 0 \text{ (no interactions present)} \\ &C_2\text{: } D \neq 0 \text{ (interactions } D\alpha_i\beta_j \text{ present)} \end{aligned} \tag{19.9}$$

we use the test statistic F^* defined in (19.8). Large values of F^* lead to conclusion C_2. The appropriate decision rule for controlling the risk of a Type I error at α is:

$$\begin{aligned} &\text{If } F^* \leq F(1 - \alpha; 1, ab - a - b), \text{ conclude } C_1 \\ &\text{If } F^* > F(1 - \alpha; 1, ab - a - b), \text{ conclude } C_2 \end{aligned} \tag{19.10}$$

The power of this test is unknown, but it appears that if interactions of approximately the type postulated in (19.3) are present, the test is effective in detecting these. The test is usually called the *Tukey one degree of freedom test.*

Example. We shall conduct the Tukey test for our earlier insurance premium example. The data are presented in Table 19.2a. First, we obtain the components of $SSAB$:

$$\begin{aligned} \sum\sum(\bar{Y}_{i.} - \bar{Y}_{..})(\bar{Y}_{.j} - \bar{Y}_{..})Y_{ij} &= (120 - 175)(190 - 175)(140) + \cdots \\ &\quad + (210 - 175)(160 - 175)(200) = -13{,}500 \end{aligned}$$

$$\sum(\bar{Y}_{i.} - \bar{Y}_{..})^2 = \frac{SSA}{2} = \frac{9{,}300}{2} = 4{,}650$$

$$\sum(\bar{Y}_{.j} - \bar{Y}_{..})^2 = \frac{SSB}{3} = \frac{1{,}350}{3} = 450$$

Hence, the interaction sum of squares is:

$$SSAB = \frac{(-13{,}500)^2}{(4{,}650)(450)} = 87.1$$

We found earlier in Table 19.2b that $SSE = 100$; hence we have by (19.7):

$$SSPE = 100 - 87.1 = 12.9$$

Finally, we obtain the test statistic by (19.8):

$$F^* = \frac{87.1}{1} \div \frac{12.9}{(3)(2) - 3 - 2} = 6.8$$

Assuming the level of significance is to be controlled at .10, we require $F(.90; 1, 1) = 39.9$. Since $F^* = 6.8 < 39.9$, we conclude that region and size of city do not interact. The use of the no-interaction model for the data in Table 19.2 therefore appears to be reasonable.

Comments

1. If the Tukey test indicates the presence of interaction effects, some simple transformations such as a square root or logarithmic transformation may be tried to see if the interactions can be removed or made unimportant. (See the discussion in Chapter 17 in this connection.)

2. If one or both factors are quantitative, a test for interaction effects can be obtained by regression methods, which will have more degrees of freedom associated with the interaction sum of squares than the Tukey test. These regression tests for interaction effects were discussed earlier. For example, model (9.6) illustrates a case where one independent variable is quantitative and the regression relation is linear. Model (7.15) is a second-order model where both independent variables are quantitative. For both of these models, the regression test for interaction effects is a simple test whether the interaction regression coefficients are zero.

19.2 UNEQUAL SAMPLE SIZES

Unequal sample sizes for the treatments are not infrequent with survey data. For instance, a market research analyst may wish to study the effect of temperature and precipitation on sales of a product, from data for the 20 largest metropolitan areas in the United States. In this type of uncontrolled situation, it may easily happen that the number of cities in the various temperature–precipitation classes are not constant.

One may also encounter unequal frequencies in experimental work. For instance, an experimenter may seek to have the same number of cases for each treatment, but for a variety of reasons (e.g., illness of subject, incomplete records) ends up with unequal sample sizes.

Notation

Our notation remains the same as before, except that the sample size for the treatment consisting of the ith level of factor A and the jth level of factor B will be denoted by n_{ij}. The total number of observations for the ith level of A is:

(19.11a) $$n_{i.} = \sum_j n_{ij}$$

for the jth level of B is:

(19.11b) $$n_{.j} = \sum_i n_{ij}$$

and for the total study is:

(19.11c) $$n_T = \sum_i \sum_j n_{ij}$$

Proportional Frequencies

Occasionally, the unequal sample sizes follow a proportional pattern. This is illustrated in Table 19.3. Here, 250 men and 250 women are to be included in a study on subliminal advertising. Fifty of each sex are to be exposed to one of three subliminal advertisements about a company's product, and their change in attitude measured. Since major emphasis is to be placed on comparing each subliminal advertisement with the control (no subliminal advertisement), 100 men and 100 women are assigned to the control advertisement. Note that the following relation holds here:

$$n_{ij} = \frac{n_{i.}n_{.j}}{n_T} \tag{19.12}$$

The condition (19.12) implies that the sample sizes in any two rows (or columns) are proportional. This is called a case of *proportional frequencies.*

TABLE 19.3

Illustration of Proportional Frequencies

	Men	*Women*	*Total*
Advertisement A	50	50	100
Advertisement B	50	50	100
Advertisement C	50	50	100
Control	100	100	200
Total	250	250	500

When the frequencies are proportional, the regular analysis of variance model (17.21), which includes an interaction term, can be employed. Further, the ordinary analysis of variance can be utilized. The definitional formulas in (17.31c) and (17.32a, b, c) are appropriate for obtaining the various sums of squares. The means required for these formulas are defined as follows:

$$\bar{Y}_{ij.} = \frac{Y_{ij.}}{n_{ij}} \tag{19.13a}$$

$$\bar{Y}_{i..} = \frac{Y_{i..}}{n_{i.}} \tag{19.13b}$$

$$\bar{Y}_{.j.} = \frac{Y_{.j.}}{n_{.j}} \tag{19.13c}$$

$$\bar{Y}_{...} = \frac{Y_{...}}{n_T} \tag{19.13d}$$

Computational formulas for the sums of squares with proportional frequencies, suitable for hand computations, are:

$$SSTO = \sum_i \sum_j \sum_k Y_{ijk}^2 - \frac{Y_{...}^2}{n_T} \tag{19.14a}$$

$$SSE = \sum_i \sum_j \sum_k Y_{ijk}^2 - \sum_i \sum_j \frac{Y_{ij.}^2}{n_{ij}} \tag{19.14b}$$

$$SSA = \sum_i \frac{Y_{i..}^2}{n_{i.}} - \frac{Y_{...}^2}{n_T} \tag{19.14c}$$

$$SSB = \sum_j \frac{Y_{.j.}^2}{n_{.j}} - \frac{Y_{...}^2}{n_T} \tag{19.14d}$$

$$SSAB = SSTO - SSE - SSA - SSB \tag{19.14e}$$

In the analysis of factor effects, it is often desirable to consider the unweighted factor level means $\mu_{i.}$ and $\mu_{.j}$ as defined in (17.2) and (17.1) respectively. These unweighted means can be estimated unbiasedly by:

$$\hat{\mu}_{i.} = \frac{\sum_j \bar{Y}_{ij.}}{b} \tag{19.15a}$$

$$\hat{\mu}_{.j} = \frac{\sum_i \bar{Y}_{ij.}}{a} \tag{19.15b}$$

Note that these estimators are not the equivalents of $\bar{Y}_{i..}$ and $\bar{Y}_{.j.}$ respectively, because of the unequal sample sizes. Confidence intervals for the factor effects can be obtained in the usual way, from the general definition of a contrast. Nothing new is encountered in the use of the Scheffé and Bonferroni multiple comparison procedures; the Tukey method cannot be employed because of the unequal sample sizes.

General Case of Unequal Frequencies

Exact Analysis of Variance. When the treatment sample sizes are unequal and not proportional, the analysis of variance for factor effects becomes complex. The least squares equations are no longer of a simple structure, yielding direct and easy solutions. Hence the regular analysis of variance calculations are inappropriate. Furthermore, the component sums of squares in the analysis of variance are no longer orthogonal, that is, they do not sum to $SSTO$. The easiest way to obtain the proper sums of squares for testing factor effects and interactions is through the regression approach, discussed in Section 19.4.

Approximate Analysis of Variance. If the sample sizes n_{ij} do not differ too much (some statisticians say not more than by the ratio 2 to 1, with most n_{ij} agreeing more closely) and if no n_{ij} is zero, an approximate analysis of variance, called the *method of unweighted means*, may be utilized. This approximate procedure is also used sometimes when the n_{ij} do differ substantially but only a quick first approximation to the factor effects is desired.

The procedure is simple. An analysis of variance is conducted, using the $\bar{Y}_{ij.}$ as if they were single observations for each treatment. *SSA*, *SSB*, and *SSAB* are thus calculated in the usual way, except that each treatment has only one "observation" $\bar{Y}_{ij.}$. We know that the "observation" $\bar{Y}_{ij.}$ has variance σ^2/n_{ij}. Hence the average variance of the "observations" $\bar{Y}_{ij.}$ is:

$$\frac{\sum_i \sum_j \frac{\sigma^2}{n_{ij}}}{ab} = \frac{\sigma^2}{ab} \sum_i \sum_j \frac{1}{n_{ij}} \tag{19.16}$$

The variance σ^2 is estimated by *MSE*, whether frequencies are equal or unequal:

$$MSE = \frac{\sum_i \sum_j \sum_k (Y_{ijk} - \bar{Y}_{ij.})^2}{n_T - ab} \tag{19.17}$$

Hence, the estimated average variance of the "observations" is:

$$\frac{MSE}{ab} \sum_i \sum_j \frac{1}{n_{ij}} \tag{19.18}$$

This estimate is then used for the error variance in the analysis of variance of the "observations" $\bar{Y}_{ij.}$. The analysis of variance thus is:

Source of Variation	*df*	*MS*
Factor *A*	$a-1$	*MSA*
Factor *B*	$b-1$	*MSB*
AB interactions	$(a-1)(b-1)$	*MSAB*
Error	$n_T - ab$	$\frac{MSE}{ab} \sum \sum \frac{1}{n_{ij}}$

Another method of approximation has been developed by Federer and Zelen (Ref. 19.1). It is more exact, though somewhat more complicated, than the method of unweighted means.

Multiple Comparison Procedures. No new problems arise in making multiple comparisons when the sample sizes are unequal. If it is desired to estimate a number of contrasts involving the $\hat{\mu}_{i.}$ as defined in (19.15a), the estimated contrasts are of the type:

$$\hat{L} = \sum c_i \hat{\mu}_{i.} \qquad \text{where } \sum c_i = 0 \tag{19.19}$$

Now $\hat{\mu}_{i.}$ is a simple average of b $\bar{Y}_{ij.}$ terms, so that we have:

$$\hat{L} = \sum_i c_i \left[\sum_j \frac{\bar{Y}_{ij.}}{b} \right] \tag{19.19a}$$

Since the $\bar{Y}_{ij.}$ terms are independent, we obtain:

$$\sigma^2(\hat{L}) = \sum_i c_i^2 \left[\frac{1}{b^2} \sum_j \frac{\sigma^2}{n_{ij}} \right] = \frac{\sigma^2}{b^2} \sum_i \sum_j \frac{c_i^2}{n_{ij}} \tag{19.20}$$

The variance σ^2 is estimated as usual by MSE in (19.17) so that the estimated variance of $\hat{L}$ is:

$$s^2(\hat{L}) = \frac{MSE}{b^2} \sum_i \sum_j \frac{c_i^2}{n_{ij}} \tag{19.21}$$

The Scheffé multiple comparison confidence intervals in (18.28) can then be used with $\hat{L}$ and $s(\hat{L})$ as defined in (19.19) and (19.21) respectively, and the Scheffé multiple defined by:

$$S^2 = (a-1)F(1-\alpha; a-1, n_T - ab) \tag{19.22}$$

If the number of contrasts is small, the Bonferroni intervals can be used with:

$$B = t(1-\alpha/2s; n_T - ab) \tag{19.23}$$

Similarly, the Scheffé and Bonferroni multiple comparison procedures are applicable for contrasts involving the $\hat{\mu}_{.j}$. Both methods can also be used for other types of contrasts involving the treatment means $\bar{Y}_{ij.}$, when the presence of interactions calls for analyses not based on factor level means.

19.3 MODELS II (RANDOM EFFECTS) AND III (MIXED EFFECTS) FOR TWO-FACTOR STUDIES

Random Effects Model

Consider an investigation of the effects of machine operators (factor A) and machines (factor B) on the number of pieces produced in a day. Five operators and three machines are used in the study. Yet the inferences are not to be confined to the particular five operators and three machines participating in the study, but rather are to pertain to all operators and all machines available to the company. Here a random effects model (model II) would be appropriate for the two-factor study, since each of the two sets of factor levels may be considered as a sample from a population (all operators, all machines) about which inferences are to be drawn.

In a random effects model for a two-factor study, we assume analogously to the random effects model for a one-factor study that both the factor A effects α_i and the factor B effects β_j are random variables, selected independently. Further, it will be assumed that the interaction effects $(\alpha\beta)_{ij}$ are independent random variables. Thus, the random effects model for a two-factor study with equal sample sizes n is:

(19.24) $$Y_{ijk} = \mu_{..} + \alpha_i + \beta_j + (\alpha\beta)_{ij} + \varepsilon_{ijk}$$

where:

$\mu_{..}$ is a constant

α_i, β_j, $(\alpha\beta)_{ij}$ are independent normal random variables with expectation zero and respective variances σ_α^2, σ_β^2, $\sigma_{\alpha\beta}^2$

ε_{ijk} are independent $N(0, \sigma^2)$, and independent of α_i, β_j, and $(\alpha\beta)_{ij}$

$i = 1, \ldots, a; j = 1, \ldots, b; k = 1, \ldots, n$

Meaning of Model. We shall explain the meaning of the terms in the random effects model (19.24) with reference to our production example involving the two factors machine operators and machines. Model (19.24) states that the effects of operators on output per day are normally distributed with zero mean and variance σ_α^2. Similarly, the effects of machines are normally distributed with zero mean and variance σ_β^2. The effect of operator i in the study (selected at random from the population of operators) is α_i. Similarly, the effect of machine j in the study (selected at random from the population of machines) is β_j. If μ_{ij} is the mean output per day for the operator i–machine j combination, the interaction $(\alpha\beta)_{ij}$ is defined in the usual way: $(\alpha\beta)_{ij} = \mu_{ij} - (\mu_{..} + \alpha_i + \beta_j)$. Surprising as it may seem, the interaction term $(\alpha\beta)_{ij}$ turns out to be independent of α_i and β_j when all random terms in the model are assumed to be normally distributed. Hence, the mean output for the operator i–machine j combination, namely $\mu_{ij} = \mu_{..} + \alpha_i + \beta_j + (\alpha\beta)_{ij}$, may be viewed in the random effects model as the result of independent selections of α_i, β_j, and $(\alpha\beta)_{ij}$ from three different normal distributions.

Note

We should caution that a random effects model only be used if the factor levels of the different factors do indeed represent random samples from populations of interest.

Mixed Effects Model

When one of the two factors involves fixed effects while the other involves random effects, a mixed effects model (model III) is appropriate. An instance where this model may be appropriate is an investigation of the effects of four different training materials (factor A) and five instructors (factor B) upon learning in a company training program. The four levels for training materials may be considered fixed, since interest centers in these particular training materials. On the other hand, the levels for instructors may be viewed as random since inferences are to be made about a population of instructors of which the five used in the study are a sample.

When factor A has fixed effects and factor B random effects, the α_i effects are constants and the β_j effects are random variables. The interaction effects $(\alpha\beta)_{ij}$ are also random variables because the factor B levels are random.

For this case where A is the fixed effects factor, B the random effects factor, and equal sample sizes n are selected for each treatment, a relatively simple mixed effects model is:

$$Y_{ijk} = \mu_{..} + \alpha_i + \beta_j + (\alpha\beta)_{ij} + \varepsilon_{ijk} \tag{19.25}$$

where:

$\mu_{..}$ is a constant

α_i are constants subject to the restriction $\sum \alpha_i = 0$

β_j are independent $N(0, \sigma_\beta^2)$

$(\alpha\beta)_{ij}$ are $N\left(0, \dfrac{a-1}{a} \sigma_{\alpha\beta}^2\right)$, subject to the restriction $\sum (\alpha\beta)_{ij} = 0$ for all j

ε_{ijk} are independent $N(0, \sigma^2)$, and independent of the β_j and $(\alpha\beta)_{ij}$

β_j and $(\alpha\beta)_{ij}$ are independent

$i = 1, \ldots, a; j = 1, \ldots, b; k = 1, \ldots, n$

More complex mixed effects models have also been formulated.

Comments

1. In model (19.25), any two interaction terms $(\alpha\beta)_{ij}$ and $(\alpha\beta)_{i'j'}$ are independent unless both refer to the same random factor B level. In the latter case, the covariance is assumed to be $\sigma((\alpha\beta)_{ij}, (\alpha\beta)_{i'j}) = -\dfrac{1}{a} \sigma_{\alpha\beta}^2$.

2. We noted that $\sum_i (\alpha\beta)_{ij} = 0$ for all j, since all factor A levels are included in the study. However, $\sum_j (\alpha\beta)_{ij}$ will ordinarily not equal zero.

Analysis of Variance

For both the mixed and random effects models for a two-factor study, the analysis of variance calculations for sums of squares are identical to those for a fixed effects model. Thus, formulas (17.31)–(17.33) are entirely applicable for models II and III. Similarly, the degrees of freedom and mean square determinations are exactly the same as for the fixed effects model, as shown in Table 17.9. The random and mixed effects models depart from the fixed effects model only in the expected mean squares and the consequent choice of the appropriate test statistic.

Expected Mean Squares

The expected mean squares for the random and mixed effects models can be worked out by utilizing the properties of the model and applying the usual expectation theorems. They are shown in Table 19.4, together with those

TABLE 19.4

Expected Mean Squares in Two-Factor Studies

Mean Square	*df*	*Fixed Effects* *(A and B fixed)*	*Random Effects* *(A and B random)*	*Mixed Effects* *(A fixed, B random)*
MSA	$a-1$	$\sigma^2 + nb\dfrac{\sum \alpha_i^2}{a-1}$	$\sigma^2 + nb\sigma_\alpha^2 + n\sigma_{\alpha\beta}^2$	$\sigma^2 + nb\dfrac{\sum \alpha_i^2}{a-1} + n\sigma_{\alpha\beta}^2$
MSB	$b-1$	$\sigma^2 + na\dfrac{\sum \beta_j^2}{b-1}$	$\sigma^2 + na\sigma_\beta^2 + n\sigma_{\alpha\beta}^2$	$\sigma^2 + na\sigma_\beta^2$
$MSAB$	$(a-1)(b-1)$	$\sigma^2 + n\dfrac{\sum\sum (\alpha\beta)_{ij}^2}{(a-1)(b-1)}$	$\sigma^2 + n\sigma_{\alpha\beta}^2$	$\sigma^2 + n\sigma_{\alpha\beta}^2$
MSE	$(n-1)ab$	σ^2	σ^2	σ^2

for the fixed effects model. Because the derivations are tedious, we now take up some simple rules for finding the expected mean squares for random and mixed effects models. Indeed, these rules apply also for the fixed effects model.

Rules for Finding Expected Mean Squares. The simple rules to be presented are appropriate for finding expected mean squares for random, mixed, and fixed effects models when the sample size is constant for all treatments. The rules apply for any number of factors studied in an investigation.

An expected mean square is a sum of variance components, each multiplied by a coefficient. The first rule tells us which variance components to include. To simplify notation, we shall initially use σ_α^2, σ_β^2, $\sigma_{\alpha\beta}^2$, etc. for denoting the variance components whether the factors are fixed or random.

(19.26*a*)

1. Always include σ^2 for the error term.
2. List the set of variance components in the model whose subscripts contain all the factors included in the desired expected mean square.
3. From the list in step 2, delete any interaction variance component for which the additional factors beyond those in the desired expected mean square are not all random.

This rule is illustrated in Table 19.5 for various types of two-factor studies. Note that in case 2, $\sigma_{\alpha\beta}^2$ is deleted because factor B, which is not included in MSA, is fixed.

TABLE 19.5

Illustration of Rule (19.26a)

1. $E(MSA)$, A fixed, B random		2. $E(MSA)$, A random, B fixed	
Step 1	σ^2	Step 1	σ^2
Step 2	σ_α^2, $\sigma_{\alpha\beta}^2$	Step 2	σ_α^2, $\sigma_{\alpha\beta}^2$
Step 3	delete none	Step 3	delete $\sigma_{\alpha\beta}^2$
3. $E(MSAB)$, A fixed, B random		**4. $E(MSAB)$, A and B random**	
Step 1	σ^2	Step 1	σ^2
Step 2	$\sigma_{\alpha\beta}^2$	Step 2	$\sigma_{\alpha\beta}^2$
Step 3	delete none	Step 3	delete none

The coefficients of the variance components to be included in the expected mean square may be determined from the following rule:

(19.26b)

1. The coefficient of σ^2 is always 1.
2. The coefficient of any other variance component is n times the product of the number of factor levels of the factors that do *not* appear in the subscript of the variance component.

To illustrate this rule, if σ_α^2 appears in the expected mean square in a two-factor study, its coefficient is nb, where b is the number of levels of factor B which is not included in the subscript of σ_α^2. Similarly, if $\sigma_{\alpha\beta}^2$ appears in the expected mean square in a two-factor study, its coefficient is simply n, since there are no other factors that are not included in its subscript.

Let us now obtain $E(MSA)$ for the case where factors A and B are both random. Rule (19.26a) indicates that σ^2, σ_α^2, and $\sigma_{\alpha\beta}^2$ are to be included ($\sigma_{\alpha\beta}^2$ is not deleted since B is a random factor). Rule (19.26b) then states that the respective coefficients are 1, nb, and n. Hence, we have:

$$E(MSA) = \sigma^2 + nb\sigma_\alpha^2 + n\sigma_{\alpha\beta}^2$$

just as stated in Table 19.4.

At this point, one last step may have to be taken. If any of the main effect variance components refers to a fixed factor, it should be replaced in the final expected mean square by the corresponding sum of squared effects divided by the degrees of freedom. For instance, if A is a fixed factor, σ_α^2 should be replaced by $\sum \alpha_i^2/(a-1)$. Similarly, an interaction component involving only fixed factors should be replaced by $\sum\sum (\alpha\beta)_{ij}^2/(a-1)(b-1)$.

Note

The reason for step 3 in rule (19.26a) is that if any factor beyond those included in the desired expected mean square has fixed levels, the sum of the interaction terms over that factor is zero. Hence that variance component does not appear in the expected mean square.

Illustrative Derivation. We illustrate a derivation of an expected mean square by finding $E(MSA)$ for the two-factor random effects model (19.24). We wish to find:

$$E(MSA) = E\left[\frac{nb\sum(\bar{Y}_{i..} - \bar{Y}_{...})^2}{a-1}\right]$$

Now:

$$\begin{aligned} Y_{i..} &= \sum_j \sum_k [\mu_{..} + \alpha_i + \beta_j + (\alpha\beta)_{ij} + \varepsilon_{ijk}] \\ &= nb\mu_{..} + nb\alpha_i + n\sum_j \beta_j + n\sum_j (\alpha\beta)_{ij} + \sum_j \sum_k \varepsilon_{ijk} \end{aligned}$$

We obtain then:

$$\bar{Y}_{i..} = \frac{Y_{i..}}{nb} = \mu_{..} + \alpha_i + \bar{\beta}_. + \overline{(\alpha\beta)}_{i.} + \bar{\varepsilon}_{i..} \tag{19.27}$$

where the bars as usual indicate means over the subscripts shown by dots. Similarly, we find:

$$\bar{Y}_{...} = \mu_{..} + \bar{\alpha}_. + \bar{\beta}_. + \overline{(\alpha\beta)}_{..} + \bar{\varepsilon}_{...} \tag{19.27a}$$

Hence we have:

$$\bar{Y}_{i..} - \bar{Y}_{...} = (\alpha_i - \bar{\alpha}_.) + (\overline{(\alpha\beta)}_{i.} - \overline{(\alpha\beta)}_{..}) + (\bar{\varepsilon}_{i..} - \bar{\varepsilon}_{...}) \tag{19.28}$$

Squaring each side of (19.28) and summing, we obtain:

$$(19.29) \qquad \sum_i (\bar{Y}_{i..} - \bar{Y}_{...})^2 = \sum_i (\alpha_i - \bar{\alpha}_.)^2 + \sum_i ((\overline{\alpha\beta})_{i.} - (\overline{\alpha\beta})_{..})^2 + \sum_i (\bar{\varepsilon}_{i..} - \bar{\varepsilon}_{...})^2 + \text{cross-product terms}$$

To find $E[\sum(\bar{Y}_{i..} - \bar{Y}_{...})^2]$, we need to take expectations of each term on the right. The cross-product terms drop out because of the independence of α_i, $(\alpha\beta)_{ij}$, and ε_{ijk} and the fact that each of these random variables has zero expectation. Each of the remaining terms can be thought of as the numerator of a sample variance of a observations. We know from the unbiasedness of a sample variance that:

$$(19.30) \qquad E\left[\sum_{i=1}^{a} (Y_i - \bar{Y})^2\right] = (a-1)\sigma^2(Y_i)$$

Hence:

$$(19.31) \qquad E\left[\sum_i (\alpha_i - \bar{\alpha}_.)^2\right] = (a-1)\sigma_\alpha^2$$

because $\sigma^2(\alpha_i) = \sigma_\alpha^2$. Similarly, we find:

$$(19.32) \qquad E\left[\sum_i (\bar{\varepsilon}_{i..} - \bar{\varepsilon}_{...})^2\right] = (a-1)\frac{\sigma^2}{bn}$$

since $\sigma^2(\bar{\varepsilon}_{i..}) = \sigma^2/bn$, and:

$$(19.33) \qquad E\left[\sum_i ((\overline{\alpha\beta})_{i.} - (\overline{\alpha\beta})_{..})^2\right] = (a-1)\frac{\sigma_{\alpha\beta}^2}{b}$$

Using (19.31)–(19.33), we find:

$$(19.34) \qquad E[\sum(\bar{Y}_{i..} - \bar{Y}_{...})^2] = (a-1)\sigma_\alpha^2 + (a-1)\frac{\sigma_{\alpha\beta}^2}{b} + (a-1)\frac{\sigma^2}{bn}$$

and:

$$(19.35) \qquad E(MSA) = \frac{nb}{a-1} E[\sum(\bar{Y}_{i..} - \bar{Y}_{...})^2] = nb\sigma_\alpha^2 + n\sigma_{\alpha\beta}^2 + \sigma^2$$

as shown in Table 19.4.

Construction of Test Statistics

As usual, the test statistic is constructed by comparing two mean squares which have the properties:

1. Under C_1, both have the same expectation.
2. Under C_2, the numerator mean square has a larger expectation than the denominator mean square.

It can be shown that such a test statistic follows the F distribution if C_1 holds. The decision rule is constructed in the ordinary fashion, with large values of the test statistic leading to C_2.

For instance, to test for the presence of factor A effects in the random effects model (19.24) namely:

$$C_1: \sigma_\alpha^2 = 0 \qquad C_2: \sigma_\alpha^2 \neq 0 \tag{19.36}$$

we see from Table 19.4 that MSA and $MSAB$ both have the same expectation if $\sigma_\alpha^2 = 0$, that is, if factor A has no effect. If $\sigma_\alpha^2 \neq 0$, $E(MSA)$ is greater than $E(MSAB)$. Hence the appropriate test statistic is:

$$F^* = \frac{MSA}{MSAB} \tag{19.37}$$

and the decision rule for controlling the Type I error at α is:

$$\begin{aligned} &\text{If } F^* \leq F(1-\alpha; a-1, (a-1)(b-1)), \text{ conclude } C_1 \\ &\text{If } F^* > F(1-\alpha; a-1, (a-1)(b-1)), \text{ conclude } C_2 \end{aligned} \tag{19.38}$$

Note that the denominator for testing for factor A effects in the random effects model is $MSAB$, whereas it is MSE in the fixed effects model.

We summarize the appropriate test statistics for mixed and random effects models in Table 19.6. For comparison purposes, we also present the test statistics for the fixed effects model there. As may be seen from Table 19.6 in a number of instances the denominator of the test statistic for mixed and random effects models differs from that for the fixed effects model. Hence it is important that the expected mean squares be obtained when random or mixed models are utilized, so that the appropriate test statistics can be determined.

TABLE 19.6

Test Statistics for Mixed and Random Effects Models

Test for Presence of Effects of—	*Fixed Effects Model (A and B fixed)*	*Random Effects Model (A and B random)*	*Mixed Effects Model (A fixed, B random)*
Factor A	MSA/MSE	$MSA/MSAB$	$MSA/MSAB$
Factor B	MSB/MSE	$MSB/MSAB$	MSB/MSE
AB interactions	$MSAB/MSE$	$MSAB/MSE$	$MSAB/MSE$

Example. We return to our earlier mixed effects example of four different training materials (factor A, fixed) and five instructors (factor B, random). Four classes were tested for each training material–instructor combination.

TABLE 19.7

ANOVA Table for Mixed Effects Model Training Example (A fixed, B random, $a = 4$, $b = 5$, $n = 4$)

Source of Variation	*SS*	*df*	*MS*	F^*
Factor A (training materials–fixed)	42	3	14.0	$14.0/3.9 = 3.59$
Factor B (instructors–random)	54	4	13.5	$13.5/2.1 = 6.43$
AB interactions	47	12	3.9	$3.9/2.1 = 1.86$
Error	126	60	2.1	
Total	269	79		

$F(.95; 3, 12) = 3.49$ $F(.95; 4, 60) = 2.53$
$F(.95; 12, 60) = 1.92$

The analysis of variance as obtained from a computer run is shown in Table 19.7; the original data are not presented. To test whether or not training materials and instructors interact:

$$C_1: \sigma_{\alpha\beta}^2 = 0$$
$$C_2: \sigma_{\alpha\beta}^2 \neq 0$$

we utilize according to Table 19.6 the test statistic:

$$F^* = \frac{MSAB}{MSE}$$

Using the results from Table 19.7, we obtain:

$$F^* = \frac{3.9}{2.1} = 1.86$$

For a level of significance of .05, we require $F(.95; 12, 60) = 1.92$. Since $F^* = 1.86 < 1.92$, we conclude that training materials and instructors do not interact.

The test statistics for testing training material effects and instructor effects are shown in Table 19.7. We find, by comparing the test statistics with the percentiles of the F distribution shown at the bottom of Table 19.7 for level of significance .05, that both training materials and instructors differ in effectiveness.

Note

If only one observation per treatment ($n = 1$) is obtained, it will be recalled from Section 19.1 that no exact tests are possible with the fixed effects two-factor model unless one can assume that there are no AB interactions. The reason is that no estimate MSE of σ^2 can be obtained in that case. Table 19.6 indicates that exact tests for both factor A and factor B effects are possible with the random effects model when $n = 1$ without any restrictive assumptions about the interactions. This is because $MSAB$ is the appropriate denominator of the test statistic here, and $MSAB$

can be determined regardless of sample size. With the mixed effects model where factor A is the fixed factor, the presence of factor A effects can also be tested when $n = 1$ without the need for assuming all interactions are zero. However, an exact test for factor B effects would require the assumption that all interactions are zero.

Estimation of Variance Components

For random factors that have significant effects, we often would like to estimate the magnitude of the variance component. Unbiased point estimators can readily be derived, using linear combinations of the expected mean squares in Table 19.4. With the random effects model, for instance, σ_α^2 can be estimated by noting that:

$$E(MSA) - E(MSAB) = \sigma^2 + nb\sigma_\alpha^2 + n\sigma_{\alpha\beta}^2 - \sigma^2 - n\sigma_{\alpha\beta}^2 = nb\sigma_\alpha^2$$

Hence, an unbiased point estimator of σ_α^2 is:

$$s_\alpha^2 = \frac{MSA - MSAB}{nb} \tag{19.39}$$

Example. In the training illustration of Table 19.7, random effects factor B (instructors) had significant effects. To estimate σ_β^2, we utilize the expected mean squares in Table 19.4 for the mixed model with factor A fixed and B random, and determine the unbiased point estimator to be:

$$s_\beta^2 = \frac{MSB - MSE}{na} \tag{19.40}$$

Substituting, we obtain:

$$s_\beta^2 = \frac{13.5 - 2.1}{16} = .71$$

Estimation of Fixed Effects in Mixed Model

If the fixed effects factor in a mixed two-factor model has real effects, one usually desires to make some estimates of these effects in the form of contrasts. Our earlier discussion for the two-factor fixed effects model in Chapter 18 is applicable here, with the principal change occurring in the estimated variance of the contrast. For a fixed effects model, this estimated variance involves MSE, as shown in (18.8). When dealing with a mixed model, however, the appropriate mean square to be used in the estimated variance formula is no longer MSE. A simple rule tells us which mean square is appropriate, namely that which is used in the denominator of the test statistic for testing the presence of the fixed factor under consideration. For instance, with the mixed model (19.25) where A is the fixed factor, $MSAB$ is the appropriate mean square (see Table 19.6). The degrees of freedom in constructing the confidence interval are those associated with the mean square utilized for estimating the variance of the contrast.

Example. In our training illustration of Table 19.7, we wish to estimate the difference in the mean amount of learning with training materials 1 and 2, using a 95 percent confidence interval. Suppose that we have:

$$\bar{Y}_{1..} = 43.1 \qquad \bar{Y}_{2..} = 40.8$$

Hence, our point estimate of $L = \mu_{1.} - \mu_{2.}$ is:

$$\hat{L} = \bar{Y}_{1..} - \bar{Y}_{2..} = 43.1 - 40.8 = 2.3 \tag{19.41}$$

The estimated variance is, using formula (18.8) with MSE replaced by $MSAB$:

$$s^2(\hat{L}) = \frac{MSAB}{bn}[(1)^2 + (-1)^2] = \frac{2MSAB}{bn} \tag{19.42}$$

For our example, we have:

$$s^2(\hat{L}) = \frac{2(3.9)}{20} = .39$$

of $s(\hat{L}) = .62$. There are 12 degrees of freedom associated with $MSAB$, hence we require $t(.975,12) = 2.179$. Our confidence interval therefore is:

$$2.3 - (2.179)(.62) \le \mu_{1.} - \mu_{2.} \le 2.3 + (2.179)(.62)$$
$$.9 \le \mu_{1.} - \mu_{2.} \le 3.7$$

Thus we conclude that training material 1 is more effective than training material 2.

Multiple Comparison Procedures. Multiple comparison procedures can be utilized for mixed models as for the fixed effects model. In the estimated variance of the contrast, MSE will simply need to be replaced by the appropriate mean square. For example, suppose we wish to obtain all pairwise comparisons between the different training materials cited in Table 19.7 by means of the Tukey method. We would calculate $s^2(D)$ from (18.13b), replacing MSE by $MSAB$:

$$s^2(D) = \frac{2MSAB}{bn} \tag{19.43}$$

The T multiple in (18.13c) now would be:

$$T = \frac{1}{\sqrt{2}}q(1 - \alpha; a, (a-1)(b-1)) \tag{19.43a}$$

where $(a-1)(b-1)$ replaces $(n-1)ab$ as the degrees of freedom associated with the mean square utilized. With specific reference to our illustration in Table 19.7, we would utilize for constructing 95 percent family confidence coefficient intervals for all pairwise comparisons between training materials:

$$q(.95; 4, 12) = 4.20$$

and:

$$T = \frac{1}{\sqrt{2}}(4.20) = 2.97$$

19.4 REGRESSION APPROACH TO TWO-FACTOR ANALYSIS OF VARIANCE

Development of Regression Model

The development of a regression model for two-factor analysis of variance involves a direct extension of the regression model for single-factor studies. Consider a two-factor study with factor A at three levels and factor B at two levels. The two-factor fixed effects analysis of variance model is:

$$Y_{ijk} = \mu_{..} + \alpha_i + \beta_j + (\alpha\beta)_{ij} + \varepsilon_{ijk} \qquad i = 1, 2, 3; j = 1, 2; k = 1, \ldots, n \tag{19.44}$$

In formulating the corresponding regression model, we shall use two indicator variables for factor A:

$$X_1 = \begin{matrix} 1 \text{ if observation from level 1 for factor } A \\ 0 \text{ otherwise} \end{matrix} \qquad X_2 = \begin{matrix} 1 \text{ if observation from level 2 for factor } A \\ 0 \text{ otherwise} \end{matrix} \tag{19.45a}$$

and one indicator variable for factor B:

$$X_3 = \begin{matrix} 1 \text{ if observation from level 1 for factor } B \\ 0 \text{ otherwise} \end{matrix} \tag{19.45b}$$

The interaction terms, as we know from Chapter 7, are represented by cross-product terms between the variables for the two factors:

$$X_1X_3 \qquad X_2X_3$$

The regression model counterpart to the analysis of variance model (19.44) can therefore be expressed as follows, using γ to denote the regression coefficients because β here denotes a factor B effect:

$$Y_{ijk} = \gamma_0 + \underbrace{\gamma_1 X_{ijk1} + \gamma_2 X_{ijk2}}_{A \text{ main effect}} + \underbrace{\gamma_3 X_{ijk3}}_{B \text{ main effect}} + \underbrace{\gamma_4 X_{ijk1} X_{ijk3} + \gamma_5 X_{ijk2} X_{ijk3}}_{AB \text{ interaction effect}} + \varepsilon_{ijk} \tag{19.46}$$

$$i = 1, 2, 3; j = 1, 2; k = 1, \ldots, n$$

Here X_{ijk1} is the value of X_1 for the kth observation when factor A is at the ith level and factor B is at the jth level, and similarly for the other X variables.

The correspondences between the ANOVA and regression model parameters are as follows:

(19.47a) $\gamma_0 = \mu_{..} + \alpha_3 + \beta_2 + (\alpha\beta)_{32} = \mu_{32}$

(19.47b) $\gamma_1 = \alpha_1 - \alpha_3 + (\alpha\beta)_{12} - (\alpha\beta)_{32} = \mu_{12} - \mu_{32}$

(19.47c) $\gamma_2 = \alpha_2 - \alpha_3 + (\alpha\beta)_{22} - (\alpha\beta)_{32} = \mu_{22} - \mu_{32}$

(19.47d) $\gamma_3 = \beta_1 - \beta_2 + (\alpha\beta)_{31} - (\alpha\beta)_{32} = \mu_{31} - \mu_{32}$

(19.47e) $\gamma_4 = (\alpha\beta)_{11} - (\alpha\beta)_{12} - (\alpha\beta)_{31} + (\alpha\beta)_{32} = \mu_{11} - \mu_{12} - \mu_{31} + \mu_{32}$

(19.47f) $\gamma_5 = (\alpha\beta)_{21} - (\alpha\beta)_{22} - (\alpha\beta)_{31} + (\alpha\beta)_{32} = \mu_{21} - \mu_{22} - \mu_{31} + \mu_{32}$

These correspondences are easy to establish. Suppose an observation is from the third level of factor A and the second level of factor B. According to the ANOVA model (19.44), we have:

(19.48) $$Y_{32k} = \mu_{..} + \alpha_3 + \beta_2 + (\alpha\beta)_{32} + \varepsilon_{32k}$$

For the regression model (19.46), the values of the independent variables for this observation are:

$$X_1 = 0 \qquad X_2 = 0 \qquad X_3 = 0 \qquad X_1X_3 = 0 \qquad X_2X_3 = 0$$

Hence, the regression model yields for this observation:

(19.49) $$\begin{aligned} Y_{32k} &= \gamma_0 + \gamma_1(0) + \gamma_2(0) + \gamma_3(0) + \gamma_4(0) + \gamma_5(0) + \varepsilon_{32k} \\ &= \gamma_0 + \varepsilon_{32k} \end{aligned}$$

Equating (19.48) and (19.49), we find:

$$\gamma_0 = \mu_{..} + \alpha_3 + \beta_2 + (\alpha\beta)_{32} = \mu_{32}$$

The other correspondences are established in a similar fashion.

Once the regression model has been defined, tests concerning factor effects and interactions are carried out in a fashion similar to that for a single-factor model. The same is true for estimation of factor effects and interactions.

Example

We illustrate the use of the regression approach for tests and estimation of factor effects and interactions by returning to the Castle Bakery example concerning the effects of display height and width on sales of the bakery's bread. Here, $a = 3$, $b = 2$, and $n = 2$. The regression model we shall employ for this example is based on the response function:

(19.50) $$E(Y) = \gamma_0 + \underbrace{\gamma_1 X_1 + \gamma_2 X_2}_{A \text{ main effect}} + \underbrace{\gamma_3 X_3}_{B \text{ main effect}} + \underbrace{\gamma_4 X_1X_3 + \gamma_5 X_2X_3}_{AB \text{ interaction effect}} \qquad \text{Full model}$$

where:

$$X_1 = \begin{matrix} 1 \text{ if observation for display height 1} \\ 0 \text{ otherwise} \end{matrix}$$

$$X_2 = \begin{matrix} 1 \text{ if observation for display height 2} \\ 0 \text{ otherwise} \end{matrix}$$

$$X_3 = \begin{matrix} 1 \text{ if observation for display width 1} \\ 0 \text{ otherwise} \end{matrix}$$

$$\gamma_0 = \mu_{32}$$

$$\gamma_1 = \mu_{12} - \mu_{32}$$

$$\gamma_2 = \mu_{22} - \mu_{32}$$

$$\gamma_3 = \mu_{31} - \mu_{32}$$

$$\gamma_4 = \mu_{11} - \mu_{12} - \mu_{31} + \mu_{32}$$

$$\gamma_5 = \mu_{21} - \mu_{22} - \mu_{31} + \mu_{32}$$

The observation vector **Y** and the **X** matrix for this example are shown in Table 19.8, based on the data in Table 17.7. A fit by a multiple regression

TABLE 19.8

Data Matrices for Model (19.50) Based on Data in Table 17.7

$$\mathbf{Y} = \begin{bmatrix} Y_{111} = 47 \\ Y_{112} = 43 \\ Y_{121} = 46 \\ Y_{122} = 40 \\ Y_{211} = 62 \\ Y_{212} = 68 \\ Y_{221} = 67 \\ Y_{222} = 71 \\ Y_{311} = 41 \\ Y_{312} = 39 \\ Y_{321} = 42 \\ Y_{322} = 46 \end{bmatrix} \qquad \mathbf{X} = \begin{bmatrix} 1 & 1 & 0 & 1 & 1 & 0 \\ 1 & 1 & 0 & 1 & 1 & 0 \\ 1 & 1 & 0 & 0 & 0 & 0 \\ 1 & 1 & 0 & 0 & 0 & 0 \\ 1 & 0 & 1 & 1 & 0 & 1 \\ 1 & 0 & 1 & 1 & 0 & 1 \\ 1 & 0 & 1 & 0 & 0 & 0 \\ 1 & 0 & 1 & 0 & 0 & 0 \\ 1 & 0 & 0 & 1 & 0 & 0 \\ 1 & 0 & 0 & 1 & 0 & 0 \\ 1 & 0 & 0 & 0 & 0 & 0 \\ 1 & 0 & 0 & 0 & 0 & 0 \end{bmatrix}$$

(Columns of **X** after the first: X_1, X_2, X_3, X_1X_3, X_2X_3.)

computer program yielded the results in Table 19.9. We illustrate the test for factor A effects:

$$(19.51) \qquad \begin{aligned} &C_1: \alpha_1 = \alpha_2 = \alpha_3 = 0 \\ &C_2: \text{not all } \alpha_i = 0 \end{aligned}$$

TABLE 19.9

Computer Output for Castle Bakery Multiple Regression Example

(a) Regression Coefficients

$g_0 = 44.0$	$g_2 = 25.0$	$g_4 = 6.0$
$g_1 = -1.0$	$g_3 = -4.0$	$g_5 = .0$

(b) Analysis of Variance

Source of Variation	*SS*	*df*	*MS*
Regression	$SSR = 1{,}580.0$	5	316.0
Error	$SSE = 62.0$	6	10.3
Total	$SSTO = 1{,}642.0$	11	

(c) Estimated Variance-Covariance Matrix for Regression Coefficients

	g_0	g_1	g_2	g_3	g_4	g_5
g_0	5.1667					
g_1	−5.1667	10.3333				
g_2	−5.1667	5.1667	10.3333			
g_3	−5.1667	5.1667	5.1667	10.3333		
g_4	5.1667	−10.3333	−5.1667	−10.3333	20.6667	
g_5	5.1667	−5.1667	−10.3333	−10.3333	10.3333	20.6667

C_1 can be expressed also in the following alternative forms:

(19.51a)
$$\mu_{1.} = \mu_{2.} = \mu_{3.}$$

(19.51b)
$$\mu_{1.} - \mu_{3.} = 0 \qquad \mu_{2.} - \mu_{3.} = 0$$

(19.51c)
$$\frac{\mu_{11} + \mu_{12}}{2} - \frac{\mu_{31} + \mu_{32}}{2} = 0 \qquad \frac{\mu_{21} + \mu_{22}}{2} - \frac{\mu_{31} + \mu_{32}}{2} = 0$$

(19.51d)
$$\mu_{11} + \mu_{12} - \mu_{31} - \mu_{32} = 0 \qquad \mu_{21} + \mu_{22} - \mu_{31} - \mu_{32} = 0$$

In terms of the regression coefficient correspondences defined in (19.50), the C_1 statement in (19.51d) can be expressed as follows:

$$\gamma_4 + 2\gamma_1 = 0 \qquad 2\gamma_2 + \gamma_5 = 0$$

or:

(19.52)
$$\gamma_4 = -2\gamma_1 \qquad \gamma_5 = -2\gamma_2$$

Hence the reduced regression model under C_1 becomes:

$$E(Y) = \gamma_0 + \gamma_1 X_1 + \gamma_2 X_2 + \gamma_3 X_3 + (-2\gamma_1)X_1 X_3 + (-2\gamma_2)X_2 X_3$$

or:

(19.53) $$E(Y) = \gamma_0 + \gamma_1(X_1 - 2X_1X_3) + \gamma_2(X_2 - 2X_2X_3) + \gamma_3 X_3 \qquad \text{Reduced model}$$

A fit of this regression model yielded the following ANOVA table:

Source of Variation	*SS*	*df*
Regression	$SSR = 36.0$	3
Error	$SSE = 1{,}606.0$	8
Total	$SSTO = 1{,}642.0$	11

Thus, $SSE(R) = 1{,}606.0$. From Table 19.9b, we know that $SSE(F) = 62.0$. Hence, the test statistic is:

$$F^* = \frac{SSE(R) - SSE(F)}{8 - 6} \div \frac{SSE(F)}{6}$$

$$= \frac{1{,}606 - 62}{8 - 6} \div \frac{62}{6}$$

$$= \frac{772}{10.3} = 75.0$$

which is identical to the test statistic on page 581 obtained by the analysis of variance.

As an illustration of estimating factor effects with the regression approach, consider the estimation of:

(19.54) $$\mu_{1.} - \mu_{3.} = \frac{\mu_{11} + \mu_{12}}{2} - \frac{\mu_{31} + \mu_{32}}{2}$$

We know from our correspondences in (19.50) that:

(19.54a) $$\mu_{1.} - \mu_{3.} = \tfrac{1}{2}(\gamma_4 + 2\gamma_1)$$

Hence, an unbiased estimator of $\mu_{1.} - \mu_{3.}$ is:

(19.55) $$\tfrac{1}{2}(g_4 + 2g_1)$$

From Table 19.9a, we obtain as an estimator of $\mu_{1.} - \mu_{3.}$ for our example:

$$\tfrac{1}{2}[6.0 + (2)(-1.0)] = 2.0$$

which is precisely the same as the result found on page 593. The estimated variance of this estimator can be obtained by utilizing the estimated variance-covariance matrix for the regression coefficients in Table 19.9c.

Comments

1. Once the analysis of variance has been obtained with the regression approach, it is usually easier to estimate factor level effects directly by making use of the estimated factor level means $\bar{Y}_{i..}$ or $\bar{Y}_{.j.}$. The use of the regression coefficients, as in (19.55), is cumbersome.

2. Unequal frequencies with a two-factor model present no problem with the regression approach. For each observation in the study, one simply determines the appropriate values of the indicator variables and places them in the **X** matrix. Table 19.10 illustrates the **Y** and **X** matrices for our Castle Bakery example if the observations Y_{112} and Y_{321} had been lost. Note that these matrices are the same as in Table 19.8 except for the deletion of the rows for the two lost observations. Once the observation vector **Y** and the **X** matrix have been specified, the fitting of the regression model with a computer program is straightforward. Testing and estimation of effects with the regression approach proceed in the same fashion whether the frequencies are equal or unequal. Remember, though, that sums of squares are no longer additive and that the Tukey method for multiple comparisons is not applicable when the sample sizes are unequal.

TABLE 19.10

Data Matrices for Model (19.50) If Y_{112} and Y_{321} Missing

$$
\mathbf{Y} = \begin{bmatrix} Y_{111}=47 \\ Y_{121}=46 \\ Y_{122}=40 \\ Y_{211}=62 \\ Y_{212}=68 \\ Y_{221}=67 \\ Y_{222}=71 \\ Y_{311}=41 \\ Y_{312}=39 \\ Y_{322}=46 \end{bmatrix}
\qquad
\mathbf{X} = \begin{bmatrix} 1 & 1 & 0 & 1 & 1 & 0 \\ 1 & 1 & 0 & 0 & 0 & 0 \\ 1 & 1 & 0 & 0 & 0 & 0 \\ 1 & 0 & 1 & 1 & 0 & 1 \\ 1 & 0 & 1 & 1 & 0 & 1 \\ 1 & 0 & 1 & 0 & 0 & 0 \\ 1 & 0 & 1 & 0 & 0 & 0 \\ 1 & 0 & 0 & 1 & 0 & 0 \\ 1 & 0 & 0 & 1 & 0 & 0 \\ 1 & 0 & 0 & 0 & 0 & 0 \end{bmatrix}
$$

(Columns of **X** after the first: X_1, X_2, X_3, X_1X_3, X_2X_3.)

3. If the factors have a large number of levels, and/or if a large number of factors are included in the study, a big computer may be necessary to fit the regression model. Suppose, for instance, that two factors with eight levels each are included in a study. We then require γ_0, seven regression coefficients for the indicator variables for A effects, seven regression coefficients for the indicator variables for B effects, and 49 regression coefficients for the AB interaction terms, or altogether 64 parameters. Some computer regression packages cannot handle this many independent variables.

4. Other versions of indicator variables are available which make it easier to test and estimate factor effects by the regression approach than 0, 1 indicator variables. The improved ease of testing and estimation comes about because of simpler correspondences between the regression model parameters and the analysis

of variance model parameters. Consider again the earlier illustration where factor A is at three levels and factor B is at two levels. Let us define the indicator variables as follows:

(19.56)
$$X_1 = \begin{cases} 1 & \text{if observation from level 1 for factor } A \\ -1 & \text{if observation from level 3 for factor } A \\ 0 & \text{otherwise} \end{cases}$$
$$X_2 = \begin{cases} 1 & \text{if observation from level 2 for factor } A \\ -1 & \text{if observation from level 3 for factor } A \\ 0 & \text{otherwise} \end{cases}$$
$$X_3 = \begin{cases} 1 & \text{if observation from level 1 for factor } B \\ -1 & \text{if observation from level 2 for factor } B \end{cases}$$

Interaction terms will still be represented by X_1X_3 and X_2X_3. The correspondences between the parameters of the ANOVA model (19.44) and the parameters of the regression model (19.46), with the X's defined as in (19.56), are then simply:

(19.57a) $$\gamma_0 = \mu_{..}$$

(19.57b) $$\gamma_1 = \alpha_1 = \mu_{1.} - \mu_{..}$$

(19.57c) $$\gamma_2 = \alpha_2 = \mu_{2.} - \mu_{..}$$

(19.57d) $$\gamma_3 = \beta_1 = \mu_{.1} - \mu_{..}$$

(19.57e) $$\gamma_4 = (\alpha\beta)_{11} = \mu_{11} - \mu_{1.} - \mu_{.1} + \mu_{..}$$

(19.57f) $$\gamma_5 = (\alpha\beta)_{21} = \mu_{21} - \mu_{2.} - \mu_{.1} + \mu_{..}$$

These correspondences show easily how to obtain the reduced models for testing factor effects and interactions. To test for factor A effects, $\gamma_1 = 0$ and $\gamma_2 = 0$. To test for factor B effects, $\gamma_3 = 0$. To test for interaction effects, $\gamma_4 = 0$ and $\gamma_5 = 0$. Likewise, the correspondences show readily how to estimate factor effects by means of the regression coefficients.

When the analysis of variance model contains interaction terms, use of indicator variables (19.56) is easiest for testing and analysis. When there are no interaction terms in the model, use of the regular 0, 1 indicator variables is easiest.

5. Ordinarily, regular analysis of variance techniques are to be preferred over the regression approach. For complex situations, however, such as when the number of observations are unequal, the regression approach often is the preferred method of analysis.

PROBLEMS

19.1. Suppose the analysis of variance model (17.21) were to be employed with $n = 1$ observation for each factor combination. How many degrees of freedom would be associated with SSE in (17.31c)? What does this imply?

19.2. A researcher investigated whether brainstorming works better for larger groups than smaller ones by studying four group sizes (2 persons, 3 persons, 4 persons, 5 persons). He used four teams of advertising executives, one for each of the group sizes, and likewise used four teams of research personnel, one for each group size. He gave each group the problem: How can the

United States attract more tourists? and he allowed the group 30 minutes to generate ideas. The variable of interest was the number of ideas proposed. The results were:

	Size of Group			
Type of Group	2	3	4	5
Advertising executives	18	24	31	30
Research personnel	15	19	25	27

a) State the model you will employ to analyze the effects of group size and type of group on the number of ideas generated.
b) Test whether the factors have any main effects. Use a level of significance of .01 for each test.
c) Give an upper bound for the family level of significance.
d) Analyze the factor effects. Explain the reasons why you chose your particular analytical approach and summarize your findings.
e) Would the model you employed in part (a) have been appropriate if the dependent variable had been number of ideas generated per person?

19.3. Refer to Problem 19.2. Conduct the Tukey test to determine whether the additive model (19.1) is appropriate. Use a level of significance of .01. If the additive model were not appropriate, what might you do?

19.4. A college computer service operates four resource centers on the campus. Each contains one time sharing terminal and one each of two types of electronic calculators. During a recent week, the machines were in use for the following numbers of hours:

	Location			
Type of Equipment	W	X	Y	Z
Time sharing terminal	65.3	68.2	45.4	61.1
Electronic calculator A	39.1	38.4	21.0	37.2
Electronic calculator B	28.4	27.7	12.3	28.7

a) State the model you will employ to analyze the effects of resource center location and type of equipment on machine usage.
b) Test whether the two factors have main effects. Use a level of significance of .025 for each test.
c) Give an upper bound for the family level of significance.
d) Analyze the factor effects. Explain the reasons why you chose your particular analytical approach and summarize your findings.

19.5. Refer to Problem 19.4. Conduct the Tukey test to determine whether the additive model (19.1) is appropriate. Use a level of significance of .01. If the additive model were not appropriate, what might you do?

19.6. A survey analyst possessed data for a two-factor study where the treatment cells contained unequal sample sizes. He used a table of random digits to eliminate randomly enough observations so that equal sample sizes existed in all cells. He then analyzed the data by the usual two-factor methods. Comment on this procedure.

19.7. Refer to Problem 17.9. Suppose that the observations $Y_{115} = +1$, $Y_{222} = +3$, and $Y_{224} = +6$ were missing. (Note that Y_{115} is subject 5 for high eye contact, male subject; Y_{222} is subject 2 for low eye contact, female subject; etc.) Use the method of unweighted means to obtain an approximate analysis of variance. Test for main effects and interactions, using a level of significance of .05 each time.

19.8. a) Find the expected mean squares for a two-factor study where factor *A* has random effects and factor *B* fixed effects. Use rule (19.26).
b) Develop the appropriate test statistics.

19.9. A survey statistician has commented: "I am rather suspicious of uses of random and mixed effects ANOVA models. Seldom are the factor levels chosen by a random mechanism from a known population." Comment.

19.10. Establish the correspondence in (19.47b).

19.11. Refer to the full model (19.50) for the Castle Bakery example. Develop the reduced model for testing for factor *B* effects.

19.12. Refer to Problem 17.10.
a) Set up the regression model counterpart to the ANOVA model. Use the type of indicator variables in (19.56).
b) State the correspondences between the regression model parameters and the ANOVA model parameters.
c) Obtain the statistics for testing for the presence of main effects and interactions by means of the regression approach. Are they the same as those obtained in Problem 17.10b?
d) Obtain an interval estimate by the regression approach for the contrast:

$$L = \mu_{1.} - \mu_{2.}$$

where $\mu_{1.}$ is the mean response for small-scale systems experience only and $\mu_{2.}$ is the mean response for large-scale systems experience. Use a 95 percent confidence coefficient.

19.13. Refer to Problem 19.7.
a) Set up the regression model counterpart to the ANOVA model. Use the type of indicator variables in (19.56).
b) State the correspondences between the regression model parameters and the ANOVA model parameters.
c) Obtain the statistics for testing for the presence of main effects and interactions by means of the regression approach.
d) Conduct the tests, each with a level of significance of .01.
e) Obtain an interval estimate by the regression approach for the contrast:

$$L = \mu_{1.} - \mu_{2.}$$

where $\mu_{1.}$ is the mean response for high eye contact, etc. Use a 97.5 percent confidence coefficient.

CITED REFERENCE

19.1. Federer, W. T., and Zelen, M. "Analysis of Multifactor Classifications with Unequal Numbers of Observations," *Biometrics*, Vol. 22 (1966), pp. 525–52.

20

Multifactor Studies

When three or more factors are studied simultaneously, the model and analysis employed are straightforward extensions of the two-factor case. We shall illustrate the nature of the extensions with reference to the three-factor case. Ordinarily, computer packages will be utilized for performing the needed calculations for multifactor studies involving three or more factors. For completeness, however, we shall present the necessary computational formulas for three-factor studies.

20.1 MODEL I (FIXED EFFECTS) FOR THREE-FACTOR STUDIES

Notation

Three factors, A, B, and C, are investigated at a, b, and c levels respectively. The mean response for the treatment when factor A is at the ith level ($i = 1, \ldots, a$), factor B is at the jth level ($j = 1, \ldots, b$), and factor C is at the kth level ($k = 1, \ldots, c$) is denoted by μ_{ijk}. The number of observations for each treatment is assumed to be constant, denoted by n. We assume $n \geq 2$.

The mean response when A is at the ith level and B is at the jth level is denoted by $\mu_{ij.}$, and the like. We define:

(20.1a)
$$\mu_{ij.} = \frac{\sum_k \mu_{ijk}}{c}$$

(20.1b)
$$\mu_{i.k} = \frac{\sum_j \mu_{ijk}}{b}$$

(20.1c)
$$\mu_{.jk} = \frac{\sum_i \mu_{ijk}}{a}$$

The mean response when A is at the ith level is denoted by $\mu_{i..}$, and similarly for the other factor level means. We define:

(20.2a)
$$\mu_{i..} = \frac{\sum_j \sum_k \mu_{ijk}}{bc}$$

(20.2b)
$$\mu_{.j.} = \frac{\sum_i \sum_k \mu_{ijk}}{ac}$$

(20.2c)
$$\mu_{..k} = \frac{\sum_i \sum_j \mu_{ijk}}{ab}$$

Finally, the overall mean response is:

(20.3)
$$\mu_{...} = \frac{\sum_i \sum_j \sum_k \mu_{ijk}}{abc}$$

Example

To illustrate the meaning of the model terms for a three-factor analysis of variance model, we shall consider a study of the effects of sex, age, and intelligence level of college graduates on learning of a complex task. Sex is factor A and has $a = 2$ levels (male, female). Age is factor B and is defined in terms of $b = 3$ levels (young, middle, old). Finally, intelligence is factor C and is defined in terms of $c = 2$ levels (high I.Q., normal I.Q.). Table 20.1 shows the treatment means for all factor level combinations, as well as their notational representations. Also shown in Table 20.1 are the various means of the μ_{ijk}. We shall refer repeatedly to this illustration as we explain the model terms for a three-factor study.

Specific Effects

Specific effects are defined correspondingly to the two-factor case. In our illustration, the specific effect of male sex for old persons of high I.Q., denoted by $\alpha_{1(31)}$, is:

$$\alpha_{1(31)} = \mu_{131} - \mu_{.31} = 18 - 16 = 2$$

This specific effect indicates that the mean learning time for old men of high I.Q. is two minutes higher than the mean learning time for all old persons of high I.Q. We thus define the *specific effect* of factor A at the ith level, when B is at the jth level and C at the kth level, as:

(20.4a)
$$\alpha_{i(jk)} = \mu_{ijk} - \mu_{.jk}$$

TABLE 20.1

Mean Learning Times according to Sex, Age, and Intelligence (in minutes)

Factor A—Sex	Intelligence (factor C) and Age (factor B)											
	$k=1$ *High I.Q.*				$k=2$ *Normal I.Q.*				*Average*			
	$j=1$ *Young*	$j=2$ *Middle*	$j=3$ *Old*	*Average*	$j=1$ *Young*	$j=2$ *Middle*	$j=3$ *Old*	*Average*	$j=1$ *Young*	$j=2$ *Middle*	$j=3$ *Old*	*Average*
$i=1$ Male	9 (μ_{111})	12 (μ_{121})	18 (μ_{131})	13 ($\mu_{1.1}$)	19 (μ_{112})	20 (μ_{122})	21 (μ_{132})	20 ($\mu_{1.2}$)	14 ($\mu_{11.}$)	16 ($\mu_{12.}$)	19.5 ($\mu_{13.}$)	16.5 ($\mu_{1..}$)
$i=2$ Female	9 (μ_{211})	10 (μ_{221})	14 (μ_{231})	11 ($\mu_{2.1}$)	19 (μ_{212})	20 (μ_{222})	21 (μ_{232})	20 ($\mu_{2.2}$)	14 ($\mu_{21.}$)	15 ($\mu_{22.}$)	17.5 ($\mu_{23.}$)	15.5 ($\mu_{2..}$)
Average	9 ($\mu_{.11}$)	11 ($\mu_{.21}$)	16 ($\mu_{.31}$)	12 ($\mu_{..1}$)	19 ($\mu_{.12}$)	20 ($\mu_{.22}$)	21 ($\mu_{.32}$)	20 ($\mu_{..2}$)	14 ($\mu_{.1.}$)	15.5 ($\mu_{.2.}$)	18.5 ($\mu_{.3.}$)	16 ($\mu_{...}$)

Similarly, we define the specific effects of factor B and factor C levels as follows:

(20.4b)
$$\beta_{j(ik)} = \mu_{ijk} - \mu_{i.k}$$

(20.4c)
$$\gamma_{k(ij)} = \mu_{ijk} - \mu_{ij.}$$

Main Effects

Like in the two-factor case, the main effect of a factor level in a three-factor study is an average of the specific effects for that factor level. Thus, in our illustration the main effect of male sex, denoted by α_1, is:

$$\alpha_1 = \frac{(9-9)+(12-11)+(18-16)+(19-19)+(20-20)+(21-21)}{6}$$

$$= .5$$

or symbolically:

$$\alpha_1 = \frac{\alpha_{1(11)} + \alpha_{1(21)} + \alpha_{1(31)} + \alpha_{1(12)} + \alpha_{1(22)} + \alpha_{1(32)}}{6}$$

It follows at once from the definition of the specific effects in (20.4a) that:

$$\alpha_1 = \mu_{1..} - \mu_{...}$$

For our illustration, we have:

$$\alpha_1 = \mu_{1..} - \mu_{...} = 16.5 - 16 = .5$$

which is the same result we obtained by averaging the specific effects.

Thus, the *main effect* of the ith level of factor A is defined:

(20.5a)
$$\alpha_i = \frac{\sum_j \sum_k \alpha_{i(jk)}}{bc} = \mu_{i..} - \mu_{...}$$

Similarly, we define the main effect of the jth level of B:

(20.5b)
$$\beta_j = \frac{\sum_i \sum_k \beta_{j(ik)}}{ac} = \mu_{.j.} - \mu_{...}$$

and the main effect of the kth level of C:

(20.5c)
$$\gamma_k = \frac{\sum_i \sum_j \gamma_{k(ij)}}{ab} = \mu_{..k} - \mu_{...}$$

It follows from these definitions that the sums of the main effects are zero:

(20.6)
$$\sum_i \alpha_i = \sum_j \beta_j = \sum_k \gamma_k = 0$$

Specific Two-Factor Interactions

In a two-factor study, the interaction of the ith level of factor A with the jth level of factor B is defined as the difference between μ_{ij} and the value $\mu_{..} + \alpha_i + \beta_j$ which would be expected if the two factors were additive, in other words:

$$(\alpha\beta)_{ij} = \mu_{ij} - (\mu_{..} + \alpha_i + \beta_j)$$

An algebraically identical definition is:

$$(\alpha\beta)_{ij} = \mu_{ij} - \mu_{i.} - \mu_{.j} + \mu_{..}$$

In a three-factor study, the *specific two-factor interaction* of the ith level of factor A with the jth level of factor B when factor C is at the kth level, denoted by $(\alpha\beta)_{ij(k)}$, is defined in identical fashion:

(20.7a) $$(\alpha\beta)_{ij(k)} = \mu_{ijk} - \mu_{i.k} - \mu_{.jk} + \mu_{..k}$$

The third subscript simply shows that factor C is at the kth level, Thus, $(\alpha\beta)_{ij(k)}$ has the interpretation of a two-factor interaction in a two-factor study with the restriction that the interrelations between factors A and B are only considered when factor C is at the kth level.

For our example in Table 20.1, we have, for instance:

$$(\alpha\beta)_{11(1)} = 9 - 13 - 9 + 12 = -1$$

This interaction is the ordinary two-factor interaction between sex at the level male and age at the level young, when intelligence is at the high level.

In similar fashion, we define the specific interactions:

(20.7b) $$(\alpha\gamma)_{ik(j)} = \mu_{ijk} - \mu_{ij.} - \mu_{.jk} + \mu_{.j.}$$

(20.7c) $$(\beta\gamma)_{jk(i)} = \mu_{ijk} - \mu_{ij.} - \mu_{i.k} + \mu_{i..}$$

Main Two-Factor Interactions

The *main two-factor interaction* between factor A at the ith level and factor B at the jth level is simply the average of the specific interactions $(\alpha\beta)_{ij(k)}$ over all levels of factor C. Denoted as before by $(\alpha\beta)_{ij}$, it is defined:

(20.8a) $$(\alpha\beta)_{ij} = \frac{\sum_k (\alpha\beta)_{ij(k)}}{c}$$

For our illustration in Table 20.1, we have for instance:

$$(\alpha\beta)_{11} = \frac{(9 - 13 - 9 + 12) + (19 - 20 - 19 + 20)}{2} = -.5$$

It may readily be shown that this main interaction can also be expressed as follows:

$$(\alpha\beta)_{ij} = \mu_{ij.} - \mu_{i..} - \mu_{.j.} + \mu_{...} \tag{20.8b}$$

Thus, we have for our illustration:

$$(\alpha\beta)_{11} = 14 - 16.5 - 14 + 16 = -.5$$

which is the same result as obtained before.

Formula (20.8b) indicates that the two-factor interaction $(\alpha\beta)_{ij}$ in a three-factor study may be viewed as the equivalent of that in a two-factor study except that all means are averaged over factor C. Often, however, it will be more useful to interpret $(\alpha\beta)_{ij}$ from the definitional form (20.8a) as the average of the specific interactions $(\alpha\beta)_{ij(k)}$. The latter approach is particularly helpful in recognizing that factors A and B may interact even though $(\alpha\beta)_{ij}$ is zero for all combinations of i and j. This can occur if the specific interactions $(\alpha\beta)_{ij(k)}$ are not zero but just happen to average out to zero for all i, j combinations. As long as we remember that the $(\alpha\beta)_{ij}$ are averages of specific interactions between A and B at different levels of factor C, we can be alert to the fact that $(\alpha\beta)_{ij} = 0$ does not necessarily imply that all specific interactions $(\alpha\beta)_{ij(k)}$ are zero.

In corresponding fashion, we define the AC and BC main interactions:

$$(\alpha\gamma)_{ik} = \frac{\sum_j (\alpha\gamma)_{ik(j)}}{b} = \mu_{i.k} - \mu_{i..} - \mu_{..k} + \mu_{...} \tag{20.8c}$$

$$(\beta\gamma)_{jk} = \frac{\sum_i (\beta\gamma)_{jk(i)}}{a} = \mu_{.jk} - \mu_{.j.} - \mu_{..k} + \mu_{...} \tag{20.8d}$$

The main interactions $(\alpha\beta)_{ij}$, $(\alpha\gamma)_{ik}$, and $(\beta\gamma)_{jk}$ are usually called *two-factor interactions* or *first-order interactions.* It can readily be shown that the sums of the first-order interactions over each subscript are zero:

$$\sum_i (\alpha\beta)_{ij} = \sum_j (\alpha\beta)_{ij} = 0 \tag{20.9a}$$

$$\sum_i (\alpha\gamma)_{ik} = \sum_k (\alpha\gamma)_{ik} = 0 \tag{20.9b}$$

$$\sum_j (\beta\gamma)_{jk} = \sum_k (\beta\gamma)_{jk} = 0 \tag{20.9c}$$

Three-Factor Interactions

Just as in a two-factor study, where the interaction between the ith level of factor A and the jth level of factor B is defined as the difference between the treatment mean μ_{ij} and the value that would be expected if the factors were additive, so in a three-factor study the three-factor interaction $(\alpha\beta\gamma)_{ijk}$ is defined as the difference between the treatment mean μ_{ijk} and the value that would be expected if main effects and first-order interactions were sufficient to account for all factor effects. The value that would be expected from main effects and first-order interactions when A is at the ith level, B at the jth level, and C at the kth level, is:

(20.10) $$\mu_{...} + \alpha_i + \beta_j + \gamma_k + (\alpha\beta)_{ij} + (\alpha\gamma)_{ik} + (\beta\gamma)_{jk}$$

Hence, the *three-factor interaction* $(\alpha\beta\gamma)_{ijk}$, also called the *second-order interaction*, is defined as:

(20.11a) $$(\alpha\beta\gamma)_{ijk} = \mu_{ijk} - [\mu_{...} + \alpha_i + \beta_j + \gamma_k + (\alpha\beta)_{ij} + (\alpha\gamma)_{ik} + (\beta\gamma)_{jk}]$$

or equivalently:

(20.11b) $$(\alpha\beta\gamma)_{ijk} = \mu_{ijk} - \mu_{ij.} - \mu_{i.k} - \mu_{.jk} + \mu_{i..} + \mu_{.j.} + \mu_{..k} - \mu_{...}$$

From the definition of the three-factor interactions, it follows that they sum to zero when added over any index:

(20.12) $$\sum_i (\alpha\beta\gamma)_{ijk} = \sum_j (\alpha\beta\gamma)_{ijk} = \sum_k (\alpha\beta\gamma)_{ijk} = 0$$

If *all* $(\alpha\beta\gamma)_{ijk}$ are zero, we say that there are no three-factor interactions between factors A, B, and C. If some $(\alpha\beta\gamma)_{ijk}$ are not zero, we say that three-factor interactions are present.

Let us find the three-factor interaction $(\alpha\beta\gamma)_{111}$ for our example in Table 20.1. We require the following terms:

$$\mu_{...} = 16 \qquad (\alpha\beta)_{11} = 14 - 16.5 - 14 + 16 = -.5$$
$$\alpha_1 = 16.5 - 16 = .5 \qquad (\alpha\gamma)_{11} = 13 - 16.5 - 12 + 16 = .5$$
$$\beta_1 = 14 - 16 = -2 \qquad (\beta\gamma)_{11} = 9 - 14 - 12 + 16 = -1$$
$$\gamma_1 = 12 - 16 = -4 \qquad \mu_{111} = 9$$

Hence, we have:

$$(\alpha\beta\gamma)_{111} = 9 - (16 + .5 - 2 - 4 - .5 + .5 - 1) = -.5$$

Since $(\alpha\beta\gamma)_{111}$ is not zero, we know at once that three-factor interactions are present in this example.

Note

The second-order interaction $(\alpha\beta\gamma)_{ijk}$ can also be expressed as the difference between the specific two-factor interaction $(\alpha\beta)_{ij(k)}$ and the average two-factor interaction $(\alpha\beta)_{ij}$:

(20.13a) $$(\alpha\beta\gamma)_{ijk} = (\alpha\beta)_{ij(k)} - (\alpha\beta)_{ij}$$

Thus in our illustration, we found $(\alpha\beta)_{11(1)} = -1$ and $(\alpha\beta)_{11} = -.5$, so that $(\alpha\beta\gamma)_{111} = -1 - (-.5) = -.5$, as we saw above.

Because of the symmetrical structure of three-factor interactions, they can also be expressed in the following ways:

(20.13b) $$(\alpha\beta\gamma)_{ijk} = (\alpha\gamma)_{ik(j)} - (\alpha\gamma)_{ik}$$

(20.13c) $$(\alpha\beta\gamma)_{ijk} = (\beta\gamma)_{jk(i)} - (\beta\gamma)_{jk}$$

Significance of Three-Factor Interactions

The presence of three-factor interactions implies that at least some of the specific two-factor interactions for any two factors differ, depending on the level of the third factor. This can be seen from (20.13), for if the specific interactions were equal, the deviations of the specific interactions from the average would all be zero. Thus, if three-factor interactions are present, the interactions between any two factors need be studied separately for each level of the third factor.

Admittedly, this explanation of the significance of three-factor interactions is somewhat abstruse. To shed more light on the significance of three-factor interactions, we shall examine some examples by means of tables and graphs.

Example 1. Table 20.2 contains the specific sex-age interactions for each intelligence group for our illustration in Table 20.1. Note that these specific interactions differ for the two intelligence groups, implying the presence of three-factor interactions.

TABLE 20.2

Specific Sex-Age Interactions by I.Q. Level

	$j=1$ *Young*	$j=2$ *Middle*	$j=3$ *Old*
High I.Q. ($k=1$):			
$i=1$ Male	-1	0	$+1$
$i=2$ Female	$+1$	0	-1
Normal I.Q. ($k=2$):			
$i=1$ Male	0	0	0
$i=2$ Female	0	0	0

Example 2. Figure 20.1 illustrates a case where there are A, B, and C main effects but no interactions of any kind. The AB response curves are plotted against C in Figure 20.1a. Here, the main A effects are reflected by the slopes of the AB curves not being zero. The main B effects are reflected by the differences in the heights of the two AB curves within each panel, and the main C effects are reflected by the corresponding curves in the two panels being at different heights.

The absence of AB interactions is shown by the parallel response curves in each panel. We know from our discussion of two-factor analysis that parallel response curves imply absence of interactions. Here, the parallel response curves within each panel imply the absence of specific AB interactions, that is, that all specific AB interactions $(\alpha\beta)_{ij(k)} = 0$. This in turn implies:

FIGURE 20.1

A, *B*, and *C* Main Effects and No Interactions

(a)

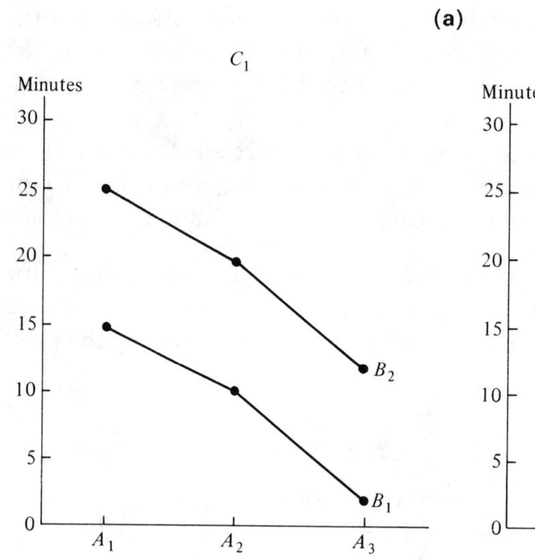

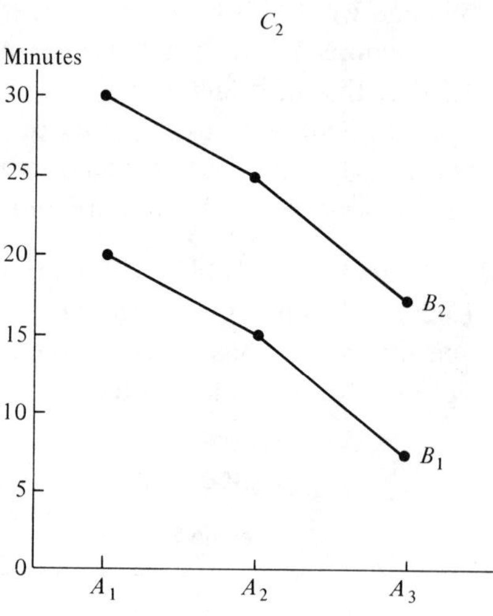

(b)

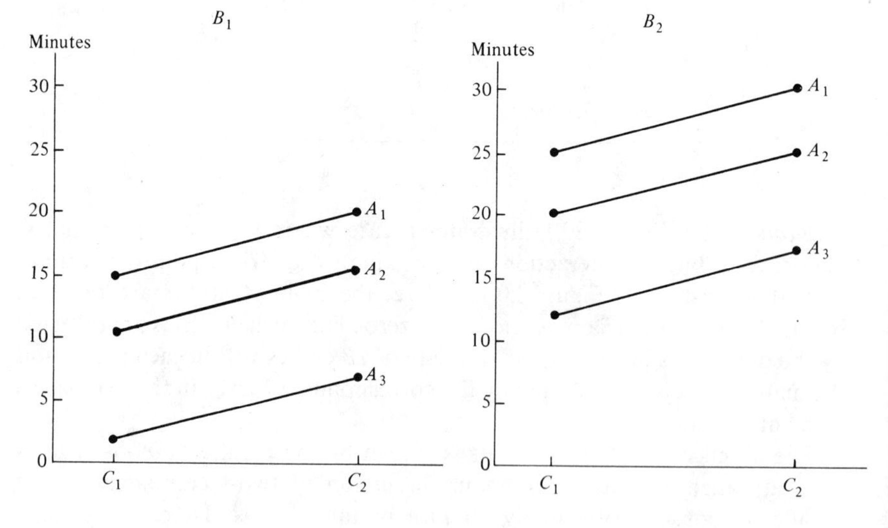

1. The main AB interactions $(\alpha\beta)_{ij}$ are zero. This follows from (20.8a), since the averages of specific interactions which are all zero must also be zero.
2. The ABC interactions $(\alpha\beta\gamma)_{ijk}$ are zero. This follows at once from the previous point and from (20.13a).

The absences of BC and AC interactions in this example become evident when the BC and AC response curves are plotted against A and B respectively. Thus, in Figure 20.1b, the same data are shown with the AC response curves plotted against B. Note that these curves are parallel in each panel, implying that the main AC interactions are zero.

Example 3. Figure 20.2 illustrates a case where there are A, B, and C main effects and AB interactions, but no other interactions. Note that the two AB response curves in each panel of Figure 20.2a are no longer parallel, indicating the presence of AB interactions. However, the upper curves in the

FIGURE 20.2

A, *B*, and *C* Main Effects and *AB* Interactions

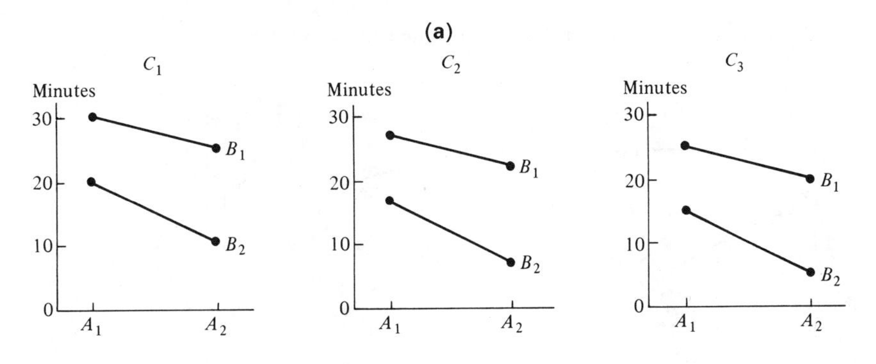

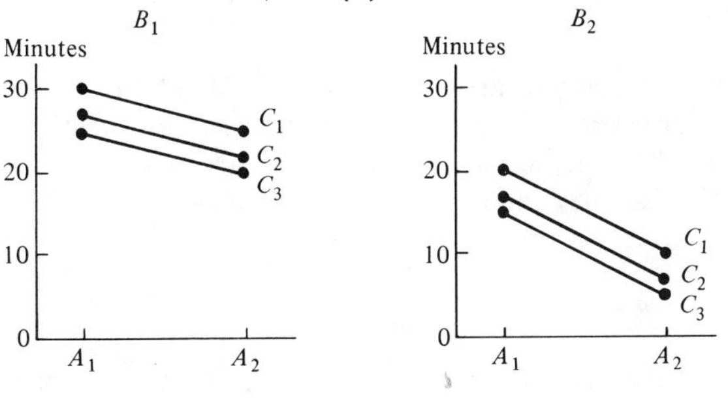

three panels are parallel, as are the lower curves. This implies that if AC curves are plotted against B, the AC response curves within each panel will be parallel. This is shown in Figure 20.2b, which contains the same data as in Figure 20.2a. The parallelism of the AC response curves within each panel in Figure 20.2b in turn implies by the earlier reasoning that there are no AC interactions, as well as no ABC interactions, that is, all $(\alpha\gamma)_{ik} = 0$ and all $(\alpha\beta\gamma)_{ijk} = 0$.

Thus, if the AB (AC, BC) response curves corresponding to any given level of factor C (B, A) are parallel for all levels of factor C (B, A), even though the response curves are not parallel within a panel, it follows that there are no three-factor interactions.

Example 4. It does not follow, however, that lack of parallelism either within or between panels implies the presence of three-factor interactions. Figure 20.3 portrays a case where main A, B, C effects and main two-factor interactions AB and AC are present, but no three-factor interactions exist. Yet there are no parallel response curves either within or between panels.

FIGURE 20.3

A, B, and *C* Main Effects and *AB* and *AC* Interactions

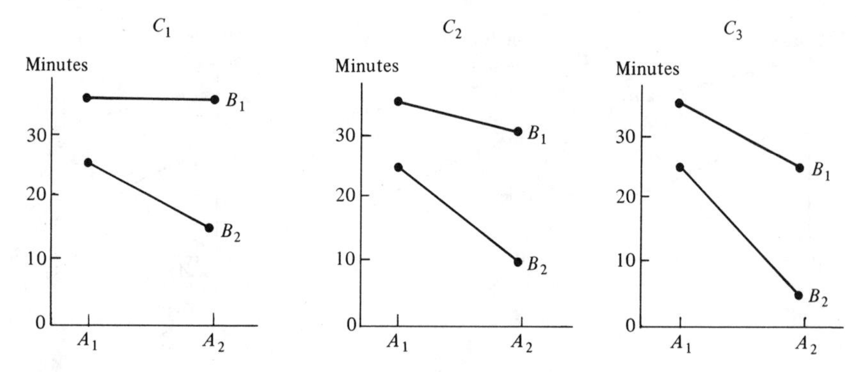

Note

If three-factor interactions are difficult to understand, higher-order interactions such as four-factor interactions are yet more abstruse. Fortunately, it is often found in practice that these higher-order interactions are quite small or nonexistent. When this is the case, they can be disregarded in the analysis of factor effects.

Model

Let Y_{ijkm} be the mth observation ($m = 1, \ldots, n$) for the treatment consisting of the ith level of A ($i = 1, ., \ldots a$), the jth level of B ($j = 1, \ldots, b$), and the kth level of C ($k = 1, \ldots, c$). Thus, the total number of observations in the study is:

(20.14)
$$n_T = nabc$$

Neglecting the factorial structure of the study, the fixed effects ANOVA model for a three-factor study is:

(20.15)
$$Y_{ijkm} = \mu_{ijk} + \varepsilon_{ijkm}$$

where:

μ_{ijk} are parameters
ε_{ijkm} are independent $N(0, \sigma^2)$
$i = 1, \ldots, a; j = 1, \ldots, b; k = 1, \ldots, c; m = 1, \ldots, n$

We incorporate the factorial structure into the model by expressing the mean μ_{ijk} in terms of the various factor effects. From the three-factor interaction definition (20.11a), we have:

(20.16)
$$\mu_{ijk} = \mu_{\ldots} + \alpha_i + \beta_j + \gamma_k + (\alpha\beta)_{ij} + (\alpha\gamma)_{ik} + (\beta\gamma)_{jk} + (\alpha\beta\gamma)_{ijk}$$

Hence, the fixed effects ANOVA model for a three-factor study is:

(20.17)
$$Y_{ijkm} = \mu_{\ldots} + \alpha_i + \beta_j + \gamma_k + (\alpha\beta)_{ij} + (\alpha\gamma)_{ik} + (\beta\gamma)_{jk} + (\alpha\beta\gamma)_{ijk} + \varepsilon_{ijkm}$$

where:

ε_{ijkm} are independent $N(0, \sigma^2)$
$\alpha_i, \beta_j, \gamma_k, (\alpha\beta)_{ij}, (\alpha\gamma)_{ik}, (\beta\gamma)_{jk}, (\alpha\beta\gamma)_{ijk}$ are constants subject to the restrictions:

$$\sum_i \alpha_i = \sum_j \beta_j = \sum_k \gamma_k = 0$$

$$\sum_i (\alpha\beta)_{ij} = \sum_j (\alpha\beta)_{ij} = \sum_i (\alpha\gamma)_{ik} = \sum_k (\alpha\gamma)_{ik} = \sum_j (\beta\gamma)_{jk} = \sum_k (\beta\gamma)_{jk} = 0$$

$$\sum_i (\alpha\beta\gamma)_{ijk} = \sum_j (\alpha\beta\gamma)_{ijk} = \sum_k (\alpha\beta\gamma)_{ijk} = 0$$

20.2 ANALYSIS OF VARIANCE

Sample Notation

The notation for sample totals and means is a straightforward extension of that for two-factor studies. As usual, a dot in the subscript indicates aggregation or averaging over the index represented by the dot. We have:

(20.18a) $$Y_{ijk.} = \sum_m Y_{ijkm} \qquad \bar{Y}_{ijk.} = \frac{Y_{ijk.}}{n}$$

(20.18b) $$Y_{ij..} = \sum_k \sum_m Y_{ijkm} \qquad \bar{Y}_{ij..} = \frac{Y_{ij..}}{cn}$$

(20.18c) $$Y_{i.k.} = \sum_j \sum_m Y_{ijkm} \qquad \bar{Y}_{i.k.} = \frac{Y_{i.k.}}{bn}$$

(20.18d) $$Y_{.jk.} = \sum_i \sum_m Y_{ijkm} \qquad \bar{Y}_{.jk.} = \frac{Y_{.jk.}}{an}$$

(20.18e) $$Y_{i...} = \sum_j \sum_k \sum_m Y_{ijkm} \qquad \bar{Y}_{i...} = \frac{Y_{i...}}{bcn}$$

(20.18f) $$Y_{.j..} = \sum_i \sum_k \sum_m Y_{ijkm} \qquad \bar{Y}_{.j..} = \frac{Y_{.j..}}{acn}$$

(20.18g) $$Y_{..k.} = \sum_i \sum_j \sum_m Y_{ijkm} \qquad \bar{Y}_{..k.} = \frac{Y_{..k.}}{abn}$$

(20.18h) $$Y_{....} = \sum_i \sum_j \sum_k \sum_m Y_{ijkm} \qquad \bar{Y}_{....} = \frac{Y_{....}}{abcn}$$

Table 20.3 illustrates this notation for a simple example. Here there are three factors, namely age of subject, education of subject, and size of firm in which subject is employed. Each factor has two levels, and there are two replications of the experiment. Tables 20.3a, b, and c show respectively the observations, totals, and means, together with the corresponding notation.

Breakdown of Total Sum of Squares

Neglecting the factorial structure of the study and simply considering it to contain *abc* treatments, we obtain the usual breakdown of the total sum of squares:

(20.19) $$SSTO = SSTR + SSE$$

where:

(20.19a) $$SSTO = \sum_i \sum_j \sum_k \sum_m (Y_{ijkm} - \bar{Y}_{....})^2$$

(20.19b) $$SSTR = n \sum_i \sum_j \sum_k (\bar{Y}_{ijk.} - \bar{Y}_{....})^2$$

(20.19c) $$SSE = \sum_i \sum_j \sum_k \sum_m (Y_{ijkm} - \bar{Y}_{ijk.})^2$$

TABLE 20.3

Three-Factor Study on Facility of Problem Solving

(a) Observations

	Size of Firm	
	Small ($k=1$)	*Large* ($k=2$)
Young age ($j=1$):		
College education ($i=1$)	6 (Y_{1111})	13 (Y_{1121})
	10 (Y_{1112})	15 (Y_{1122})
No college ($i=2$)	10 (Y_{2111})	15 (Y_{2121})
	10 (Y_{2112})	17 (Y_{2122})
Older age ($j=2$):		
College education ($i=1$)	5 (Y_{1211})	11 (Y_{1221})
	7 (Y_{1212})	13 (Y_{1222})
No college ($i=2$)	6 (Y_{2211})	9 (Y_{2221})
	10 (Y_{2212})	11 (Y_{2222})

(b) Totals

	$k=1$	$k=2$	*All k*
$j=1$:			
$i=1$	16 ($Y_{111.}$)	28 ($Y_{112.}$)	44 ($Y_{11..}$)
$i=2$	20 ($Y_{211.}$)	32 ($Y_{212.}$)	52 ($Y_{21..}$)
All i	36 ($Y_{.11.}$)	60 ($Y_{.12.}$)	96 ($Y_{.1..}$)
$j=2$:			
$i=1$	12 ($Y_{121.}$)	24 ($Y_{122.}$)	36 ($Y_{12..}$)
$i=2$	16 ($Y_{221.}$)	20 ($Y_{222.}$)	36 ($Y_{22..}$)
All i	28 ($Y_{.21.}$)	44 ($Y_{.22.}$)	72 ($Y_{.2..}$)
All j:			
$i=1$	28 ($Y_{1.1.}$)	52 ($Y_{1.2.}$)	80 ($Y_{1...}$)
$i=2$	36 ($Y_{2.1.}$)	52 ($Y_{2.2.}$)	88 ($Y_{2...}$)
All i	64 ($Y_{..1.}$)	104 ($Y_{..2.}$)	168 ($Y_{....}$)

(c) *Averages*

	$k=1$	$k=2$	*All k*
$j=1$:			
$i=1$	8 ($\bar{Y}_{111.}$)	14 ($\bar{Y}_{112.}$)	11 ($\bar{Y}_{11..}$)
$i=2$	10 ($\bar{Y}_{211.}$)	16 ($\bar{Y}_{212.}$)	13 ($\bar{Y}_{21..}$)
All i	9 ($\bar{Y}_{.11.}$)	15 ($\bar{Y}_{.12.}$)	12 ($\bar{Y}_{.1..}$)
$j=2$:			
$i=1$	6 ($\bar{Y}_{121.}$)	12 ($\bar{Y}_{122.}$)	9 ($\bar{Y}_{12..}$)
$i=2$	8 ($\bar{Y}_{221.}$)	10 ($\bar{Y}_{222.}$)	9 ($\bar{Y}_{22..}$)
All i	7 ($\bar{Y}_{.21.}$)	11 ($\bar{Y}_{.22.}$)	9 ($\bar{Y}_{.2..}$)
All j:			
$i=1$	7 ($\bar{Y}_{1.1.}$)	13 ($\bar{Y}_{1.2.}$)	10 ($\bar{Y}_{1...}$)
$i=2$	9 ($\bar{Y}_{2.1.}$)	13 ($\bar{Y}_{2.2.}$)	11 ($\bar{Y}_{2...}$)
All i	8 ($\bar{Y}_{..1.}$)	13 ($\bar{Y}_{..2.}$)	10.5 ($\bar{Y}_{....}$)

Consider now the treatment mean deviation $\bar{Y}_{ijk.} - \bar{Y}_{....}$, which appears in *SSTR*. This can be decomposed in terms of the least-squares estimators of the main effects, two-factor interactions, and a three-factor interaction:

$$\underbrace{\bar{Y}_{ijk.} - \bar{Y}_{....}}_{\substack{\text{Treatment}\\\text{mean}\\\text{deviation}}} = \underbrace{\bar{Y}_{i...} - \bar{Y}_{....}}_{A\text{ main effect}} + \underbrace{\bar{Y}_{.j..} - \bar{Y}_{....}}_{B\text{ main effect}} + \underbrace{\bar{Y}_{..k.} - \bar{Y}_{....}}_{C\text{ main effect}}$$

$$+ \underbrace{\bar{Y}_{ij..} - \bar{Y}_{i...} - \bar{Y}_{.j..} + \bar{Y}_{....}}_{AB\text{ interaction effect}}$$

$$+ \underbrace{\bar{Y}_{i.k.} - \bar{Y}_{i...} - \bar{Y}_{..k.} + \bar{Y}_{....}}_{AC\text{ interaction effect}} + \underbrace{\bar{Y}_{.jk.} - \bar{Y}_{.j..} - \bar{Y}_{..k.} + \bar{Y}_{....}}_{BC\text{ interaction effect}}$$

$$+ \underbrace{\bar{Y}_{ijk.} - \bar{Y}_{ij..} - \bar{Y}_{i.k.} - \bar{Y}_{.jk.} + \bar{Y}_{i...} + \bar{Y}_{.j..} + \bar{Y}_{..k.} - \bar{Y}_{....}}_{ABC\text{ interaction effect}}$$

If we square each side and sum over i, j, k, and m, all cross-product terms drop out and we obtain:

(20.20) $$SSTR = SSA + SSB + SSC + SSAB + SSAC + SSBC + SSABC$$

where:

(20.20a) $$SSA = nbc \sum_i (\bar{Y}_{i...} - \bar{Y}_{....})^2$$

(20.20b) $$SSB = nac \sum_j (\bar{Y}_{.j..} - \bar{Y}_{....})^2$$

(20.20c) $$SSC = nab \sum_k (\bar{Y}_{..k.} - \bar{Y}_{....})^2$$

(20.20d) $$SSAB = nc \sum_i \sum_j (\bar{Y}_{ij..} - \bar{Y}_{i...} - \bar{Y}_{.j..} + \bar{Y}_{....})^2$$

(20.20e) $$SSAC = nb \sum_i \sum_k (\bar{Y}_{i.k.} - \bar{Y}_{i...} - \bar{Y}_{..k.} + \bar{Y}_{....})^2$$

(20.20f) $$SSBC = na \sum_j \sum_k (\bar{Y}_{.jk.} - \bar{Y}_{.j..} - \bar{Y}_{..k.} + \bar{Y}_{....})^2$$

(20.20g) $$SSABC = n \sum_i \sum_j \sum_k (\bar{Y}_{ijk.} - \bar{Y}_{ij..} - \bar{Y}_{i.k.} - \bar{Y}_{.jk.} + \bar{Y}_{i...} + \bar{Y}_{.j..} + \bar{Y}_{..k.} - \bar{Y}_{....})^2$$

Combining (20.19) and (20.20), we have thus established that:

(20.21) $$SSTO = SSA + SSB + SSC + SSAB + SSAC + SSBC + SSABC + SSE$$

SSA, *SSB*, and *SSC* are the usual main effect sums of squares. For instance, the larger (absolutely) are the estimated main B effects $(\bar{Y}_{.j..} - \bar{Y}_{....})$, the larger will be *SSB*.

SSAB, *SSAC*, and *SSBC* are the usual two-factor interaction sums of squares. For instance, the larger (absolutely) are the estimated *AB* interactions $(\bar{Y}_{ij..} - \bar{Y}_{i...} - \bar{Y}_{.j..} + \bar{Y}_{....})$, the larger will be *SSAB*.

Finally, *SSABC* is the three-factor interaction sum of squares. The larger (absolutely) are these estimated three-factor interactions, the larger will be *SSABC*.

Computational Formulas. When calculations are to be done by hand, the use of the definitional formulas previously given is cumbersome. The computational formulas for the three-factor case follow the pattern of the two-factor formulas:

(20.22a) $$SSTO = \sum_i \sum_j \sum_k \sum_m Y_{ijkm}^2 - \frac{Y_{....}^2}{nabc}$$

(20.22b) $$SSE = \sum_i \sum_j \sum_k \sum_m Y_{ijkm}^2 - \sum_i \sum_j \sum_k \frac{Y_{ijk.}^2}{n}$$

(20.22c) $$SSA = \frac{\sum_i Y_{i...}^2}{nbc} - \frac{Y_{....}^2}{nabc}$$

(20.22d) $$SSB = \frac{\sum_j Y_{.j..}^2}{nac} - \frac{Y_{....}^2}{nabc}$$

(20.22e) $$SSC = \frac{\sum_k Y_{..k.}^2}{nab} - \frac{Y_{....}^2}{nabc}$$

The two-factor interaction sum of squares *SSAB* can be obtained by working with the means $\bar{Y}_{ij..}$ and treating these *ab* means as constituting a two-factor study. The "treatment sum of squares" for this two-factor study, which we will denote by *SSTR*(*A*, *B*), is as usual:

(20.23) $$SSTR(A, B) = nc \sum_i \sum_j (\bar{Y}_{ij..} - \bar{Y}_{....})^2 = \frac{\sum_i \sum_j Y_{ij..}^2}{nc} - \frac{Y_{....}^2}{nabc}$$

It can be shown that:

(20.24) $$SSTR(A, B) = SSA + SSB + SSAB$$

Hence, we can find *SSAB* by subtraction:

(20.25a) $$SSAB = SSTR(A, B) - SSA - SSB$$

Similarly, we find *SSAC* and *SSBC* as follows:

(20.25b) $$SSAC = SSTR(A, C) - SSA - SSC$$

(20.25c) $$SSBC = SSTR(B, C) - SSB - SSC$$

where:

$$SSTR(A, C) = \frac{\sum_i \sum_k Y_{i.k.}^2}{nb} - \frac{Y_{....}^2}{nabc} \tag{20.25d}$$

$$SSTR(B, C) = \frac{\sum_j \sum_k Y_{.jk.}^2}{na} - \frac{Y_{....}^2}{nabc} \tag{20.25e}$$

The three-factor interaction sum of squares is obtained by subtraction:

$$SSABC = SSTO - SSE - SSA - SSB - SSC - SSAB - SSAC - SSBC \tag{20.26}$$

Note

The computational formulas above can be extended readily if more than three factors are studied simultaneously. The total sum of squares, the error sum of squares, and the main effect sums of squares present no difficulty. Two-factor interaction sums of squares are obtained by aggregating over all other factors, calculating the "treatment sum of squares" for this two-factor situation, and subtracting the main effect sums of squares for these two factors. Three-factor interaction sums of squares are obtained by aggregating over all other factors, calculating the "treatment sum of squares" for this three-factor situation, and then subtracting all lower-order sums of squares. Four-factor and still higher-order interaction sums of squares are obtained by successive extensions of this procedure.

Degrees of Freedom and Mean Squares

Table 20.4 contains the ANOVA table for the three-factor fixed effects model (20.17). The degrees of freedom for main effect and two-factor interaction sums of squares correspond to those for two-factor studies. The number of degrees of freedom associated with $SSABC$ is obtained by subtraction, and corresponds to the number of independent linear relations among all the interaction terms $(\alpha\beta\gamma)_{ijk}$.

The expected mean squares are obtained in the same way as in our earlier derivations. Note from Table 20.4 that MSA, MSB, MSC, $MSAB$, $MSAC$, $MSBC$, and $MSABC$ all have expectations equal to σ^2 if there are no factor effects of the type reflected by the mean square. If there are such effects, each mean square has an expectation exceeding σ^2. As usual, $E(MSE) = \sigma^2$ always. Hence, the test for factor effects consists of comparing the appropriate mean square against MSE, with large values of the ratio indicating the presence of factor effects.

Tests for Factor Effects

The various tests for factor effects all follow the same pattern; we illustrate them with the test for three-factor interactions. The alternatives are:

$$\begin{aligned} &C_1\text{: all } (\alpha\beta\gamma)_{ijk} = 0 \\ &C_2\text{: not all } (\alpha\beta\gamma)_{ijk} \text{ are zero} \end{aligned} \tag{20.27a}$$

TABLE 20.4

ANOVA Table for Three-Factor Fixed Effects Study

Source of Variation	*SS*	*df*	*MS*	*E(MS)*
Between treatments	$SSTR$	$abc-1$	$MSTR$	$\sigma^2 + \frac{n \sum \sum \sum (\mu_{ijk} - \mu_{...})^2}{abc-1}$
Factor A	SSA	$a-1$	MSA	$\sigma^2 + \frac{bcn}{a-1} \sum \alpha_i^2$
Factor B	SSB	$b-1$	MSB	$\sigma^2 + \frac{acn}{b-1} \sum \beta_j^2$
Factor C	SSC	$c-1$	MSC	$\sigma^2 + \frac{abn}{c-1} \sum \gamma_k^2$
AB interactions	$SSAB$	$(a-1)(b-1)$	$MSAB$	$\sigma^2 + \frac{cn}{(a-1)(b-1)} \sum \sum (\alpha\beta)_{ij}^2$
AC interactions	$SSAC$	$(a-1)(c-1)$	$MSAC$	$\sigma^2 + \frac{bn}{(a-1)(c-1)} \sum \sum (\alpha\gamma)_{ik}^2$
BC interactions	$SSBC$	$(b-1)(c-1)$	$MSBC$	$\sigma^2 + \frac{an}{(b-1)(c-1)} \sum \sum (\beta\gamma)_{jk}^2$
ABC interactions	$SSABC$	$(a-1)(b-1)(c-1)$	$MSABC$	$\sigma^2 + \frac{n}{(a-1)(b-1)(c-1)} \sum \sum \sum (\alpha\beta\gamma)_{ijk}^2$
Error	SSE	$abc(n-1)$	MSE	σ^2
Total	$SSTO$	$abcn-1$		

The appropriate test statistic is:

(20.27b)
$$F^* = \frac{MSABC}{MSE}$$

If C_1 holds, F^* follows the F distribution with $(a-1)(b-1)(c-1)$ degrees of freedom for the numerator and $abc(n-1)$ degrees of freedom for the denominator. Hence the decision rule to control the Type I error at α is:

(20.27c)
$$\begin{aligned}&\text{If } F^* \leq F(1-\alpha; (a-1)(b-1)(c-1), (n-1)abc), \text{ conclude } C_1\\&\text{If } F^* > F(1-\alpha; (a-1)(b-1)(c-1), (n-1)abc), \text{ conclude } C_2\end{aligned}$$

Table 20.5 contains the test statistics and appropriate percentiles of the F distribution for the various possible tests for a three-factor study.

TABLE 20.5

Test Statistics for Three-Factor Fixed Effects Study

Alternatives	*Test Statistic*	*Percentile*
C_1: all $\alpha_i = 0$ C_2: not all $\alpha_i = 0$	$F^* = \frac{MSA}{MSE}$	$F(1-\alpha; a-1, (n-1)abc)$
C_1: all $\beta_j = 0$ C_2: not all $\beta_j = 0$	$F^* = \frac{MSB}{MSE}$	$F(1-\alpha; b-1, (n-1)abc)$
C_1: all $\gamma_k = 0$ C_2: not all $\gamma_k = 0$	$F^* = \frac{MSC}{MSE}$	$F(1-\alpha; c-1, (n-1)abc)$
C_1: all $(\alpha\beta)_{ij} = 0$ C_2: not all $(\alpha\beta)_{ij} = 0$	$F^* = \frac{MSAB}{MSE}$	$F(1-\alpha; (a-1)(b-1), (n-1)abc)$
C_1: all $(\alpha\gamma)_{ik} = 0$ C_2: not all $(\alpha\gamma)_{ik} = 0$	$F^* = \frac{MSAC}{MSE}$	$F(1-\alpha; (a-1)(c-1), (n-1)abc)$
C_1: all $(\beta\gamma)_{jk} = 0$ C_2: not all $(\beta\gamma)_{jk} = 0$	$F^* = \frac{MSBC}{MSE}$	$F(1-\alpha; (b-1)(c-1), (n-1)abc)$
C_1: all $(\alpha\beta\gamma)_{ijk} = 0$ C_2: not all $(\alpha\beta\gamma)_{ijk} = 0$	$F^* = \frac{MSABC}{MSE}$	$F(1-\alpha; (a-1)(b-1)(c-1), (n-1)abc)$

Comments

1. The power of the factor effects tests can be obtained from Table A–8 in the manner described for one-factor and two-factor studies. The noncentrality parameter ϕ for a given test is defined as follows:

(20.28)
$$\phi = \frac{1}{\sigma}\left[\frac{\text{numerator of second term in } E(MS) \text{ in Table 20.4}}{\text{denominator of second term in } E(MS) \text{ plus 1}}\right]^{1/2}$$

Thus, for testing the existence of three-factor interactions, we have:

$$\phi = \frac{1}{\sigma}\left[\frac{n \sum\sum\sum (\alpha\beta\gamma)_{ijk}^2}{(a-1)(b-1)(c-1)+1}\right]^{1/2}$$

2. The Kimball inequality (see p. 582) for the family level of significance α for the combined set of seven tests on main effects, two-factor interactions, and three-factor interactions is:

$$\alpha < 1 - (1 - \alpha_1)(1 - \alpha_2) \cdots (1 - \alpha_7) \tag{20.29}$$

where α_i is the level of significance for the ith test.

3. If the three-factor interactions (and also perhaps some sets of two-factor interactions) are zero, the question arises whether the corresponding sums of squares should be pooled with the error sum of squares. Our earlier discussion on pooling is applicable here also.

4. If there is only one observation per treatment in a three-factor fixed effects study, analysis of variance tests can only be conducted if it is possible to assume that some interactions are zero. Usually, the interactions most likely to be zero are the three-factor interactions. If it is possible to assume that all three-factor interactions are zero, $MSABC$ has expectation σ^2 and is used as the error mean square MSE. All mean squares are calculated in the usual manner, except that $n = 1$.

20.3 ANALYSIS OF FACTOR EFFECTS

No new problems are encountered in the analysis of factor effects for fixed effects three-factor studies.

Contrasts among Factor Level Means

Suppose that a contrast involving the factor level means $\mu_{i..}$ is to be estimated:

$$L = \sum c_i \mu_{i..} \qquad \text{where } \sum c_i = 0 \tag{20.30}$$

The unbiased estimator of L we shall employ is:

$$\hat{L} = \sum c_i \bar{Y}_{i...} \tag{20.31}$$

The variance of this estimator is:

$$\sigma^2(\hat{L}) = \sum c_i^2 \sigma^2(\bar{Y}_{i...}) = \frac{\sigma^2}{nbc} \sum c_i^2 \tag{20.32}$$

This result follows because each $\bar{Y}_{i...}$ is based on nbc observations, and all $\bar{Y}_{i...}$ are independent because of the independence of the error terms ε_{ijkm}.

The estimated variance of $\hat{L}$ is:

$$s^2(\hat{L}) = \frac{MSE}{nbc} \sum c_i^2 \tag{20.33}$$

and the $1 - \alpha$ confidence interval for L is:

$$\hat{L} - t(1 - \alpha/2; (n-1)abc)s(\hat{L}) \leq L \leq \hat{L} + t(1 - \alpha/2; (n-1)abc)s(\hat{L}) \tag{20.34}$$

If a number of contrasts among the factor level means $\mu_{i..}$ are to be estimated and assurance is to be provided by a family confidence coefficient, the t multiple in (20.34) is simply replaced by the T, S, or B multiple defined as follows:

(20.35a) Tukey procedure (for pairwise comparisons) $$T = \frac{1}{\sqrt{2}} q(1 - \alpha; a, (n-1)abc)$$

(20.35b) Scheffé procedure $$S^2 = (a-1)F(1 - \alpha; a-1, (n-1)abc)$$

(20.35c) Bonferroni procedure $$B = t(1 - \alpha/2s; (n-1)abc)$$

Contrasts based on the factor level means $\mu_{.j.}$ or $\mu_{..k}$ are estimated in corresponding fashion.

Contrasts among Treatment Means

When interactions are present, contrasts among the treatment means μ_{ijk} may be desired. Let, as usual, L denote such a constrast:

(20.36) $$L = \sum \sum \sum c_{ijk} \mu_{ijk} \quad \text{where } \sum \sum \sum c_{ijk} = 0$$

An unbiased estimator of L is:

(20.37) $$\hat{L} = \sum \sum \sum c_{ijk} \bar{Y}_{ijk.}$$

for which the estimated variance is:

(20.38) $$s^2(\hat{L}) = \frac{MSE}{n} \sum \sum \sum c_{ijk}^2$$

The $1 - \alpha$ confidence interval for L in (20.34) is appropriate, with $\hat{L}$ and $s(\hat{L})$ defined in (20.37) and (20.38) respectively. For multiple comparisons, the t multiple should be replaced by the T, S, or B multiple defined as follows:

(20.39a) Tukey procedure (for pairwise comparisons) $$T = \frac{1}{\sqrt{2}} q(1 - \alpha; abc, (n-1)abc)$$

(20.39b) Scheffé procedure $$S^2 = (abc-1)F(1 - \alpha; abc-1, (n-1)abc)$$

(20.39c) Bonferroni procedure $$B = t(1 - \alpha/2s; (n-1)abc)$$

20.4 EXAMPLE OF THREE-FACTOR STUDY

This example illustrates a typical three-factor study in which the ANOVA calculations are obtained by a computer package, and a sequence of testing and estimation procedures is employed to answer questions of interest.

A marketing research consultant, wishing to evaluate the effects of certain factors of interest on the quality of work performed under contract by independent marketing research agencies, classified a large number of such agencies by these factors, as follows:

Factor		*Levels of Factor*
A: Fee level	i:	1. High
		2. Average
		3. Low
B: Scope	j:	1. Performs all work in house
		2. Subcontracts out some work
C: Field-work control	k:	1. Has local supervisors
		2. Has only traveling supervisors

Thus there were 12 combinations of factor levels or treatments. The consultant then selected four agencies at random from the agencies falling into each combination group, and rated the quality of their contract work according to a formal system in which points were awarded for various attributes. The ratings are shown in Table 20.6. Model (20.17) was employed.

TABLE 20.6

Data for Marketing Research Contractor Example

	Fee Schedule (factor A)		
Scope (factor B) and Control (factor C)	*High* ($i=1$)	*Average* ($i=2$)	*Low* ($i=3$)
All work in house ($j=1$):			
Has local supervisors ($k=1$)	58	56	59
	62	59	61
	61	64	58
	59	57	62
Traveling supervisors only ($k=2$)	54	45	52
	52	51	49
	53	46	47
	50	53	45
Subcontracts out ($j=2$):			
Has local supervisors ($k=1$)	63	62	63
	55	59	56
	66	61	57
	57	56	62
Traveling supervisors only ($k=2$)	50	48	55
	53	53	45
	51	51	54
	60	47	53

Data Input

Since all factor level combinations were included in the study, the consultant utilized a "complete factorial ANOVA" program for his preliminary calculations. Figure 20.4 shows the data of Table 20.6 rearranged into the

FIGURE 20.4

Input Data for Marketing Research Contractor Example

Replication			
1st	58	56	59
	63	62	63
	54	45	52
	50	48	55
2d	62	59	61
	55	59	56
	52	51	49
	53	53	45
3d	61	64	58
	66	61	57
	53	46	47
	51	51	54
4th	59	57	62
	57	56	62
	50	53	45
	60	47	53

particular data input format required by the consultant's program. Note that the observations are entered one replication at a time, in the following format, which is repeated in turn for each replication:

$$\begin{array}{ccc} A_1B_1C_1 & A_2B_1C_1 & A_3B_1C_1 \\ A_1B_2C_1 & A_2B_2C_1 & A_3B_2C_1 \\ \\ A_1B_1C_2 & A_2B_1C_2 & A_3B_1C_2 \\ A_1B_2C_2 & A_2B_2C_2 & A_3B_2C_2 \end{array}$$

Analysis of Variance

Output. The computer printout obtained by the consultant is given in Figure 20.5. The output format here takes the form of a matrix. The rows pertain, in turn, to the following sources of variation: factors A, B, C; two-factor interactions AB, AC, BC; three-factor interactions ABC; error; and total. The columns of the output matrix show, in turn: source identification, *df*, *SS*, and *MS*. We have annotated the output for convenience.

FIGURE 20.5

Computer Output for Marketing Research Contractor Example

COMPLETE FACTORIAL ANOVA

Source		df	SS	MS
1	A	2	43.16666667	21.58333333
10	B	1	12	12
100	C	1	972	972
11	AB	2	1.5	0.75
101	AC	2	18.5	9.25
110	BC	1	10.08333333	10.08333333
111	ABC	2	7.166666667	3.583333333
0	ERROR	36	421.5	11.70833333
	TOTAL	47	1485.916667	

The analyst used this output to test for the various factor effects. He decided to employ $\alpha = .01$ in each test since his chief concern here was to control, at a low level, the risk of concluding that an effect exists when it does not. The alternative conclusions, form of F^*, and percentile point in the decision rule for each test were as shown in Table 20.5.

Test for Three-Factor Interactions. The first test was for three-factor interactions. The decision rule was:

$$\text{If } F^* \leq F(.99; 2, 36) = 5.25, \text{ conclude } C_1$$

$$\text{If } F^* > F(.99; 2. 36) = 5.25, \text{ conclude } C_2$$

and the F^* statistic obtained from Figure 20.5 was:

$$F^* = \frac{MSABC}{MSE} = \frac{3.583}{11.708} = .306$$

Hence the analyst concluded that all ABC interactions are zero.

Tests for Main Two-Factor Interactions. The consultant next tested for main two-factor interactions. In the test for AB interactions, the decision rule was:

$$\text{If } F^* \leq F(.99; 2, 36) = 5.25, \text{ conclude } C_1$$
$$\text{If } F^* > F(.99; 2, 36) = 5.25, \text{ conclude } C_2$$

and the test statistic is seen to be:

$$F^* = \frac{MSAB}{MSE} = \frac{.75}{11.708} = .064$$

Hence the analyst concluded that all AB interactions are zero.

The tests for AC interactions and BC interactions proceeded similarly. We obtain:

$$F^* = \frac{MSAC}{MSE} = \frac{9.250}{11.708} = .790 < F(.99; 2, 36) = 5.25$$

Conclusion: All AC interactions are zero

$$F^* = \frac{MSBC}{MSE} = \frac{10.083}{11.708} = .861 < F(.99; 1, 36) = 7.40$$

Conclusion: All BC interactions are zero

Thus the consultant concluded that all first-order interactions are zero.

Tests for Main Effects. Attention next turned to testing for A, B, and C main effects. In testing for factor A main effects, the decision rule was:

$$\text{If } F^* \leq F(.99; 2, 36) = 5.25, \text{ conclude } C_1$$
$$\text{If } F^* > F(.99; 2, 36) = 5.25, \text{ conclude } C_2$$

and the test statistic is seen to be:

$$F^* = \frac{MSA}{MSE} = \frac{21.583}{11.708} = 1.84$$

Hence the conclusion reached was that all A main effects are zero.

The tests for factor B main effects and factor C main effects proceeded similarly. We obtain:

$$F^* = \frac{MSB}{MSE} = \frac{12.000}{11.708} = 1.02 < F(.99; 1, 36) = 7.40$$

Conclusion: All B main effects are zero

$$F^* = \frac{MSC}{MSE} = \frac{972.000}{11.708} = 83.02 > F(.99; 1, 36) = 7.40$$

Conclusion: Not all C main effects are zero

Hence the consultant concluded that factor A main effects and factor B main effects are all zero, but factor C main effects are not all zero.

Family of Tests. The seven separate F tests for factor effects taken as a family indicated to the consultant that:

1. There are no three-factor interactions;
2. There are no two-factor interactions;
3. There are no main effects as to fee (factor A) and scope (factor B), but there are main effects as to control (factor C).

The upper bound for the family level of significance α for this set of tests is, by the Kimball inequality (20.29):

$$\alpha \leq 1 - (.99)^7 = .068$$

This set of test results, with family level of significance $\alpha \leq .068$, was most useful to the consultant. The final step in his analysis now was to estimate the difference between the factor level means for factor C.

Estimation of Difference between Factor Level Means

Since factor C has only two levels, a single comparison is required:

$$L = \mu_{..1} - \mu_{..2}$$

which is estimated by:

$$\hat{L} = \bar{Y}_{..1.} - \bar{Y}_{..2.}$$

From Table 20.6, we obtain:

$$\hat{L} = 59.7 - 50.7 = 9.0$$

The analyst used confidence interval (20.34), with a 95 percent confidence coefficient. Hence he required:

$$t(.975; 36) = 2.03$$

$$s^2(\hat{L}) = \frac{MSE}{nab}[(1)^2 + (-1)^2]$$

$$= \frac{11.708}{24}(2) = .9757$$

$$s(\hat{L}) = .988$$

The confidence interval therefore is:

$$7.0 = 9.0 - (2.03)(.988) \leq \mu_{..1} - \mu_{..2} \leq 9.0 + (2.03)(.988) = 11.0$$

Thus the consultant estimates, with confidence coefficient .95, that the mean rating of agencies using local supervisors is somewhere between 7 and 11 points higher than that of agencies using only traveling supervisors.

20.5 MODELS II AND III FOR THREE-FACTOR STUDIES

The principal problem with random and mixed models, as we saw for two-factor models, is the determination of the appropriate expected mean squares. Once these are known, the proper test statistics and confidence intervals can be constructed readily. Rule (19.26) for finding the expected mean squares for random and mixed models applies for any number of factors. We shall illustrate how it is applied to three-factor investigations.

Random Effects Model

In a study of the effects of operators, machines, and batches of raw material on daily output, all three factors may be considered to have random effects. The model then would be:

(20.40)
$$Y_{ijkm} = \mu_{...} + \alpha_i + \beta_j + \gamma_k + (\alpha\beta)_{ij} + (\alpha\gamma)_{ik} + (\beta\gamma)_{jk} + (\alpha\beta\gamma)_{ijk} + \varepsilon_{ijkm}$$

where:

$\mu_{...}$ is a constant

$\alpha_i, \beta_j, \gamma_k, (\alpha\beta)_{ij}, (\alpha\gamma)_{ik}, (\beta\gamma)_{jk}, (\alpha\beta\gamma)_{ijk}, \varepsilon_{ijkm}$ are independent normal random variables with expectation zero and respective variances $\sigma_\alpha^2, \sigma_\beta^2, \sigma_\gamma^2, \sigma_{\alpha\beta}^2, \sigma_{\alpha\gamma}^2, \sigma_{\beta\gamma}^2, \sigma_{\alpha\beta\gamma}^2, \sigma^2$

$i = 1, \ldots, a; j = 1, \ldots, b; k = 1, \ldots, c; m = 1, \ldots, n$

We illustrate how to find $E(MSA)$ for this model. Applying rule (19.26a) we obtain:

Step 1	σ^2
Step 2	$\sigma_\alpha^2, \sigma_{\alpha\beta}^2, \sigma_{\alpha\gamma}^2, \sigma_{\alpha\beta\gamma}^2$
Step 3	No terms in previous step deleted

Rule (19.26b) indicates that the coefficients of these terms are 1, nbc, nc, nb, and n respectively. Hence we have:

$$E(MSA) = \sigma^2 + nbc\sigma_\alpha^2 + nc\sigma_{\alpha\beta}^2 + nb\sigma_{\alpha\gamma}^2 + n\sigma_{\alpha\beta\gamma}^2$$

Table 20.7 contains the expected mean squares for all components of the ANOVA table for the random effects model (20.40).

Mixed Effects Model

Suppose that in a three-factor study, factors B and C have random effects while factor A has fixed effects. The model then is:

(20.41)
$$Y_{ijkm} = \mu_{...} + \alpha_i + \beta_j + \gamma_k + (\alpha\beta)_{ij} + (\alpha\gamma)_{ik} + (\beta\gamma)_{jk} + (\alpha\beta\gamma)_{ijk} + \varepsilon_{ijkm}$$

TABLE 20.7

Expected Mean Squares in Random Effects Three-Factor Study

Mean Square	*df*	*Expected Mean Square*
MSA	$a-1$	$\sigma^2 + nbc\sigma_\alpha^2 + nc\sigma_{\alpha\beta}^2 + nb\sigma_{\alpha\gamma}^2 + n\sigma_{\alpha\beta\gamma}^2$
MSB	$b-1$	$\sigma^2 + nac\sigma_\beta^2 + nc\sigma_{\alpha\beta}^2 + na\sigma_{\beta\gamma}^2 + n\sigma_{\alpha\beta\gamma}^2$
MSC	$c-1$	$\sigma^2 + nab\sigma_\gamma^2 + nb\sigma_{\alpha\gamma}^2 + na\sigma_{\beta\gamma}^2 + n\sigma_{\alpha\beta\gamma}^2$
$MSAB$	$(a-1)(b-1)$	$\sigma^2 + nc\sigma_{\alpha\beta}^2 + n\sigma_{\alpha\beta\gamma}^2$
$MSAC$	$(a-1)(c-1)$	$\sigma^2 + nb\sigma_{\alpha\gamma}^2 + n\sigma_{\alpha\beta\gamma}^2$
$MSBC$	$(b-1)(c-1)$	$\sigma^2 + na\sigma_{\beta\gamma}^2 + n\sigma_{\alpha\beta\gamma}^2$
$MSABC$	$(a-1)(b-1)(c-1)$	$\sigma^2 + n\sigma_{\alpha\beta\gamma}^2$
MSE	$(n-1)abc$	σ^2

where:

$\mu_{...}$ is a constant

α_i are constants

$\beta_j, \gamma_k, (\alpha\beta)_{ij}, (\alpha\gamma)_{ik}, (\beta\gamma)_{jk}, (\alpha\beta\gamma)_{ijk}$ are normal random variables with expectation zero

ε_{ijkm} are independent $N(0, \sigma^2)$, and are independent of the other random components

$$\sum_i \alpha_i = \sum_i (\alpha\beta)_{ij} = \sum_i (\alpha\gamma)_{ik} = \sum_i (\alpha\beta\gamma)_{ijk} = 0$$

$$i = 1, \ldots, a; j = 1, \ldots, b; k = 1, \ldots, c; m = 1, \ldots, n$$

Note that all interaction terms in this model are random, since at least one of the factors contained in each is a random effects factor. Note also that all sums of effects involving the fixed factor are zero over the fixed factor levels. Various correlations exist between the random effects terms, which we shall not detail.

The expected mean squares are obtained in accordance with rule (19.26). We illustrate the determination of $E(MSBC)$. Rule (19.26a) yields:

Step 1 σ^2
Step 2 $\sigma_{\beta\gamma}^2, \sigma_{\alpha\beta\gamma}^2$
Step 3 Delete $\sigma_{\alpha\beta\gamma}^2$ since A is a fixed effects factor

The coefficients of the components are, according to rule (19.26b), 1 and na respectively. Hence, we have:

$$E(MSBC) = \sigma^2 + na\sigma_{\beta\gamma}^2$$

Table 20.8 contains all the expected mean squares for the mixed effects model (20.41). Note that the components referring to fixed effects have been restated in terms of sums of squared effects divided by degrees of freedom.

Other mixed effects models and their expected mean squares can be developed in similar fashion.

TABLE 20.8

Expected Mean Squares in Mixed Effects Three-Factor Study (*A* fixed, *B* and *C* random)

Mean Square	*df*	*Expected Mean Square*
MSA	$a-1$	$\sigma^2 + nbc\dfrac{\sum \alpha_i^2}{a-1} + nc\sigma_{\alpha\beta}^2 + nb\sigma_{\alpha\gamma}^2 + n\sigma_{\alpha\beta\gamma}^2$
MSB	$b-1$	$\sigma^2 + nac\sigma_\beta^2 + na\sigma_{\beta\gamma}^2$
MSC	$c-1$	$\sigma^2 + nab\sigma_\gamma^2 + na\sigma_{\beta\gamma}^2$
MSAB	$(a-1)(b-1)$	$\sigma^2 + nc\sigma_{\alpha\beta}^2 + n\sigma_{\alpha\beta\gamma}^2$
MSAC	$(a-1)(c-1)$	$\sigma^2 + nb\sigma_{\alpha\gamma}^2 + n\sigma_{\alpha\beta\gamma}^2$
MSBC	$(b-1)(c-1)$	$\sigma^2 + na\sigma_{\beta\gamma}^2$
MSABC	$(a-1)(b-1)(c-1)$	$\sigma^2 + n\sigma_{\alpha\beta\gamma}^2$
MSE	$(n-1)abc$	σ^2

Appropriate Test Statistics

Once the expected mean squares have been obtained, we seek to determine the appropriate F^* statistic for a given test. This can often be found for random and mixed models, but sometimes it cannot.

Exact F Test. Suppose we wish to determine whether or not *BC* interactions are present in the random effects model for Table 20.7. We see easily from the expected mean squares column that the appropriate test statistic is $MSBC/MSABC$. If we wish to study the same question for the mixed effects model for Table 20.8, we are again able to find an appropriate test statistic, but this time it is $MSBC/MSE$. We thus see that the two test statistics are not the same, even though the same factor effects are being studied, because of the differences between the two models.

Approximate F Test. It is not always possible to find appropriate test statistics for mixed and random effects models. For instance, one cannot directly test for the presence of factor *A* effects in the random effects model for Table 20.7. Note from this table that there is no expected mean square which consists of the components of $E(MSA)$ except for the $nbc\sigma_\alpha^2$ term.

There are two ways out this difficulty. It may be possible to assume that certain interactions are zero, and then proceed in the usual way with an exact *F* test. For example, to test for factor *A* effects in the random effects model for Table 20.7, it may be possible to assume that $\sigma_{\alpha\gamma}^2 = 0$ (indeed, this can be tested with $MSAC/MSABC$). If this assumption is appropriate one can use the test statistic $MSA/MSAB$ to test for factor *A* effects.

Often, however, one may not know whether certain interactions are zero. In that case, an approximate *F* test may be employed which utilizes a *pseudo F* or *quasi F* statistic. The basic idea is to develop a linear combination of mean squares which has the same expectation, when C_1 holds, as the factor effects mean square. Let us express this linear combination as follows:

$$a_1 MS_1 + a_2 MS_2 + \cdots + a_h MS_h \tag{20.42}$$

where the a's are constants. It can be shown that the approximate number of degrees of freedom associated with the linear combination (20.42) is:

$$df \cong \frac{(a_1 MS_1 + a_2 MS_2 + \cdots + a_h MS_h)^2}{\frac{(a_1 MS_1)^2}{df_1} + \frac{(a_2 MS_2)^2}{df_2} + \cdots + \frac{(a_h MS_h)^2}{df_h}} \tag{20.43}$$

where df_i denotes the degrees of freedom associated with MS_i. The test statistic is then set up in the usual way, and follows approximately the F distribution when C_1 holds.

We illustrate this procedure for testing factor A effects in the random effects model for Table 20.7:

$$\begin{aligned} &C_1: \sigma_\alpha^2 = 0 \\ &C_2: \sigma_\alpha^2 \neq 0 \end{aligned} \tag{20.44}$$

Note from Table 20.7 that:

$$E(MSAB) + E(MSAC) - E(MSABC) = \sigma^2 + nc\sigma_{\alpha\beta}^2 + nb\sigma_{\alpha\gamma}^2 + n\sigma_{\alpha\beta\gamma}^2 \tag{20.45}$$

This equals precisely $E(MSA)$ when $\sigma_\alpha^2 = 0$. Hence, the suggested test statistic is:

$$F^{**} = \frac{MSA}{MSAB + MSAC - MSABC} \tag{20.46}$$

where we denote the test statistic as F^{**} as a reminder that a pseudo F test is involved.

Table 20.9 contains the analysis of variance for a study of operators, machines, and batches, each of which is assumed to be a random effects

TABLE 20.9

ANOVA Table for Random Effects Three-Factor Study ($a = 3$, $b = 2$, $c = 5$, $n = 3$)

Source of Variation	*SS*	*df*	*MS*
Factor A (operators)	17	2	8.5
Factor B (machines)	4	1	4.0
Factor C (batches)	25	4	6.2
AB	5	2	2.5
AC	32	8	4.0
BC	12	4	3.0
ABC	12	8	1.5
Error	138	60	2.3
Total	245	89	

factor. To test whether operators (factor A) have an effect, we use the test statistic (20.46):

$$F^{**} = \frac{8.5}{2.5 + 4.0 - 1.5} = \frac{8.5}{5.0} = 1.7$$

The approximate number of degrees of freedom associated with the denominator is:

$$df \cong \frac{(5.0)^2}{\frac{(2.5)^2}{2} + \frac{(4.0)^2}{8} + \frac{(-1.5)^2}{8}} = 4.6$$

Usually, the degrees of freedom will not turn out to be an integer. One can then either interpolate in the F distribution table, or round to the nearest integer. We shall round. For a level of significance of .05, we require $F(.95; 2, 5) = 5.79$. Since $F^{**} = 1.7 < 5.79$, we conclude C_1, that operators do not have an effect on daily output. This test, it should be remembered, is an approximate one but can be quite useful if employed with caution.

Estimation of Effects

No new problems arise in the development of unbiased estimators of variance components for random effects factors or in the estimation of contrasts for fixed effects factors in mixed models, when three or more factors are studied at one time. Confidence limits for contrasts among the factor levels of a fixed effects factor are obtained by using the mean square utilized in the denominator of the test statistic for examining the presence of effects for that factor. The degrees of freedom are those associated with the mean square utilized.

PROBLEMS

20.1. Refer to Table 20.1 containing the mean responses μ_{ijk} for a three-factor study.
   a) What is the specific effect of high I.Q. for young males?
   b) What is the specific effect of young age for females with normal I.Q.?
   c) What are the main effects of age?
   d) What is the specific interaction effect of young age and normal I.Q. for male persons? For female persons?
   e) What is the main interaction effect of young age and normal I.Q.?
   f) What is the interaction effect of young age, normal I.Q., and female sex?

20.2. Refer to Table 20.1. Plot the mean responses μ_{ijk} in one or more arrangements so that you can analyze the nature of the main effects and interaction effects. Summarize the findings of your analysis.

20.3. Refer to Figure 20.3. Plot the mean responses shown there, in the form of BC response curves against A. Does this plot bring out any information on main effects and interactions not as readily seen from the plot in Figure 20.3?

20.4. Derive the sum of squares breakdown in (20.24).

20.5. For the three-factor fixed effects ANOVA model (20.17), what is the noncentrality parameter ϕ for testing main A effects? For testing main AB interactions?

20.6. State the fixed effects ANOVA model for a three-factor study with $n = 1$ if it can be assumed that all three-factor interactions are zero. Show the ANOVA table for this case.

20.7. Refer to Problem 17.10. In each "type of experience–amount of experience" group, the programmers were equally divided between men and women. The first two observations in each group are those for men; the second two those for women.

a) State the model you will use to analyze the effects of amount of experience, type of experience, and sex of programmer on the forecasting error.
b) Test whether the factors have any main effects and whether they interact. Use a level of significance of .01 for each test. Summarize your results.
c) Give an upper bound for the family level of significance.
d) Analyze the results of the study. Explain the reasons why you chose your particular analytical approach and summarize your findings.

20.8. A wine distributor wished to study the effect of wine bottle labeling on the perception of the quality of the wine by ordinary wine drinkers. He used two domestic and two foreign white wines. One of each was a high quality wine, the other a low quality wine. Thirty-two subjects were employed. Each subject was shown a glass of wine and the label, then he was asked to taste the wine, and finally he was requested to rate the quality of the wine on a 20-point scale (0—extremely poor, 20—outstanding). Half of the subjects were shown the correct label, while for the other half the label was switched as to the country of origin. For instance, a person in the switched label group might be given the high quality domestic wine and shown the high quality foreign wine label. All treatment assignments were made at random. The results were as follows:

	Domestic Wine		*Foreign Wine*	
	Low Quality	*High Quality*	*Low Quality*	*High Quality*
Correct label:				
	12	16	8	17
	11	15	11	16
	9	19	10	18
	10	18	12	17
Switched label:				
	14	16	6	17
	13	18	10	19
	17	17	9	18
	15	15	8	18

A fixed effects model is to be employed.

a) State the model you will use to analyze the effects of origin of wine, quality of wine, and label on the perceived quality.

b) Test whether the factors have any main effects and whether they interact. Use a level of significance of .05 for each test. Summarize your results.

c) Analyze the factor effects. Explain the reasons why you chose your particular analytical approach, and summarize your findings.

20.9. In a three-factor study, factor A has random effects while factors B and C have fixed effects.

a) State the model for this case.

b) Find the expected mean squares for this model by using rule (19.26).

c) Set up the appropriate test statistics for testing the various main factor effects and interactions. Did you need to use any approximate F tests?

d) If you wished to estimate a contrast among the factor level means for factor B, what would be the appropriate mean square to use for the estimated variance?

20.10. In a three-factor study, factors A and B have random effects, while factor C has fixed effects.

a) State the model for this case.

b) Find the expected mean squares for this model by using rule (19.26).

c) Set up the appropriate test statistics for testing the various main factor effects and interactions. Did you need to use any approximate F tests?

d) If you wished to estimate a contrast among the factor level means for factor A, what would be the appropriate mean square to use for the estimated variance?

20.11. Refer to Table 20.9. All three factors in this study have random effects.

a) Test whether machines (factor B) have an effect. Use a level of significance of .01.

b) Test whether AB interactions are present. Use a level of significance of .01.

c) If you wished to estimate a contrast among the factor level means for factor A, what would be the appropriate mean square to use for the estimated variance? What would be the appropriate number of degrees of freedom?

part V

Experimental Designs

21

Completely Randomized Designs

FORMAL EXPERIMENTATION is being widely employed in the social sciences. It has come of age somewhat more recently in business and economics, but a wide variety of uses now are found in these fields also. One example is an experiment to investigate the best level of aggregation of company data furnished by a management information system to middle management. Another is an experiment on the effect of a guaranteed annual income on the consumption behavior of families. The latter experiment is designed by dividing a group of low-income families at random into two halves, one of which receives income supplements up to a guaranteed annual income, while the other half receives no supplements. The consumption behavior of each group of families is then observed.

Much of our earlier discussion has been concerned with the analysis of investigations, both formal experiments and surveys. In this and the following chapters, we focus largely on the *design* of experimental studies, as distinct from the *analysis* of results. In particular, we shall concentrate on those elements of experimental design where statistical methods have made the greatest contributions.

In this chapter, we shall first consider the major elements of any experimental design, and then shall take up completely randomized designs. In succeeding chapters, other widely used experimental designs will be discussed.

21.1 ELEMENTS OF EXPERIMENTAL DESIGN

The *design of an experiment* refers to the structure of the experiment, with particular reference to:

1. The set of treatments included in the study.

2. The set of experimental units included in the study.
3. The rules and procedures by which the treatments are assigned to the experimental units (or vice versa).
4. The measurements that are made on the experimental units after the treatments have been applied.

Statistical designs for experiments are concerned with the rules and procedures whereby treatments are assigned to the experimental units. Statistical methodology also makes contributions to the other elements of experimental design, but we shall dwell chiefly on how to assign the treatments to experimental units so that efficient use is made of the experimental units. First, however, we shall discuss briefly the other key elements of experimental design.

Treatments

A treatment, we know, is a factor level in a single-factor study or a combination of factor levels in a multifactor study. Three principal problems in experimental design pertaining to treatments are: (1) choice of treatments to be studied, (2) definition of each treatment, and (3) need for a control treatment.

Choice of Treatments. The choice of treatments to be included in an experimental investigation is basically a matter to be decided by the investigator. A few general comments may, however, be appropriate. In a scientific investigation, the treatments included should be able to provide some insights into the mechanism underlying the phenomenon under study. Initial experiments should not attempt to study this mechanism in full detail, but rather should seek to find the principal factors involved and obtain indications of the magnitudes of their effects. Subsequent experiments can then be conducted to provide more detailed findings.

Even if the investigation is not of scientific but rather of practical interest, inclusion of treatments to provide some explanation of the results is often desirable. Suppose a company is considering the purchase of tool machines from a new maker and wishes to compare this new type of machine with the present ones by means of an experiment. Suppose further that the new machines are of a much larger size than the present ones. Under these circumstances, the company may wish to include a third type of machine in the experiment, namely one corresponding in size to the present ones, but from the new maker. By doing so, the company will get information as to whether any differences that might be observed between the present machines and the proposed new ones are connected with the maker, the size of the machine, or both.

In earlier chapters, we discussed the usefulness of factorial studies for investigating the effects of several factors simultaneously. It is often desirable to include some factors in an investigation even though the main interest is not in them, simply because the chief factors of interest may interact in special ways with these additional factors.

Definition of Treatment. The definition of a treatment can be a difficult problem. Consider an experiment to study whether BASIC or FORTRAN is a better programming language to teach in an introductory statistics course. Some teachers will prefer BASIC, others FORTRAN. Should the treatments then be defined as the programming language taught by instructors who prefer that language? If so, differences in findings may be due to differences between the two groups of instructors. Should the definition of a treatment not include the instructor, and instructors be randomized, with some being forced to teach a language they do not prefer? Or should instructor preference be a second factor, with each instructor teaching both languages? Problems of this kind need careful resolution so that the results of the study will be useful.

Control Treatment. A control treatment is needed in some experiments, but not in all. A control treatment consists of applying the identical procedures to experimental units that are used with the other treatments, except for the effect under investigation. In a study of food additives, for instance, a treatment may consist of a portion of a vegetable containing a particular additive which is served to a consumer in a particular experimental setting in the laboratory. A control treatment here would consist of a portion of the same vegetable served to a consumer in the identical experimental setting except that no food additive has been used.

A control treatment is required when the general effectiveness of the treatments under study is not known, or when the general effectiveness of the treatments is known but is not consistent under all conditions. In our food additives illustration, suppose it is known that food additive A is highly effective in enhancing the tastiness of vegetables and it is desired to see if additives B and C are equally effective or possibly even more effective. In that case, a standard of comparison is available and no control treatment is required. On the other hand, suppose there is no knowledge about the general effectiveness of the three additives, and the following results are obtained (ratings can range between 0 and 60):

Additive	*Mean Rating*
A	39
B	37
C	41

Assume that the sample sizes are large so that negligible random error is present. In the absence of a standard of comparison, one would not know here whether each of the three additives is effective or whether none of the additives is effective.

It is crucial that the control treatment be conducted in the identical experimental setting as the other treatments. In our food additives illustration, for instance, a survey of consumers at home, in which persons are asked to rate the general tastiness of the vegetable (without any additive) on the same scale as in the experiment, would not qualify as a control treatment. Such a survey might yield a mean rating of 22, suggesting that the three additives substantially increase the tastiness of the vegetable. This conclusion, however, could be grossly misleading. If the control treatment actually were incorporated into the experiment so that consumers are given portions of the vegetable with no additive in the laboratory setting, the mean rating for the control treatment might be 40. This result would imply that none of the three additives is effective in enhancing the tastiness of the vegetable. The reason for the higher mean rating in the laboratory setting could be a "halo" effect connected with the experimental procedures. Possibly, foods served in the experimental setting taste better than at home, or perhaps consumers try to oblige by giving higher ratings when they participate in an experimental study. Thus, only a control treatment incorporated into the experiment can serve as the proper standard of comparison.

Experimental Units

Definition of Experimental Unit. We may define an experimental unit as the smallest subunit of the experimental material such that any two different experimental units may receive different treatments. Suppose that two incentive pay systems are studied in two plants each, with the plant assignments made at random, and observations are then made on the productivity of a sample of employees in each plant. Here, the experimental unit is the plant, not the employee, since all employees in a given plant are assigned the same incentive pay system in the experiment.

Size of Experimental Unit. An important problem in designing an experiment sometimes is the size of the experimental unit. In our example above, should the experimental unit be an individual employee, a shift, or a plant? The theory of hierarchical designs can be helpful in considering the question of size of the experimental unit. A different aspect of size occurs in studies of sales and similar phenomena. Suppose that we are interested in measuring the effectiveness of five different television commercials in terms of sales during a period of time subsequent to their showing. Should the length of time be one week, two weeks, one month, or some other time period? Clearly, the purposes of the study will need to govern the length of time which makes up the experimental unit here.

Choice of Experimental Units. Representativeness of experimental units is another important consideration in the design of experiments. Consider a study of management behavior with different communications networks. A university investigator may be tempted to use students as subjects because of their ready availability. If, however, information is desired about the behavior of businessmen, the students may not be representative experimental units. It hardly needs to be stated that an investigator should make every effort to obtain representative experimental units. Conversely, one should be cautious in extending results of an investigation to groups for which the experimental units are not representative. Thus if the communications network study cited above *did* use students, one should not automatically assume that the findings are relevant to businessmen.

There is another facet to the selection of experimental units. The more similar they are, the smaller will be the experimental errors (i.e., the smaller the variance of the random error component ε) and the more precise the experimental results will be. Thus, in a learning experiment, the use of persons of the same age, intelligence, economic and social background, and the like will tend to lead to smaller experimental errors than if a more heterogeneous group of subjects was used. However the more homogeneous are the experimental units, the smaller is the range for which the experimental results are valid. For instance, findings for persons in one age group may not be valid for persons in other age groups. Thus, to make the conclusions broadly valid, one should vary the characteristics of experimental units, but the cost of this is less precision in the experimental results. Blocking designs, to be discussed in later chapters, can be employed to have one's cake and eat it too, namely to have sufficient variability between experimental units for a wide range of validity and yet achieve high precision due to small experimental errors.

Measurements

The measurements to be made on the experimental units after the treatments have been applied constitute the values of the dependent variable. The analyst will not only need to decide in advance which characteristics to measure, but also how to measure them. Important scaling problems may arise in the measurement process. The observations may include readings on variables that are concerned not so much with the phenomenon directly under study as with factors which may help to explain observed effects or which may help to reduce experimental errors. For instance, we mentioned earlier a study of five television commercials for which sales in an ensuing period are observed. Suppose the experimental unit is a metropolitan area. The observations taken in each area during the period may include readings on the weather to help reduce the experimental errors by the method of covariance (Chapter 22).

The measurement process ideally should produce measurements that are unbiased and precise. Often, extensive research prior to the experiment is required to develop a satisfactory measurement process.

Assignment of Treatments to Experimental Units

Once the treatments and the experimental units have been specified, rules and procedures are required for assigning the treatments to the experimental units (or vice versa). Randomization is always used in some stage of the assignment process in a statistically designed experiment. We shall discuss this facet of experimental design in Section 21.3.

21.2 CONTRIBUTIONS OF STATISTICS TO EXPERIMENTATION

Statistics has made a number of important contributions to experimentation. We consider briefly four major ones.

Factorial Experiments

This contribution was considered in Chapter 17. There we noted that multifactor investigations permit the analysis of a number of factors with the same precision as if the entire experiment had been devoted to the study of only one factor. In addition, a single factorial experiment provides information on interaction effects while the classical one-factor-at-a-time approach requires a series of experiments for doing so.

Replication

Replication refers to the repetition of an experiment. Consider an experiment consisting of three treatments. The assignment of three experimental units at random, one to each treatment, constitutes one replication of the experiment. The assignment of an additional three experimental units at random to the three treatments constitutes a second replication, and so on.

We need to observe at once that not all repetitions are replications. Suppose two incentive pay plans are being investigated and two plants are used in the study, with one plant assigned to each incentive plan. Assume now that 10 employees in each plant are selected and their productivity measured. For purposes of comparing the incentive pay plans, the plants are the experimental units so that there is only one replication (one plant for each plan), not 10 (the number of employees studied in each plant). Indeed, with only one replication here, plant effects and incentive pay plan effects are confounded and cannot be disentangled. Selecting additional employees will not enable one to disentangle the incentive pay plan effects from the plant effects; only repetition of the experiment in additional plants (i.e., replicating the experiment) will permit this.

Replication makes it possible to assess the mean square error required for testing the presence of treatment effects or for establishing confidence interval estimates of these effects, as we have seen in earlier chapters. Replication also plays a second role, namely it permits control over the precision of the estimates or the power of the tests through manipulation of the replication (sample) size. Again, we have observed this in earlier chapters.

Randomization

Randomization in experiments is a relatively recent idea, first introduced by the famous British statistician R. A. Fisher. In the past, treatments had been assigned to experimental units either on a systematic or on a subjective basis. With both of these procedures, serious biases can arise. Consider, for instance, an experiment using 10 employees and 2 treatments. If the first five employees on the payroll listing are assigned treatment 1 and the next five treatment 2, bias could be introduced in the comparison of the two treatments. Suppose that the payroll listing is by seniority and that this variable is related to the phenomenon under study, say productivity. A comparison of treatments 1 and 2 then reflects not only differences between the two treatments but also differences in the amount of experience between the two groups of employees. This potential bias may be so transparent that no good experimenter would use the type of systematic assignment just described. Nevertheless, there may be many other sources of bias that are not so apparent.

Subjective assignments of treatments to experimental units can also lead to bias, as when an experimenter subconsciously tends to assign one treatment to highly extrovert subjects and the other treatment to less extrovert subjects.

With randomization, the treatments are assigned to experimental units at random, by means of a table of random digits. (The mechanics will be discussed shortly.) Randomization tends to average out between the treatments whatever systematic effects may be present, apparent or hidden, so that comparisons between treatments measure only the pure treatment effects. Thus, randomization tends to eliminate the influence of extraneous factors not under the direct control of the experimenter. Cochran and Cox (Ref. 21.1, p. 8) have likened randomization to an insurance policy in that it is a precaution against biases that may or may not occur.

Randomization is appropriate not only for the assignment of treatments to experimental units but also for any other phase of the experiment where systematic effects not under the control of the experimenter may be present. For instance, consider an experiment in which 5 treatments (alternative methods of measuring subjective probability) and 20 subjects are used. Only one subject can be run per day, thus four weeks are required to complete the experiment. In this type of situation, it usually is highly desirable to determine the order of the treatments by means of random numbers since a variety of systematic time effects could be present. The experimenter may with time improve his explanation of the methods of measuring subjective probability, there may be a streak of extremely hot weather during a week, and the like. With these possible time effects, a systematic assignment of one treatment per week could lead to seriously biased results. Randomization, on the other hand, will tend to average out whatever systematic effects are present, whether anticipated or not.

Advice on randomizations needed in addition to the assignment of treatments to experimental units can only be general. Certainly, randomization should be employed whenever the consequences of systematic effects could be serious. In our illustration on methods of measuring subjective probability, suppose that two observers conduct the experiment. Randomization of observers to treatment-experimental unit combinations would then be highly desirable, since large differences between observers are known to occur in this kind of situation. If the seriousness of the consequences is not known, the safe course is to randomize when feasible and not too costly. If randomization cannot be easily carried out and no serious consequences of systematic effects are anticipated, the experimenter may be willing to forego randomization. He must then realize, however, that the validity of the treatment comparisons depends on the absence of serious systematic effects.

Comments

1. One may view the implications of randomization in a somewhat different fashion than that presented so far. The random errors of experimental units that are adjacent in time or space are often correlated, not independent, as a result of various systematic effects over time or space. Randomization does not erase this correlation pattern but, by making it equally likely that any two treatments are adjacent, tends to eliminate the correlations between treatments with increasing replications. Thus, randomization makes it reasonable to analyze the data as though the model random error terms are independent, an assumption which has been made in almost all models discussed so far.

2. Once in a while, randomization may provide a pattern that makes the experimenter uneasy, such as running the four experimental units for treatment 1 first and then running the four experimental units for treatment 2. This is not a likely occurrence, but one that can take place. Some solutions have been suggested for this problem, but none provides a final answer. In practice, the experimenter typically will discard a randomization sequence that has apparent dangers of systematic effects for his particular experiment, and select another randomization.

3. Randomization also can provide the basis for making inferences without requiring that the error terms ε are independent $N(0, \sigma^2)$. We illustrate this for a simple one-factor experiment, consisting of two treatments and three replications. In this experiment, the treatments were assigned to the experimental units at random. Suppose the data are:

Treatment 1	*Treatment 2*
3	6
9	2
4	10

We assume now the following simple model (it can be generalized):

$$Y_{ij} = \begin{pmatrix} \text{A quantity depending on} \\ \text{the experimental unit} \end{pmatrix} + \begin{pmatrix} \text{A quantity depending} \\ \text{on the treatment} \end{pmatrix} \tag{21.1}$$

Both the quantities for the experimental units and the quantities for the treatments are viewed as fixed. The randomness in the model arises solely from the random assignment of treatments to experimental units. Suppose now that the two treatment effects are equal. In that case, it would have been just as likely that we observed the numbers 3, 9, 4 for treatment 2, and 6, 2, 10 for treatment 1, since the treatments are assigned to experimental units at random. In fact, if the two treatment effects are equal, any division of the six observations into two groups of three is equally likely. Thus in the list of all possible arrangements, all are equally likely if no treatment effects are present:

Treatment 1	*Treatment 2*
3, 9, 4	6, 2, 10
3, 9, 6	4, 2, 10
3, 9, 2	6, 4, 10
etc.	etc.

We then view the problem of comparing treatments as a one-way analysis of variance, and calculate $F^* = MSTR/MSE$ for each arrangement. We thereby obtain the exact sampling distribution of F^* when the two treatment effects are equal. Both empirical and theoretical studies have shown that the sampling distribution so obtained is distributed approximately as the F distribution, provided the sample sizes are not very small. Thus, randomization alone can justify the F test as a good approximate test, without requiring any normality assumption.

Local Control

The fourth contribution of statistics to experimental design is the concept of local control, which often is considered statistical design proper. Local control is intended to reduce experimental errors and make the experiment more powerful by suitable restrictions on the randomization of treatments to experimental units. Consider again the study of five methods of measuring subjective probability, to be conducted over a period of four weeks. The thought may have occurred in our earlier discussion that complete randomization might not provide full balance of all treatments within the four-week period. Would it not be better if we required that each week contain each of the five treatments once? If a substantial time effect is likely, it would indeed be desirable to use this form of restricted randomization, called *blocking*. Thus, we would randomize the order of treatments subject to the restriction that each treatment occur once in each week. It will be seen later that if a time effect is present, blocking will lead to more precise results than complete randomization.

Let us consider this same example from a slightly different point of view. With unrestricted randomization, the five observations in a single replication of the experiment will differ among themselves because of treatment effects, because of time effects (since the treatments may end up in any of the four weeks), and so on. If we require that each of the five treatments be conducted in each week, then a week's observations constitute a replication. Within such a replication, the observations will differ again because of treatment effects and a variety of other causes, but not because of any time effects from one week to another. The only effect of time that is left is that within a week, which may be anticipated to be substantially smaller than that between weeks. Thus, blocking by week will reduce the experimental errors when a time effect is present, and in this way make the experiment more powerful.

There are many other methods available for restricting the randomization of treatments to experimental units in order to improve the efficiency of the experimental design. The most important of these will be discussed in subsequent chapters.

21.3 COMPLETELY RANDOMIZED DESIGNS

Description

The simplest statistical design for an experiment is the *completely randomized design*. With this design, treatments are assigned to experimental units (or vice versa) completely at random. This complete randomization provides that every experimental unit has an equal chance to receive any one of the treatments, also that all combinations of experimental units assigned to the different treatments are equally likely. We shall discuss the mechanics of accomplishing this randomization shortly.

A completely randomized design is useful when the experimental units are quite homogeneous. It also must be used when the experimental units are heterogeneous and no information is available for "blocking" the experimental units into more homogeneous groups and then applying restricted randomization. A completely randomized design may at times also be used with heterogeneous experimental units if covariance analysis (Chapter 22) is utilized to reduce the experimental errors.

Advantages and Disadvantages

The completely randomized design has a number of important advantages:

1. It is very flexible. It can accommodate any number of treatments and any number of replications.
2. The sample sizes can be varied from treatment to treatment. This might be desirable when a control treatment is included in the experiment. (While the power of tests is maximized with equal sample sizes, comparisons between treatments and control can be made more precise by assigning a greater number of observations to the control.)
3. The statistical analysis of the data is easy even when the sample sizes in a single-factor study are unequal.
4. The number of degrees of freedom associated with the experimental error mean square is larger than for any restricted randomization design.
5. Missing observations (e.g., due to sickness of a subject, loss of a questionnaire) create no problems in the analysis of single-factor studies. Neither does a dropped treatment, as when it is realized after the experiment is concluded that the intended deception in a treatment worked improperly.
6. The model requires fewer assumptions than those for other designs.

The chief disadvantage of the completely randomized design is its inefficiency when the experimental units are heterogeneous. Restricted randomization designs can utilize knowledge about the sources of variation in the observations (e.g., time effects, observer effects) to reduce the experimental errors and thereby lead to more precise results. Unless the experiment is a very small-scale one, the larger number of degrees of freedom for the experimental error mean square with the completely randomized design will often not compensate for the reduction in experimental errors which can be obtained by restricted randomization when the experimental units are heterogeneous.

Note

In business and economics, the experimental units frequently are a person, a family, a store, a town, a metropolitan area. These are typically highly heterogeneous experimental units. Hence, blocking and restricted randomization are often employed to reduce the experimental errors which would be encountered with a completely randomized design.

How to Randomize

Use of Table of Random Permutations. All randomizations for experimental designs require that a series of objects (treatments, rows, columns, etc.) be placed in a random order. For a completely randomized design, the situation might be thus:

$$\text{Treatments: } T_1, T_2, T_3$$
$$\text{Sample sizes: } n_1 = 3, n_2 = 2, n_3 = 3$$

Here eight treatment assignments need be made to eight experimental units. These treatment assignments are listed in arbitrary order:

$$T_1 \quad T_1 \quad T_1 \quad T_2 \quad T_2 \quad T_3 \quad T_3 \quad T_3$$

To randomize the treatments to the experimental units, we number the latter from 1 to 8. Now we go to Table A–13, which contains random permutations of 9 (we use these since a table of 8 is not available). This table has been obtained by selecting permutations of the digits 1 to 9 at random, thus making each permutation equally probable. Suppose we pick the random permutation in the upper left corner:

$$5 \quad 4 \quad 9 \quad 7 \quad 1 \quad 6 \quad 8 \quad 3 \quad 2$$

We drop the number 9 since we have only eight experimental units, and obtain the following random assignment:

Experimental unit:	5	4	7	1	6	8	3	2
Treatment:	T_1	T_1	T_1	T_2	T_2	T_3	T_3	T_3

Thus, experimental units 5, 4, and 7 receive treatment T_1, and so on.

If more than 9 but less than 17 experimental units had been involved, we could have used Table A–14, which contains random permutations of 16. The tables by Moses and Oakford (Ref. 21.2) are a good source of random permutations. They contain random permutations of 9, 16, 20, 30, 50, 100, 200, 500, and 1,000.

Use of Table of Random Digits. If a table of random permutations is not available, it is not difficult to generate a random permutation from a table of random digits. We illustrate how to do this for a random permutation of 6. First, we list the digits in order:

$$1 \quad 2 \quad 3 \quad 4 \quad 5 \quad 6$$

Next, we select six random numbers (3 digit numbers are used here to make the probability of a tie small), and enter them over the digits 1 to 6. Suppose we use Table A–15, proceeding down from the upper left corner. We obtain:

132	212	990	001	605	912
1	2	3	4	5	6

Finally, we rearrange the pairs of numbers in ascending sequence for the random numbers:

001	132	212	605	912	990
4	1	2	5	6	3

Thus we obtain the random permutation:

$$4 \quad 1 \quad 2 \quad 5 \quad 6 \quad 3$$

If a tie should be found between two random numbers, a selection of another random digit can be used to break the tie.

It can be shown that the procedure just described makes all permutations equally likely. Thus, in the absence of a table of random permutations, a table of random digits can be used quite readily to obtain random permutations.

Additional Randomizations. Finally, we wish to illustrate how to make additional randomizations besides that of treatments to experimental units. Suppose that in our previous illustration with three treatments and eight experimental units, the experimenter also would like to order the experiment at random over time. We would then select another random permutation of 8 for the time order of the experimental units, say:

Random time order:	5	1	3	6	4	7	2	8
Experimental unit:	1	2	3	4	5	6	7	8

This permutation indicates that experimental unit 1 comes in time position 5, experimental unit 2 in time position 1, etc. The complete set of assignments for our earlier example would be as follows:

Time position:	1	2	3	4	5	6	7	8
Experimental unit:	2	7	3	5	1	4	6	8
Treatment:	T_3	T_1	T_3	T_1	T_2	T_1	T_2	T_3

Comments

1. We have not explicitly indicated whether the treatments to be randomized are for a single-factor or multifactor experiment. The reason is that the randomization procedure for a completely randomized design is unaffected. Thus, in a two-factor study with each factor at two levels, there are simply four treatments T_1, T_2, T_3, and T_4 for purposes of randomization.

2. Should one of the factors in a multifactor experiment be a classification factor (e.g., age of subject, sales volume of store), we will consider such an experiment not to be a completely randomized one, since the randomization of treatments to experimental units is restricted.

Models

The models which we considered in Parts III and IV for single-factor and multifactor investigations are appropriate for completely randomized designs for experimental data. Hence, there is no need to discuss models in any detail. Recall that for a single-factor study with a completely randomized design, the model is of the form:

$$Y_{ij} = \mu_. + \tau_j + \varepsilon_{ij} \tag{21.2}$$

The τ_j are fixed or random according to the nature of the factor levels.

In a two-factor study with a completely randomized design, the model is of the form:

$$Y_{ijk} = \mu_{..} + \alpha_i + \beta_j + (\alpha\beta)_{ij} + \varepsilon_{ijk} \tag{21.3}$$

The main effects α_i and β_j and the interactions $(\alpha\beta)_{ij}$ may be fixed or random depending on the nature of the factors.

Design Considerations

The major statistical question in designing a completely randomized experiment concerns the appropriate sample sizes for the treatments. We discussed this subject in Chapters 15 and 18, and presented various approaches there whereby the necessary sample sizes can be determined. This discussion applies in its entirety to planning sample sizes for completely randomized designs. Hence, there is no need for further discussion of this topic here.

Analysis of Results

The analysis of single-factor and multifactor studies discussed in Parts III and IV is entirely applicable for experimental data based on completely randomized designs. Hence, we present no further examples of analysis of such results at this point.

Evaluation of Aptness of Model

Finally, we mention that the discussion of evaluating the aptness of analysis of variance models in Chapter 15 applies directly to completely randomized experimental designs.

PROBLEMS

21.1. In a study of the effectiveness of subliminal advertising, subjects are brought to a studio and shown a film with one of three types of subliminal advertising for a product. Each subject's attitudes toward the product before and after the film viewing are measured.
- a) In this type of situation, should a control treatment be employed?
- b) Explain precisely what would be the nature of the control treatment for this study.

21.2. Give an example of an experimental study where a repetition is not a replication.

21.3. A random permutation of 12 is desired.
- a) Use Table A-14 to obtain a random permutation of 12.
- b) Suppose a table of random permutations is not available. Use a table of random digits (Table A–15) to obtain a random permutation of 12.

21.4. Four treatments are to be studied in a completely randomized design, and 16 experimental units are available. Each treatment is to be assigned to four experimental units selected at random. There are four observers; each is to be assigned at random to one experimental unit for each treatment. Use a table of random permutations or random digits to make the treatment and observer assignments.

21.5. Refer to Problem 13.11. Explain how you would make the treatment assignments at random in this single-factor study. Use a table of random digits to obtain assignments.

21.6. Refer to Problem 17.9.
- a) Explain how you would make the treatment assignments at random in this two-factor study. Use a table of random digits to obtain assignments.
- b) Did you randomize the factor levels for each factor?

21.7. Refer to Problem 17.10.
- a) Is any randomization of treatment assignments called for in this study?
- b) Would you consider this study to be experimental in nature?

21.8. How does the randomization of treatment assignments in a two-factor study differ when both factors are experimental factors and when only one factor is an experimental factor?

CITED REFERENCES

21.1. Cochran, William G., and Cox, Gertrude M. *Experimental Designs*. 2d ed. New York: John Wiley & Sons, Inc., 1957.

21.2. Moses, Lincoln E., and Oakford, Robert V. *Tables of Random Permutations*. Stanford: Stanford University Press, 1963.

22

Analysis of Covariance for Completely Randomized Designs

WITH COMPLETELY randomized designs, the experimental errors may be quite large due to heterogeneity of the experimental units. One possibility for reducing the experimental errors without utilizing blocking designs, which are discussed in later chapters, is to employ analysis of covariance. We considered covariance models briefly in Chapter 9, and saw there that they are a special type of regression model. In this chapter, we shall first consider the basic ideas underlying the use of covariance models for completely randomized designs. Then, we shall discuss the major analysis problems with covariance models in the framework of regression. Finally, we shall indicate how analysis of covariance may be viewed as a modification of analysis of variance.

22.1 BASIC IDEAS

How Covariance Analysis Reduces Experimental Errors

Blocking can be a highly effective device for reducing experimental errors. In blocking designs, the experimental units are divided into groups such that each group is more homogeneous than the set of experimental units as a whole. Thus, in testing reactions to four different scents that may be used in a new deodorant, the experimental subjects may be grouped by age so that differences in scent preferences related to age are removed from the experimental errors.

Often, however, blocking is not feasible. Covariance analysis may then be helpful to reduce experimental errors. Consider an experiment in which the effects of three different films promoting travel in a state are studied. A subject receives an initial questionnaire to elicit information about his or her attitudes toward the state. The subject is then shown a five-minute film, and immediately afterwards is questioned about the film, about desire to travel in the state, and so on. In this type of experiment, it would be highly desirable to block the subjects according to familiarity with and attitudes toward the state before randomizing the treatment assignments. Blocking is not possible here, however, since only one subject can be studied at a time, and information about the subject's initial attitudes is not obtained until the time the treatment assignment has to be made.

In this type of situation, covariance analysis can be utilized. To see why it might be highly effective, consider Figure 22.1a. Here are plotted the desire-to-travel scores, obtained after the promotional films were shown to five subjects each in a completely randomized design. Three different symbols are used to distinguish the different treatments. It is evident from Figure 22.1a that the experimental errors, as shown by the scatter around the treatment means $\bar{Y}_{.j}$, are fairly large.

Suppose now that we were to utilize the subjects' initial attitude scores, and plot the desire-to-travel scores (obtained after exposure to the film) against the initial attitude scores. This is done in Figure 22.1b. It happens there that the regression relations are linear, but this need not be so. Notice that the scatter around the treatment regression lines is much less than the scatter in Figure 22.1a around the treatment means $\bar{Y}_{.j}$. The reason, of course, is that in this instance the desire-to-travel scores are highly related to the initial attitude scores. The relatively large scatter in Figure 22.1a reflects the large experimental errors that would be encountered with analysis of variance for this completely randomized design. The smaller scatter in Figure 22.1b reflects the smaller experimental errors that would be involved in an analysis of covariance for this same completely randomized design.

Covariance analysis, it is thus seen, utilizes the relationship between the dependent variable (desire-to-travel score, in our example) and an independent quantitative variable for which observations are available (pre-experiment attitude score, in our example) in order to reduce the experimental errors and make the experiment a more powerful one for studying treatment effects. The use of regression relationships in covariance analysis may be thought of as an indirect means of controlling experimental errors.

Concomitant Variable

In covariance analysis terminology, the independent quantitative variable is called the *concomitant variable*. Clearly, the choice of concomitant variable is an important one. If this variable has no relation to the dependent variable, nothing is to be gained by covariance analysis, and one might as well use the simpler analysis of variance. Concomitant variables frequently used with

FIGURE 22.1

Illustration of Experimental Error Reduction by Covariance Analysis

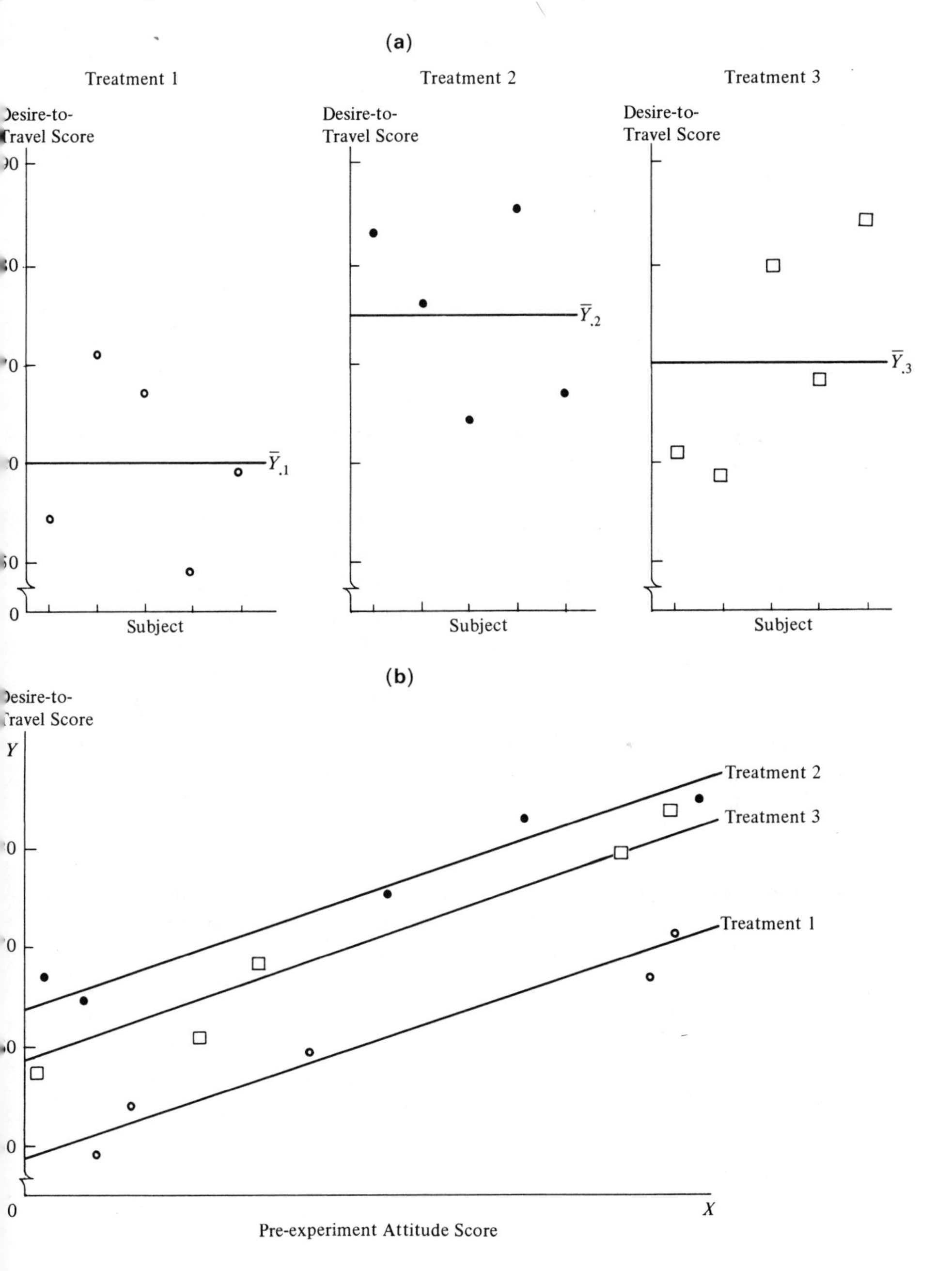

human subjects include pre-experiment attitudes, age, I.Q., and aptitude. When stores are used as experimental units, the concomitant variable might be last period's sales or number of employees.

Choice of Concomitant Variable. There is one problem in selecting the concomitant variable that is unique to covariance analysis. For clear interpretations of the results, the concomitant variable should be observed before the experiment, or if observed during the experiment, it should not be influenced by the treatments in any way. A pre-experiment attitude score meets this requirement. Also, if a subject's age is ascertained during the experiment, it would be reasonable in many instances to expect that his response about age will not be affected by the treatment. The reason for this requirement can be readily seen from the following example. A company was conducting a training school for engineers to teach them accounting and budgeting principles. Two teaching methods were used, and engineers were assigned at random to one of the two. At the end of the program, a score was obtained for each engineer reflecting the amount of learning. The analyst decided to use as the concomitant variable in covariance analysis the amount of time devoted to study (which the engineers were required to record). After conducting the analysis of covariance, the analyst found that training method had virtually no effect. He was baffled by this until it was pointed out to him that the amount of study time probably was also affected by the treatments, and analysis indeed confirmed this. One of the training methods involved computer-assisted learning which appealed to the engineers so that they spent more study time and also learned more. In other words, both learning score and amount of study time were dependent upon the treatment in this case.

When the concomitant variable is affected by the treatments, covariance analysis will remove some (or much) of the effect which the treatments had on the dependent variable, so that an uncritical analysis may be badly misleading. Great care should be taken in the analysis when the covariance technique is employed with a concomitant variable affected by treatments.

Figure 22.2 shows the data for the experiment involving the training of engineers. Treatment 1 is the one using computer-assisted learning. Note that most persons with this treatment devoted large amounts of time to study. On the other hand, persons receiving treatment 2 tended to devote smaller amounts of time to study. As a result, the observations for the two treatments tend to be bunched over different intervals on the X scale.

Contrast this situation with the one seen in Figure 22.1b for the experiment on promotional films. Figure 22.1b illustrates how the concomitant variable observations should be scattered if treatments have no effect on the concomitant variable. Because of the randomization procedure employed in a completely randomized design, any subject has equal probability of being assigned to any one treatment. Hence the distribution of subjects along the X scale by pre-experiment attitude scores (or any other attribute not affected by treatments) should be roughly similar for all treatments, subject only to chance variation.

FIGURE 22.2

Illustration of Treatments Affecting the Concomitant Variable

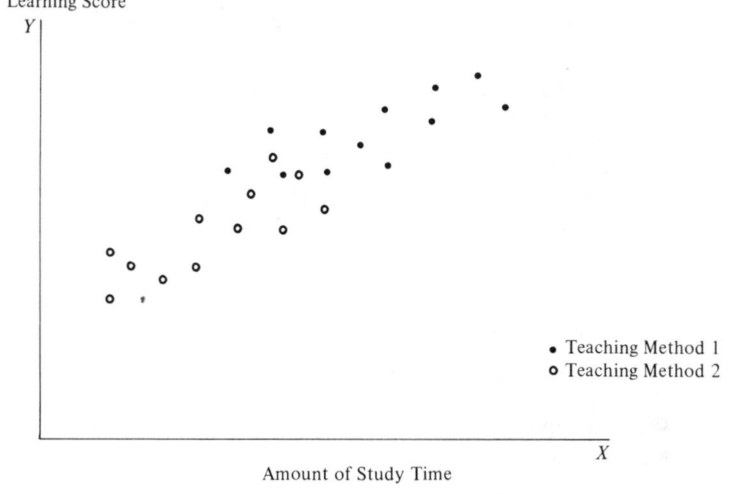

22.2 COVARIANCE MODEL

With a completely randomized design, n_j experimental units are assigned randomly to the jth treatment and the total number of observations is $n_T = \sum n_j$. The observation on the dependent variable for the ith experimental unit assigned to treatment j is denoted Y_{ij}, as usual. Let X_{ij} be the value of the concomitant variable for this experimental unit.

The fixed effects analysis of variance model (13.6) is:

(22.1) $$Y_{ij} = \mu_{.} + \tau_j + \varepsilon_{ij}$$

The covariance model starts with this analysis of variance model and simply adds another term (or several), reflecting the relationship between the concomitant and dependent variables. Usually, a linear relation is utilized as a first approximation:

(22.2) $$Y_{ij} = \mu_{.} + \tau_j + \gamma X_{ij} + \varepsilon_{ij}$$

Here γ is a regression coefficient for the relation between Y and X. The constant $\mu_{.}$ now is no longer an overall mean. We can, however, make the constant an overall mean, and incidentally simplify some computations, if we express the concomitant variable as a deviation from the overall mean $\bar{X}_{..}$. The resulting model is the usual covariance model for a completely randomized design with fixed treatment effects:

(22.3) $$Y_{ij} = \mu_{.} + \tau_j + \gamma(X_{ij} - \bar{X}_{..}) + \varepsilon_{ij}$$

where:

$\mu_{.}$ is an overall mean
τ_j are the fixed treatment effects subject to the restriction $\sum n_j \tau_j = 0$
γ is a regression coefficient for the relation between Y and X
X_{ij} are constants
ε_{ij} are independent $N(0, \sigma^2)$
$i = 1, \ldots, n_j; j = 1, \ldots, r$

Model (22.3) corresponds to the analysis of variance model (22.1) except for the added term $\gamma(X_{ij} - \bar{X}_{..})$ to reflect the relationship between Y and X. Note that the concomitant observations X_{ij} are assumed to be constants. Since ε_{ij} is the only random variable on the right side of (22.3), it follows at once that:

(22.4a) $$E(Y_{ij}) = \mu_{.} + \tau_j + \gamma(X_{ij} - \bar{X}_{..})$$

(22.4b) $$\sigma^2(Y_{ij}) = \sigma^2$$

In view of the independence of the ε_{ij}, the Y_{ij} are also independent. Hence, an alternate statement of model (22.3) is:

(22.5) $$Y_{ij} \text{ are independent } N(\mu_{ij}, \sigma^2)$$

where:

$$\mu_{ij} = \mu_{.} + \tau_j + \gamma(X_{ij} - \bar{X}_{..})$$
$$\sum n_j \tau_j = 0$$

Properties of Model

Some of the properties of the covariance model (22.3) are identical to those of the analysis of variance model (22.1). For instance, the error terms ε_{ij} are independent and have constant variance. There are also some new properties, and we discuss these now.

Comparisons of Treatment Effects. With the analysis of variance model, all observations for the jth treatment have the same mean response (μ_j). This is not so with the covariance model, since the mean response here depends on the treatment and also on the value of the concomitant variable X_{ij} for the experimental unit. Thus, the expected response for the jth treatment with covariance model (22.3) is given by a regression line:

(22.6) $$\mu_{ij} = \mu_{.} + \tau_j + \gamma(X_{ij} - \bar{X}_{..})$$

which indicates, for any value of X, the mean response with treatment j. Figure 22.3 illustrates for an experiment with three treatments how these treatment regression lines might appear. Note that $\mu_{.} + \tau_j$ is the ordinate of the line for the jth treatment when $X - \bar{X}_{..} = 0$, that is, when $X = \bar{X}_{..}$, and that γ is the slope of each line. Since all treatment regression lines have the same slope, they are parallel.

FIGURE 22.3

Treatment Regression Lines with Covariance Model (22.3)

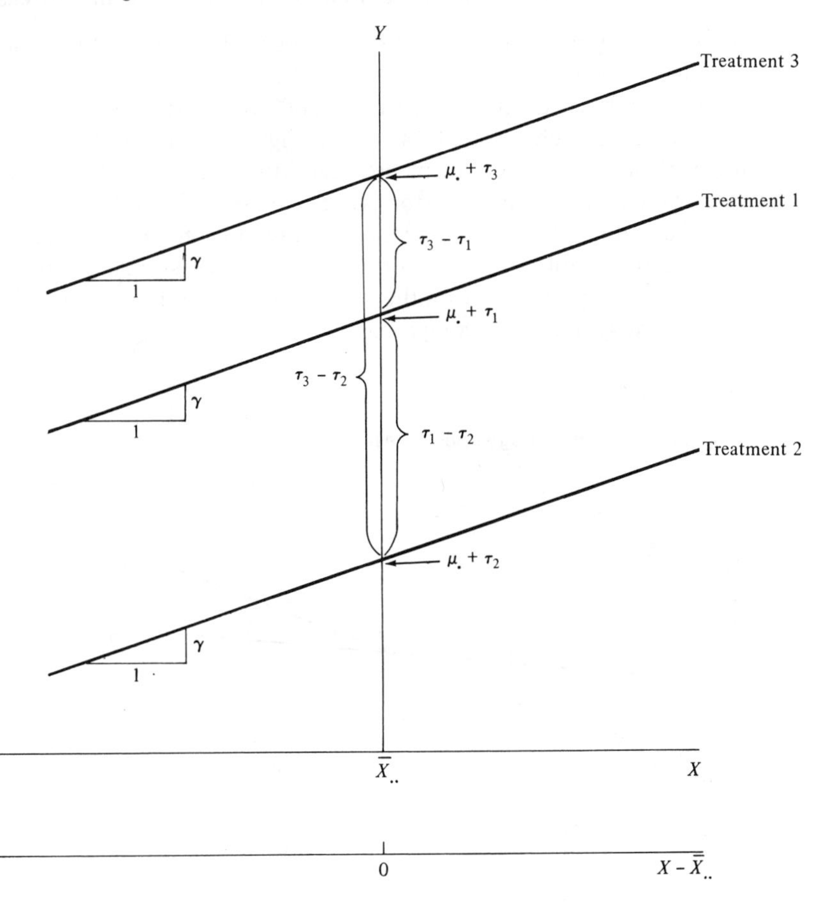

While we no longer can speak of *the* mean response with the jth treatment since it varies with X, we can still measure the effect of any treatment compared with any other by a single number. In Figure 22.3, for instance, treatment 1 leads to a higher mean response than treatment 2, no matter what is the value of X. The difference between the two mean responses is the same for all values of X, since the slopes of the regression lines are equal. Hence we can measure the difference at any convenient X, say at $X = \bar{X}_{..}$:

$$\mu_. + \tau_1 - (\mu_. + \tau_2) = \tau_1 - \tau_2 \tag{22.7}$$

Thus, $\tau_1 - \tau_2$ measures how much higher the mean response is with treatment 1 than with treatment 2, for any value of X. We can compare any other two treatments similarly. It follows directly from this discussion that when all treatments have the same mean responses for each X (i.e., the treatments have no differential effects), the treatment regression lines must be identical and hence $\tau_1 - \tau_2 = 0$, $\tau_1 - \tau_3 = 0$, etc. Indeed, all τ_j are zero in that case.

Constancy of Slopes. The assumption in model (22.3) that all treatment regression lines have the same slope is a crucial one. Without it, one cannot summarize the difference between the effects of two treatments by a single number based on main effects, such as $\tau_2 - \tau_1$. Figure 22.4 illustrates the case of nonparallel slopes for two treatments. Here, treatment 1 leads to higher mean responses than treatment 2 for some values of X, and the reverse holds for other values of X. When the treatment regression lines interact with the concomitant variable X in the form of nonparallel slopes, covariance analysis is not appropriate. Instead, the treatment regression lines should be estimated separately, and then compared.

FIGURE 22.4

Nonparallel Treatment Regression Lines

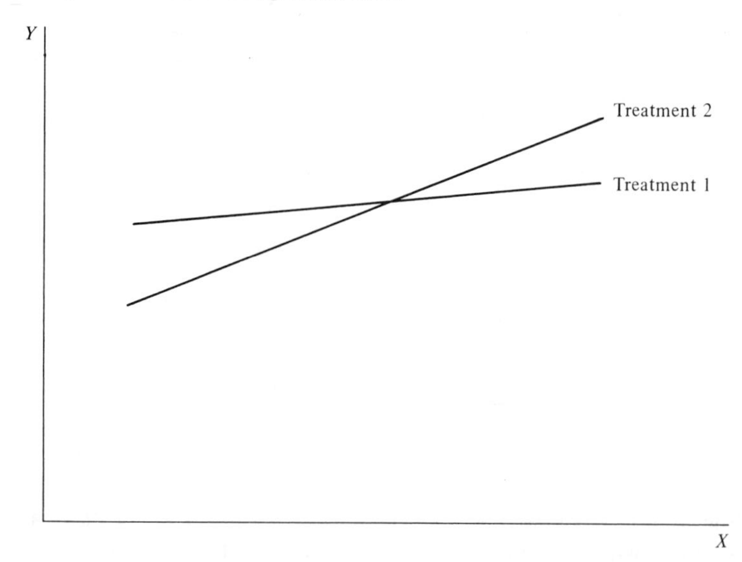

Constant X's. Model (22.3) assumes that the observations on the concomitant variable are constants. At times, it might be more reasonable to consider the concomitant observations as random variables. In that case, if one is willing to interpret model (22.3) as a conditional one, applying for any X values that might be observed, the covariance analysis to be presented is still appropriate.

Linearity of Relation. The linear relation between Y and X assumed in model (22.3) is not essential to covariance analysis. Any other relation could be used. For instance, the model would be as follows for a quadratic relation:

$$Y_{ij} = \mu_{.} + \tau_j + \gamma_1(X_{ij} - \bar{X}_{..}) + \gamma_2(X_{ij} - \bar{X}_{..})^2 + \varepsilon_{ij} \tag{22.8}$$

Linearity of the relation leads to simpler analysis, and is often a sufficiently good approximation to provide meaningful results. If a linear relation is not a good approximation, however, a more adequate description of the relation should be utilized in the model.

One Concomitant Variable. Model (22.3) uses a single concomitant variable. This is often sufficient to reduce the experimental errors substantially. However, the model can be extended in a straightforward fashion to include two or more concomitant variables. The model for two concomitant variables, X_1 and X_2, would be:

$$Y_{ij} = \mu_{.} + \tau_j + \gamma_1(X_{ij1} - \bar{X}_{..1}) + \gamma_2(X_{ij2} - \bar{X}_{..2}) + \varepsilon_{ij} \tag{22.9}$$

Questions of Interest

Inferences. The key statistical inferences of interest in covariance analysis are the same as with analysis of variance models, namely whether the treatments have any effect, and if so what these effects are. Testing for treatment effects involves the same alternatives as for analysis of variance models:

$$\begin{aligned} &C_1\colon \tau_1 = \tau_2 = \cdots = \tau_r = 0 \\ &C_2\colon \text{not all } \tau_j \text{ are zero} \end{aligned} \tag{22.10}$$

If the treatment effects differ, one usually wishes to investigate the nature of these effects. Pairwise comparisons of treatment effects $\tau_j - \tau_{j'}$ (the vertical distance between the two treatment regression lines) may be made, or more general contrasts of the τ_j's may be utilized. The nature of the regression relationship between Y and X sometimes is of interest, but usually the concomitant variable X is only employed to help reduce the experimental errors.

Appropriateness of Model. A number of questions concerning the appropriateness of the covariance model may also be of interest. These may be answered by means of statistical methods previously discussed. Specifically, these questions pertain to:

1. Normality of error terms.
2. Equality of error variances for different treatments.
3. Equality of slopes for the different treatment regression lines.
4. Linearity of regression. (Because there are typically no replications at the different X levels, this is often examined via quadratic regression and testing whether the curvature coefficient is zero.)

Note

One is usually not concerned with whether the regression coefficient γ is zero, that is, whether there is indeed a relation between Y and X. If there is no relation, no bias results in the covariance analysis. The experimental error mean square would simply be the same as for analysis of variance (allowing for sampling variation), and one degree of freedom for the error mean square would be lost.

Regression Formulation

The easiest way to estimate the parameters of the covariance model (22.3) and analyze questions of interest is through the regression approach. (If the reader has not yet read Sections 9.1–9.3, he should do so before proceeding further.)

We shall denote $X_{ij} - \bar{X}_{..}$ in model (22.3) by Z_{ij}:

(22.11) $$Z_{ij} = X_{ij} - \bar{X}_{..}$$

Further, we shall use $r - 1$ indicator variables for the r treatments:

(22.12)
$$X_1 = \begin{matrix} 1 \text{ if observation from treatment } 1 \\ 0 \text{ otherwise} \end{matrix}$$
$$\vdots \qquad \vdots$$
$$X_{r-1} = \begin{matrix} 1 \text{ if observation from treatment } r-1 \\ 0 \text{ otherwise} \end{matrix}$$

We can then express the covariance model (22.3) as follows:

(22.13) $$Y_{ij} = \beta_0 + \beta_1 X_{ij1} + \cdots + \beta_{r-1} X_{ij,r-1} + \beta_r Z_{ij} + \varepsilon_{ij}$$

The correspondences between the parameters in the regression model (22.13) and the parameters in the covariance analysis model (22.3) are as follows:

(22.14)
$$\beta_0 = \mu_. + \tau_r$$
$$\beta_j = \tau_j - \tau_r \qquad j = 1, \ldots, r-1$$
$$\beta_r = \gamma$$

These correspondences are established in the usual fashion. For instance, if an experimental unit received treatment r, the analysis of covariance model (22.3) would yield:

(22.15) $$Y_{ir} = \mu_. + \tau_r + \gamma Z_{ir} + \varepsilon_{ir}$$

With the regression model (22.13), the X variables would have the values:

$$X_1 = 0 \qquad X_2 = 0 \qquad \ldots \qquad X_{r-1} = 0$$

Hence, we obtain with the regression model:

$$Y_{ir} = \beta_0 + \beta_r Z_{ir} + \varepsilon_{ir} \tag{22.16}$$

Equating (22.15) and (22.16), we obtain:

$$\beta_0 = \mu_. + \tau_r$$
$$\beta_r = \gamma$$

The other correspondences are obtained in similar fashion.

It is evident from the correspondences in (22.14) that tests concerning the treatment effects τ_j can be carried out as tests on the regression coefficients β_j $(j = 1, \ldots, r - 1)$. Similarly, estimates of the treatment effects can be obtained in terms of estimates of the regression coefficients. Since there are no new basic principles involved, we shall illustrate the usual tests and estimation procedures in terms of an example.

22.3 EXAMPLE

A company wished to study the effects of three different types of promotions on sales of its crackers. The three promotions were:

Treatment 1—Sampling of product by customers in store and regular shelf space
Treatment 2—Additional shelf space in regular location
Treatment 3—Special display shelves at ends of aisle in addition to regular shelf space

Fifteen stores were selected as the experimental units. Each store was randomly assigned one of the promotion types, with five stores assigned to each type of promotion. Other relevant conditions under the control of the company, such as price and advertising, were kept the same for all stores in the experiment. Data on the number of cases of the product sold during the promotional period, denoted by Y, are presented in Table 22.1a, as are also data on the sales of the product in the preceding period, denoted by X. Sales of the preceding period are to be used as the concomitant variable.

Development of Model

Figure 22.5 presents the data of Table 22.1a. Linear regression and parallel slopes for the treatment regression lines appear to be reasonable. (Later we shall illustrate a formal statistical test for parallel slopes of the treatment regression lines.) Hence, the following regression model was employed:

$$Y_{ij} = \beta_0 + \beta_1 X_{ij1} + \beta_2 X_{ij2} + \beta_3 Z_{ij} + \varepsilon_{ij} \qquad \text{Full model} \tag{22.17}$$

TABLE 22.1

Data for Cracker Promotion Example (number of cases sold)

(a) Data

Experimental Unit	Treatment 1		Treatment 2		Treatment 3	
	Y	X	Y	X	Y	X
1	38	21	43	34	24	23
2	39	26	38	26	32	29
3	36	22	38	29	31	30
4	45	28	27	18	21	16
5	33	19	34	25	28	29

(b) Means, Totals, and Sums of Squares

$\bar{Y}_{.1} = 38.2$ $\bar{Y}_{.2} = 36.0$ $\bar{Y}_{.3} = 27.2$ $\bar{Y}_{..} = 33.8$
$\bar{X}_{.1} = 23.2$ $\bar{X}_{.2} = 26.4$ $\bar{X}_{.3} = 25.4$ $\bar{X}_{..} = 25.0$

Treatment	$\sum_i Y_{ij}$	$\sum_i Y_{ij}^2$	$\sum_i X_{ij}$	$\sum_i X_{ij}^2$	$\sum_i X_{ij} Y_{ij}$
1	191	7,375	116	2,746	4,491
2	180	6,622	132	3,622	4,888
3	136	3,786	127	3,367	3,558
Total	507	17,783	375	9,735	12,937

FIGURE 22.5

Scatter Plot of Cracker Sales

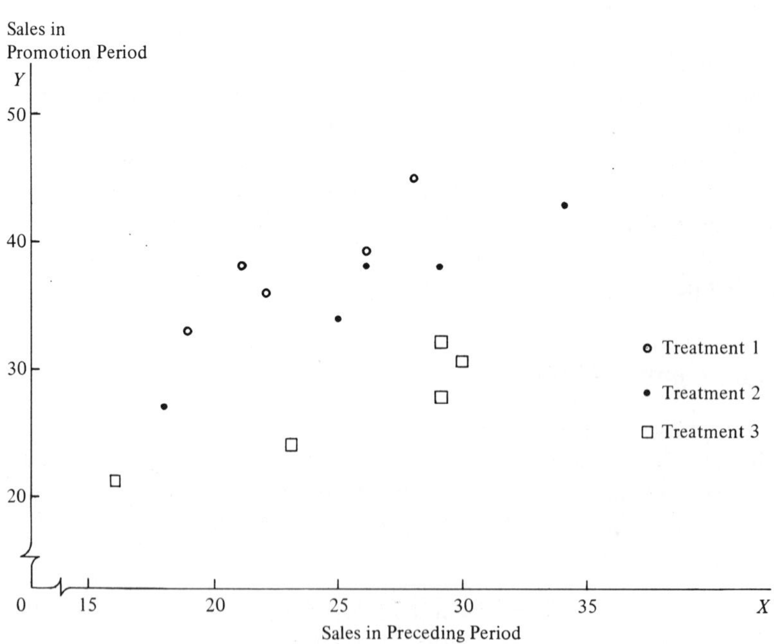

where:

$$X_1 = \begin{matrix} 1 \text{ if observation from treatment 1} \\ 0 \text{ otherwise} \end{matrix}$$

$$X_2 = \begin{matrix} 1 \text{ if observation from treatment 2} \\ 0 \text{ otherwise} \end{matrix}$$

$$Z_{ij} = X_{ij} - \bar{X}_{..}$$

$$\beta_0 = \mu_. + \tau_3$$

$$\beta_1 = \tau_1 - \tau_3$$

$$\beta_2 = \tau_2 - \tau_3$$

$$\beta_3 = \gamma$$

The **Y** observation vector and the **X** matrix for the data in Table 22.1a are given in Table 22.2. A computer run of a multiple regression package led to the results summarized in Table 22.3.

TABLE 22.2

Data Matrices for Cracker Promotion Example—Covariance Model (22.17)

$$\mathbf{Y} = \begin{bmatrix} 38 \\ 39 \\ 36 \\ 45 \\ 33 \\ 43 \\ 38 \\ 38 \\ 27 \\ 34 \\ 24 \\ 32 \\ 31 \\ 21 \\ 28 \end{bmatrix} \qquad \mathbf{X} = \begin{bmatrix} & X_1 & X_2 & Z \\ 1 & 1 & 0 & 21-25=-4 \\ 1 & 1 & 0 & 26-25=1 \\ 1 & 1 & 0 & 22-25=-3 \\ 1 & 1 & 0 & 28-25=3 \\ 1 & 1 & 0 & 19-25=-6 \\ 1 & 0 & 1 & 34-25=9 \\ 1 & 0 & 1 & 26-25=1 \\ 1 & 0 & 1 & 29-25=4 \\ 1 & 0 & 1 & 18-25=-7 \\ 1 & 0 & 1 & 25-25=0 \\ 1 & 0 & 0 & 23-25=-2 \\ 1 & 0 & 0 & 29-25=4 \\ 1 & 0 & 0 & 30-25=5 \\ 1 & 0 & 0 & 16-25=-9 \\ 1 & 0 & 0 & 29-25=4 \end{bmatrix}$$

Test for Treatment Effects

To test whether or not the three cracker promotions differ in effectiveness, we follow the general procedures for testing a linear model discussed in Chapter 3. The alternatives are:

$$\begin{aligned} &C_1: \tau_1 = \tau_2 = \tau_3 = 0 \\ &C_2: \text{not all } \tau_j \text{ are zero} \end{aligned} \tag{22.18}$$

TABLE 22.3

Computer Output for Cracker Promotion Example—Covariance Model (22.17)

(a) Regression Coefficients

$b_0 = 26.841$	$b_2 = 7.901$
$b_1 = 12.977$	$b_3 = .899$

(b) Analysis of Variance

Source of Variation	*SS*	*df*	*MS*
Regression	$SSR = 607.829$	3	$MSR = 202.610$
Error	$SSE = 38.571$	11	$MSE = 3.506$
Total	$SSTO = 646.400$	14	

(c) Estimated Variance-Covariance Matrix for Regression Coefficients

	b_0	b_1	b_2	b_3
b_0	.7030			
b_1	−.7106	1.4535		
b_2	−.6971	.6781	1.4131	
b_3	−.00421	.0232	−.0105	.0105

Using the correspondences in (22.17), we can restate these alternatives in terms of the regression parameters:

(22.18a)
$$C_1: \beta_1 = \beta_2 = 0$$
$$C_2: \text{not both } \beta_1 \text{ and } \beta_2 \text{ are zero}$$

Hence, the reduced model under C_1 is:

(22.19)
$$Y_{ij} = \beta_0 + \beta_3 Z_{ij} + \varepsilon_{ij} \qquad \text{Reduced model}$$

where now $\beta_0 = \mu_.$ and all other terms are the same as before. Note, indeed, that model (22.19) is just a simple linear regression model where none of the parameters vary for the different treatments:

(22.19a)
$$Y_{ij} = \mu_. + \gamma(X_{ij} - \bar{X}_{..}) + \varepsilon_{ij}$$

The data matrices for this reduced model are shown in Table 22.4. A computer run, based on these data, led to the following results:

Source of Variation	*SS*	*df*
Regression	$SSR = 190.678$	1
Error	$SSE = 455.722$	13
Total	$SSTO = 646.400$	14

TABLE 22.4

Data Matrices for Cracker Promotion Example—Reduced Model (22.19)

$$\mathbf{Y} = \begin{bmatrix} 38 \\ 39 \\ 36 \\ 45 \\ 33 \\ 43 \\ 38 \\ 38 \\ 27 \\ 34 \\ 24 \\ 32 \\ 31 \\ 21 \\ 28 \end{bmatrix} \qquad \mathbf{X} = \begin{bmatrix} 1 & -4 \\ 1 & 1 \\ 1 & -3 \\ 1 & 3 \\ 1 & -6 \\ 1 & 9 \\ 1 & 1 \\ 1 & 4 \\ 1 & -7 \\ 1 & 0 \\ 1 & -2 \\ 1 & 4 \\ 1 & 5 \\ 1 & -9 \\ 1 & 4 \end{bmatrix}$$

(The second column of $\mathbf{X}$ is headed Z.)

Thus, $SSE(R) = 455.722$. We know from Table 22.3 that the error sum of squares for the full model is $SSE(F) = 38.571$. Hence, test statistic (3.66) here is:

$$F^* = \frac{SSE(R) - SSE(F)}{(n_T - 2) - [n_T - (r + 1)]} \div \frac{SSE(F)}{n_T - r - 1}$$

$$= \frac{455.72 - 38.57}{13 - 11} \div \frac{38.57}{11} = 59.5$$

The level of significance is to be controlled at .05; hence we require:

$$F(.95; 2, 11) = 3.98$$

The decision rule therefore is:

$$\text{If } F^* \leq 3.98, \text{ conclude } C_1$$
$$\text{If } F^* > 3.98, \text{ conclude } C_2$$

Since $F^* = 59.5 > 3.98$, we conclude C_2, that the three cracker promotions differ in sales effectiveness.

Estimation of Treatment Effects

Since treatment effects were found to be present, the analyst wished to investigate them. We noted earlier that a comparison of two treatments involves $\tau_j - \tau_{j'}$, the vertical distance between the two treatment regression lines. The counterpart of $\tau_j - \tau_{j'}$ in the regression model (22.17) may be obtained from the associated correspondences:

$$\begin{aligned} \tau_1 - \tau_2 &= \beta_1 - \beta_2 \\ \tau_1 - \tau_3 &= \beta_1 \\ \tau_2 - \tau_3 &= \beta_2 \end{aligned} \tag{22.20}$$

Hence, we know at once that the estimators of these comparisons, and their variances, are as follows:

(22.21)

Comparison	*Estimator*	*Variance of Estimator*
$\tau_1 - \tau_2$	$b_1 - b_2$	$\sigma^2(b_1) + \sigma^2(b_2) - 2\sigma(b_1, b_2)$
$\tau_1 - \tau_3$	b_1	$\sigma^2(b_1)$
$\tau_2 - \tau_3$	b_2	$\sigma^2(b_2)$

Table 22.3 furnishes us the needed estimated regression coefficients as well as their estimated variances and covariances. We obtain from there:

(22.21a)

Comparison	*Estimate*	*Estimated Variance*
$\tau_1 - \tau_2$	12.977 − 7.901 = 5.076	1.4535 + 1.4131 − 2(.6781) = 1.5104
$\tau_1 - \tau_3$	12.977	1.4535
$\tau_2 - \tau_3$	7.901	1.4131

If a single interval estimate is to be constructed, the t distribution with $n_T - r - 1$ degrees of freedom would be used. (The degrees of freedom are those associated with MSE in the full covariance model.) Usually, however, a family of interval estimates is desired. In that case, the Scheffé multiple comparison procedure may be employed with the S multiple defined by:

$$S^2 = (r - 1)F(1 - \alpha; r - 1, n_T - r - 1) \tag{22.22}$$

or the Bonferroni method may be employed with the B multiple:

$$B = t(1 - \alpha/2s; n_T - r - 1) \tag{22.23}$$

where s is the number of statements in the family. The Tukey method is not appropriate for covariance analysis.

In the case at hand, the analyst wished to obtain all pairwise comparisons with a 95 percent family confidence coefficient. The analyst used the Scheffé procedure because he anticipated making some additional estimates of contrasts. He required therefore:

$$S^2 = (3 - 1)F(.95; 2, 11) = 2(3.98) = 7.96$$

or:

$$S = 2.82$$

Using the results in (22.21a), the confidence intervals for all treatment comparisons, with a 95 percent family confidence coefficient, then are:

$$1.61 = 5.076 - 2.82\sqrt{1.5104} \leq \tau_1 - \tau_2 \leq 5.076 + 2.82\sqrt{1.5104} = 8.54$$
$$9.58 = 12.977 - 2.82\sqrt{1.4535} \leq \tau_1 - \tau_3 \leq 12.977 + 2.82\sqrt{1.4535} = 16.38$$
$$4.55 = 7.901 - 2.82\sqrt{1.4131} \leq \tau_2 - \tau_3 \leq 7.901 + 2.82\sqrt{1.4131} = 11.25$$

These results indicate clearly that sampling in store (treatment 1) is significantly better for stimulating cracker sales than either of the two shelf promotions, and that increasing the regular shelf space (treatment 2) is superior to additional displays at the end of the aisle (treatment 3).

Comments

1. Occasionally, more general contrasts among treatment effects than pairwise comparisons are desired. No new problems arise either in the use of the t distribution for a single contrast or in the use of the Scheffé or Bonferroni procedures for multiple comparisons. For instance, if the analyst had desired in our cracker promotion example to compare the treatment effect for sampling in the store (treatment 1) with the two treatments involving shelf displays (treatments 2 and 3), he would have been interested in the constrast:

$$L = \tau_1 - \frac{\tau_2 + \tau_3}{2} \tag{22.24}$$

From the correspondences in (22.17), we see that the appropriate estimator in terms of the regression coefficients is:

$$\hat{L} = b_1 - \frac{1}{2} b_2 \tag{22.25}$$

The variance of this estimator, from basic probability theory, is:

$$\sigma^2(\hat{L}) = \sigma^2(b_1) + \frac{1}{4} \sigma^2(b_2) - \sigma(b_1, b_2) \tag{22.26}$$

2. It may be of interest to estimate the mean response of the jth treatment for some value of X. Frequently $X = \bar{X}_{..}$ is considered to be a "typical" value of X. We know from Figure 22.3 that at $X = \bar{X}_{..}$, the mean response for the jth treatment is the intercept of the treatment regression line, $\mu_. + \tau_j$. An estimator of $\mu_. + \tau_j$ can be readily developed. For our cracker promotion example, we obtain the following estimators and their variances:

(22.27)

Mean Response at $X = \bar{X}_{..}$	*Estimator*	*Variance*
$\mu_. + \tau_1$	$b_0 + b_1$	$\sigma^2(b_0) + \sigma^2(b_1) + 2\sigma(b_0, b_1)$
$\mu_. + \tau_2$	$b_0 + b_2$	$\sigma^2(b_0) + \sigma^2(b_2) + 2\sigma(b_0, b_2)$
$\mu_. + \tau_3$	b_0	$\sigma^2(b_0)$

Use of the results in Table 22.3 leads to the following estimates:

Treatment	*Estimated Mean Response at* $\bar{X}_{..}$	*Estimated Variance*
1	$26.841 + 12.977 = 39.818$	$.7030 + 1.4535 + 2(-.7106) = .7353$
2	$26.841 + 7.901 = 34.742$	$.7030 + 1.4131 + 2(-.6971) = .7219$
3	26.841	.7030

3. Computational formulas for estimating treatment comparisons and general contrasts, which can be used in case a computer with a multiple regression program is not available, will be presented in the next section.

Test for Parallel Slopes

An important assumption made in covariance analysis is that all treatment regression lines have the same slope γ. The analyst who conducted the cracker promotion study indeed tested this assumption before proceeding with the analysis discussed earlier. We know from Chapter 9 that model (22.17) can be generalized to allow for different slopes for the treatments by introducing interaction terms. Specifically, interaction variables X_1Z and X_2Z will be required. Thus, the generalized model is:

$$Y_{ij} = \beta_0 + \beta_1 X_{ij1} + \beta_2 X_{ij2} + \beta_3 Z_{ij} + \beta_4 X_{ij1}Z_{ij} + \beta_5 X_{ij2}Z_{ij} + \varepsilon_{ij} \qquad \text{Generalized model} \tag{22.28}$$

Table 22.5 contains the data matrices for this generalized model for our cracker promotion example. Note that the **X** matrix for the generalized model differs from the **X** matrix for the covariance model (22.17) simply by the addition of the X_1Z and X_2Z columns. A fit of model (22.28) by a computer multiple regression package yielded the following ANOVA sums of squares:

Source of Variation	*SS*	*df*
Regression	$SSR = 614.879$	5
Error	$SSE = 31.521$	9
Total	$SSTO = 646.400$	14

TABLE 22.5

Data Matrices for Cracker Promotion Example—Generalized Model (22.28)

$$\mathbf{Y} = \begin{bmatrix} 38 \\ 39 \\ 36 \\ 45 \\ 33 \\ 43 \\ 38 \\ 38 \\ 27 \\ 34 \\ 24 \\ 32 \\ 31 \\ 21 \\ 28 \end{bmatrix} \qquad \mathbf{X} = \begin{bmatrix} 1 & 1 & 0 & -4 & -4 & 0 \\ 1 & 1 & 0 & 1 & 1 & 0 \\ 1 & 1 & 0 & -3 & -3 & 0 \\ 1 & 1 & 0 & 3 & 3 & 0 \\ 1 & 1 & 0 & -6 & -6 & 0 \\ 1 & 0 & 1 & 9 & 0 & 9 \\ 1 & 0 & 1 & 1 & 0 & 1 \\ 1 & 0 & 1 & 4 & 0 & 4 \\ 1 & 0 & 1 & -7 & 0 & -7 \\ 1 & 0 & 1 & 0 & 0 & 0 \\ 1 & 0 & 0 & -2 & 0 & 0 \\ 1 & 0 & 0 & 4 & 0 & 0 \\ 1 & 0 & 0 & 5 & 0 & 0 \\ 1 & 0 & 0 & -9 & 0 & 0 \\ 1 & 0 & 0 & 4 & 0 & 0 \end{bmatrix}$$

(Columns of **X** after the first: X_1, X_2, Z, X_1Z, X_2Z.)

Remember that SSE obtained by fitting the generalized model (22.28) is the equivalent of fitting separate regression lines for each treatment and summing these error sums of squares.

To test for parallel slopes is equivalent to testing for no interactions in the generalized model (22.28):

(22.29)
$$C_1: \beta_4 = \beta_5 = 0$$
$$C_2: \text{not both } \beta_4 \text{ and } \beta_5 \text{ are zero}$$

We now need to recognize that the generalized model (22.28) here is the "full" model and the full model (22.17) is now the "reduced" model. Hence we have:

$$SSE(F) = 31.521 \qquad SSE(R) = 38.571$$

Thus, the test statistic (3.66) becomes here:

$$F^* = \frac{38.571 - 31.521}{11 - 9} \div \frac{31.521}{9} = 1.01$$

For a level of significance of .05, we require $F(.95; 2, 9) = 4.26$. Since $F^* = 1.01 < 4.26$, we conclude C_1, that the three treatment regression lines have the same slope.

22.4 COVARIANCE ANALYSIS AS MODIFICATION OF ANALYSIS OF VARIANCE

In previous sections, we explained the fundamental ideas underlying covariance analysis by viewing the model as a regression model. We saw that the statistical questions which arise in covariance analysis fit readily in the regression framework. When analysis of covariance was first developed, however, electronic computers did not exist and it was not feasible to do the covariance analysis computations by means of the regression approach. Instead, computational formulas were developed which exploit the fact that the indicator variables take on the values 0 and 1 in certain structures. These computational formulas may appear formidable, but computations by hand via these formulas are much simpler than hand computations via the regression approach. To explain the rationale of these computational formulas, analysis of covariance is usually viewed as a process which begins with the regular analysis of variance and *adjusts* this analysis for the concomitant variable. We shall now examine this adjustment approach to covariance analysis and demonstrate its equivalence to the regression approach.

Analysis of Covariance Table

Analysis of Variance for X and XY. The first idea to be considered when viewing covariance analysis as an adjustment of regular analysis of variance is that the decomposition of the total sum of squares in (13.40):

(22.30)
$$SSTO_Y = SSTR_Y + SSE_Y$$

can also be carried out for the variable X and the product XY. Note that the subscript Y has been added in (22.30) to make clear that this decomposition refers to the variable Y. It will be recalled that:

$$SSTO_Y = \sum_i \sum_j (Y_{ij} - \bar{Y}_{..})^2 = \sum_i \sum_j Y_{ij}^2 - \frac{Y_{..}^2}{n_T} \tag{22.31a}$$

$$SSTR_Y = \sum_j n_j(\bar{Y}_{.j} - \bar{Y}_{..})^2 = \sum_j \frac{Y_{.j}^2}{n_j} - \frac{Y_{..}^2}{n_T} \tag{22.31b}$$

$$SSE_Y = \sum_i \sum_j (Y_{ij} - \bar{Y}_{.j})^2 = \sum_i \sum_j Y_{ij}^2 - \sum_j \frac{Y_{.j}^2}{n_j} \tag{22.31c}$$

The same analysis of variance for the variable X is:

$$SSTO_X = \sum_i \sum_j (X_{ij} - \bar{X}_{..})^2 = \sum_i \sum_j X_{ij}^2 - \frac{X_{..}^2}{n_T} \tag{22.32a}$$

$$SSTR_X = \sum_j n_j(\bar{X}_{.j} - \bar{X}_{..})^2 = \sum_j \frac{X_{.j}^2}{n_j} - \frac{X_{..}^2}{n_T} \tag{22.32b}$$

$$SSE_X = \sum_i \sum_j (X_{ij} - \bar{X}_{.j})^2 = \sum_i \sum_j X_{ij}^2 - \sum_j \frac{X_{.j}^2}{n_j} \tag{22.32c}$$

where the notation for the X's corresponds exactly to that for the Y's.

The analysis of variance for the products XY starts with the definition of *total sum of products* ($SPTO$):

$$SPTO = \sum_i \sum_j (X_{ij} - \bar{X}_{..})(Y_{ij} - \bar{Y}_{..}) = \sum_i \sum_j X_{ij} Y_{ij} - \frac{X_{..} Y_{..}}{n_T} \tag{22.33a}$$

To see that the total sum of products is related to the total sum of squares for Y or X, note that if X_{ij} is replaced by Y_{ij} in (22.33a), we obtain $SSTO_Y$, and if Y_{ij} is replaced by X_{ij}, we obtain $SSTO_X$. The two components of the total sum of products are the *treatment sum of products* ($SPTR$) and the *error sum of products* (SPE):

$$SPTR = \sum_j n_j(\bar{X}_{.j} - \bar{X}_{..})(\bar{Y}_{.j} - \bar{Y}_{..}) = \sum_j \frac{X_{.j} Y_{.j}}{n_j} - \frac{X_{..} Y_{..}}{n_T} \tag{22.33b}$$

$$SPE = \sum_i \sum_j (X_{ij} - \bar{X}_{.j})(Y_{ij} - \bar{Y}_{.j}) = \sum_i \sum_j X_{ij} Y_{ij} - \sum_j \frac{X_{.j} Y_{.j}}{n_j} \tag{22.33c}$$

Unlike sums of squares, *sums of products may be negative.*

Example. Table 22.6 contains the analyses of variance for Y, X, and XY for our cracker promotions example in Table 22.1a. We illustrate two of the calculations, using the results presented in Table 22.1b:

$$SPTR = \frac{1}{5}[(116)(191) + (132)(180) + (127)(136)] - \frac{(375)(507)}{15}$$

$$= 12{,}637.6 - 12{,}675.0 = -37.4$$

$$SSE_X = 9{,}735 - \frac{1}{5}[(116)^2 + (132)^2 + (127)^2] = 9{,}735 - 9{,}401.8 = 333.2$$

TABLE 22.6

Analyses of Variance for Y, X, and XY for Cracker Promotion Example

Source of Variation	*Sums of Squares or Products*			
	Y	*X*	*XY*	*df*
Treatments	338.8	26.8	−37.4	2
Error	307.6	333.2	299.4	12
Total	646.4	360.0	262.0	14

Adjusted Analysis of Variance for Y. We are now ready to adjust the analysis of variance for Y to obtain the analysis of covariance. The rationale of the adjustment can best be seen from simple regression analysis. There we found that the error sum of squares (2.21):

$$SSE = \sum (Y_i - \hat{Y}_i)^2 = \sum (Y_i - b_0 - b_1 X_i)^2 \tag{22.34}$$

could be expressed in the algebraically equivalent form (2.24b):

$$SSE = \sum (Y_i - \bar{Y})^2 - \frac{[\sum (X_i - \bar{X})(Y_i - \bar{Y})]^2}{\sum (X_i - \bar{X})^2} \tag{22.34a}$$

which can be rewritten as follows:

$$SSE = \sum (Y_i - \bar{Y})^2 - b_1 \sum (X_i - \bar{X})(Y_i - \bar{Y}) \tag{22.34b}$$

In terms of our analysis of variance notation (which uses double subscripts and $\bar{X}_{..}$ and $\bar{Y}_{..}$ for $\bar{X}$ and $\bar{Y}$), the regression SSE can therefore be expressed in either of the following ways:

$$SSE = SSTO_Y - \frac{(SPTO)^2}{SSTO_X} \tag{22.35a}$$

$$SSE = SSTO_Y - b_1(SPTO) \tag{22.35b}$$

Hence, in the analysis of covariance the total sum of squares for Y, adjusted for the linear relation to X, would be obtained as follows:

$$SSTO(\text{adj.}) = SSTO_Y - \frac{(SPTO)^2}{SSTO_X} \tag{22.36a}$$

where $SSTO(\text{adj.})$ stands for the *adjusted total sum of squares for Y*.

It may then be argued by analogy that:

$$SSE(\text{adj.}) = SSE_Y - \frac{(SPE)^2}{SSE_X} \tag{22.36b}$$

where SSE(adj.) stands for the *adjusted error sum of squares for Y.*

Finally, we obtain by subtraction:

$$SSTR(\text{adj.}) = SSTO(\text{adj.}) - SSE(\text{adj.}) \tag{22.36c}$$

where $SSTR$(adj.) stands for the *adjusted treatment sum of squares for Y.* Note carefully that $SSTR$(adj.) is obtained by subtraction and not by an analogous adjustment. The reason for this will become clear later.

TABLE 22.7

Covariance Analysis for Completely Randomized Design

Source of Variation	Sums of Squares or Products: Y	X	XY	df
Treatments	$SSTR_Y$	$SSTR_X$	$SPTR$	$r-1$
Error	SSE_Y	SSE_X	SPE	n_T-r
Total	$SSTO_Y$	$SSTO_X$	$SPTO$	n_T-1

Source of Variation	Adjusted SS	Adjusted df	Adjusted MS
Treatments	$SSTR$(adj.)	$r-1$	$MSTR$(adj.)
Error	SSE(adj.)	n_T-r-1	MSE(adj.)
Total	$SSTO$(adj.)	n_T-2	

Table 22.7 contains the general covariance analysis table for a completely randomized design. First are presented the sums of squares and products. Then the adjusted sums of squares are given. When these are divided by the degrees of freedom, the adjusted mean squares are obtained. Note that there is one less degree of freedom for $SSTO$(adj.) and for SSE(adj.) than with the analysis of variance model. The reason is that the regression coefficient γ for the concomitant variable had to be estimated.

Example. For our cracker promotion example, we find, using (22.36) and the results in Table 22.6:

$$SSTO(\text{adj.}) = 646.4 - \frac{(262)^2}{360} = 455.722$$

$$SSE(\text{adj.}) = 307.6 - \frac{(299.4)^2}{333.2} = 38.571$$

$$SSTR(\text{adj.}) = 455.722 - 38.571 = 417.151$$

These results are presented in an analysis of covariance table in Table 22.8, which also contains the adjusted degrees of freedom and the adjusted mean squares. Note there is one less degree of freedom for the adjusted total sum of squares and for the adjusted error sum of squares than in the analysis of variance (Table 22.6).

TABLE 22.8

Covariance Analysis Table for Cracker Promotion Example

Source of Variation	*Adjusted SS*	*Adjusted df*	*Adjusted MS*
Treatments	417.151	2	208.576
Error	38.571	11	3.506
Total	455.722	13	

Test for Treatment Effects

The test for treatment effects:

(22.37a)
$$\begin{aligned} &C_1: \tau_1 = \tau_2 = \cdots = \tau_r = 0 \\ &C_2: \text{not all } \tau_j \text{ are zero} \end{aligned}$$

is then based on the usual test statistic:

(22.37b)
$$F^* = \frac{MSTR(\text{adj.})}{MSE(\text{adj.})}$$

If C_1 holds, F^* follows the $F(r-1, n_T - r - 1)$ distribution. Hence the decision rule for controlling the level of significance at α is:

(22.37c)
$$\begin{aligned} &\text{If } F^* \leq F(1-\alpha; r-1, n_T - r - 1), \text{ conclude } C_1 \\ &\text{If } F^* > F(1-\alpha; r-1, n_T - r - 1), \text{ conclude } C_2 \end{aligned}$$

Example. For our cracker promotion example, we have from Table 22.8:

$$F^* = \frac{MSTR(\text{adj.})}{MSE(\text{adj.})} = \frac{208.576}{3.506} = 59.5$$

which, of course, is the same value as we obtained on page 699 with the regression approach. If $\alpha = .05$, we require $F(.95; 2, 11) = 3.98$. Since $F^* = 59.5 > 3.98$, we conclude that the three promotions had different effects on the sales of crackers.

Reconciliation of Two Approaches. Figure 22.6 summarizes the relations between the regression and adjustment approaches to covariance analysis. The equivalence of $SSE(R)$ with the regression approach and $SSTO(\text{adj.})$ with

FIGURE 22.6

Reconciliation between Regression and Adjustment Approaches to Covariance Analysis

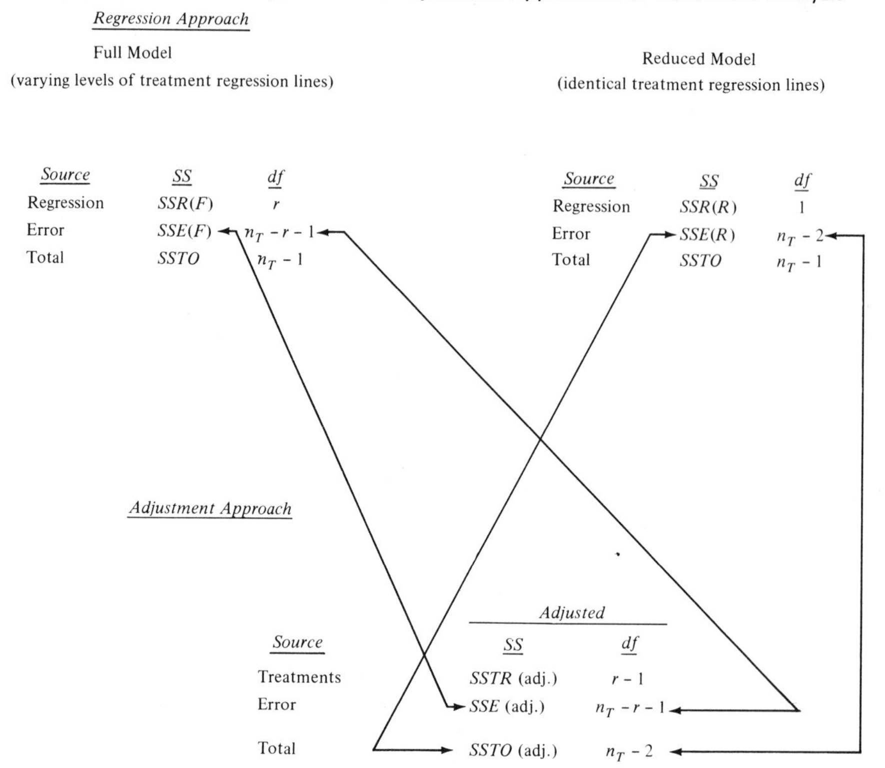

the adjustment approach is easy to see. When we fit the reduced model (22.19) with the regression approach:

$$Y_{ij} = \mu_. + \gamma(X_{ij} - \bar{X}_{..}) + \varepsilon_{ij}$$

we obtain the usual least squares estimator (2.10a) for the slope γ:

$$g = \frac{\sum_i \sum_j (X_{ij} - \bar{X}_{..})(Y_{ij} - \bar{Y}_{..})}{\sum_i \sum_j (X_{ij} - \bar{X}_{..})^2} = \frac{SPTO}{SSTO_X} \tag{22.38}$$

and the usual error sum of squares (2.24b):

$$SSE(R) = \sum_i \sum_j (Y_{ij} - \bar{Y}_{..})^2 - \frac{\left[\sum_i \sum_j (X_{ij} - \bar{X}_{..})(Y_{ij} - \bar{Y}_{..})\right]^2}{\sum_i \sum_j (X_{ij} - \bar{X}_{..})^2} \tag{22.39}$$

FIGURE 22.7

Schematic Representation of Residuals for $SSE(R)$ and $SSE(F)$

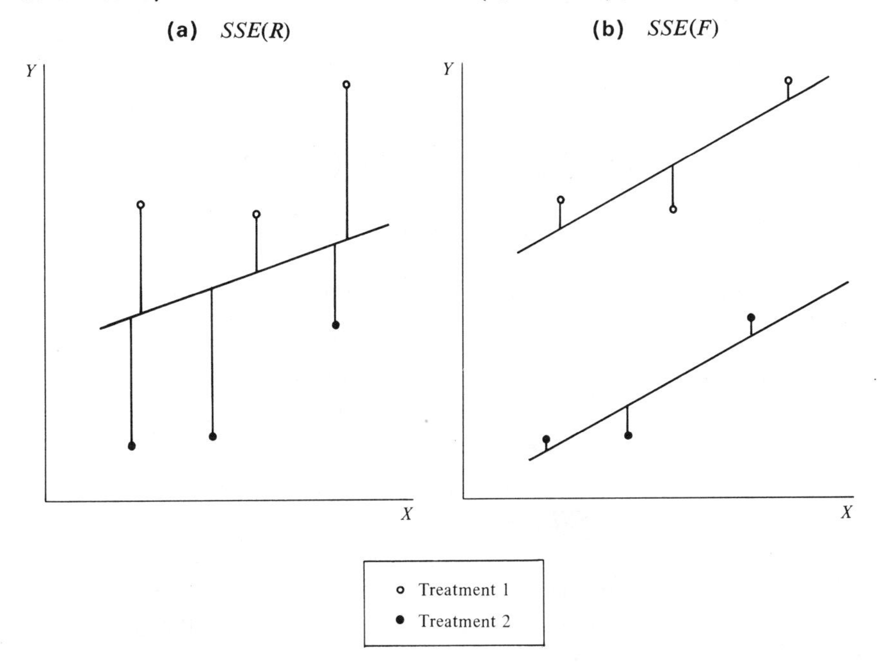

Figure 22.7a schematically illustrates the residuals entering into $SSE(R)$ when two treatments are in an experiment.

Expressed in our notation, $SSE(R)$ becomes:

$$SSE(R) = SSTO_Y - \frac{(SPTO)^2}{SSTO_X} \tag{22.39a}$$

which is the definition of $SSTO$(adj.) in formula (22.36a). Thus, $SSTO$(adj.) is simply the error sum of squares when a linear regression is fitted to the entire set of experimental data.

When the full covariance model (22.3):

$$Y_{ij} = \mu_. + \tau_j + \gamma(X_{ij} - \bar{X}_{..}) + \varepsilon_{ij}$$

is fitted to the experimental data, allowing different intercepts $\mu_. + \tau_j$ for the treatment regression lines but requiring a common slope γ, it can be shown that the least squares estimator of this common slope is:

$$g_w = \frac{\sum_i \sum_j (X_{ij} - \bar{X}_{.j})(Y_{ij} - \bar{Y}_{.j})}{\sum_i \sum_j (X_{ij} - \bar{X}_{.j})^2} = \frac{SPE}{SSE_X} \tag{22.40}$$

and that the least squares estimator of $\mu_. + \tau_j$ is:

$$\bar{Y}_{.j} - g_w(\bar{X}_{.j} - \bar{X}_{..}) \tag{22.41}$$

Hence, the error sum of squares for the jth treatment is:

$$\text{(22.42)} \qquad \sum_i \{Y_{ij} - [\bar{Y}_{.j} - g_w(\bar{X}_{.j} - \bar{X}_{..})] - g_w(X_{ij} - \bar{X}_{..})\}^2 = \sum_i [(Y_{ij} - \bar{Y}_{.j}) - g_w(X_{ij} - \bar{X}_{.j})]^2$$

Summing these error sums of squares over all treatments, we obtain $SSE(F)$:

$$\text{(22.43)} \qquad SSE(F) = \sum_i \sum_j [(Y_{ij} - \bar{Y}_{.j}) - g_w(X_{ij} - \bar{X}_{.j})]^2$$

Figure 22.7b illustrates the residuals entering into $SSE(F)$ for the case of two treatments.

Expanding out the expression for $SSE(F)$ in (22.43) and simplifying, we obtain:

$$\text{(22.43a)} \qquad SSE(F) = \sum_i \sum_j (Y_{ij} - \bar{Y}_{.j})^2 - g_w \sum_i \sum_j (X_{ij} - \bar{X}_{.j})(Y_{ij} - \bar{Y}_{.j})$$

But in our new notation, this expression becomes:

$$\text{(22.43b)} \qquad SSE(F) = SSE_Y - \frac{SPE}{SSE_X} SPE = SSE_Y - \frac{(SPE)^2}{SSE_X}$$

which is the definition of SSE(adj.) in (22.36b). Thus, SSE(adj.) is simply the error sum of squares when separate regression lines, each with the same slope, are fitted to the treatments.

$SSTR$(adj.) is obtained as the difference $SSTO$(adj.) $-$ SSE(adj.), just as the numerator of the test statistic with the general linear test approach is based on the difference $SSE(R) - SSE(F)$.

Comments

1. One can obtain an indication of the effectiveness of the analysis of covariance in reducing experimental errors by comparing MSE(adj.) for covariance analysis with MSE for regular analysis of variance. For our cracker promotion example, we know from Table 22.8 that MSE(adj.) $= 3.51$. We can also see from Table 22.6 that the error mean square for regular analysis of variance would have been:

$$MSE = \frac{307.6}{12} = 26.63$$

Hence, in this case covariance analysis reduced the experimental error variance by about 87 percent, a substantial reduction.

2. Covariance analysis and analysis of variance need not lead to the same conclusions about the treatment effects. For instance, analysis of variance might not indicate any treatment effects whereas covariance analysis with smaller experimental errors could show significant treatment effects. Ordinarily, of course, one should decide in advance which of the two analyses is to be used.

3. The estimator g_w of the common slope γ can be considered as an average of the separately estimated treatment regression line slopes g_j. If we were fitting a

separate regression line for each treatment, the estimated slope g_j for the jth treatment would be given by the method of least squares as follows:

$$g_j = \frac{\sum_i (X_{ij} - \bar{X}_{.j})(Y_{ij} - \bar{Y}_{.j})}{\sum_i (X_{ij} - \bar{X}_{.j})^2} \tag{22.44}$$

A weighted average of the g_j, using $\sum_i (X_{ij} - \bar{X}_{.j})^2$ as weights, gives us precisely g_w as defined in (22.40):

$$\frac{\sum_j \left[\sum_i (X_{ij} - \bar{X}_{.j})^2 \right] g_j}{\sum_j \left[\sum_i (X_{ij} - \bar{X}_{.j})^2 \right]} = \frac{\sum_i \sum_j (X_{ij} - \bar{X}_{.j})(Y_{ij} - \bar{Y}_{.j})}{\sum_i \sum_j (X_{ij} - \bar{X}_{.j})^2} = g_w$$

Thus, g_w may be thought of as an average within-treatments regression slope.

For our cracker promotion example, the average within-treatments regression slope is:

$$g_w = \frac{SPE}{SSE_X} = \frac{299.4}{333.2} = .8986$$

and the slope when a single regression line is fitted to all data is:

$$g = \frac{SPTO}{SSTO_X} = \frac{262}{360} = .7278$$

Adjusted Treatment Means

In analysis of variance, the treatment mean $\bar{Y}_{.j}$ is an estimate of the mean response with the jth treatment. In analysis of covariance, many writers speak of the need to *adjust* these $\bar{Y}_{.j}$'s to make them comparable with respect to the concomitant variable since the X values usually will not be the same for all treatments. The adjustment takes the form:

$$\bar{Y}_{.j}(\text{adj.}) = \bar{Y}_{.j} - g_w(\bar{X}_{.j} - \bar{X}_{..}) \tag{22.45}$$

The rationale of the adjustment can be seen from Figure 22.8. This figure contains the points $(\bar{X}_{.j}, \bar{Y}_{.j})$ for the three treatments in our cracker promotion example. Through each of these points, a regression line with the average within-treatment slope $g_w = .8986$ is drawn. The adjusted treatment mean $\bar{Y}_{.j}(\text{adj.})$ is simply the ordinate of its regression line at $X = \bar{X}_{..}$. In this way, the treatments are said to be made comparable with respect to X. For our example, the adjusted treatment means are obtained as follows:

Treatment	$\bar{Y}_{.j}$	$\bar{X}_{.j}$	$\bar{X}_{..}$	g_w	$g_w(\bar{X}_{.j} - \bar{X}_{..})$	$\bar{Y}_{.j}(\text{adj.})$
1	38.2	23.2	25	.8986	−1.62	39.82
2	36.0	26.4	25	.8986	1.26	34.74
3	27.2	25.4	25	.8986	.36	26.84

FIGURE 22.8

Representation of Adjusted Treatment Means for Cracker Promotion Example

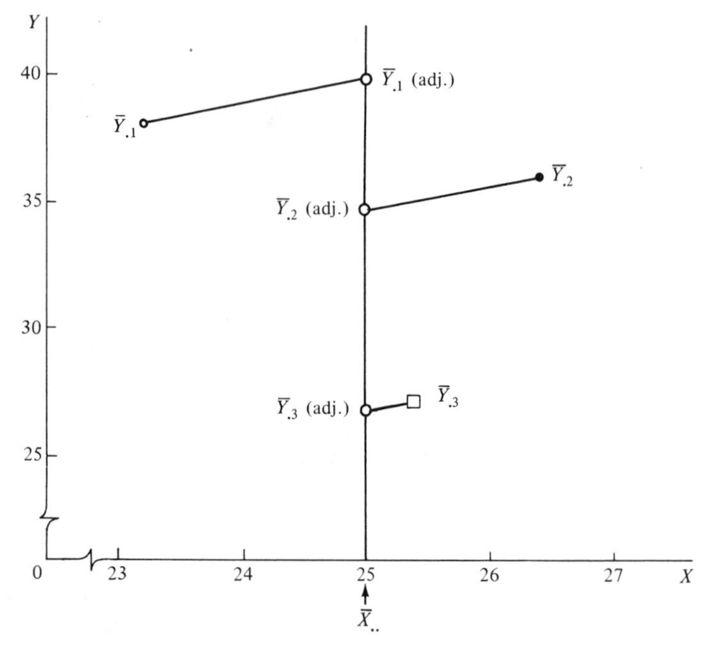

A comparison with the results on page 701 indicates that these adjusted treatment means are simply estimates of the intercepts of the treatment regression lines. In other words, $\bar{Y}_{.j}(\text{adj.})$ is simply an estimator of $\mu_{.} + \tau_j$.

The variance of $\bar{Y}_{.j}(\text{adj.})$ can be shown to be:

$$\sigma^2(\bar{Y}_{.j}(\text{adj.})) = \sigma^2\left[\frac{1}{n_j} + \frac{(\bar{X}_{.j} - \bar{X}_{..})^2}{SSE_X}\right] \tag{22.46}$$

and an unbiased estimator of this variance is:

$$s^2(\bar{Y}_{.j}(\text{adj.})) = MSE(\text{adj.})\left[\frac{1}{n_j} + \frac{(\bar{X}_{.j} - \bar{X}_{..})^2}{SSE_X}\right] \tag{22.47}$$

For instance, the estimated variance of the adjusted mean for treatment 1 in our cracker promotion example is:

$$s^2(\bar{Y}_{.1}(\text{adj.})) = 3.506\left[\frac{1}{5} + \frac{(23.2 - 25)^2}{333.2}\right] = .735$$

which is the same result obtained on page 701 by the regression approach.

Comparisons among Adjusted Treatment Means. Since $\bar{Y}_{.j}$(adj.) is an estimator of $\mu_{.} + \tau_j$, a pairwise comparison $\bar{Y}_{.j}$(adj.) $- \bar{Y}_{.j'}$(adj.) is an estimator of $\tau_j - \tau_{j'}$. For our example, we would thus estimate $\tau_1 - \tau_3$ from:

$$\bar{Y}_{.1}(\text{adj.}) - \bar{Y}_{.3}(\text{adj.}) = 39.82 - 26.84 = 12.98$$

which is the same result we obtained on page 700. The variance of the difference of two adjusted treatment means is:

$$\sigma^2(\bar{Y}_{.j}(\text{adj.}) - \bar{Y}_{.j'}(\text{adj.})) = \sigma^2\left[\frac{1}{n_j} + \frac{1}{n_{j'}} + \frac{(\bar{X}_{.j} - \bar{X}_{.j'})^2}{SSE_X}\right] \tag{22.48}$$

and this is estimated by:

$$s^2(\bar{Y}_{.j}(\text{adj.}) - \bar{Y}_{.j'}(\text{adj.})) = MSE(\text{adj.})\left[\frac{1}{n_j} + \frac{1}{n_{j'}} + \frac{(\bar{X}_{.j} - \bar{X}_{.j'})^2}{SSE_X}\right] \tag{22.49}$$

For our example, we obtain:

$$s^2(\bar{Y}_{.1}(\text{adj.}) - \bar{Y}_{.3}(\text{adj.})) = 3.506\left[\frac{1}{5} + \frac{1}{5} + \frac{(23.2 - 25.4)^2}{333.2}\right] = 1.453$$

which, of course, is the same estimated variance as was obtained via the regression approach on page 700.

The use of the t distribution for a single interval estimate of a contrast among the adjusted treatment means and of the Scheffé and Bonferroni methods for multiple comparisons is identical to their use with the regression approach. An estimator of a contrast among the adjusted treatment means is defined as follows:

$$\hat{L} = \sum c_j[\bar{Y}_{.j}(\text{adj.})] = \sum c_j \bar{Y}_{.j} - g_w \sum c_j \bar{X}_{.j} \qquad \text{where } \sum c_j = 0 \tag{22.50}$$

Its estimated variance is:

$$s^2(\hat{L}) = MSE(\text{adj.})\left[\sum \frac{c_j^2}{n_j} + \frac{(\sum c_j \bar{X}_{.j})^2}{SSE_X}\right] \tag{22.51}$$

A confidence interval for a single comparison is:

$$\hat{L} - t(1 - \alpha/2; n_T - r - 1)s(\hat{L}) \leq L \leq \hat{L} + t(1 - \alpha/2; n_T - r - 1)s(\hat{L}) \tag{22.52}$$

For multiple comparisons, the t multiple is replaced by either the S or B multiples defined in (22.22) and (22.23) respectively.

22.5 FACTORIAL EXPERIMENTS

We have until now considered the case of a completely randomized design with r treatments, where the treatments refer to a single factor. Covariance analysis can also be employed when the r treatments are defined in terms of two or more factors. We illustrate now the use of covariance analysis in a two-factor experiment with a completely randomized design.

Covariance Model

The analysis of variance model for a two-factor experiment with fixed effects was given in (17.21):

$$Y_{ijk} = \mu_{..} + \alpha_i + \beta_j + (\alpha\beta)_{ij} + \varepsilon_{ijk} \tag{22.53}$$
$$i = 1, \ldots, a; j = 1, \ldots, b; k = 1, \ldots, n$$

where α_i is the main effect of factor A at the ith level, β_j is the main effect of factor B at the jth level, and $(\alpha\beta)_{ij}$ is the interaction effect when factor A is at the ith level and factor B is at the jth level. The covariance model for a two-factor experiment, assuming the relation between Y and the concomitant variable X is linear, is:

$$Y_{ijk} = \mu_{..} + \alpha_i + \beta_j + (\alpha\beta)_{ij} + \gamma(X_{ijk} - \bar{X}_{...}) + \varepsilon_{ijk} \tag{22.54}$$
$$i = 1, \ldots, a; j = 1, \ldots, b; k = 1, \ldots, n$$

Regression Approach

To illustrate the regression approach to covariance analysis for a two-factor experiment in a completely randomized design, suppose that both factors A and B are at two levels and have fixed effects. We shall use the modified indicator variable coding of (19.56) to simplify analyses. The regression model counterpart to (22.54) then is, using δ to denote a regression coefficient:

$$Y_{ijk} = \delta_0 + \delta_1 X_{ijk1} + \delta_2 X_{ijk2} + \delta_3 X_{ijk1} X_{ijk2} + \delta_4 Z_{ijk} + \varepsilon_{ijk} \tag{22.55}$$

where:

$$X_{ijk1} = \begin{matrix} 1 \text{ if observation from level 1 for factor } A \\ -1 \text{ if observation from level 2 for factor } A \end{matrix}$$

$$X_{ijk2} = \begin{matrix} 1 \text{ if observation from level 1 for factor } B \\ -1 \text{ if observation from level 2 for factor } B \end{matrix}$$

$$Z_{ijk} = X_{ijk} - \bar{X}_{...}$$
$$\delta_0 = \mu_{..}$$
$$\delta_1 = \alpha_1$$
$$\delta_2 = \beta_1$$
$$\delta_3 = (\alpha\beta)_{11}$$
$$\delta_4 = \gamma$$

Testing for factor A effects requires that $\delta_1 = 0$ in the reduced model. Correspondingly, $\delta_2 = 0$ in the reduced model when testing for factor B effects, and $\delta_3 = 0$ in the reduced model when testing for AB interactions.

Estimation of factor A and B main effects can easily be done in terms of comparisons among the regression coefficients. The use of the Scheffé and

Bonferroni multiple comparison procedures presents no new problems. For instance, the S multiple for multiple comparisons among the factor A means is defined as follows:

$$S^2 = (a-1)F(1-\alpha; a-1, n_T - ab - 1) \tag{22.56}$$

and the B multiple would be given by (22.23) with $r = ab$.

Adjustment Approach

In applying the adjustment approach to covariance analysis for a two-factor experiment in a completely randomized design, we again require analyses of variance for Y, X, and XY. The analysis of variance for Y is given in (17.32) and (17.33). The analysis of variance for X is identical, with X_{ij} replacing Y_{ij}. Table 22.9 contains these analyses of variance for Y and X.

TABLE 22.9

Analyses of Variance for Y, X, and XY—Two-Factor Experiment in a Completely Randomized Design

Source of Variation	*Sums of Squares or Products*			
	Y	X	XY	*df*
Factor A	SSA_Y	SSA_X	SPA	$a-1$
Factor B	SSB_Y	SSB_X	SPB	$b-1$
AB interactions	$SSAB_Y$	$SSAB_X$	$SPAB$	$(a-1)(b-1)$
Error	SSE_Y	SSE_X	SPE	$ab(n-1)$
Total	$SSTO_Y$	$SSTO_X$	$SPTO$	$abn-1$

Note that subscripts for Y and X are used for identification.

The analysis of variance for products XY is as follows:

$$SPTO = \sum_i \sum_j \sum_k (X_{ijk} - \bar{X}_{...})(Y_{ijk} - \bar{Y}_{...}) = \sum_i \sum_j \sum_k X_{ijk} Y_{ijk} - \frac{X_{...} Y_{...}}{abn} \tag{22.57a}$$

$$SPA = bn \sum_i (\bar{X}_{i..} - \bar{X}_{...})(\bar{Y}_{i..} - \bar{Y}_{...}) = \frac{\sum_i X_{i..} Y_{i..}}{bn} - \frac{X_{...} Y_{...}}{abn} \tag{22.57b}$$

$$SPB = an \sum_j (\bar{X}_{.j.} - \bar{X}_{...})(\bar{Y}_{.j.} - \bar{Y}_{...}) = \frac{\sum_j X_{.j.} Y_{.j.}}{an} - \frac{X_{...} Y_{...}}{abn} \tag{22.57c}$$

$$SPE = \sum_i \sum_j \sum_k (X_{ijk} - \bar{X}_{ij.})(Y_{ijk} - \bar{Y}_{ij.}) = SPTO - SPTR \tag{22.57d}$$

$$SPAB = n \sum_i \sum_j (\bar{X}_{ij.} - \bar{X}_{i..} - \bar{X}_{.j.} + \bar{X}_{...})(\bar{Y}_{ij.} - \bar{Y}_{i..} - \bar{Y}_{.j.} + \bar{Y}_{...}) = SPTR - SPA - SPB \tag{22.57e}$$

where:

(22.57f) $$SPTR = n \sum_i \sum_j (\bar{X}_{ij.} - \bar{X}_{...})(\bar{Y}_{ij.} - \bar{Y}_{...}) = \frac{\sum_i \sum_j X_{ij.} Y_{ij.}}{n} - \frac{X_{...} Y_{...}}{abn}$$

Table 22.9 also contains this analysis of variance for products XY.

The adjustment process ignores all components other than the error component and the component for which the test is to be made, and then proceeds to make adjustments in a fashion analogous to a one-factor experiment in a completely randomized design. Thus, to test for factor A effects:

(22.58) $$\begin{aligned} &C_1: \alpha_1 = \alpha_2 = \cdots = \alpha_a = 0 \\ &C_2: \text{not all } \alpha_i \text{ are zero} \end{aligned}$$

we extract the lines for factor A and error in Table 22.9. This is done in Table 22.10a. We then derive the adjusted sums of squares in the usual fashion:

(22.59a) $$SS(A + E; \text{adj.}) = (SSA_Y + SSE_Y) - \frac{(SPA + SPE)^2}{SSA_X + SSE_X}$$

(22.59b) $$SSE(\text{adj.}) = SSE_Y - \frac{(SPE)^2}{SSE_X}$$

(22.59c) $$SSA(\text{adj.}) = SS(A + E; \text{adj.}) - SSE(\text{adj.})$$

TABLE 22.10

Covariance Analysis for Testing Factor A Effects in a Two-Factor Experiment with a Completely Randomized Design

(a) Analyses of Variance

Source of Variation	*Sums of Squares or Products* Y	X	XY	*df*
Factor A	SSA_Y	SSA_X	SPA	$a - 1$
Error	SSE_Y	SSE_X	SPE	$ab(n - 1)$
Sum	$SSA_Y + SSE_Y$	$SSA_X + SSE_X$	$SPA + SPE$	$a + ab(n - 1) - 1$

(b) Analysis of Covariance

Source of Variation	*Adjusted SS*	*Adjusted df*	*Adjusted MS*
Factor A	$SSA(\text{adj.})$	$a - 1$	$MSA(\text{adj.})$
Error	$SSE(\text{adj.})$	$ab(n - 1) - 1$	$MSE(\text{adj.})$
Sum	$SS(A + E; \text{adj.})$	$a + ab(n - 1) - 2$	

Table 22.10b contains the covariance analysis for testing factor A effects. The degrees of freedom for error and the sum are reduced by one to take account of the concomitant variable X, and the adjusted degrees of freedom for factor A effects are obtained as a residual.

The statistic for testing the alternatives in (22.58) is, as usual:

$$F^* = \frac{MSA(\text{adj.})}{MSE(\text{adj.})} \tag{22.60}$$

If C_1 holds, F^* follows the $F(a-1, ab(n-1)-1)$ distribution.

22.6 ADDITIONAL CONSIDERATIONS FOR THE USE OF COVARIANCE ANALYSIS

Use of Differences

In a variety of experiments, a pre-experiment observation X and a post-experiment observation Y on the same variable under study are available for each experimental unit. For instance, X may be the score for a subject's attitude toward a company prior to reading its annual report, and Y may be the score after reading the report. In this situation, an obvious alternative to covariance analysis is to do an analysis of variance on the difference $Y - X$. Sometimes, $Y - X$ is called an *index of response*, because it makes one observation out of two separate ones.

If the slope of the treatment regression lines is $\gamma = 1$, analysis of covariance and analysis of variance on $Y - X$ are essentially equivalent. When $\gamma = 1$, the covariance model (22.2) becomes:

$$Y_{ij} = \mu_{.} + \tau_j + X_{ij} + \varepsilon_{ij} \tag{22.61}$$

which can be written as a regular analysis of variance model:

$$Y_{ij} - X_{ij} = \mu_{.} + \tau_j + \varepsilon_{ij} \tag{22.61a}$$

Thus, if a unit change in X leads to about the same change in Y, it makes sense to perform an analysis of variance on $Y - X$, rather than to use covariance analysis, because analysis of variance is much easier. If the regression slope is not near 1, however, covariance analysis may be substantially more effective than use of the difference $Y - X$.

In our earlier cracker promotion example, use of $Y - X$ would have been effective. It would have involved an error mean square (see Table 22.6):

$$\frac{SSE_Y + SSE_X - 2SPE}{12} = \frac{307.6 + 333.2 - 2(299.4)}{12} = 3.50$$

which is practically the same as the error mean square for covariance analysis, $MSE(\text{adj.}) = 3.51$. Recall that the regression slope in our example was close to 1 ($g_w = .8986$), hence the approximate equivalence of the two procedures.

Correction for Bias

The suggestion is sometimes made that analysis of covariance can be helpful in correcting for bias when survey data, rather than experimental data, are at hand. With survey data, the groups under study may differ substantially with respect to a concomitant variable, and this may bias the comparisons of the groups. Consider, for instance, a study in which attitudes toward no-fault automobile insurance were compared for persons who are risk averse and persons who are risk seeking. It was found that many persons in the risk averse group tended to be older (50 to 70 years old), while many persons in the risk seeking group tended to be younger (20 to 40 years old). In this type of situation, some would advise that covariance analysis, with age as the concomitant variable, be employed to help remove any bias that may be in the survey data because the two age groups differ so much.

Even though there is great appeal in the idea of removing bias in survey data, covariance analysis should be used with caution for this purpose. In the first place, the adjusted means may require substantial extrapolation of the regression lines to a region where there are no or only few data points (in our example, to near 45 years). It may well be that the regression relationship used in the covariance analysis is not appropriate for substantial extrapolation. In the second place, the treatment variable may be dependent on the concomitant variable (or vice versa) which could affect the proper conclusions to be drawn.

Interest in Nature of Treatment Effects

Covariance analysis is sometimes employed for the principal purpose of shedding more light on the nature of the treatment effects, rather than merely for increasing the precision of the analysis. For instance, a market researcher in a study of the effects of three different advertisements on the maximum price consumers are willing to pay for a new type of home siding may use covariance analysis, with value of the consumer's home as the independent variable. The reason is because he is truly interested in the relation for each advertisement between home value and maximum price. Reduction of error variance in this instance may be a secondary consideration for the analyst in using covariance analysis.

As in all regression analyses, care must be used in drawing inferences about the causal nature of the relation between the concomitant variable and the dependent variable. In our advertising example, it might well be that value of a consumer's home is largely influenced by income. If this were so, the relation between value of the consumer's home and maximum price he is willing to pay may actually be largely a reflection of a more underlying relation between income and maximum price.

PROBLEMS

22.1. A student's reaction to the instructor's statement that covariance analysis is inappropriate when the treatment regression lines do not have the same slope was as follows: "It seems to me that this is ducking a real-world problem. If the treatment slopes are different, just use a covariance model which allows for different treatment slopes." Evaluate this reaction.

22.2. A survey analyst has remarked: "When covariance analysis is used with survey data, there is the danger that the treatment variable may be related to the concomitant variable." What is the nature of the problem? Does this same problem exist when the treatments are randomly assigned to the experimental units?

22.3. Portray in graphic form the nature of the covariance model (22.3) when there are three treatments and the parameters have the values: $\mu_{.} = 190$, $\tau_1 = -2$, $\tau_2 = +5$, $\tau_3 = -3$, $\gamma = 4$, $\bar{X}_{..} = 72$, $\sigma = 3$. Show several distributions of Y for each treatment.

22.4. Refer to the cracker promotion study on page 695. A student stated, in discussing this case: "Strictly speaking, you can't conclude anything about whether the three promotions differ in effectiveness because there was no control. The preceding period does not qualify as a control because it might have differed from the promotion period due to seasonal or other unique factors." Comment.

22.5. Refer to the cracker promotion study on page 700, where three pairwise comparisons of treatment effects were made by the Scheffé procedure.

a) What would be the value of the Bonferroni multiple here for estimating the three comparisons?

b) Did the analyst have to pay a substantial price for using the Scheffé procedure which permits him to make additional estimates without modifying the present ones?

22.6. A manufacturer of felt-tip markers investigated by an experiment whether a proposed new display is more effective than the present counter display. Fifteen drugstores of similar characteristics, which featured the current display, were chosen for the study. They were assigned at random in equal numbers to one of the following three treatments: (1) present counter display continues in stationery area; (2) new carousel display in stationery area; (3) two carousel displays, one in stationery area, the other in the front of the store. Sales with the present display were recorded for a three-week period, then the new displays were set up in the 10 stores receiving the new display, and sales were observed for the next three-week period. The sales were (in dollars):

	Store				
	1	2	3	4	5
Treatment 1:					
1st 3 weeks	67	74	52	68	92
2d 3 weeks	43	58	38	57	70
Treatment 2:					
1st 3 weeks	81	69	73	77	80
2d 3 weeks	75	73	78	74	82
Treatment 3:					
1st 3 weeks	42	65	81	73	69
2d 3 weeks	49	63	84	75	76

a) Using sales in the first three weeks as a concomitant variable, fit the covariance model (22.3). (If you use the regression approach, state the regression model and the correspondences between the two sets of parameters.)
b) Test whether or not treatment effects are present. Use a level of significance of .05.
c) Analyze the nature of the treatment effects. Explain the reasons why you chose your particular analytical approach, and summarize your findings.
d) How much wider would your confidence intervals in part (c) have been if you had simply analyzed sales in the second three weeks without using any concomitant variable?

22.7. Refer to Problem 22.6.
a) Obtain the residuals and make appropriate residual plots to analyze whether model (22.3) is apt for this case. Summarize your findings.
b) Conduct a formal test to determine whether or not the three treatment regression slopes are equal. Use $\alpha = .01$. What do you conclude?
c) Could you conduct a formal test here as to whether the regression functions are linear?

22.8. Refer to Problem 13.11. The annual sales volumes (in hundred thousand dollars) of the six dealers for each of the four owners were as follows:

	Owner			
Dealer	A	B	C	D
1	2.0	1.6	1.4	1.3
2	.9	2.2	2.1	1.7
3	1.8	2.1	1.5	1.8
4	1.9	1.8	1.7	2.1
5	1.2	1.1	1.8	1.0
6	2.1	1.9	1.9	1.5

The analyst wished to investigate whether it would be helpful to use the dealer's sales volume as a concomitant variable.
a) Fit the covariance model (22.3). (If you use the regression approach, state the regression model and the correspondences between the two sets of parameters.)
b) How much smaller is the error mean square with the covariance model (22.3) than with the analysis of variance model? Does this appear to be a worthwhile reduction?
c) Obtain the residuals and make appropriate residual plots to analyze whether model (22.3) is apt here. Summarize your findings.
d) Conduct a formal test to determine whether or not the four treatment regression slopes are equal. Use a level of significance of .05. What do you conclude?

22.9. Refer to Problem 17.9.
a) Why would it not make sense to use the experimenter's age as a concomitant variable?
b) The ages of the personnel officials were as follows:

Personnel Official	High Eye Contact		Low Eye Contact	
	Male	*Female*	*Male*	*Female*
1	32	28	58	51
2	49	43	27	29
3	29	29	43	43
4	53	51	30	31
5	36	30	31	27

Use these ages as a concomitant variable, and fit covariance model (22.54). (If you use the regression approach, state the regression model and the correspondences between the two sets of parameters and employ the type of indicator variables in (19.56).)

c) Obtain the residuals and make appropriate residual plots to analyze whether model (22.54) is apt here. Summarize your findings.

d) Conduct a formal test to determine whether or not the treatment regression slopes are equal. Use $\alpha = .01$. What do you conclude?

22.10. Refer to Problem 22.9. Assume that model (22.54) is appropriate.

a) Test whether the factors have any main effects and whether they interact. Use a level of significance of .01 for each test.

b) Analyze the factor effects, and describe your principal findings.

22.11. Refer to Problem 22.6. Suppose we had used the differences between the sales in the two three-week periods as the dependent variable.

a) Obtain the analysis of variance and test for treatment effects using $\alpha = .05$.

b) How effective is the use of differences in this case, compared to the use of the first three weeks' sales as the concomitant variable with covariance model (22.3)?

23

Randomized Block Designs

IN THIS CHAPTER, we consider the most widely used design that employs local control to reduce experimental errors—the randomized block design.

23.1 BASIC ELEMENTS

Description of Design

A randomized block design is a restricted randomization design in which the experimental units are first sorted into homogeneous groups, called *blocks*, and the treatments are then assigned at random within the blocks. Consider, for instance, an experiment on the effect of four levels of newspaper advertising saturation on sales volume. The experimental unit is a city, and 16 cities are available for the study. Size of city usually is highly correlated with the dependent variable, sales volume. Hence it is desirable to block the 16 cities into four groups of four cities each, according to population size. Thus, the four largest cities will constitute block 1, and so on. Within each block, the four treatments are then assigned at random to the four cities, and the assignments from one block to another are made independently.

As another example, consider an experiment on the effects of three different incentive pay schemes on employee productivity of electronic assemblies. The experimental unit is an employee, and 30 employees are available for the study. Since productivity in this situation is highly correlated with manual dexterity, it is desirable to block the 30 employees into 10 groups of 3, according to manual dexterity. Thus, the three employees with the highest manual dexterity ratings are grouped into one block, and so on for the other employees. Within each block, the three incentive pay schemes are then assigned randomly to the three employees.

As these examples imply, the key objective in blocking the experimental units is to make them as homogeneous as possible within blocks with respect to the dependent variable under study, and to make the different blocks as heterogeneous as possible with respect to the dependent variable. The design in which each treatment is included in each block is called a *randomized complete block design*. Often, we shall drop the term "complete" because the context makes it clear that all treatments are included in each block.

Comments

1. In a complete block design, each block constitutes a replication of the experiment. For that reason, it is highly desirable that the experimental units within a block be processed together whenever this will help to reduce experimental errors. As an example, an experimenter may tend to make changes in his experimental techniques over time (e.g., in the administration of the experiment to subjects) without being aware of it. Consecutive processing of the experimental units block by block will tend to keep such sources of variation out of the variation within blocks, and thereby make the experimental results more precise.

2. In factorial experiments, some of the factors of interest are often characteristics of the experimental unit, such as sex, age, and amount of experience on the job. Even though these factors are not introduced to reduce experimental errors, but rather are included for their intrinsic interest, we shall nevertheless consider such experiments to be randomized block designs since the randomization of treatments to experimental units is restricted by the nature of the classification factors considered.

Criteria for Blocking

As noted earlier, the purpose of blocking is to sort experimental units into groups within each of which the elements are homogeneous with respect to the dependent variable, such that the differences between groups are as great as possible. To help recognize some of the characteristics of experimental units which are fruitful criteria for blocking, we need a more precise definition of an experimental unit than that given in Chapter 21. All elements of the experimental situation which are not included in the definition of a treatment need to be assigned to the definition of an experimental unit. Suppose the treatment in an experiment consists of a portion of a vegetable containing a particular additive served in the laboratory. The experimental unit might then be defined as a housewife of a given age, processed by a given observer on a specified day during a particular part of the day, and served food from a given batch of cooked vegetable. Still other elements of the experimental setting might be included in the definition of the experimental unit, and should be if they could be the cause of material variability in the observations.

A full definition of the experimental unit such as the one just given suggests two types of blocking criteria:

1. Characteristics associated with the unit—for persons: sex, age, income, intelligence, education, job experience, attitudes, etc.; for geographic areas: population size, average income, etc.
2. Characteristics associated with the experimental setting—observer, time of processing, machine, batch of material, measuring instrument, etc.

Use of time as a blocking variable frequently captures a number of different sources of variability, such as learning by observer, changes in equipment, and drifts in environmental conditions (e.g., weather). Blocking by observers often eliminates a substantial amount of interobserver variability; similarly, blocking by batches of material frequently is very effective.

There is no need to use only a single blocking criterion; several may be employed if the experimental errors can be reduced substantially thereby. We shall consider later the use of more than one blocking criterion, such as when a block consists of subjects in a given age group processed by a particular observer.

To design effective randomized block experiments requires the ability to select blocking variables which will reduce the experimental errors. Often, past experience in the subject matter field enables the experimenter to select good blocking variables. If some experiments have been run in the past in which blocking has been employed, these results can be analyzed to determine the effectiveness of the blocking variables. We shall discuss an appropriate method of analysis for doing this later. In the absence of any information on potential blocking variables, uniformity trials can be run where all experimental units are assigned the same treatment. From these trials, information can be obtained on the effectiveness of different variables for blocking.

Note

There is another blocking variable often used in social science research which has not been mentioned yet, namely the subject himself. With the subject as a complete block, all treatments are given to every subject. Such designs are often called *repeated measures designs*. Since they involve some special problems, we will discuss them separately later on.

Advantages and Disadvantages

The advantages of a randomized complete block design are:

1. It can, with effective grouping, provide substantially more precise results than a completely randomized design of comparable size.
2. It can accommodate any number of treatments and replications.

3. Different treatments need not have equal sample sizes. For instance, if the control is to have twice as large a sample size as each of three treatments, blocks of size five would be used; three units in a block are then assigned at random to the three treatments and two to the control.
4. The statistical analysis is relatively simple.
5. If an entire treatment or a block needs to be dropped from the analysis for some reason, such as spoiled results, the analysis is not complicated thereby.
6. Variability in experimental units can be deliberately introduced to widen the range of validity of the experimental results without sacrificing the precision of the results.

Disadvantages include:

1. Missing observations within a block cause additional calculations. This is a particularly serious problem if there are many missing observations.
2. The degrees of freedom for experimental error are not as large as with a completely randomized design. One degree of freedom is lost for each block after the first.
3. More assumptions are required for the model (e.g., no interactions between treatments and blocks, constant variance from block to block) than for a completely randomized design model.

How to Randomize

The randomization procedure for a randomized block design is perfectly straightforward. Within each block a random permutation is used to assign treatments to experimental units, just as in a completely randomized design. Independent permutations are selected for the several blocks.

Illustration

In an experiment on decision making, businessmen were exposed to one of three methods of quantifying the maximum risk premium they would be willing to pay to avoid uncertainty. The three methods are the utility method, the worry method, and the comparison method. After using the method assigned to him, a subject was asked to state his degree of confidence in the method of quantifying the risk premium, on a scale from 0 (no confidence) to 20 (highest confidence).

Fifteen subjects were used in the study. They were grouped into five blocks of three businessmen, according to age. Block 1 contained the three oldest businessmen, and so on. The design layout, after five independent random permutations of 3 were employed, was as shown in Table 23.1. Table 23.2 contains the results of the experiment.

TABLE 23.1

Layout for Randomized Block Design for Risk Premium Example

Block	Experimental Unit 1	2	3
Block 1 (oldest businessmen)	C	W	U
2	C	U	W
3	U	W	C
4	W	U	C
5 (youngest businessmen)	W	C	U

C: Comparison method
W: Worry method
U: Utility method

TABLE 23.2

Risk Premium Experiment Results

Block	*Confidence Rating for—* *Utility Method*	*Worry Method*	*Comparison Method*	*Average*
1 (oldest)	1	5	8	4.7
2	2	8	14	8.0
3	7	9	16	10.7
4	6	13	18	12.3
5 (youngest)	12	14	17	14.3
Average	5.6	9.8	14.6	10.0

23.2 MODEL

Table 23.2 is similar in appearance to Table 19.2a, which shows the data for a two-factor study with one observation in each cell. In fact, one may think of a randomized complete block design as corresponding to a two-factor study (blocks and treatments are the factors), with one observation in each cell. As we noted in Section 19.1, we must be able to assume that there are no interactions between the two factors when there is only one observation in each cell and the factors have fixed effects.

Thus, the usual model for a randomized complete block design, when both the block and treatment effects are fixed and there are n blocks (replications) and r treatments, is of the form:

$$Y_{ij} = \mu_{..} + \rho_i + \tau_j + \varepsilon_{ij} \tag{23.1}$$

where:

$\mu_{..}$ is a constant

ρ_i are constants for the block (row) effects, subject to the restriction $\sum \rho_i = 0$

τ_j are constants for the treatment effects, subject to the restriction $\sum \tau_j = 0$

ε_{ij} are independent $N(0, \sigma^2)$

$i = 1, \ldots, n; j = 1, \ldots, r$

This model is identical to that in (19.1) for the two-factor, no-interaction model, except that we now use ρ_i for the block effect, and n to designate the total number of blocks. Note that Y_{ij} here stands for the observation for the jth treatment in the ith block.

Comments

1. When the experimental units are grouped according to specified categories, such as into particular age groups, income groups, and order-of-processing groups, the block effects ρ_i are usually considered to be fixed. Sometimes the block effects are viewed as random. For instance, when observers or subjects are used as blocks, the particular observers or subjects in the study may be considered to be a sample from a population of observers or subjects. The case of random block effects will be taken up later.

2. If the treatment effects are random, the only changes in model (23.1) are that the τ_j now represent independent normal variables with expectation zero and variance σ_τ^2, and that the τ_j are independent of the ε_{ij}.

3. The additive model (23.1) implies that the expected values of observations in different blocks for the same treatment may differ (e.g., older businessmen may tend to have lower confidence ratings for any of the methods of quantifying the risk premium than younger businessmen), but the treatment effects (e.g., how much higher the confidence rating for one method is over that for another) are the same for all blocks. We shall consider the possibility of interactions between blocks and treatments later.

23.3 ANALYSIS OF VARIANCE

The analysis of variance for a randomized complete block design is identical to that for a two-factor, no-interaction model with one observation per cell, as described in Section 19.1:

$$SSBL = r \sum_i (\bar{Y}_{i.} - \bar{Y}_{..})^2 = \sum_i \frac{Y_{i.}^2}{r} - \frac{Y_{..}^2}{rn} \tag{23.2a}$$

$$SSTR = n \sum_j (\bar{Y}_{.j} - \bar{Y}_{..})^2 = \sum_j \frac{Y_{.j}^2}{n} - \frac{Y_{..}^2}{rn} \tag{23.2b}$$

$$SSE = \sum_i \sum_j (Y_{ij} - \bar{Y}_{i.} - \bar{Y}_{.j} + \bar{Y}_{..})^2 \tag{23.2c}$$

$$= \sum_i \sum_j Y_{ij}^2 - \sum_i \frac{Y_{i.}^2}{r} - \sum_j \frac{Y_{.j}^2}{n} + \frac{Y_{..}^2}{rn}$$

Here, $SSBL$ is the *sum of squares for blocks* and $SSTR$ and SSE as usual are the treatment and error sums of squares respectively; rn is the total number of observations in the experiment.

A summary of the analysis of variance, including the expected mean squares for both fixed and random treatment effects, is given in Table 23.3. The expected mean squares can be derived readily from rule (19.26). Note that there are no interaction terms in the model, hence the expected mean squares contain only σ^2 and, as appropriate, the treatment or block effects term. Also note that the variance component for treatment effects needs to be replaced by the corresponding sum of squared effects divided by degrees of freedom when the treatment effects are fixed.

TABLE 23.3

ANOVA Table for Randomized Complete Block Design, Block Effects Fixed

				$E(MS)$	
Source of Variation	*SS*	*df*	*MS*	*Treatments Fixed*	*Treatments Random*
Blocks	$SSBL$	$n-1$	$MSBL$	$\sigma^2 + \frac{r \sum \rho_i^2}{n-1}$	$\sigma^2 + \frac{r \sum \rho_i^2}{n-1}$
Treatments	$SSTR$	$r-1$	$MSTR$	$\sigma^2 + \frac{n \sum \tau_j^2}{r-1}$	$\sigma^2 + n\sigma_\tau^2$
Error	SSE	$(n-1)(r-1)$	MSE	σ^2	σ^2
Total	$SSTO$	$nr-1$			

As the $E(MS)$ column in Table 23.3 indicates, the test for treatment effects:

	Fixed Treatment Effects	*Random Treatment Effects*
(23.3a)	C_1: all $\tau_j = 0$	$C_1: \sigma_\tau^2 = 0$
	C_2: not all τ_j are zero	$C_2: \sigma_\tau^2 \neq 0$

uses the same test statistic whether the treatment effects are fixed or random:

$$F^* = \frac{MSTR}{MSE} \tag{23.3b}$$

and the decision rule for controlling the Type I error at α is:

$$\begin{aligned} &\text{If } F^* \leq F(1-\alpha; r-1, (n-1)(r-1)), \text{ conclude } C_1 \\ &\text{If } F^* > F(1-\alpha; r-1, (n-1)(r-1)), \text{ conclude } C_2 \end{aligned} \tag{23.3c}$$

Example

Table 23.4 contains the analysis of variance for our risk premium example in Table 23.2. The calculations are straightforward, and were carried out by a computer package. To test for treatment effects:

$$C_1: \tau_1 = \tau_2 = \tau_3 = 0$$
$$C_2: \text{not all } \tau_j \text{ are zero}$$

we use the results of Table 23.4:

$$F^* = \frac{MSTR}{MSE} = \frac{101.4}{2.99} = 33.9$$

For a level of significance of .01, we require $F(.99; 2, 8) = 8.65$. Since $F^* = 33.9 > 8.65$, we conclude that the mean confidence ratings for the three methods differ.

TABLE 23.4

ANOVA Table for Randomized Complete Block Design Risk Premium Example of Table 23.2

Source of Variation	*SS*	*df*	*MS*
Blocks	171.3	4	42.8
Methods for risk premium specification	202.8	2	101.4
Error	23.9	8	2.99
Total	398.0	14	

Comments

1. Sometimes one may also wish to conduct a test for block effects:

(23.4a)
$$C_1: \text{all } \rho_i = 0$$
$$C_2: \text{not all } \rho_i \text{ are zero}$$

Usually, however, the treatments are of primary interest, and blocks are chiefly the means for reducing experimental errors. Table 23.3 indicates that the test for block effects uses the test statistic:

(23.4b)
$$F^* = \frac{MSBL}{MSE}$$

For our risk premium example, this test statistic is:

$$F^* = \frac{42.8}{2.99} = 14.3$$

For a level of significance of .01, we require $F(.99; 4, 8) = 7.01$. We conclude that the mean confidence ratings (averaged over treatments) differ for the various blocks.

Since blocks correspond to a classification factor, one needs to be careful in interpreting the implications of block effects. In our example, for instance, the block effects might not be due to age, even though age was the grouping variable. Education could be the pivotal independent variable, the effect by age arising if older businessmen have less formal education than younger businessmen.

2. The power of the F test for treatment effects for a randomized complete block design involves the same noncentrality parameter as for a completely randomized design. Formula (13.70a) gives the appropriate measure. Despite the same form of the noncentrality parameter, the two designs generally lead to different power levels even when based on the same sample sizes, for two reasons. First, the experimental error variance σ^2 will differ for the two designs. Second, the degrees of freedom associated with the denominator of the F^* statistic differ for the two designs.

3. If only two treatments are investigated in a randomized complete block design, it can readily be shown that the F test for treatment effects in (23.3b) is equivalent to the two-sided t test for paired observations based on (1.62).

23.4 ANALYSIS OF TREATMENT EFFECTS

Once the block effects have been isolated through the analysis of variance, the analysis of treatment effects proceeds as in the completely randomized design. If desired, $SSTR$ can be decomposed into components of interest along the lines of Section 14.1. More frequently, the analysis of treatment effects proceeds directly by estimating one or more contrasts. MSE obtained from (23.2c) is the appropriate variance term to be used in the estimated variance of the contrast, since it is the denominator of the F^* statistic for testing treatment effects. The multiple for the estimated standard deviation of the contrast is as follows:

(23.5a) Single comparison $\quad t(1 - \alpha/2; (n-1)(r-1))$

(23.5b) Tukey procedure (for pairwise comparisons) $\quad T = \frac{1}{\sqrt{2}} q(1 - \alpha; r, (n-1)(r-1))$

(23.5c) Scheffé procedure $\quad S^2 = (r-1)F(1 - \alpha; r-1, (n-1)(r-1))$

(23.5d) Bonferroni procedure $\quad B = t(1 - \alpha/2s; (n-1)(r-1))$

Example

The researcher who conducted the risk premium study wished to obtain all pairwise comparisons with a 95 percent family confidence coefficient. Using (14.36b) and the results in Table 23.4, we obtain:

$$s^2(D) = \frac{2MSE}{n} = \frac{2(2.99)}{5} = 1.20$$

Remember that each treatment mean $\bar{Y}_{.j}$ consists of n observations (one from each of n blocks). Using (23.5b), we find for a 95 percent family confidence coefficient:

$$T = \frac{1}{\sqrt{2}} q(.95; 3, 8) = \frac{1}{\sqrt{2}}(4.04) = 2.86$$

Hence:

$$Ts(D) = 2.86\sqrt{1.20} = 3.13$$

Thus, we obtain for the pairwise comparisons (see Table 23.2 for the $\bar{Y}_{.j}$):

$$1.7 = (14.6 - 9.8) - 3.1 \leq \mu_{.3} - \mu_{.2} \leq (14.6 - 9.8) + 3.1 = 7.9$$
$$5.9 = (14.6 - 5.6) - 3.1 \leq \mu_{.3} - \mu_{.1} \leq (14.6 - 5.6) + 3.1 = 12.1$$
$$1.1 = (9.8 - 5.6) - 3.1 \leq \mu_{.2} - \mu_{.1} \leq (9.8 - 5.6) + 3.1 = 7.3$$

Here $\mu_{.1}$ is the mean confidence rating, averaged over all blocks, for the utility method, and $\mu_{.2}$ and $\mu_{.3}$ are the mean confidence ratings for the worry and comparison methods respectively.

We conclude that the comparison method has a substantially greater mean confidence rating than the worry method, which in turn has a higher mean confidence rating than the utility method. The confidence coefficient of .95 applies to this entire set of comparisons.

23.5 FACTORIAL TREATMENTS

If the treatments in a randomized block design are combinations of different factor levels, one can simply write the model showing the factor effects in place of the treatment effect. For a two-factor study, we have:

$$Y_{ijk} = \mu_{...} + \rho_i + \alpha_j + \beta_k + (\alpha\beta)_{jk} + \varepsilon_{ijk} \tag{23.6}$$

where the terms in the model have the usual meaning and (j, k) identifies the treatment.

In the analysis of variance, we proceed as always by decomposing the treatment sum of squares *SSTR* into sums of squares for factor effects and interactions. This is shown in Table 23.5 for a two-factor study, the factors having a and b levels respectively. Thus, the total number of treatments r here

TABLE 23.5

ANOVA Table for a Two-Factor Study in a Randomized Complete Block Design—Model (23.6)

Source of Variation	*SS*	*df*	*MS*
Blocks	*SSBL*	$n - 1$	*MSBL*
Treatments	*SSTR*	$r - 1$	*MSTR*
Factor *A*	*SSA*	$a - 1$	*MSA*
Factor *B*	*SSB*	$b - 1$	*MSB*
AB interactions	*SSAB*	$(a - 1)(b - 1)$	*MSAB*
Error	*SSE*	$(n - 1)(r - 1)$	*MSE*
Total	*SSTO*	$nr - 1$	

Note: $r = ab$

equals ab. The decomposition is done in the usual fashion, as explained in Section 17.4, utilizing the relation in (17.32):

$$SSTR = SSA + SSB + SSAB$$

Formulas (17.32a, b, c) and their alternate versions (17.33c, d, f) are appropriate for calculating the component sums of squares, remembering that (i, j) subscripts are there used to identify the treatments in terms of the factor combinations. Tests for factor effects are conducted as usual, and no new problems are encountered in the estimation of factor effects.

Note

Model (23.6) assumes no interactions between treatments and blocks. Specifically, it implies that all block–factor A interactions (denoted by $BL.A$) are zero, and similarly that all $BL.B$ and $BL.AB$ interactions are zero. One can make a less restrictive analysis by assuming only that the $BL.AB$ interactions are zero. To see this, consider the layout for a two-factor study ($a = 2$, $b = 2$) in $n = 3$ blocks, as shown in Table 23.6. Here Y_{ijk} denotes the observation in the ith block of the (j, k) factor combination. Note that this layout corresponds to the three-factor layout in Table 20.3, but

TABLE 23.6

Layout for a Two-Factor Study in a Randomized Complete Block Design

	A_1		A_2	
	B_1	B_2	B_1	B_2
Block 1	Y_{111}	Y_{112}	Y_{121}	Y_{122}
2	Y_{211}	Y_{212}	Y_{221}	Y_{222}
3	Y_{311}	Y_{312}	Y_{321}	Y_{322}

with only one observation per cell. Table 20.4 contains the general analysis of variance for a three-factor study, and it can be seen from there that if all $BL.AB$ interactions ($BL.AB$ corresponds to ABC in Table 20.4) are zero, the $BL.AB$ interaction sum of squares is an unbiased estimator of σ^2, the experimental error variance. Hence, we can conduct all tests and make all desired estimates with a factorial experiment in a complete block design by only assuming that the $BL.AB$ interactions are zero. The price of the less restrictive assumptions is fewer degrees of freedom for the experimental error.

The model for this less restrictive case is:

$$Y_{ijk} = \mu_{...} + \rho_i + \alpha_j + \beta_k + (\alpha\beta)_{jk} + (\rho\alpha)_{ij} + (\rho\beta)_{ik} + \varepsilon_{ijk} \tag{23.7}$$

where the model terms have the usual meaning. The factor effects and interactions may be fixed or random, depending on the nature of the factors. The analysis of

TABLE 23.7

ANOVA Table for a Two-Factor Study in a Randomized Complete Block Design—Model (23.7)

Source of Variation	*SS*	*df*	*MS*
Blocks (*BL*)	*SSBL*	$n-1$	*MSBL*
Factor *A*	*SSA*	$a-1$	*MSA*
Factor *B*	*SSB*	$b-1$	*MSB*
AB interactions	*SSAB*	$(a-1)(b-1)$	*MSAB*
BL.A interactions	*SSBL.A*	$(n-1)(a-1)$	*MSBL.A*
BL.B interactions	*SSBL.B*	$(n-1)(b-1)$	*MSBL.B*
Error	*SSE*	$(n-1)(a-1)(b-1)$	*MSE*
Total	*SSTO*	$nab-1$	

variance for this model is given in Table 23.7. The sums of squares may be calculated using formulas (20.20a–g) or their alternate computational versions. In using these formulas, remember that n in these formulas (number of observations per cell) now is 1, and that the number of levels for blocks (corresponding to factor C) is n.

23.6 0, 1 RESPONSES

When the responses in a randomized block experiment are either 1 or 0 (buy or not buy, succeed in task or fail), a chi-square test may be used for assessing the presence of treatment effects. To test whether treatment effects are present:

$$C_1: \text{all } \tau_j = 0 \qquad C_2: \text{not all } \tau_j \text{ are zero} \tag{23.8}$$

the following test statistic due to Cochran may be used:

$$X_C^2 = SSTR \div \frac{SSTR + SSE}{n(r-1)} \tag{23.9}$$

which reduces to:

$$X_C^2 = \frac{(r-1)\left[r \sum_j Y_{.j}^2 - Y_{..}^2\right]}{rY_{..} - \sum_i Y_{i.}^2} \tag{23.9a}$$

where the usual notation is employed.

The number of 1's in each block may differ because of block-to-block differences. Given the number of 1's in each block, if all treatments have the same effect, all permutations of 0's and 1's within a block are equally likely. It can then be shown that when C_1 holds, X_C^2 is distributed approximately as χ^2 with $r-1$ degrees of freedom provided the number of blocks is not very small. Large values of X_C^2 lead to conclusion C_2.

Example

Table 23.8 contains data for an experiment in which 15 teams were grouped into 5 blocks of 3 teams each, according to a criterion on the creativity of the team. Within each block the teams were assigned at random to one of three training methods, and upon completion of training each team was given the same complex task to perform. Success in the task was coded 1, failure 0.

TABLE 23.8

0,1 Response Experiment in Randomized Block Design

	Instruction Method			
Block	1	2	3	*Total*
1 (high creativity)	1	1	1	3
2	1	0	1	2
3	1	0	0	1
4	0	1	0	1
5 (low creativity)	1	0	1	2
Total	4	2	3	9

$\sum Y_{.j}^2 = 4^2 + 2^2 + 3^2 = 29$
$\sum Y_{i.}^2 = 3^2 + 2^2 + 1^2 + 1^2 + 2^2 = 19$

To test whether training methods have differential effects on successful performance, we use the test statistic version (23.9a) and obtain:

$$X_C^2 = \frac{2[3(29) - (9)^2]}{3(9) - 19} = 1.5$$

Suppose the level of significance is to be controlled at .05. We require then $\chi^2(.95; 2) = 5.99$. Since $X_C^2 = 1.5 < 5.99$, we conclude that instructional methods do not differ in effectiveness.

23.7 IMPLEMENTATION CONSIDERATIONS

Necessary Number of Replications

The planning of necessary sample size for a randomized complete block design is very similar to that for a completely randomized design. One may determine the needed number of blocks n either to obtain specified protection against making Type I and II errors or to obtain specified precision for key contrasts among the treatment means. With either approach, it is necessary to assess in advance the magnitude of the experimental error variance σ^2.

Power Approach. The same charts as for the completely randomized design (Table A–10) may be used, provided the number of treatments and blocks are not very small, specifically provided that $r(n-1) \geq 20$. The development in Section 15.1 may be followed directly.

We illustrate the determination of sample size for the experiment on confidence ratings for three methods of measuring the risk premium. Suppose the number of replications had not yet been determined and the experimenter desired the following risk protections:

1. Type I error is to be controlled at .05.
2. If any two treatment means differ by 3 or more rating points, the risk of concluding that there are no treatment effects should not exceed .20.

The experimenter anticipates that the experimental error standard deviation when businessmen are grouped by age will be approximately $\sigma = 2$.

Thus, the specifications can be summarized as follows:

$$r = 3 \qquad \alpha = .05 \qquad d = 3$$
$$\beta = .20 \text{ or Power} = .80 \qquad \sigma = 2$$

Using (15.4) we find:

$$\phi' = \frac{d}{\sigma}\sqrt{\frac{1}{2r}} = \frac{3}{2}\sqrt{\frac{1}{(2)(3)}} = .6$$

Entering the chart for $r = 3$ in Table A–10 and using $\alpha = .05$ and $P = .80$, we find for $\phi' = .6$ that $n = 10$ approximately. Thus, the experimenter requires approximately 10 blocks of 3 businessmen each in order to obtain the desired protection against incorrect decisions.

Estimation Approach. If the experimenter wishes to determine the number of replications n by means of the estimation approach, he need simply calculate the anticipated standard deviations of key contrasts, and modify the replication size iteratively until the desired precision is attained. Often, he would use a multiple comparison procedure for encompassing the different estimates under a family confidence coefficient.

With reference to our risk premium illustration, suppose the Tukey procedure is to be used for all pairwise comparisons, with a 95 percent family confidence coefficient. Using $n = 10$ as a starting point, and assuming that $\sigma = 2$ approximately, the anticipated variance of any pairwise difference is:

$$\sigma^2(D) = \frac{2\sigma^2}{n} = \frac{2(4)}{10} = .8$$

or $\sigma(D) = .89$. Further:

$$T = \frac{1}{\sqrt{2}}q(.95; r, (n-1)(r-1)) = \frac{1}{\sqrt{2}}q(.95; 3, 18) = \frac{1}{\sqrt{2}}(3.61) = 2.55$$

Thus, the anticipated half-width of the confidence interval is $T\sigma(D) = 2.55(.89) = 2.3$. If this precision is not adequate, a larger number of blocks should be tried next. If the precision is greater than necessary, a smaller number of blocks should be used in the next iteration.

Examination of Aptness of Model

Since the importance of examining the aptness of the model for a given set of data has been mentioned many times earlier and since the techniques of examination are similar, we shall make only a few points of special relevance to randomized block designs here.

Residual Plots. Some of the chief ways in which the data may not fit the randomized block model (23.1) are:

1. Unequal error variability by blocks
2. Unequal error variability by treatments
3. Time effects
4. Block-treatment interactions

Use of residual plots in connection with points 2 and 3 has been considered in Section 15.4 with reference to a completely randomized design, but the discussion there applies also to the residuals of a randomized block design:

$$e_{ij} = Y_{ij} - \overline{Y}_{i.} - \overline{Y}_{.j} + \overline{Y}_{..} \tag{23.10}$$

We simply add here that if treatments do have unequal error variability in a randomized complete block design, one can always estimate differences between any two treatments by working with the differences between the paired observations, $Y_{ij} - Y_{ij'}$, which are unaffected by the unequal treatment variances.

Unequal error variability by blocks can be studied by plotting the residuals for each block, as shown in Figure 23.1. The residual plot in Figure 23.1 is suggestive of increasing error variability with increasing block number. If, for instance, the blocks were processed in block number order, some modifications in procedures may have taken place leading to larger experimental errors over time. Tests concerning the equality of variances, such as those described in Section 15.6, may be employed for a more formal determination,

FIGURE 23.1

Residual Plots Suggesting Unequal Error Variances by Blocks

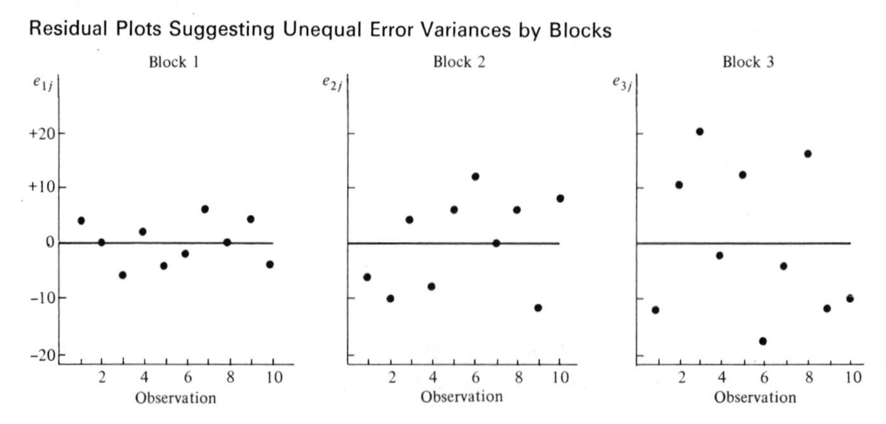

provided that the sample sizes are reasonably large so that the residuals can be treated as if they were independent.

Interactions between treatments and blocks are not easy to detect from residual plots. Figure 23.2 contains the residuals for an experiment with two treatments run in four blocks. The reversal in pattern of the residuals is suggestive of an interaction effect. There are, however, many other possible types of interaction patterns which would appear very much different from that in Figure 23.2.

FIGURE 23.2

Residual Plots Suggesting Block–Treatment Interactions

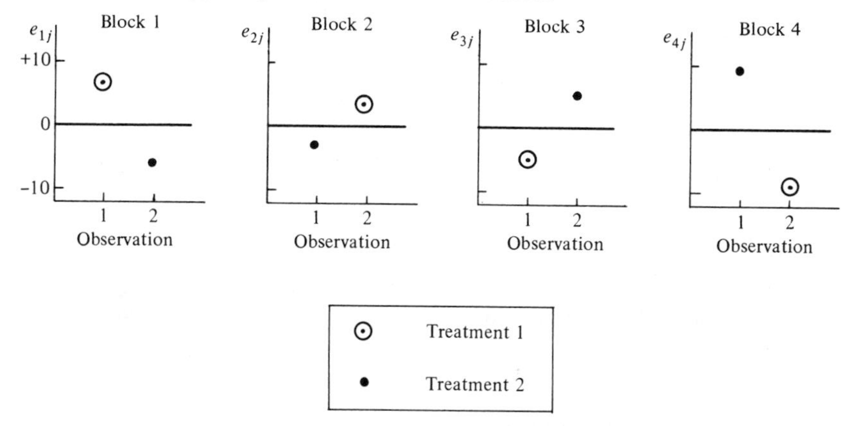

Tukey Test for Additivity. The Tukey test for additivity, discussed in Section 19.1, may be employed for a formal test of possible interaction effects between blocks and treatments. We illustrate this test for the risk premium example data in Table 23.2. The interaction sum of squares is given by (19.6):

$$SSBL.TR = \frac{\left[\sum_i \sum_j (\bar{Y}_{i.} - \bar{Y}_{..})(\bar{Y}_{.j} - \bar{Y}_{..})Y_{ij}\right]^2}{\sum_i (\bar{Y}_{i.} - \bar{Y}_{..})^2 \sum_j (\bar{Y}_{.j} - \bar{Y}_{..})^2}$$

Using the data in Table 23.2, we find the numerator:

$$[(4.7 - 10)(5.6 - 10)(1) + \cdots + (14.3 - 10)(14.6 - 10)(17)]^2 = 615.04$$

and from the results in Table 23.4 and formulas (23.2a) and (23.2b), we obtain the two terms in the denominator:

$$\sum (\bar{Y}_{i.} - \bar{Y}_{..})^2 = \frac{SSBL}{r} = \frac{171.3}{3} = 57.1$$

$$\sum (\bar{Y}_{.j} - \bar{Y}_{..})^2 = \frac{SSTR}{n} = \frac{202.8}{5} = 40.56$$

Hence:

$$SSBL.TR = \frac{615.04}{(57.1)(40.56)} = .27$$

We know from Table 23.4 that $SSE = 23.9$. Hence the pure error sum of squares (19.7) is:

$$SSPE = SSE - SSBL.TR = 23.9 - .27 = 23.63$$

and the test statistic (19.8) is:

$$F^* = \frac{SSBL.TR}{1} \div \frac{SSPE}{rn - r - n}$$

$$= \frac{.27}{1} \div \frac{23.63}{7} = .08$$

For a level of significance of .05, we need $F(.95; 1, 7) = 5.59$. Since $F^* = .08 < 5.59$, we conclude that no block–treatment interaction effects are present.

Note

If interaction effects are present, transformations on the data should be attempted to remove at least the important interaction effects. The discussion in Section 17.2 is relevant to this point.

Efficiency of Blocking Variable

Once a randomized complete block experiment has been run, it is often desired to estimate the efficiency of the blocking variable employed for guidance in future experimentation.

Let σ_b^2 stand for the experimental error variance for the randomized block design. Up to this point, we have used σ^2 for this error variance, but now that we will compare two designs we need to be more specific. Let σ_r^2 denote the experimental error variance for a completely randomized design. The relative efficiency of blocking, compared to a completely randomized design, is then defined as follows:

$$E = \frac{\sigma_r^2}{\sigma_b^2} \tag{23.11}$$

The measure E indicates how much larger the replications need be with a completely randomized design as compared to a randomized block design in order that the variance of any estimated treatment contrast is the same for the two designs.

We know that MSE for the randomized block design is an unbiased estimator of σ_b^2. The question is how to estimate σ_r^2 from the data for the randomized block design. Since the same experimental units are involved in either case and there are assumed to be no interactions between treatments and blocks, it can be shown that an unbiased estimator of σ_r^2 is:

$$s_r^2 = \frac{(n-1)MSBL + n(r-1)MSE}{nr-1} \tag{23.12}$$

Hence, we estimate E as follows:

$$\hat{E} = \frac{s_r^2}{MSE} = \frac{(n-1)MSBL + n(r-1)MSE}{(nr-1)MSE} \tag{23.13}$$

Since the number of degrees of freedom for experimental error for a randomized block design is not as great as for a completely randomized one, E overstates the efficiency a little because it considers only the error variances. Several modified measures of efficiency have been suggested to take this overstatement into account. Unless the degrees of freedom for experimental error with both designs are very small, these modifications have little effect. One frequently used modification, applicable for assessing any design relative to another, is:

$$\hat{E}' = \frac{(df_2 + 1)(df_1 + 3)}{(df_2 + 3)(df_1 + 1)}\hat{E} \tag{23.14}$$

where df_1 denotes the degrees of freedom for the experimental error in the base design (completely randomized design, in our case) and df_2 denotes the degrees of freedom for the experimental error in the design whose efficiency is being assessed (randomized block design, in our case).

Example. We shall evaluate the efficiency of blocking by age of businessman in our risk premium example. Placing the appropriate results from Table 23.4 in the efficiency measure (23.13), we obtain:

$$\hat{E} = \frac{4(42.8) + 5(2)(2.99)}{(14)(2.99)} = 4.8$$

Thus, we would have required almost five times as many replications per treatment with a completely randomized design to achieve the same variance of any estimated contrast as is obtained with blocking by age. Clearly blocking by age was highly effective here.

If we had used the modified efficiency measure (23.14), we would have found:

$$\hat{E}' = \frac{(8+1)(12+3)}{(8+3)(12+1)}(4.8) = 4.5$$

This result, of course, does not differ greatly from that obtained by using (23.13).

Note

The efficiency measure $\hat{E}$ in (23.13) equals 1 if $MSBL = MSE$, is greater than 1 if $MSBL > MSE$, and is less than 1 if $MSBL < MSE$. Thus, good blocking is achieved when $MSBL$ exceeds MSE by a considerable margin.

Missing Observations

There are, unfortunately, occasions when one or several observations in a randomized complete block design are "missing"—a subject may have been sick, a record may have been mislaid, a treatment may have been applied incorrectly in one instance. Such missing observations destroy the symmetry (orthogonality) of the complete block design, and make the usual ANOVA calculations inappropriate.

Regression Approach. The regression approach to the analysis of randomized block designs, which is discussed in Section 23.12, is appropriate whether or not there are any missing observations. If indeed several observations are missing, the regression approach often will be the preferred one.

Yates Method. If, however, only one or perhaps two observations are missing, a simplified alternate procedure due to Yates is available which leads approximately or exactly to the correct least squares solutions. With the Yates method, we obtain one or more pseudo-observations and then proceed with the ANOVA calculations as if all observations were at hand. The degrees of freedom do need to be modified for the missing data.

Suppose a single observation, namely Y_{ij}, were missing. The pseudo-observation will be denoted X_{ij}. The Yates method requires that pseudo-observation X_{ij} have a residual equal to zero, which minimizes the error sum of squares. Let $Y'_{i.}$, $Y'_{.j}$, and $Y'_{..}$ be the totals based on the actual observations (to be called the *unaugmented* data), and let $Y_{i.}$, $Y_{.j}$, $Y_{..}$ be the totals for the data including the pseudo-observation (to be called the *augmented* data). The expectation of X_{ij} with the complete block design is $\mu_{..} + \rho_i + \tau_j$, and the estimate of this quantity based on the augmented data is:

$$\bar{Y}_{..} + (\bar{Y}_{i.} - \bar{Y}_{..}) + (\bar{Y}_{.j} - \bar{Y}_{..}) = \bar{Y}_{i.} + \bar{Y}_{.j} - \bar{Y}_{..}$$

Since the residual for X_{ij} must be zero, we have:

$$e_{ij} = X_{ij} - \bar{Y}_{i.} - \bar{Y}_{.j} + \bar{Y}_{..} = 0 \tag{23.15}$$

We can express the means in (23.15) in terms of the unaugmented data totals and the pseudo-observation as follows:

$$X_{ij} - \frac{Y'_{i.} + X_{ij}}{r} - \frac{Y'_{.j} + X_{ij}}{n} + \frac{Y'_{..} + X_{ij}}{nr} = 0 \tag{23.15a}$$

which can be readily solved for X_{ij}:

$$X_{ij} = \frac{nY'_{i.} + rY'_{.j} - Y'_{..}}{(n-1)(r-1)} \tag{23.16}$$

Once the pseudo-observation X_{ij} has been obtained, we proceed in the ANOVA calculations as if there were no missing observation.

The Yates method has the following properties:

1. The estimates of the treatment effects τ_j and the block effects ρ_i are identical to the correct least squares estimates.
2. SSE is identical to the correct least squares quantity.
3. $SSTR$ is slightly biased upward as compared to the correct least squares quantity. The bias is negligible unless there are only a few observations, in which case the bias can be removed as follows:

$$SSTR(\text{unbiased}) = SSTO' - SSE - SSBL' \tag{23.17}$$

where $SSTO'$ and $SSBL'$ are the sums of squares calculated from the unaugmented data and SSE is obtained from the augmented data.

4. Ordinary confidence intervals for differences between treatment means are only approximately correct, but this can be allowed for when it matters. For example, if a single observation is missing and it comes from treatment j, the variance for the comparison $D = \bar{Y}_{.j} - \bar{Y}_{.j'}$ is:

$$\sigma^2(D) = \sigma^2\left[\frac{2}{n} + \frac{r}{n(n-1)(r-1)}\right] \tag{23.18}$$

For $n = 3$ and $r = 3$, (23.18) yields:

$$\sigma^2(D) = \sigma^2\left[\frac{2}{3} + \frac{3}{3(2)(2)}\right] = \frac{11}{12}\sigma^2$$

If we recognize that $n_j = 2$ and $n_{j'} = 3$ and calculate the ordinary variance of a difference between two means, we obtain:

$$\sigma^2(D) = \frac{\sigma^2}{2} + \frac{\sigma^2}{3} = \frac{10}{12}\sigma^2$$

Thus, for all practical purposes we can analyze treatment differences when observations are missing by using the augmented data for calculating the treatment means $\bar{Y}_{.j}$, and in the variance of the difference between two treatment means recognize the actual number of observations:

$$\sigma^2(D) \cong \frac{\sigma^2}{n_j} + \frac{\sigma^2}{n_{j'}} \tag{23.19}$$

An estimate of (23.19) is, as usual:

$$s^2(D) = MSE\left[\frac{1}{n_j} + \frac{1}{n_{j'}}\right] \tag{23.20}$$

This approximate procedure is satisfactory provided the total number of observations is not very small and the number of missing observations is not large. In setting up confidence intervals, the degrees of freedom associated with MSE are those excluding the missing observations.

TABLE 23.9

Missing Observation Illustration ($r = 3$, $n = 3$)

(a) Unaugmented Data

Block	Treatment 1	2	3	Total
1	Missing	10	9	19
2	11	10	7	28
3	6	4	3	13
Total	17	24	19	60

(b) Augmented Data

Block	Treatment 1	2	3	Total
1	12	10	9	31
2	11	10	7	28
3	6	4	3	13
Total	29	24	19	72

Example. We illustrate the Yates method with the data in Table 23.9a. Here $n = 3$, $r = 3$, and Y_{11} is missing. From Table 23.9a, we obtain:

$$Y'_{1.} = 19 \qquad Y'_{.1} = 17 \qquad Y'_{..} = 60$$

Hence the pseudo-observation is by (23.16):

$$X_{11} = \frac{3(19) + 3(17) - 60}{(3-1)(3-1)} = 12$$

The pseudo-observation 12 has been placed in Table 23.9b. The analysis of variance is then calculated in the usual way from the augmented data, and is shown in Table 23.10. Note carefully that there are a total of 7 degrees of freedom since there are only 8 observations. Also, *SSE* has associated with it

TABLE 23.10

ANOVA Table for Missing Data Example

Source of Variation	*SS*	*df*	*MS*
Blocks	62	2	31
Treatments	16.67	2	8.3
Error	1.33	3	.44
Total	80.0	7	

3 degrees of freedom; there would have been four without the missing observation. Thus, *for each missing observation, there is one less degree of freedom associated with SSE.*

To obtain the unbiased treatment sum of squares, we use the data from Table 23.9a to find:

$$SSTO' = (10)^2 + (9)^2 + \cdots + (3)^2 - \frac{(60)^2}{8} = 62$$

$$SSBL' = \frac{(19)^2}{2} + \frac{(28)^2}{3} + \frac{(13)^2}{3} - \frac{(60)^2}{8} = 48.17$$

Since we know from Table 23.10 that $SSE = 1.33$, we obtain:

$$SSTR(\text{unbiased}) = 62 - 1.33 - 48.17 = 12.50$$

Note that the unbiased treatment sum of squares is somewhat smaller than $SSTR$ in Table 23.10. With more observations, the bias would have been much smaller.

The test for treatment effects is conducted as usual. Table 23.10 indicates that $MSE = .44$. Using our earlier $SSTR$(unbiased) result, we find:

$$MSTR(\text{unbiased}) = \frac{12.50}{2} = 6.25$$

Hence:

$$F^* = \frac{MSTR(\text{unbiased})}{MSE} = \frac{6.25}{.44} = 14.2$$

For $\alpha = .05$, we need $F(.95; 2, 3) = 9.55$. Since $F^* = 14.2 > 9.55$, we conclude that differential treatment effects are present.

Comments

1. If more than one observation is missing, iteration can be used with the Yates procedure for determining the pseudo-observations. For instance, if two observations are missing, one begins by assigning any reasonable value to one of the two missing observations. Then, formula (23.16) is used to get a tentative pseudo-observation for the second missing value. That, in turn, is used to revise the pseudo-observation for the first missing value, and so on. If several observations are missing, the regression approach may be preferred.

2. The procedure for deriving formula (23.16) for the pseudo-observation in a randomized block design is applicable to other designs also. One simply sets the residual equal to zero, after having expressed the estimated mean in the form appropriate for the model employed, and solves for the pseudo-observation.

3. Missing observations are lost forever. Pseudo-observations are solely a device for obtaining results that the exact least squares procedure would provide based on the available data.

4. Formula (23.17) for obtaining an unbiased treatment sum of squares is appropriate whether one or several observations are missing.

23.8 RANDOM BLOCK EFFECTS

When blocks can be considered a random sample from a population (e.g., a population of observers, batches, machines), either an additive or a non-additive (i.e., interaction) model can be employed.

Additive Model

The additive model for random block effects and fixed treatment effects is analogous to (23.1):

$$Y_{ij} = \mu_{..} + \rho_i + \tau_j + \varepsilon_{ij} \tag{23.21}$$

where:

$\mu_{..}$ is a constant

ρ_i are independent $N(0, \sigma_\rho^2)$

τ_j are constants subject to the restriction $\sum \tau_j = 0$

ε_{ij} are independent $N(0, \sigma^2)$, and independent of the ρ_i

$i = 1, \ldots, n; j = 1, \ldots, r$

TABLE 23.11

ANOVA for Randomized Complete Block Design—Block Effects Random, Treatment Effects Fixed

Source of Variation	SS	df	MS	E(MS) Additive Model	E(MS) Interaction Model
Blocks	$SSBL$	$n-1$	$MSBL$	$\sigma^2 + r\sigma_\rho^2$	$\sigma^2 + r\sigma_\rho^2$
Treatments	$SSTR$	$r-1$	$MSTR$	$\sigma^2 + \frac{n}{r-1}\sum \tau_j^2$	$\sigma^2 + \sigma_{\rho\tau}^2 + \frac{n}{r-1}\sum \tau_j^2$
Residual	SSE	$(n-1)(r-1)$	MSE	σ^2	$\sigma^2 + \sigma_{\rho\tau}^2$
Total	$SSTO$	$nr-1$			

Table 23.11 contains the analysis of variance for this model. The sums of squares are the same as in (23.2) for the fixed block effects model. The $E(MS)$ are obtained as usual by applying rule (19.26). The statistic for testing for treatment effects is $F^* = MSTR/MSE$, as may be seen from the $E(MS)$ column in Table 23.11. Thus the test statistic is the same whether block effects are fixed or random. Confidence intervals for treatment contrasts present no new problems.

Interaction Model

When block effects are random, one need not use the restrictive model (23.21) which assumes that there are no block–treatment interactions. Instead, a model allowing for interactions between blocks and treatments may be employed:

$$Y_{ij} = \mu_{..} + \rho_i + \tau_j + (\rho\tau)_{ij} + \varepsilon_{ij} \tag{23.22}$$

where the terms have the same meaning as in (23.21) with the following additions:

$(\rho\tau)_{ij}$ are $N\left(0, \dfrac{r-1}{r}\sigma^2_{\rho\tau}\right)$, subject to the restriction $\sum_j (\rho\tau)_{ij} = 0$ for all i

$(\rho\tau)_{ij}$ are independent of the ε_{ij} and of the ρ_i

The sums of squares and degrees of freedom are the same as for the no-interaction model. The difference in the two models shows its effect in the expected mean squares. Table 23.11 also contains the ANOVA for the interaction model. The expected mean squares are obtained by applying rule (19.26) in the usual fashion (see comment 2 below in this connection). Again, the test statistic for examining whether treatment effects are present is $F^* = MSTR/MSE$. Thus, from the point of view of testing for treatment effects or estimating treatment contrasts, there is no difference when block effects are random whether one employs the additive or nonadditive model.

Comments

1. Table 23.11 indicates that the principal difference between the two models is that no exact test for block effects is possible with the interaction model, whereas an exact test can be obtained with the additive model. This distinction is unimportant whenever the blocks are not of intrinsic interest but serve merely to reduce experimental errors.

2. To be precise, *SSE* for the interaction model should be labeled *SSBL.TR* in Table 23.11, and *MSE* should be labeled *MSBL.TR*. The reason is, as can be seen from Table 23.11, that the expected value of the mean square is not the experimental error variance σ^2 for this model. However, the sum of squares $\sum\sum(Y_{ij} - \bar{Y}_{i.} - \bar{Y}_{.j} + \bar{Y}_{..})^2$ is the same as for the additive model where, when it is divided by the degrees of freedom, it does estimate σ^2. Hence we wished to avoid introducing another label for the same term. For purposes of obtaining $E(MSE)$ for the interaction model, however, we need to recognize *MSE* as an interaction mean square.

23.9 REPEATED MEASURES DESIGNS

We mentioned earlier that at times a subject (or store, plant, city, etc.) is given each of the treatments under study. In that case, the subject serves as a block, and the experimental units within a block must be viewed as the different instances when a treatment can be applied to the subject. The treatments are then assigned at random to these instances, that is, the order of the treatments is randomized for each subject independently. Social scientists often call such a randomized complete block design a *repeated measures design*. Table 23.12 contains the layout for a repeated measures design with five subjects and four treatments, after independent randomizations of the treatment orders.

TABLE 23.12

Layout for Repeated Measures Design

	Treatment Order			
	1	2	3	4
Subject 1	T_4	T_3	T_2	T_1
2	T_3	T_4	T_1	T_2
3	T_4	T_3	T_1	T_2
4	T_2	T_1	T_4	T_3
5	T_1	T_2	T_4	T_3

Advantages and Disadvantages

A principal advantage of this type of design is that all sources of variability between subjects are excluded from the experimental errors; only variation within subjects enters the experimental errors. Thus, one may view each subject as serving as his own control. Another advantage of a repeated measures design is that it economizes on subjects. This is particularly important when only a few subjects (stores, plants, etc.) can be utilized for the experiment. Also, when interest is in the effects of a treatment over time, as when one is concerned with the shape of the learning curve for different instructional methods, it is usually desirable to observe the same subject at different points in time rather than observing different subjects at each specified point in time.

Repeated measures designs have a serious potential disadvantage, however, namely that there may be several types of interference. One type of interference is connected with the position in the treatment order. For instance, in evaluating five different advertisements, subjects may tend to give higher (or lower) ratings for advertisements shown toward the end of the sequence than at the beginning. Another type of interference is connected with the preceding treatment or treatments. For instance, in evaluating five different soup recipes, a bland recipe may get a higher (or lower) rating when preceded by a highly spiced recipe than when preceded by a blander recipe. This type of interference is called a *carry-over effect.*

Various steps can be taken to minimize the danger of interference. Randomization of the treatment orders for each subject independently will make it more reasonable to analyze the data as if the error terms are independent. Allowing sufficient time between treatments is often an effective means of reducing carry-over effects. It may be desirable at times to balance the order of treatment presentations and sometimes even the number of times each treatment is preceded by any other treatment. Latin square designs and cross-over designs (discussed in Chapter 24) are helpful to this end.

Models

In repeated measures designs, the subjects are usually considered to be a random sample from a population, and either models (23.21) or (23.22) are employed when the treatment effects are fixed. Should the subjects not be viewed as a sample but of intrinsic interest in themselves, model (23.1) would be used when the treatment effects are fixed. The analysis of repeated measures designs follows the lines presented earlier since such designs are simply randomized complete block designs.

Friedman Test for Ranked Data

In some repeated measures designs, subjects are asked to rank the treatments. For instance, a subject may be asked to rank four different colored papers according to the vividness of the red color. Again, a subject may be asked to rank five coffee sweeteners according to his taste preferences. When the observations Y_{ij} in a randomized block design are ranks, the Friedman test may be used to study whether or not differential treatment effects exist. The test statistic is (using the regular notation):

$$X_r^2 = SSTR \div \frac{SSTR + SSE}{n(r-1)} \tag{23.23}$$

which can be reduced to (when no ties are present):

$$X_r^2 = \frac{12}{nr(r+1)} \sum_j Y_{.j}^2 - 3n(r+1) \tag{23.23a}$$

If there are no treatment differences, all ranking permutations for a subject are assumed to be equally likely and the statistic X_r^2 will be distributed approximately as χ^2 with $r - 1$ degrees of freedom, provided the number of subjects and/or treatments is not too small. Large values of the test statistic lead to the conclusion that the treatments have unequal effects. If the number of subjects and/or treatments is small, tables such as those in Reference 23.1 may be used for an exact test.

TABLE 23.13

Ranked Data for Coffee Sweeteners in a Repeated Measures Design

	Sweetener				
Subject	A	B	C	D	E
1	5	1	2	4	3
2	4	2	1	5	3
3	3	2	1	4	5
4	5	2	3	4	1
5	4	1	2	3	5
6	4	1	3	5	2
Total	25	9	12	25	19

$\sum Y_{.j}^2 = (25)^2 + (9)^2 + (12)^2 + (25)^2 + (19)^2 = 1{,}836$

Example. Table 23.13 contains an illustration of ranked data in a repeated measures design. Six subjects were asked to rank five coffee sweeteners according to their taste preferences, with rank 1 assigned to the most preferred sweetener. We have:

$$X_r^2 = \frac{12}{6(5)(6)}(1{,}836) - 3(6)(6) = 14.4$$

For a level of significance of .05, we need $\chi^2(.95; 4) = 9.49$. Since $X_r^2 = 14.4 > 9.49$, we conclude that the five sweeteners are not equally liked. Analysis should now be undertaken as to which sweeteners are the preferred ones and why. Inspection of the data in Table 23.13 strongly suggests that sweeteners B and C are the preferred ones.

Comments

1. X_r^2 is related to Kendall's coefficient of concordance W in the following way:

$$W = \frac{X_r^2}{n(r-1)} = \frac{SSTR}{SSTR + SSE} \tag{23.24}$$

The coefficient of concordance W is a measure of the agreement of the rankings of the n subjects. It equals 1 if there is perfect agreement, and equals 0 if there is no agreement, that is, if all treatments receive the same mean ranking. For our example in Table 23.13, W is:

$$W = \frac{14.4}{6(4)} = .60$$

indicating a fair amount of agreement between the subjects.

2. The test statistic X_r^2 may also be used for randomized complete block designs when different experimental units are assigned the different treatments. The observations within a block would simply be ranked, and formula (23.23) then used. This would be desirable if the observations are far from normally distributed and transformations are of little help.

23.10 GENERALIZED RANDOMIZED BLOCK DESIGNS

When block effects are fixed, use of a no-interaction model in the presence of interactions between blocks and treatments has the effect of reducing the power of the test and increasing the width of interval estimates of treatment effects, thus making the experiment less sensitive. In addition, there are occasions when one is interested in the nature of the interactions between blocks and treatments and would like to obtain estimates of these. It is possible to use a design which permits an interaction term in the model even when the block effects are fixed, and which allows the nature of the interaction effects to be investigated. This design is called the *generalized randomized block design*. It is the same as a complete randomized block design except that d experimental units are assigned to each treatment within a block. This design increases the size of a block from r units for a randomized complete block design to dr units. This increase in block size often has the effect of increasing experimental errors when the number of experimental units is fixed. In the social sciences, however, increasing the size of the block moderately may cause little loss in efficiency. For instance, having one block of 10 persons aged 20–29 instead of two blocks of five persons of ages 20–24 and 25–29 respectively will for many types of experiments involve little loss of efficiency.

As we shall demonstrate by an example, a generalized randomized block design is analyzed like an ordinary multifactor study where blocks are one factor. Hence no new problems are encountered with a generalized randomized block design in testing for treatment effects or in estimating them.

Example

Table 23.14 contains the data for a two-factor experiment in which the effects of motivation (factor A: high level, low level) and distraction (factor B: high level, low level) on the time required to complete a task were studied, using eight men and eight women. Two men were assigned at random to each treatment, and independently two women were assigned at random to each treatment. Here sex is the blocking variable. Each block contains eight persons, with two randomly assigned to each treatment within the block. The layout in Table 23.14 corresponds to the layout in Table 20.3a for a three-factor experiment; to stress the correspondence, we have placed the blocks in

TABLE 23.14

Two-Factor Study in Generalized Randomized Block Design with $d = 2$ (observations are times for task completion)

	Block (sex)	
	Male	*Female*
High motivation:		
High distraction	12	3
	8	11
Low distraction	7	5
	5	9
Low motivation:		
High distraction	14	11
	16	9
Low distraction	15	10
	13	14

columns rather than in rows as usual. Since blocks, motivation levels, and distraction levels are considered to be fixed effects, we utilize the fixed effects three-factor model (20.17), with notation modified to fit the present context:

$$Y_{ijkm} = \mu_{...} + \rho_i + \alpha_j + \beta_k + (\rho\alpha)_{ij} + (\rho\beta)_{ik} + (\alpha\beta)_{jk} + (\rho\alpha\beta)_{ijk} + \varepsilon_{ijkm} \tag{23.25}$$

where:

$\mu_{...}$ is a constant

ρ_i, α_j, and β_k are constants subject to the restrictions $\sum \rho_i = \sum \alpha_j = \sum \beta_k = 0$

$(\rho\alpha)_{ij}$, $(\rho\beta)_{ik}$, $(\alpha\beta)_{jk}$, $(\rho\alpha\beta)_{ijk}$ are constants subject to the restrictions that the sums over any subscript are zero

ε_{ijkm} are independent $N(0, \sigma^2)$

$i = 1, \ldots, n; j = 1, \ldots, a; k = 1, \ldots, b; m = 1, \ldots, d$

The analysis of variance for model (23.25) is the ordinary three-factor ANOVA of Table 20.4, with slight modifications in notation. A computer package was employed to obtain the analysis of variance for the data in Table 23.14, and the results are shown in Table 23.15. We know from Table 20.4 that all test statistics use MSE in the denominator. These F^* statistics are shown in Table 23.15. For $\alpha = .01$, we require $F(.99; 1, 8) = 11.3$ for each of the tests. It is evident from the results in Table 23.15 that blocks do not interact with the factorial treatments and that among the two factors, only motivation affects the time required to complete the task. At this point, further analysis of the motivation effects would be indicated.

TABLE 23.15

Analysis of Variance for Task Completion Example in Table 23.14 ($n=2$, $a=2$, $b=2$, $d=2$)

Source of Variation	SS	df	MS	F*
Blocks (sex)	$SSBL = 20.25$	$n-1=1$	$MSBL = 20.25$	2.53
Factor A (motivation)	$SSA = 110.25$	$a-1=1$	$MSA = 110.25$	13.78
Factor B (distraction)	$SSB = 2.25$	$b-1=1$	$MSB = 2.25$	.28
$BL.A$ interactions	$SSBL.A = 6.25$	$(n-1)(a-1)=1$	$MSBL.A = 6.25$	.78
$BL.B$ interactions	$SSBL.B = 12.25$	$(n-1)(b-1)=1$	$MSBL.B = 12.25$	1.53
AB interactions	$SSAB = 6.25$	$(a-1)(b-1)=1$	$MSAB = 6.25$	.78
$BL.AB$ interactions	$SSBL.AB = .25$	$(n-1)(a-1)(b-1)=1$	$MSBL.AB = .25$	.03
Error	$SSE = 64.00$	$(d-1)nab=8$	$MSE = 8.00$	
Total	$SSTO = 221.75$	$dnab-1=15$		

23.11 USE OF MORE THAN ONE BLOCKING VARIABLE

Sometimes, substantial reduction in the experimental errors can only be obtained by utilizing more than one variable for determining blocks. For instance, both age and sex might be needed for designating blocks:

Block	*Characteristics of Experimental Units*
1	Male, aged 20–29
2	Female, aged 20–29
3	Male, aged 30–39
etc.	etc.

As another example, both observer and day of treatment application may be helpful as blocking variables:

Block	*Characteristics of Experimental Units*
1	Observer 1, day 1
2	Observer 2, day 1
3	Observer 1, day 2
etc.	etc.

Unless one wishes to study the separate effects of each of the blocking variables, no new problems arise when the blocks are defined by two or more variables. The n blocks are simply treated as ordinary blocks, and the usual block sum of squares is calculated.

If the effect of each of the blocking variables is to be isolated and the blocks are defined in a complete factorial fashion (e.g., nine blocks are used when three observers and three days are the blocking variables), the analysis simply treats each of the blocking variables as a factor and utilizes the methods developed in Chapter 20.

A problem that sometimes arises when two or more blocking variables are to be used is the large number of blocks called for. Suppose an experiment is to be conducted where the experimental units are stores. In order to reduce the experimental errors to a reasonable level, it would be desirable to group the stores into six sales volume classes and also into six location classes (suburban shopping center, suburban other, etc.). Thirty-six blocks result from combining these two blocking variables. If six treatments are to be studied, 216 stores would be required for the experiment. Often, use of this many stores would be much too costly. Latin square designs, to be discussed in Chapter 24, permit in this type of case the use of a much smaller number of replications while still preserving the full benefits of error variance reduction by using both blocking variables in six classes each.

23.12 REGRESSION APPROACH TO RANDOMIZED BLOCK DESIGNS

Model (23.1) for a randomized block design with both block and treatment effects fixed:

$$Y_{ij} = \mu_{..} + \rho_i + \tau_j + \varepsilon_{ij} \qquad i = 1, \ldots, n; j = 1, \ldots, r \tag{23.26}$$

can be expressed in straightforward fashion in the form of a regression model with indicator variables. Since there are no interaction terms in model (23.26), use of 0, 1 indicator variables is easiest. The regression model for the risk premium example in Table 23.2, with $r = 3$ treatments and $n = 5$ blocks, can be written as follows:

$$Y_{ij} = \beta_0 + \underbrace{\beta_1 X_{ij1} + \beta_2 X_{ij2} + \beta_3 X_{ij3} + \beta_4 X_{ij4}}_{\text{Block effect}} + \underbrace{\beta_5 X_{ij5} + \beta_6 X_{ij6}}_{\text{Treatment effect}} + \varepsilon_{ij} \tag{23.27}$$

where:

$$X_1 = \begin{cases} 1 & \text{if experimental unit from block 1} \\ 0 & \text{otherwise} \end{cases}$$

X_2, X_3, X_4 are defined similarly

$$X_5 = \begin{cases} 1 & \text{if experimental unit received treatment 1} \\ 0 & \text{otherwise} \end{cases}$$

$$X_6 = \begin{cases} 1 & \text{if experimental unit received treatment 2} \\ 0 & \text{otherwise} \end{cases}$$

The correspondences between the regression coefficients in model (23.27) and the parameters in the analysis of variance model (23.26) are as follows:

$$\begin{aligned} \beta_0 &= \mu_{..} + \rho_5 + \tau_3 & \beta_4 &= \rho_4 - \rho_5 \\ \beta_1 &= \rho_1 - \rho_5 & \beta_5 &= \tau_1 - \tau_3 \\ \beta_2 &= \rho_2 - \rho_5 & \beta_6 &= \tau_2 - \tau_3 \\ \beta_3 &= \rho_3 - \rho_5 & & \end{aligned} \tag{23.27a}$$

The **Y** observation vector and **X** matrix for the risk premium example are shown in Table 23.16. Note, for instance, that for observation Y_{12}, the indicator variables have the values:

$$X_1 = 1 \quad X_2 = 0 \quad X_3 = 0 \quad X_4 = 0 \quad X_5 = 0 \quad X_6 = 1$$

Hence:

$$\begin{aligned} Y_{12} &= \beta_0 + \beta_1 + \beta_6 + \varepsilon_{12} \\ &= (\mu_{..} + \rho_5 + \tau_3) + (\rho_1 - \rho_5) + (\tau_2 - \tau_3) + \varepsilon_{12} \\ &= \mu_{..} + \rho_1 + \tau_2 + \varepsilon_{12} \end{aligned}$$

TABLE 23.16

Data Matrices for Model (23.27) Based on Risk Premium Data in Table 23.2

$$\mathbf{Y} = \begin{bmatrix} Y_{11} = 1 \\ Y_{12} = 5 \\ Y_{13} = 8 \\ Y_{21} = 2 \\ Y_{22} = 8 \\ Y_{23} = 14 \\ Y_{31} = 7 \\ Y_{32} = 9 \\ Y_{33} = 16 \\ Y_{41} = 6 \\ Y_{42} = 13 \\ Y_{43} = 18 \\ Y_{51} = 12 \\ Y_{52} = 14 \\ Y_{53} = 17 \end{bmatrix} \qquad \mathbf{X} = \overset{\begin{matrix} X_1 & X_2 & X_3 & X_4 & X_5 & X_6 \end{matrix}}{\begin{bmatrix} 1 & 1 & 0 & 0 & 0 & 1 & 0 \\ 1 & 1 & 0 & 0 & 0 & 0 & 1 \\ 1 & 1 & 0 & 0 & 0 & 0 & 0 \\ 1 & 0 & 1 & 0 & 0 & 1 & 0 \\ 1 & 0 & 1 & 0 & 0 & 0 & 1 \\ 1 & 0 & 1 & 0 & 0 & 0 & 0 \\ 1 & 0 & 0 & 1 & 0 & 1 & 0 \\ 1 & 0 & 0 & 1 & 0 & 0 & 1 \\ 1 & 0 & 0 & 1 & 0 & 0 & 0 \\ 1 & 0 & 0 & 0 & 1 & 1 & 0 \\ 1 & 0 & 0 & 0 & 1 & 0 & 1 \\ 1 & 0 & 0 & 0 & 1 & 0 & 0 \\ 1 & 0 & 0 & 0 & 0 & 1 & 0 \\ 1 & 0 & 0 & 0 & 0 & 0 & 1 \\ 1 & 0 & 0 & 0 & 0 & 0 & 0 \end{bmatrix}}$$

which is precisely the expression for Y_{12} according to the analysis of variance model (23.26).

Tests and estimation of treatment effects with the regression approach follow easily, in view of the simple correspondences in (23.27a).

Note

One or more missing observations in a randomized block design cause no difficulties with the regression approach. We simply set up the regression model for the observations available, and then fit the model to the data.

23.13 COVARIANCE ANALYSIS FOR RANDOMIZED BLOCK DESIGNS

Covariance analysis can be employed to further reduce the experimental errors in a randomized block design. The extension is a straightforward one from covariance analysis for a completely randomized design.

Model

The usual randomized block design model was given in (23.1):

(23.28)
$$Y_{ij} = \mu_{..} + \rho_i + \tau_j + \varepsilon_{ij}$$
$$i = 1, \ldots, n; j = 1, \ldots, r$$

The covariance model for a randomized block design is obtained by simply adding a term (or several terms) for the relation between the dependent variable Y and the concomitant variable X. Assuming this relation can be described by a linear function, we obtain:

$$Y_{ij} = \mu_{..} + \rho_i + \tau_j + \gamma(X_{ij} - \bar{X}_{..}) + \varepsilon_{ij} \qquad (23.29)$$
$$i = 1, \ldots, n; j = 1, \ldots, r$$

Regression Approach

The regression approach to the covariance model (23.29) involves no new principles. As in Chapter 22, we shall denote $X_{ij} - \bar{X}_{..}$ in model (23.29) by Z_{ij}:

$$Z_{ij} = X_{ij} - \bar{X}_{..} \qquad (23.30)$$

Further, we shall use 0, 1 indicator variables for the block and treatment effects.

Suppose in a randomized complete block design study, $n = 4$ blocks and $r = 3$ treatments are utilized. The regression model counterpart to covariance model (23.29) then is:

$$Y_{ij} = \beta_0 + \beta_1 X_{ij1} + \beta_2 X_{ij2} + \beta_3 X_{ij3} + \beta_4 X_{ij4} + \beta_5 X_{ij5} + \beta_6 Z_{ij} + \varepsilon_{ij} \qquad \text{Full model} \qquad (23.31)$$

$X_1 =$ 1 if experimental unit from block 1; 0 otherwise

$X_2 =$ 1 if experimental unit from block 2; 0 otherwise

$X_3 =$ 1 if experimental unit from block 3; 0 otherwise

$X_4 =$ 1 if experimental unit received treatment 1; 0 otherwise

$X_5 =$ 1 if experimental unit received treatment 2; 0 otherwise

$$\beta_0 = \mu_{..} + \rho_4 + \tau_3$$
$$\beta_1 = \rho_1 - \rho_4$$
$$\beta_2 = \rho_2 - \rho_4$$
$$\beta_3 = \rho_3 - \rho_4$$
$$\beta_4 = \tau_1 - \tau_3$$
$$\beta_5 = \tau_2 - \tau_3$$
$$\beta_6 = \gamma$$

To test for treatment effects:

(23.32)

$$C_1: \tau_1 = \tau_2 = \tau_3 = 0$$
$$C_2: \text{not all } \tau_j \text{ are zero}$$

we would also need to fit the reduced model under C_1:

(23.33) $$Y_{ij} = \beta_0 + \beta_1 X_{ij1} + \beta_2 X_{ij2} + \beta_3 X_{ij3} + \beta_6 Z_{ij} + \varepsilon_{ij} \qquad \text{Reduced model}$$

The test for treatment effects would then be conducted in the usual way.

If we wish to compare two treatment effects by the regression approach, we utilize the correspondences in (23.31) to obtain an unbiased estimator. For estimating $\tau_1 - \tau_2$, for example, the estimator is:

(23.34) $$b_4 - b_5$$

The variance of this estimator is:

(23.35) $$\sigma^2(b_4 - b_5) = \sigma^2(b_4) + \sigma^2(b_5) - 2\sigma(b_4, b_5)$$

The estimated variance-covariance matrix for the regression coefficients from the computer printout can then be used to estimate this variance.

Adjustment Approach

If the adjustment approach to covariance analysis is to be employed, we begin as before by obtaining analyses of variance for Y, X, and XY. The analysis of variance for Y is given in (23.2), and that for X would be identical, with X_{ij} replacing Y_{ij}. Table 23.17 contains these analyses of variance. Note that subscripts for X and Y are used for identification.

TABLE 23.17

Analyses of Variance for Y, X, and XY—Randomized Block Design

Source of Variation	*Sums of Squares or Products*			*df*
	Y	X	XY	
Blocks	$SSBL_Y$	$SSBL_X$	$SPBL$	$n-1$
Treatments	$SSTR_Y$	$SSTR_X$	$SPTR$	$r-1$
Error	SSE_Y	SSE_X	SPE	$(n-1)(r-1)$
Total	$SSTO_Y$	$SSTO_X$	$SPTO$	$nr-1$

The analysis of variance for products proceeds as follows:

23.36a) $$SPTO = \sum_i \sum_j (X_{ij} - \bar{X}_{..})(Y_{ij} - \bar{Y}_{..}) = \sum_i \sum_j X_{ij} Y_{ij} - \frac{X_{..} Y_{..}}{nr}$$

23.36b) $$SPBL = r \sum_i (\bar{X}_{i.} - \bar{X}_{..})(\bar{Y}_{i.} - \bar{Y}_{..}) = \frac{\sum_i X_{i.} Y_{i.}}{r} - \frac{X_{..} Y_{..}}{nr}$$

(23.36c) $$SPTR = n \sum_j (\bar{X}_{.j} - \bar{X}_{..})(\bar{Y}_{.j} - \bar{Y}_{..}) = \frac{\sum_j X_{.j} Y_{.j}}{n} - \frac{X_{..} Y_{..}}{nr}$$

(23.36d) $$SPE = \sum_i \sum_j (X_{ij} - \bar{X}_{i.} - \bar{X}_{.j} + \bar{X}_{..})(Y_{ij} - \bar{Y}_{i.} - \bar{Y}_{.j} + \bar{Y}_{..})$$
$$= SPTO - SPBL - SPTR$$

Table 23.17 also contains this analysis of variance for XY.

The adjustment process now ignores the line for block effects in Table 23.17, as if this effect has thereby been eliminated, and proceeds in a manner analogous to a completely randomized design. Table 23.18a contains the

TABLE 23.18

Covariance Analysis for Testing Treatment Effects in a Randomized Block Design

(a) Analyses of Variance

Source of Variation	*Sums of Squares or Products* Y	X	XY	df
Treatments	$SSTR_Y$	$SSTR_X$	$SPTR$	$r-1$
Error	SSE_Y	SSE_X	SPE	$(n-1)(r-1)$
Sum	$SSTR_Y + SSE_Y$	$SSTR_X + SSE_X$	$SPTR + SPE$	$n(r-1)$

(b) Analysis of Covariance

Source of Variation	Adjusted SS	Adjusted df	Adjusted MS
Treatments	$SSTR$(adj.)	$r-1$	$MSTR$(adj.)
Error	SSE(adj.)	$(n-1)(r-1)-1$	MSE(adj.)
Sum	$SS(TR + E$; adj.)	$n(r-1)-1$	

analyses of variance without the block effects line. Note that the total line is called "sum," since it excludes the block effects. The adjustment process is now entirely analogous to that for a completely randomized design. We obtain for the "sum" line:

(23.37a) $$SS(TR + E; \text{adj.}) = (SSTR_Y + SSE_Y) - \frac{(SPTR + SPE)^2}{SSTR_X + SSE_X}$$

which corresponds to (22.36a) for the total line for a completely randomized design. The adjusted error sum of squares is as before in (22.36b):

(23.37b) $$SSE(\text{adj.}) = SSE_Y - \frac{(SPE)^2}{SSE_X}$$

Finally, the adjusted treatment sum of squares is obtained as a remainder, as in (22.36c):

$$SSTR(\text{adj.}) = SS(TR + E; \text{adj.}) - SSE(\text{adj.}) \tag{23.37c}$$

Table 23.18b contains the covariance analysis for a randomized block design. Note that the degrees of freedom for error and the sum are reduced by one on account of the introduction of the concomitant variable X. The adjusted degrees of freedom for treatments are obtained as a remainder.

A test for treatment effects uses the usual statistic:

$$F^* = \frac{MSTR(\text{adj.})}{MSE(\text{adj.})} \tag{23.38}$$

which, if C_1 holds, is distributed as $F(r - 1, (n - 1)(r - 1) - 1)$.

Blocking as Alternative to Covariance Analysis

At times, a choice exists between: (1) a completely randomized design with covariance analysis used to reduce the experimental errors, and (2) a randomized block design, with the blocks formed by means of the concomitant variable. Generally, the latter alternative is to be preferred. There are several reasons for this:

1. If the regression is linear, randomized block designs and covariance analysis are about equally efficient. If the regression is not linear but covariance analysis with a linear relationship is utilized, covariance analysis with a completely randomized design will tend to be not as effective as a randomized block design.
2. Computations with randomized block designs are simpler than those with covariance analysis.
3. Randomized block designs are essentially free of assumptions about the nature of the relationship between the blocking variable and the dependent variable, while covariance analysis assumes a definite form of relationship.

A drawback of randomized block designs is that somewhat fewer degrees of freedom are available for experimental error than with covariance analysis for a completely randomized design. However, in all but small-scale experiments, this difference in degrees of freedom has little effect on the precision of the estimates.

PROBLEMS

23.1. A student commented in a discussion group: "Random permutations are used to assign treatments to experimental units with a randomized block design just as with a completely randomized design. Hence there is no basic difference between these two designs." Comment.

23.2. a) What might be some useful blocking variables for an experiment about the effects of different price levels on sales of a product, using stores as experimental units?

b) What might be some useful blocking variables for an experiment about the effects of different flight crew schedules on the morale of the crews, using flight crews as experimental units?

23.3. Five treatments are studied in an experiment with a randomized complete block design, using four blocks. Use a table of random permutations to obtain the assignments of treatments to experimental units.

23.4. Two treatments and a control are studied in an experiment with a randomized block design. Five blocks are employed, each containing four experimental units. In each block, each treatment is to be assigned to one experimental unit, and the control is to be assigned to two experimental units. Use a table of random permutations to obtain the assignments of treatments to experimental units.

23.5. Show that when two treatments are studied in a randomized complete block design, the F test statistic for treatment effects is equivalent to the two-sided t test statistic for paired observations based on (1.62).

23.6. An accounting firm, prior to introducing in the firm widespread training in statistical sampling for auditing, tested three training methods: (1) study at home with programmed training materials; (2) training sessions at local offices conducted by local staff; (3) training sessions in Chicago conducted by national staff. Thirty auditors were grouped into 10 blocks of 3, according to time elapsed since college graduation, and the auditors in each block were randomly assigned to the 3 training methods. At the end of the training, each auditor was asked to analyze a complex case involving statistical applications; a proficiency measure based on this analysis was obtained for each auditor. The results were (block 1 consists of auditors graduated most recently, block 10 consists of those graduated most distantly):

	Training Method		
Block	1	2	3
1	73	81	92
2	76	79	89
3	75	76	87
4	74	77	90
5	76	71	88
6	73	75	85
7	68	72	88
8	64	71	82
9	65	73	81
10	62	69	78

a) Why do you think the blocking variable "time elapsed since college graduation" was employed?

b) State the model you will use to analyze the effects of different training methods.

c) Test whether the training methods differ in effectiveness. Use a level of significance of .05. What do you conclude?

d) Analyze the comparative effectiveness of the three training methods. Explain the reasons why you chose your particular analytical approach, and summarize your findings.

23.7. Refer to Problem 23.6.

a) Obtain the residuals and analyze them to determine the appropriateness of the model you employed in Problem 23.6b. Summarize your findings.

b) Conduct the Tukey test for additivity of block and treatment effects, using a level of significance of .01. What do you conclude?

23.8. Refer to Problem 23.6.

a) How effective was use of the blocking variable "time elapsed since graduation," as compared to a completely randomized design?

b) Another accounting firm wishes to conduct the same experiment with some of its auditors, using the same design and model. How many blocks would you recommend that this firm employ if it wishes to make all pairwise treatment comparisons with precision ± 1.5, with a 99 percent family confidence coefficient?

23.9. A computer company, to test the efficiency of its new pocket electronic calculator, selected six engineers who were proficient in the use of both this calculator and a slide rule and asked them to work out two problems on both the electronic calculator and the slide rule. One of the problems was statistical in nature, the other was an engineering problem. The order of the four calculations was randomized independently for each engineer. The length of time (in minutes) required to solve each problem was observed. The results were:

	Statistical Problem		*Engineering Problem*	
Engineer	*Electronic Calculator*	*Slide Rule*	*Electronic Calculator*	*Slide Rule*
Jones	3.1	7.4	2.3	5.1
Williams	3.8	8.1	2.8	5.3
Adams	3.0	7.6	2.0	4.9
Dixon	3.5	7.9	2.7	5.5
Erickson	3.3	6.9	2.5	5.2
Maynes	3.6	7.8	2.4	4.9

Since the six engineers chosen for the experiment were those who were proficient in the use of both the calculator and the slide rule, and were not selected by a random mechanism from a defined population, the company does not wish to regard them as a sample from a population of engineers. Model (23.6) is to be employed.

a) What assumptions are involved in using this model, and which of these are you most concerned about?

b) Analyze the results of this study. Explain the reasons why you chose your particular analytical approach, and summarize your findings.

23.10. Refer to Problem 23.9.

a) Obtain the residuals and analyze them to determine the appropriateness of model (23.6). Summarize your findings. What additional information would have been helpful?

b) Conduct the Tukey test for additivity of block and treatment effects, using a level of significance of .01. What do you conclude?

23.11. Refer to Problem 23.9. Another computer company wishes to conduct the same experiment with some of its engineers and its pocket electronic calculator, using the same experimental design and model. How many engineers should this company use, assuming that the principal objective is to estimate the difference in expected time between the two computing aids for each problem within $\pm.2$ minutes, with a 98 percent family confidence coefficient?

23.12. Refer to Problem 23.6. Suppose that observation $Y_{23} = 89$ were missing.

a) Use the Yates method to test whether the training methods differ in effectiveness. Use a level of significance of .05. What do you conclude?

b) Compare the mean proficiency for training methods 2 and 3, using a 95 percent confidence interval.

23.13. A consumer research organization showed five different advertisements to eight subjects, and asked each to rank them in order of truthfulness. A rank of 1 denotes the most truthful. The results were:

	Advertisement				
Subject	A	B	C	D	E
1	3	1	2	5	4
2	4	2	1	3	5
3	4	2	3	1	5
4	3	1	2	5	4
5	4	1	2	5	3
6	4	2	1	3	5
7	4	1	2	3	5
8	5	1	3	2	4

a) Do the subjects perceive the five advertisements as having equal truthfulness? Conduct a formal test, using a level of significance of .05. What do you conclude?

b) The organization's staff had information available besides that in the advertisements, and ranked the advertisements as follows:

Advertisement:	A	B	C	D	E
Rank:	3	2	1	5	4

Analyze the data, and summarize your findings.

23.14. A social scientist, after learning about generalized randomized block designs, asked: "Why would anyone use a randomized complete block design which requires the assumption that block and treatment effects do not interact, when this assumption can be avoided with a generalized randomized block design?" Comment.

23.15. Refer to the motivation-distraction study on page 749.

a) Using the data in Table 23.14, verify the analysis of variance in Table 23.15.

b) Estimate the difference in mean effects for the two motivation levels, using a 90 percent confidence interval.

23.16. Refer to Problem 23.6. The accounting firm repeated the experiment with another group of 30 auditors, but this time grouped them into five blocks of six each. In each block, each treatment was randomly assigned to two auditors. The results were:

	Training Method		
Block	1	2	3
1	74	84	94
	71	78	95
2	73	75	93
	69	83	98
3	75	81	89
	67	74	86
4	68	73	84
	70	75	85
5	64	71	81
	61	74	74

a) State the model you will use to analyze the effects of the different training methods.
b) Fit the model and obtain the residuals. Analyze them to determine the appropriateness of the model you employed. Summarize your findings.
c) Assuming that your model is appropriate, analyze the results of the study and state your findings. Explain why you chose your particular analytical approach.

23.17. Refer to Problem 23.9.
a) Set up the regression model counterpart to the ANOVA model. State the correspondences between the two sets of parameters.
b) Test whether the mean times for the two problems are the same, using the regression approach and a level of significance of .05.
c) Obtain an interval estimate by the regression approach for the difference between the mean times for the two problems. Use a confidence coefficient of 99 percent.

23.18. Refer to Problem 23.6. Suppose that observations $Y_{31} = 75$ and $Y_{82} = 71$ were missing.
a) Use the regression approach to test whether the training methods differ in effectiveness; employ $\alpha = .05$. State the regression model and the correspondences between the two sets of parameters. What do you conclude?
b) Compare the mean proficiency for training methods 1 and 2 by means of the regression approach, using a 99 percent confidence interval.

23.19. Refer to Problem 23.6. The analyst wished to examine whether use of a concomitant variable would help to reduce the experimental error significantly.
a) Would you expect the auditor's age to be a helpful concomitant variable here?

b) State the regression model formulation you would employ for this study if the auditor's statistical proficiency score obtained prior to the experiment is used as the concomitant variable. What are the correspondences between the regression model parameters and those for the covariance model?

c) The pre-training statistical proficiency scores for the auditors were as follows:

	Training Method		
Block	1	2	3
1	93	98	91
2	94	93	94
3	89	91	92
4	86	84	90
5	78	76	84
6	75	74	78
7	79	76	72
8	71	69	64
9	74	71	70
10	63	68	64

Test whether the training methods differ in effectiveness, using the concomitant variable and a level of significance of .05. What do you conclude?

d) Analyze the comparative effectiveness of the three training methods, using the concomitant variable, and state your findings.

e) Is the pre-training statistical proficiency score a helpful concomitant variable here? Would it be a more helpful concomitant variable if no blocking had been employed?

CITED REFERENCE

23.1. Owen, Donald B. *Handbook of Statistical Tables*. Reading, Mass.: Addison-Wesley Publishing Co., Inc., 1962.

24

Latin Square Designs

IN THIS CHAPTER, we consider a widely used design which employs two blocking variables to reduce experimental errors—the latin square design.

24.1 BASIC ELEMENTS

Complete and Incomplete Block Designs

We saw in Section 23.11 that two blocking variables can be used simultaneously in randomized complete block designs to eliminate, from the experimental errors, the variation associated with each of the blocking variables. Thus the blocking variables might be age of subject and income of subject, with a block containing subjects in a given age group and income group.

The full use of two blocking variables in a complete block design may, however, have the disadvantage of requiring too many experimental units. For instance, if the age and income variables in our earlier illustration have six classes each, 36 blocks would be required. If six treatments were to be studied, 216 subjects would be needed for the experiment. Cost considerations may not permit the use of this many experimental units, yet precision and range of validity considerations may require the simultaneous use of two blocking variables, each with six classes, in order to reduce the experimental error variance sufficiently and to have a reasonable variety of experimental subjects. In this type of situation, an *incomplete block design* may be helpful. In such a design, all 36 blocks in our example would still be used, but now each block would not contain all six treatments.

Latin Square Designs

Taking incomplete block designs to the extreme in our example, the number of experimental units is minimized, given the employment of 36 blocks, if only one treatment is run in each block. This extreme case, where each block contains only one treatment, is the type of situation for which a latin square design is appropriate. Table 24.1 provides an illustration of the difference between complete and incomplete block designs for the example considered. Column 1 shows the complete block design for this case, while columns 2 and 3 illustrate incomplete block designs, with three and one treatments in each block respectively.

There is another reason, besides economy, why latin squares with only one treatment per block are used, namely that blocks sometimes cannot contain more than one treatment. Consider the repeated measures design discussed in the last chapter, where each subject receives every treatment. It was stressed there that the order of treatments be randomized in case interference effects between the different treatments were to be present. If indeed interference effects due to the order position of the treatments are anticipated, it may be desirable to use the order position as another blocking variable. Thus, "subject" would be one blocking variable and "order position of the treatments" a second blocking variable. Blocks would then be defined as follows:

Block 1: Subject 1, position 1
Block 2: Subject 1, position 2
⋮ ⋮
Block 7: Subject 2, position 1
etc. etc.

Notice that the blocks so defined can contain only one treatment, since the order position refers to the place of a single treatment in the sequence of treatments for a subject.

Description of Latin Square Designs

Let A, B, C represent three treatments; it is conventional with latin square designs to use Latin letters for the treatments. Suppose that day of week (Monday, Tuesday, Wednesday) and operator (1, 2, 3) are to be used as blocking variables. A latin square design might then be shown as follows:

	Operator		
Day	1	2	3
Monday	*B*	*A*	*C*
Tuesday	*A*	*C*	*B*
Wednesday	*C*	*B*	*A*

TABLE 24.1

Complete and Incomplete Block Designs

Block Description	(1) *Complete Block Design*	(2) *Incomplete Block Design (three treatments per block)*	(3) *Incomplete Block Design (one treatment per block)*
Age under 25, income under $4,000	$T_1, T_2, T_3, T_4, T_5, T_6$	T_1, T_3, T_5	T_2
Age under 25, income $4,000–6,999	$T_1, T_2, T_3, T_4, T_5, T_6$	T_2, T_4, T_6	T_5
⋮	⋮	⋮	⋮
Age 25–34, income under $4,000	$T_1, T_2, T_3, T_4, T_5, T_6$	T_2, T_4, T_5	T_3
⋮	⋮	⋮	⋮
Age 25–34, income over $20,000	$T_1, T_2, T_3, T_4, T_5, T_6$	T_3, T_4, T_6	T_2
etc.	etc.	etc.	etc.

Operator 1 would run treatment *B* on Monday, treatment *A* on Tuesday, and treatment *C* on Wednesday, and so on for the other operators. Note that each operator runs each treatment, and that all treatments are run on each day.

A latin square design thus has the following features:

1. There are r treatments.
2. There are two blocking variables, each containing r classes.
3. Each row and each column in the design square contains all treatments; that is, each class of each blocking variable constitutes a replication.

Advantages and Disadvantages of Latin Square Designs

Advantages of a latin square design include:

1. The use of two blocking variables often permits greater reductions in experimental errors than can be obtained with either blocking variable alone.
2. Treatment effects can be studied from a small-scale experiment. This is particularly helpful in preliminary or pilot studies.
3. It is often helpful in repeated measures experiments to take into account the order effect of treatments by means of a latin square design.

Disadvantages of a latin square design are:

1. The number of classes of each blocking variable must equal the number of treatments. This leads to a very small number of degrees of freedom for experimental error when only a few treatments are studied. On the other hand, when many treatments are studied, the degrees of freedom for experimental error may be larger than necessary.
2. The assumptions of the model are restrictive (e.g., that there are no interactions between either blocking variable and treatments, and also none between the two blocking variables).
3. The two blocking variables cannot have different numbers of classes.
4. The randomization required is somewhat more complex than that for earlier designs considered.

Because of the limitations on the degrees of freedom for experimental error just described, latin squares are rarely used when more than eight treatments are being investigated. For the same reason, when there are few treatments, say four or less, additional replications are usually required when a latin square design is employed.

Randomization of Latin Square Design

There exist many latin squares for a given number of treatments. Suppose that the number of treatments is $r = 3$; some possible latin square designs are (we omit the row and column blocking variable labels):

A	B	C		A	C	B		B	A	C
B	C	A		B	A	C		C	B	A
C	A	B		C	B	A		A	C	B

For $r = 3$, there are altogether 12 different possible arrangements. This number increases rapidly as the number of treatments gets larger; for $r = 5$, there are 161,280 possible arrangements.

The objective of randomization is to select one of all the possible latin squares for the given number of treatments r, such that each square has an equal probability of being selected. Clearly, it is not generally feasible to list all possible latin squares so that one can be selected at random.

Instead, we utilize *standard latin squares*, which are latin squares in which the elements of the first row and the first column are arranged alphabetically. The latin square on the left is a standard latin square. Table A–16 contains all the standard squares for $r = 3$ and 4, and a single selected standard square for $r = 5$, 6, 7, 8, and 9. The randomization procedure usually employed is as follows:

1. For $r = 3$, independently arrange the rows and columns at random.
2. For $r = 4$, select one of the standard squares at random. Then, independently arrange its rows and columns at random.
3. For $r = 5$ and higher, independently arrange the rows, columns, and treatments of the given standard square at random.

It can be shown that this procedure selects one latin square at random from all possible squares for $r = 3$ and 4. For $r = 5$ or higher, the randomization procedure is not based on all possible latin squares, but rather on very large and suitable subsets thereof.

Example 1. In our earlier illustration, the blocking variables were day (Monday, Tuesday, Wednesday) and operator (1, 2, 3). Thus, we can show the borders of the square as follows:

	Operator		
Day	1	2	3
Monday			
Tuesday			
Wednesday			

The standard latin square for $r = 3$ is:

	A	B	C
Step 1	B	C	A
	C	A	B

We select now a random permutation of 3 to arrange the rows. Suppose it is: 2, 3, 1. Thus, row 2 in the standard square comes first, row 3 second, and row 1 third. We obtain:

	Original Row Number			
	2	*B*	*C*	*A*
Step 2	3	*C*	*A*	*B*
	1	*A*	*B*	*C*

We then select independently another permutation of 3 to rearrange the columns. Suppose it is: 2, 1, 3. Column 2 in the square of step 2 now becomes the first column, and so on. We obtain:

	Step 2 *column number*:	2	1	3
		C	*B*	*A*
Step 3		*A*	*C*	*B*
		B	*A*	*C*

The selected design therefore is:

	Operator		
Day	1	2	3
Monday	*C*	*B*	*A*
Tuesday	*A*	*C*	*B*
Wednesday	*B*	*A*	*C*

Example 2. Consider an experiment about the effects of five different types of background music (A, B, C, D, E) on the productivity of bank tellers. A given type of music is played for one day and the productivity is observed. Day of the week and week of the experimental period are to be the two blocking variables. The borders of the latin square design therefore are:

	Day				
Week	*M*	*T*	*W*	*Th*	*F*
1					
2					
3					
4					
5					

We start with the standard latin square of Table A–16 for $r = 5$:

	A	*B*	*C*	*D*	*E*
	B	*A*	*E*	*C*	*D*
Step 1	*C*	*D*	*A*	*E*	*B*
	D	*E*	*B*	*A*	*C*
	E	*C*	*D*	*B*	*A*

Next we permute the rows randomly, using a random permutation of 5, say: 4, 2, 3, 1, 5. We obtain then:

	D	*E*	*B*	*A*	*C*
	B	*A*	*E*	*C*	*D*
Step 2	*C*	*D*	*A*	*E*	*B*
	A	*B*	*C*	*D*	*E*
	E	*C*	*D*	*B*	*A*

We now permute the columns randomly, using an independent random permutation of 5, say: 2, 4, 1, 5, 3. We obtain:

	E	*A*	*D*	*C*	*B*
	A	*C*	*B*	*D*	*E*
Step 3	*D*	*E*	*C*	*B*	*A*
	B	*D*	*A*	*E*	*C*
	C	*B*	*E*	*A*	*D*

Finally, we need to permute the treatment labels randomly. Note that the treatment designations *A*, *B*, *C*, *D*, *E* remain fixed. We simply wish to randomize the cells in which the specific treatments occur, given the pattern obtained in step 3. Assume the correspondence:

1	2	3	4	5
A	*B*	*C*	*D*	*E*

and that an independent random selection of a permutation of 5 yields: 3, 5, 2, 1, 4. We obtain then:

Present cell designation:	*A*	*B*	*C*	*D*	*E*
New cell designation:	*C*	*E*	*B*	*A*	*D*

Thus, every cell designated *A* in step 3 now is designated *C*, and so on. Hence, the latin square design to be used is:

	Day				
Week	*M*	*T*	*W*	*Th*	*F*
1	*D*	*C*	*A*	*B*	*E*
2	*C*	*B*	*E*	*A*	*D*
3	*A*	*D*	*B*	*E*	*C*
4	*E*	*A*	*C*	*D*	*B*
5	*B*	*E*	*D*	*C*	*A*

Note

For $r = 3$ and 4, it would be sufficient to randomize all r rows and the last $r - 1$ columns. It is equally good to randomize all rows and columns, the procedure suggested here.

Example

Let us return to the experiment cited above on the effects of different types of background music on the productivity of bank tellers. The treatments were defined as various combinations of tempo of music (slow, medium, fast) and style of music (instrumental and vocal, instrumental only). The treatments and Latin letter designations were as follows:

Treatment	*Latin Letter Designation*	*Tempo and Style of Music*
1	*A*	Slow, instrumental and vocal
2	*B*	Medium, instrumental and vocal
3	*C*	Fast, instrumental and vocal
4	*D*	Medium, instrumental only
5	*E*	Fast, instrumental only

Table 24.2 contains the results of this experiment. The treatment in each cell is shown in parentheses. Note that in this study, the experimental unit is a working day for the crew of bank tellers, and that the productivity data pertain to the performance of the entire crew. Let $Y_{ij(k)}$ denote the observation in the cell defined by the ith class of the row blocking variable and the jth class of the column blocking variable. The subscript k indicates the treatment assigned to this cell by the particular latin square design employed; it is shown in parentheses since it is dependent on (i, j). As soon as the (i, j) cell combination is specified, the treatment is determined for the particular latin square design employed. Thus, $Y_{12(3)} = 17$ is the productivity on Tuesday of the first week, and Table 24.2 indicates that the type of music on that day was C.

TABLE 24.2

Latin Square Design Background Music Study (productivity of crew—data coded)

	Day					
Week	*M*	*T*	*W*	*Th*	*F*	*Total*
1	18 (*D*)	17 (*C*)	14 (*A*)	21 (*B*)	17 (*E*)	$Y_{1..} = 87$
2	13 (*C*)	34 (*B*)	21 (*E*)	16 (*A*)	15 (*D*)	$Y_{2..} = 99$
3	7 (*A*)	29 (*D*)	32 (*B*)	27 (*E*)	13 (*C*)	$Y_{3..} = 108$
4	17 (*E*)	13 (*A*)	24 (*C*)	31 (*D*)	25 (*B*)	$Y_{4..} = 110$
5	21 (*B*)	26 (*E*)	26 (*D*)	31 (*C*)	7 (*A*)	$Y_{5..} = 111$
Total	$Y_{.1.} = 76$	$Y_{.2.} = 119$	$Y_{.3.} = 117$	$Y_{.4.} = 126$	$Y_{.5.} = 77$	$Y_{...} = 515$

$$Y_{..1} = 7 + 13 + 14 + 16 + 7 = 57 \qquad \bar{Y}_{..1} = 11.4$$
$$Y_{..2} = 21 + 34 + 32 + 21 + 25 = 133 \qquad \bar{Y}_{..2} = 26.6$$
$$Y_{..3} = 13 + 17 + 24 + 31 + 13 = 98 \qquad \bar{Y}_{..3} = 19.6$$
$$Y_{..4} = 18 + 29 + 26 + 31 + 15 = 119 \qquad \bar{Y}_{..4} = 23.8$$
$$Y_{..5} = 17 + 26 + 21 + 27 + 17 = 108 \qquad \bar{Y}_{..5} = 21.6$$

24.2 MODEL

A latin square design involves the effect of the row blocking variable, denoted by ρ_i, the effect of the column blocking variable, denoted by κ_j, and the treatment effect, denoted by τ_k. It is necessary to assume that no interactions exist between these three variables in order that the latin square design can be analyzed. Thus the model employed must be an additive one. For the case of fixed treatment and block effects, the model is:

(24.1)
$$Y_{ij(k)} = \mu_{...} + \rho_i + \kappa_j + \tau_k + \varepsilon_{ij(k)}$$

where:

$\mu_{...}$ is a constant
ρ_i, κ_j, τ_k are constants subject to the restrictions $\sum \rho_i = \sum \kappa_j = \sum \tau_k = 0$
$\varepsilon_{ij(k)}$ are independent $N(0, \sigma^2)$
$i = 1, \ldots, r; j = 1, \ldots, r; k = 1, \ldots, r$

Note again that the number of classes for each of the two blocking variables is the same as the number of treatments; also note that the total number of observations is r^2.

Comments

1. Sometimes, one or both of the blocking variable effects are viewed as random, as when the blocking variables refer to subjects, observers, machines, etc. We shall consider the case of random blocking variable effects in Section 24.8.

2. If the treatment effects are random, the only change in model (24.1) would be that the τ_k now are independent $N(0, \sigma_\tau^2)$, and are independent of the $\varepsilon_{ij(k)}$.

3. The aptness of the additive model (24.1) needs to be carefully explored. A test for this purpose is explained in Section 24.6.

24.3 ANALYSIS OF VARIANCE

We shall employ the usual notation for row, column, and treatment totals and means:

(24.2a)
$$Y_{i..} = \sum_j Y_{ij(k)} \qquad \bar{Y}_{i..} = \frac{Y_{i..}}{r}$$

(24.2b)
$$Y_{.j.} = \sum_i Y_{ij(k)} \qquad \bar{Y}_{.j.} = \frac{Y_{.j.}}{r}$$

(24.2c)
$$Y_{..k} = \sum_{i,j} Y_{ij(k)} \qquad \bar{Y}_{..k} = \frac{Y_{..k}}{r}$$

The overall total and mean are denoted as always:

(24.2d)
$$Y_{...} = \sum_i \sum_j Y_{ij(k)} \qquad \bar{Y}_{...} = \frac{Y_{...}}{r^2}$$

Note the redundancy of any one of the three subscripts, arising from the fact that the treatment is uniquely determined by the row and column specifications for the latin square utilized. The various totals for our background music example are shown in Table 24.2.

To obtain the analysis of variance, we express the total deviation $Y_{ij(k)} - \bar{Y}_{...}$ as follows:

$$(24.3) \qquad \underbrace{Y_{ij(k)} - \bar{Y}_{...}}_{\text{Total deviation}} = \underbrace{\bar{Y}_{i..} - \bar{Y}_{...}}_{\text{Row effect}} + \underbrace{\bar{Y}_{.j.} - \bar{Y}_{...}}_{\text{Column effect}} + \underbrace{\bar{Y}_{..k} - \bar{Y}_{...}}_{\text{Treatment effect}} + \underbrace{Y_{ij(k)} - \bar{Y}_{i..} - \bar{Y}_{.j.} - \bar{Y}_{..k} + 2\bar{Y}_{...}}_{\text{Residual}}$$

Squaring (24.3) and then summing over all observations, we find that all cross-product terms drop out and we obtain:

$$(24.4) \qquad SSTO = SSROW + SSCOL + SSTR + SSE$$

where:

$$(24.4a) \qquad SSTO = \sum_i \sum_j (Y_{ij(k)} - \bar{Y}_{...})^2$$

$$(24.4b) \qquad SSROW = r \sum_i (\bar{Y}_{i..} - \bar{Y}_{...})^2$$

$$(24.4c) \qquad SSCOL = r \sum_j (\bar{Y}_{.j.} - \bar{Y}_{...})^2$$

$$(24.4d) \qquad SSTR = r \sum_k (\bar{Y}_{..k} - \bar{Y}_{...})^2$$

$$(24.4e) \qquad SSE = \sum_i \sum_j (Y_{ij(k)} - \bar{Y}_{i..} - \bar{Y}_{.j.} - \bar{Y}_{..k} + 2\bar{Y}_{...})^2$$

$SSROW$ is the *row sum of squares*. The more the row means $\bar{Y}_{i..}$ differ, the larger is $SSROW$. Similarly, $SSCOL$ is the *column sum of squares*, and measures the variability of the column means $\bar{Y}_{.j.}$. $SSTR$ and SSE denote, as always, the treatment sum of squares and error sum of squares respectively.

Formulas for these sums of squares suitable for hand computations are as follows:

$$(24.5a) \qquad SSTO = \sum_i \sum_j Y_{ij(k)}^2 - \frac{Y_{...}^2}{r^2}$$

$$(24.5b) \qquad SSROW = \frac{\sum_i Y_{i..}^2}{r} - \frac{Y_{...}^2}{r^2}$$

$$(24.5c) \qquad SSCOL = \frac{\sum_j Y_{.j.}^2}{r} - \frac{Y_{...}^2}{r^2}$$

$$(24.5d) \qquad SSTR = \frac{\sum_k Y_{..k}^2}{r} - \frac{Y_{...}^2}{r^2}$$

$$(24.5e) \qquad SSE = SSTO - SSROW - SSCOL - SSTR$$

The complete analysis of variance table for a latin square design is given in Table 24.3. There are r^2 observations, and hence $SSTO$ has $r^2 - 1$ degrees of freedom associated with it. Since there are r classes for the row and column blocking variables each, and also r treatments, each of the corresponding sums of squares has $r - 1$ degrees of freedom associated with it. The number of degrees of freedom associated with SSE is the remainder, namely $(r^2 - 1) - 3(r - 1) = (r - 1)(r - 2)$. Note that the addition of a second blocking variable has reduced the number of degrees of freedom for SSE from $(r - 1)^2$ for a randomized complete block design based on the same number of experimental units to $(r - 1)(r - 2)$, a reduction of $r - 1$ degrees of freedom.

The $E(MS)$ column in Table 24.3 can be obtained by using rule (19.26), remembering that any one of the three subscripts is redundant and hence should not be considered in rule (19.26b) for obtaining the coefficients of the variance components. Also, n in rule (19.26b) corresponds to 1 here, since there is only one observation per cell.

To test for treatment effects:

(24.6a)
$$C_1: \text{all } \tau_k = 0$$
$$C_2: \text{not all } \tau_k \text{ are zero}$$

we see from the $E(MS)$ column in Table 24.3 that we use the usual test statistic:

(24.6b)
$$F^* = \frac{MSTR}{MSE}$$

The appropriate decision rule is:

(24.6c)
$$\text{If } F^* \leq F(1 - \alpha; r - 1, (r - 1)(r - 2)), \text{ conclude } C_1$$
$$\text{If } F^* > F(1 - \alpha; r - 1, (r - 1)(r - 2)), \text{ conclude } C_2$$

Example

The analysis of variance calculations for the background music data in Table 24.2 are as follows, using the computational formulas (24.5):

$$SSTO = (18)^2 + (17)^2 + (14)^2 + \cdots + (7)^2 - \frac{(515)^2}{25} = 1{,}412$$

$$SSROW = \frac{1}{5}[(87)^2 + (99)^2 + (108)^2 + (110)^2 + (111)^2] - \frac{(515)^2}{25} = 82$$

$$SSCOL = \frac{1}{5}[(76)^2 + (119)^2 + (117)^2 + (126)^2 + (77)^2] - \frac{(515)^2}{25} = 477.2$$

$$SSTR = \frac{1}{5}[(57)^2 + (133)^2 + (98)^2 + (119)^2 + (108)^2] - \frac{(515)^2}{25} = 664.4$$

$$SSE = 1{,}412 - 82 - 477.2 - 664.4 = 188.4$$

TABLE 24.3

ANOVA Table for Latin Square Design—Model (24.1)

Source of Variation	*SS*	*df*	*MS*	*E(MS)*
Row blocking variable	$SSROW$	$r-1$	$MSROW = \frac{SSROW}{r-1}$	$\sigma^2 + r\frac{\sum \rho_i^2}{r-1}$
Column blocking variable	$SSCOL$	$r-1$	$MSCOL = \frac{SSCOL}{r-1}$	$\sigma^2 + r\frac{\sum \kappa_j^2}{r-1}$
Treatments	$SSTR$	$r-1$	$MSTR = \frac{SSTR}{r-1}$	$\sigma^2 + r\frac{\sum \tau_k^2}{r-1}$
Error	SSE	$(r-1)(r-2)$	$MSE = \frac{SSE}{(r-1)(r-2)}$	σ^2
Total	$SSTO$	r^2-1		

TABLE 24.4

ANOVA Table for Background Music Example

Source of Variation	SS	df	MS
Weeks	82.0	4	20.5
Days within week	477.2	4	119.3
Type of music	664.4	4	166.1
Error	188.4	12	15.7
Total	1,412.0	24	

The complete analysis of variance table for the background music example is shown in Table 24.4.

To test for treatment effects:

$$C_1: \tau_1 = \tau_2 = \tau_3 = \tau_4 = \tau_5 = 0$$
$$C_2: \text{not all } \tau_k \text{ are zero}$$

we find from Table 24.4:

$$F^* = \frac{MSTR}{MSE} = \frac{166.1}{15.7} = 10.6$$

Assuming we are to control the risk of making a Type I error at .01, we require $F(.99; 4, 12) = 5.41$. Since $F^* = 10.6 > 5.41$, we conclude that the various types of background music have differential effects on the productivity of the bank tellers.

Comments

1. At times, it may be desired to test for the presence of blocking variable effects. The $E(MS)$ column in Table 24.3 indicates this is done in the usual way for model (24.1). The statistic for testing row blocking variable effects is:

$$F^* = \frac{MSROW}{MSE} \tag{24.7a}$$

and that for testing column blocking variable effects is:

$$F^* = \frac{MSCOL}{MSE} \tag{24.7b}$$

For instance, to test in our background music example whether productivity varies according to the day of the week, we use the test statistic $F^* = 119.3/15.7 = 7.6$. For a level of significance of .01, we need $F(.99; 4, 12) = 5.41$. Hence we would conclude that there is variation in productivity (averaged over all treatments and weeks) within a week. Since blocking variables correspond to a classification factor, the interpretation of blocking variable effects must be done with care.

2. The power of the F test for treatment effects in a latin square design involves the noncentrality parameter:

$$\phi = \frac{1}{\sigma}\sqrt{\sum \tau_k^2} \tag{24.8}$$

with degrees of freedom $(r-1)$ for numerator and $(r-1)(r-2)$ for denominator. Other than these modifications, no new problems are encountered in obtaining the power of the test for treatment effects in a latin square design.

3. If the treatment effects were random, the alternatives to be considered would be:

$$\begin{aligned} &C_1: \sigma_\tau^2 = 0 \\ &C_2: \sigma_\tau^2 \neq 0 \end{aligned} \tag{24.9}$$

but the test statistic and decision rule would be the same as in (24.6) for the fixed treatment effects case.

24.4 ANALYSIS OF TREATMENT EFFECTS

If differential treatment effects are found by the analysis of variance, one will usually wish to estimate some contrasts involving the treatment effects, often utilizing multiple comparison procedures. MSE obtained from (24.4e) is the appropriate mean square to be used in the estimated variance of the contrast, and the multiples for the estimated standard deviation of the contrast are as follows:

(24.10a)	Single comparison	$t(1-\alpha/2; (r-1)(r-2))$
(24.10b)	Tukey procedure (for pairwise comparisons)	$T = \frac{1}{\sqrt{2}} q(1-\alpha; r, (r-1)(r-2))$
(24.10c)	Scheffé procedure	$S^2 = (r-1)F(1-\alpha; r-1, (r-1)(r-2))$
(24.10d)	Bonferroni procedure	$B = t(1-\alpha/2s; (r-1)(r-2))$

Example

In our background music example, pairwise comparisons between the different kinds of music were desired, with a family confidence coefficient of .90. The analyst used the Tukey procedure. Substituting into (14.21) with $n_j = n_{j'} = r$ and using the results from Table 24.4, he obtained:

$$s^2(D) = \frac{2MSE}{r} = \frac{2(15.7)}{5} = 6.28$$

or:

$$s(D) = 2.51$$

Remember that each treatment mean $\bar{Y}_{..k}$ is based on five observations here. Next, the analyst found the T multiple in (24.10b):

$$T = \frac{1}{\sqrt{2}} q(.90; 5, 12) = \frac{1}{\sqrt{2}}(3.92) = 2.77$$

so that:

$$Ts(D) = 2.77(2.51) = 7.0$$

Using the treatment means in Table 24.2, the pairwise comparisons were then obtained. For instance, we have:

$$8.2 = (26.6 - 11.4) - 7.0 \le \mu_{..2} - \mu_{..1} \le (26.6 - 11.4) + 7.0 = 22.2$$

Here $\mu_{..1}$ is the mean productivity for treatment 1 averaged over all weeks and days of the week, and $\mu_{..2}$ has the corresponding meaning for treatment 2. The entire set of pairwise comparisons is as follows:

$$+8.2 \le \mu_{..2} - \mu_{..1} \le +22.2 \qquad +5.4 \le \mu_{..4} - \mu_{..1} \le +19.4$$
$$-14.0 \le \mu_{..3} - \mu_{..2} \le -.05 \qquad -9.2 \le \mu_{..5} - \mu_{..4} \le +4.8$$
$$+1.2 \le \mu_{..3} - \mu_{..1} \le +15.2 \qquad -5.0 \le \mu_{..5} - \mu_{..3} \le +9.0$$
$$-2.8 \le \mu_{..4} - \mu_{..3} \le +11.2 \qquad -12.0 \le \mu_{..5} - \mu_{..2} \le +2.0$$
$$-9.8 \le \mu_{..4} - \mu_{..2} \le +4.2 \qquad +3.2 \le \mu_{..5} - \mu_{..1} \le +17.2$$

These pairwise differences led to the following conclusions by the analyst, with family confidence coefficient of 90 percent:

1. Treatment 2 yields greater mean productivity than treatments 1 or 3.
2. Treatments 3, 4, and 5 all yield greater mean productivity than treatment 1.
3. No pairwise differences in mean productivity for treatments 2, 4, and 5 are evident.
4. No pairwise differences in mean productivity for treatments 3, 4, and 5 are evident.

This is to say, the most promising treatment appears to be mixed instrumental–vocal music in medium tempo ($k = 2$). There is clear evidence that it is better than instrumental–vocal music in slow tempo ($k = 1$) or instrumental–vocal music in fast tempo ($k = 3$). The point estimates suggest it also is better than solely instrumental music in medium ($k = 4$) or fast ($k = 5$) tempo, but the experimental evidence on these latter two comparisons is indecisive.

24.5 FACTORIAL TREATMENTS

If the treatments in a latin square design are factorial in nature, the treatment sum of squares $SSTR$ is decomposed in the usual manner. For a two-factor experiment, involving factors A and B, we would have:

(24.11)
$$SSTR = SSA + SSB + SSAB$$

Estimates of the factor effects can be made readily since they are simply contrasts among the treatment means.

Example

Suppose the background music study mentioned earlier had consisted of four treatments to investigate the effects of loudness of music (soft, loud) and type of music (semipopular, popular) on productivity of bank tellers. The treatments are defined as follows:

T_1—soft, semipopular

T_2—loud, semipopular

T_3—soft, popular

T_4—loud, popular

The blocking variables for the latin square design are day of week (Monday, Tuesday, Wednesday, Thursday) and week (1, 2, 3, 4).

The analysis of variance for this latin square design experiment is shown in Table 24.5. The effect of loudness is reflected by the contrast:

(24.12)
$$L_1 = \frac{\mu_{..1} + \mu_{..3}}{2} - \frac{\mu_{..2} + \mu_{..4}}{2}$$

where $\mu_{..k}$ is the mean productivity for the kth treatment averaged over all weeks and days of the week. The point estimator of this contrast is:

(24.12a)
$$\hat{L}_1 = \frac{\bar{Y}_{..1} + \bar{Y}_{..3}}{2} - \frac{\bar{Y}_{..2} + \bar{Y}_{..4}}{2}$$

TABLE 24.5

ANOVA Table for Latin Square Design—Factorial Treatments ($r = 4$, $a = 2$, $b = 2$)

Source of Variation	*SS*	*df*	*MS*
Weeks	*SSROW*	3	*MSROW*
Days within week	*SSCOL*	3	*MSCOL*
Treatments	*SSTR*	3	*MSTR*
Loudness of music (*A*)	*SSA*	1	*MSA*
Type of music (*B*)	*SSB*	1	*MSB*
AB interactions	*SSAB*	1	*MSAB*
Error	*SSE*	6	*MSE*
Total	*SSTO*	15	

Similarly, the effect of type of music is reflected by the contrast:

(24.13)
$$L_2 = \frac{\mu_{..1} + \mu_{..2}}{2} - \frac{\mu_{..3} + \mu_{..4}}{2}$$

A point estimator of this contrast is:

(24.13a)
$$\hat{L}_2 = \frac{\bar{Y}_{..1} + \bar{Y}_{..2}}{2} - \frac{\bar{Y}_{..3} + \bar{Y}_{..4}}{2}$$

Confidence intervals for these contrasts are obtained in the usual fashion, employing the multiples in (24.10).

24.6 IMPLEMENTATION CONSIDERATIONS

Examination of Aptness of Model

The use of residuals for examining the aptness of a model has been discussed for other designs, and the basic points made earlier apply also to latin square designs.

Tukey Test for Additivity. A key question concerning the aptness of the latin square design model (24.1) is whether the effects of blocking variables and treatments are indeed additive. If nonadditivity is present in the data, transformations of the data should be studied to see if it can thereby be eliminated. The use of a model assuming additivity when in fact the effects are nonadditive will lower the level of significance and power of the test for treatment effects, or widen the confidence limits, thus making the experiment less sensitive.

The Tukey test for additivity in a randomized complete block design, discussed in Section 23.7, has been extended to latin square designs. For completeness, we outline the steps required for the Tukey test for additivity in a latin square experiment:

1. For each cell, obtain the fitted value:

(24.14a)
$$\hat{Y}_{ij(k)} = \bar{Y}_{i..} + \bar{Y}_{.j.} + \bar{Y}_{..k} - 2\bar{Y}_{...}$$

2. Find the residual for each cell:

(24.14b)
$$e_{ij(k)} = Y_{ij(k)} - \hat{Y}_{ij(k)}$$

Check that the residuals sum to zero over every row, column, and treatment.

3. Calculate SSE for the latin square. This can be done through (24.5e), or through (24.4e) which can be expressed as follows:

(24.14c)
$$SSE = \sum_i \sum_j e_{ij(k)}^2$$

4. Calculate for each cell:

(24.14d)
$$U_{ij(k)} = (\hat{Y}_{ij(k)} - \bar{Y}_{...})^2$$

5. Calculate:

$$N = \sum_i \sum_j e_{ij(k)} U_{ij(k)} \tag{24.14e}$$

6. Treating the $U_{ij(k)}$ as the observations in a latin square, obtain the error sum of squares, denoted by $SSE(U)$, using (24.4e) or (24.5e), with the Y's replaced by the U's.

7. The lack of fit sum of squares, associated with nonadditivity, is:

$$SSLF = \frac{N^2}{SSE(U)} \tag{24.14f}$$

This sum of squares has one degree of freedom associated with it.

8. The pure error sum of squares is:

$$SSPE = SSE - SSLF \tag{24.14g}$$

It has $(r-1)(r-2)-1$ degrees of freedom associated with it.

9. The test statistic is:

$$F^* = \frac{SSLF}{1} \div \frac{SSPE}{(r-1)(r-2)-1} \tag{24.14h}$$

This test statistic follows the $F(1, (r-1)(r-2)-1)$ distribution if no interactions exist. Large values of F^* lead to the conclusion that an additive model is not appropriate.

Efficiency of Blocking Variables

The efficiency of a latin square design can be assessed relative to a completely randomized design or relative to a randomized complete block design. The efficiency relative to a completely randomized design is defined by:

$$E_1 = \frac{\sigma_r^2}{\sigma_L^2} \tag{24.15a}$$

where σ_r^2 and σ_L^2 are the experimental error variances with a completely randomized design and a latin square design respectively. The efficiency relative to a randomized complete block design can be measured in two ways, depending on whether the row or the column blocking variable is used in the randomized block design:

$$E_2 = \frac{\sigma_{br}^2}{\sigma_L^2} \tag{24.15b}$$

$$E_3 = \frac{\sigma_{bc}^2}{\sigma_L^2} \tag{24.15c}$$

where σ_{br}^2 and σ_{bc}^2 are the experimental error variances with a randomized block design if the row blocking variable and the column blocking variable is utilized respectively.

If the treatment mean $\bar{Y}_{..k}$ containing the missing observation is to be compared with any other treatment mean $\bar{Y}_{..k'}$, the correct variance of the difference is:

(24.21)
$$\sigma^2(D) = \sigma^2\left[\frac{2}{r} + \frac{1}{(r-1)(r-2)}\right]$$

This variance is estimated by:

(24.22)
$$s^2(D) = MSE\left[\frac{2}{r} + \frac{1}{(r-1)(r-2)}\right]$$

24.7 ADDITIONAL REPLICATIONS WITH LATIN SQUARE DESIGNS

Need for Additional Replications

A latin square design, as noted earlier, provides r replications for each treatment. Power and/or estimation considerations may indicate that this is too few replications, particularly when r is small, say 3, 4, or 5. Two basic methods are available for increasing the number of replications with a latin square design, namely replications within cells and additional latin squares. We consider each in turn.

Replications Within Cells

This method of increasing the replications per treatment is feasible when two or more experimental units can be obtained for each cell defined by the row and column blocking variables. Consider, for instance, an experiment in which I.Q. (low, normal, high) and age (young, middle, old) are the blocking variables. In this type of situation, it is possible to obtain two or more experimental subjects for each cell, and each of the subjects in a cell will then receive the treatment assigned to that cell by the latin square employed. Suppose that n experimental units are available for each cell, and let $Y_{ij(k)m}$ denote the observation for the mth unit ($m = 1, \ldots, n$) in the (i, j) cell for which the assigned treatment is k. For the additive fixed effects model (24.1), with replications in cells, the analysis of variance is straightforward as shown in Table 24.6. The treatment, row, and column sums of squares are respectively:

(24.23a)
$$SSTR = rn \sum_k (\bar{Y}_{..k.} - \bar{Y}_{....})^2 = \frac{\sum_k Y_{..k.}^2}{rn} - \frac{Y_{....}^2}{r^2 n}$$

(24.23b)
$$SSROW = rn \sum_i (\bar{Y}_{i...} - \bar{Y}_{....})^2 = \frac{\sum_i Y_{i...}^2}{rn} - \frac{Y_{....}^2}{r^2 n}$$

(24.23c)
$$SSCOL = rn \sum_j (\bar{Y}_{.j..} - \bar{Y}_{....})^2 = \frac{\sum_j Y_{.j..}^2}{rn} - \frac{Y_{....}^2}{r^2 n}$$

TABLE 24.6

ANOVA Table for Latin Square Design with n Replications per Cell

Source of Variation	SS	df	MS
Row blocking variable	$SSROW$	$r-1$	$MSROW$
Column blocking variable	$SSCOL$	$r-1$	$MSCOL$
Treatments	$SSTR$	$r-1$	$MSTR$
Error	SSE	$nr^2 - 3r + 2$	MSE
Total	$SSTO$	$nr^2 - 1$	

The total sum of squares as usual is:

$$SSTO = \sum_i \sum_j \sum_m (Y_{ij(k)m} - \bar{Y}_{....})^2 = \sum_i \sum_j \sum_m Y_{ij(k)m}^2 - \frac{Y_{....}^2}{r^2 n} \tag{24.23d}$$

while SSE is obtained as a remainder:

$$SSE = SSTO - SSROW - SSCOL - SSTR \tag{24.23e}$$

The degrees of freedom for row, column, and treatment sums of squares are unchanged, while those associated with SSE are increased from $(r-1)(r-2)$ to $nr^2 - 3r + 2$, an increase of $(n-1)r^2$ degrees of freedom.

Test for Additivity. When there are n replications within a cell for a latin square, it is possible to obtain a pure error measure irrespective of the correctness of the additive model (24.1). The pure error sum of squares is obtained in the usual manner:

$$SSPE = \sum_i \sum_j \sum_m (Y_{ij(k)m} - \bar{Y}_{ij(k).})^2 = \sum_i \sum_j \sum_m Y_{ij(k)m}^2 - \sum_i \sum_j \frac{Y_{ij(k).}^2}{n} \tag{24.24}$$

It has associated with it $(n-1)r^2$ degrees of freedom, representing the increase in the degrees of freedom for SSE from having n observations in each cell of the latin square. The difference between SSE and $SSPE$ is a reflection of the lack of fit of the additive model:

$$SSLF = SSE - SSPE \tag{24.25}$$

and has associated with it $(r-1)(r-2)$ degrees of freedom. Let:

$$MSPE = \frac{SSPE}{(n-1)r^2} \tag{24.26a}$$

$$MSLF = \frac{SSLF}{(r-1)(r-2)} \tag{24.26b}$$

Then:

$$F^* = \frac{MSLF}{MSPE} \tag{24.27}$$

can be used to test whether or not the additive model is appropriate.

It can be shown that if the additive model is appropriate, F^* follows the $F((r-1)(r-2), (n-1)r^2)$ distribution, and that large values of F^* lead to the conclusion that the additive model is not appropriate. Table 24.7 contains the analysis of variance breaking down SSE into the $SSLF$ and $SSPE$ components.

TABLE 24.7

ANOVA Table to Test Additivity when n Replications per Latin Square Cell

Source of Variation	*SS*	*df*	*MS*
Row blocking variable	$SSROW$	$r-1$	
Column blocking variable	$SSCOL$	$r-1$	
Treatments	$SSTR$	$r-1$	
Error	SSE	nr^2-3r+2	
Lack of fit	$SSLF$	$(r-1)(r-2)$	$MSLF$
Pure error	$SSPE$	$(n-1)r^2$	$MSPE$
Total	$SSTO$	nr^2-1	

Example. A state university is undertaking a prototype retraining program designed to teach general automobile repair skills to persons who have been displaced from their previous manual occupations. Table 24.8 shows the results of an experiment to evaluate the effects of three different incentive methods on achievement scores made by participants in the program. The blocking variables are I.Q. and age of subject. Part (a) of the table contains the achievement scores, while part (b) contains the cell totals.

In testing for additivity we obtain:

$$SSTO = (19)^2 + (16)^2 + (20)^2 + \cdots + (4)^2 - \frac{(288)^2}{18} = 5{,}206 - 4{,}608 = 598$$

$$SSROW = \frac{1}{6}[(125)^2 + (103)^2 + (60)^2] - 4{,}608 = 364.3$$

$$SSCOL = \frac{1}{6}[(105)^2 + (98)^2 + (85)^2] - 4{,}608 = 34.3$$

$$SSTR = \frac{1}{6}[(96)^2 + (75)^2 + (117)^2] - 4{,}608 = 147$$

$$SSE = 598 - 364.3 - 34.3 - 147 = 52.4$$

$$SSPE = (19)^2 + (16)^2 + (20)^2 + \cdots + (4)^2 - \frac{1}{2}[(35)^2 + (44)^2 + \cdots + (11)^2] = 36$$

TABLE 24.8

Example of Latin Square Design with Two Replications per Cell in Retraining Program Experiment

(a) Data

I.Q.	*Age* Young	Middle	Old
High	(*B*) 19, 16	(*A*) 20, 24	(*C*) 25, 21
Normal	(*C*) 24, 22	(*B*) 14, 15	(*A*) 14, 14
Low	(*A*) 10, 14	(*C*) 12, 13	(*B*) 7, 4

(b) Cell Totals

I.Q.	*Age* Young	Middle	Old	Total
High	35	44	46	125
Normal	46	29	28	103
Low	24	25	11	60
Total	105	98	85	288

$Y_{..1.} = 96 \quad Y_{..2.} = 75 \quad Y_{..3.} = 117$

To test the appropriateness of the additive model, we find:

$$SSLF = SSE - SSPE = 52.4 - 36 = 16.4$$

$$MSLF = \frac{SSLF}{(r-1)(r-2)} = \frac{16.4}{(2)(1)} = 8.2$$

$$MSPE = \frac{SSPE}{(n-1)r^2} = \frac{36}{1(3)^2} = 4$$

The test statistic (24.27) here is:

$$F^* = \frac{MSLF}{MSPE} = \frac{8.2}{4} = 2.05$$

For a level of significance .05, we need $F(.95; 2, 9) = 4.26$. Since $F^* = 2.05 < 4.26$, we conclude that the additive model (24.1) is appropriate. The analysis of treatment effects can now proceed as usual, based on the additive model (24.1).

Additional Latin Squares

At times, it is not possible to obtain additional experimental units within a cell. This is the case, for instance, in the background music example of Table 24.2, where only one type of music can be played in one day. When it is not possible to replicate within cells, additional replications for each treatment frequently can be obtained by adding one or more latin squares to one of the blocking variables. In our background music example of Table 24.2, the experiment could be run for another five weeks. In an experiment using plant crews as experimental units and employing as blocking variables plant shift (morning, afternoon, evening) and production department (A, B, C), additional replications can be obtained by running the experiment in other production departments.

The layout for the background music example of Table 24.2, when run over another five weeks, is shown in Table 24.9. The second latin square, and additional ones when required, is selected independently of the first.

TABLE 24.9

Two Latin Squares Design for Background Music Example of Table 24.2

		Day				
Square	*Week*	*M*	*T*	*W*	*Th*	*F*
	1	*D*	*C*	*A*	*B*	*E*
	2	*C*	*B*	*E*	*A*	*D*
1	3	*A*	*D*	*B*	*E*	*C*
	4	*E*	*A*	*C*	*D*	*B*
	5	*B*	*E*	*D*	*C*	*A*
	6	*E*	*D*	*C*	*A*	*B*
	7	*B*	*A*	*E*	*D*	*C*
2	8	*D*	*C*	*A*	*B*	*E*
	9	*A*	*E*	*B*	*C*	*D*
	10	*C*	*B*	*D*	*E*	*A*

Frequently, the additional squares may be viewed as classes of a third blocking variable. For instance, in our background music example of Table 24.9, the two latin squares may be considered to refer to the blocking variable "time period." The first five weeks may be viewed as time period 1, and the second five weeks as time period 2. As another example, in the experiment with plant crews mentioned previously, the production departments for the first latin square may be on an hourly rate, while the departments for the second latin square may be on incentive pay, as shown in Table 24.10. Thus, with additional latin squares, one can in effect introduce a third blocking variable. As a consequence, the variation associated with the third blocking

TABLE 24.10

Two Latin Squares Design for Experiment with Plant Crews

Square	*Production Department*	*Shift*		
		Morning	*Afternoon*	*Evening*
	1	*C*	*B*	*A*
1 (hourly pay)	2	*B*	*A*	*C*
	3	*A*	*C*	*B*
	4	*A*	*B*	*C*
2 (incentive pay)	5	*C*	*A*	*B*
	6	*B*	*C*	*A*

variable can be removed from the experimental errors. In addition, one can study the interactions between the third blocking variable and the other variables and need not assume that these do not exist.

Table 24.11 shows an analysis of variance for n independent squares. This analysis is appropriate when (1) the n latin squares have the same rows and columns, and (2) the model permits third blocking variable effects and interactions between the third blocking variable and the other variables. These conditions might be appropriate for our background music example of Table 24.9. There, it may at times be appropriate to consider the rows to refer to the position of the week within the five-week period, and the columns refer to the position of the day within a week. Note in Table 24.11 that besides the row, column, and treatment sums of squares, there are third blocking variable effects, interactions involving the third blocking variable, and a term reflecting experimental error.

TABLE 24.11

ANOVA Table when n Latin Squares with Same Rows and Columns Are Employed and Interactions Involving Third Blocking Variable Are Permitted

Source of Variation	*SS*	*df*
Row blocking variable (*ROW*)	$SSROW$	$r-1$
Column blocking variable (*COL*)	$SSCOL$	$r-1$
Treatments (*TR*)	$SSTR$	$r-1$
Third blocking variable (3)	$SS3$	$n-1$
Third-row interactions (3.*ROW*)	$SS3.ROW$	$(n-1)(r-1)$
Third-column interactions (3.*COL*)	$SS3.COL$	$(n-1)(r-1)$
Third-treatment interactions (3.*TR*)	$SS3.TR$	$(n-1)(r-1)$
Error (*E*)	SSE	$n(r-1)(r-2)$
Total	$SSTO$	nr^2-1

Replications in Repeated Measures Studies

We noted earlier that a latin square design is highly suitable for a repeated measures study when there are r treatments and r subjects. If additional replications are needed, however, replications within cells cannot be used since a cell pertains to an individual subject. Instead, cross-over designs or independent latin squares may be used.

Cross-over Designs. These designs, also called *change-over designs*, are often useful when a latin square is to be used in a repeated measures study to balance the order positions of treatments, yet more subjects are required than called for by a single latin square. With this type of design, the subjects are randomly assigned to the different treatment order patterns given by a latin square (several latin squares may be used at times). Consider an experiment in which treatments A, B, C are to be administered to each subject, and the three treatment order patterns are given by the latin square:

	Order Position		
Pattern	1	2	3
1	A	B	C
2	B	C	A
3	C	A	B

Suppose that $3n$ subjects are available for the study. Then n subjects will be assigned at random to each of the three order patterns in a cross-over design. Note that this design is a mixture of randomized blocks (subjects are blocks) and latin square (order patterns form a latin square).

Assuming that an additive model is appropriate, the analysis of variance of a cross-over design is straightforward; the total sum of squares is decomposed into components for blocks (subjects), order positions, treatments, and error. Table 24.12 contains the analysis of variance table for the general case of r treatments and n subjects per order pattern in a cross-over design. Computational formulas for the sums of squares in Table 24.12 follow the usual

TABLE 24.12

ANOVA Table for Cross-over Design

Source of Variation	*SS*	*df*	*MS*
Subjects (S)	SSS	$rn - 1$	MSS
Order positions (O)	SSO	$r - 1$	MSO
Treatments (TR)	$SSTR$	$r - 1$	$MSTR$
Error (E)	SSE	$(r - 1)(nr - 2)$	MSE
Total	$SSTO$	$nr^2 - 1$	

pattern. Let $Y_{ij(k)m}$ denote the observation for the mth unit (e.g., subject) for the ith treatment order pattern which in period j assigned treatment k; $i = 1, \ldots, r; j = 1, \ldots, r; k = 1, \ldots, r; m = 1, \ldots, n$. We then have:

(24.28a) $$SSTO = \sum_i \sum_j \sum_m Y_{ij(k)m}^2 - \frac{Y_{....}^2}{r^2 n}$$

(24.28b) $$SSS = \frac{\sum_i \sum_m Y_{i..m}^2}{r} - \frac{Y_{....}^2}{r^2 n}$$

(24.28c) $$SSO = \frac{\sum_j Y_{.j..}^2}{nr} - \frac{Y_{....}^2}{r^2 n}$$

(24.28d) $$SSTR = \frac{\sum_k Y_{..k.}^2}{nr} - \frac{Y_{....}^2}{r^2 n}$$

(24.28e) $$SSE = SSTO - SSS - SSO - SSTR$$

where SSS is the *sum of squares for subjects*, SSO is the *sum of squares for order positions*, and the other sums of squares are defined as usual.

Example. Table 24.13a contains data for a study of the effects of three different displays on the sale of apples, using the cross-over design. Six stores were used, and two were assigned at random to each of the three treatment

TABLE 24.13

Cross-over Design Apple Sales Example

(a) Data (coded)

	Two-Week Period		
Store	1	2	3
1	9 (*B*)	12 (*C*)	15 (*A*)
2	12 (*A*)	14 (*B*)	3 (*C*)
3	13 (*A*)	14 (*B*)	3 (*C*)
4	7 (*C*)	18 (*A*)	6 (*B*)
5	5 (*C*)	20 (*A*)	4 (*B*)
6	4 (*B*)	12 (*C*)	9 (*A*)

(b) Analysis of Variance

Source of Variation	*SS*	*df*	*MS*
Stores	21.3	5	4.26
Order positions	233.3	2	116.7
Displays	189.0	2	94.5
Error	20.4	8	2.55
Total	464.0	17	

order patterns shown. Each display was kept for two weeks, and the observed variable was sales per 100 customers. Table 24.13b contains the analysis of variance. The sums of squares were obtained from a computer run.

To test for treatment effects, we use:

$$F^* = \frac{MSTR}{MSE} = \frac{94.5}{2.55} = 37.1$$

For $\alpha = .05$, we require $F(.95; 2, 8) = 4.46$. Since $F^* = 37.1 > 4.46$, we conclude that there are differential sales effects for the three displays. Tests for order position effects and store effects were also carried out. They indicated that order position effects were present, but no store effects. Order position effects here are associated with the three time periods in which the displays were studied, and may reflect seasonal effects as well as the results of special events, such as unusually hot weather in one period.

Use of Independent Latin Squares. If the order position effects are not approximately constant for all subjects (stores, etc.), a cross-over design is not fully effective. It may then be preferable to place the subjects into homogeneous groups with respect to the order position effects and use independent latin squares for each group. Suppose that four treatments are to be administered to eight subjects each, four males and four females, and that the experimenter expects the fatigue effect to be strong for females but only mild for males. The use of two independent latin squares, one for male subjects and the other for female subjects, may then be advisable.

Carry-over Effects. If carry-over effects from one treatment to another are anticipated, that is, if not only the order position but also the preceding treatment has an effect, one may balance out these carry-over effects by choosing a latin square in which every treatment follows every other treatment an equal number of times. For $r = 4$, an example of such a latin square is:

	Period			
Subject	1	2	3	4
1	A	B	D	C
2	B	C	A	D
3	C	D	B	A
4	D	A	C	B

Note that treatment A follows each of the other treatments once, and similarly for the other treatments. This design is suited when the carry-over effects do not persist for more than one following treatment.

When r is odd, the sequence balance can be obtained by using a pair of latin squares with the property that the treatment sequences in one are reversed in the other square. Indeed, even when r is even, it is usually desirable to use a pair of such squares so that the degrees of freedom associated with the error mean square are reasonably large. Such a design is sometimes called a *double cross-over design.* This type of design retains the advantages of employing two blocking variables in a latin square, while enabling the experimenter also to balance and measure the carry-over effects.

For our earlier apple display illustration in which three displays were studied in six stores, the two latin squares might be as shown in Table 24.14. The stores should first be placed into two homogeneous groups and these should then be assigned to the two latin squares.

TABLE 24.14

Illustration of a Double Cross-over Design

		Two-Week Period		
Square	*Store*	1	2	3
	1	*A*	*B*	*C*
1	2	*B*	*C*	*A*
	3	*C*	*A*	*B*
	4	*A*	*C*	*B*
2	5	*B*	*A*	*C*
	6	*C*	*B*	*A*

24.8 RANDOM BLOCKING VARIABLE EFFECTS

If the row and/or column blocking variable in a latin square design has classes which should be viewed as random selections from a population, the fixed effects model (24.1) no longer is applicable.

Both Blocking Variables Random

Consider the case where the row blocking variable is subject and the column blocking variable observer, and both the subjects and the observers included in the study are viewed as random samples from relevant populations. In that case, the additive model for latin square designs is, assuming treatment effects are fixed:

(24.29)
$$Y_{ij(k)} = \mu_{...} + \rho_i + \kappa_j + \tau_k + \varepsilon_{ij(k)}$$

where:

$\mu_{...}$ is a constant
ρ_i are independent $N(0, \sigma_\rho^2)$
κ_j are independent $N(0, \sigma_\kappa^2)$
τ_k are constants subject to the restriction $\sum \tau_k = 0$
$\varepsilon_{ij(k)}$ are independent $N(0, \sigma^2)$
ρ_i, κ_j, and $\varepsilon_{ij(k)}$ are independent
$i = 1, \ldots, r; j = 1, \ldots, r; k = 1, \ldots, r$

The analysis of variance is the same as for the fixed blocking variable effects model. The expected mean squares are obtained by replacing, in Table 24.3, the sums of squared effects divided by degrees of freedom by variance terms. Consequently, all tests and estimates of treatment effects are conducted as for fixed blocking variable effects.

One Blocking Variable Random, Other Fixed

When one of the blocking variables has random effects while the other has fixed effects, the additive model becomes a mixture of models (24.1) and (24.29). Again, there will be no change in the analysis of variance or in tests or estimates of treatment effects. An instance where this mixed model would be appropriate is a repeated measures study where the row blocking variable is subject and the column blocking variable is order of treatments.

24.9 YOUDEN AND GRAECO-LATIN SQUARES

When it is not possible to use a latin square design because the number of column classes is less than the number of row classes, a *Youden square design* may be helpful. Consider a repeated measures study involving four treatments and four subjects (row blocking variable). Suppose now that a subject can be given only three treatments because of serious fatigue effects. Hence the column blocking variable (order position of treatments) can have only three classes. A Youden square design suitable for this occasion is shown in Table 24.15. Note that this layout would become a latin square with the addition

TABLE 24.15

Illustration of a Youden Square Design

	Order Position of Treatment		
Subject	1	2	3
1	A	B	C
2	D	A	B
3	C	D	A
4	B	C	D

of the column (D, C, B, A). Also note that each treatment occurs once in each order position, and that every pair of treatments appears together an equal number of times within subjects. These are characteristics present in all Youden squares. The analysis of Youden square designs is more complex than that of latin squares because not all treatments are run in each class of the row blocking variable. A reference, such as Reference 24.1, should be consulted if a Youden square design is to be used.

TABLE 24.16

Illustration of a Graeco-Latin Square Design

Row Blocking Variable	Column Blocking Variable			
	1	2	3	4
1	$\alpha:A$	$\beta:B$	$\gamma:C$	$\delta:D$
2	$\beta:C$	$\alpha:D$	$\delta:A$	$\gamma:B$
3	$\delta:B$	$\gamma:A$	$\beta:D$	$\alpha:C$
4	$\gamma:D$	$\delta:C$	$\alpha:B$	$\beta:A$

A *graeco-latin square design* is an extension of a latin square design when three blocking variables are to be used simultaneously. Table 24.16 illustrates a graeco-latin square design for $r = 4$. The symbols α, β, γ, δ represent the four levels of the third blocking variable. Thus, the cell corresponding to the first class of each of the three blocking variables is to receive treatment A, and so on. Note that the levels of the third blocking variable appear once in each row and once in each column, and they appear once only with each treatment. Graeco-latin square designs are used much less frequently in practice than the other designs we have discussed.

24.10 REGRESSION APPROACH TO LATIN SQUARE DESIGNS

Model (24.1) for a latin square design with fixed blocking and treatment effects:

$$Y_{ij(k)} = \mu_{...} + \rho_i + \kappa_j + \tau_k + \varepsilon_{ij(k)} \qquad (24.30)$$
$$i = 1, \ldots, r; j = 1, \ldots, r; k = 1, \ldots, r$$

can be expressed easily in the form of a regression model with indicator variables. Since no interaction terms are present, we shall use 0, 1 indicator variables. The regression model for the background music example in Table 24.2, with $r = 5$, can be expressed as follows:

$$(24.31) \quad Y_{ij(k)} = \beta_0 + \underbrace{\beta_1 X_{ij(k)1} + \beta_2 X_{ij(k)2} + \beta_3 X_{ij(k)3} + \beta_4 X_{ij(k)4}}_{\text{Row blocking effect}}$$
$$+ \underbrace{\beta_5 X_{ij(k)5} + \beta_6 X_{ij(k)6} + \beta_7 X_{ij(k)7} + \beta_8 X_{ij(k)8}}_{\text{Column blocking effect}}$$
$$+ \underbrace{\beta_9 X_{ij(k)9} + \beta_{10} X_{ij(k)10} + \beta_{11} X_{ij(k)11} + \beta_{12} X_{ij(k)12}}_{\text{Treatment effect}}$$
$$+ \varepsilon_{ij(k)}$$

where:

$$X_1 = \begin{matrix} 1 \text{ if experimental unit from row blocking class 1} \\ 0 \text{ otherwise} \end{matrix}$$

X_2, X_3, X_4 are defined similarly

$$X_5 = \begin{matrix} 1 \text{ if experimental unit from column blocking class 1} \\ 0 \text{ otherwise} \end{matrix}$$

X_6, X_7, X_8 are defined similarly

$$X_9 = \begin{matrix} 1 \text{ if experimental unit received treatment 1} \\ 0 \text{ otherwise} \end{matrix}$$

X_{10}, X_{11}, X_{12} are defined similarly

The correspondences between the regression model (24.31) and the ANOVA model (24.30) with $r = 5$ are:

(24.32)

$$\beta_0 = \mu_{...} + \rho_5 + \kappa_5 + \tau_5$$

$$\beta_1 = \rho_1 - \rho_5$$

β_2, β_3, β_4 are defined similarly

$$\beta_5 = \kappa_1 - \kappa_5$$

β_6, β_7, β_8 are defined similarly

$$\beta_9 = \tau_1 - \tau_5$$

β_{10}, β_{11}, β_{12} are defined similarly

The **Y** observation vector and the **X** matrix for the background music example in Table 24.2 are shown in Table 24.17. Note, for instance, that for observation $Y_{12(3)}$, $X_1 = 1$, $X_6 = 1$, $X_{11} = 1$ and all other X's are zero. Hence we have:

$$\begin{aligned} Y_{12(3)} &= \beta_0 + \beta_1 + \beta_6 + \beta_{11} + \varepsilon_{12(3)} \\ &= (\mu_{...} + \rho_5 + \kappa_5 + \tau_5) + (\rho_1 - \rho_5) + (\kappa_2 - \kappa_5) + (\tau_3 - \tau_5) + \varepsilon_{12(3)} \\ &= \mu_{...} + \rho_1 + \kappa_2 + \tau_3 + \varepsilon_{12(3)} \end{aligned}$$

which is the appropriate expression for $Y_{12(3)}$ according to the ANOVA model (24.30) for the particular latin square used in the example.

Tests and estimation of treatment effects with the regression model are conducted in the usual fashion.

Note

One or several missing observations in a latin square design cause no difficulties with the regression approach. We just set up the regression model for the available observations, and then fit the model to the data.

TABLE 24.17

Data Matrices for Model (24.31) Based on the Background Music Data in Table 24.2

$$
\mathbf{Y}=\begin{bmatrix}
Y_{11(4)}=18\\
Y_{12(3)}=17\\
Y_{13(1)}=14\\
Y_{14(2)}=21\\
Y_{15(5)}=17\\
Y_{21(3)}=13\\
Y_{22(2)}=34\\
Y_{23(5)}=21\\
Y_{24(1)}=16\\
Y_{25(4)}=15\\
Y_{31(1)}=7\\
Y_{32(4)}=29\\
Y_{33(2)}=32\\
Y_{34(5)}=27\\
Y_{35(3)}=13\\
Y_{41(5)}=17\\
Y_{42(1)}=13\\
Y_{43(3)}=24\\
Y_{44(4)}=31\\
Y_{45(2)}=25\\
Y_{51(2)}=21\\
Y_{52(5)}=26\\
Y_{53(4)}=26\\
Y_{54(3)}=31\\
Y_{55(1)}=7
\end{bmatrix}
\qquad
\begin{matrix} & X_1 & X_2 & X_3 & X_4 & X_5 & X_6 & X_7 & X_8 & X_9 & X_{10} & X_{11} & X_{12} \end{matrix}
$$

$$
\mathbf{X}=\begin{bmatrix}
1&1&0&0&0&1&0&0&0&0&0&0&1\\
1&1&0&0&0&0&1&0&0&0&0&1&0\\
1&1&0&0&0&0&0&1&0&1&0&0&0\\
1&1&0&0&0&0&0&0&1&0&1&0&0\\
1&1&0&0&0&0&0&0&0&0&0&0&0\\
1&0&1&0&0&1&0&0&0&0&0&1&0\\
1&0&1&0&0&0&1&0&0&0&1&0&0\\
1&0&1&0&0&0&0&1&0&0&0&0&0\\
1&0&1&0&0&0&0&0&1&1&0&0&0\\
1&0&1&0&0&0&0&0&0&0&0&0&1\\
1&0&0&1&0&1&0&0&0&1&0&0&0\\
1&0&0&1&0&0&1&0&0&0&0&0&1\\
1&0&0&1&0&0&0&1&0&0&1&0&0\\
1&0&0&1&0&0&0&0&1&0&0&0&0\\
1&0&0&1&0&0&0&0&0&0&0&1&0\\
1&0&0&0&1&1&0&0&0&0&0&0&0\\
1&0&0&0&1&0&1&0&0&1&0&0&0\\
1&0&0&0&1&0&0&1&0&0&0&1&0\\
1&0&0&0&1&0&0&0&1&0&0&0&1\\
1&0&0&0&1&0&0&0&0&0&1&0&0\\
1&0&0&0&0&1&0&0&0&0&1&0&0\\
1&0&0&0&0&0&1&0&0&0&0&0&0\\
1&0&0&0&0&0&0&1&0&0&0&0&1\\
1&0&0&0&0&0&0&0&1&0&0&1&0\\
1&0&0&0&0&0&0&0&0&1&0&0&0
\end{bmatrix}
$$

PROBLEMS

24.1. A behavioral scientist explained why he uses latin square designs so frequently: "Many times, we require the use of repeated measures designs because variability between human subjects is so great. Since an order effect may be present in this situation, we employ latin square designs to eliminate any bias due to order effects." Comment.

24.2. a) Use Table A–13 to select randomly a 3 by 3 latin square. Show all steps.

b) Use Table A–13 to select randomly a 6 by 6 latin square. Show all steps.

24.3. A manufacturer conducted a small pilot study of the effect of the price of one of its products on sales of this product in hardware stores. Since it might be confusing to customers if prices were switched repeatedly within a store, only one price was used for any one store during the six-month study period. Sixteen stores were employed in the study. To reduce experimental error, they were chosen so that there would be one store for each "sales volume–geographic location" class. The four price levels (A—\$1.79, B—\$1.69, C—\$1.59, D—\$1.49) were assigned to the stores according to the latin square design shown below. Sales during the six-month period, in hundred dollars, were as follows:

Sales Volume Class	Geographic Location Class: Northeast	Northwest	Southeast	Southwest
1 (smallest)	1.1 (B)	1.5 (C)	1.0 (A)	1.7 (D)
2	1.4 (A)	1.9 (D)	1.6 (B)	1.5 (C)
3	2.8 (C)	2.2 (B)	2.7 (D)	2.1 (A)
4 (largest)	3.4 (D)	2.5 (A)	2.9 (C)	2.7 (B)

a) State the model you will use to analyze the effects of different price levels.
b) Test whether price level affects sales. Use a level of significance of .05. What do you conclude?
c) Analyze the nature of the price effect on sales. Explain the reasons why you chose your particular analytical approach, and summarize your findings.

24.4. Refer to Problem 24.3.

a) Obtain the residuals and analyze them to determine the appropriateness of the model you employed in Problem 24.3a. Summarize your findings.
b) Conduct the Tukey test for additivity, using a level of significance of .01. What do you conclude?

24.5. Refer to Problem 24.3. How effective were the two blocking variables in reducing experimental errors? Would a randomized block design have been adequate here?

24.6. A management information systems consultant conducted a small-scale study of five different daily summary reports (A—greatest amount of details; B; C; D; E—least amount of details). He used five sales executives. Each was given one type of daily report for a month and then was asked to rate its helpfulness on a 25-point scale (0—no help; 25—extremely helpful). Over a five-month period, each executive received each type of report for one month according to the latin square design shown below. The results were:

Executive	Month: March	April	May	June	July
Harrison	21 (D)	8 (A)	17 (C)	9 (B)	16 (E)
Smith	5 (A)	10 (E)	3 (B)	12 (C)	15 (D)
Carmichael	20 (C)	10 (B)	15 (E)	21 (D)	12 (A)
Loeb	4 (B)	15 (D)	3 (A)	9 (E)	10 (C)
Munch	17 (E)	16 (C)	20 (D)	8 (A)	11 (B)

The consultant wishes to use a fixed effects model.

a) State the model you will use to analyze the effects of the different daily summary reports.

b) Test whether the five types of reports differ in effectiveness. Use a level of significance of .025.

c) Analyze the comparative effectiveness of the five types of reports. Explain why you chose your particular analytical approach, and summarize your findings.

24.7. Refer to Problem 24.6.

a) Obtain the residuals and analyze them to determine the appropriateness of the model you employed in Problem 24.6a. State your findings.

b) Conduct the Tukey test for additivity, using $\alpha = .01$. What do you conclude?

c) How effective was the use of the repeated measures design here?

24.8. Refer to Problem 24.3. Suppose that observation $Y_{23(2)} = 1.6$ were missing.

a) Use the Yates method to test whether price level affects sales, with a level of significance of .05. What do you conclude?

b) Compare the expected sales for the \$1.79 and \$1.69 price levels, using a 90 percent confidence interval.

24.9. A study was undertaken to determine whether the volume of sound of a television commercial affects recall, and whether this varies by product. Thirty-two subjects were chosen, two each for 16 groups defined according to age (class 1—youngest; 2; 3; 4—oldest) and amount of education (class 1—lowest education level; 2; 3; 4—highest education level). Each subject was exposed to one of four television commercial showings (*A*—high volume, product X; *B*—low volume, product X; *C*—high volume, product Y; *D*—low volume, product Y), according to the latin square design shown below. Two different commercials were involved, one for each product. During the following week, the subjects were asked to mention everything they could remember about the advertisement. Scores were based on the number of learning points mentioned, suitably standardized. The results were as follows:

	Education Level			
Age Class	1	2	3	4
1	(*D*) 83 86	(*A*) 64 69	(*C*) 78 75	(*B*) 76 75
2	(*B*) 70 73	(*C*) 81 74	(*A*) 64 60	(*D*) 87 81
3	(*C*) 67 74	(*B*) 67 62	(*D*) 76 81	(*A*) 64 57
4	(*A*) 56 60	(*D*) 72 67	(*B*) 63 64	(*C*) 64 66

a) Fit the fixed effects model (24.1) modified to allow for factorial treatments, and obtain the residuals. Analyze the residuals to determine the appropriateness of the model. State your findings.

b) Conduct a formal test whether the effects of blocking variables and treatments are additive. Use a level of significance of .01.

24.10. Refer to Problem 24.9. Assume that the model employed in Problem 24.9a is appropriate. Analyze the results of the study. Explain why you chose your particular analytical approach, and summarize your findings.

24.11. Refer to the apple sales experiment on page 791.

a) Verify the analysis of variance in Table 24.13b for the data in Table 24.13a.

b) Conduct tests for the presence of order and store effects, using a level of significance of .05 for each. State your findings.

c) Analyze the display effects, and report your findings.

24.12. Refer to Problem 24.3.

a) Set up the regression model counterpart to the ANOVA model. State the correspondences between the two sets of parameters.

b) Test by means of the regression approach whether price level affects sales. Use a level of significance of .05.

c) Obtain an interval estimate by the regression approach for the difference in expected sales for the $1.49 and $1.59 price levels, with a confidence coefficient of 95 percent.

24.13. Refer to Problem 24.6. Suppose observations $Y_{11(4)} = 21$ and $Y_{45(3)} = 10$ were missing.

a) Use the regression approach to test whether the five types of reports differ in effectiveness; employ $\alpha = .025$. State the regression model and the correspondences between the two sets of parameters. What do you conclude?

b) Compare the mean helpfulness of reports A and D by means of the regression approach, using a 90 percent confidence interval.

CITED REFERENCE

24.1. Cochran, William G., and Cox, Gertrude M. *Experimental Designs*. 2d ed. New York: John Wiley & Sons, Inc., 1957.

Appendix Tables

TABLE A–1

Cumulative Probabilities of the Standard Normal Distribution

Entry is area $1 - \alpha$ under the standard normal curve from $-\infty$ to $z(1 - \alpha)$

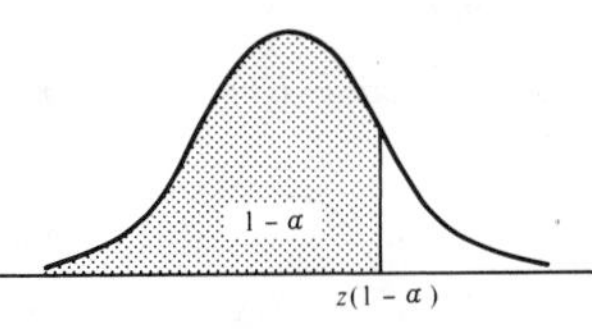

z	.00	.01	.02	.03	.04	.05	.06	.07	.08	.09
.0	.5000	.5040	.5080	.5120	.5160	.5199	.5239	.5279	.5319	.5359
.1	.5398	.5438	.5478	.5517	.5557	.5596	.5636	.5675	.5714	.5753
.2	.5793	.5832	.5871	.5910	.5948	.5987	.6026	.6064	.6103	.6141
.3	.6179	.6217	.6255	.6293	.6331	.6368	.6406	.6443	.6480	.6517
.4	.6554	.6591	.6628	.6664	.6700	.6736	.6772	.6808	.6844	.6879
.5	.6915	.6950	.6985	.7019	.7054	.7088	.7123	.7157	.7190	.7224
.6	.7257	.7291	.7324	.7357	.7389	.7422	.7454	.7486	.7517	.7549
.7	.7580	.7611	.7642	.7673	.7704	.7734	.7764	.7794	.7823	.7852
.8	.7881	.7910	.7939	.7967	.7995	.8023	.8051	.8078	.8106	.8133
.9	.8159	.8186	.8212	.8238	.8264	.8289	.8315	.8340	.8365	.8389
1.0	.8413	.8438	.8461	.8485	.8508	.8531	.8554	.8577	.8599	.8621
1.1	.8643	.8665	.8686	.8708	.8729	.8749	.8770	.8790	.8810	.8830
1.2	.8849	.8869	.8888	.8907	.8925	.8944	.8962	.8980	.8997	.9015
1.3	.9032	.9049	.9066	.9082	.9099	.9115	.9131	.9147	.9162	.9177
1.4	.9192	.9207	.9222	.9236	.9251	.9265	.9279	.9292	.9306	.9319
1.5	.9332	.9345	.9357	.9370	.9382	.9394	.9406	.9418	.9429	.9441
1.6	.9452	.9463	.9474	.9484	.9495	.9505	.9515	.9525	.9535	.9545
1.7	.9554	.9564	.9573	.9582	.9591	.9599	.9608	.9616	.9625	.9633
1.8	.9641	.9649	.9656	.9664	.9671	.9678	.9686	.9693	.9699	.9706
1.9	.9713	.9719	.9726	.9732	.9738	.9744	.9750	.9756	.9761	.9767
2.0	.9772	.9778	.9783	.9788	.9793	.9798	.9803	.9808	.9812	.9817
2.1	.9821	.9826	.9830	.9834	.9838	.9842	.9846	.9850	.9854	.9857
2.2	.9861	.9864	.9868	.9871	.9875	.9878	.9881	.9884	.9887	.9890
2.3	.9893	.9896	.9898	.9901	.9904	.9906	.9909	.9911	.9913	.9916
2.4	.9918	.9920	.9922	.9925	.9927	.9929	.9931	.9932	.9934	.9936
2.5	.9938	.9940	.9941	.9943	.9945	.9946	.9948	.9949	.9951	.9952
2.6	.9953	.9955	.9956	.9957	.9959	.9960	.9961	.9962	.9963	.9964
2.7	.9965	.9966	.9967	.9968	.9969	.9970	.9971	.9972	.9973	.9974
2.8	.9974	.9975	.9976	.9977	.9977	.9978	.9979	.9979	.9980	.9981
2.9	.9981	.9982	.9982	.9983	.9984	.9984	.9985	.9985	.9986	.9986
3.0	.9987	.9987	.9987	.9988	.9988	.9989	.9989	.9989	.9990	.9990
3.1	.9990	.9991	.9991	.9991	.9992	.9992	.9992	.9992	.9993	.9993
3.2	.9993	.9993	.9994	.9994	.9994	.9994	.9994	.9995	.9995	.9995
3.3	.9995	.9995	.9995	.9996	.9996	.9996	.9996	.9996	.9996	.9997
3.4	.9997	.9997	.9997	.9997	.9997	.9997	.9997	.9997	.9997	.9998

Selected Percentiles

Cumulative probability $1 - \alpha$:	.90	.95	.975	.98	.99	.995	.999
$z(1 - \alpha)$:	1.282	1.645	1.960	2.054	2.326	2.576	3.090
Two Sides	.80%	.90	.95	.96	.98		

TABLE A–2

Percentiles of the t Distribution

Entry is $t(1-\alpha;\nu)$ where $P\{t(\nu) \le t(1-\alpha;\nu)\} = 1-\alpha$

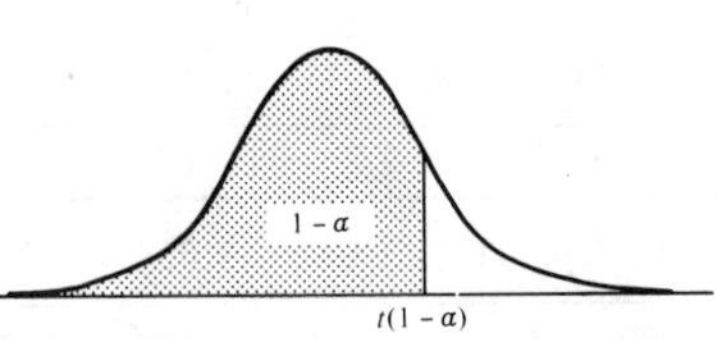

ν	.55	.60	.65	.70	.75	.80	.85
1	0.158	0.325	0.510	0.727	1.000	1.376	1.963
2	0.142	0.289	0.445	0.617	0.816	1.061	1.386
3	0.137	0.277	0.424	0.584	0.765	0.978	1.250
4	0.134	0.271	0.414	0.569	0.741	0.941	1.190
5	0.132	0.267	0.408	0.559	0.727	0.920	1.156
6	0.131	0.265	0.404	0.553	0.718	0.906	1.134
7	0.130	0.263	0.402	0.549	0.711	0.896	1.119
8	0.130	0.262	0.399	0.546	0.706	0.889	1.108
9	0.129	0.261	0.398	0.543	0.703	0.883	1.100
10	0.129	0.260	0.397	0.542	0.700	0.879	1.093
11	0.129	0.260	0.396	0.540	0.697	0.876	1.088
12	0.128	0.259	0.395	0.539	0.695	0.873	1.083
13	0.128	0.259	0.394	0.538	0.694	0.870	1.079
14	0.128	0.258	0.393	0.537	0.692	0.868	1.076
15	0.128	0.258	0.393	0.536	0.691	0.866	1.074
16	0.128	0.258	0.392	0.535	0.690	0.865	1.071
17	0.128	0.257	0.392	0.534	0.689	0.863	1.069
18	0.127	0.257	0.392	0.534	0.688	0.862	1.067
19	0.127	0.257	0.391	0.533	0.688	0.861	1.066
20	0.127	0.257	0.391	0.533	0.687	0.860	1.064
21	0.127	0.257	0.391	0.532	0.686	0.859	1.063
22	0.127	0.256	0.390	0.532	0.686	0.858	1.061
23	0.127	0.256	0.390	0.532	0.685	0.858	1.060
24	0.127	0.256	0.390	0.531	0.685	0.857	1.059
25	0.127	0.256	0.390	0.531	0.684	0.856	1.058
26	0.127	0.256	0.390	0.531	0.684	0.856	1.058
27	0.127	0.256	0.389	0.531	0.684	0.855	1.057
28	0.127	0.256	0.389	0.530	0.683	0.855	1.056
29	0.127	0.256	0.389	0.530	0.683	0.854	1.055
30	0.127	0.256	0.389	0.530	0.683	0.854	1.055
40	0.126	0.255	0.388	0.529	0.681	0.851	1.050
60	0.126	0.254	0.387	0.527	0.679	0.848	1.046
120	0.126	0.254	0.386	0.526	0.677	0.845	1.041
∞	0.126	0.253	0.385	0.524	0.674	0.842	1.036

TABLE A–2 (*continued*)

Percentiles of the t Distribution

ν	$1-\alpha$					
	.90	.95	.975	.99	.995	.9995
1	3.078	6.314	12.706	31.821	63.657	636.619
2	1.886	2.920	4.303	6.965	9.925	31.598
3	1.638	2.353	3.182	4.541	5.841	12.924
4	1.533	2.132	2.776	3.747	4.604	8.610
5	1.476	2.015	2.571	3.365	4.032	6.869
6	1.440	1.943	2.447	3.143	3.707	5.959
7	1.415	1.895	2.365	2.998	3.499	5.408
8	1.397	1.860	2.306	2.896	3.355	5.041
9	1.383	1.833	2.262	2.821	3.250	4.781
10	1.372	1.812	2.228	2.764	3.169	4.587
11	1.363	1.796	2.201	2.718	3.106	4.437
12	1.356	1.782	2.179	2.681	3.055	4.318
13	1.350	1.771	2.160	2.650	3.012	4.221
14	1.345	1.761	2.145	2.624	2.977	4.140
15	1.341	1.753	2.131	2.602	2.947	4.073
16	1.337	1.746	2.120	2.583	2.921	4.015
17	1.333	1.740	2.110	2.567	2.898	3.965
18	1.330	1.734	2.101	2.552	2.878	3.922
19	1.328	1.729	2.093	2.539	2.861	3.883
20	1.325	1.725	2.086	2.528	2.845	3.850
21	1.323	1.721	2.080	2.518	2.831	3.819
22	1.321	1.717	2.074	2.508	2.819	3.792
23	1.319	1.714	2.069	2.500	2.807	3.767
24	1.318	1.711	2.064	2.492	2.797	3.745
25	1.316	1.708	2.060	2.485	2.787	3.725
26	1.315	1.706	2.056	2.479	2.779	3.707
27	1.314	1.703	2.052	2.473	2.771	3.690
28	1.313	1.701	2.048	2.467	2.763	3.674
29	1.311	1.699	2.045	2.462	2.756	3.659
30	1.310	1.697	2.042	2.457	2.750	3.646
40	1.303	1.684	2.021	2.423	2.704	3.551
60	1.296	1.671	2.000	2.390	2.660	3.460
120	1.289	1.658	1.980	2.358	2.617	3.373
∞	1.282	1.645	1.960	2.326	2.576	3.291

Source: Table A–2 is taken from Table III of Fisher and Yates: *Statistical Tables for Biological, Agricultural and Medical Research*, 6th ed., 1963, published by Oliver and Boyd, Edinburgh, and by permission of the authors and publishers.

TABLE A–3

Percentiles of the χ^2 Distribution

Entry is $\chi^2(1-\alpha;\nu)$ where $P\{\chi^2(\nu) \leq \chi^2(1-\alpha;\nu)\} = 1-\alpha$

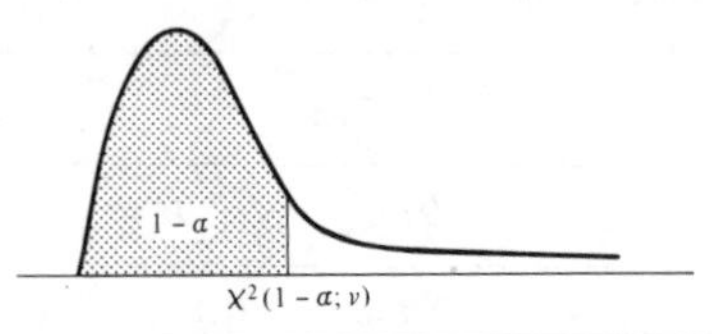

ν	.005	.010	.025	.050	.100	.900	.950	.975	.990	.995
	$1-\alpha$									
1	0.0^4393	0.0^3157	0.0^3982	0.0^2393	0.0158	2.71	3.84	5.02	6.63	7.88
2	0.0100	0.0201	0.0506	0.103	0.211	4.61	5.99	7.38	9.21	10.60
3	0.072	0.115	0.216	0.352	0.584	6.25	7.81	9.35	11.34	12.84
4	0.207	0.297	0.484	0.711	1.064	7.78	9.49	11.14	13.28	14.86
5	0.412	0.554	0.831	1.145	1.61	9.24	11.07	12.83	15.09	16.75
6	0.676	0.872	1.24	1.64	2.20	10.64	12.59	14.45	16.81	18.55
7	0.989	1.24	1.69	2.17	2.83	12.02	14.07	16.01	18.48	20.28
8	1.34	1.65	2.18	2.73	3.49	13.36	15.51	17.53	20.09	21.96
9	1.73	2.09	2.70	3.33	4.17	14.68	16.92	19.02	21.67	23.59
10	2.16	2.56	3.25	3.94	4.87	15.99	18.31	20.48	23.21	25.19
11	2.60	3.05	3.82	4.57	5.58	17.28	19.68	21.92	24.73	26.76
12	3.07	3.57	4.40	5.23	6.30	18.55	21.03	23.34	26.22	28.30
13	3.57	4.11	5.01	5.89	7.04	19.81	22.36	24.74	27.69	29.82
14	4.07	4.66	5.63	6.57	7.79	21.06	23.68	26.12	29.14	31.32
15	4.60	5.23	6.26	7.26	8.55	22.31	25.00	27.49	30.58	32.80
16	5.14	5.81	6.91	7.96	9.31	23.54	26.30	28.85	32.00	34.27
17	5.70	6.41	7.56	8.67	10.09	24.77	27.59	30.19	33.41	35.72
18	6.26	7.01	8.23	9.39	10.86	25.99	28.87	31.53	34.81	37.16
19	6.84	7.63	8.91	10.12	11.65	27.20	30.14	32.85	36.19	38.58
20	7.43	8.26	9.59	10.85	12.44	28.41	31.41	34.17	37.57	40.00
21	8.03	8.90	10.28	11.59	13.24	29.62	32.67	35.48	38.93	41.40
22	8.64	9.54	10.98	12.34	14.04	30.81	33.92	36.78	40.29	42.80
23	9.26	10.20	11.69	13.09	14.85	32.01	35.17	38.08	41.64	44.18
24	9.89	10.86	12.40	13.85	15.66	33.20	36.42	39.36	42.98	45.56
25	10.52	11.52	13.12	14.61	16.47	34.38	37.65	40.65	44.31	46.93
26	11.16	12.20	13.84	15.38	17.29	35.56	38.89	41.92	45.64	48.29
27	11.81	12.88	14.57	16.15	18.11	36.74	40.11	43.19	46.96	49.64
28	12.46	13.56	15.31	16.93	18.94	37.92	41.34	44.46	48.28	50.99
29	13.12	14.26	16.05	17.71	19.77	39.09	42.56	45.72	49.59	52.34
30	13.79	14.95	16.79	18.49	20.60	40.26	43.77	46.98	50.89	53.67
40	20.71	22.16	24.43	26.51	29.05	51.81	55.76	59.34	63.69	66.77
50	27.99	29.71	32.36	34.76	37.69	63.17	67.50	71.42	76.15	79.49
60	35.53	37.48	40.48	43.19	46.46	74.40	79.08	83.30	88.38	91.95
70	43.28	45.44	48.76	51.74	55.33	85.53	90.53	95.02	100.4	104.2
80	51.17	53.54	57.15	60.39	64.28	96.58	101.9	106.6	112.3	116.3
90	59.20	61.75	65.65	69.13	73.29	107.6	113.1	118.1	124.1	128.3
100	67.33	70.06	74.22	77.93	82.36	118.5	124.3	129.6	135.8	140.2

Source: Reprinted, with permission, from E. S. Pearson and C. M. Thompson, "Table of Percentage Points of the Chi-Square Distribution," *Biometrika*, Vol. 32 (1941), pp. 188–89.

TABLE A–4

Percentiles of the F Distribution

Entry is $F(1-\alpha; \nu_1, \nu_2)$ where $P\{F(\nu_1, \nu_2) \le F(1-\alpha; \nu_1, \nu_2)\} = 1-\alpha$

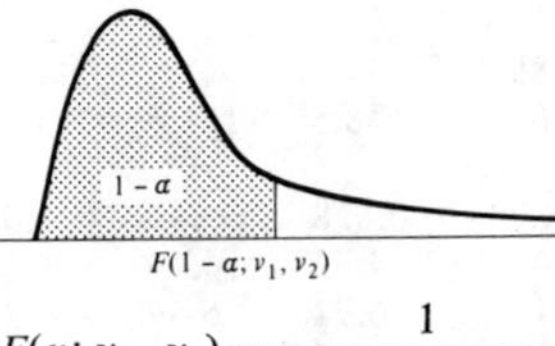

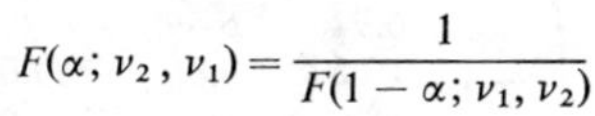

$$F(\alpha; \nu_2, \nu_1) = \frac{1}{F(1-\alpha; \nu_1, \nu_2)}$$

TABLE A–4 (*continued*)

Percentiles of the *F* Distribution

ν_2	$1-\alpha$	ν_1 = 1	2	3	4	5	6	7	8	9
1	.50	1.00	1.50	1.71	1.82	1.89	1.94	1.98	2.00	2.03
	.90	39.9	49.5	53.6	55.8	57.2	58.2	58.9	59.4	59.9
	.95	161	200	216	225	230	234	237	239	241
	.975	648	800	864	900	922	937	948	957	963
	.99	4,052	5,000	5,403	5,625	5,764	5,859	5,928	5,981	6,022
	.995	16,211	20,000	21,615	22,500	23,056	23,437	23,715	23,925	24,091
	.999	405,280	500,000	540,380	562,500	576,400	585,940	592,870	598,140	602,280
2	.50	0.667	1.00	1.13	1.21	1.25	1.28	1.30	1.32	1.33
	.90	8.53	9.00	9.16	9.24	9.29	9.33	9.35	9.37	9.38
	.95	18.5	19.0	19.2	19.2	19.3	19.3	19.4	19.4	19.4
	.975	38.5	39.0	39.2	39.2	39.3	39.3	39.4	39.4	39.4
	.99	98.5	99.0	99.2	99.2	99.3	99.3	99.4	99.4	99.4
	.995	199	199	199	199	199	199	199	199	199
	.999	998.5	999.0	999.2	999.2	999.3	999.3	999.4	999.4	999.4
3	.50	0.585	0.881	1.00	1.06	1.10	1.13	1.15	1.16	1.17
	.90	5.54	5.46	5.39	5.34	5.31	5.28	5.27	5.25	5.24
	.95	10.1	9.55	9.28	9.12	9.01	8.94	8.89	8.85	8.81
	.975	17.4	16.0	15.4	15.1	14.9	14.7	14.6	14.5	14.5
	.99	34.1	30.8	29.5	28.7	28.2	27.9	27.7	27.5	27.3
	.995	55.6	49.8	47.5	46.2	45.4	44.8	44.4	44.1	43.9
	.999	167.0	148.5	141.1	137.1	134.6	132.8	131.6	130.6	129.9
4	.50	0.549	0.828	0.941	1.00	1.04	1.06	1.08	1.09	1.10
	.90	4.54	4.32	4.19	4.11	4.05	4.01	3.98	3.95	3.94
	.95	7.71	6.94	6.59	6.39	6.26	6.16	6.09	6.04	6.00
	.975	12.2	10.6	9.98	9.60	9.36	9.20	9.07	8.98	8.90
	.99	21.2	18.0	16.7	16.0	15.5	15.2	15.0	14.8	14.7
	.995	31.3	26.3	24.3	23.2	22.5	22.0	21.6	21.4	21.1
	.999	74.1	61.2	56.2	53.4	51.7	50.5	49.7	49.0	48.5
5	.50	0.528	0.799	0.907	0.965	1.00	1.02	1.04	1.05	1.06
	.90	4.06	3.78	3.62	3.52	3.45	3.40	3.37	3.34	3.32
	.95	6.61	5.79	5.41	5.19	5.05	4.95	4.88	4.82	4.77
	.975	10.0	8.43	7.76	7.39	7.15	6.98	6.85	6.76	6.68
	.99	16.3	13.3	12.1	11.4	11.0	10.7	10.5	10.3	10.2
	.995	22.8	18.3	16.5	15.6	14.9	14.5	14.2	14.0	13.8
	.999	47.2	37.1	33.2	31.1	29.8	28.8	28.2	27.6	27.2
6	.50	0.515	0.780	0.886	0.942	0.977	1.00	1.02	1.03	1.04
	.90	3.78	3.46	3.29	3.18	3.11	3.05	3.01	2.98	2.96
	.95	5.99	5.14	4.76	4.53	4.39	4.28	4.21	4.15	4.10
	.975	8.81	7.26	6.60	6.23	5.99	5.82	5.70	5.60	5.52
	.99	13.7	10.9	9.78	9.15	8.75	8.47	8.26	8.10	7.98
	.995	18.6	14.5	12.9	12.0	11.5	11.1	10.8	10.6	10.4
	.999	35.5	27.0	23.7	21.9	20.8	20.0	19.5	19.0	18.7
7	.50	0.506	0.767	0.871	0.926	0.960	0.983	1.00	1.01	1.02
	.90	3.59	3.26	3.07	2.96	2.88	2.83	2.78	2.75	2.72
	.95	5.59	4.74	4.35	4.12	3.97	3.87	3.79	3.73	3.68
	.975	8.07	6.54	5.89	5.52	5.29	5.12	4.99	4.90	4.82
	.99	12.2	9.55	8.45	7.85	7.46	7.19	6.99	6.84	6.72
	.995	16.2	12.4	10.9	10.1	9.52	9.16	8.89	8.68	8.51
	.999	29.2	21.7	18.8	17.2	16.2	15.5	15.0	14.6	14.3

TABLE A–4 (*continued*)

Percentiles of the F Distribution

ν_2	$1-\alpha$	ν_1 = 10	12	15	20	24	30	60	120	∞
1	.50	2.04	2.07	2.09	2.12	2.13	2.15	2.17	2.18	2.20
	.90	60.2	60.7	61.2	61.7	62.0	62.3	62.8	63.1	63.3
	.95	242	244	246	248	249	250	252	253	254
	.975	969	977	985	993	997	1,001	1,010	1,014	1,018
	.99	6,056	6,106	6,157	6,209	6,235	6,261	6,313	6,339	6,366
	.995	24,224	24,426	24,630	24,836	24,940	25,044	25,253	25,359	25,464
	.999	605,620	610,670	615,760	620,910	623,500	626,100	631,340	633,970	636,620
2	.50	1.34	1.36	1.38	1.39	1.40	1.41	1.43	1.43	1.44
	.90	9.39	9.41	9.42	9.44	9.45	9.46	9.47	9.48	9.49
	.95	19.4	19.4	19.4	19.4	19.5	19.5	19.5	19.5	19.5
	.975	39.4	39.4	39.4	39.4	39.5	39.5	39.5	39.5	39.5
	.99	99.4	99.4	99.4	99.4	99.5	99.5	99.5	99.5	99.5
	.995	199	199	199	199	199	199	199	199	200
	.999	999.4	999.4	999.4	999.4	999.5	999.5	999.5	999.5	999.5
3	.50	1.18	1.20	1.21	1.23	1.23	1.24	1.25	1.26	1.27
	.90	5.23	5.22	5.20	5.18	5.18	5.17	5.15	5.14	5.13
	.95	8.79	8.74	8.70	8.66	8.64	8.62	8.57	8.55	8.53
	.975	14.4	14.3	14.3	14.2	14.1	14.1	14.0	13.9	13.9
	.99	27.2	27.1	26.9	26.7	26.6	26.5	26.3	26.2	26.1
	.995	43.7	43.4	43.1	42.8	42.6	42.5	42.1	42.0	41.8
	.999	129.2	128.3	127.4	126.4	125.9	125.4	124.5	124.0	123.5
4	.50	1.11	1.13	1.14	1.15	1.16	1.16	1.18	1.18	1.19
	.90	3.92	3.90	3.87	3.84	3.83	3.82	3.79	3.78	3.76
	.95	5.96	5.91	5.86	5.80	5.77	5.75	5.69	5.66	5.63
	.975	8.84	8.75	8.66	8.56	8.51	8.46	8.36	8.31	8.26
	.99	14.5	14.4	14.2	14.0	13.9	13.8	13.7	13.6	13.5
	.995	21.0	20.7	20.4	20.2	20.0	19.9	19.6	19.5	19.3
	.999	48.1	47.4	46.8	46.1	45.8	45.4	44.7	44.4	44.1
5	.50	1.07	1.09	1.10	1.11	1.12	1.12	1.14	1.14	1.15
	.90	3.30	3.27	3.24	3.21	3.19	3.17	3.14	3.12	3.11
	.95	4.74	4.68	4.62	4.56	4.53	4.50	4.43	4.40	4.37
	.975	6.62	6.52	6.43	6.33	6.28	6.23	6.12	6.07	6.02
	.99	10.1	9.89	9.72	9.55	9.47	9.38	9.20	9.11	9.02
	.995	13.6	13.4	13.1	12.9	12.8	12.7	12.4	12.3	12.1
	.999	26.9	26.4	25.9	25.4	25.1	24.9	24.3	24.1	23.8
6	.50	1.05	1.06	1.07	1.08	1.09	1.10	1.11	1.12	1.12
	.90	2.94	2.90	2.87	2.84	2.82	2.80	2.76	2.74	2.72
	.95	4.06	4.00	3.94	3.87	3.84	3.81	3.74	3.70	3.67
	.975	5.46	5.37	5.27	5.17	5.12	5.07	4.96	4.90	4.85
	.99	7.87	7.72	7.56	7.40	7.31	7.23	7.06	6.97	6.88
	.995	10.2	10.0	9.81	9.59	9.47	9.36	9.12	9.00	8.88
	.999	18.4	18.0	17.6	17.1	16.9	16.7	16.2	16.0	15.7
7	.50	1.03	1.04	1.05	1.07	1.07	1.08	1.09	1.10	1.10
	.90	2.70	2.67	2.63	2.59	2.58	2.56	2.51	2.49	2.47
	.95	3.64	3.57	3.51	3.44	3.41	3.38	3.30	3.27	3.23
	.975	4.76	4.67	4.57	4.47	4.42	4.36	4.25	4.20	4.14
	.99	6.62	6.47	6.31	6.16	6.07	5.99	5.82	5.74	5.65
	.995	8.38	8.18	7.97	7.75	7.65	7.53	7.31	7.19	7.08
	.999	14.1	13.7	13.3	12.9	12.7	12.5	12.1	11.9	11.7

TABLE A–4 (*continued*)

Percentiles of the F Distribution

ν_2	$1-\alpha$	ν_1 = 1	2	3	4	5	6	7	8	9
8	.50	0.499	0.757	0.860	0.915	0.948	0.971	0.988	1.00	1.01
	.90	3.46	3.11	2.92	2.81	2.73	2.67	2.62	2.59	2.56
	.95	5.32	4.46	4.07	3.84	3.69	3.58	3.50	3.44	3.39
	.975	7.57	6.06	5.42	5.05	4.82	4.65	4.53	4.43	4.36
	.99	11.3	8.65	7.59	7.01	6.63	6.37	6.18	6.03	5.91
	.995	14.7	11.0	9.60	8.81	8.30	7.95	7.69	7.50	7.34
	.999	25.4	18.5	15.8	14.4	13.5	12.9	12.4	12.0	11.8
9	.50	0.494	0.749	0.852	0.906	0.939	0.962	0.978	0.990	1.00
	.90	3.36	3.01	2.81	2.69	2.61	2.55	2.51	2.47	2.44
	.95	5.12	4.26	3.86	3.63	3.48	3.37	3.29	3.23	3.18
	.975	7.21	5.71	5.08	4.72	4.48	4.32	4.20	4.10	4.03
	.99	10.6	8.02	6.99	6.42	6.06	5.80	5.61	5.47	5.35
	.995	13.6	10.1	8.72	7.96	7.47	7.13	6.88	6.69	6.54
	.999	22.9	16.4	13.9	12.6	11.7	11.1	10.7	10.4	10.1
10	.50	0.490	0.743	0.845	0.899	0.932	0.954	0.971	0.983	0.992
	.90	3.29	2.92	2.73	2.61	2.52	2.46	2.41	2.38	2.35
	.95	4.96	4.10	3.71	3.48	3.33	3.22	3.14	3.07	3.02
	.975	6.94	5.46	4.83	4.47	4.24	4.07	3.95	3.85	3.78
	.99	10.0	7.56	6.55	5.99	5.64	5.39	5.20	5.06	4.94
	.995	12.8	9.43	8.08	7.34	6.87	6.54	6.30	6.12	5.97
	.999	21.0	14.9	12.6	11.3	10.5	9.93	9.52	9.20	8.96
12	.50	0.484	0.735	0.835	0.888	0.921	0.943	0.959	0.972	0.981
	.90	3.18	2.81	2.61	2.48	2.39	2.33	2.28	2.24	2.21
	.95	4.75	3.89	3.49	3.26	3.11	3.00	2.91	2.85	2.80
	.975	6.55	5.10	4.47	4.12	3.89	3.73	3.61	3.51	3.44
	.99	9.33	6.93	5.95	5.41	5.06	4.82	4.64	4.50	4.39
	.995	11.8	8.51	7.23	6.52	6.07	5.76	5.52	5.35	5.20
	.999	18.6	13.0	10.8	9.63	8.89	8.38	8.00	7.71	7.48
15	.50	0.478	0.726	0.826	0.878	0.911	0.933	0.949	0.960	0.970
	.90	3.07	2.70	2.49	2.36	2.27	2.21	2.16	2.12	2.09
	.95	4.54	3.68	3.29	3.06	2.90	2.79	2.71	2.64	2.59
	.975	6.20	4.77	4.15	3.80	3.58	3.41	3.29	3.20	3.12
	.99	8.68	6.36	5.42	4.89	4.56	4.32	4.14	4.00	3.89
	.995	10.8	7.70	6.48	5.80	5.37	5.07	4.85	4.67	4.54
	.999	16.6	11.3	9.34	8.25	7.57	7.09	6.74	6.47	6.26
20	.50	0.472	0.718	0.816	0.868	0.900	0.922	0.938	0.950	0.959
	.90	2.97	2.59	2.38	2.25	2.16	2.09	2.04	2.00	1.96
	.95	4.35	3.49	3.10	2.87	2.71	2.60	2.51	2.45	2.39
	.975	5.87	4.46	3.86	3.51	3.29	3.13	3.01	2.91	2.84
	.99	8.10	5.85	4.94	4.43	4.10	3.87	3.70	3.56	3.46
	.995	9.94	6.99	5.82	5.17	4.76	4.47	4.26	4.09	3.96
	.999	14.8	9.95	8.10	7.10	6.46	6.02	5.69	5.44	5.24
24	.50	0.469	0.714	0.812	0.863	0.895	0.917	0.932	0.944	0.953
	.90	2.93	2.54	2.33	2.19	2.10	2.04	1.98	1.94	1.91
	.95	4.26	3.40	3.01	2.78	2.62	2.51	2.42	2.36	2.30
	.975	5.72	4.32	3.72	3.38	3.15	2.99	2.87	2.78	2.70
	.99	7.82	5.61	4.72	4.22	3.90	3.67	3.50	3.36	3.26
	.995	9.55	6.66	5.52	4.89	4.49	4.20	3.99	3.83	3.69
	.999	14.0	9.34	7.55	6.59	5.98	5.55	5.23	4.99	4.80

(over)

TABLE A–4 (*continued*)

Percentiles of the F Distribution

ν_2	$1-\alpha$	ν_1 = 10	12	15	20	24	30	60	120	∞
8	.50	1.02	1.03	1.04	1.05	1.06	1.07	1.08	1.08	1.09
	.90	2.54	2.50	2.46	2.42	2.40	2.38	2.34	2.32	2.29
	.95	3.35	3.28	3.22	3.15	3.12	3.08	3.01	2.97	2.93
	.975	4.30	4.20	4.10	4.00	3.95	3.89	3.78	3.73	3.67
	.99	5.81	5.67	5.52	5.36	5.28	5.20	5.03	4.95	4.86
	.995	7.21	7.01	6.81	6.61	6.50	6.40	6.18	6.06	5.95
	.999	11.5	11.2	10.8	10.5	10.3	10.1	9.73	9.53	9.33
9	.50	1.01	1.02	1.03	1.04	1.05	1.05	1.07	1.07	1.08
	.90	2.42	2.38	2.34	2.30	2.28	2.25	2.21	2.18	2.16
	.95	3.14	3.07	3.01	2.94	2.90	2.86	2.79	2.75	2.71
	.975	3.96	3.87	3.77	3.67	3.61	3.56	3.45	3.39	3.33
	.99	5.26	5.11	4.96	4.81	4.73	4.65	4.48	4.40	4.31
	.995	6.42	6.23	6.03	5.83	5.73	5.62	5.41	5.30	5.19
	.999	9.89	9.57	9.24	8.90	8.72	8.55	8.19	8.00	7.81
10	.50	1.00	1.01	1.02	1.03	1.04	1.05	1.06	1.06	1.07
	.90	2.32	2.28	2.24	2.20	2.18	2.16	2.11	2.08	2.06
	.95	2.98	2.91	2.84	2.77	2.74	2.70	2.62	2.58	2.54
	.975	3.72	3.62	3.52	3.42	3.37	3.31	3.20	3.14	3.08
	.99	4.85	4.71	4.56	4.41	4.33	4.25	4.08	4.00	3.91
	.995	5.85	5.66	5.47	5.27	5.17	5.07	4.86	4.75	4.64
	.999	8.75	8.45	8.13	7.80	7.64	7.47	7.12	6.94	6.76
12	.50	0.989	1.00	1.01	1.02	1.03	1.03	1.05	1.05	1.06
	.90	2.19	2.15	2.10	2.06	2.04	2.01	1.96	1.93	1.90
	.95	2.75	2.69	2.62	2.54	2.51	2.47	2.38	2.34	2.30
	.975	3.37	3.28	3.18	3.07	3.02	2.96	2.85	2.79	2.72
	.99	4.30	4.16	4.01	3.86	3.78	3.70	3.54	3.45	3.36
	.995	5.09	4.91	4.72	4.53	4.43	4.33	4.12	4.01	3.90
	.999	7.29	7.00	6.71	6.40	6.25	6.09	5.76	5.59	5.42
15	.50	0.977	0.989	1.00	1.01	1.02	1.02	1.03	1.04	1.05
	.90	2.06	2.02	1.97	1.92	1.90	1.87	1.82	1.79	1.76
	.95	2.54	2.48	2.40	2.33	2.29	2.25	2.16	2.11	2.07
	.975	3.06	2.96	2.86	2.76	2.70	2.64	2.52	2.46	2.40
	.99	3.80	3.67	3.52	3.37	3.29	3.21	3.05	2.96	2.87
	.995	4.42	4.25	4.07	3.88	3.79	3.69	3.48	3.37	3.26
	.999	6.08	5.81	5.54	5.25	5.10	4.95	4.64	4.48	4.31
20	.50	0.966	0.977	0.989	1.00	1.01	1.01	1.02	1.03	1.03
	.90	1.94	1.89	1.84	1.79	1.77	1.74	1.68	1.64	1.61
	.95	2.35	2.28	2.20	2.12	2.08	2.04	1.95	1.90	1.84
	.975	2.77	2.68	2.57	2.46	2.41	2.35	2.22	2.16	2.09
	.99	3.37	3.23	3.09	2.94	2.86	2.78	2.61	2.52	2.42
	.995	3.85	3.68	3.50	3.32	3.22	3.12	2.92	2.81	2.69
	.999	5.08	4.82	4.56	4.29	4.15	4.00	3.70	3.54	3.38
24	.50	0.961	0.972	0.983	0.994	1.00	1.01	1.02	1.02	1.03
	.90	1.88	1.83	1.78	1.73	1.70	1.67	1.61	1.57	1.53
	.95	2.25	2.18	2.11	2.03	1.98	1.94	1.84	1.79	1.73
	.975	2.64	2.54	2.44	2.33	2.27	2.21	2.08	2.01	1.94
	.99	3.17	3.03	2.89	2.74	2.66	2.58	2.40	2.31	2.21
	.995	3.59	3.42	3.25	3.06	2.97	2.87	2.66	2.55	2.43
	.999	4.64	4.39	4.14	3.87	3.74	3.59	3.29	3.14	2.97

TABLE A–4 (*continued*)

Percentiles of the F Distribution

ν_2	$1-\alpha$	ν_1 = 1	2	3	4	5	6	7	8	9
30	.50	0.466	0.709	0.807	0.858	0.890	0.912	0.927	0.939	0.948
	.90	2.88	2.49	2.28	2.14	2.05	1.98	1.93	1.88	1.85
	.95	4.17	3.32	2.92	2.69	2.53	2.42	2.33	2.27	2.21
	.975	5.57	4.18	3.59	3.25	3.03	2.87	2.75	2.65	2.57
	.99	7.56	5.39	4.51	4.02	3.70	3.47	3.30	3.17	3.07
	.995	9.18	6.35	5.24	4.62	4.23	3.95	3.74	3.58	3.45
	.999	13.3	8.77	7.05	6.12	5.53	5.12	4.82	4.58	4.39
60	.50	0.461	0.701	0.798	0.849	0.880	0.901	0.917	0.928	0.937
	.90	2.79	2.39	2.18	2.04	1.95	1.87	1.82	1.77	1.74
	.95	4.00	3.15	2.76	2.53	2.37	2.25	2.17	2.10	2.04
	.975	5.29	3.93	3.34	3.01	2.79	2.63	2.51	2.41	2.33
	.99	7.08	4.98	4.13	3.65	3.34	3.12	2.95	2.82	2.72
	.995	8.49	5.80	4.73	4.14	3.76	3.49	3.29	3.13	3.01
	.999	12.0	7.77	6.17	5.31	4.76	4.37	4.09	3.86	3.69
120	.50	0.458	0.697	0.793	0.844	0.875	0.896	0.912	0.923	0.932
	.90	2.75	2.35	2.13	1.99	1.90	1.82	1.77	1.72	1.68
	.95	3.92	3.07	2.68	2.45	2.29	2.18	2.09	2.02	1.96
	.975	5.15	3.80	3.23	2.89	2.67	2.52	2.39	2.30	2.22
	.99	6.85	4.79	3.95	3.48	3.17	2.96	2.79	2.66	2.56
	.995	8.18	5.54	4.50	3.92	3.55	3.28	3.09	2.93	2.81
	.999	11.4	7.32	5.78	4.95	4.42	4.04	3.77	3.55	3.38
∞	.50	0.455	0.693	0.789	0.839	0.870	0.891	0.907	0.918	0.927
	.90	2.71	2.30	2.08	1.94	1.85	1.77	1.72	1.67	1.63
	.95	3.84	3.00	2.60	2.37	2.21	2.10	2.01	1.94	1.88
	.975	5.02	3.69	3.12	2.79	2.57	2.41	2.29	2.19	2.11
	.99	6.63	4.61	3.78	3.32	3.02	2.80	2.64	2.51	2.41
	.995	7.88	5.30	4.28	3.72	3.35	3.09	2.90	2.74	2.62
	.999	10.8	6.91	5.42	4.62	4.10	3.74	3.47	3.27	3.10

TABLE A–4 (*continued*)

Percentiles of the F Distribution

		ν_1								
ν_2	$1-\alpha$	10	12	15	20	24	30	60	120	∞
30	.50	0.955	0.966	0.978	0.989	0.994	1.00	1.01	1.02	1.02
	.90	1.82	1.77	1.72	1.67	1.64	1.61	1.54	1.50	1.46
	.95	2.16	2.09	2.01	1.93	1.89	1.84	1.74	1.68	1.62
	.975	2.51	2.41	2.31	2.20	2.14	2.07	1.94	1.87	1.79
	.99	2.98	2.84	2.70	2.55	2.47	2.39	2.21	2.11	2.01
	.995	3.34	3.18	3.01	2.82	2.73	2.63	2.42	2.30	2.18
	.999	4.24	4.00	3.75	3.49	3.36	3.22	2.92	2.76	2.59
60	.50	0.945	0.956	0.967	0.978	0.983	0.989	1.00	1.01	1.01
	.90	1.71	1.66	1.60	1.54	1.51	1.48	1.40	1.35	1.29
	.95	1.99	1.92	1.84	1.75	1.70	1.65	1.53	1.47	1.39
	.975	2.27	2.17	2.06	1.94	1.88	1.82	1.67	1.58	1.48
	.99	2.63	2.50	2.35	2.20	2.12	2.03	1.84	1.73	1.60
	.995	2.90	2.74	2.57	2.39	2.29	2.19	1.96	1.83	1.69
	.999	3.54	3.32	3.08	2.83	2.69	2.55	2.25	2.08	1.89
120	.50	0.939	0.950	0.961	0.972	0.978	0.983	0.994	1.00	1.01
	.90	1.65	1.60	1.55	1.48	1.45	1.41	1.32	1.26	1.19
	.95	1.91	1.83	1.75	1.66	1.61	1.55	1.43	1.35	1.25
	.975	2.16	2.05	1.95	1.82	1.76	1.69	1.53	1.43	1.31
	.99	2.47	2.34	2.19	2.03	1.95	1.86	1.66	1.53	1.38
	.995	2.71	2.54	2.37	2.19	2.09	1.98	1.75	1.61	1.43
	.999	3.24	3.02	2.78	2.53	2.40	2.26	1.95	1.77	1.54
∞	.50	0.934	0.945	0.956	0.967	0.972	0.978	0.989	0.994	1.00
	.90	1.60	1.55	1.49	1.42	1.38	1.34	1.24	1.17	1.00
	.95	1.83	1.75	1.67	1.57	1.52	1.46	1.32	1.22	1.00
	.975	2.05	1.94	1.83	1.71	1.64	1.57	1.39	1.27	1.00
	.99	2.32	2.18	2.04	1.88	1.79	1.70	1.47	1.32	1.00
	.995	2.52	2.36	2.19	2.00	1.90	1.79	1.53	1.36	1.00
	.999	2.96	2.74	2.51	2.27	2.13	1.99	1.66	1.45	1.00

Source: Reprinted from Table 5 of Pearson and Hartley, *Biometrika Tables for Statisticians*, Volume 2, 1972, published by the Cambridge University Press, on behalf of The Biometrika Society, by permission of the authors and publishers.

TABLE A–5

Power Function for Two-Sided t Test

$\alpha = .05$

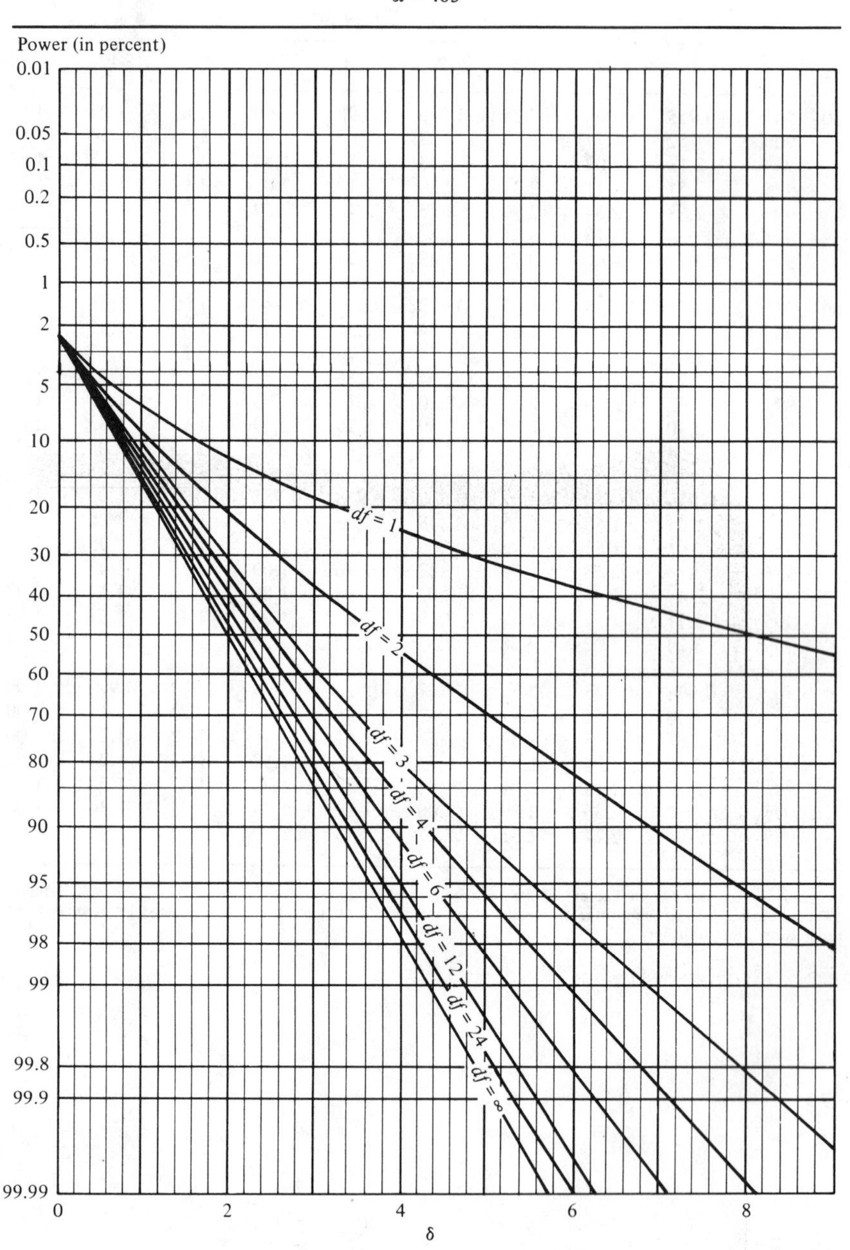

TABLE A–5 (*continued*)

Power Function for Two-Sided t Test

$\alpha = .01$

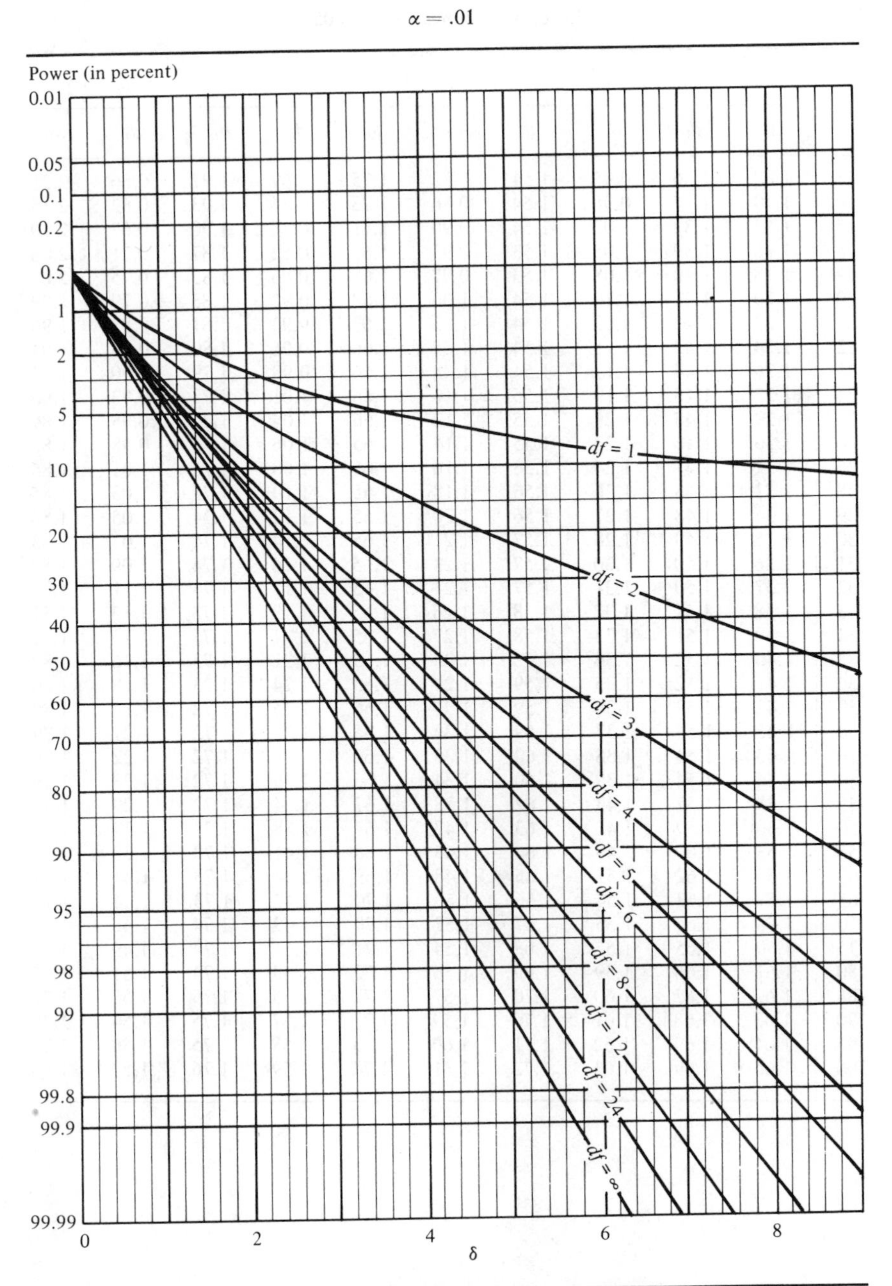

Source: Reprinted, with permission, from D. B. Owen, *Handbook of Statistical Tables* (Reading, Mass.: Addison-Wesley Publishing Co., Inc., 1962), pp. 32, 34. Courtesy of U.S. Atomic Energy Commission.

TABLE A–6

Durbin-Watson Test Bounds

Level of Significance $\alpha = .05$

	$p-1=1$		$p-1=2$		$p-1=3$		$p-1=4$		$p-1=5$	
n	d_L	d_U	d_L	d_U	d_L	d_U	d_L	d_U	d_L	d_U
15	1.08	1.36	0.95	1.54	0.82	1.75	0.69	1.97	0.56	2.21
16	1.10	1.37	0.98	1.54	0.86	1.73	0.74	1.93	0.62	2.15
17	1.13	1.38	1.02	1.54	0.90	1.71	0.78	1.90	0.67	2.10
18	1.16	1.39	1.05	1.53	0.93	1.69	0.82	1.87	0.71	2.06
19	1.18	1.40	1.08	1.53	0.97	1.68	0.86	1.85	0.75	2.02
20	1.20	1.41	1.10	1.54	1.00	1.68	0.90	1.83	0.79	1.99
21	1.22	1.42	1.13	1.54	1.03	1.67	0.93	1.81	0.83	1.96
22	1.24	1.43	1.15	1.54	1.05	1.66	0.96	1.80	0.86	1.94
23	1.26	1.44	1.17	1.54	1.08	1.66	0.99	1.79	0.90	1.92
24	1.27	1.45	1.19	1.55	1.10	1.66	1.01	1.78	0.93	1.90
25	1.29	1.45	1.21	1.55	1.12	1.66	1.04	1.77	0.95	1.89
26	1.30	1.46	1.22	1.55	1.14	1.65	1.06	1.76	0.98	1.88
27	1.32	1.47	1.24	1.56	1.16	1.65	1.08	1.76	1.01	1.86
28	1.33	1.48	1.26	1.56	1.18	1.65	1.10	1.75	1.03	1.85
29	1.34	1.48	1.27	1.56	1.20	1.65	1.12	1.74	1.05	1.84
30	1.35	1.49	1.28	1.57	1.21	1.65	1.14	1.74	1.07	1.83
31	1.36	1.50	1.30	1.57	1.23	1.65	1.16	1.74	1.09	1.83
32	1.37	1.50	1.31	1.57	1.24	1.65	1.18	1.73	1.11	1.82
33	1.38	1.51	1.32	1.58	1.26	1.65	1.19	1.73	1.13	1.81
34	1.39	1.51	1.33	1.58	1.27	1.65	1.21	1.73	1.15	1.81
35	1.40	1.52	1.34	1.58	1.28	1.65	1.22	1.73	1.16	1.80
36	1.41	1.52	1.35	1.59	1.29	1.65	1.24	1.73	1.18	1.80
37	1.42	1.53	1.36	1.59	1.31	1.66	1.25	1.72	1.19	1.80
38	1.43	1.54	1.37	1.59	1.32	1.66	1.26	1.72	1.21	1.79
39	1.43	1.54	1.38	1.60	1.33	1.66	1.27	1.72	1.22	1.79
40	1.44	1.54	1.39	1.60	1.34	1.66	1.29	1.72	1.23	1.79
45	1.48	1.57	1.43	1.62	1.38	1.67	1.34	1.72	1.29	1.78
50	1.50	1.59	1.46	1.63	1.42	1.67	1.38	1.72	1.34	1.77
55	1.53	1.60	1.49	1.64	1.45	1.68	1.41	1.72	1.38	1.77
60	1.55	1.62	1.51	1.65	1.48	1.69	1.44	1.73	1.41	1.77
65	1.57	1.63	1.54	1.66	1.50	1.70	1.47	1.73	1.44	1.77
70	1.58	1.64	1.55	1.67	1.52	1.70	1.49	1.74	1.46	1.77
75	1.60	1.65	1.57	1.68	1.54	1.71	1.51	1.74	1.49	1.77
80	1.61	1.66	1.59	1.69	1.56	1.72	1.53	1.74	1.51	1.77
85	1.62	1.67	1.60	1.70	1.57	1.72	1.55	1.75	1.52	1.77
90	1.63	1.68	1.61	1.70	1.59	1.73	1.57	1.75	1.54	1.78
95	1.64	1.69	1.62	1.71	1.60	1.73	1.58	1.75	1.56	1.78
100	1.65	1.69	1.63	1.72	1.61	1.74	1.59	1.76	1.57	1.78

TABLE A–6 (*continued*)

Durbin-Watson Test Bounds

Level of Significance $\alpha = .01$

	$p-1=1$		$p-1=2$		$p-1=3$		$p-1=4$		$p-1=5$	
n	d_L	d_U	d_L	d_U	d_L	d_U	d_L	d_U	d_L	d_U
15	0.81	1.07	0.70	1.25	0.59	1.46	0.49	1.70	0.39	1.96
16	0.84	1.09	0.74	1.25	0.63	1.44	0.53	1.66	0.44	1.90
17	0.87	1.10	0.77	1.25	0.67	1.43	0.57	1.63	0.48	1.85
18	0.90	1.12	0.80	1.26	0.71	1.42	0.61	1.60	0.52	1.80
19	0.93	1.13	0.83	1.26	0.74	1.41	0.65	1.58	0.56	1.77
20	0.95	1.15	0.86	1.27	0.77	1.41	0.68	1.57	0.60	1.74
21	0.97	1.16	0.89	1.27	0.80	1.41	0.72	1.55	0.63	1.71
22	1.00	1.17	0.91	1.28	0.83	1.40	0.75	1.54	0.66	1.69
23	1.02	1.19	0.94	1.29	0.86	1.40	0.77	1.53	0.70	1.67
24	1.04	1.20	0.96	1.30	0.88	1.41	0.80	1.53	0.72	1.66
25	1.05	1.21	0.98	1.30	0.90	1.41	0.83	1.52	0.75	1.65
26	1.07	1.22	1.00	1.31	0.93	1.41	0.85	1.52	0.78	1.64
27	1.09	1.23	1.02	1.32	0.95	1.41	0.88	1.51	0.81	1.63
28	1.10	1.24	1.04	1.32	0.97	1.41	0.90	1.51	0.83	1.62
29	1.12	1.25	1.05	1.33	0.99	1.42	0.92	1.51	0.85	1.61
30	1.13	1.26	1.07	1.34	1.01	1.42	0.94	1.51	0.88	1.61
31	1.15	1.27	1.08	1.34	1.02	1.42	0.96	1.51	0.90	1.60
32	1.16	1.28	1.10	1.35	1.04	1.43	0.98	1.51	0.92	1.60
33	1.17	1.29	1.11	1.36	1.05	1.43	1.00	1.51	0.94	1.59
34	1.18	1.30	1.13	1.36	1.07	1.43	1.01	1.51	0.95	1.59
35	1.19	1.31	1.14	1.37	1.08	1.44	1.03	1.51	0.97	1.59
36	1.21	1.32	1.15	1.38	1.10	1.44	1.04	1.51	0.99	1.59
37	1.22	1.32	1.16	1.38	1.11	1.45	1.06	1.51	1.00	1.59
38	1.23	1.33	1.18	1.39	1.12	1.45	1.07	1.52	1.02	1.58
39	1.24	1.34	1.19	1.39	1.14	1.45	1.09	1.52	1.03	1.58
40	1.25	1.34	1.20	1.40	1.15	1.46	1.10	1.52	1.05	1.58
45	1.29	1.38	1.24	1.42	1.20	1.48	1.16	1.53	1.11	1.58
50	1.32	1.40	1.28	1.45	1.24	1.49	1.20	1.54	1.16	1.59
55	1.36	1.43	1.32	1.47	1.28	1.51	1.25	1.55	1.21	1.59
60	1.38	1.45	1.35	1.48	1.32	1.52	1.28	1.56	1.25	1.60
65	1.41	1.47	1.38	1.50	1.35	1.53	1.31	1.57	1.28	1.61
70	1.43	1.49	1.40	1.52	1.37	1.55	1.34	1.58	1.31	1.61
75	1.45	1.50	1.42	1.53	1.39	1.56	1.37	1.59	1.34	1.62
80	1.47	1.52	1.44	1.54	1.42	1.57	1.39	1.60	1.36	1.62
85	1.48	1.53	1.46	1.55	1.43	1.58	1.41	1.60	1.39	1.63
90	1.50	1.54	1.47	1.56	1.45	1.59	1.43	1.61	1.41	1.64
95	1.51	1.55	1.49	1.57	1.47	1.60	1.45	1.62	1.42	1.64
100	1.52	1.56	1.50	1.58	1.48	1.60	1.46	1.63	1.44	1.65

Source: Reprinted, with permission, from J. Durbin and G. S. Watson, "Testing for Serial Correlation in Least Squares Regression. II," *Biometrika*, Vol. 38 (1951), pp. 159–78.

TABLE A–7

Table of z' Transformation of Correlation Coefficient

r ρ	z' ζ	r ρ	z' ζ	r ρ	z' ζ	r ρ	z' ζ
.00	.0000	.25	.2554	.50	.5493	.75	.973
.01	.0100	.26	.2661	.51	.5627	.76	.996
.02	.0200	.27	.2769	.52	.5763	.77	1.020
.03	.0300	.28	.2877	.53	.5901	.78	1.045
.04	.0400	.29	.2986	.54	.6042	.79	1.071
.05	.0500	.30	.3095	.55	.6184	.80	1.099
.06	.0601	.31	.3205	.56	.6328	.81	1.127
.07	.0701	.32	.3316	.57	.6475	.82	1.157
.08	.0802	.33	.3428	.58	.6625	.83	1.188
.09	.0902	.34	.3541	.59	.6777	.84	1.221
.10	.1003	.35	.3654	.60	.6931	.85	1.256
.11	.1104	.36	.3769	.61	.7089	.86	1.293
.12	.1206	.37	.3884	.62	.7250	.87	1.333
.13	.1307	.38	.4001	.63	.7414	.88	1.376
.14	.1409	.39	.4118	.64	.7582	.89	1.422
.15	.1511	.40	.4236	.65	.7753	.90	1.472
.16	.1614	.41	.4356	.66	.7928	.91	1.528
.17	.1717	.42	.4477	.67	.8107	.92	1.589
.18	.1820	.43	.4599	.68	.8291	.93	1.658
.19	.1923	.44	.4722	.69	.8480	.94	1.738
.20	.2027	.45	.4847	.70	.8673	.95	1.832
.21	.2132	.46	.4973	.71	.8872	.96	1.946
.22	.2237	.47	.5101	.72	.9076	.97	2.092
.23	.2342	.48	.5230	.73	.9287	.98	2.298
.24	.2448	.49	.5361	.74	.9505	.99	2.647

Source: Abridged from Table 14 of Pearson and Hartley, *Biometrika Tables for Statisticians*, Volume 1, 1966, published by the Cambridge University Press, on behalf of The Biometrika Society, by permission of the authors and publishers.

TABLE A–8

Power Function for Analysis of Variance (fixed effects model)

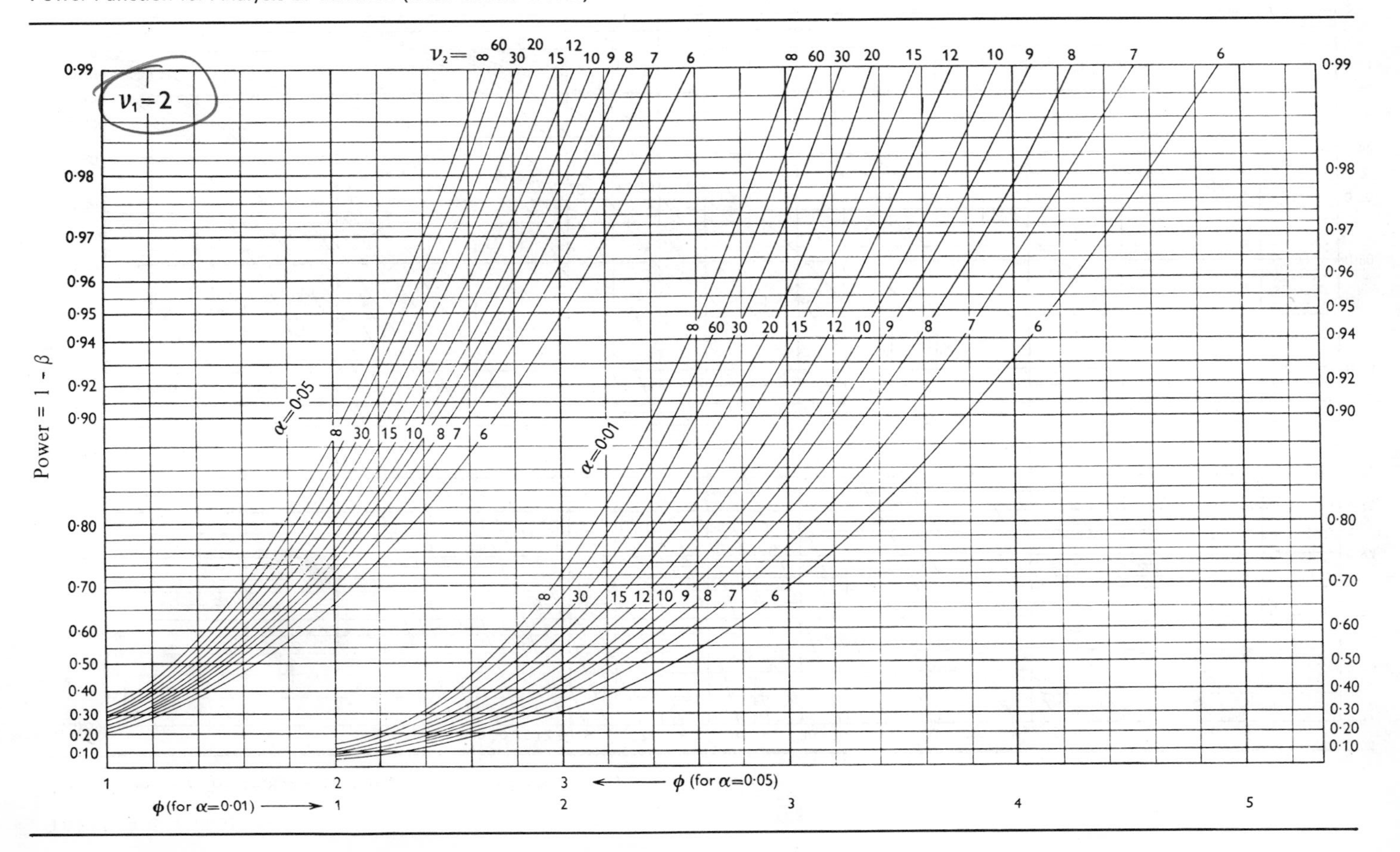

TABLE A–8 (*continued*)

Power Function for Analysis of Variance (fixed effects model)

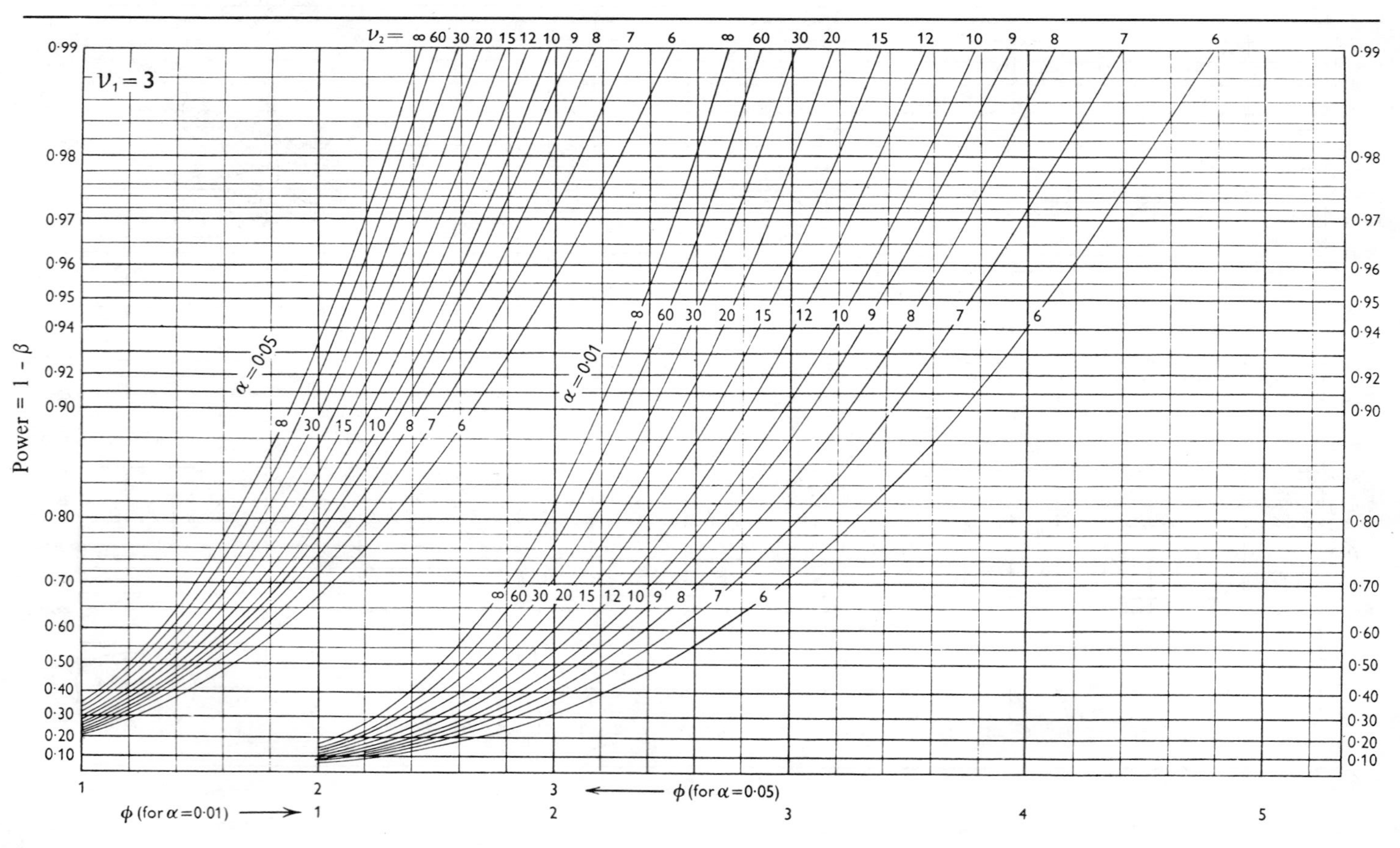

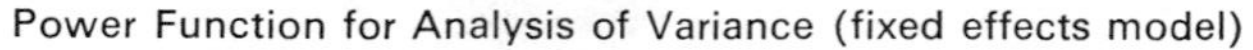

TABLE A–8 (*continued*)

Power Function for Analysis of Variance (fixed effects model)

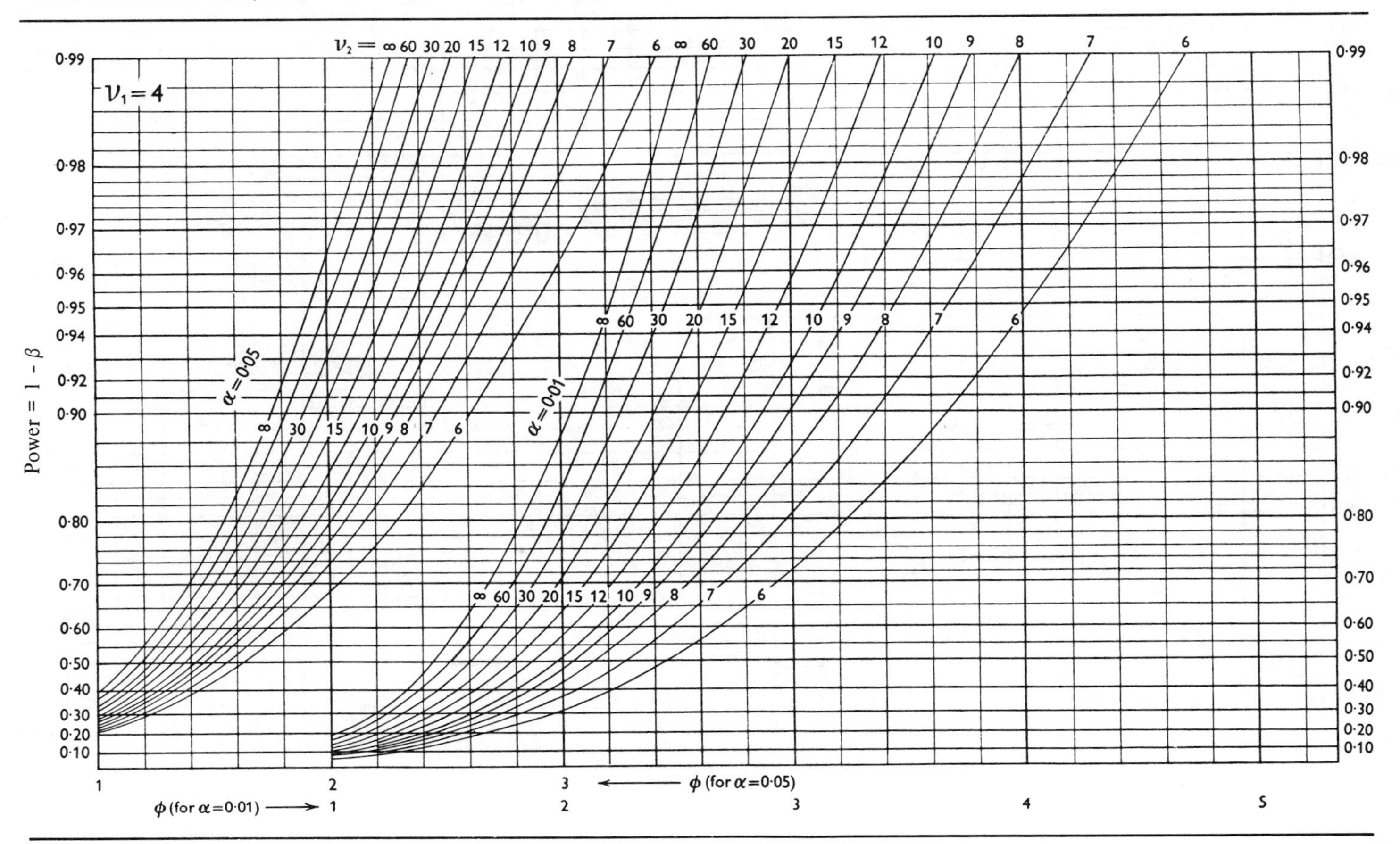

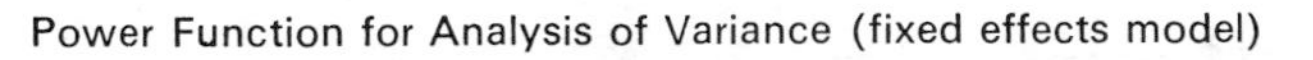

TABLE A–8 (*continued*)

Power Function for Analysis of Variance (fixed effects model)

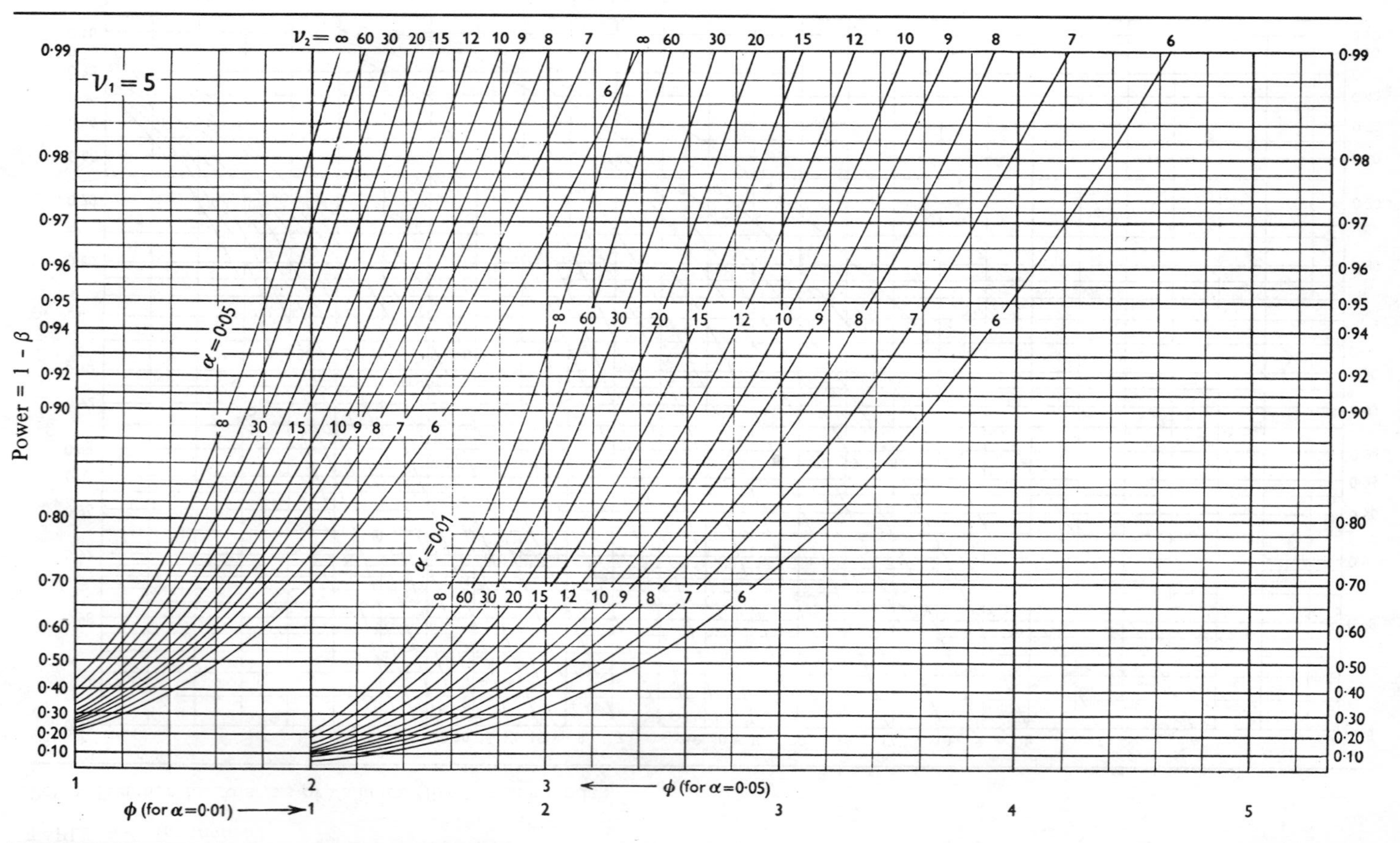

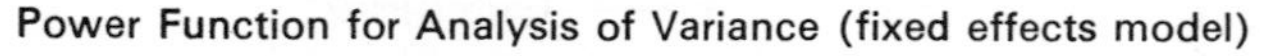

TABLE A–8 (*continued*)

Power Function for Analysis of Variance (fixed effects model)

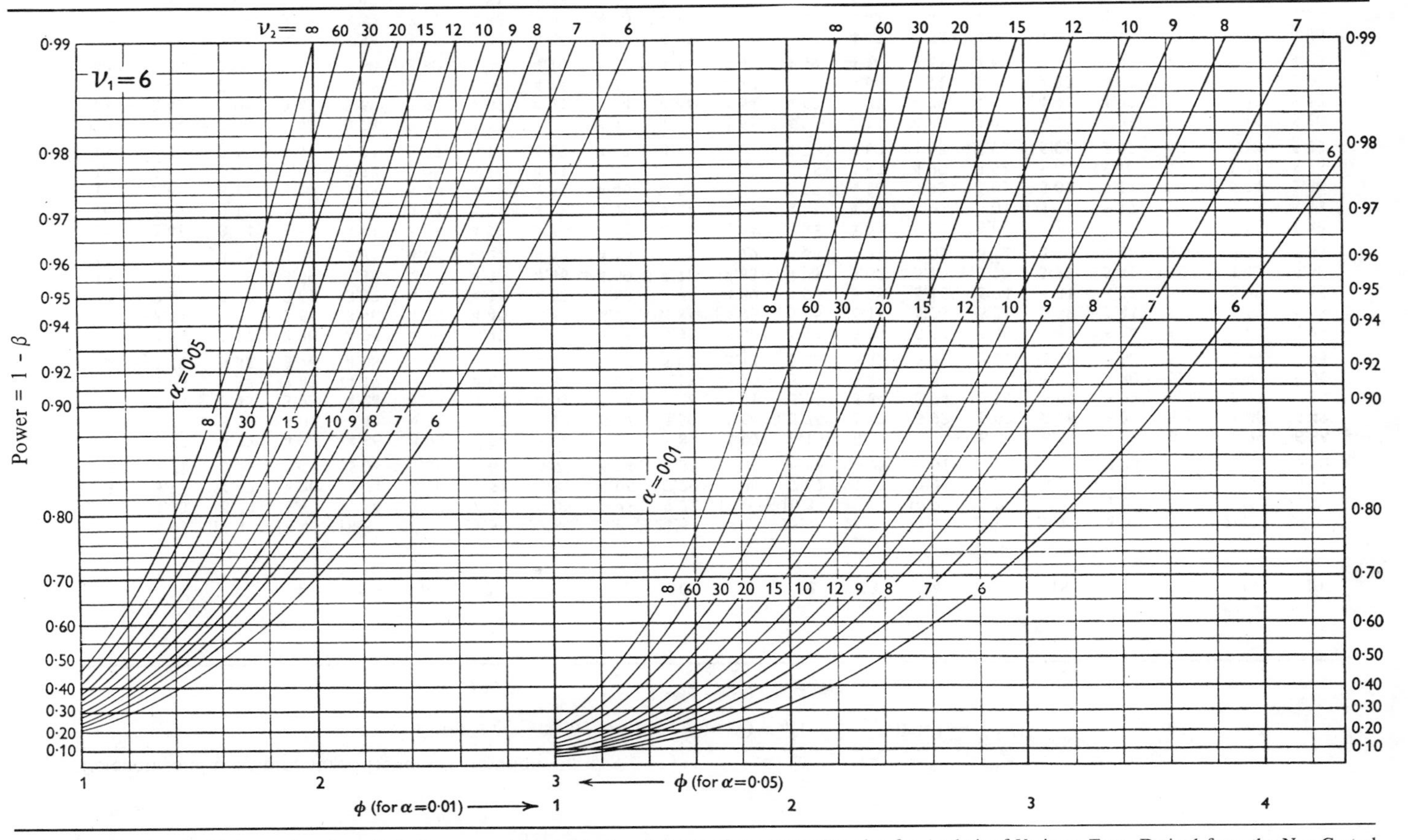

Source: Reprinted, with permission, from E. S. Pearson and H. O. Hartley, "Charts of the Power Function for Analysis of Variance Tests, Derived from the Non-Central *F*-Distribution," *Biometrika*, Vol. 38 (1951), pp. 112–30.

TABLE A–9

Percentiles of the Studentized Range Distribution

Entry is $q(1-\alpha; r, \nu)$ where $P\{q(r, \nu) \le q(1-\alpha; r, \nu)\} = 1-\alpha$

$1-\alpha = .90$

ν	r = 2	3	4	5	6	7	8	9	10	11	12	13	14	15	16	17	18	19	20
1	8.93	13.4	16.4	18.5	20.2	21.5	22.6	23.6	24.5	25.2	25.9	26.5	27.1	27.6	28.1	28.5	29.0	29.3	29.7
2	4.13	5.73	6.77	7.54	8.14	8.63	9.05	9.41	9.72	10.0	10.3	10.5	10.7	10.9	11.1	11.2	11.4	11.5	11.7
3	3.33	4.47	5.20	5.74	6.16	6.51	6.81	7.06	7.29	7.49	7.67	7.83	7.98	8.12	8.25	8.37	8.48	8.58	8.68
4	3.01	3.98	4.59	5.03	5.39	5.68	5.93	6.14	6.33	6.49	6.65	6.78	6.91	7.02	7.13	7.23	7.33	7.41	7.50
5	2.85	3.72	4.26	4.66	4.98	5.24	5.46	5.65	5.82	5.97	6.10	6.22	6.34	6.44	6.54	6.63	6.71	6.79	6.86
6	2.75	3.56	4.07	4.44	4.73	4.97	5.17	5.34	5.50	5.64	5.76	5.87	5.98	6.07	6.16	6.25	6.32	6.40	6.47
7	2.68	3.45	3.93	4.28	4.55	4.78	4.97	5.14	5.28	5.41	5.53	5.64	5.74	5.83	5.91	5.99	6.06	6.13	6.19
8	2.63	3.37	3.83	4.17	4.43	4.65	4.83	4.99	5.13	5.25	5.36	5.46	5.56	5.64	5.72	5.80	5.87	5.93	6.00
9	2.59	3.32	3.76	4.08	4.34	4.54	4.72	4.87	5.01	5.13	5.23	5.33	5.42	5.51	5.58	5.66	5.72	5.79	5.85
10	2.56	3.27	3.70	4.02	4.26	4.47	4.64	4.78	4.91	5.03	5.13	5.23	5.32	5.40	5.47	5.54	5.61	5.67	5.73
11	2.54	3.23	3.66	3.96	4.20	4.40	4.57	4.71	4.84	4.95	5.05	5.15	5.23	5.31	5.38	5.45	5.51	5.57	5.63
12	2.52	3.20	3.62	3.92	4.16	4.35	4.51	4.65	4.78	4.89	4.99	5.08	5.16	5.24	5.31	5.37	5.44	5.49	5.55
13	2.50	3.18	3.59	3.88	4.12	4.30	4.46	4.60	4.72	4.83	4.93	5.02	5.10	5.18	5.25	5.31	5.37	5.43	5.48
14	2.49	3.16	3.56	3.85	4.08	4.27	4.42	4.56	4.68	4.79	4.88	4.97	5.05	5.12	5.19	5.26	5.32	5.37	5.43
15	2.48	3.14	3.54	3.83	4.05	4.23	4.39	4.52	4.64	4.75	4.84	4.93	5.01	5.08	5.15	5.21	5.27	5.32	5.38
16	2.47	3.12	3.52	3.80	4.03	4.21	4.36	4.49	4.61	4.71	4.81	4.89	4.97	5.04	5.11	5.17	5.23	5.28	5.33
17	2.46	3.11	3.50	3.78	4.00	4.18	4.33	4.46	4.58	4.68	4.77	4.86	4.93	5.01	5.07	5.13	5.19	5.24	5.30
18	2.45	3.10	3.49	3.77	3.98	4.16	4.31	4.44	4.55	4.65	4.75	4.83	4.90	4.98	5.04	5.10	5.16	5.21	5.26
19	2.45	3.09	3.47	3.75	3.97	4.14	4.29	4.42	4.53	4.63	4.72	4.80	4.88	4.95	5.01	5.07	5.13	5.18	5.23
20	2.44	3.08	3.46	3.74	3.95	4.12	4.27	4.40	4.51	4.61	4.70	4.78	4.85	4.92	4.99	5.05	5.10	5.16	5.20
24	2.42	3.05	3.42	3.69	3.90	4.07	4.21	4.34	4.44	4.54	4.63	4.71	4.78	4.85	4.91	4.97	5.02	5.07	5.12
30	2.40	3.02	3.39	3.65	3.85	4.02	4.16	4.28	4.38	4.47	4.56	4.64	4.71	4.77	4.83	4.89	4.94	4.99	5.03
40	2.38	2.99	3.35	3.60	3.80	3.96	4.10	4.21	4.32	4.41	4.49	4.56	4.63	4.69	4.75	4.81	4.86	4.90	4.95
60	2.36	2.96	3.31	3.56	3.75	3.91	4.04	4.16	4.25	4.34	4.42	4.49	4.56	4.62	4.67	4.73	4.78	4.82	4.86
120	2.34	2.93	3.28	3.52	3.71	3.86	3.99	4.10	4.19	4.28	4.35	4.42	4.48	4.54	4.60	4.65	4.69	4.74	4.78
∞	2.33	2.90	3.24	3.48	3.66	3.81	3.93	4.04	4.13	4.21	4.28	4.35	4.41	4.47	4.52	4.57	4.61	4.65	4.69

TABLE A–9 (*continued*)

Percentiles of the Studentized Range Distribution

$1-\alpha=.95$

	r																		
	2	3	4	5	6	7	8	9	10	11	12	13	14	15	16	17	18	19	20
1	18.0	27.0	32.8	37.1	40.4	43.1	45.4	47.4	49.1	50.6	52.0	53.2	54.3	55.4	56.3	57.2	58.0	58.8	59.6
2	6.08	8.33	9.80	10.9	11.7	12.4	13.0	13.5	14.0	14.4	14.7	15.1	15.4	15.7	15.9	16.1	16.4	16.6	16.8
3	4.50	5.91	6.82	7.50	8.04	8.48	8.85	9.18	9.46	9.72	9.95	10.2	10.3	10.5	10.7	10.8	11.0	11.1	11.2
4	3.93	5.04	5.76	6.29	6.71	7.05	7.35	7.60	7.83	8.03	8.21	8.37	8.52	8.66	8.79	8.91	9.03	9.13	9.23
5	3.64	4.60	5.22	5.67	6.03	6.33	6.58	6.80	6.99	7.17	7.32	7.47	7.60	7.72	7.83	7.93	8.03	8.12	8.21
6	3.46	4.34	4.90	5.30	5.63	5.90	6.12	6.32	6.49	6.65	6.79	6.92	7.03	7.14	7.24	7.34	7.43	7.51	7.59
7	3.34	4.16	4.68	5.06	5.36	5.61	5.82	6.00	6.16	6.30	6.43	6.55	6.66	6.76	6.85	6.94	7.02	7.10	7.17
8	3.26	4.04	4.53	4.89	5.17	5.40	5.60	5.77	5.92	6.05	6.18	6.29	6.39	6.48	6.57	6.65	6.73	6.80	6.87
9	3.20	3.95	4.41	4.76	5.02	5.24	5.43	5.59	5.74	5.87	5.98	6.09	6.19	6.28	6.36	6.44	6.51	6.58	6.64
10	3.15	3.88	4.33	4.65	4.91	5.12	5.30	5.46	5.60	5.72	5.83	5.93	6.03	6.11	6.19	6.27	6.34	6.40	6.47
11	3.11	3.82	4.26	4.57	4.82	5.03	5.20	5.35	5.49	5.61	5.71	5.81	5.90	5.98	6.06	6.13	6.20	6.27	6.33
12	3.08	3.77	4.20	4.51	4.75	4.95	5.12	5.27	5.39	5.51	5.61	5.71	5.80	5.88	5.95	6.02	6.09	6.15	6.21
13	3.06	3.73	4.15	4.45	4.69	4.88	5.05	5.19	5.32	5.43	5.53	5.63	5.71	5.79	5.86	5.93	5.99	6.05	6.11
14	3.03	3.70	4.11	4.41	4.64	4.83	4.99	5.13	5.25	5.36	5.46	5.55	5.64	5.71	5.79	5.85	5.91	5.97	6.03
15	3.01	3.67	4.08	4.37	4.59	4.78	4.94	5.08	5.20	5.31	5.40	5.49	5.57	5.65	5.72	5.78	5.85	5.90	5.96
16	3.00	3.65	4.05	4.33	4.56	4.74	4.90	5.03	5.15	5.26	5.35	5.44	5.52	5.59	5.66	5.73	5.79	5.84	5.90
17	2.98	3.63	4.02	4.30	4.52	4.70	4.86	4.99	5.11	5.21	5.31	5.39	5.47	5.54	5.61	5.67	5.73	5.79	5.84
18	2.97	3.61	4.00	4.28	4.49	4.67	4.82	4.96	5.07	5.17	5.27	5.35	5.43	5.50	5.57	5.63	5.69	5.74	5.79
19	2.96	3.59	3.98	4.25	4.47	4.65	4.79	4.92	5.04	5.14	5.23	5.31	5.39	5.46	5.53	5.59	5.65	5.70	5.75
20	2.95	3.58	3.96	4.23	4.45	4.62	4.77	4.90	5.01	5.11	5.20	5.28	5.36	5.43	5.49	5.55	5.61	5.66	5.71
24	2.92	3.53	3.90	4.17	4.37	4.54	4.68	4.81	4.92	5.01	5.10	5.18	5.25	5.32	5.38	5.44	5.49	5.55	5.59
30	2.89	3.49	3.85	4.10	4.30	4.46	4.60	4.72	4.82	4.92	5.00	5.08	5.15	5.21	5.27	5.33	5.38	5.43	5.47
40	2.86	3.44	3.79	4.04	4.23	4.39	4.52	4.63	4.73	4.82	4.90	4.98	5.04	5.11	5.16	5.22	5.27	5.31	5.36
60	2.83	3.40	3.74	3.98	4.16	4.31	4.44	4.55	4.65	4.73	4.81	4.88	4.94	5.00	5.06	5.11	5.15	5.20	5.24
120	2.80	3.36	3.68	3.92	4.10	4.24	4.36	4.47	4.56	4.64	4.71	4.78	4.84	4.90	4.95	5.00	5.04	5.09	5.13
∞	2.77	3.31	3.63	3.86	4.03	4.17	4.29	4.39	4.47	4.55	4.62	4.68	4.74	4.80	4.85	4.89	4.93	4.97	5.01

TABLE A–9 (*continued*)

Percentiles of the Studentized Range Distribution

$1 - \alpha = .99$

ν	r = 2	3	4	5	6	7	8	9	10	11	12	13	14	15	16	17	18	19	20
1	90.0	135	164	186	202	216	227	237	246	253	260	266	272	277	282	286	290	294	298
2	14.0	19.0	22.3	24.7	26.6	28.2	29.5	30.7	31.7	32.6	33.4	34.1	34.8	35.4	36.0	36.5	37.0	37.5	37.9
3	8.26	10.6	12.2	13.3	14.2	15.0	15.6	16.2	16.7	17.1	17.5	17.9	18.2	18.5	18.8	19.1	19.3	19.5	19.8
4	6.51	8.12	9.17	9.96	10.6	11.1	11.5	11.9	12.3	12.6	12.8	13.1	13.3	13.5	13.7	13.9	14.1	14.2	14.4
5	5.70	6.97	7.80	8.42	8.91	9.32	9.67	9.97	10.2	10.5	10.7	10.9	11.1	11.2	11.4	11.6	11.7	11.8	11.9
6	5.24	6.33	7.03	7.56	7.97	8.32	8.61	8.87	9.10	9.30	9.49	9.65	9.81	9.95	10.1	10.2	10.3	10.4	10.5
7	4.95	5.92	6.54	7.01	7.37	7.68	7.94	8.17	8.37	8.55	8.71	8.86	9.00	9.12	9.24	9.35	9.46	9.55	9.65
8	4.74	5.63	6.20	6.63	6.96	7.24	7.47	7.68	7.87	8.03	8.18	8.31	8.44	8.55	8.66	8.76	8.85	8.94	9.03
9	4.60	5.43	5.96	6.35	6.66	6.91	7.13	7.32	7.49	7.65	7.78	7.91	8.03	8.13	8.23	8.32	8.41	8.49	8.57
10	4.48	5.27	5.77	6.14	6.43	6.67	6.87	7.05	7.21	7.36	7.48	7.60	7.71	7.81	7.91	7.99	8.07	8.15	8.22
11	4.39	5.14	5.62	5.97	6.25	6.48	6.67	6.84	6.99	7.13	7.25	7.36	7.46	7.56	7.65	7.73	7.81	7.88	7.95
12	4.32	5.04	5.50	5.84	6.10	6.32	6.51	6.67	6.81	6.94	7.06	7.17	7.26	7.36	7.44	7.52	7.59	7.66	7.73
13	4.26	4.96	5.40	5.73	5.98	6.19	6.37	6.53	6.67	6.79	6.90	7.01	7.10	7.19	7.27	7.34	7.42	7.48	7.55
14	4.21	4.89	5.32	5.63	5.88	6.08	6.26	6.41	6.54	6.66	6.77	6.87	6.96	7.05	7.12	7.20	7.27	7.33	7.39
15	4.17	4.83	5.25	5.56	5.80	5.99	6.16	6.31	6.44	6.55	6.66	6.76	6.84	6.93	7.00	7.07	7.14	7.20	7.26
16	4.13	4.78	5.19	5.49	5.72	5.92	6.08	6.22	6.35	6.46	6.56	6.66	6.74	6.82	6.90	6.97	7.03	7.09	7.15
17	4.10	4.74	5.14	5.43	5.66	5.85	6.01	6.15	6.27	6.38	6.48	6.57	6.66	6.73	6.80	6.87	6.94	7.00	7.05
18	4.07	4.70	5.09	5.38	5.60	5.79	5.94	6.08	6.20	6.31	6.41	6.50	6.58	6.65	6.72	6.79	6.85	6.91	6.96
19	4.05	4.67	5.05	5.33	5.55	5.73	5.89	6.02	6.14	6.25	6.34	6.43	6.51	6.58	6.65	6.72	6.78	6.84	6.89
20	4.02	4.64	5.02	5.29	5.51	5.69	5.84	5.97	6.09	6.19	6.29	6.37	6.45	6.52	6.59	6.65	6.71	6.76	6.82
24	3.96	4.54	4.91	5.17	5.37	5.54	5.69	5.81	5.92	6.02	6.11	6.19	6.26	6.33	6.39	6.45	6.51	6.56	6.61
30	3.89	4.45	4.80	5.05	5.24	5.40	5.54	5.65	5.76	5.85	5.93	6.01	6.08	6.14	6.20	6.26	6.31	6.36	6.41
40	3.82	4.37	4.70	4.93	5.11	5.27	5.39	5.50	5.60	5.69	5.77	5.84	5.90	5.96	6.02	6.07	6.12	6.17	6.21
60	3.76	4.28	4.60	4.82	4.99	5.13	5.25	5.36	5.45	5.53	5.60	5.67	5.73	5.79	5.84	5.89	5.93	5.98	6.02
120	3.70	4.20	4.50	4.71	4.87	5.01	5.12	5.21	5.30	5.38	5.44	5.51	5.56	5.61	5.66	5.71	5.75	5.79	5.83
∞	3.64	4.12	4.40	4.60	4.76	4.88	4.99	5.08	5.16	5.23	5.29	5.35	5.40	5.45	5.49	5.54	5.57	5.61	5.65

Source: Reprinted, with permission, from Henry Scheffé, *The Analysis of Variance* (New York: John Wiley & Sons, Inc., 1959), pp. 434–36.

TABLE A–10

Charts for Determining Sample Size for Analysis of Variance (fixed effects model)

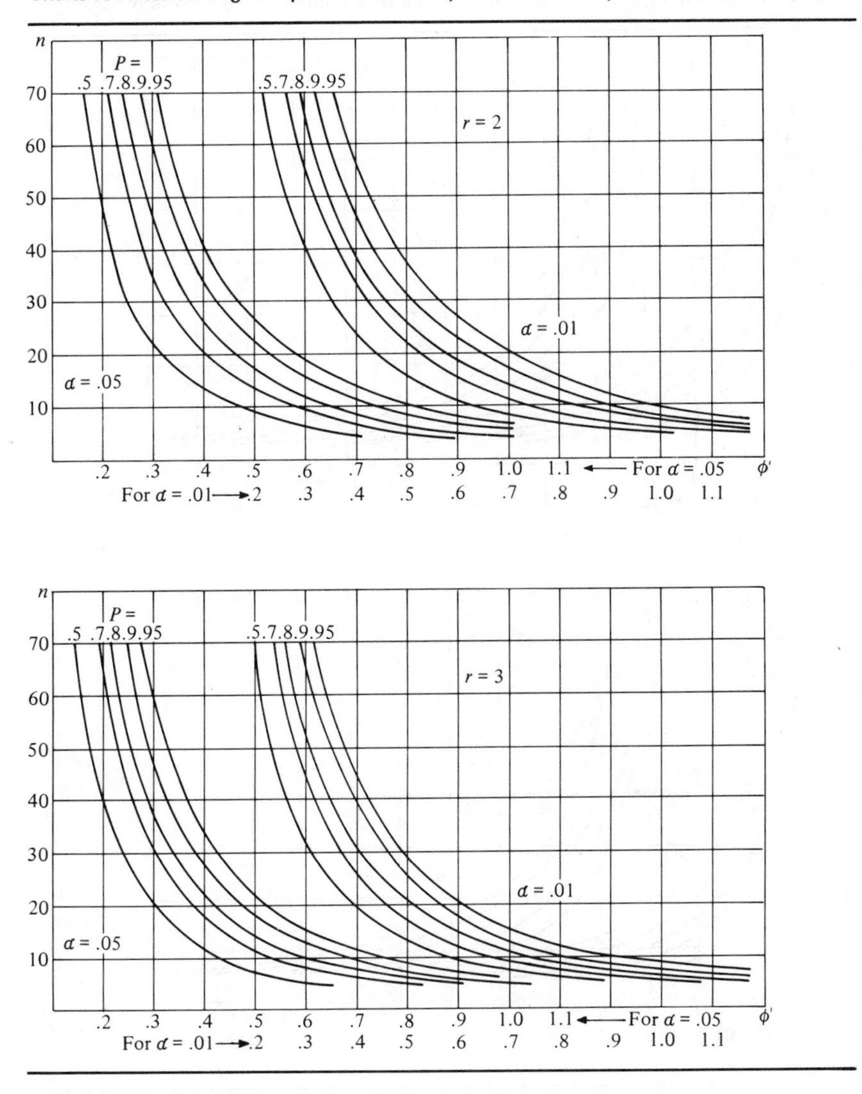

TABLE A–10 (*continued*)

Charts for Determining Sample Size for Analysis of Variance (fixed effects model)

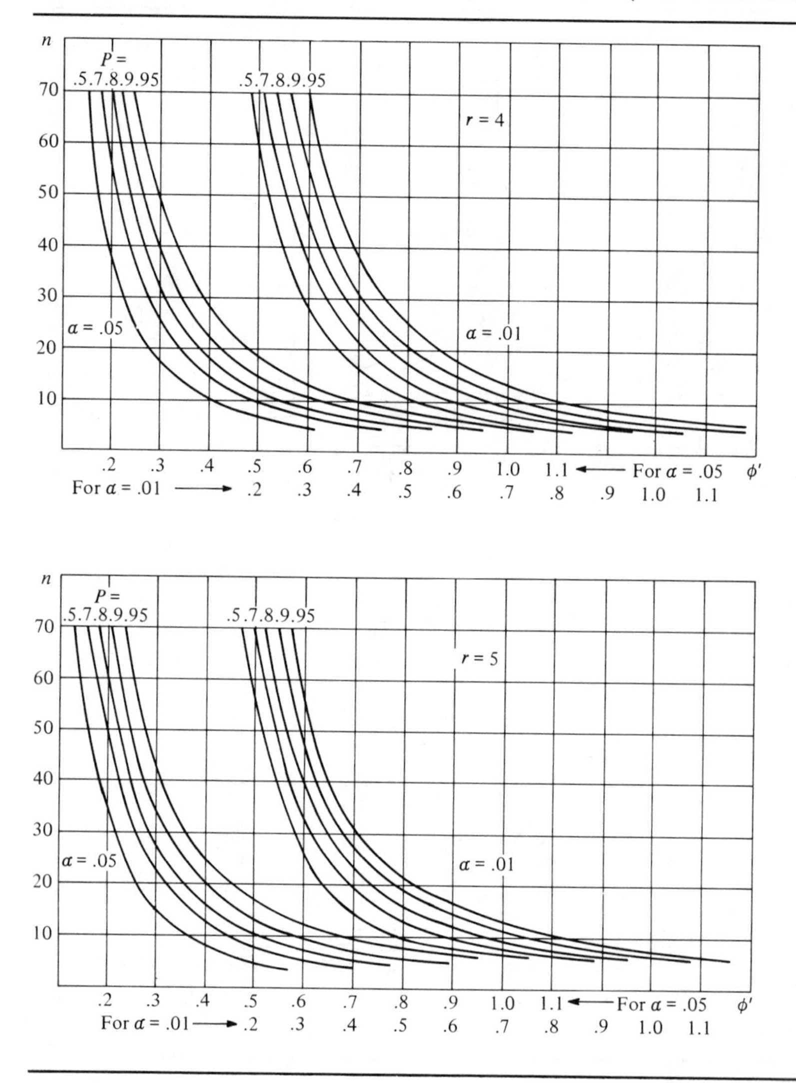

Source: Reprinted, with permission, from L. S. Feldt and M. W. Mahmoud, "Power Function Charts for Specifying Numbers of Observations in Analyses of Variance of Fixed Effects," *The Annals of Mathematical Statistics*, Vol. 29 (1958), pp. 871–77.

TABLE A–11

Table of $\lambda\sqrt{n}/\sigma$ for Determining Sample Size to Find "Best" of r Population Means

Number of Populations (r)	*Probability of Correct Identification* $(1 - \alpha)$		
	.90	.95	.99
2	1.8124	2.3262	3.2900
3	2.2302	2.7101	3.6173
4	2.4516	2.9162	3.7970
5	2.5997	3.0552	3.9196
6	2.7100	3.1591	4.0121
7	2.7972	3.2417	4.0861
8	2.8691	3.3099	4.1475
9	2.9301	3.3679	4.1999
10	2.9829	3.4182	4.2456

Source: Reprinted, with permission, from R. E. Bechhofer, "A Single-Sample Multiple Decision Procedure for Ranking Means of Normal Populations with Known Variances," *The Annals of Mathematical Statistics*, Vol. 25 (1954), pp. 16–39.

TABLE A–12

Percentiles of H Statistic Distribution

Entry is $H(1-\alpha; r, n)$ where $P\{H \leq H(1-\alpha; r, n)\} = 1-\alpha$

$1-\alpha = .95$

n	r = 2	3	4	5	6	7	8	9	10	11	12
3	39.0	87.5	142	202	266	333	403	475	550	626	704
4	15.4	27.8	39.2	50.7	62.0	72.9	83.5	93.9	104	114	124
5	9.60	15.5	20.6	25.2	29.5	33.6	37.5	41.1	44.6	48.0	51.4
6	7.15	10.8	13.7	16.3	18.7	20.8	22.9	24.7	26.5	28.2	29.9
7	5.82	8.38	10.4	12.1	13.7	15.0	16.3	17.5	18.6	19.7	20.7
8	4.99	6.94	8.44	9.70	10.8	11.8	12.7	13.5	14.3	15.1	15.8
9	4.43	6.00	7.18	8.12	9.03	9.78	10.5	11.1	11.7	12.2	12.7
10	4.03	5.34	6.31	7.11	7.80	8.41	8.95	9.45	9.91	10.3	10.7
11	3.72	4.85	5.67	6.34	6.92	7.42	7.87	8.28	8.66	9.01	9.34
13	3.28	4.16	4.79	5.30	5.72	6.09	6.42	6.72	7.00	7.25	7.48
16	2.86	3.54	4.01	4.37	4.68	4.95	5.19	5.40	5.59	5.77	5.93
21	2.46	2.95	3.29	3.54	3.76	3.94	4.10	4.24	4.37	4.49	4.59
31	2.07	2.40	2.61	2.78	2.91	3.02	3.12	3.21	3.29	3.36	3.39
61	1.67	1.85	1.96	2.04	2.11	2.17	2.22	2.26	2.30	2.33	2.36
∞	1.00	1.00	1.00	1.00	1.00	1.00	1.00	1.00	1.00	1.00	1.00

$1-\alpha = .99$

n	r = 2	3	4	5	6	7	8	9	10	11	12
3	199	448	729	1,036	1,362	1,705	2,063	2,432	2,813	3,204	3,605
4	47.5	85	120	151	184	216	249	281	310	337	361
5	23.2	37	49	59	69	79	89	97	106	113	120
6	14.9	22	28	33	38	42	46	50	54	57	60
7	11.1	15.5	19.1	22	25	27	30	32	34	36	37
8	8.89	12.1	14.5	16.5	18.4	20	22	23	24	26	27
9	7.50	9.9	11.7	13.2	14.5	15.8	16.9	17.9	18.9	19.8	21
10	6.54	8.5	9.9	11.1	12.1	13.1	13.9	14.7	15.3	16.0	16.6
11	5.85	7.4	8.6	9.6	10.4	11.1	11.8	12.4	12.9	13.4	13.9
13	4.91	6.1	6.9	7.6	8.2	8.7	9.1	9.5	9.9	10.2	10.6
16	4.07	4.9	5.5	6.0	6.4	6.7	7.1	7.3	7.5	7.8	8.0
21	3.32	3.8	4.3	4.6	4.9	5.1	5.3	5.5	5.6	5.8	5.9
31	2.63	3.0	3.3	3.4	3.6	3.7	3.8	3.9	4.0	4.1	4.2
61	1.96	2.2	2.3	2.4	2.4	2.5	2.5	2.6	2.6	2.7	2.7
∞	1.00	1.0	1.0	1.0	1.0	1.0	1.0	1.0	1.0	1.0	1.0

Source: Reprinted, with permission, from H. A. David, "Upper 5 and 1% Points of the Maximum *F*-Ratio," *Biometrika*, Vol. 39 (1952), pp. 422–24.

TABLE A–13

Table of Random Permutations of 9

5 5 6 7 1	4 3 3 7 3	8 7 4 6 3	9 7 4 9 4	9 2 2 8 8	2 7 9 3 5	8 3 1 9 4
4 1 2 8 2	7 1 1 2 9	9 5 7 8 2	8 9 3 6 6	1 7 7 2 4	4 8 5 7 3	3 7 4 5 6
9 3 3 2 9	8 8 8 4 5	2 4 6 1 6	3 6 7 7 8	7 4 4 7 1	7 3 2 8 6	6 1 2 2 2
7 9 7 4 3	5 5 2 9 2	1 6 5 3 5	7 8 5 1 9	5 1 9 1 3	6 5 1 4 9	2 9 8 7 8
1 6 9 6 5	6 9 4 3 6	4 3 9 2 9	5 1 8 2 3	8 3 3 3 2	8 9 6 1 2	4 5 7 6 9
6 4 4 3 6	2 4 6 8 1	7 9 3 4 1	6 2 6 4 2	2 9 8 5 9	9 2 4 2 8	9 6 9 8 1
8 7 8 1 7	1 2 5 6 8	3 1 2 9 8	4 4 1 8 7	6 5 1 6 7	5 4 3 5 1	1 4 3 1 7
3 2 1 9 4	3 6 7 5 7	6 8 8 7 7	2 5 9 5 1	3 8 5 4 6	3 6 7 9 4	5 2 5 4 5
2 8 5 5 8	9 7 9 1 4	5 2 1 5 4	1 3 2 3 5	4 6 6 9 5	1 1 8 6 7	7 8 6 3 3
7 4 6 1 5	9 2 2 2 9	2 8 1 7 3	2 4 2 1 9	2 4 8 3 1	2 6 5 4 8	8 4 9 4 2
9 3 8 3 2	1 1 1 9 8	9 4 9 5 4	8 8 8 8 6	7 7 5 4 6	5 3 2 7 6	9 3 8 2 1
1 6 3 4 7	6 5 8 4 5	6 1 7 1 9	5 2 5 6 3	8 5 7 5 5	6 9 9 8 1	3 6 7 9 7
6 8 2 8 4	4 8 7 8 6	5 7 5 4 5	9 6 7 5 8	5 9 9 7 7	8 5 3 3 5	6 9 4 6 9
4 1 4 7 8	2 3 9 3 4	4 2 2 3 6	4 7 4 2 5	6 3 3 6 9	1 7 8 5 4	4 5 2 1 4
2 9 1 9 3	7 9 6 6 2	1 6 4 6 1	7 9 9 7 4	1 8 4 1 8	9 2 7 9 3	1 8 3 5 5
5 5 5 5 1	3 7 4 7 7	8 5 8 9 2	1 5 1 3 2	9 6 2 8 4	3 8 1 1 9	5 7 1 3 3
8 2 9 2 9	8 6 5 5 3	7 9 6 8 8	3 1 6 9 7	4 1 6 9 3	4 4 6 6 2	7 2 6 8 8
3 7 7 6 6	5 4 3 1 1	3 3 3 2 7	6 3 3 4 1	3 2 1 2 2	7 1 4 2 7	2 1 5 7 6
9 7 7 5 5	9 9 9 3 8	9 8 6 1 7	5 8 6 1 2	1 9 8 3 3	3 1 7 7 3	7 6 6 5 5
3 8 1 7 2	6 2 7 1 6	4 1 3 4 2	3 6 2 4 3	2 6 1 2 8	8 8 6 2 7	8 9 7 4 7
4 3 4 2 7	7 3 1 7 2	1 5 4 8 6	6 2 1 6 1	7 8 5 1 7	5 9 1 3 6	3 1 2 3 1
5 9 2 8 3	3 7 5 8 9	2 9 1 7 1	2 3 8 3 4	3 5 9 9 9	7 2 3 4 1	5 7 1 7 8
1 6 5 1 1	5 6 4 4 1	7 3 7 2 3	4 7 3 8 8	9 3 2 5 6	6 6 9 5 9	9 8 9 1 2
6 2 8 3 6	8 4 6 2 5	5 2 2 6 8	9 1 7 5 6	4 7 4 6 4	1 7 4 6 4	1 2 8 8 6
2 4 9 6 4	1 8 3 5 4	3 6 5 9 4	8 5 9 7 9	8 1 6 8 1	4 5 5 9 5	2 4 5 9 4
8 5 6 9 9	2 5 2 6 7	8 7 8 3 9	1 9 4 2 5	6 4 7 4 5	2 3 2 8 2	6 3 3 2 3
7 1 3 4 8	4 1 8 9 3	6 4 9 5 5	7 4 5 9 7	5 2 3 7 2	9 4 8 1 8	4 5 4 6 9
7 4 9 8 7	9 7 1 7 1	9 2 3 8 7	7 8 5 3 5	5 1 6 4 9	7 8 6 1 8	2 9 7 3 4
5 6 1 1 2	6 4 6 1 4	5 9 1 2 8	2 4 6 8 7	7 3 7 6 1	5 1 7 4 1	9 3 4 7 7
4 9 3 5 6	1 1 8 4 8	3 5 4 9 3	3 6 1 2 3	2 6 8 7 7	4 5 3 8 5	8 5 9 5 1
3 3 2 2 8	5 2 3 2 2	7 3 8 6 9	4 1 8 6 1	1 9 2 3 6	3 9 5 7 7	1 2 8 1 2
2 1 4 9 4	4 6 2 8 3	2 7 6 5 1	5 7 3 1 2	9 8 4 1 3	6 3 1 2 9	6 1 5 8 8
9 7 5 4 5	3 9 7 9 9	1 4 2 3 4	6 9 7 4 4	3 2 5 2 2	8 4 2 6 3	5 6 3 6 3
6 2 6 3 9	8 8 5 5 5	8 6 7 7 2	9 3 4 5 8	8 7 9 9 4	9 2 4 9 4	4 8 1 2 9
8 5 8 7 1	2 3 9 3 7	4 1 5 1 5	8 5 9 7 6	4 5 3 5 8	1 6 8 5 2	3 4 6 4 5
1 8 7 6 3	7 5 4 6 6	6 8 9 4 6	1 2 2 9 9	6 4 1 8 5	2 7 9 3 6	7 7 2 9 6
8 4 6 8 6	2 1 9 9 7	2 2 1 8 9	5 1 9 2 4	5 2 6 2 8	1 6 8 8 3	8 1 9 4 1
9 9 4 5 8	4 4 8 7 8	8 7 5 9 7	3 6 4 7 7	3 8 5 3 6	4 4 6 7 7	6 6 8 7 8
6 6 3 1 1	6 8 3 1 9	7 5 7 5 5	6 5 1 8 5	2 4 3 8 2	5 1 4 3 6	4 9 7 8 6
7 3 7 7 2	7 3 6 2 2	3 8 9 4 6	4 7 2 6 9	7 9 7 4 1	3 8 2 6 5	3 5 3 1 4
2 8 9 3 4	1 5 5 5 1	5 4 3 6 4	7 8 7 5 3	9 5 8 6 5	8 2 7 9 2	5 3 4 3 5
3 7 2 6 9	8 6 4 6 3	4 1 8 2 1	1 9 6 4 8	4 7 2 1 3	6 3 5 5 1	2 2 6 9 9
5 1 8 4 5	9 9 1 8 4	1 9 4 3 2	8 2 8 9 6	6 3 4 9 9	2 7 1 2 4	9 8 2 6 2
4 5 5 2 7	3 2 7 3 6	9 3 2 1 8	9 3 5 1 2	1 6 9 7 7	9 5 9 1 8	7 7 1 5 7
1 2 1 9 3	5 7 2 4 5	6 6 6 7 3	2 4 3 3 1	8 1 1 5 4	7 9 3 4 9	1 4 5 2 3

Source: Reprinted, with permission, from William G. Cochran and Gertrude M. Cox, *Experimental Designs* (2d. ed.; New York: John Wiley & Sons, Inc., 1957), p. 577.

TABLE A–14

Table of Random Permutations of 16

7	12	15	15	1	2	7	16	10	2	14	15	7	13	13	10	6	1	8	10
13	3	8	16	7	10	11	10	13	5	11	7	13	16	7	7	5	13	2	14
3	1	4	5	14	13	3	14	9	13	13	2	9	15	6	2	8	4	5	8
11	8	16	14	15	6	2	6	2	16	8	5	12	3	9	13	4	3	10	4
14	9	1	6	3	9	14	13	8	6	5	8	14	7	3	15	13	11	4	7
2	16	10	13	5	5	13	2	11	7	3	12	5	14	12	16	2	2	9	15
4	6	13	7	2	15	1	9	1	4	7	10	6	9	11	9	7	6	16	11
6	14	6	10	4	14	4	15	3	3	4	16	2	6	5	1	12	10	6	9
10	15	2	1	13	12	16	3	4	8	10	1	15	5	14	12	14	12	3	2
12	10	7	12	9	11	9	8	12	14	15	4	11	8	16	8	9	14	14	1
15	7	5	2	10	7	8	12	6	15	6	13	16	12	15	4	11	8	12	6
16	2	11	8	8	8	15	5	16	1	1	9	8	1	8	14	16	5	13	5
9	13	14	3	6	4	10	11	5	12	9	3	10	4	4	3	10	9	1	3
8	11	9	4	11	3	12	7	7	10	12	14	3	10	1	6	15	16	15	12
1	5	12	11	16	16	5	4	14	9	16	11	1	2	10	5	1	15	7	13
5	4	3	9	12	1	6	1	15	11	2	6	4	11	2	11	3	7	11	16

11	8	16	5	5	13	1	13	2	16	14	12	9	8	7	5	13	3	13	3
2	2	8	8	14	16	4	3	8	11	10	14	15	1	2	11	4	5	15	9
6	13	2	13	6	5	9	15	11	10	12	6	16	15	16	9	10	12	16	15
14	12	4	16	16	11	14	10	5	12	3	3	12	14	15	13	6	4	1	16
8	6	3	9	4	10	6	4	16	2	2	9	8	16	4	6	5	15	7	8
9	15	12	10	3	2	12	6	1	15	4	13	7	7	9	12	14	8	8	11
3	10	11	12	13	12	5	11	7	8	9	5	14	11	10	1	3	13	3	5
16	1	13	14	8	14	15	5	3	7	11	15	6	12	5	7	11	1	14	4
1	14	14	2	9	15	16	14	6	14	7	8	3	13	11	8	7	7	12	7
4	4	6	4	12	3	11	8	15	9	8	1	13	6	3	3	15	9	9	12
15	5	1	11	10	6	3	7	10	5	5	11	10	10	12	15	16	14	5	2
5	3	5	6	7	7	13	2	14	3	16	4	5	5	13	4	9	16	2	6
12	7	15	15	15	9	8	12	12	13	15	10	1	4	6	16	2	6	11	1
10	11	10	3	2	4	2	1	4	6	6	7	11	9	14	10	8	11	4	13
7	9	7	7	11	1	7	16	13	1	13	2	4	2	1	2	12	2	10	14
13	16	9	1	1	8	10	9	9	4	1	16	2	3	8	14	1	10	6	10

1	6	7	4	8	6	5	2	8	15	4	6	6	1	4	5	7	13	2	10
9	15	11	3	11	15	9	10	1	3	8	2	15	7	9	8	16	1	14	3
10	16	4	5	12	9	16	11	7	1	7	16	11	8	3	3	12	2	3	4
4	14	1	9	5	5	4	13	6	8	15	5	12	5	7	16	5	11	8	1
7	3	13	14	15	2	1	14	16	5	14	9	2	16	1	12	6	14	4	13
16	11	2	1	14	16	6	9	3	4	16	14	3	15	11	11	3	9	12	5
3	10	16	16	13	7	13	1	11	14	9	10	16	2	10	2	10	7	10	16
11	13	9	13	4	13	8	3	5	13	10	12	5	12	5	14	13	16	5	6
15	2	3	12	9	12	2	4	13	10	3	13	14	4	2	1	14	8	6	12
14	1	14	6	10	1	3	12	4	2	2	4	13	3	16	9	9	3	7	14
13	12	5	11	3	11	15	8	2	7	11	7	8	14	6	4	4	4	15	11
12	5	10	7	2	14	7	15	14	16	13	1	9	10	12	10	11	10	9	8
8	9	8	10	6	4	11	7	10	11	6	8	4	9	8	15	8	6	11	9
2	7	6	2	1	8	10	6	15	12	1	11	7	11	13	6	1	15	13	15
6	4	15	8	16	10	14	16	9	6	12	3	10	6	14	7	2	12	16	7
5	8	12	15	7	3	12	5	12	9	5	15	1	13	15	13	15	5	1	2

13	4	10	4	16	13	16	13	5	3	6	14	1	16	8	7	2	3	3	12
5	14	4	6	8	2	15	1	13	14	16	4	15	4	3	12	12	1	4	7
2	2	2	15	14	16	9	12	16	6	10	15	14	9	10	1	14	8	8	16
7	12	15	8	12	3	5	14	7	12	5	13	16	1	7	5	11	2	9	3
6	9	7	14	9	14	10	11	15	11	12	1	12	12	14	16	3	11	11	8
14	5	16	7	10	8	11	8	14	13	7	11	6	3	11	4	4	6	6	9
15	11	8	9	7	12	8	7	1	15	9	3	3	7	13	11	10	4	5	1
11	6	6	1	4	1	3	16	12	5	4	9	13	13	6	8	15	9	1	14
4	10	3	16	2	11	7	9	6	9	1	8	4	11	5	2	16	10	12	4
1	8	1	13	1	15	4	4	11	4	2	16	5	8	1	9	5	12	16	6
9	7	14	2	6	4	14	10	9	8	15	10	7	10	9	10	6	14	10	11
12	1	9	10	15	5	2	15	10	2	14	2	8	2	4	13	8	5	15	5
3	3	12	11	5	9	6	6	3	10	13	12	9	6	2	15	7	15	7	13
10	15	11	5	13	7	12	5	2	7	11	5	10	15	12	3	1	13	13	10
8	13	13	3	3	10	13	2	4	1	8	6	11	14	15	6	9	16	2	2
16	16	5	12	11	6	1	3	8	16	3	7	2	5	16	14	13	7	14	15

Source: Reprinted, with permission, from William G. Cochran and Gertrude M. Cox, *Experimental Designs* (2d. ed.; New York: John Wiley & Sons, Inc., 1957), p. 583.

TABLE A–15

Table of Random Digits

Line	(1)–(5)	(6)–(10)	(11)–(15)	(16)–(20)	(21)–(25)	(26)–(30)	(31)–(35)
101	13284	16834	74151	92027	24670	36665	00770
102	21224	00370	30420	03883	94648	89428	41583
103	99052	47887	81085	64933	66279	80432	65793
104	00199	50993	98603	38452	87890	94624	69721
105	60578	06483	28733	37867	07936	98710	98539
106	91240	18312	17441	01929	18163	69201	31211
107	97458	14229	12063	59611	32249	90466	33216
108	35249	38646	34475	72417	60514	69257	12489
109	38980	46600	11759	11900	46743	27860	77940
110	10750	52745	38749	87365	58959	53731	89295
111	36247	27850	73958	20673	37800	63835	71051
112	70994	66986	99744	72438	01174	42159	11392
113	99638	94702	11463	18148	81386	80431	90628
114	72055	15774	43857	99805	10419	76939	25993
115	24038	65541	85788	55835	38835	59399	13790
116	74976	14631	35908	28221	39470	91548	12854
117	35553	71628	70189	26436	63407	91178	90348
118	35676	12797	51434	82976	42010	26344	92920
119	74815	67523	72985	23183	02446	63594	98924
120	45246	88048	65173	50989	91060	89894	36036
121	76509	47069	86378	41797	11910	49672	88575
122	19689	90332	04315	21358	97248	11188	39062
123	42751	35318	97513	61537	54955	08159	00337
124	11946	22681	45045	13964	57517	59419	58045
125	96518	48688	20996	11090	48396	57177	83867
126	35726	58643	76869	84622	39098	36083	72505
127	39737	42750	48968	70536	84864	64952	38404
128	97025	66492	56177	04049	80312	48028	26408
129	62814	08075	09788	56350	76787	51591	54509
130	25578	22950	15227	83291	41737	59599	96191
131	68763	69576	88991	49662	46704	63362	56625
132	17900	00813	64361	60725	88974	61005	99709
133	71944	60227	63551	71109	05624	43836	58254
134	54684	93691	85132	64399	29182	44324	14491
135	25946	27623	11258	65204	52832	50880	22273
136	01353	39318	44961	44972	91766	90262	56073
137	99083	88191	27662	99113	57174	35571	99884
138	52021	45406	37945	75234	24327	86978	22644
139	78755	47744	43776	83098	03225	14281	83637
140	25282	69106	59180	16257	22810	43609	12224
141	11959	94202	02743	86847	79725	51811	12998
142	11644	13792	98190	01424	30078	28197	55583
143	06307	97912	68110	59812	95448	43244	31262
144	76285	75714	89585	99296	52640	46518	55486
145	55322	07598	39600	60866	63007	20007	66819
146	78017	90928	90220	92503	83375	26986	74399
147	44768	43342	20696	26331	43140	69744	82928
148	25100	19336	14605	86603	51680	97678	24261
149	83612	46623	62876	85197	07824	91392	58317
150	41347	81666	82961	60413	71020	83658	02415

Source: *Table of* 105,000 *Random Decimal Digits*, Interstate Commerce Commission, Bureau of Transport Economics and Statistics, 1949.

TABLE A–16

Selected Standard Latin Squares

3 × 3

A	B	C
B	C	A
C	A	B

4 × 4

1

A	B	C	D
B	A	D	C
C	D	B	A
D	C	A	B

2

A	B	C	D
B	C	D	A
C	D	A	B
D	A	B	C

3

A	B	C	D
B	D	A	C
C	A	D	B
D	C	B	A

4

A	B	C	D
B	A	D	C
C	D	A	B
D	C	B	A

5 × 5

A	B	C	D	E
B	A	E	C	D
C	D	A	E	B
D	E	B	A	C
E	C	D	B	A

6 × 6

A	B	C	D	E	F
B	F	D	C	A	E
C	D	E	F	B	A
D	A	F	E	C	B
E	C	A	B	F	D
F	E	B	A	D	C

7 × 7

A	B	C	D	E	F	G
B	C	D	E	F	G	A
C	D	E	F	G	A	B
D	E	F	G	A	B	C
E	F	G	A	B	C	D
F	G	A	B	C	D	E
G	A	B	C	D	E	F

8 × 8

A	B	C	D	E	F	G	H
B	C	D	E	F	G	H	A
C	D	E	F	G	H	A	B
D	E	F	G	H	A	B	C
E	F	G	H	A	B	C	D
F	G	H	A	B	C	D	E
G	H	A	B	C	D	E	F
H	A	B	C	D	E	F	G

9 × 9

A	B	C	D	E	F	G	H	I
B	C	D	E	F	G	H	I	A
C	D	E	F	G	H	I	A	B
D	E	F	G	H	I	A	B	C
E	F	G	H	I	A	B	C	D
F	G	H	I	A	B	C	D	E
G	H	I	A	B	C	D	E	F
H	I	A	B	C	D	E	F	G
I	A	B	C	D	E	F	G	H

Index

Index

E

F

G

H

I

J–K

L

M

N

O

P

This book has been set in 10 and 9 point Times New Roman, leaded 2 points. Chapter and part numbers are in 30 point Univers Bold #693. Part titles are in 24 point Univers Medium # 689 and chapter titles are in 18 point Univers Medium #689. The size of the type page is 27x46½ picas.